TOPICS IN MATRIX ANALYSIS

Topics in matrix analysis

ROGER A. HORN
University of Utah

CHARLES R. JOHNSON
College of William and Mary

CAMBRIDGE
UNIVERSITY PRESS

CAMBRIDGE UNIVERSITY PRESS
Cambridge, New York, Melbourne, Madrid, Cape Town, Singapore,
São Paulo, Delhi, Dubai, Tokyo

Cambridge University Press
32 Avenue of the Americas, New York, NY 10013-2473, USA

www.cambridge.org
Information on this title: www.cambridge.org/9780521467131

First published 1991
First paperback edition (with corrections) 1994
10th printing 2008

A catalog record for this publication is available from the British Library

ISBN 978-0-521-30587-7 Hardback
ISBN 978-0-521-46713-1 Paperback

Transferred to digital printing 2010

Contents

Preface

This volume is a sequel to the previously published *Matrix Analysis* and includes development of further topics that support applications of matrix theory. We refer the reader to the preface of the prior volume for many general comments that apply here also. We adopt the notation and referencing conventions of that volume and make specific reference to it [HJ] as needed.

Matrix Analysis developed the topics of broadest utility in the connection of matrix theory to other subjects and for modern research in the subject. The current volume develops a further set of slightly more specialized topics in the same spirit. These are: the field of values (or classical numerical range), matrix stability and inertia (including *M*-matrices), singular values and associated inequalities, matrix equations and Kronecker products, Hadamard (or entrywise) products of matrices, and several ways in which matrices and functions interact. Each of these topics is an area of active current research, and several of them do not yet enjoy a broad exposition elsewhere.

Though this book should serve as a reference for these topics, the exposition is designed for use in an advanced course. Chapters include motivational background, discussion, relations to other topics, and literature references. Most sections include exercises in the development as well as many problems that reinforce or extend the subject under discussion. There are, of course, other matrix analysis topics not developed here that warrant attention. Some of these already enjoy useful expositions; for example, totally positive matrices are discussed in [And] and [Kar].

We have included many exercises and over 650 problems because we feel they are essential to the development of an understanding of the subject and its implications. The exercises occur throughout the text as part of the development of each section; they are generally elementary and of immediate use in understanding the concepts. We recommend that the reader work at least a broad selection of these. Problems are listed (in no particular order) at the end of sections; they cover a range of difficulties and types (from theoretical to computational) and they may extend the topic, develop special aspects, or suggest alternate proofs of major ideas. In order to enhance the utility of the book as a reference, many problems have hints; these are collected in a separate section following Chapter 6. The results of some problems are referred to in other problems or in the text itself. We cannot overemphasize the importance of the reader's active involvement in carrying out the exercises and solving problems.

As in the prior volume, a broad list of related books and major surveys is given prior to the index, and references to this list are given via mnemonic code in square brackets. Readers may find the reference list of independent utility.

We appreciate the assistance of our colleagues and students who have offered helpful suggestions or commented on the manuscripts that preceded publication of this volume. They include M. Bakonyi, W. Barrett, O. Chan, C. Cullen, M. Cusick, J. Dietrich, S. H. Friedberg, S. Gabriel, F. Hall, C.-K. Li, M. Lundquist, R. Mathias, D. Merino, R. Merris, P. Nylen, A. Sourour, G. W. Stewart, R. C. Thompson, P. van Dooren, and E. M. E. Wermuth.

The authors wish to maintain the utility of this volume to the community and welcome communication from readers of errors or omissions that they find. Such communications will be rewarded with a current copy of all known errata.

R. A. H.

C. R. J.

Preface to the Second Printing

We have corrected all known errata in the first printing, polished the exposition of a few points, noted the resolution of several conjectures, and added some items to the notation list and index. It is a pleasure to acknowledge helpful comments from our colleagues T. Ando, R. Bhatia, S. Friedberg, D. Jesperson, B. Kroschel, I. Lewkowicz, C.-K. Li, R. Loewy, J. Miao, and F. Uhlig.

Chapter 1

The field of values

1.0 Introduction

Like the spectrum (or set of eigenvalues) $\sigma(\cdot)$, the field of values $F(\cdot)$ is a set of complex numbers naturally associated with a given n-by-n matrix A:

$$F(A) \equiv \{x^* A x : x \in \mathbb{C}^n, \, x^* x = 1\}$$

The spectrum of a matrix is a discrete point set; while the field of values can be a continuum, it is always a compact convex set. Like the spectrum, the field of values is a set that can be used to learn something about the matrix, and it can often give information that the spectrum alone cannot give. The eigenvalues of Hermitian and normal matrices have especially pleasant properties, and the field of values captures certain aspects of this nice structure for general matrices.

1.0.1 Subadditivity and eigenvalues of sums

If only the eigenvalues $\sigma(A)$ and $\sigma(B)$ are known about two n-by-n matrices A and B, remarkably little can be said about $\sigma(A + B)$, the eigenvalues of the sum. Of course, $\text{tr}(A + B) = \text{tr}\,A + \text{tr}\,B$, so the sum of all the eigenvalues of $A + B$ is the sum of all the eigenvalues of A plus the sum of all the eigenvalues of B. But beyond this, nothing can be said about the eigenvalues of $A + B$ without more information about A and B. For example, even if all the eigenvalues of two n-by-n matrices A and B are known and fixed, the spectral radius of $A + B$ [the largest absolute value of an eigenvalue of $A + B$, denoted by $\rho(A + B)$] can be arbitrarily large (see Problem 1). On the other hand, if A and B are normal, then much can be said about the

1

eigenvalues of $A + B$; for example, $\rho(A + B) \leq \rho(A) + \rho(B)$ in this case. Sums of matrices do arise in practice, and two relevant properties of the field of values $F(\cdot)$ are:

(a)   The field of values is subadditive: $F(A + B) \subset F(A) + F(B)$, where the set sum has the natural definition of sums of all possible pairs, one from each; and

(b)   The eigenvalues of a matrix lie inside its field of values: $\sigma(A) \subset F(A)$.

Combining these two properties yields the inclusions

$$\sigma(A + B) \subset F(A + B) \subset F(A) + F(B)$$

so if the two fields of values $F(A)$ and $F(B)$ are known, something can be said about the spectrum of the sum.

1.0.2   An application from the numerical solution of partial differential equations

Suppose that $A = [a_{ij}] \in M_n(\mathbb{R})$ satisfies

(a) A is tridiagonal ($a_{ij} = 0$ for $|i - j| > 1$), and

(b) $a_{i,i+1} a_{i+1,i} < 0$ for $i = 1, ..., n - 1$.

Matrices of this type arise in the numerical solution of partial differential equations and in the analysis of dynamical systems arising in mathematical biology. In both cases, knowledge about the real parts of the eigenvalues of A is important. It turns out that rather good information about the eigenvalues of such a matrix can be obtained easily using the field of values $F(\cdot)$.

1.0.2.1   Fact: For any eigenvalue λ of a matrix A of the type indicated, we have

$$\min_{1 \leq i \leq n} a_{ii} \leq \operatorname{Re} \lambda \leq \max_{1 \leq i \leq n} a_{ii}$$

A proof of this fact is fairly simple using some properties of the field of values to be developed in Section (1.2). First, choose a diagonal matrix D

with positive diagonal entries such that $D^{-1}AD \equiv \hat{A} = [\hat{a}_{ij}]$ satisfies $\hat{a}_{ji} = -\hat{a}_{ij}$ for $j \neq i$. The matrix $D \equiv \text{diag}(d_1,..., d_n)$ defined by

$$d_1 = 1, \text{ and } d_i = \left|\frac{a_{i,i-1}}{a_{i-1,i}}\right|^{\frac{1}{2}} d_{i-1}, \ d_i > 0, \ i = 2,..., n$$

will do. Since $\hat{A}$ and A are similar, their eigenvalues are the same. We then have

$$\text{Re } \sigma(A) = \text{Re } \sigma(\hat{A}) \subset \text{Re } F(\hat{A}) = F(\tfrac{1}{2}(\hat{A} + \hat{A}^T))$$
$$= F(\text{diag}(a_{11},..., a_{nn}))$$
$$= \text{Convex hull of } \{a_{11},..., a_{nn}\} = [\min_i a_{ii}, \max_i a_{ii}]$$

The first inclusion follows from the spectral containment property (1.2.6), the next equality follows from the projection property (1.2.5), the next equality follows from the special form achieved for $\hat{A}$, and the last equality follows from the normality property (1.2.9) and the fact that the eigenvalues of a diagonal matrix are its diagonal entries. Since the real part of each eigenvalue $\lambda \in \sigma(A)$ is a convex combination of the main diagonal entries a_{ii}, $i = 1,..., n$, the asserted inequalities are clear and the proof is complete.

1.0.3   Stability analysis

In an analysis of the stability of an equilibrium in a dynamical system governed by a system of differential equations, it is important to know if the real part of every eigenvalue of a certain matrix A is negative. Such a matrix is called *stable*. In order to avoid juggling negative signs, we often work with *positive stable* matrices (all eigenvalues have positive real parts). Obviously, A is positive stable if and only if $-A$ is stable. An important sufficient condition for a matrix to be positive stable is the following fact.

1.0.3.1   Fact: Let $A \in M_n$. If $A + A^*$ is positive definite, then A is positive stable.

This is another application of properties of the field of values $F(\cdot)$ to be developed in Section (1.2). By the spectral containment property (1.2.6), $\text{Re } \sigma(A) \subset \text{Re } F(A)$, and, by the projection property (1.2.5), $\text{Re } F(A) =$

$F(\frac{1}{2}(A + A^*))$. But, since $A + A^*$ is positive definite, so is $\frac{1}{2}(A + A^*)$, and hence, by the normality property (1.2.9), $F(\frac{1}{2}(A + A^*))$ is contained in the positive real axis. Thus, each eigenvalue of A has a positive real part, and A is positive stable.

Actually, more is true. If $A + A^*$ is positive definite, and if $P \in M_n$ is any positive definite matrix, then PA is positive stable because

$$(P^{\frac{1}{2}})^{-1}[PA]P^{\frac{1}{2}} = P^{\frac{1}{2}}AP^{\frac{1}{2}}, \text{ and}$$

$$P^{\frac{1}{2}}AP^{\frac{1}{2}} + (P^{\frac{1}{2}}AP^{\frac{1}{2}})^* = P^{\frac{1}{2}}(A + A^*)P^{\frac{1}{2}}$$

where $P^{\frac{1}{2}}$ is the unique (Hermitian) positive definite square root of P. Since congruence preserves positive definiteness, the eigenvalues of PA have positive real parts for the same reason as A. Lyapunov's theorem (2.2.1) shows that all positive stable matrices arise in this way.

1.0.4   An approximation problem

Suppose we wish to approximate a given matrix $A \in M_n$ by a complex multiple of a Hermitian matrix of rank at most one, as closely as possible in the Frobenius norm $\|\cdot\|_2$. This is the problem

$$\text{minimize } \|A - cxx^*\|_2^2 \text{ for } x \in \mathbb{C}^n \text{ with } x^*x = 1 \text{ and } c \in \mathbb{C} \qquad (1.0.4.1)$$

Since the inner product $[A,B] \equiv \text{tr } AB^*$ generates the Frobenius norm, we have

$$\|A - cxx^*\|_2^2 = [A - cxx^*, A - cxx^*]$$

$$= \|A\|_2^2 - 2 \text{ Re } \bar{c}[A,xx^*] + |c|^2$$

which, for a given unit vector x, is minimized by $c = [A,xx^*]$. Substitution of this value into (1.0.4.1) transforms our problem into

$$\text{minimize } (\|A\|_2^2 - |[A,xx^*]|^2) \text{ for } x \in \mathbb{C}^n \text{ with } x^*x = 1$$

or, equivalently,

$$\text{maximize } |[A,xx^*]| \text{ for } x \in \mathbb{C}^n \text{ with } x^*x = 1$$

A vector x_0 that solves the latter problem (we are maximizing a continuous function on a compact set) will yield a rank one solution $[A, x_0 x_0^*] x_0 x_0^*$ to our original problem. Since $[A, xx^*] = \text{tr } Axx^* = x^* Ax$, we are led naturally to finding a unit vector x such that the point $x^* Ax$ in the field of values $F(A)$ has maximum distance from the origin. The absolute value of such a point is called the *numerical radius* of A [often denoted by $r(A)$] by analogy with the *spectral radius* [often denoted by $\rho(A)$], which is the absolute value of a point in the spectrum $\sigma(A)$ that is at maximum distance from the origin.

Problems

1. Consider the real matrices

$$A = \begin{bmatrix} 1-\alpha & 1 \\ \alpha(1-\alpha)-1 & \alpha \end{bmatrix} \text{ and } B = \begin{bmatrix} 1+\alpha & 1 \\ -\alpha(1+\alpha)-1 & -\alpha \end{bmatrix}$$

Show that $\sigma(A)$ and $\sigma(B)$ are independent of the value of $\alpha \in \mathbb{R}$. What are they? What is $\sigma(A+B)$? Show that $\rho(A+B)$ is unbounded as $\alpha \to \infty$.

2. In contrast to Problem 1, show that if $A, B \in M_n$ are normal, then $\rho(A+B) \le \rho(A) + \rho(B)$.

3. Show that "$<$" in (1.0.2(b)) may be replaced by "$\le$," the main diagonal entries a_{ii} may be complex, and Fact (1.0.2.1) still holds if a_{ii} is replaced by $\text{Re } a_{ii}$.

4. Show that the problem of approximating a given $A \in M_n$ by a positive semidefinite rank one matrix with spectral radius one can be solved if one can find a unit vector x such that the point $x^* Ax$ in $F(A)$ is furthest to the right in the complex plane, that is, $\text{Re } x^* Ax$ is maximized.

1.1   Definitions

In this section we define the field of values and certain related objects.

1.1.1   Definition. The *field of values* of $A \in M_n$ is

$$F(A) \equiv \{x^* Ax : x \in \mathbb{C}^n, x^* x = 1\}$$

Thus, $F(\cdot)$ is a function from M_n into subsets of the complex plane.

$F(A)$ is just the normalized locus of the Hermitian form associated with A. The field of values is often called the *numerical range*, especially in the context of its analog for operators on infinite dimensional spaces.

Exercise. Show that $F(I) = \{1\}$ and $F(\alpha I) = \{\alpha\}$ for all $\alpha \in \mathbb{C}$. Show that $F\begin{bmatrix} 1 & 0 \\ 0 & 0 \end{bmatrix}$ is the closed unit interval $[0,1]$, and $F\begin{bmatrix} 0 & 2 \\ 0 & 0 \end{bmatrix}$ is the closed unit disc $\{z \in \mathbb{C}: |z| \leq 1\}$.

The field of values $F(A)$ may also be thought of as the image of the surface of the Euclidean unit ball in $\mathbb{C}^n$ (a compact set) under the continuous transformation $x \to x^* A x$. As such, $F(A)$ is a compact (and hence bounded) set in $\mathbb{C}$. An unbounded analog of $F(\cdot)$ is also of interest.

1.1.2   Definition. The *angular field of values* is

$$F'(A) \equiv \{x^* A x: x \in \mathbb{C}^n, x \neq 0\}$$

Exercise. Show that $F'(A)$ is determined geometrically by $F(A)$; every open ray from the origin that intersects $F(A)$ in a point other than the origin is in $F'(A)$, and $0 \in F'(A)$ if and only if $0 \in F(A)$. Draw a typical picture of an $F(A)$ and $F'(A)$ assuming that $0 \notin F(A)$.

It will become clear that $F'(A)$ is an angular sector of the complex plane that is anchored at the origin (possibly the entire complex plane). The angular opening of this sector is of interest.

1.1.3   Definition. The *field angle* $\Theta \equiv \Theta(A) \equiv \Theta(F'(A)) \equiv \Theta(F(A))$ of $A \in M_n$ is defined as follows:

(a)   If 0 is an interior point of $F(A)$, then $\Theta(A) \equiv 2\pi$.
(b)   If 0 is on the boundary of $F(A)$ and there is a (unique) tangent to the boundary of $F(A)$ at 0, then $\Theta(A) \equiv \pi$.
(c)   If $F(A)$ is contained in a line through the origin, $\Theta(A) \equiv 0$.
(d)   Otherwise, consider the two different support lines of $F(A)$ that go through the origin, and let $\Theta(A)$ be the angle subtended by these two lines at the origin. If $0 \notin F(A)$, these support lines will be uniquely determined; if 0 is on the boundary of $F(A)$, choose the two support lines that give the minimum angle.

We shall see that $F(A)$ is a compact convex set for every $A \in M_n$, so this informal definition of the field angle makes sense. The field angle is just the angular opening of the smallest angular sector that includes $F(A)$, that is, the angular opening of the sector $F'(A)$.

Finally, the size of the bounded set $F(A)$ is of interest. We measure its size in terms of the radius of the smallest circle centered at the origin that contains $F(A)$.

1.1.4   Definition. The *numerical radius* of $A \in M_n$ is

$$r(A) \equiv \max \{|z| : z \in F(A)\}$$

The numerical radius is a vector norm on matrices that is not a matrix norm (see Section (5.7) of [HJ]).

Problems

1. Show that among the vectors entering into the definition of $F(A)$, only vectors with real nonnegative first coordinate need be considered.

2. Show that both $F(A)$ and $F'(A)$ are simply connected for any $A \in M_n$.

3. Show that for each $0 \le \theta \le \pi$, there is an $A \in M_2$ with $\Theta(A) = \theta$. Is $\Theta(A) = 3\pi/2$ possible?

4. Why is the "max" in (1.1.4) attained?

5. Show that the following alternative definition of $F(A)$ is equivalent to the one given:

$$F(A) \equiv \{x^* A x / x^* x : x \in \mathbb{C}^n \text{ and } x \ne 0\}$$

Thus, $F(\cdot)$ is a normalized version of $F'(\cdot)$.

6. Determine $F\begin{bmatrix} 1 & 1 \\ 1 & 0 \end{bmatrix}$, $F\begin{bmatrix} 1 & 1 \\ 0 & 1 \end{bmatrix}$, and $F\begin{bmatrix} 1 & 1 \\ 1 & 1 \end{bmatrix}$.

7. If $A \in M_n$ and $\alpha \in F(A)$, show that there is a unitary matrix $U \in M_n$ such that α is the 1,1 entry of $U^* A U$.

8. Determine as many different possible types of sets as you can that can be an $F'(A)$.

9. Show that $F(A^*) = \overline{F(A)}$ and $F'(A^*) = \overline{F'(A)}$ for all $A \in M_n$.

10. Show that all of the main diagonal entries and eigenvalues of a given $A \in M_n$ are in its field of values $F(A)$.

1.2   Basic properties of the field of values

As a function from M_n into subsets of $\mathbb{C}$, the field of values $F(\cdot)$ has many useful functional properties, most of which are easily established. We catalog many of these properties here for reference and later use. The important property of convexity is left for discussion in the next section.

The sum or product of two subsets of $\mathbb{C}$, or of a subset of $\mathbb{C}$ and a scalar, has the usual algebraic meaning. For example, if $S, T \subset \mathbb{C}$, then $S + T \equiv \{s + t: s \in S, t \in T\}$.

1.2.1   Property: *Compactness.* For all $A \in M_n$,

$F(A)$ is a compact subset of $\mathbb{C}$

Proof: The set $F(A)$ is the range of the continuous function $x \to x^* A x$ over the domain $\{x: x \in \mathbb{C}^n, x^* x = 1\}$, the surface of the Euclidean unit ball, which is a compact set. Since the continuous image of a compact set is compact, it follows that $F(A)$ is compact. ☐

1.2.2   Property: *Convexity.* For all $A \in M_n$,

$F(A)$ is a convex subset of $\mathbb{C}$

The next section of this chapter is reserved for a proof of this fundamental fact, known as the *Toeplitz-Hausdorff theorem*. At this point, it is clear that $F(A)$ must be a *connected* set since it is the continuous image of a connected set.

Exercise. If A is a diagonal matrix, show that $F(A)$ is the convex hull of the diagonal entries (the eigenvalues) of A.

The field of values of a matrix is changed in a simple way by adding a scalar multiple of the identity to it or by multiplying it by a scalar.

1.2.3   Property: *Translation*. For all $A \in M_n$ and $\alpha \in \mathbb{C}$,

$$F(A + \alpha I) = F(A) + \alpha$$

Proof: We have $F(A + \alpha I) = \{x^*(A + \alpha I)x: x^*x = 1\} = \{x^*Ax + \alpha x^*x: x^*x = 1\} = \{x^*Ax + \alpha: x^*x = 1\} = \{x^*Ax: x^*x = 1\} + \alpha = F(A) + \alpha$. ☐

1.2.4   Property: *Scalar multiplication*. For all $A \in M_n$ and $\alpha \in \mathbb{C}$,

$$F(\alpha A) = \alpha F(A)$$

Exercise. Prove property (1.2.4) by the same method used in the proof of (1.2.3).

For $A \in M_n$, $H(A) \equiv \frac{1}{2}(A + A^*)$ denotes the *Hermitian part of A* and $S(A) \equiv \frac{1}{2}(A - A^*)$ denotes the *skew-Hermitian part of A*; notice that $A = H(A) + S(A)$ and that $H(A)$ and $iS(A)$ are both Hermitian. Just as taking the real part of a complex number projects it onto the real axis, taking the Hermitian part of a matrix projects its field of values onto the real axis. This simple fact helps in locating the field of values, since, as we shall see, it is relatively easy to deal with the field of values of a Hermitian matrix. For a set $S \subset \mathbb{C}$, we interpret Re S as $\{\text{Re } s: s \in S\}$, the projection of S onto the real axis.

1.2.5   Property: *Projection*. For all $A \in M_n$,

$$F(H(A)) = \text{Re } F(A)$$

Proof: We calculate $x^*H(A)x = x^*\frac{1}{2}(A + A^*)x = \frac{1}{2}(x^*Ax + x^*A^*x) = \frac{1}{2}(x^*Ax + (x^*Ax)^*) = \frac{1}{2}(x^*Ax + \overline{x^*Ax}) = \text{Re } x^*Ax$. Thus, each point in $F(H(A))$ is of the form Re z for some $z \in F(A)$ and vice versa. ☐

We denote the open upper half-plane of $\mathbb{C}$ by $UHP \equiv \{z \in \mathbb{C}: \text{Im } z > 0\}$, the open left half-plane of $\mathbb{C}$ by $LHP \equiv \{z \in \mathbb{C}: \text{Re } z < 0\}$, the open right half-plane of $\mathbb{C}$ by $RHP \equiv \{z \in \mathbb{C}: \text{Re } z > 0\}$, and the closed right half-plane of $\mathbb{C}$ by $RHP_0 \equiv \{z \in \mathbb{C}: \text{Re } z \geq 0\}$. The projection property gives a simple indication of when $F(A) \subset RHP$ or RHP_0 in terms of positive definiteness or positive semidefiniteness.

1.2.5a   **Property**: *Positive definite indicator function.* Let $A \in M_n$. Then $F(A) \subset RHP$ if and only if $A + A^*$ is positive definite

1.2.5b   **Property**: *Positive semidefinite indicator function.* Let $A \in M_n$. Then

$$F(A) \subset RHP_0 \text{ if and only if } A + A^* \text{ is positive semidefinite}$$

Exercise. Prove (1.2.5a) and (1.2.5b) (the proofs are essentially the same) using (1.2.5) and the definition of positive definite and semidefinite (see Chapter 7 of [HJ]).

The point set of eigenvalues of $A \in M_n$ is denoted by $\sigma(A)$, the *spectrum* of A. A very important property of the field of values is that it includes the eigenvalues of A.

1.2.6   **Property**: *Spectral containment.* For all $A \in M_n$,

$$\sigma(A) \subset F(A)$$

Proof: Suppose that $\lambda \in \sigma(A)$. Then there exists some nonzero $x \in \mathbb{C}^n$, which we may take to be a unit vector, for which $Ax = \lambda x$ and hence $\lambda = \lambda x^* x = x^*(\lambda x) = x^* A x \in F(A)$. $\square$

Exercise. Use the spectral containment property (1.2.6) to show that the eigenvalues of a positive definite matrix are positive real numbers.

Exercise. Use the spectral containment property (1.2.6) to show that the eigenvalues of $\begin{bmatrix} 0 & 1 \\ -1 & 0 \end{bmatrix}$ are imaginary.

The following property underlies the fact that the numerical radius is a vector norm on matrices and is an important reason why the field of values is so useful.

1.2.7   **Property**: *Subadditivity.* For all $A, B \in M_n$,

$$F(A + B) \subset F(A) + F(B)$$

Proof: $F(A + B) = \{x^*(A + B)x : x \in \mathbb{C}^n, x^* x = 1\} = \{x^* A x + x^* B x : x \in \mathbb{C}^n,$

$x^*x = 1\} \subset \{x^*Ax: \ x \in \mathbb{C}^n, \ x^*x = 1\} + \{y^*By: \ y \in \mathbb{C}^n, \ y^*y = 1\} = F(A) + F(B).$ ☐

Exercise. Use (1.2.7) to show that the numerical radius $r(\cdot)$ satisfies the triangle inequality on M_n.

Another important property of the field of values is its invariance under unitary similarity.

1.2.8   Property: *Unitary similarity invariance.* For all A, $U \in M_n$ with U unitary,

$$F(U^*AU) = F(A)$$

Proof: Since a unitary transformation leaves invariant the surface of the Euclidean unit ball, the complex numbers that comprise the sets $F(U^*AU)$ and $F(A)$ are the same. If $x \in \mathbb{C}^n$ and $x^*x = 1$, we have $x^*(U^*AU)x = y^*Ay \in F(A)$, where $y = Ux$, so $y^*y = x^*U^*Ux = x^*x = 1$. Thus, $F(U^*AU) \subset F(A)$. The reverse containment is obtained similarly. ☐

The unitary similarity invariance property allows us to determine the field of values of a normal matrix. Recall that, for a set S contained in a real or complex vector space, $\text{Co}(S)$ denotes the convex hull of S, which is the set of all convex combinations of finitely many points of S. Alternatively, $\text{Co}(S)$ can be characterized as the intersection of all convex sets containing S, so it is the "smallest" closed convex set containing S.

1.2.9   Property: *Normality.* If $A \in M_n$ is normal, then

$$F(A) = \text{Co}(\sigma(A))$$

Proof: If A is normal, then $A = U^*\Lambda U$, where $\Lambda = \text{diag}(\lambda_1,...,\lambda_n)$ is diagonal and U is unitary. By the unitary similarity invariance property (1.2.8), $F(A) = F(\Lambda)$ and, since

$$x^*\Lambda x = \sum_{i=1}^n \bar{x}_i x_i \lambda_i = \sum_{i=1}^n |x_i|^2 \lambda_i$$

$F(\Lambda)$ is just the set of all convex combinations of the diagonal entries of Λ ($x^*x = 1$ implies $\Sigma_i |x_i|^2 = 1$ and $|x_i|^2 \geq 0$). Since the diagonal entries of Λ are the eigenvalues of A, this means that $F(A) = \text{Co}(\sigma(A))$.   ▯

Exercise. Show that if H is Hermitian, $F(H)$ is a closed real line segment whose endpoints are the largest and smallest eigenvalues of A.

Exercise. Show that the field of values of a normal matrix is always a polygon whose vertices are eigenvalues of A. If $A \in M_n$, how many sides may $F(A)$ have? If A is unitary, show that $F(A)$ is a polygon inscribed in the unit circle.

Exercise. Show that $\text{Co}(\sigma(A)) \subset F(A)$ for all $A \in M_n$.

Exercise. If $A, B \in M_n$, show that

$$\sigma(A + B) \subset F(A) + F(B)$$

If A and B are normal, show that

$$\sigma(A + B) \subset \text{Co}(\sigma(A)) + \text{Co}(\sigma(B))$$

The next two properties have to do with fields of values of matrices that are built up from or extracted from other matrices in certain ways. Recall that for $A \in M_{n_1}$ and $B \in M_{n_2}$, the *direct sum* of A and B is the matrix

$$A \oplus B \equiv \begin{bmatrix} A & 0 \\ 0 & B \end{bmatrix} \in M_{n_1 + n_2}$$

If $J \subset \{1, 2,..., n\}$ is an index set and if $A \in M_n$, then $A(J)$ denotes the principal submatrix of A contained in the rows and columns indicated by J.

1.2.10   Property: *Direct sums*. For all $A \in M_{n_1}$ and $B \in M_{n_2}$,

$$F(A \oplus B) = \text{Co}(F(A) \cup F(B))$$

Proof: Note that $A \oplus B \in M_{n_1 + n_2}$. Partition any given unit vector $z \in \mathbb{C}^{n_1 + n_2}$ as $z = \begin{bmatrix} x \\ y \end{bmatrix}$, where $x \in \mathbb{C}^{n_1}$ and $y \in \mathbb{C}^{n_2}$ Then $z^*(A \oplus B)z = x^*Ax + y^*By$. If $y^*y = 1$ then $x = 0$ and $z^*(A \oplus B)z = y^*By \in F(B)$, so $F(A \oplus B) \supset F(B)$. By a similar argument when $x^*x = 1$, $F(A \oplus B) \supset F(A)$ and hence

$F(A \bullet B) \supset F(A) \cup F(B)$. But since $F(A \bullet B)$ is convex, it follows that $F(A \bullet B) \supset \mathrm{Co}(F(A) \cup F(B))$ (see Problem 21).

To prove the reverse containment, let $z = \begin{bmatrix} x \\ y \end{bmatrix} \in \mathbb{C}^{n_1 + n_2}$ be a unit vector again. If $x^* x = 0$, then $y^* y = 1$ and $z^*(A \bullet B)z = y^* By \in F(B) \subset \mathrm{Co}(F(A) \cup F(B))$. The argument is analogous if $y^* y = 0$. Now suppose that both x and y are nonzero and write

$$z^*(A \bullet B)z = x^* Ax + y^* By = x^* x \left[\frac{x^* Ax}{x^* x} \right] + y^* y \left[\frac{y^* By}{y^* y} \right]$$

Since $x^* x + y^* y = z^* z = 1$, this last expression is a convex combination of

$$\frac{x^* Ax}{x^* x} \in F(A) \quad \text{and} \quad \frac{y^* By}{y^* y} \in F(B)$$

and we have $F(A \bullet B) \subset \mathrm{Co}(F(A) \cup F(B))$. ☐

1.2.11 Property: *Submatrix inclusion.* For all $A \in M_n$ and index sets $J \subset \{1, ..., n\}$,

$$F(A(J)) \subset F(A)$$

Proof: Suppose that $J = \{j_1, ..., j_k\}$, with $1 \leq j_1 < j_2 < \cdots < j_k \leq n$, and suppose $x \in \mathbb{C}^k$ satisfies $x^* x = 1$. We may insert zero entries into appropriate locations in x to produce a vector $\hat{x} \in \mathbb{C}^n$ such that $\hat{x}_{j_i} = x_i$, $i = 1, 2, ..., k$, and $\hat{x}_j = 0$ for all other indices j. A calculation then shows that $x^* A(J)x = \hat{x}^* A\hat{x}$ and $\hat{x}^* \hat{x} = 1$, which verifies the asserted inclusion. ☐

1.2.12 Property: *Congruence and the angular field of values.* Let $A \in M_n$ and suppose that $C \in M_n$ is nonsingular. Then

$$F'(C^* AC) = F'(A)$$

Proof: Let $x \in \mathbb{C}^n$ be a nonzero vector, so that $x^* C^* ACx = y^* Ay$, where $y \equiv Cx \neq 0$. Thus, $F'(C^* AC) \subset F'(A)$. In the same way, one shows that $F'(A) \subset F'(C^* AC)$ since $A = (C^{-1})^* C^* AC(C^{-1})$. ☐

Problems

1.   For what $A \in M_n$ does $F(A)$ consist of a single point? Could $F(A)$ consist of k distinct points for finite $k > 1$?

2.   State and prove results corresponding to (1.2.5) and (1.2.5a,b) about projecting the field of values onto the imaginary axis and about when $F(A)$ is in the upper half-plane.

3.   Show that $F(A) + F(B)$ is not the same as $F(A + B)$ in general. Why not?

4.   If $A, B \in M_n$, is $F(AB) \subset F(A)F(B)$? Prove or give a counterexample.

5.   If $A, B \in M_n$, is $F'(AB) \subset F'(A)F'(B)$? Is $\Theta(AB) \leq \Theta(A) + \Theta(B)$? Prove or give a counterexample.

6.   If $A, B \in M_n$ are normal with $\sigma(A) = \{\alpha_1,..., \alpha_n\}$ and $\sigma(B) = \{\beta_1,..., \beta_n\}$, show that $\sigma(A + B) \subset \mathrm{Co}(\{\alpha_i + \beta_j: i, j = 1,..., n\})$. If $0 \leq \alpha \leq 1$ and if $U, V \in M_n$ are unitary, show that $\rho(\alpha U + (1 - \alpha)V) \leq \alpha\rho(U) + (1 - \alpha)\rho(V) = 1$, where $\rho(\cdot)$ denotes the spectral radius.

7.   If $A, B \in M_n$ are Hermitian with ordered eigenvalues $\alpha_1 \leq \cdots \leq \alpha_n$ and $\beta_1 \leq \cdots \leq \beta_n$, respectively, use a field of values argument with the subadditivity property (1.2.7) to show that $\alpha_1 + \beta_1 \leq \gamma_1$ and $\gamma_n \leq \alpha_n + \beta_n$, where $\gamma_1 \leq \cdots \leq \gamma_n$ are the ordered eigenvalues of $C = A + B$. What can you say if equality holds in either of the inequalities? Compare with the conclusions and proof of Weyl's theorem (4.3.1) in [HJ].

8.   Which convex subsets of $\mathbb{C}$ are fields of values? Show that the class of convex subsets of $\mathbb{C}$ that are fields of values is closed under the operation of taking convex hulls of finite unions of the sets. Show that any convex polygon and any disc are in this class.

9.   According to property (1.2.8), if two matrices are unitarily similar, they have the same field of values. Although the converse is true when $n = 2$ [see Problem 18 in Section (1.3)], it is not true in general. Construct two matrices that have the same size and the same field of values, but are not unitarily similar. A complete characterization of all matrices of a given size with a given field of values is unknown.

10. According to (1.2.9), if $A \in M_n$ is normal, then its field of values is the convex hull of its spectrum. Show that the converse is not true by consid-

ering the matrix $A \equiv \text{diag}(1, i, -1, -i) \oplus \begin{bmatrix} 0 & 1 \\ 0 & 0 \end{bmatrix} \in M_6$. Show that $F(A) = \text{Co}(\sigma(A))$, but that A is not normal. Construct a counterexample of the form $A = \text{diag}(\lambda_1, \lambda_2, \lambda_3) \oplus \begin{bmatrix} 0 & 1 \\ 0 & 0 \end{bmatrix} \in M_5$. Why doesn't this kind of example work for M_4? Is there some other kind of counterexample in M_4? See (1.6.9).

11. Let z be a complex number with modulus 1, and let

$$A \equiv \begin{bmatrix} 0 & 1 & 0 \\ 0 & 0 & 1 \\ z & 0 & 0 \end{bmatrix}$$

Show that $F(A)$ is the closed equilateral triangle whose vertices are the cube roots of z. More generally, what is $F(A)$ if $A = [a_{ij}] \in M_n$ has all $a_{i,i+1} = 1$, $a_{n1} = z$, and all other $a_{ij} = 0$?

12. If $A \in M_n$ and $F(A)$ is a real line segment, show that A is Hermitian.

13. Give an example of a matrix A and an index set J for which equality occurs in (1.2.11). Can you characterize the cases of equality?

14. What is the geometric relationship between $F'(C^*AC)$ and $F'(A)$ if: (a) $C \in M_n$ is singular, or (b) if C is n-by-k and rank $C = k$?

15. What is the relationship between $F(C^*AC)$ and $F(A)$ when $C \in M_n$ is nonsingular but not unitary? Compare with Sylvester's law of inertia, Theorem (4.5.8) in [HJ].

16. Properties (1.2.8) and (1.2.11) are special cases of a more general property. Let $A \in M_n$, $k \leq n$, and $P \in M_{n,k}$ be given. If $P^*P = I \in M_k$, then P is called an *isometry* and P^*AP is called an *isometric projection of A*. Notice that an isometry $P \in M_{n,k}$ is unitary if and only if $k = n$. If A' is a principal submatrix of A, show how to construct an isometry P such that $A' = P^*AP$. Prove the following statement and explain how it includes both (1.2.8) and (1.2.11):

1.2.13   **Property**: *Isometric projection.* For all $A \in M_n$ and $P \in M_{n,k}$ with $k \leq n$ and $P^*P = I$, $F(P^*AP) \subset F(A)$, and $F(P^*AP) = F(A)$ when $k = n$.

17. It is natural to inquire whether there are any nonunitary cases in which the containment in (1.2.13) is an equality. Let $A \in M_n$ be given, let $P \in M_{n,k}$ be a given isometry with $k < n$, and suppose the column space of P contains

all the columns of both A and A^*. Show that $F'(A) = F'(P^*AP) \cup \{0\}$ and $F(A) = \text{Co}(F(P^*AP) \cup \{0\})$. If, in addition, $0 \in F(P^*AP)$, show that $F(A) = F(P^*AP)$; consider $A = \begin{bmatrix} 1 & 1 \\ 1 & 1 \end{bmatrix}$ and $P = \begin{bmatrix} 1 \\ 1 \end{bmatrix}/\sqrt{2}$ to show that if $0 \notin F(P^*AP)$, then the containment $F(P^*AP) \subset F(A)$ can be strict.

18. Let $A \in M_n$ and let $P \in M_{n,k}$ be an isometry with $k \le n$. Consider $A = \begin{bmatrix} 1 & 0 \\ 0 & 1 \end{bmatrix}$ and $P = \begin{bmatrix} 1 \\ 1 \end{bmatrix}/\sqrt{2}$ to show that one can have $F(P^*AP) = F(A)$ even if the column space of P does not contain the columns of A and A^*. Suppose $U \in M_n$ is unitary and $UAU^* = \Delta$ is upper triangular with $\Delta = \Delta_1 \oplus \cdots \oplus \Delta_p$ in which each Δ_i is upper triangular and some of the matrices Δ_i are diagonal. Describe how to construct an isometry P from selected columns of U so that $F(A) = F(P^*AP)$. Apply your construction to $A = I$, to $A = \text{diag}(1,2,3) \in M_3$, and to a normal $A \in M_n$ and discuss.

19. Suppose that $A \in M_n$ is given and that $G \in M_n$ is positive definite. Show that $\sigma(A) \subset F'(GA)$ and, therefore, that $\sigma(A) \subset \cap\{F'(GA): G \text{ is positive definite}\}$.

20. Consider the two matrices

$$A = \begin{bmatrix} 0 & 0 & 0 & 1 \\ 0 & 0 & 1 & 0 \\ 0 & 0 & 0 & 0 \\ 1 & 0 & 0 & 0 \end{bmatrix} \quad \text{and} \quad B = \begin{bmatrix} 0 & 0 & 0 & 1 \\ 0 & 0 & 1 & 0 \\ 0 & 1 & 0 & 0 \\ 1 & 0 & 0 & 0 \end{bmatrix}$$

Precisely determine $F(A)$, $F(B)$, and $F(\alpha A + (1-\alpha)B)$, $0 \le \alpha \le 1$.

21. Let S be a given subset of a complex vector space, and let S_1 and S_2 be given subsets of S. Show that $S \supset \text{Co}(S_1 \cup S_2)$ if S is convex, but that this statement need not be correct if S is not convex. Notice that this general principle was used in the proof of the direct sum property (1.2.10).

22. Let $A \in M_n$ be nonsingular. Show that $F(A) \subset RHP$ if and only if $F(A^{-1}) \subset RHP$, or, equivalently, $H(A)$ is positive definite if and only if $H(A^{-1})$ is positive definite.

23. Let $A \in M_n$, and suppose λ is an eigenvalue of A.
   (a) If λ is real, show that $\lambda \in F(H(A))$ and hence $|\lambda| \le \rho(H(A)) = ||| H(A) |||_2$, where $\rho(\cdot)$ and $|||\cdot|||_2$ denote the spectral radius and spectral norm, respectively.

(b) If λ is purely imaginary, show that $\lambda \in F(S(A))$ and hence $|\lambda| \leq \rho(S(A)) = \| S(A)) \|_2$.

24. Suppose all the entries of a given $A \in M_n$ are nonnegative. Then the nonnegative real number $\rho(A)$ is always an eigenvalue of A [HJ, Theorem (8.3.1)]. Use Problem 23 (a) to show that $\rho(A) \leq \rho(H(A)) = \frac{1}{2}\rho(A + A^T)$ whenever A is a square matrix with nonnegative entries. See Problem 23 (n) in Section (1.5) for a better inequality involving the numerical radius of A.

25. We have seen that the field of values of $\begin{bmatrix} 0 & 1 \\ 0 & 0 \end{bmatrix}$ is a closed disc of radius $\frac{1}{2}$ centered at the origin; this result has a useful generalization.

(a) Let $B \in M_{m,n}$ be given, let $A = \begin{bmatrix} 0 & B \\ 0 & 0 \end{bmatrix} \in M_{m+n}$, and let $\sigma_1(B)$ denote the largest singular value (the spectral norm) of B. Show that $F(A) = \{x^* By : x \in \mathbb{C}^m,\ y \in \mathbb{C}^n,\ \|x\|_2^2 + \|y\|_2^2 = 1\}$, and conclude that $F(A)$ is a closed disc of radius $\sigma_1(B)/2$ centered at the origin. In particular, $r(A) = \sigma_1(B)/2$.

(b) Consider $A = \begin{bmatrix} 0 & B \\ 0 & 0 \end{bmatrix} \in M_3$ with $B \equiv \begin{bmatrix} 1 & 1 \\ 1 & 1 \end{bmatrix} \in M_2$. Show that $\sigma_1(B) = 2$. Consider $e^T A e$ with $e = [1, 1, 1]^T$ to show that $r(A) \geq 4/3 > \sigma_1(B)/2$. Does this contradict (a)?

26. Let $A \in M_n$ be given. Show that the following are equivalent:

(a) $r(A) \leq 1$.

(b) $\rho(H(e^{i\theta}A)) \leq 1$ for all $\theta \in \mathbb{R}$.

(c) $\lambda_{max}(H(e^{i\theta}A)) \leq 1$ for all $\theta \in \mathbb{R}$.

(d) $\| H(e^{i\theta}A) \|_2 \leq 1$ for all $\theta \in \mathbb{R}$.

1.3   Convexity

In this section we prove the fundamental convexity property (1.2.2) of the field of values and discuss several important consequences of that convexity. We shall make use of several basic properties exhibited in the previous section. Our proof contains several useful observations and consists of three parts:

1.   Reduction of the problem to the 2–by–2 case;

2.   Use of various basic properties to transform the general 2–by–2 case to 2–by–2 matrices of special form; and

3.   Demonstration of convexity of the field of values for the special 2-by-2 form.

See Problems 7 and 10 for two different proofs that do not involve reduction to the 2-by-2 case. There are other proofs in the literature that are based upon more advanced concepts from other branches of mathematics.

Reduction to the 2–by–2 case

In order to show that a given set $S \subset \mathbb{C}$ is convex, it is sufficient to show that $\alpha s + (1 - \alpha)t \in S$ whenever $0 \le \alpha \le 1$ and $s, t \in S$. Thus, for a given $A \in M_n$, $F(A)$ is convex if $\alpha x^* A x + (1 - \alpha)y^* A y \in F(A)$ whenever $0 \le \alpha \le 1$ and $x, y \in \mathbb{C}^n$ satisfy $x^* x = y^* y = 1$. It suffices to prove this only in the 2-by-2 case because we need to consider only convex combinations associated with *pairs* of vectors. For each given pair of vectors $x, y \in \mathbb{C}^n$, there is a unitary matrix U and vectors $v, w \in \mathbb{C}^n$ such that $x = Uv$, $y = Uw$, and all entries of v and w after the first two are equal to zero (see Problem 1). Using this transformation, we have

$$\alpha x^* A x + (1 - \alpha)y^* A y = \alpha v^* U^* A U v + (1 - \alpha)w^* U^* A U w$$

$$= \alpha v^* B v + (1 - \alpha)w^* B w$$

$$= \alpha \xi^* B(\{1,2\})\xi + (1 - \alpha)\eta^* B(\{1,2\})\eta$$

where $B \equiv U^* A U$, $B(\{1,2\})$ is the upper left 2-by-2 principal submatrix of B, and $\xi, \eta \in \mathbb{C}^2$ consist of the first two entries of v and w, respectively. Thus, it suffices to show that the field of values of any 2-by-2 matrix is convex. This reduction is possible because of the unitary similarity invariance property (1.2.8) of the field of values.

Sufficiency of a special 2–by–2 form

We prove next that in order to show that $F(A)$ is convex for every matrix $A \in M_2$, it suffices to demonstrate that $F\begin{bmatrix} 0 & a \\ b & 0 \end{bmatrix}$ is convex for any $a, b \ge 0$. The following observation is useful.

1.3.1   Lemma. For each $A \in M_2$, there is a unitary $U \in M_2$ such that the two main diagonal entries of $U^* A U$ are equal.

Proof: We may suppose without loss of generality that tr $A = 0$ since we may replace A by $A - (\frac{1}{2}\mathrm{tr}\, A)I$. We need to show that A is unitarily similar to a matrix whose diagonal entries are both equal to zero. To show this, it is sufficient to show that there is a nonzero vector $w \in \mathbb{C}^2$ such that $w^* A w = 0$. Such a vector may be normalized and used as the first column of a unitary matrix W, and a calculation reveals that the 1,1 entry of $W^* A W$ is zero; the 2,2 entry of $W^* A W$ must also be zero since the trace is zero. Construct the vector w as follows: Since A has eigenvalues $\pm \alpha$ for some complex number α, let x be a normalized eigenvector associated with $-\alpha$ and let y be a normalized eigenvector associated with $+\alpha$. If $\alpha = 0$, just take $w = x$. If $\alpha \neq 0$, x and y are independent and the vector $w = e^{i\theta}x + y$ is nonzero for all $\theta \in \mathbb{R}$. A calculation shows that $w^* A w = \alpha(e^{-i\theta}x^*y - e^{i\theta}y^*x) = 2i\alpha\,\mathrm{Im}(e^{-i\theta}x^*y)$. Now choose θ so that $e^{-i\theta}x^*y$ is real. □

We now use Lemma (1.3.1) together with several of the properties given in the previous section to reduce the question of convexity in the 2-by-2 case to consideration of the stated special form. If $A \in M_2$ is given, apply the translation property (1.2.3) to conclude that $F(A)$ is convex if and only if $F(A + \alpha I)$ is convex. If we choose $\alpha = -\frac{1}{2}\mathrm{tr}\, A$, we may suppose without loss of generality that our matrix has trace 0. According to (1.3.1) and the unitary similarity invariance property (1.2.8), we may further suppose that both main diagonal entries of our matrix are 0.

Thus, we may assume that the given matrix has the form $\begin{bmatrix} 0 & c \\ d & 0 \end{bmatrix}$ for some $c, d \in \mathbb{C}$. Now we can use the unitary similarity invariance property (1.2.8) and a diagonal unitary matrix to show that we may consider

$$\begin{bmatrix} 1 & 0 \\ 0 & e^{-i\theta} \end{bmatrix} \begin{bmatrix} 0 & c \\ d & 0 \end{bmatrix} \begin{bmatrix} 1 & 0 \\ 0 & e^{i\theta} \end{bmatrix} = \begin{bmatrix} 0 & ce^{i\theta} \\ de^{-i\theta} & 0 \end{bmatrix}$$

for any $\theta \in \mathbb{R}$. If $c = |c|e^{i\theta_1}$ and $d = |d|e^{i\theta_2}$, and if we choose $\theta = \frac{1}{2}(\theta_2 - \theta_1)$, the latter matrix becomes $e^{i\varphi}\begin{bmatrix} 0 & |c| \\ |d| & 0 \end{bmatrix}$ with $\varphi = \frac{1}{2}(\theta_1 + \theta_2)$.

Thus, it suffices to consider a matrix of the form $e^{i\varphi}\begin{bmatrix} 0 & a \\ b & 0 \end{bmatrix}$ with $\varphi \in \mathbb{R}$ and $a, b \geq 0$. Finally, by the scalar multiplication property (1.2.4), we need to consider only the special form

$$\begin{bmatrix} 0 & a \\ b & 0 \end{bmatrix}, \qquad a, b \geq 0 \tag{1.3.2}$$

That is, we have shown that the field of values of every 2-by-2 complex matrix is convex if the field of values of every matrix of the special form (1.3.2) is convex.

Convexity of the field of values of the special 2-by-2 form

1.3.3   Lemma. If $A \in M_2$ has the form (1.3.2), then $F(A)$ is an ellipse (with its interior) centered at the origin. Its minor axis is along the imaginary axis and has length $|a - b|$. Its major axis is along the real axis and has length $a + b$. Its foci are at $\pm\sqrt{ab}$, which are the eigenvalues of A.

Proof: Without loss of generality, we assume $a \geq b \geq 0$. Since $z^* A z = (e^{i\theta}z)^* A(e^{i\theta}z)$ for any $\theta \in \mathbb{R}$, to determine $F(A)$, it suffices to consider $z^* A z$ for unit vectors z whose first component is real and nonnegative. Thus, we consider the 2-vector $z = [t, e^{i\theta}(1 - t^2)^{\frac{1}{2}}]^T$ for $0 \leq t \leq 1$ and $0 \leq \theta \leq 2\pi$. A calculation shows that

$$z^* A z = t(1 - t^2)^{\frac{1}{2}}[(a + b)\cos\theta + i(a - b)\sin\theta]$$

As θ varies from 0 to 2π, the point $(a + b)\cos\theta + i(a - b)\sin\theta$ traces out a possibly degenerate ellipse $\mathcal{E}$ centered at the origin; the major axis extends from $-(a + b)$ to $(a + b)$ on the real axis and the minor axis extends from $i(b - a)$ to $i(a - b)$ on the imaginary axis in the complex plane. As t varies from 0 to 1, the factor $t(1 - t^2)^{\frac{1}{2}}$ varies from 0 to $\frac{1}{2}$ and back to 0, ensuring that every point in the interior of the ellipse $\frac{1}{2}\mathcal{E}$ is attained and verifying that $F(A)$ is the asserted ellipse with its interior, which is convex. The two foci of the ellipse $\frac{1}{2}\mathcal{E}$ are located on the major axis at distance $[\frac{1}{4}(a + b)^2 - \frac{1}{4}(a - b)^2]^{\frac{1}{2}} = \pm\sqrt{ab}$ from the center. This completes the argument to prove the convexity property (1.2.2). $\square$

There are many important consequences of convexity of the field of values. One immediate consequence is that Lemma (1.3.1) holds for matrices of any size, not just for 2-by-2 matrices.

1.3.4   Theorem. For each $A \in M_n$ there is a unitary matrix $U \in M_n$ such that all the diagonal entries of $U^* A U$ have the same value $\mathrm{tr}(A)/n$.

Proof: Without loss of generality, we may suppose that $\mathrm{tr}\, A = 0$, since we may replace A by $A - [\mathrm{tr}(A)/n]I$. We proceed by induction to show that A is unitarily similar to a matrix with all zero main diagonal entries. We know from Lemma (1.3.1) that this is true for $n = 2$, so let $n \geq 3$ and suppose that the assertion has been proved for all matrices of all orders less than n. We have

$$0 = \frac{1}{n}\operatorname{tr} A = \frac{1}{n}\lambda_1 + \frac{1}{n}\lambda_2 + \cdots + \frac{1}{n}\lambda_n = 0$$

and this is a convex combination of the eigenvalues λ_i of A. Since each λ_i is in $F(A)$, and since $F(A)$ is convex, we conclude that $0 \in F(A)$. If $x \in \mathbb{C}^n$ is a unit vector such that $x^*Ax = 0$, let $W = [x \; w_2 \; \ldots \; w_n] \in M_n$ be a unitary matrix whose first column is x. One computes that

$$W^*AW = \begin{bmatrix} 0 & z^* \\ \zeta & \hat{A} \end{bmatrix}, \; z, \zeta \in \mathbb{C}^{n-1}, \; \hat{A} \in M_{n-1}$$

But $0 = \operatorname{tr} A = \operatorname{tr} W^*AW = \operatorname{tr} \hat{A} = 0$, and so by the induction hypothesis there is some unitary $\hat{V} \in M_{n-1}$ such that all the main diagonal entries of $\hat{V}^*\hat{A}\hat{V}$ are zero. Define the unitary direct sum

$$V = \begin{bmatrix} 1 & 0 \\ 0 & \hat{V} \end{bmatrix} \in M_n$$

and compute

$$(WV)^*A(WV) = V^*W^*AWV = \begin{bmatrix} 0 & z^*\hat{V} \\ \hat{V}\zeta & \hat{V}^*\hat{A}\hat{V} \end{bmatrix}$$

which has a zero main diagonal by construction. $\Box$

A different proof of (1.3.4) using compactness of the set of unitary matrices is given in Problem 3 of Section (2.2) of [HJ].

Another important, and very useful, consequence of convexity of the field of values is the following *rotation property* of a matrix whose field of values does not contain the point 0.

1.3.5   Theorem. Let $A \in M_n$ be given. There exists a real number θ such that the Hermitian matrix $H(e^{i\theta}A) = \frac{1}{2}[e^{i\theta}A + e^{-i\theta}A^*]$ is positive definite if and only if $0 \notin F(A)$.

Proof: If $H(e^{i\theta}A)$ is positive definite for some $\theta \in \mathbb{R}$, then $F(e^{i\theta}A) \subset RHP$ by

(1.2.5a), so $0 \notin F(e^{i\theta}A)$ and hence $0 \notin F(A)$ by (1.2.4). Conversely, suppose $0 \notin F(A)$. By the separating hyperplane theorem (see Appendix B of [HJ]), there is a line L in the plane such that each of the two nonintersecting compact convex sets $\{0\}$ and $F(A)$ lies entirely within exactly one of the two open half-planes determined by L. The coordinate axes may now be rotated so that the line L is carried into a vertical line in the right half-plane with $F(A)$ strictly to the right of it, that is, for some $\theta \in \mathbb{R}$, $F(e^{i\theta}A) = e^{i\theta}F(A) \subset RHP$, so $H(e^{i\theta}A)$ is positive definite by (1.2.5a). $\square$

Some useful information can be extracted from a careful examination of the steps we have taken to transform a given matrix $A \in M_2$ to the special form (1.3.2). The first step was a translation $A \to A - (\frac{1}{2}\mathrm{tr}\, A)I \equiv A_0$ to achieve $\mathrm{tr}\, A_0 = 0$. The second step was a unitary similarity $A_0 \to UA_0U^* \equiv A_1$ to make both diagonal entries of A_1 zero. The third step was another unitary similarity $A_1 \to VA_1V^* \equiv A_2$ to put A_2 into the form

$$A_2 = e^{i\varphi}\begin{bmatrix} 0 & a \\ b & 0 \end{bmatrix} \text{ with } a,\, b \geq 0 \text{ and } \varphi \in \mathbb{R}$$

The last step was a unitary rotation $A_2 \to e^{-i\varphi}A_2 \equiv A_3$ to achieve the special form (1.3.2). Since the field of values of A_3 is an ellipse (possibly degenerate, that is, a point or line segment) centered at the origin with its major axis along the real axis and its foci at $\pm\sqrt{ab}$, the eigenvalues of A_3, the field of values of A_2 is also an ellipse centered at the origin, but its major axis is tilted at an angle φ to the real axis. A line through the two eigenvalues of A_2, $\pm e^{i\varphi}\sqrt{ab}$, which are the foci of the ellipse, contains the major axis of the ellipse; if $ab = 0$, the ellipse is a circle (possibly degenerate), so any diameter is a major axis. Since A_1 and A_0 are achieved from A_2 by successive unitary similarities, each of which leaves the eigenvalues and field of values invariant, we have $F(A_0) = F(A_1) = F(A_2)$. Finally,

$$F(A) = F(A_0 + [\tfrac{1}{2}\mathrm{tr}\, A]I) = F(A_0) + \tfrac{1}{2}\mathrm{tr}\, A = F(A_2) + \tfrac{1}{2}\mathrm{tr}\, A$$

a shift that moves both eigenvalues by $\frac{1}{2}\mathrm{tr}\, A$, so we conclude that the field of values of any matrix $A \in M_2$ is an ellipse (possibly degenerate) with center at the point $\frac{1}{2}\mathrm{tr}\, A$. The major axis of this ellipse lies on a line through the two eigenvalues of A, which are the foci of the ellipse; if the two eigenvalues coincide, the ellipse is a circle or a point.

According to Lemma (1.3.3), the ellipse $F(A)$ is degenerate if and only if $A_3 = \begin{bmatrix} 0 & a \\ b & 0 \end{bmatrix}$ has $a = b$. Notice that $A_3^* A_3 = \begin{bmatrix} b^2 & 0 \\ 0 & a^2 \end{bmatrix}$ and $A_3 A_3^* = \begin{bmatrix} a^2 & 0 \\ 0 & b^2 \end{bmatrix}$, so $a = b$ if and only if A_3 is normal. But A can be recovered from A_3 by a nonzero scalar multiplication, two unitary similarities, and a translation, each of which preserves both normality and nonnormality. Thus, A_3 is normal if and only if A is normal, and we conclude that for $A \in M_2$, the ellipse $F(A)$ is degenerate if and only if A is normal.

The eigenvalues of A_3 are located at the foci on the major axis of $F(A_3)$ at a distance of $\sqrt{ab}$ from the center, and the length of the semimajor axis is $\frac{1}{2}(a + b)$ by (1.3.3). Thus, $\frac{1}{2}(a + b) - \sqrt{ab} = \frac{1}{2}(\sqrt{a} - \sqrt{b})^2 \geq 0$ with equality if and only if $a = b$, that is, if and only if A is normal. We conclude that the eigenvalues of a nonnormal $A \in M_2$ always lie in the interior of $F(A)$.

For $A \in M_2$, the parameters of the ellipse $F(A)$ (even if degenerate) can be computed easily using (1.3.3) if one observes that a and b are the singular values of $A_3 = \begin{bmatrix} 0 & a \\ b & 0 \end{bmatrix}$, that is, the square roots of the eigenvalues of $A_3^* A_3$, and the singular values of A_3 are invariant under pre- or post-multiplication by any unitary matrix. Thus, the singular values $\sigma_1 \geq \sigma_2 \geq 0$ of $A_0 = A - (\frac{1}{2}\text{tr } A)I$ are the same as those of A_3. The length of the major axis of $F(A)$ is $a + b = \sigma_1 + \sigma_2$, the length of the minor axis is $|a - b| = \sigma_1 - \sigma_2$, and the distance of the foci from the center is $[(a + b)^2 - (a - b)^2]^{\frac{1}{2}}/4 = \sqrt{ab} = \sqrt{\sigma_1 \sigma_2} = |\det A_3|^{\frac{1}{2}} = |\det A_0|^{\frac{1}{2}}$. Moreover, $\sigma_1^2 + \sigma_2^2 = \text{tr } A_0^* A_0$ (the sum of the squares of the moduli of the entries of A_0), so $\sigma_1 \pm \sigma_2 = [\sigma_1^2 + \sigma_2^2 \pm 2\sigma_1\sigma_2]^{\frac{1}{2}} = [\text{tr } A_0^* A_0 \pm 2|\det A_0|]^{\frac{1}{2}}$. We summarize these observations for convenient reference in the following theorem.

1.3.6 **Theorem.** Let $A \in M_2$ be given, and set $A_0 \equiv A - (\frac{1}{2}\text{tr } A)I$. Then

(a) The field of values $F(A)$ is a closed ellipse (with interior, possibly degenerate).

(b) The center of the ellipse $F(A)$ is at the point $\frac{1}{2}\text{tr } A$. The length of the major axis is $[\text{tr } A_0^* A_0 + 2|\det A_0|]^{\frac{1}{2}}$; the length of the minor axis is $[\text{tr } A_0^* A_0 - 2|\det A_0|]^{\frac{1}{2}}$; the distance of the foci from the center is $|\det A_0|^{\frac{1}{2}}$. The major axis lies on a line passing through the two eigenvalues of A, which are the foci of $F(A)$; these two eigenvalues coincide if and only if the ellipse is a circle (possibly a point).

(c) $F(A)$ is a closed line segment if and only if A is normal; it is a single point if and only if A is a scalar matrix.

(d) $F(A)$ is a nondegenerate ellipse (with interior) if and only if A is not normal, and in this event the eigenvalues of A are interior points of $F(A)$.

Problems

1. Let $x, y \in \mathbb{C}^n$ be two given vectors. Show how to construct vectors $v, w \in \mathbb{C}^n$ and a unitary matrix $U \in M_n$ such that $x = Uv$, $y = Uw$, and all entries of v and w after the first two are zero.

2. Verify all the calculations in the proof of (1.3.1).

3. Sketch the field of values of a matrix of the form (1.3.2), with $a \geq 0$, $b \geq 0$.

4. Use (1.3.3) to show that the field of values of $\begin{bmatrix} 1 & 1 \\ 0 & 0 \end{bmatrix}$ is a closed ellipse (with interior) with foci at 0 and 1, major axis of length $\sqrt{2}$, and minor axis of length 1. Verify these assertions using Theorem (1.3.6).

5. Show that if $A \in M_n(\mathbb{R})$ then $F(A)$ is symmetric with respect to the real axis.

6. If $x_1, \ldots, x_k \in \mathbb{C}^n$ are given orthonormal vectors, let $P \equiv [x_1 \ldots x_k] \in M_{n,k}$ and observe that $P^*P = I \in M_k$. If $A \in M_n$, show that $F(P^*AP) \subset F(A)$ and that $x_i^*Ax_i \in F(P^*AP)$ for $i = 1, \ldots, k$. Use this fact for $k = 2$ to give an alternate reduction of the question of convexity of $F(A)$ to the 2-by-2 case.

7. Let $A \in M_n$ be given. Provide details for the following proof of the convexity of $F(A)$ that does not use a reduction to the 2-by-2 case. There is nothing to prove if $F(A)$ is a single point. Pick any two distinct points in $F(A)$. Using (1.2.3) and (1.2.4), there is no loss of generality to assume that these two points are 0, $a \in \mathbb{R}$, $a > 0$. We must show that the line segment joining 0 and a lies in $F(A)$. Write A as $A = H + iK$, in which $H = H(A)$ and $K = -iS(A) \equiv -i\tfrac{1}{2}(A - A^*)$, so that H and K are Hermitian. Assume further, without loss of generality after using a unitary similarity, that K is diagonal. Let $x, y \in \mathbb{C}^n$ satisfy $x^*Ax = 0$, $x^*x = 1$, $y^*Ay = a$, and $y^*y = 1$, and let $x_j = |x_j| e^{i\theta_j}$, $y_j = |y_j| e^{i\varphi_j}$, $j = 1, \ldots, n$. Note that $x^*Hx = x^*Kx = y^*Ky = 0$ and $y^*Hy = a$. Define $z(t) \in \mathbb{C}^n$, $0 \leq t \leq 1$, by

$$
z_j(t) = \begin{cases}
|x_j|\,e^{i(1-3t)\theta_j}, & 0 \le t \le 1/3 \\
\left[(2-3t)|x_j|^2 + (3t-1)|y_j|^2\right]^{\frac{1}{2}}, & 1/3 < t < 2/3 \\
|y_j|\,e^{i(3t-2)\varphi_j}, & 2/3 \le t \le 1
\end{cases}
$$

Verify that $z^*(t)z(t) = 1$ and $z^*(t)Kz(t) = 0$, $0 \le t \le 1$, and note that $z^*(t)Az(t) = z^*(t)Hz(t)$ is real and equal to 0 at $t = 0$, is equal to a at $t = 1$, and is a continuous function of t for $0 \le t \le 1$. Conclude that the line segment joining 0 and a lies in $F(A)$, which, therefore, is convex.

8. If $A \in M_n$ is such that $0 \in F(A)$, show that the Euclidean unit sphere in the definition of $F(A)$ may be replaced by the Euclidean unit ball, that is, show that $F(A) = \{x^* Ax : x \in \mathbf{C}^n, x^*x \le 1\}$. What if $0 \notin F(A)$?

9. Let $J_n(0) \in M_n$ be the n-by-n nilpotent Jordan block

$$
J_n(0) \equiv \begin{bmatrix} 0 & 1 & & 0 \\ & 0 & 1 & \\ & & 0 & \ddots \\ & & & \ddots & 1 \\ 0 & & & & 0 \end{bmatrix}
$$

Show that $F(J_n(0))$ is a disc centered at the origin with radius $\rho(\mathrm{H}(J_n(0)))$. Use this to show that $F(J_n(0))$ is strictly contained in the unit disc. If $D \in M_n$ is diagonal, show that $F(DJ_n(0))$ is also a disc centered at the origin with radius $\rho(\mathrm{H}(DJ_n(0))) = \rho(\mathrm{H}(|D|J_n(0)))$. See Problem 29 in Section (1.5) for a stronger result.

10. Let $A \in M_n$ be given. Provide details for the following proof of the convexity of $F(A)$ that does not use an explicit reduction to the 2-by-2 case. If $F(A)$ is not a single point, pick any two distinct points $a, b \in F(A)$, and let c be any given point on the open line segment between a and b. We may assume that $c = 0$, $a, b \in \mathbf{R}$, and $a < 0 < b$. Let $x, y \in \mathbf{C}^n$ be unit vectors such that $x^* Ax = a$, $y^* Ay = b$, so x and y are independent. Consider $z(t,\theta) \equiv e^{i\theta}x + ty$, where t and θ are real parameters to be determined. Show that $f(t,\theta) \equiv z(t,\theta)^* Az(t,\theta) = bt^2 + \alpha(\theta)t + a$, where $\alpha(\theta) \equiv e^{-i\theta}x^* Ay + e^{i\theta}y^* Ax$. If $y^* Ax - x^T \bar{A}y = re^{i\varphi}$ with $r \ge 0$ and $\varphi \in \mathbf{R}$, then $\alpha(\theta)$ is real for $\theta = -\varphi$, and $f(t_0, -\varphi) = 0$ for $t_0 = [-\alpha(-\varphi) + [\alpha(-\varphi)^2 - 4ab]^{\frac{1}{2}}]/2b \in \mathbf{R}$. Then $z(t_0, -\varphi) \ne 0$ and $0 = z^* Az \in F(A)$ for $z \equiv z(t_0, -\varphi)/\|z(t_0, -\varphi)\|_2$.

11. Let $A \in M_2$ be given. Show that A is normal (and hence $F(A)$ is a line segment) if and only if $A_0 \equiv A - (\frac{1}{2}\mathrm{tr}\, A)I$ is a scalar multiple of a unitary matrix, that is, $A_0^* A_0 = cI$ for some $c \geq 0$.

12. Give an example of a normal matrix $A \in M_n$, $n \geq 3$, such that $F(A)$ is not a line segment. Why can't this happen in M_2?

13. Let $A \in M_n$ be given. If $F(A)$ is a line segment or a point, show that A must be normal.

14. Consider the upper triangular matrix $A = \begin{bmatrix} \lambda_1 & \beta \\ 0 & \lambda_2 \end{bmatrix} \in M_2$. Show that the length of the major axis of the ellipse $F(A)$ is $[|\lambda_1 - \lambda_2|^2 + |\beta|^2]^{\frac{1}{2}}$ and the length of the minor axis is $|\beta|$. Where is the center? Where are the foci? In particular, conclude that the eigenvalues of A are interior points of $F(A)$ if and only if $\beta \neq 0$.

15. Let $A \in M_n$ be given with $0 \notin F(A)$. Provide the details for the following proof that $F(A)$ lies in an open half-plane determined by some line through the origin, and explain how this result may be used as an alternative to the argument involving the separating hyperplane theorem in the proof of (1.3.5). For $z \in F(A)$, consider the function $g(z) \equiv z/|z| = \exp(i \arg z)$. Since $g(\cdot)$ is a continuous function on the compact connected set $F(A)$, its range $g(F(A)) \equiv R$ is a compact connected subset of the unit circle. Thus, R is a closed arc whose length must be strictly less than π since $F(A)$ is convex and $0 \notin F(A)$.

16. Let $A \in M_n$ be given. Ignoring for the moment that we know the field of values $F(A)$ is convex, $F(A)$ is obviously nonempty, closed, and bounded, so its complement $F(A)^c$ has an unbounded component and possibly some bounded components. The *outer boundary* of $F(A)$ is the intersection of $F(A)$ with the closure of the unbounded component of $F(A)^c$. Provide details for the following proof (due to Toeplitz) that the outer boundary of $F(A)$ is a convex curve. For any given $\theta \in [0, 2\pi]$, let $e^{i\theta} A = H + iK$ with Hermitian $H, K \in M_n$. Let $\lambda_n(H)$ denote the algebraically largest eigenvalue of H, let $S_\theta \equiv \{x \in \mathbb{C}^n : x \neq 0, Hx = \lambda_n(H)x\}$, and suppose $\dim(S_\theta) = k \geq 1$. Then the intersection of $F(e^{i\theta} A)$ with the vertical line $\operatorname{Re} z = \lambda_n(H)$ is the set $\lambda_n(H) + i\{x^* K x : x \in S_\theta, \|x\|_2 = 1\}$, which is a single point if $k = 1$ and can be a finite interval if $k > 1$ (this is the convexity property of the field of values for a Hermitian matrix, which follows simply from the spectral theorem). Conclude, by varying θ, that the outer boundary of $F(A)$ is a convex curve, which may contain straight line segments. Why doesn't this prove that $F(A)$ is convex?

17. (a) Let $A \in M_2$ be given. Use Schur's unitary triangularization theorem (Theorem (2.3.1) in [HJ]) and a diagonal unitary similarity to show that A is unitarily similar to an upper triangular matrix of the form

$$\begin{bmatrix} \lambda_1 & \alpha(A) \\ 0 & \lambda_2 \end{bmatrix}$$

where $\alpha(A) \geq 0$. Show that tr $A^*A = |\lambda_1|^2 + |\lambda_2|^2 + \alpha(A)^2$ and that $\alpha(A)$ is a unitary similarity invariant, that is, $\alpha(A) = \alpha(UAU^*)$ for any unitary $U \in M_2$.
(b) If $A, B \in M_2$ have the same eigenvalues, show that A is unitarily similar to B if and only if tr $A^*A = $ tr B^*B.
(c) Show that $A, B \in M_2$ have the same eigenvalues if and only if tr $A = $ tr B and tr $A^2 = $ tr B^2.
(d) Conclude that two given matrices $A, B \in M_2$ are unitarily similar if and only if tr $A = $ tr B, tr $A^2 = $ tr B^2, and tr $A^*A = $ tr B^*B.

18. Use Theorem (1.3.6) and the preceding problem to show that two given 2-by-2 complex or real matrices are unitarily similar if and only if their fields of values are identical. Consider $A_t = \text{diag}(0,1,t)$ for $t \in [0,1]$ to show that nonsimilar 3-by-3 Hermitian matrices can have the same fields of values.

19. Let $A = [a_{ij}] \in M_2$ and suppose that $F(A) \subset UHP_0 = \{z \in \mathbb{C}: \text{Im } z \geq 0\}$. Use Theorem (1.3.6) to show that if either (a) a_{11} and a_{22} are real, or (b) tr A is real, then A is Hermitian.

Notes and Further Readings. The convexity of the field of values (1.2.2) was first discussed in O. Toeplitz, Das algebraische Analogon zu einem Satze von Fejér, *Math. Zeit.* 2 (1918), 187-197, and F. Hausdorff, Das Wertvorrat einer Bilinearform, *Math. Zeit.* 3 (1919), 314-316. Toeplitz showed that the outer boundary of $F(A)$ is a convex curve, but left open the question of whether the interior of this curve is completely filled out with points of $F(A)$; see Problem 16 for Toeplitz's elegant proof. He also proved the inequality $||| A |||_2 \leq 2r(A)$ between the spectral norm and the numerical radius; see Problem 21 in Section (5.7) of [HJ]. In a paper dated six months after Toeplitz's (the respective dates of their papers were May 22 and November 28, 1918), Hausdorff rose to the challenge and gave a short proof, similar to the argument outlined in Problem 7, that $F(A)$ is actually a convex set. There are many other proofs, besides the elementary one given in this section, such as that of W. Donoghue, On the Numerical Range of a Bounded Operator, *Mich. Math. J.* 4 (1957), 261-263. The result of Theorem (1.3.4) was first noted by W. V. Parker in Sets of Complex Numbers Associated with a Matrix, *Duke Math. J.* 15 (1948), 711-715. The modification of Hausdorff's original convexity proof outlined in Problem 7 was given by Donald Robinson; the proof outlined in Problem 10 is due to Roy Mathias.

The fact that the field of values of a square complex matrix is convex has an immediate extension to the infinite-dimensional case. If T is a bounded linear operator on a complex Hilbert space $\mathcal{X}$ with inner product $<\cdot,\cdot>$, then its field of values (often called the *numerical range*) is $F(T) \equiv \{<Tx,x>: x \in \mathcal{X}$ and $<x,x> = 1\}$. One can show that $F(T)$ is convex by reducing to the two-dimensional case, just as we did in the proof in this section.

1.4   Axiomatization

It is natural to ask (for both practical and aesthetic reasons) whether the list of properties of $F(A)$ given in Section (1.2) is, in some sense, complete. Since special cases and corollary properties may be of interest, it may be that no finite list is truly complete; but a mathematically precise version of the completeness question is whether or not, among the properties given thus far, there is a subset that characterizes the field of values. If so, then further properties, and possibly some already noted, would be corollary to a set of characterizing properties, and the mathematical utility of the field of values would be captured by these properties. This does not mean that it is not useful to write down properties beyond a characterizing set. Some of the most applicable properties do follow, if tediously, from others.

1.4.1   Example. Spectral containment (1.2.6) follows from compactness (1.2.1), translation (1.2.3), scalar multiplication (1.2.4), unitary invariance (1.2.8), and submatrix inclusion (1.2.11) in the sense that any set-valued function on M_n that has these five properties also satisfies (1.2.6). If $A \in M_n$ and if $\beta \in \sigma(A)$, then for some unitary $U \in M_n$ the matrix U^*AU is upper triangular with β in the 1,1 position. Then by (1.2.8) and (1.2.11) it is enough to show that $\beta \in F([\beta])$, and (because of (1.2.3)) it suffices to show that $0 \in F([0])$; here we think of $[\beta]$ and $[0]$ as members of M_1 However, because of (1.2.4), $F([0]) = \alpha F([0])$ for any α, and there are only two nonempty subsets of the complex plane possessing this property: $\{0\}$ and the entire plane. The latter is precluded by (1.2.1), and hence (1.2.6) follows.

Exercise. Show that (1.2.8) and (1.2.10) together imply (1.2.9).

The main result of this section is that there is a subset of the properties already mentioned that characterizes $F(\cdot)$ as a function from M_n into subsets of $\mathbb{C}$.

1.4.2   Theorem. Properties (1.2.1–4 and 5b) characterize the field of values. That is, the usual field of values $F(\cdot)$ is the only complex set-valued function on M_n such that

(a) $F(A)$ is compact (1.2.1) and convex (1.2.2) for all $A \in M_n$;

(b) $F(A + \alpha I) = F(A) + \alpha$ (1.2.3) and $F(\alpha A) = \alpha F(A)$ (1.2.4) for all $\alpha \in \mathbb{C}$ and all $A \in M_n$; and

(c) $F(A)$ is a subset of the closed right half-plane if and only if $A + A^*$ is positive semidefinite (1.2.5b).

Proof: Suppose $F_1(\cdot)$ and $F_2(\cdot)$ are two given complex set-valued functions on M_n that satisfy the five cited functional properties. Let $A \in M_n$ be given. We first show that $F_1(A) \subset F_2(A)$. Suppose, to the contrary, that $\beta \in F_1(A)$ and $\beta \notin F_2(A)$ for some complex number β. Then because of (1.2.1) and (1.2.2), there is a straight line L in the complex plane that has the point β strictly on one side of it and the set $F_2(A)$ on the other side (by the separating hyperplane theorem for convex sets; see Appendix B of [HJ]). The plane may be rotated and translated so that the imaginary axis coincides with L and β lies in the open left half-plane. That is, there exist complex numbers $\alpha_1 \neq 0$ and α_2 such that $\operatorname{Re}(\alpha_1 \beta + \alpha_2) < 0$ while $\alpha_1 F_2(A) + \alpha_2$ is contained in the closed right half-plane. However, $\alpha_1 F_2(A) + \alpha_2 = F_2(\alpha_1 A + \alpha_2 I)$ because of (1.2.3) and (1.2.4), so $\operatorname{Re} \beta' < 0$ while $F_2(A')$ lies in the closed right half-plane, where $A' \equiv \alpha_1 A + \alpha_2 I$ and $\beta' \equiv \alpha_1 \beta + \alpha_2 \in F_1(A')$. Then, by (1.2.5b), $A' + A'^*$ is positive semidefinite. This, however, contradicts the fact that $F_1(A')$ is not contained in the closed right half-plane, and we conclude that $F_1(A) \subset F_2(A)$.

Reversing the roles of $F_1(\cdot)$ and $F_2(\cdot)$ shows that $F_2(A) \subset F_1(A)$ as well. Thus, $F_1(A) = F_2(A)$ for all $A \in M_n$. Since the usual field of values $F(\cdot)$ satisfies the five stated properties (1.2.1–4 and 5b), we obtain the desired conclusion by taking $F_1(A) \equiv F(A)$. $\Box$

Problems

1.   Show that no four of the five properties cited in Theorem (1.4.2) are sufficient to characterize $F(\cdot)$; that is, each subset of four of the five properties in the theorem is satisfied by some complex set-valued function other than $F(\cdot)$.

2.   Show that (1.2.2-4 and 5a) also characterize $F(\cdot)$, and that no subset of these four properties is sufficient to characterize $F(\cdot)$.

3.   Determine other characterizing sets of properties. Can you find one that does not contain (1.2.2)?

4.   Show that the complex set-valued function $F(\cdot) \equiv \text{Co}(\sigma(\cdot))$ is characterized by the four properties (1.2.1-4) together with the fifth property "$F(A)$ is contained in the closed right half-plane if and only if all the eigenvalues of A have nonnegative real parts" (A is *positive semistable*).

5.   Give some other complex set-valued functions on M_n that satisfy (1.2.7).

Further Reading. This section is based upon C. R. Johnson, Functional Characterization of the Field of Values and the Convex Hull of the Spectrum, *Proc. Amer. Math. Soc.* 61 (1976), 201-204.

1.5   Location of the field of values

Thus far we have said little about where the field of values $F(A)$ sits in the complex plane, although it is clear that knowledge of its location could be useful for applications such as those mentioned in (1.0.2-4). In this section, we give a Geršgorin-type inclusion region for $F(A)$ and some observations that facilitate its numerical determination.

   Because the eigenvalues of a matrix depend continuously upon its entries and because the eigenvalues of a diagonal matrix are the diagonal entries, it is not too surprising that there is a spectral location result such as Geršgorin's theorem (6.1.1) in [HJ]. We argue by analogy that because the set $F(A)$ depends continuously upon the entries of A and because the field of values of a diagonal matrix is the convex hull of the diagonal entries, there ought to be some sort of inclusion region for $F(A)$ that, like the Geršgorin discs, depends in a simple way on the entries of A. We next present such an inclusion region.

1.5.1   Definition. Let $A = [a_{ij}] \in M_n$, let the deleted absolute row and column sums of A be denoted by

$$R'_i(A) = \sum_{\substack{j=1 \\ j \neq i}}^n |a_{ij}|, \quad i = 1,\dots, n$$

and

$$C'_j(A) = \sum_{\substack{i=1 \\ i \neq j}}^n |a_{ij}|, \quad j = 1,\dots, n$$

respectively, and let

$$g_i(A) = \tfrac{1}{2}[R'_i(A) + C'_i(A)], \; i = 1,\dots, n$$

be the average of the i th deleted absolute row and column sums of A. Define the complex set-valued function $G_F(\cdot)$ on M_n by

$$G_F(A) \equiv \mathrm{Co}\left[\bigcup_{i=1}^n \{z: |z - a_{ii}| \leq g_i(A)\} \right]$$

Recall Geršgorin's theorem about the spectrum $\sigma(A)$ of $A \in M_n$, which says that

$$\sigma(A) \subset G(A) \equiv \bigcup_{i=1}^n \{z: |z - a_{ii}| \leq R'_i(A)\}$$

and that

$$\sigma(A) \subset G(A^T) \equiv \bigcup_{j=1}^n \{z: |z - a_{jj}| \leq C'_j(A)\}$$

Note that the Geršgorin regions $G(A)$ and $G(A^T)$ are unions of circular discs with centers at the diagonal entries of A, while the set function $G_F(A)$ is the convex hull of some circular discs with the same centers, but with radii that are the average of the radii leading to $G(A)$ and $G(A^T)$, respectively. The Geršgorin regions $G(A)$ and $G(A^T)$ need not be convex, but $G_F(A)$ is convex by definition.

Exercise. Show that the set-valued function $G_F(\cdot)$ satisfies the properties (1.2.1-4) of the field of values. Which additional properties of the field of values function $F(\cdot)$ does $G_F(\cdot)$ share?

Exercise. Show that $R'_i(A) = C'_i(A^T)$ and use the triangle inequality to show that $R'_i(\mathrm{H}(A)) = R'_i(\frac{1}{2}[A + A^*]) \le g_i(A)$ for all $i = 1,\dots, n$.

Our Geršgorin-type inclusion result for the field of values is the following:

1.5.2   Theorem. For any $A \in M_n$, $F(A) \subset G_F(A)$.

Proof: The demonstration has three steps. Let *RHP* denote the open right half-plane $\{z \in \mathbb{C}: \operatorname{Re} z > 0\}$. We first show that if $G_F(A) \subset RHP$, then $F(A) \subset RHP$. If $G_F(A) \subset RHP$, then $\operatorname{Re} a_{ii} > g_i(A)$. Let $\mathrm{H}(A) = \frac{1}{2}(A + A^*) \equiv B = [b_{ij}]$. Since $R'_i(A^*) = C'_i(A)$ and $R'_i(B) \le g_i(A)$ (by the preceding exercise), it follows that $b_{ii} = \operatorname{Re} a_{ii} > g_i(A) \ge R'_i(B)$. In particular, $G(B) \subset RHP$. Since $\sigma(B) \subset G(B)$ by Geršgorin's theorem, we have $\sigma(B) \subset RHP$. But since B is Hermitian, $F(\mathrm{H}(A)) = F(B) = \mathrm{Co}(\sigma(B)) \subset RHP$ and hence $F(A) \subset RHP$ by (1.2.5).

We next show that if $0 \notin G_F(A)$, then $0 \notin F(A)$. Suppose $0 \notin G_F(A)$. Since $G_F(A)$ is convex, there is some $\theta \in [0, 2\pi)$ such that $G_F(e^{i\theta}A) = e^{i\theta}G_F(A) \subset RHP$. As we showed in the first step, this means that $F(e^{i\theta}A) \subset RHP$, and, since $F(A) = e^{-i\theta}F(e^{i\theta}A)$, it follows that $0 \notin F(A)$.

Finally, if $\alpha \notin G_F(A)$, then $0 \notin G_F(A - \alpha I)$ since the set function $G_F(\cdot)$ satisfies the translation property. By what we have just shown, it follows that $0 \notin F(A - \alpha I)$ and hence $\alpha \notin F(A)$, so $F(A) \subset G_F(A)$. ⬚

A simple bound for the numerical radius $r(A)$ (1.1.4) follows directly from Theorem (1.5.2).

1.5.3   **Corollary.** For all $A \in M_n$,

$$r(A) \leq \max_{1 \leq i \leq n} \tfrac{1}{2} \sum_{j=1}^{n} (|a_{ij}| + |a_{ji}|)$$

It follows immediately from (1.5.3) that

$$r(A) \leq \tfrac{1}{2} \left[\max_{1 \leq i \leq n} \sum_{j=1}^{n} |a_{ij}| + \max_{1 \leq i \leq n} \sum_{j=1}^{n} |a_{ji}| \right]$$

and the right-hand side of this inequality is just the average of the maximum absolute row and column sum matrix norms, $\||A\||_1$ and $\||A\||_\infty$ (see Section (5.6) of [HJ]).

1.5.4   **Corollary.** For all $A \in M_n$, $r(A) \leq \tfrac{1}{2}(\||A\||_1 + \||A\||_\infty)$.

A norm on matrices is called *spectrally dominant* if, for all $A \in M_n$, it is an upper bound for the spectral radius $\rho(A)$. It is apparent from the spectral containment property (1.2.6) that the numerical radius $r(\cdot)$ is spectrally dominant.

1.5.5   **Corollary.** For all $A \in M_n$, $\rho(A) \leq r(A) \leq \tfrac{1}{2}(\||A\||_1 + \||A\||_\infty)$.

We next discuss a procedure for determining and plotting $F(A)$ numerically. Because the set $F(A)$ is convex and compact, it suffices to determine the boundary of $F(A)$, which we denote by $\partial F(A)$. The general strategy is to calculate many well-spaced points on $\partial F(A)$ and support lines of $F(A)$ at these points. The convex hull of these boundary points is then a convex polygonal approximation to $F(A)$ that is contained in $F(A)$, while the intersection of the half-spaces determined by the support lines is a convex polygonal approximation to $F(A)$ that contains $F(A)$. The area of the region between these two convex polygonal approximations may be thought of as a measure of how well either one approximates $F(A)$. Furthermore, if the boundary points and support lines are produced as one traverses $\partial F(A)$ in one direction, it is easy to plot these two approximating polygons. A pictorial summary of this general scheme is given in Figure (1.5.5.1). The points q_i are at intersections of consecutive support lines and, therefore, are the vertices of the external approximating polygon.

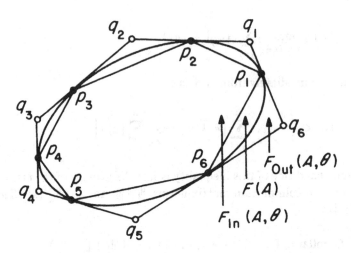

Figure 1.5.5.1

The purpose of the next few observations is to show how to produce boundary points and support lines around $\partial F(A)$.

From (1.2.4) it follows that

$$e^{-i\theta}F(e^{i\theta}A) = F(A) \qquad\qquad (1.5.6)$$

for every $A \in M_n$ and all $\theta \in [0,2\pi)$. Furthermore, we have the following lemma.

1.5.7   Lemma. If $x \in \mathbb{C}^n$, $x^*x = 1$, and $A \in M_n$, the following three conditions are equivalent:

(a) $\operatorname{Re} x^*Ax = \max \{\operatorname{Re} \alpha\colon \alpha \in F(A)\}$

(b) $x^*H(A)x = \max \{r\colon r \in F(H(A))\}$

(c) $H(A)x = \lambda_{max}(H(A))x$

where $\lambda_{max}(B)$ denotes the algebraically largest eigenvalue of the Hermitian matrix B.

Proof: The equivalence of (a) and (b) follows from the calculation Re x^*Ax $= \frac{1}{2}(x^*Ax + x^*A^*x) = x^*H(A)x$ and the projection property (1.2.5). If $\{y_1,..., y_n\}$ is an orthonormal set of eigenvectors of the Hermitian matrix $H(A)$ and if $H(A)y_j = \lambda_j y_j$, then x may be written as

$$x = \sum_{j=1}^{n} c_j y_j, \text{ with } \sum_{j=1}^{n} \bar{c}_j c_j = 1$$

since $x^*x = 1$. Thus,

$$x^*H(A)x = \sum_{j=1}^{n} \bar{c}_j c_j \lambda_j$$

from which the equivalence of (b) and (c) is immediately deduced. $\quad\square$

It follows from the lemma that

$$\max\{\text{Re } \alpha: \alpha \in F(A)\} = \max\{r: r \in F(H(A))\} = \lambda_{max}(H(A)) \quad (1.5.8)$$

This means that the furthest point to the right in $F(H(A))$ is the real part of the furthest point to the right in $F(A)$, which is $\lambda_{max}(H(A))$. A unit vector yielding any one of these values also yields the others.

Lemma (1.5.7) shows that, if we compute $\lambda_{max}(H(A))$ and an associated unit eigenvector x, we obtain a boundary point x^*Ax of $F(A)$ and a support line $\{\lambda_{max}(H(A)) + ti: t \in \mathbb{R}\}$ of the convex set $F(A)$ at this boundary point; see Figure (1.5.8.1).

Using (1.5.6), however, one can obtain as many such boundary points and support lines as desired by rotating $F(A)$ and carrying out the required eigenvalue–eigenvector calculation. For an angle $\theta \in [0, 2\pi)$, we define

$$\lambda_\theta \equiv \lambda_{max}(H(e^{i\theta}A)) \quad (1.5.9)$$

and let $x_\theta \in \mathbb{C}^n$ be an associated unit eigenvector

$$H(e^{i\theta}A)x_\theta = \lambda_\theta x_\theta, \quad x_\theta^*x_\theta = 1 \quad (1.5.10)$$

We denote

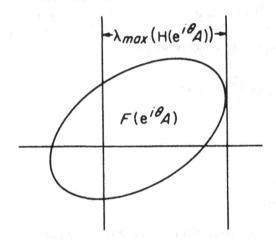

Figure 1.5.8.1

$$L_\theta \equiv \text{the line } \{e^{-i\theta}(\lambda_0 + ti): t \in \mathbb{R}\}$$

and denote the half-plane determined by the line L_0 by

$$H_\theta \equiv \text{the half-plane } e^{-i\theta}\{z: \text{Re } z \le \lambda_0\}$$

Based upon (1.5.6), (1.5.7), and the preceding discussion, we then have the following:

1.5.11   Theorem. For each $A \in M_n$ and each $\theta \in [0,2\pi)$, the complex number $p_\theta \equiv x_\theta^* A x_\theta$ is a boundary point of $F(A)$. The line L_θ is a support line for $F(A)$, with $p_\theta \in L_\theta \cap F(A)$ and $F(A) \subset H_\theta$ for all $\theta \in [0,2\pi)$.

Because $F(A)$ is convex, it is geometrically clear that each extreme point of $F(A)$ occurs as a p_θ and that for any $\alpha \notin F(A)$ there is an L_θ separating $F(A)$ and α, that is, $\alpha \notin H_\theta$. Thus, we may represent $F(A)$ in the following way:

1.5.12   Theorem. For all $A \in M_n$,

$$F(A) = \text{Co}(\{p_\theta : 0 \le \theta < 2\pi\}) = \bigcap_{0 \le \theta < 2\pi} H_\theta$$

Since it is not possible to compute infinitely many points p_θ and lines L_θ, we must be content with a discrete analog of (1.5.12) with equalities replaced by set containments. Let Θ denote a set of angular mesh points, $\Theta = \{\theta_1, \theta_2, ..., \theta_k\}$, where $0 \le \theta_1 < \theta_2 < \cdots < \theta_k < 2\pi$.

1.5.13 **Definition.** Let $A \in M_n$ be given, let a finite set of angular mesh points $\Theta = \{0 \le \theta_1 < \cdots < \theta_k < 2\pi\}$ be given, let $\{p_{\theta_i}\}$ be the associated set of boundary points of $F(A)$ given by (1.5.11), and let $\{H_{\theta_i}\}$ be the half-spaces associated with the support lines L_{θ_i} for $F(A)$ at the points p_{θ_i}. Then we define

$$F_{In}(A,\Theta) \equiv \text{Co}(\{p_{\theta_1}, ..., p_{\theta_k}\}), \text{ and}$$

$$F_{Out}(A,\Theta) \equiv H_{\theta_1} \cap \cdots \cap H_{\theta_k}$$

These are the constructive inner and outer approximating sets for $F(A)$, as illustrated in Figure (1.5.5.1).

1.5.14 **Theorem.** For every $A \in M_n$ and every angular mesh Θ,

$$F_{In}(A,\Theta) \subset F(A) \subset F_{Out}(A,\Theta)$$

The set $F_{Out}(A,\Theta)$ is most useful as an outer estimate if the angular mesh points θ_j are sufficiently numerous and well spaced that the set $\cap \{H_{\theta_i} : 1 \le i \le k\}$ is bounded (which we assume henceforth). In this case, it is also simply determined. Let q_{θ_i} denote the (finite) intersection point of L_{θ_i} and $L_{\theta_{i+1}}$, where $i = 1, ..., k$ and $i = k+1$ is identified with $i = 1$. The existence of these intersection points is equivalent to the assumption that $F_{Out}(A,\Theta)$ is bounded, in which case we have the following simple alternate representation of $F_{Out}(A,\Theta)$:

$$F_{Out}(A,\Theta) = \bigcap_{1 \le i \le k} H_{\theta_i} = \text{Co}(\{q_{\theta_1}, ..., q_{\theta_k}\}) \qquad (1.5.15)$$

Because of the ordering of the angular mesh points θ_j, the points p_{θ_j} and q_{θ_j}

occur consecutively around $\partial F(A)$ for $j = 1,\ldots, k$, and $\partial F_{In}(A,\Theta)$ is just the union of the k line segments $[p_{\theta_1},p_{\theta_2}],\ldots, [p_{\theta_{k-1}},p_{\theta_k}], [p_{\theta_k},p_{\theta_1}]$ while $\partial F_{Out}(A,\Theta)$ consists of the k line segments $[q_{\theta_1},q_{\theta_2}], \ldots, [q_{\theta_{k-1}},q_{\theta_k}], [q_{\theta_k},q_{\theta_1}]$. Thus, each approximating set is easily plotted, and the difference of their areas (or some other measure of their set difference), which is easily calculated (see Problem 10), may be taken as a measure of the closeness of the approximation. If the approximation is not sufficiently close, a finer angular mesh may be used; because of (1.5.12), such approximations can be made arbitrarily close to $F(A)$. It is interesting to note that $F(A)$ (which contains $\sigma(A)$ for all $A \in M_n$), may be approximated arbitrarily closely with only a series of Hermitian eigenvalue–eigenvector computations.

These procedures allow us to calculate the numerical radius $r(A)$ as well. The following result is an immediate consequence of Theorem (1.5.14).

1.5.16   Corollary. For each $A \in M_n$ and every angular mesh Θ,

$$\max_{1 \le i \le k} |p_{\theta_i}| \le r(A) \le \max_{1 \le i \le k} |q_{\theta_i}|$$

Recall from (1.2.11) that $F[A(i')] \subset F(A)$ for $i = 1,\ldots, n$, where $A(i')$ denotes the principal submatrix of $A \in M_n$ formed by deleting row and column i. It follows from (1.2.2) that

$$Co\left[\bigcup_{i=1}^{n} F[A(i')] \right] \subset F(A)$$

A natural question to ask is: How much of the right-hand side does the left-hand side fill up? The answer is "all of it in the limit," as the dimension goes to infinity. In order to describe this fact conveniently, we define Area(S) for a convex subset S of the complex plane to be the conventional area unless S is a line segment (possibly a point), in which case Area(S) is understood to be the length of the line segment (possibly zero). Furthermore, in the following area ratio we take $0/0$ to be 1.

1.5.17   Theorem. For $A \in M_n$ with $n \ge 2$, let $A(i') \in M_{n-1}$ denote the principal submatrix of A obtained by deleting row and column i from A. There exists a sequence of constants $c_2, c_3,\ldots \in [0,1]$ such that for any $A \in M_n$,

$$\frac{\text{Area}\left[\text{Co}\left[\overset{n}{\underset{i=1}{\cup}} F[A(i')]\right]\right]}{\text{Area}[F(A)]} \geq c_n, \qquad n = 2, 3, \ldots$$

and $\lim c_n = 1$ as $n \to \infty$. The constants

$$c_n \equiv \frac{n-2}{7n - 2 - 6[n(n-2)]^{\frac{1}{2}}} \tag{1.5.18a}$$

satisfy these conditions, and so do the constants

$$c'_n \equiv \frac{2n-5}{2n+7} \tag{1.5.18b}$$

Problems

1. Show by example that there is no general containment relation between any two of the sets $\text{Co}(G(A))$, $\text{Co}(G(A^T))$, and $G_F(A)$.

2. Show that neither $r(A) \leq ||| A |||_1$ nor $r(A) \leq ||| A |||_\infty$ holds in general. Comment in light of (1.5.4).

3. Show that $A = \begin{bmatrix} 0 & 2 \\ 0 & 0 \end{bmatrix}$ is a case of equality in Theorem (1.5.2). What are other cases of equality?

4. Show that $F(A) \subset \cap\{G_F(U^*AU) : U \in M_n \text{ is unitary}\}$. Is this containment an equality for all $A \in M_n$? Prove or give a counterexample.

5. Show that (1.5.2) is the best possible inclusion region based solely on the $3n$ pieces of information a_{ii}, $R'_i(A)$, and $C'_i(A)$, $i = 1, \ldots, n$. Let $s(x,y)$ be a given function on $[0,\infty) \times [0,\infty)$, and define

$$G_s(A) = \text{Co}\left[\overset{n}{\underset{i=1}{\cup}} \{z : |z - a_{ii}| \leq s(R'_i(A), C'_i(A))\}\right]$$

If $F(A) \subset G_s(A)$ for all $A \in M_n$, show that $s(x,y) \geq \frac{1}{2}(x + y)$.

6. Let $A = [a_{ij}] \in M_n$ with $a_{ij} = 1$ if exactly one of i, j equals 1, and $a_{ij} = 0$ otherwise. Let $B = [b_{ij}] \in M_n$ with $b_{ij} = 0$ if $i = j$ and $b_{ij} = 1$ if $i \neq j$. Compare $G_F(A)$ and $G_F(B)$ with $F(A)$ and $F(B)$ to give examples of how imprecisely $G_F(\cdot)$ can estimate $F(\cdot)$.

7.   Suppose the diagonal entries of $A \in M_n$ are positive. If A is row diagonally dominant *or* column diagonally dominant, then A is positive stable. Show that DA is positive stable for all positive diagonal D. Show that if A is row diagonally dominant *and* column diagonally dominant, then PA is positive stable for all positive definite matrices P. See (1.0.3).

8.   A real square matrix with nonnegative entries and all row and column sums equal to $+1$ is said to be *doubly stochastic* (see Section (8.7) of [HJ]). Let $A \in M_n$ be doubly stochastic.

   (a) Use (1.5.2) to show that $F(A)$ is contained in the unit disc. Is there any smaller disc that contains $F(A)$ for all doubly stochastic $A \in M_n$?

   (b) Let $P \in M_n$ be a permutation matrix. Observe that P is doubly stochastic and describe $F(P)$ explicitly in terms of the eigenvalues of P; where are they?

   (c) Now use Birkhoff's theorem (8.7.1) in [HJ], (1.2.4), and (1.2.7) to give another proof that $F(A)$ is contained in the unit disc.

   (d) From what you have proved about $F(A)$, what can you say about the location of the eigenvalues of a doubly stochastic matrix? How does this compare with what you already know from the Perron-Frobenius theorem (8.3.1) in [HJ]? See Problems 18-21 for a sharper statement about $\sigma(A)$.

9.   Show that the points q_{0_j} in (1.5.15) are given by

$$q_{0_j} = e^{-i\theta_j} \left[\lambda_{0_j} + i \frac{\lambda_{0_j} \cos \delta_j - \lambda_{0_{j+1}}}{\sin \delta_j} \right]$$

where $\delta_j \equiv \theta_{j+1} - \theta_j$.

10. If P is a k-sided polygon in the complex plane with vertices $c_1, \ldots, c_k$ given in counterclockwise order, show that the area of P may be written as

$$\text{Area}(P) = \tfrac{1}{2} \text{Im} \left[\bar{c}_1 c_2 + \bar{c}_2 c_3 + \cdots + \bar{c}_{k-1} c_k + \bar{c}_k c_1 \right]$$

Use this identity to formulate the accuracy estimate

$$\text{Area}(F_{Out}(A, \Theta)) - \text{Area}(F_{In}(A, \Theta))$$

in terms of the points p_{0_i} and q_{0_i}. What other measures of accuracy might be calculated easily?

11. If $A \in M_n$, show that the numerical radius of A is given by

$$r(A) = \max_{0 \le \theta < 2\pi} |p_\theta| = \max_{0 \le \theta < 2\pi} |\lambda_\theta| = \max_{0 \le \theta < 2\pi} \lambda_\theta$$

12. If $A \in M_n$, show that $F(A)$ is inscribed in a rectangle with vertical sides parallel to the imaginary axis going through the smallest and largest eigenvalues of $H(A)$, respectively, and with horizontal sides parallel to the real axis going through i times the smallest and largest eigenvalues, respectively, of $-iS(A)$. In particular, conclude that $\max\{\operatorname{Re}\lambda\colon \lambda \in \sigma(A)\} \le \max\{\lambda\colon \lambda \in \sigma(H(A))\}$.

13. For $A = \begin{bmatrix} -1 & 2 \\ 0 & 1 \end{bmatrix}$ and angular mesh $\Theta = \{0, \pi/3, 2\pi/3, \pi, 4\pi/3, 5\pi/3\}$, explicitly determine and sketch $F_{In}(A,\Theta)$ and $F_{Out}(A,\Theta)$.

14. If $A \in M_n$ is normal, describe a simpler procedure for plotting $F(A)$.

15. Let $A \in M_n$ have constant main diagonal, and let $A_i \in M_{n-1}$ denote the principal submatrix formed by deleting row and column i from A. Show that

$$\operatorname{Area}\left[\operatorname{Co}\left[\bigcup_{i=1}^{n} F(A_i)\right]\right] / \operatorname{Area}(F(A)) \ge \frac{n-2}{n}$$

that is, a possible choice of the constants c_n in Theorem (1.5.17) in this special case is $c_n = (n-2)/n$.

16. Let $A \in M_n$ be a given Hermitian matrix. If A is positive definite, show that every principal submatrix $A_i \in M_{n-1}$ (formed by deleting row i and column i from A) is also positive definite. Show by example that the simple converse of this statement is false: Each A_i can be positive definite and A indefinite. Now let α denote the smallest and β the largest of all the eigenvalues of the A_i, $i = 1,\dots, n$, and let $\{c_k\}_{k=1}^{\infty}$ be any sequence of constants satisfying the conditions in Theorem (1.5.17). Show that if all A_i are positive definite and $(\alpha/\beta) > 1 - c_n$, then A is positive definite.

17. Consider the constants c_n and c_n' defined in (1.5.18a,b). Show that the sequence $\{c_n'\}$ increases monotonically to 1, $c_n \ge c_n'$ for $n = 2, 3,\dots,$ $\lim c_n/c_n' = 1$ as $n \to \infty$, and the approximation $c_n \sim c_n'$ as $n \to \infty$ is quite good since $c_n - c_n' = 9/(2n^3) + O(1/n^4)$.

The following four problems improve Problem 8's simple estimate that the field of values of a doubly stochastic matrix is contained in the unit disc.

18. Let $m \geq 2$ and let $Q = [q_{ij}] \in M_m$ be a *cyclic permutation matrix*, that is, $q_{i,i+1} = 1$ for $i = 1,..., m-1$, $q_{m,1} = 1$, and all other $q_{ij} = 0$. What does this have to do with a cyclic permutation of the integers $\{1,..., m\}$? Show that $F(Q)$ is the convex hull of the m th roots of unity $e^{2\pi i k/m}$, $k = 1,..., m$, which is contained in the angular wedge L_m in the complex plane with vertex at $z = 1$ and passing through $e^{\pm 2\pi i/m}$:

$$L_m \equiv \left\{ z = 1 + r e^{i\theta} : r \geq 0, \; |\theta - \pi| \leq \frac{m-2}{2m}\pi \right\} \qquad (1.5.19)$$

Sketch L_m. Notice that the directed graph of Q contains a *simple cycle of length* m, that is, a directed path that begins and ends at the same node, which occurs exactly twice in the path, and no other node occurs more than once in the path; see (6.2.12) in [HJ].

19. For any $B \in M_n$, define $\mu(B) \equiv$ the length of a maximum-length simple cycle in the directed graph of B. Notice that $0 \leq \mu(B) \leq n$, and $\mu(B) = n$ if and only if B is irreducible. Now let $B_1, B_2,..., B_k \in M_n$ be nonnegative matrices, and let $\alpha_1, \alpha_2,..., \alpha_k > 0$. Show that $\mu(\alpha_1 B_1 + \cdots + \alpha_k B_k) \geq \max \{\mu(B_1),..., \mu(B_k)\}$.

20. Let $P \in M_n$ be a given permutation matrix. Since any permutation of the integers $\{1,..., n\}$ is a product of cyclic permutations, explain why P is permutation similar (hence, unitarily similar) to a direct sum of cyclic permutation matrices $Q_1,..., Q_k$, $Q_i \in M_{n_i}$, $1 \leq n_i \leq n$, $n_1 + \cdots + n_k = n$, where $\mu(P) = \max \{n_1,..., n_k\}$. Use (1.2.10) to show that $F(P)$ is contained in the wedge $L_{\mu(P)}$ defined by (1.5.19).

21. Let $A \in M_n$ be a given doubly stochastic matrix. Use Birkhoff's theorem (8.7.1) in [HJ] to write A as a finite positive convex combination of permutation matrices, that is,

$$A = \sum_i \alpha_i P_i, \quad \text{all } \alpha_i > 0, \quad \sum_i \alpha_i = 1$$

Although there may be many ways to express A as such a sum, show that $\mu(A) \geq \max_i \mu(P_i)$ always, and hence we have the uniform upper bound $F(P_i) \subset L_{\mu(A)} \subset L_n$ for all i. Use (1.2.4) and (1.2.7) to show that $F(A) \subset$

$L_{\mu(A)} \subset L_n$. Conclude that $F(A)$ is contained in the part of the wedge $L_{\mu(A)}$ (and L_n) that is contained in the unit disc. Sketch this set. What does this say when $n = 2$? When $\mu(A) = 2$? When A is tridiagonal? From what you have proved about $F(A)$, what can you say about $\sigma(A)$? Do not use the fact that a doubly stochastic tridiagonal matrix must be symmetric.

22. Let $A \in M_n(\mathbb{R})$ have nonnegative entries and spectral radius 1. Let $\mu(A)$ denote the length of the maximum-length simple cycle in the directed graph of A. It is a theorem of Kellogg and Stephens that $\sigma(A)$ is contained in the intersection of the unit disc and the angular wedge $L_{\mu(A)}$ defined by (1.5.19).

   (a) Compare this statement with the result of Problem 21.

   (b) If $\mu(A) = 2$, use the Kellogg-Stephens theorem to show that all the eigenvalues of A are real, and deduce that all the eigenvalues of a nonnegative tridiagonal matrix are real. See Problem 5 of Section (4.1) of [HJ] for a different proof of this fact.

   (c) Use the Kellogg-Stephens theorem to show that all the eigenvalues of a general n-by-n nonnegative matrix with spectral radius 1 are contained in the intersection of the unit disc and the wedge L_n. Sketch this set. A precise, but rather complicated, description of the set of all possible eigenvalues of a nonnegative matrix with spectral radius 1 has been given by Dmitriev, Dynkin, and Karpelevich.

23. Verify the following facts about the *numerical radius* function $r(\cdot)$: $M_n \to \mathbb{R}_+$ defined by $r(A) = \max \{ |z| : z \in F(A) \}$. We write $\rho(A)$ for the spectral radius, $\||\, A\, \||_2$ for the spectral norm, $\| A \|_2$ for the Frobenius norm, and $\||\, A\, \||_1$, $\||\, A\, \||_\infty$ for the maximum column sum and maximum row sum matrix norms, respectively. The following statements hold for all $A, B \in M_n$. See Section (5.7) of [HJ] and its problems for background information about the numerical radius and its relationship to other norms.

(a) $r(\cdot)$ is a norm on M_n but is not a matrix norm.

(b) $4r(\cdot)$ is submultiplicative with respect to the ordinary matrix product, that is, $4r(AB) \le 4r(A)4r(B)$, and 4 is the least positive constant with this property. Thus, $4r(\cdot)$ is a matrix norm on M_n.

(c) $2r(\cdot)$ is submultiplicative with respect to the Hadamard product, that is, $2r(A \circ B) \le 2r(A)2r(B)$, and 2 is the least positive constant with this property, but if either A or B is normal then $r(A \circ B) \le r(A)r(B)$; see Corollaries (1.7.24-25).

(d) $r(A^m) \le [r(A)]^m$ for all $m = 1, 2, \dots$; this is the *power inequality* for the numerical radius. Give an example to show that it is *not* always true that $r(A^{k+m}) \leq r(A^k)r(A^m)$.

(e) $\rho(A) \leq r(A)$.

(f) $r(A) = r(A^*)$.

(g) $\frac{1}{2}||| A |||_2 \leq r(A) \leq ||| A |||_2$ and both bounds are sharp.

(h) $c_n || A ||_2 \leq r(A) \leq || A ||_2$ with $c_n = (4n)^{-\frac{1}{2}}$, and the upper, but not the lower, bound is sharp. A sharp lower bound is given by $c_n = (2n)^{-\frac{1}{2}}$ if n is even and $c_n = (2n-1)^{-\frac{1}{2}}$ if n is odd.

(i) $r(A) \leq ||| A |||_2 \leq (||| A ||| \; ||| A^* |||)^{\frac{1}{2}}$ for any matrix norm $||| \cdot |||$ on M_n. In particular, $r(A) \leq ||| A |||_2 \leq (||| A |||_1 \; ||| A |||_\infty)^{\frac{1}{2}} \leq \frac{1}{2}(||| A |||_1 + ||| A |||_\infty)$, an inequality discovered by I. Schur; see Problem 21 in Section (5.6) of [HJ].

(j) $r(\hat{A}) \leq r(A)$ for any principal submatrix $\hat{A}$ of A.

(k) $r(A_1 \oplus A_2) = \max \{ r(A_1), r(A_2) \}$ for any $A_1 \in M_k$ and $A_2 \in M_m$.

(l) $r(A) \leq r(|A|)$, where $A = [a_{ij}]$ and $|A| \equiv [|a_{ij}|]$.

(m) If $A \geq 0$, that is, all the entries of A are real and nonnegative, then $\rho(A) \leq r(A) = r(\mathrm{H}(A)) = \rho(\mathrm{H}(A))$, where $\mathrm{H}(A) \equiv \frac{1}{2}(A + A^T)$ is the symmetric part of A; moreover, $r(A) = \max \{ x^T A x : x \in \mathbb{R}^n, x \geq 0, x^T x = 1 \}$.

(n) $r(A) \leq r(|A|) = \frac{1}{2}\rho(|A| + |A|^T)$, where $A = [a_{ij}]$ and $|A| \equiv [|a_{ij}|]$.

(o) $||| A^m |||_2 \leq 2\, r(A)^m$ for all $m = 1, 2, \dots$.

24. Let $A \in M_n$ be given. In the preceding problem we saw that $\rho(A) \leq r(A) \leq ||| A |||_2$, so $\rho(A) = ||| A |||_2$ implies $r(A) = ||| A |||_2$. A. Wintner proved in 1929 that the converse is also true: $r(A) = ||| A |||_2$ if and only if $\rho(A) = ||| A |||_2$. Prove this. A matrix $A \in M_n$ is said to be *radial* (some authors use the term *radialoid*) if $\rho(A) = ||| A |||_2$, or, equivalently, if $r(A) = ||| A |||_2$.

25. Provide details for the following sketch for a proof of a theorem of Ptak: Let $A \in M_n$ be given with $||| A |||_2 \leq 1$. Then $\rho(A) = 1$ if and only if $||| A^n |||_2 = 1$. *Proof:* $S \equiv \{ x \in \mathbb{C}^n : || x ||_2 = || Ax ||_2 \}$ is a subspace of $\mathbb{C}^n$ since it is the nullspace of the positive semidefinite matrix $I - A^* A$. If $||| A^n |||_2 = 1$, let $x_0 \in \mathbb{C}^n$ be a unit vector such that $|| A^n x_0 ||_2 = || x_0 ||_2$. Then $|| x_0 ||_2 = || A^n x_0 ||_2 \leq || A^{n-1} x_0 ||_2 \leq \cdots \leq || A x_0 ||_2 \leq || x_0 ||_2$, so all these inequalities are equalities and $S_0 \equiv \mathrm{Span} \{ x_0, A x_0, \dots, A^{n-1} x_0 \} \subset S$, $k \equiv \dim S_0 \geq 1$, and $A S_0 \subset S_0$

by the Cayley-Hamilton theorem. Let $U = [U_1 \ U_2] \in M_n$ be a unitary matrix such that the columns of $U_1 \in M_{n,k}$ form an orthonormal basis of S_0. Then $U^* A U$ is a block triangular matrix of the form

$$U^* A U = \begin{bmatrix} U_1^* A U_1 & * \\ 0 & * \end{bmatrix}$$

where $U_1^* A U_1$ is a Euclidean isometry on S_0 (a linear transformation from S_0 to S_0 that preserves the Euclidean length of every vector) and hence is unitary (see (2.1.4-5) in [HJ]). Thus, $1 = \rho(U_1^* A U_1) \leq \rho(U^* A U) = \rho(A) \leq ||| A |||_2 \leq 1$.

26. Modify the proof of the preceding problem to show that if $A \in M_n$ has $||| A |||_2 \leq 1$, then $\rho(A) = 1$ if and only if $||| A^m |||_2 = 1$ for some positive integer m greater than or equal to the degree of the minimal polynomial of A.

27. Let $A \in M_n$ be given. Combine the results of the preceding three problems and verify that the following conditions are equivalent:

(a) A is radial.
(b) $\rho(A) = ||| A |||_2$.
(c) $r(A) = ||| A |||_2$.
(d) $||| A^n |||_2 = (||| A |||_2)^n$.
(e) $||| A^m |||_2 = (||| A |||_2)^m$ for some integer m not less than the degree of the minimal polynomial of A.
(f) $||| A^k |||_2 = (||| A |||_2)^k$ for all $k = 1, 2, \ldots$.

In addition, one more equivalent condition follows from Problem 37(f) in Section (1.6); see Problem 38 in Section (1.6).

(g) A is unitarily similar to $||| A |||_2 (U \oplus B)$, where $U \in M_k$ is unitary, $1 \leq k \leq n$, and $B \in M_{n-k}$ has $\rho(B) < 1$ and $||| B |||_2 \leq 1$.

28. (a) If $A \in M_n$ is normal, show that A is radial. If A is radial and $n = 2$, show that A is normal. If $n \geq 3$, exhibit a radial matrix in M_n that is not normal.
(b) Show that if $A \in M_n$ is radial, then every positive integer power of A is radial, but it is possible for some positive integer power of a matrix to be radial without the matrix being radial.

29. Let $J_n(0)$ denote the n-by-n nilpotent Jordan block considered in Problem 9 in Section (1.3), where it was shown that $F(J_n(0))$ is a closed disc centered at the origin with radius $r(J_n(0)) = \rho(\mathrm{H}(J_n(0)))$. If k is a positive integer such that $2k \le n$, show that $2^{-1/k} \le r(J_n(0)) < 1$. Describe $F(J_n(0))$ as $n \to \infty$.

30. A given $A \in M_{m,n}$ is called a *contraction* if its spectral norm satisfies $||| A |||_2 \le 1$; it is a *strict contraction* if $||| A |||_2 < 1$. T. Ando has proved two useful representations for the unit ball of the numerical radius norm in M_n:

(a) $r(A) \le 1$ if and only if

$$A = (I - Z)^{\frac{1}{2}} C (I + Z)^{\frac{1}{2}} \tag{1.5.20}$$

for some contractions $C, Z \in M_n$ with Z Hermitian; the indicated square roots are the unique positive semidefinite square roots of $I \pm Z$.

(b) $r(A) \le 1$ if and only if

$$A = 2(I - C^* C)^{\frac{1}{2}} C \tag{1.5.21}$$

for some contraction $C \in M_n$.

Verify half of Ando's theorem—show that both representations (1.5.20, 21) give matrices A for which $r(A) \le 1$. See Problem 44 in Section (3.1) for an identity related to (1.5.21).

31. Use Ando's representation (1.5.20) to show the following for $A \in M_n$:

(a) $r(A^m) \le [r(A)]^m$ for all $m = 1, 2, \dots$. This is the famous *power inequality* for the numerical radius; see Problem 23(d).

(b) $r(A) \le 1$ if and only if there is a Hermitian $Z \in M_n$ such that the 2-by-2 block matrix

$$\begin{bmatrix} I - Z & A \\ A^* & I + Z \end{bmatrix}$$

is positive semidefinite.

See Lemma (3.5.12).

32. Let $p(t)$ be a polynomial, all of whose coefficients are nonnegative. Show that $r(p(A)) \le p(r(A))$ for all $A \in M_n$, and that the same inequality holds if the numerical radius $r(\cdot)$ is replaced by the spectral radius $\rho(\cdot)$ or by any vector norm $||\cdot||$ on M_n that satisfies the power inequality $||A^k|| \le$

$\|A\|^k$ for $k = 1, 2,\dots$ for all $A \in M_n$.

Notes and Further Readings. Most of this section is based on A Geršgorin Inclusion Set for the Field of Values of a Finite Matrix, *Proc. Amer. Math. Soc.* 41 (1973), 57–60, and Numerical Determination of the Field of Values of a General Complex Matrix, *SIAM J. Numer. Anal.* 15 (1978), 595–602, both by C. R. Johnson. See also C. R. Johnson, An Inclusion Region for the Field of Values of a Doubly Stochastic Matrix Based on Its Graph, *Aequationes Mathematicae* 17 (1978), 305–310. For a proof of Theorem (1.5.17), see C. R. Johnson, Numerical Ranges of Principal Submatrices, *Linear Algebra Applic.* 37 (1981), 11–22; the best possible value for the lower bound c_n in each dimension n is not yet known. For a proof of the theorem mentioned at the beginning of Problem 22, see R. B. Kellogg and A. B. Stephens, Complex Eigenvalues of a Non-Negative Matrix with a Specified Graph, *Linear Algebra Appl.* 20 (1978), 179–187. For the precise description of the set of all possible eigenvalues of n-by-n nonnegative matrices mentioned at the end of Problem 22, see N. Dmitriev and E. Dynkin, Eigenvalues of a Stochastic Matrix, *Izv. Akad. Nauk SSSR Ser. Mat.* 10 (1946), 167–184, and F. I. Karpelevich, On the Eigenvalues of a Matrix with Non-Negative Elements, *Izv. Akad. Nauk SSSR Ser. Mat.* 15 (1951), 361–383. A discussion of the bounds in Problem 23 (h) is in C. R. Johnson and C.-K. Li, Inequalities Relating Unitarily Invariant Norms and the Numerical Range, *Linear Multilinear Algebra* 23 (1988), 183–191. The theorem mentioned in Problem 24 is in A. Wintner, Zur Theorie der beschränkten Bilinearformen, *Math. Zeit.* 30 (1929), 228–282. Ptak's theorem mentioned in Problem 25 was first published in 1960; the proof we have outlined, and many interesting related results, is in V. Ptak, Lyapunov Equations and Gram Matrices, *Linear Algebra Applic.* 49 (1983), 33–55. For surveys of results about the numerical radius and field of values, generalizations, applications to finite difference methods for hyperbolic partial differential equations, and an extensive bibliography, see M. Goldberg, On Certain Finite Dimensional Numerical Ranges and Numerical Radii, *Linear Multilinear Algebra* 7 (1979), 329–342, as well as M. Goldberg and E. Tadmor, On the Numerical Radius and its Applications, *Linear Algebra Applic.* 42 (1982), 263–284. The representations (1.5.20, 21) for the unit ball of the numerical radius norm in M_n are in T. Ando, Structure of Operators with Numerical Radius One, *Acta. Sci. Math. (Szeged)* 34 (1973), 11–15.

1.6   Geometry

In Section (1.3) we saw that if $A \in M_2$, then the field of values $F(A)$ is a (possibly degenerate) ellipse with interior and that the ellipse may be determined in several ways from parameters associated with the entries of the matrix A.

Exercise. If $A = \begin{bmatrix} \lambda_1 & \beta \\ 0 & \lambda_2 \end{bmatrix} \in M_2$, show that $F(A)$ is: (a) a point if and only if $\lambda_2 = \lambda_1$ and $\beta = 0$; (b) a line segment joining λ_1 and λ_2 if and only if $\beta = 0$; (c) a circular disc of radius $\frac{1}{2}|\beta|$ if and only if $\lambda_2 = \lambda_1$; or (d) an ellipse (with interior) with foci at λ_1 and λ_2 otherwise, and λ_1 and λ_2 are interior points of the ellipse in this case.

In higher dimensions, however, a considerably richer variety of shapes is possible for the field of values. Any convex polygon is the field of values of a matrix of sufficiently high dimension; by (1.2.9), one can use a normal matrix whose eigenvalues are the vertices of the polygon. Thus, any bounded convex set can be approximated as closely as desired by the field of values of some matrix, but the dimension of the matrix may have to be large. Further interesting shapes can be pieced together using the direct sum property (1.2.10); if two sets occur, then so does their convex hull, but a direct sum of higher dimension may be required to realize it. Except for $n = 1$ and 2, it is not known which compact convex sets in $\mathbb{C}$ occur as fields of values for matrices of a fixed finite dimension n. Our purpose in this section is to present a few basic facts about the shape of $F(A)$ and its relation to $\sigma(A)$. These results elaborate upon some of the basic properties in (1.2).

The relative interior of a set in $\mathbb{C}$ is just its interior relative to the smallest dimensional space in which it sits. A compact convex set in the complex plane is either a point (with no relative interior), a nontrivial closed line segment (whose relative interior is an open line segment), or a set with a two–dimensional interior in the usual sense. What can be said about interior points of $F(A)$? If $A \in M_n$, the point $\frac{1}{n} \operatorname{tr} A$ always lies in $F(A)$ since it is a convex combination of points in $F(A)$ (the eigenvalues as well as the main diagonal entries of A), but somewhat more can be said. The key observation is that if a convex set C is contained in the closed upper half-plane and if a strict convex combination of given points in C is real, then all of these points must be real.

1.6.1   Theorem Let $A = [a_{ij}] \in M_n$ be given, let $\lambda_1, \ldots, \lambda_n$ denote the eigenvalues of A, and let $\mu_1, ..., \mu_n$ be given positive real numbers such that $\mu_1 + \cdots + \mu_n = 1$. The following are equivalent:

(a) A is not a scalar multiple of the identity.
(b) $\mu_1 a_{11} + \cdots + \mu_n a_{nn}$ is in the relative interior of $F(A)$.
(c) $\mu_1 \lambda_1 + \cdots + \mu_n \lambda_n$ is in the relative interior of $F(A)$.

In particular, we have the following dichotomy: Either A is a scalar multiple of the identity or the point $\frac{1}{n} \operatorname{tr} A$ lies in the relative interior of $F(A)$.

Proof: If A is a scalar multiple of the identity, then $F(A)$ has no relative interior, so both (b) and (c) imply (a).

Let $\zeta \equiv \mu_1 a_{11} + \cdots + \mu_n a_{nn}$, observe that $\zeta \in F(A)$, and suppose that ζ is not in the relative interior of $F(A)$. Let $B = [b_{ij}] \equiv A - \zeta I$ and observe that $\mu_1 b_{11} + \cdots + \mu_n b_{nn} = 0$, $0 \in F(B)$, and 0 is not in the relative interior of $F(B)$. Since $F(B)$ is convex, it can be rotated about the boundary point 0 so that, for some $\theta \in \mathbb{R}$, the field of values of $C = [c_{ij}] \equiv e^{i\theta} B$ lies in the closed upper half-plane $UHP \equiv \{z \in \mathbb{C} : \operatorname{Im} z \geq 0\}$. Then $0 \in F(C) \subset UHP$, 0 is not in the relative interior of $F(C)$, and $\mu_1 c_{11} + \cdots + \mu_n c_{nn} = 0$. Since each $c_{ii} \in F(C)$, $\operatorname{Im} c_{ii} \geq 0$ for all $i = 1, ..., n$. But $\mu_1 \operatorname{Im} c_{11} + \cdots + \mu_n \operatorname{Im} c_{nn} = 0$ and each $\mu_i > 0$, so all the main diagonal entries of C are real. For any given indices $i, j \in \{1, ..., n\}$, let Γ_{ij} denote the 2-by-2 principal submatrix of C obtained as the intersections of rows and columns i and j, and let λ_1, λ_2 denote its eigenvalues. Since $F(\Gamma_{ij}) \subset F(C) \subset UHP$, we have $\lambda_1, \lambda_2 \in UHP$. But $\operatorname{Im}(\lambda_1) + \operatorname{Im}(\lambda_2) = \operatorname{Im}(\lambda_1 + \lambda_2) = \operatorname{Im}(\operatorname{tr} \Gamma_{ij}) = 0$, so λ_1 and λ_2 are both real and neither is an interior point of $F(\Gamma_{ij}) \subset UHP$. It follows from Theorem (1.3.6(d)) that Γ_{ij} is normal; it is actually Hermitian since its eigenvalues are real. Since i and j are arbitrary indices, it follows that C is Hermitian and hence $F(C)$ is a real interval, which must have 0 as an endpoint since 0 is not in the relative interior of $F(C)$. Thus, every point in $F(C)$, in particular, every c_{ii} and every λ_i, has the same sign. But $\mu_1 c_{11} + \cdots + \mu_n c_{nn} = 0$ and all $\mu_i > 0$, so all $c_{ii} = 0$. Since $\lambda_1 + \cdots + \lambda_n = c_{11} + \cdots + c_{nn}$, it follows that all $\lambda_i = 0$ as well, and hence $C = 0$. We conclude that $A = \zeta I$ and hence (a) implies (b).

To show that (a) implies (c), suppose $\zeta \equiv \mu_1 \lambda_1 + \cdots + \mu_n \lambda_n$ is not in the relative interior of $F(A)$. Choose a unitary $U \in M_n$ such that $\mathcal{A} \equiv UAU^*$ is upper triangular and has main diagonal entries $\lambda_1, ..., \lambda_n$. Then $F(A) = F(\mathcal{A})$, and a strict convex combination of the main diagonal entries of $\mathcal{A}$ is not in the relative interior of its field of values. By the equivalence of (a) and (b) we conclude that $\mathcal{A} = \zeta I$ and hence $A = U^* \mathcal{A} U = \zeta I$. $\square$

We next investigate the smoothness of the boundary of the field of values $\partial F(A)$ and its relationship with $\sigma(A)$. Intuitively, a "sharp point" of a convex set S is an extreme point at which the boundary takes an abrupt turn, a boundary point where there are nonunique tangents, or a "corner."

1.6.2   Definition. Let $A \in M_n$. A point $\alpha \in \partial F(A)$ is called a *sharp point* of $F(A)$ if there are angles θ_1 and θ_2 with $0 \leq \theta_1 < \theta_2 < 2\pi$ for which

$$\text{Re } e^{i\theta}\alpha = \max \{\text{Re } \beta : \beta \in F(e^{i\theta}A)\} \text{ for all } \theta \in (\theta_1, \theta_2)$$

1.6.3   Theorem. Let $A \in M_n$. If α is a sharp point of $F(A)$, then α is an eigenvalue of A.

Proof: If α is a sharp point of $F(A)$, we know from (1.5.7) that there is a unit vector $x \in \mathbb{C}^n$ such that

$$x^* H(e^{i\theta}A)x = \lambda_{max}(H(e^{i\theta}A)) \qquad \text{for all } \theta \in (\theta_1, \theta_2)$$

from which it follows that (see (1.5.9))

$$H(e^{i\theta}A)x = \lambda_\theta x \qquad \text{for all } \theta \in (\theta_1, \theta_2) \quad (1.6.3a)$$

The vector x is independent of θ. Differentiation of (1.6.3a) with respect to θ yields

$$H(i e^{i\theta}A) x = \lambda_\theta' x$$

which is equivalent to

$$S(e^{i\theta}A) x = -i\lambda_\theta' x \qquad\qquad (1.6.3b)$$

Adding (1.6.3a) and (1.6.3b) gives $e^{i\theta}Ax = (\lambda_\theta - i\lambda_\theta')x$, or

$$Ax = e^{-i\theta}(\lambda_\theta - i\lambda_\theta')x$$

Interpreted as an evaluation at any θ in the indicated interval, this means that

$$\alpha = x^* Ax = e^{-i\theta}(\lambda_\theta - i\lambda_\theta')$$

is an eigenvalue of A.   ☐

Since each $A \in M_n$ has a finite number of eigenvalues, Theorem (1.6.3) implies that if a set is the field of values of a finite matrix, it can have only a finite number of sharp points. Moreover, if the only extreme points of $F(A)$ are sharp points, then $F(A)$ is the convex hull of some eigenvalues of A.

1.6.4   Corollary. Let $A \in M_n$ be given. Then $F(A)$ has at most n sharp points, and $F(A)$ is a convex polygon if and only if $F(A) = \mathrm{Co}(\sigma(A))$.

Although every sharp point on the boundary of $F(A)$ is an eigenvalue, not every eigenvalue on the boundary of $F(A)$ is a sharp point. Nevertheless, every such point does have a special characteristic.

1.6.5   Definition. A point $\lambda \in \sigma(A)$ is a *normal eigenvalue* for the matrix $A \in M_n$ if

(a) Every eigenvector of A corresponding to λ is orthogonal to every eigenvector of A corresponding to each eigenvalue different from λ, and

(b) The geometric multiplicity of the eigenvalue λ (the dimension of the corresponding eigenspace of A) is equal to the algebraic multiplicity of λ (as a root of the characteristic polynomial of A).

Exercise. Show that every eigenvalue of a normal matrix is a normal eigenvalue.

Exercise. If $A \in M_n$ has as many as $n-1$ normal eigenvalues, counting multiplicity, show that A is a normal matrix.

Exercise. Show that λ is a normal eigenvalue of a matrix $A \in M_n$ if and only if it is a normal eigenvalue of UAU^* for every unitary matrix $U \in M_n$, that is, the property of being a normal eigenvalue is a unitary similarity invariant.

Our main observation here is that every eigenvalue on the boundary of the field of values is a normal eigenvalue.

1.6.6   Theorem. If $A \in M_n$ and if $\alpha \in \partial F(A) \cap \sigma(A)$, then α is a normal eigenvalue of A. If m is the multiplicity of α, then A is unitarily similar to $\alpha I \oplus B$, with $I \in M_m$, $B \in M_{n-m}$, and $\alpha \notin \sigma(B)$.

Proof: If the algebraic multiplicity of α is m, then A is unitarily similar to an upper triangular matrix T (this is Schur's theorem; see (2.3.1) in [HJ]) whose first m main diagonal entries are equal to α and whose remaining diagonal entries (all different from α) are the other eigenvalues of A. Suppose there were a nonzero entry off the main diagonal in one of the first m rows of T. Then T would have a 2-by-2 principal submatrix T_0 of the form $T_0 = \begin{bmatrix} \alpha & \gamma \\ 0 & \beta \end{bmatrix}$, $\gamma \neq 0$. Since $F(T_0)$ is either a circular disc about α with radius $\tfrac{1}{2}|\gamma|$ or a nondegenerate ellipse (with interior) with foci at α and β, the point α must be in the interior of $F(T_0)$. But $F(T_0) \subset F(T) = F(A)$ by (1.2.8) and (1.2.11), which means that α is in the interior of $F(A)$. This contradiction shows that there are no nonzero off-diagonal entries in the first m rows of T, and hence $T = \alpha I \oplus B$ with $I \in M_m$ and $B \in M_{n-m}$. The remaining assertions are easily verified.   □

Exercise. Complete the details of the proof of Theorem (1.6.6).

We have already noted that the converse of the normality property (1.2.9) does not hold, but we are now in a position to understand fully the relationship between the two conditions on $A \in M_n$: A is normal, and $F(A) = \mathrm{Co}(\sigma(A))$.

1.6.7   Corollary. If $\sigma(A) \subset \partial F(A)$, then A is normal.

Proof: If all the eigenvalues of A are normal eigenvalues, then there is an orthonormal basis of eigenvectors of A, which must therefore be normal.   □

Exercise. Show that a given matrix $A \in M_n$ is normal if and only if A is unitarily similar to a direct sum $A_1 \oplus A_2 \oplus \cdots \oplus A_k$ with $\sigma(A_i) \subset \partial F(A_i)$, $i = 1, \ldots, k$.

Two further corollary facts complete our understanding of the extent to which there is a converse to (1.2.9).

1.6.8   Theorem. Let $A \in M_n$. Then $F(A) = \mathrm{Co}(\sigma(A))$ if and only if *either* A is normal *or* A is unitarily similar to a matrix of the form $\begin{bmatrix} A_1 & 0 \\ 0 & A_2 \end{bmatrix}$, where A_1 is normal and $F(A_2) \subset F(A_1)$.

1.6.9   Corollary. If $A \in M_n$ and $n \leq 4$, then A is normal if and only if

$F(A) = \text{Co}(\sigma(A))$.

Exercise. Supply a proof for (1.6.8) using (1.6.6).
Exercise. Supply a proof for (1.6.9) by considering the geometrical possibilities allowed in (1.6.8).

We wish to discuss one additional connection between normality and the field of values. If $A \in M_{m,n}$ is given and if $B = \begin{bmatrix} A & * \\ * & * \end{bmatrix}$ is a larger matrix that contains A in its upper-left corner, we say that B is a *dilation* of A; by the submatrix inclusion property (1.2.11), $F(A) \subset F(B)$ whenever $B \in M_N$ is a dilation of $A \in M_n$. One can always find a $2n$-by-$2n$ *normal* dilation of A, for example, the matrix B defined by (1.6.11), in which case $F(B) = \text{Co}(\sigma(B))$. It is clear that $F(A)$ is contained in the intersection of the fields of values of all the normal dilations of A, and it is a pleasant observation that this intersection is *exactly* $F(A)$.

1.6.10   Theorem. Let $A \in M_n$ be given. Then

$$F(A) = \bigcap \left\{ F(B) \colon B = \begin{bmatrix} A & * \\ * & * \end{bmatrix} \in M_{2n} \text{ is normal} \right\}$$

Proof: Since we have already noted that $F(A)$ is a subset of the given intersection, we need only prove the reverse inclusion. Because $F(A)$ is closed and convex, it is the intersection of all the closed half-planes that contain it, a fact used in the preceding section to develop a numerical algorithm to compute $F(A)$. Thus, it is sufficient to show that for each closed half-plane that contains $F(A)$, there is some normal matrix $B = \begin{bmatrix} A & * \\ * & * \end{bmatrix} \in M_{2n}$ such that $F(B)$ is contained in the same half-plane. By the translation and scalar multiplication properties (1.2.3-4), there is no loss of generality to assume that the given half-plane is the closed right half-plane RHP_0. Thus, we shall be done if we show that whenever $A + A^*$ is positive semidefinite, that is, $F(A) \subset RHP_0$, then A has a normal dilation $B \in M_{2n}$ such that $B + B^*$ is positive semidefinite, that is, $F(B) \subset RHP_0$. Consider

$$B \equiv \begin{bmatrix} A & A^* \\ A^* & A \end{bmatrix} \tag{1.6.11}$$

Then B is a dilation of A and

$$B^* B = \begin{bmatrix} A^* A + A A^* & A^{*2} + A^2 \\ A^2 + A^{*2} & A A^* + A^* A \end{bmatrix} = B B^*$$

so B is normal, independent of any assumption about $F(A)$. Moreover, if $A + A^*$ is positive semidefinite, then

$$B + B^* = \begin{bmatrix} A + A^* & A + A^* \\ A + A^* & A + A^* \end{bmatrix} \qquad (1.6.12)$$

is positive semidefinite, since if $y, z \in \mathbf{C}^n$ and $x \equiv \begin{bmatrix} y \\ z \end{bmatrix} \in \mathbf{C}^{2n}$, then $x^*(B + B^*)x = (y + z)^*(A + A^*)(y + z) \geq 0$. $\qquad \square$

Since the field of values of a normal matrix is easily shown to be convex and the intersection of convex sets is convex, Theorem (1.6.10) suggests a clean conceptual proof of the convexity property of the field of values (1.2.2). Unfortunately, the convexity of $F(A)$ is used in a crucial way in the proof of Theorem (1.6.10); thus, it would be very pleasant to have a different proof that does not rely on the Toeplitz-Hausdorff theorem.

The matrix defined by (1.6.11) gives a $2n$-by-$2n$ normal dilation of any given matrix $A \in M_n$. It is sometimes useful to know that one can find dilations with even more structure. For example, one can choose the dilation $B \in M_{2n}$ to be of the form $B = cV$, where $V \in M_{2n}$ is unitary and $c \equiv \max \{ ||| A |||_2, 1 \}$ (see Problem 22). Moreover, for any given integer $k \geq 1$, one can find a dilation $B \in M_{(k+1)n}$ that is a scalar multiple of a unitary matrix and has the property that $B^m = \begin{bmatrix} A^m & * \\ * & * \end{bmatrix}$ for all $m = 1, 2, ..., k$ (see Problem 25).

Problems

1.   If $A \in M_2$, show that $F(A)$ is a possibly degenerate ellipse with center at $\frac{1}{2} \mathrm{tr}\, A$ and foci at the two eigenvalues of A. Characterize the case in which $F(A)$ is a circular disc. What is the radius of this disc? Characterize the case in which $F(A)$ is a line segment. What are its endpoints? In the remaining case of an ellipse, what is the eccentricity and what is the equation of the boundary?

2.   For $A = \begin{bmatrix} a & b \\ c & d \end{bmatrix} \in M_2(\mathbb{R})$, show that $F(A)$ is an ellipse (possibly degenerate) with center at the point $\frac{1}{2}(a + d)$ on the real axis, whose vertices are the two points $\frac{1}{2}(a + d \pm [(a - d)^2 + (b + c)^2]^{\frac{1}{2}})$ on the real axis and the two

points $\frac{1}{2}(a + d \pm i|b-c|)$.   Show that the degenerate cases correspond exactly to the cases in which A is normal. Can you find similarly explicit formulae if $A \in M_2(\mathbb{C})$?

3.   If $A \in M_2$ has distinct eigenvalues, and if $A_0 \equiv A - (\frac{1}{2}\text{tr } A)I$, show that A_0 is nonsingular and the major axis of $F(A)$ lies in the direction $-(\det A_0)/|\det A_0|$ in the complex plane. What is the length of the major axis? What is the length of the minor axis?

4.   If $A \in M_2$, and if one of the eigenvalues of A appears on the main diagonal of A, show that at least one of the off-diagonal entries of A must be zero.

5.   Show that any given (possibly degenerate) ellipse (with interior) is the field of values of some matrix $A \in M_2$. Show how to construct such a matrix A for a given ellipse.

6.   Show by example that 4 is the best possible value in (1.6.9), that is, for every $n \geq 5$ show that there is a nonnormal $A \in M_n$ with $F(A) = \text{Co}(\sigma(A))$.

7.   Let $n \geq 3$ be given and let $1 \leq k \leq n$. Show that there is some $A \in M_n$ such that $F(A)$ has exactly k sharp points.

8.   If $A \in M_n$ and if α is a sharp point of $F(A)$, show that a unit vector $x \in \mathbb{C}^n$ for which $\alpha = x^*Ax$ is an eigenvector for both A and $H(A)$.

9.   Give an alternate, more geometric proof of (1.6.3) along the following lines: If $A \in M_n$ and $\lambda \in F(A)$, pick a unitary $U \in M_n$ so that λ is the 1,1 entry of U^*AU. (How may U be constructed?) Since

$$\lambda \in F[(U^*AU)(\{1,j\})]$$

it must be a sharp point of $F[(U^*AU)(\{1,j\})]$, $j = 2,..., n$, if λ is a sharp point of $F(A)$. But since $F[(U^*AU)(\{1,j\})]$ is either a circular disc, a line segment, or an ellipse (with interior), it must actually be a line segment with λ as one endpoint in order for λ to be a sharp point. But then $F[(U^*AU)(\{1,j\})]$ must be diagonal, and

$$U^*AU = [\lambda] \oplus (U^*AU)(\{2,..., n\})$$

that is, λ is an eigenvalue of A.

10.   Give an alternate proof of (1.6.3) along these lines: Let $x \in \mathbb{C}^n$ be a unit

vector and let $\lambda = x^* A x$ be a sharp point of $F(A)$. Let $y \in \mathbb{C}^n$ be independent of x. Let $H \in M_n$ denote the (Hermitian) orthogonal projection onto the two-dimensional subspace spanned by x and y, that is, let $\{v, w\}$ be an orthonormal basis of $V \equiv \mathrm{Span}\, \{x, y\}$, let $P \equiv [v \ w] \in M_{n,2}$ be an isometric projection (see Problem 16 in Section (1.2)), and let $H \equiv PP^*$, so $Hz = z$ if $z \in V$ and $Hz = 0$ if $z \perp V$. Then λ is a sharp point of $F(HAH) \subset F(A)$. Since $F(HAH)$ is a (possibly degenerate) ellipse, it must be a line segment with λ as an endpoint, so λ is an eigenvalue of HAH and of A, with eigenvector x.

11. If $A \in M_n$ and $\lambda \in \sigma(A)$, and if $x \in \mathbb{C}^n$ is an associated eigenvector, show that λ is a normal eigenvalue of A if and only if x is an eigenvector of both A and A^*. What can be said if the multiplicity of λ is greater than one?

12. If $\lambda, \mu \in \sigma(A)$ and if $\mu \neq \lambda$, show that any eigenvector of A^* associated with $\bar{\mu}$ is orthogonal to any eigenvector of A associated with λ.

13. Let $A \in M_n$. Show that if x is an eigenvector of both A and $H(A)$, then x is an eigenvector of A^*. How are the eigenvalues related?

14. Give another proof of (1.6.6) using (1.5.7) and the result of Problem 13.

15. Let $A \in M_n$. If $F(A)$ is a polygon with at least $n-1$ vertices, show that A is normal.

16. Let $A \in M_n$. If at least $n-1$ of the eigenvalues of A lie on the boundary of $F(A)$, show that A is normal. Show that the value $n-1$ is the best possible.

17. Let $A \in M_n$. (a) If $F(A)$ has finitely many extreme points, show that they are all eigenvalues of A. (b) If A is normal, show that the extreme points of $F(A)$ are eigenvalues of A.

18. Show that $A \in M_n$ is normal if and only if A is unitarily similar to a direct sum of square matrices $A_1 \oplus \cdots \oplus A_k$, $A_i \in M_{n_i}$, $i = 1,\ldots, k$, and each $F(A_i)$ is a polygon (perhaps a line segment or a point) with n_i vertices, $i = 1,\ldots, k$.

19. Let $A \in M_n$ be nonsingular and normal, with $\sigma(A) = \{r_1 e^{i\theta_1},\ldots, r_n e^{i\theta_n}\}$, all $r_j > 0$, and all $\theta_j \in [0, 2\pi)$. If A has some diagonal entry equal to 0, show that there does not exist a value t such that $t \leq \theta_i < t + \pi$ for all $i = 1, 2,\ldots, n$, that is, the eigenvalues do not project onto the unit circle inside any semicircle.

20. Give another proof that the matrix B defined by (1.6.11) is normal, and that $B + B^*$ is positive semidefinite if $A + A^*$ is positive semidefinite, as follows: Show that the $2n$-by-$2n$ matrix $U \equiv \begin{bmatrix} I & I \\ -I & I \end{bmatrix} / \sqrt{2}$ is unitary, $U^* B U = \begin{bmatrix} A - A^* & 0 \\ 0 & A + A^* \end{bmatrix}$, and $U^*(B + B^*) U = \begin{bmatrix} 0 & 0 \\ 0 & 2(A + A^*) \end{bmatrix}$. Thus, B and $B + B^*$ are unitarily similar to something obviously normal and something positive semidefinite, respectively. How is the spectrum of B determined by A? Show that $||| B |||_2 \leq 2 \, ||| A |||_2$.

21. Recall that a matrix A is a *contraction* if $||| A |||_2 \leq 1$. See Problem 30 in Section (1.5) for an important link between contractions and the field of values.

(a) If there exists a unitary dilation $V = \begin{bmatrix} A & * \\ * & * \end{bmatrix}$ of a given square complex matrix A, show that A must be a contraction.

(b) Conversely, if $A \in M_n$ is a given contraction, show that A has a unitary dilation as follows: Let $A = PU$ be a polar decomposition of A, where $U \in M_n$ is unitary and $P \equiv (AA^*)^{\frac{1}{2}}$ is positive semidefinite. Let

$$ Z \equiv \begin{bmatrix} A & (I - P^2)^{\frac{1}{2}} U \\ -(I - P^2)^{\frac{1}{2}} U & A \end{bmatrix} \in M_{2n} \qquad (1.6.13) $$

and show that Z is unitary.

(c) If $A \in M_n$ is a contraction, show that

$$ Z \equiv \begin{bmatrix} A & (I - AA^*)^{\frac{1}{2}} \\ -(I - A^* A)^{\frac{1}{2}} & A^* \end{bmatrix} \in M_{2n} \qquad (1.6.14) $$

is also a unitary dilation of A.

(d) Use the dilation Z in (1.6.14) and the argument in the proof of Theorem (1.6.10) to show that if A is a *normal* contraction, then

$$ F(A) = \bigcap \left\{ \{ F(U) : U = \begin{bmatrix} A & * \\ * & * \end{bmatrix} \in M_{2n} \text{ is unitary} \right\} $$

(e) Combine the normal dilation (1.6.11) with the result in (d) and the norm bound in Problem 20 to show that if $A \in M_n$ is a contraction with $||| A |||_2 \leq \frac{1}{2}$, then

$$ F(A) = \bigcap \left\{ F(U) : U = \begin{bmatrix} A & * \\ * & * \end{bmatrix} \in M_{4n} \text{ is unitary} \right\} $$

(f) Now consider a third unitary dilation, one that does not require the polar decomposition, the singular value decomposition, or matrix square roots. The following construction can result in a dilation of smaller size than $2n$; an analogous construction gives a complex orthogonal dilation of an arbitrary matrix (see Problem 40). Let $A \in M_n$ be a given contraction and let $\delta \equiv \mathrm{rank}(I - A^*A)$, so $0 \le \delta \le n$ with $\delta = 0$ if and only if A is unitary; δ may be thought of as a measure of how far A is from being unitary. Explain why $\mathrm{rank}(I - AA^*) = \delta$ and why there are nonsingular $X, Y \in M_n$ with

$$I - AA^* = X \begin{bmatrix} I_\delta & 0 \\ 0 & 0 \end{bmatrix} X^* \quad \text{and} \quad I - A^*A = Y^* \begin{bmatrix} 0 & 0 \\ 0 & I_\delta \end{bmatrix} Y \qquad (*)$$

and $I_\delta \in M_\delta$. Define

$$B \equiv X \begin{bmatrix} I_\delta \\ 0 \end{bmatrix} \in M_{n,\delta}, \quad C \equiv -[0 \ I_\delta] Y \in M_{\delta,n}, \text{ and}$$

$$D \equiv [0 \ I_\delta] YA^*(X^*)^{-1} \begin{bmatrix} I_\delta \\ 0 \end{bmatrix} \in M_\delta$$

and form

$$Z \equiv \begin{bmatrix} A & B \\ C & D \end{bmatrix} \in M_{n+\delta} \qquad (1.6.15)$$

Use the definitions $(*)$ and the identity $A(I - A^*A) = (I - AA^*)A$ to show that

$$X^{-1}AY^* \begin{bmatrix} 0 & 0 \\ 0 & I_\delta \end{bmatrix} = \begin{bmatrix} I_\delta & 0 \\ 0 & 0 \end{bmatrix} X^*AY^{-1} \qquad (**)$$

What do (1.6.15) and $(**)$ become when $\delta = n$, that is, when A is a strict contraction ($\|\|A\|\|_2 < 1$)? In this special case, verify that (1.6.15) gives a unitary matrix. Now consider the general case when $\delta < n$; considerable algebraic manipulation seems to be needed to show that (1.6.15) is unitary.

(g) Let n and k be given positive integers. Show that a given $A \in M_k$ is a principal submatrix of a unitary $U \in M_n$ if and only if A is a contraction and $\mathrm{rank}(I - A^*A) \le \min\{k, n-k\}$; the latter condition imposes no restriction on A if $k \le n/2$.

(h) Show that a given $A \in M_{m,n}$ has a unitary dilation if and only if A is a contraction.

22. Let $A \in M_n$ be given. Show that there exists a unitary $V \in M_{2n}$ such that $cV = \begin{bmatrix} A & * \\ * & * \end{bmatrix}$ with $c \equiv \max \{ \|\| A \|\|_2, 1 \}$. Compare with Theorem (1.6.10).

23. Consider

$$A = \begin{bmatrix} 0 & 1 \\ 0 & 0 \end{bmatrix} \in M_2 \text{ and } B_z = \begin{bmatrix} 0 & 1 & 0 \\ 0 & 0 & 1 \\ z & 0 & 0 \end{bmatrix} \in M_3$$

Show that $F(A) = \cap\{F(B_z): z \in \mathbb{C} \text{ and } |z| = 1\}$. Discuss this in light of Problem 21.

24. The dilation results in Problem 21 have a very interesting generalization. Let

$$Z \equiv \begin{bmatrix} A & Z_{12} \\ Z_{21} & Z_{22} \end{bmatrix}$$

denote any unitary dilation of a given contraction $A \in M_n$, for example, Z could be any of the dilations given by (1.6.13,14,15), and let $k \geq 1$ be a given integer. Consider the block matrix $V = [V_{ij}]_{i,j=1}^{k+1}$, in which each block V_{ij} is defined by: $V_{11} \equiv A$, $V_{1,2} \equiv Z_{12}$, $V_{k+1,1} \equiv Z_{21}$, $V_{k+1,2} \equiv Z_{22}$, and $V_{2,3} = V_{3,4} = \cdots = V_{k,k+1} \equiv I$, where this identity matrix is the same size as Z_{22}, and all other blocks are zero:

$$V \equiv \begin{bmatrix} A & Z_{12} & 0 & \cdots\cdots & 0 \\ 0 & 0 & I & 0 & \vdots \\ \vdots & \vdots & & I & \ddots & 0 \\ 0 & 0 & 0 & & \ddots & I \\ Z_{21} & Z_{22} & 0 & \cdots\cdots & 0 \end{bmatrix} \qquad (1.6.16)$$

(a) When $k = 1$, show that $V = Z$.

(b) For $k \geq 2$, show that V is unitary and that $V^m = \begin{bmatrix} A^m & * \\ * & * \end{bmatrix}$ for $m = 1, \ldots, k$, that is, each of these powers of V is a unitary dilation of the corresponding power of A.

25. Let $A \in M_n$ and let $k \geq 1$ be a given integer. Show that there exists a unitary $V \in M_{(k+1)n}$ such that $(cV)^m = \begin{bmatrix} A^m & * \\ * & * \end{bmatrix}$ for $m = 1, 2, \ldots, k$, where $c \equiv$

max $\{\||| A \||_2, 1\}$. Compare with Theorem (1.6.10).

26. Let $A \in M_n$ be a given contraction, that is, $\||| A \||_2 \leq 1$. If $p(\cdot)$ is a polynomial such that $|p(z)| \leq 1$ for all $|z| = 1$, use Problem 25 to show that $p(A)$ is a contraction.

27. Use Theorem (1.6.6) to show that the algebraic and geometric multiplicities of the Perron root (and any other eigenvalues of maximum modulus) of a doubly stochastic matrix must be equal, and show by example that this is not true of a general nonnegative matrix. Show, however, that this *is* true of a nonnegative matrix that is *either* row or column stochastic, and give an example of such a matrix whose Perron root does not lie on the boundary of its field of values.

28. Use Corollary (1.6.7) to give a shorter proof of Theorem (1.6.1). Is there any circular reasoning involved in doing so?

29. A matrix $A \in M_n$ is said to be *unitarily reducible* if there is some unitary $U \in M_n$ such that $U^*AU = A_1 \oplus A_2$ with $A_1 \in M_k$, $A_2 \in M_{n-k}$, and $1 < k < n$; A is *unitarily irreducible* if it is not unitarily reducible. If $A \in M_n$ is unitarily irreducible, show that every eigenvalue of A lies in the (two-dimensional) interior of $F(A)$ and that the boundary of $F(A)$ is a smooth curve (no sharp points).

30. Let $A \in M_n$ be given. Show that $\text{Co}(\sigma(A)) = \cap \{F(SAS^{-1}): S \in M_n$ is nonsingular$\}$. Moreover, if $J_{n_1}(\lambda_1) \oplus \cdots \oplus J_{n_k}(\lambda_k)$ is the Jordan canonical form of A (see (3.1.11) in [HJ]), show that there is some one nonsingular $S \in M_n$ such that $\text{Co}(\sigma(A)) = F(SAS^{-1})$ if and only if $n_i = 1$ for every eigenvalue λ_i that lies on the boundary of $\text{Co}(\sigma(A))$.

31. If $<\cdot,\cdot>$ is a given inner product on $\mathbb{C}^n$, the *field of values generated by* $<\cdot,\cdot>$ is defined by $F_{<\cdot,\cdot>}(A) \equiv \{<Ax,x>: x \in \mathbb{C}^n, <x,x> = 1\}$ for $A \in M_n$. Note that the usual field of values is the field of values generated by the usual Euclidean inner product $<x,y> \equiv y^*x$. Show that $\text{Co}(\sigma(A)) = \cap\{F_{<\cdot,\cdot>}(A): <\cdot,\cdot>$ is an inner product on $\mathbb{C}^n\}$.

32. Let $A \in M_n$ be given. In general, $\rho(A) \leq r(A)$, but if $\rho(A) = r(A)$, we say that A is *spectral* (the term *spectraloid* is also used). In particular, A is spectral if it is radial, that is, if either $\||| A \||_2 = \rho(A)$ or $\||| A \||_2 = r(A)$; see Problems 23, 24, and 27 in Section (1.5). If A is spectral and λ is any eigenvalue of A such that $|\lambda| = \rho(A)$, show that λ is a normal eigenvalue of A. Show that every normal matrix $A \in M_n$ is spectral and that the converse is true only for $n = 2$.

33. Use Problem 32 to show that a given $A \in M_n$ is spectral if and only if it is unitarily similar to a direct sum of the form $\lambda_1 I_{n_1} \oplus \cdots \oplus \lambda_k I_{n_k} \oplus B$, where $1 \le k \le n$, $n_1 + \cdots + n_k = n - m$, $0 \le m \le n - 1$, $B \in M_m$, $|\lambda_1| = \cdots = |\lambda_k| = \rho(A)$, $\rho(B) < \rho(A)$, and $r(B) \le r(A)$.

34. Use Problem 33 to show that a given $A \in M_n$ is spectral if and only if it is unitarily similar to $r(A)(U \oplus B)$, where $U \in M_k$ is unitary, $1 \le k \le n$, and $B \in M_{n-k}$ has $\rho(B) < \rho(A)$ and $r(B) \le r(A)$.

35. Show that a given $A \in M_n$ is spectral if and only if $r(A^k) = r(A)^k$ for all $k = 1, 2, \ldots$. Conclude that if A is spectral, then every positive integer power of A is spectral. For each $n \ge 2$, exhibit a nonspectral matrix $B \in M_n$ such that B^m is spectral for all $m \ge 2$.

36. Let $A \in M_n$ be given with $r(A) \le 1$. Show that if $\rho(A) = 1$, then $r(A^m) = 1$ for all $m = 1, 2, \ldots$. Goldberg, Tadmor, and Zwas have proved a converse: If $r(A) \le 1$, then $\rho(A) = 1$ if and only if $r(A^m) = 1$ for some positive integer m not less than the degree of the minimal polynomial of A. Use this result to establish the following:

(a) A is spectral if and only if $r(A^m) = r(A)^m$ for some positive integer m not less than the degree of the minimal polynomial of A.

(b) Show that the 3-by-3 nilpotent Jordan block

$$A = \begin{bmatrix} 0 & 1 & 0 \\ 0 & 0 & 1 \\ 0 & 0 & 0 \end{bmatrix}$$

has $\rho(A) = 0$ and $r(A^2) = r(A)^2 = \frac{1}{4}$, but A is not spectral. Explain why the bound on the exponent in the Goldberg-Tadmor-Zwas theorem is sharp.

(c) Show that A is spectral if and only if $r(A^n) = r(A)^n$.

(d) When $n = 2$, show that A is normal if and only if $r(A^2) = r(A)^2$.

Compare the Goldberg-Tadmor-Zwas theorem with Ptak's theorem on radial matrices and its generalization in Problems 25 and 26 in Section (1.5).

37. Let $A \in M_n$ be given. Combine the results of the preceding five problems and verify that the following conditions are equivalent:

(a) A is spectral.

(b) $\rho(A) = r(A)$.

(c) $r(A^n) = r(A)^n$.

(d) $r(A^m) = r(A)^m$ for some integer m not less than the degree of the minimal polynomial of A.

(e) $r(A^k) = r(A)^k$ for all $k = 1, 2,....$

(f) A is unitarily similar to $r(A)(U \oplus B)$, where $U \in M_k$ is unitary, $1 \leq k \leq n$, and $B \in M_{n-k}$ has $\rho(B) < 1$ and $r(B) \leq 1$.

Compare with the conditions for a matrix to be *radial* given in Problem 27 in Section (1.5).

38. Use Problem 37(f) to prove the assertion in Problem 27(g) in Section (1.5).

39. The main geometrical content of Theorem (1.6.6) is the following fact: Any eigenvector of A corresponding to an eigenvalue that is a boundary point of the field of values is orthogonal to any eigenvector of A corresponding to any other (different) eigenvalue. Provide details for the following alternative proof of this fact. We write $<x,y>$ for the usual inner product. (1) Let $\lambda \in \partial F(A)$, and let $\mu \neq \lambda$ be any other eigenvalue of A. Let x, y be unit vectors with $Ax = \lambda x$ and $Ay = \mu y$. Since $F(A)$ is compact and convex, and λ is a boundary point of $F(A)$, there is a supporting line for $F(A)$ that passes through λ. Show that this geometrical statement is equivalent to the existence of a nonzero $c \in \mathbb{C}$ and a real number a such that $\text{Im } c<Az,z> \geq a$ for all unit vectors z, with equality for $z = x$, that is, $\text{Im } c\lambda = a$.   (2) Let $A_1 \equiv cA - iaI$, $\lambda_1 \equiv c\lambda - ia$, and $\mu_1 \equiv c\mu - ia$.   Show that $\text{Im } <A_1z,z> \geq 0$ for all unit vectors z, λ_1 is real, $\bar{\mu}_1 \neq \lambda_1$, and $\text{Im } \mu_1 \neq 0$. (3) For arbitrary $\xi, \eta \in \mathbb{C}$, note that $\text{Im } <A_1(\xi x + \eta y),(\xi x + \eta y)> \geq 0$ and deduce that $\text{Im } [|\eta|^2 \mu_1 + (\lambda_1 - \bar{\mu}_1)\xi\bar{\eta}<x,y>] \geq 0$.   (4) Show that you may choose η so that $|\eta| = 1$ and $\text{Im } [(\lambda_1 - \bar{\mu}_1)\bar{\eta}<x,y>] = |(\lambda_1 - \bar{\mu}_1)<x,y>|$; deduce that $\text{Im } \mu_1 + |(\lambda_1 - \bar{\mu}_1)<x,y>|\xi \geq 0$ for all $\xi \in \mathbb{R}$.   (5) Conclude that $<x,y> = 0$, as desired.   *Remark:* The proof we have outlined is of interest because it does not rely on an assumption that A is a (finite-dimensional) matrix. It is valid for any bounded linear operator A on a complex Hilbert space with inner product $<\cdot,\cdot>$, and uses the fact that $F(A)$ is a compact convex set. Thus, the geometrical statement about orthogonality of eigenvectors given at the beginning of this problem is valid for any bounded linear operator A on a complex Hilbert space.

40. Although a given matrix has a unitary dilation if and only if it is a contraction (Problem 21), *every complex matrix has a complex orthogonal* *dilation.*

(a) Let $A \in M_n$ and let $\delta \equiv \text{rank}(I - A^T A)$, so $0 \leq \delta \leq n$ with $\delta = 0$ if and only if A is complex orthogonal; δ may be thought of as a measure of how far A is from being complex orthogonal. Explain why $\text{rank}(I - A A^T) = \delta$ and use Corollary (4.4.4) in [HJ] to show that there are nonsingular $X, Y \in M_n$ with

$$I - AA^T = X \begin{bmatrix} I_\delta & 0 \\ 0 & 0 \end{bmatrix} X^T \text{ and } I - A^T A = Y^T \begin{bmatrix} 0 & 0 \\ 0 & I_\delta \end{bmatrix} Y \qquad (T)$$

and $I_\delta \in M_\delta$. Define

$$B \equiv X \begin{bmatrix} I_\delta \\ 0 \end{bmatrix} \in M_{n,\delta}, \ C \equiv -[0 \ I_\delta] Y \in M_{\delta,n}, \text{ and}$$

$$D \equiv [0 \ I_\delta] Y A^T (X^T)^{-1} \begin{bmatrix} I_\delta \\ 0 \end{bmatrix} \in M_\delta$$

and form

$$Q \equiv \begin{bmatrix} A & B \\ C & D \end{bmatrix} \in M_{n+\delta} \qquad (1.6.17)$$

Use the definitions (T) and the identity $A(I - A^T A) = (I - A A^T) A$ to show that

$$X^{-1} A Y^T \begin{bmatrix} 0 & 0 \\ 0 & I_\delta \end{bmatrix} = \begin{bmatrix} I_\delta & 0 \\ 0 & 0 \end{bmatrix} X^T A Y^{-1} \qquad (TT)$$

Notice that these calculations carry over exactly to (or from) (**) in Problem 21; one just interchanges * and T. What do (1.6.17) and (TT) become when $\delta = n$? In this special case, verify that (1.6.17) gives a complex orthogonal matrix. The algebraic manipulations necessary to establish the general case are analogous to those needed to establish the general case in Problem 21.

(b) Assuming that (1.6.17) gives a complex orthogonal dilation of A, explain why the construction of (1.6.16) in Problem 24 gives a complex orthogonal dilation $P = \begin{bmatrix} A & * \\ * & * \end{bmatrix} \in M_{n+k\delta}$ such that $P^m = \begin{bmatrix} A^m & * \\ * & * \end{bmatrix}$ for $m = 1, \dots, k$.

(c) Let n and k be given positive integers. Show that a given $A \in M_k$ is a principal submatrix of a complex orthogonal $Q \in M_n$ if and only if $\mathrm{rank}(I - A^TA) \leq \min\{k, n-k\}$, which imposes no restriction on A if $k \leq n/2$. (d) Show that every $A \in M_{m,n}$ has a complex orthogonal dilation.

Notes and Further Readings. The proof given for Theorem (1.6.1) is due to Onn Chan. Theorem (1.6.3) may be found in Section I of R. Kippenhahn, Über den Wertevorrat einer Matrix, *Math. Nachr.* 6 (1951), 193-228, where a proof different from the three given here may be found (see Problems 9 and 10), as well as many additional geometric results about the field of values; however, see the comments at the end of Section (1.8) about some of the quaternion results in Section II of Kippenhahn's paper. Theorems (1.6.6) and (1.6.8) are from C. R. Johnson, Normality and the Numerical Range, *Linear Algebra Appl.* 15 (1976), 89-94, where additional related references may be found. A version of (1.6.1) and related ideas are discussed in O. Taussky, Matrices with Trace Zero, *Amer. Math. Monthly* 69 (1962), 40-42. Theorem (1.6.10) and its proof are in P. R. Halmos, Numerical Ranges and Normal Dilations, *Acta Sci. Math. (Szeged)* 25 (1964), 1-5. The constructions of unitary dilations of a given contraction given in (1.6.13) and (1.6.16) are in E. Egervary, On the Contractive Linear Transformations of n-Dimensional Vector Space, *Acta Sci. Math. (Szeged)* 15 (1953), 178-182. The constructions for (1.6.15-17) as well as sharp bounds on the sizes of unitary and complex orthogonal dilations with the power properties discussed in Problems 24 and 40 are given in R. C. Thompson and C.-C. T. Kuo, Doubly Stochastic, Unitary, Unimodular, and Complex Orthogonal Power Embeddings, *Acta. Sci. Math. (Szeged)* 44 (1982), 345-357. Generalizations and extensions of the result in Problem 26 are in K. Fan, Applications and Sharpened Forms of an Inequality of Von Neumann, pp. 113-121 of [UhGr]; $p(z)$ need not be a polynomial, but may be any analytic function on an open set containing the closed unit disc, and there are matrix analytic versions of classical results such as the Schwarz lemma, Pick's theorem, and the Koebe $\frac{1}{4}$-theorem for univalent functions. The theorem referred to in Problem 36 is proved in M. Goldberg, E. Tadmor, and G. Zwas, The Numerical Radius and Spectral Matrices, *Linear Multilinear Algebra* 2 (1975), 317-326. The argument in Problem 39 is due to Ky Fan, A Remark on Orthogonality of Eigenvectors, *Linear Multilinear Algebra* 23 (1988), 283-284.

1.7   Products of matrices

We survey here a broad range of facts relating products of matrices to the field of values. These fall into four categories:

(a)   Examples of the failure of submultiplicativity of $F(\cdot)$;
(b)   Results about the usual product when zero is not in the field of values of one of the factors;
(c)   Discussion of simultaneous diagonalization by congruence;
(d)   A brief survey of the field of values of a Hadamard product.

Examples of the failure of submultiplicativity for $F(\cdot)$

Unfortunately, even very weak multiplicative analogs of the subadditivity property (1.2.7) do not hold. We begin by noting several examples that illustrate what is *not* true; these place in proper perspective the few facts that *are* known about products.

The containment

$$F(AB) \subset F(A)F(B), \qquad A, B \in M_n \tag{1.7.1}$$

fails to hold both "angularly" and "magnitudinally."

1.7.2   Example. Let $A = \begin{bmatrix} 0 & 2 \\ 0 & 0 \end{bmatrix}$ and $B = \begin{bmatrix} 0 & 0 \\ 2 & 0 \end{bmatrix}$. Then $F(A) = F(B) = F(A)F(B) =$ the unit disc, while $F(AB)$ is the line segment joining 0 and 4. Thus, $F(AB)$ contains points much further (four times as far) from the origin than $F(A)F(B)$, so this example shows that the numerical radius $r(\cdot)$ is not a matrix norm. However, $4r(\cdot)$ is a matrix norm (see Problem 22 in Section (5.7) of [HJ]).

1.7.3   Example. Let $A = \begin{bmatrix} 1 & 0 \\ 0 & -1 \end{bmatrix}$ and $B = \begin{bmatrix} 0 & 1 \\ 1 & 0 \end{bmatrix}$. Then $F(A) = F(B) = F(A)F(B) =$ the line segment joining -1 and 1, while $F(AB)$ is the line segment joining $-i$ and i. Also, $F'(A) = F'(B) = F'(A)F'(B) =$ the real line, while $F'(AB) =$ the imaginary axis. Note that $r(AB) \leq r(A)r(B)$ in this case, but that (1.7.1) still fails—for angular reasons; $F'(AB)$ is not contained in $F'(A)F'(B)$ either.

1.7.4   Example. Let $A = \begin{bmatrix} 1 & 0 \\ 0 & i \end{bmatrix}$. Then $F(A)$ is the line segment joining

1 and i, while $F(A^2)$ is the line segment joining -1 and 1. Thus, $F(A^2)$ is not even contained in $F(A)^2$, since $0 \in F(A^2)$ and $0 \notin F(A)$. However, $\mathrm{Co}(F(A)^2) = \mathrm{Co}\{-1, i, 1\}$, so $F(A^2) \subset \mathrm{Co}(F(A)^2)$.

One reason it is not surprising that (1.7.1) does not hold in general is that $F(A)F(B)$ is not generally a convex set, whereas $F(AB)$ is always convex. But, since $F(A)F(B)$ is convex in the case of both examples (1.7.2) and (1.7.3), the inclusion

$$F(AB) \subset \mathrm{Co}(F(A)F(B)), \quad A, B \in M_n \tag{1.7.1a}$$

also fails to hold in general. Examples (1.7.2) and (1.7.3) also show that another weaker statement

$$\sigma(AB) \subset F(A)F(B) \tag{1.7.1b}$$

fails to hold, too.

Product results when zero is not in the field of values of a factor

Now that we know a few things that we can *not* prove, we turn to examining some of what *can* be said about the field of values and products of matrices. The importance of the condition that zero not be in the field of values of one of the factors should be noted at the outset. This means, for example, that the field of values may be rotated into the right half-plane, an observation that links many of the results presented.

1.7.5   Observation. If $A \in M_n$ is nonsingular, then $F'(A^{-1}) = \overline{F'(A)}$.

Proof: Employing the congruence property (1.2.12), we have $F'(A^{-1}) = F'(AA^{-1}A^*) = F'(A^*) = \overline{F'(A)}$. $\square$

Exercise. Verify that $F(A^*) = \overline{F(A)}$ and, therefore, that $F'(A^*) = \overline{F'(A)}$.

Although (1.7.1b) fails to hold, the following four facts indicate that certain limited statements can be made about the relationship between the spectrum of a product and the product of the fields of values.

1.7.6   **Theorem.** Let $A, B \in M_n$ and assume that $0 \notin F(B)$.   Then $\sigma(AB^{-1}) \subset F(A)/F(B)$.

Proof: Since $0 \notin F(B)$, we know from the spectral containment property (1.2.6) that B is nonsingular. The set ratio $F(A)/F(B)$ (which has the usual algebraic interpretation) makes sense and is bounded. If $\lambda \in \sigma(AB^{-1})$, then there is a unit vector $x \in \mathbb{C}^n$ such that $x^* AB^{-1} = \lambda x^*$. Then $x^* A = \lambda x^* B$, $x^* Ax = \lambda x^* Bx$, and $x^* Bx \neq 0$; hence $\lambda = x^* Ax/x^* Bx \in F(A)/F(B)$. $\square$

1.7.7   **Corollary.** Let $A, B \in M_n$. If B is positive semidefinite, then $\sigma(AB) \subset F(A)F(B)$.

Proof: First suppose that B is nonsingular and set $B = C^{-1}$. If $\lambda \in \sigma(AB) = \sigma(AC^{-1})$, then by (1.7.6) we must have $\lambda = a/c$ for some $a \in F(A)$ and some $c \in F(C)$. If β_{min} and β_{max} are the smallest and largest eigenvalues of B, then β_{min}^{-1} and β_{max}^{-1} are the largest and smallest eigenvalues of the positive definite matrix C, and $F(C)$ is the interval $[\beta_{max}^{-1}, \beta_{min}^{-1}]$. Therefore, $c^{-1} \in [\beta_{min}, \beta_{max}] = F(B)$ and $\lambda = ac^{-1} \in F(A)F(B)$. The case in which B is singular may be reduced to the nonsingular case by replacing B with $B + \epsilon I$, $\epsilon > 0$, and then letting $\epsilon \to 0$. $\square$

1.7.8   **Theorem.** Let $A, B \in M_n$ and assume that $0 \notin F(B)$.   Then $\sigma(AB) \subset F'(A)F'(B)$.

Proof: Since $0 \notin F(B)$, B is nonsingular and we may write $B = C^{-1}$. By (1.7.6), for each $\lambda \in \sigma(AB) = \sigma(AC^{-1})$ we have $\lambda = a/c$ for some $a \in F(A) \subset F'(A)$. For some unit vector $x \in \mathbb{C}^n$ we have $c = x^* Cx = (Cx)^* B^*(Cx) = y^* B^* y \in F'(B^*) = \overline{F'(B)}$, where we have set $y \equiv Cx$. Then $\bar{c} \in F'(B)$, $c \neq 0$, and $c^{-1} = \bar{c}/|c|^2 \in F'(B)$. Thus, $\lambda = ac^{-1} \in F'(A)F'(B)$. $\square$

1.7.9   **Theorem.** Let $A \in M_n$ and $\lambda \in \mathbb{C}$ be given with $\lambda \neq 0$. The following are equivalent:

   (a) $\lambda \in F'(A)$;
   (b) $\lambda \in \sigma(HA)$ for some positive definite $H \in M_n$; and
   (c) $\lambda \in \sigma(C^* AC)$ for some nonsingular $C \in M_n$.

Proof: Since H is positive definite if and only if $H = CC^*$ for some nonsingular $C \in M_n$, and since $\sigma(CC^* A) = \sigma(C^* AC)$, it is clear that (b) and (c) are equivalent. To show that (c) implies (a), suppose that $\lambda \in \sigma(C^*AC)$ with an associated unit eigenvector y. Then $\lambda = y^*C^*ACy = (Cy)^*A(Cy) = x^*Ax$ for $x \equiv Cy \neq 0$, and hence $\lambda \in F'(A)$. Conversely, suppose $\lambda = x^*Ax$. Then $x \neq 0$ since $\lambda \neq 0$, and we let $C_1 \in M_n$ be any nonsingular matrix whose first column is x. The 1,1 entry of $C_1^*AC_1$ is λ. Let v^T be the row vector formed by the remaining $n-1$ entries of the first row of $C_1^*AC_1$ and let $z^T = -v^T/\lambda$. Then $C_2 = \begin{bmatrix} 1 & z^T \\ 0 & I \end{bmatrix} \in M_n$ is nonsingular and $C_2^*(C_1^*AC_1)C_2 = \begin{bmatrix} \lambda & 0 \\ * & * \end{bmatrix}$. Thus, $\lambda \in \sigma(C^*AC)$, where $C = C_1C_2$ is nonsingular.   ▯

It is now possible to characterize the so-called *H-stable matrices*, that is, those $A \in M_n$ such that all eigenvalues of HA have positive real part for all positive definite matrices H.

1.7.10   Corollary. Let $A \in M_n$. Then $\sigma(HA) \subset RHP$ for all positive definite $H \in M_n$ if and only if $F(A) \subset RHP \cup \{0\}$ and A is nonsingular.

Exercise. Prove (1.7.10) using (1.7.9).

We have already seen the special role played by matrices for which zero is exterior to the field of values. Another example of their special nature is the following result, which amounts to a characterization.

1.7.11   Theorem. Let $A \in M_n$ be nonsingular. The following are equivalent:

(a) $A^{-1}A^* = B^{-1}B^*$ for some $B \in M_n$ with $0 \notin F(B)$.
(b) A is *congruent to a normal matrix.
(b′) A is *congruent to a normal matrix via a positive definite congruence.
(c) $A^{-1}A^*$ is similar to a unitary matrix.

Proof: Since $A^{-1}A^*$ is unitary if and only if $(A^{-1}A^*)^{-1} = (A^{-1}A^*)^*$, which holds if and only if $A^*A = AA^*$, we see that $A^{-1}A^*$ is unitary if and only if A is normal. The equivalence of (b) and (c) then follows directly from the calculation $S^{-1}A^{-1}A^*S = S^{-1}A^{-1}(S^*)^{-1}S^*A^*S = (S^*AS)^{-1}(S^*AS)^*$; if $S^{-1}A^{-1}A^*S$ is unitary, then S^*AS (a *congruence of A) must be normal, and conversely.

To verify that (c) implies (a), suppose that $U \equiv S^{-1}A^{-1}A^*S$ is unitary, and let $U^{\frac{1}{2}}$ be one of the (several) unitary square roots of U such that all eigenvalues of $U^{\frac{1}{2}}$ lie on an arc of the unit circle of length less than π. By (1.2.9), $0 \notin F(U^{\frac{1}{2}})$ and by (1.2.12) and (1.7.5), $0 \notin F((S^{-1})^* U^{-\frac{1}{2}} S^{-1})$, where $U^{-\frac{1}{2}} \equiv (U^{\frac{1}{2}})^{-1}$. Now calculate $A^{-1}A^* = SUS^{-1} = S(U^{-\frac{1}{2}})^{-1}(U^{-\frac{1}{2}})^* S^{-1} = S(U^{-\frac{1}{2}})^{-1} S^*(S^{-1})^*(U^{-\frac{1}{2}})^* S^{-1} = B^{-1}B^*$, with $B \equiv (S^{-1})^* U^{-\frac{1}{2}} S^{-1}$ and $0 \notin F(B)$ as (a) asserts. Assuming (a), we may suppose $0 \notin F(A)$ and then we may suppose, without loss of generality, that $\mathrm{H}(A)$ is positive definite by (1.3.5). We now show that (a) implies (b') by writing $A = H + S$, where $H \equiv \mathrm{H}(A) = \frac{1}{2}(A + A^*)$ is positive definite and $S \equiv \mathrm{S}(A) = \frac{1}{2}(A - A^*)$. If $H^{-\frac{1}{2}}$ is the inverse of the unique positive definite square root of H, we have $(H^{-\frac{1}{2}})^* A H^{-\frac{1}{2}} = H^{-\frac{1}{2}}(H + S) H^{-\frac{1}{2}} = I + H^{-\frac{1}{2}} S H^{-\frac{1}{2}}$, which is easily verified to be normal. Thus, A is congruent to a normal matrix via the positive definite congruence $\mathrm{H}(A)^{-\frac{1}{2}}$. Finally, (b') trivially implies (b).   ☐

We next relate angular information about the spectrum of A to the angular field of values of positive definite multiples of A. Compare the next result to (1.7.9).

1.7.12   Theorem. Let Γ be an open angular sector of $\mathbb{C}$ anchored at the origin, with angle not greater than π, and let $A \in M_n$. The following are equivalent:

(a) $\sigma(A) \subset \Gamma$
(b) $F'(HA) \subset \Gamma$ for some positive definite $H \in M_n$

Proof: By (3.1.13) in [HJ] there is for each $\epsilon > 0$ a nonsingular $S_\epsilon \in M_n$ such that $S_\epsilon^{-1} A S_\epsilon$ is in modified Jordan canonical form: In place of every off-diagonal 1 that occurs in the Jordan canonical form of A is ϵ. Then, for sufficiently small ϵ, $F(S_\epsilon A S_\epsilon^{-1}) \subset \Gamma$ if $\sigma(A) \subset \Gamma$. But, by (1.2.12), we have $F'(S_\epsilon^* S_\epsilon A S_\epsilon^{-1} S_\epsilon) = F'(S_\epsilon A S_\epsilon^{-1}) \subset \Gamma$. Letting $H = S_\epsilon^* S_\epsilon$ demonstrates that (a) implies (b).
     The proof that (b) implies (a) is similar. If $F'(HA) \subset \Gamma$, then $F'(H^{\frac{1}{2}} A H^{-\frac{1}{2}}) = F'(H^{-\frac{1}{2}} H A H^{-\frac{1}{2}}) \subset \Gamma$, using (1.2.12) again, where $H^{\frac{1}{2}}$ is the positive definite square root of H and $H^{-\frac{1}{2}}$ is its inverse. But $\sigma(A) = \sigma(H^{\frac{1}{2}} A H^{-\frac{1}{2}}) \subset F(H^{\frac{1}{2}} A H^{-\frac{1}{2}}) \subset F'(H^{\frac{1}{2}} A H^{-\frac{1}{2}}) \subset \Gamma$, which completes the proof. ☐

Simultaneous diagonalization by congruence

We now return to the notion of simultaneous diagonalization of matrices by congruence (*congruence), which was discussed initially in Section (4.5) of [HJ].

1.7.13   Definition. We say that $A_1, A_2,..., A_m \in M_n$ are *simultaneously diagonalizable by congruence* if there is a nonsingular matrix $C \in M_n$ such that each of $C^* A_1 C, C^* A_2 C,..., C^* A_m C$ is diagonal.

A notion that arises in the study of simultaneous diagonalization by congruence and links it to the field of values, especially those topics studied in this section, is that of zeroes of the Hermitian form $x^* Ax$.

1.7.14   Definition. A nonzero vector $x \in \mathbb{C}^n$ is said to be an *isotropic vector* for a given matrix $A \in M_n$ if $x^* Ax = 0$, and x is further said to be a *common isotropic vector* for $A_1,..., A_m \in M_n$ if $x^* A_i x = 0, i = 1,..., m$.

Two simple observations will be of use.

1.7.15   Replacement Lemma. Let $\alpha_1,..., \alpha_m \in \mathbb{C}$ and $A_1,..., A_m \in M_n$ be given. The matrices $A_1,..., A_m$ are simultaneously diagonalizable by congruence if and only if the matrices

$$\sum_{i=1}^{n} \alpha_i A_i, A_2, A_3,..., A_m, \quad \alpha_1 \neq 0$$

are simultaneously diagonalizable by congruence.

Exercise. Perform the calculation necessary to verify (1.7.15).

If $m = 2$ and if $A_1, A_2 \in M_n$ are Hermitian, we often study $A = A_1 + iA_2$ in order to determine whether A_1 and A_2 are simultaneously diagonalizable by congruence. Note that $A_1 = \mathrm{H}(A)$ and $iA_2 = \mathrm{S}(A)$.

1.7.16   Observation. Let $A_1, A_2 \in M_n$ be Hermitian. Then A_1 and A_2 are simultaneously diagonalizable by congruence if and only if $A = A_1 + iA_2$ is congruent to a normal matrix.

Proof: If A_1 and A_2 are simultaneously diagonalizable by congruence via $C \in M_n$, then $C^* AC = C^* A_1 C + iC^* A_2 C = D_1 + iD_2$, where D_1 and D_2 are real diagonal matrices. The matrix C^*AC is then diagonal and, thus, is normal.

Conversely, suppose that a given nonsingular matrix $B \in M_n$ is such that B^*AB is normal. Then there is a unitary matrix $U \in M_n$ such that $U^*B^*ABU = D$ is diagonal. Now set $C = BU$ and notice that $H(D) = H(C^*AC) = C^*H(A)C = C^*A_1C$ is diagonal and that $-iS(D) = -iS(C^*AC) = -iC^*S(A)C = C^*A_2C$ is diagonal. □

We may now apply (1.7.11) to obtain a classical sufficient condition for two Hermitian matrices to be simultaneously diagonalizable by congruence.

Exercise. If $A_1, A_2 \in M_n$ are Hermitian and if $A = A_1 + iA_2$, show that $0 \notin F(A)$ if and only if A_1 and A_2 have no common isotropic vector.

1.7.17 Theorem. If $A_1, A_2 \in M_n$ are Hermitian and have no common isotropic vector, then A_1 and A_2 are simultaneously diagonalizable by congruence.

Proof: If A_1 and A_2 have no common isotropic vector, then $0 \notin F(A)$, where $A \equiv A_1 + iA_2$. Choosing $B = A$ in (1.7.11a), we conclude from the equivalence of (1.7.11a) and (1.7.11b) that A is congruent to a normal matrix. According to (1.7.16), then, A_1 and A_2 are simultaneously diagonalizable by congruence. □

Other classical sufficient conditions for pairs of Hermitian matrices to be simultaneously diagonalizable by congruence follow directly from (1.7.17).

Exercise. Let $A_1, A_2 \in M_n$ be given. Show that the set of common isotropic vectors of A_1 and A_2 is the same as the set of common isotropic vectors of $\alpha_1 A + \alpha_2 A_2$ and A_2 if $\alpha_1 \neq 0$. Since a positive definite matrix has no isotropic vectors, conclude that A_1 and A_2 have no common isotropic vectors if a linear combination of A_1 and A_2 is positive definite.

1.7.18 Corollary. If $A_1 \in M_n$ is positive definite and if $A_2 \in M_n$ is Hermitian, then A_1 and A_2 are simultaneously diagonalizable by congruence.

1.7.19 Corollary. If $A_1, A_2 \in M_n$ are Hermitian and if there is a linear combination of A_1 and A_2 that is positive definite, then A_1 and A_2 are simultaneously diagonalizable by congruence.

Exercise. Deduce (1.7.18) and (1.7.19) from (1.7.17) using (1.7.15) in the case of (1.7.19).

Exercise. Prove (1.7.18) directly, using Sylvester's law of inertia.

Exercise. If $A_1, A_2 \in M_n$ are Hermitian, show that there is a real linear combination of A_1 and A_2 that is positive definite if and only if A_1 and A_2 have no common isotropic vector. *Hint:* Choose $0 \le \theta < 2\pi$ so that $e^{i\theta}F(A)$ ⊂ *RHP*, $A \equiv A_1 + iA_2$, and calculate $\mathrm{H}(e^{i\theta}A) = (\cos\theta)A_1 - (\sin\theta)A_2$.

1.7.20   Example. The simple example $A_1 = A_2 = \begin{bmatrix} 1 & 0 \\ 0 & -1 \end{bmatrix}$ shows that the direct converses of (1.7.17), (1.7.18), and (1.7.19) do not hold.

Using (1.7.11) again, however, we can provide a converse to (1.7.17) that emphasizes the role of the condition "$0 \notin F(A)$" in the study of simultaneous diagonalization by congruence. We conclude this discussion of simultaneous diagonalization by congruence for two Hermitian matrices by listing this converse (1.7.21e) along with other equivalent conditions.

1.7.21   Theorem. Let $A_1, A_2 \in M_n$ be Hermitian and assume that A_1 and $A \equiv A_1 + iA_2$ are nonsingular. The following are equivalent:

(a) A_1 and A_2 are simultaneously diagonalizable by congruence;
(b) A is diagonalizable by congruence;
(c) A is congruent to a normal matrix;
(d) $A^{-1}A^*$ is similar to a unitary matrix;
(e)   There is some $\hat{A} \in M_n$ with $0 \notin F(\hat{A})$, such that $A^{-1}A^* = \hat{A}^{-1}\hat{A}^*$; and
(f) $A_1^{-1}A_2$ is similar to a real diagonal matrix.

Exercise. Prove (1.7.21). These equivalences are merely a compilation of previous developments. Observation (1.7.16) shows that (a) and (c) are equivalent, independent of the nonsingularity assumptions. Show that (b) and (c) are equivalent, independent of the nonsingularity assumptions, using the theory of normal matrices. Items (c), (d), and (e) are equivalent because of (1.7.11), and the equivalence of (a) and (f) was proven in Theorem (4.5.15) in [HJ]. Notice that the assumption of nonsingularity of A_1 is necessary only because of (f) and the assumption of nonsingularity of A is necessary only because of (d) and (e).

Exercise. Many conditions for two Hermitian matrices to be simultaneously diagonalizable by congruence can be thought of as generalizations of Corollary (1.7.18). Consider (1.7.21f) and show that if A_1 is positive definite, then $A_1^{-1}A_2$ has real eigenvalues and is diagonalizable by similarity.

Extensions of the Hermitian pair case to non-Hermitian matrices and to more than two matrices are developed in Problems 18 and 19.

Hadamard products

Recall that the Hadamard product $A \circ B$ is just the entrywise product of two matrices of the same size (see Definition (5.0.1)). For comparison, we state, without proof here, some results relating $F(\cdot)$ and the Hadamard product. Tools for the first of these (1.7.22) are developed at the end of Section (4.2). The remaining four results (1.7.23-26) may be deduced from (1.7.22).

1.7.22   Theorem. If A, $N \in M_n$ and if N is normal, then $F(N \circ A) \subset Co(F(N)F(A))$.

Proof: See Corollary (4.2.17).

1.7.23   Corollary. If A, $H \in M_n$ and if H is positive semidefinite, then $F(H \circ A) \subset F(H)F(A)$.

Exercise. Deduce (1.7.23) from (1.7.22) using the facts that H is normal and $F(H)F(A)$ is convex.

1.7.24   Corollary. If A, $B \in M_n$ and if either A or B is normal, then $r(A \circ B) \le r(A)r(B)$.

Exercise. Deduce (1.7.24) from (1.7.22).

1.7.25   Corollary. If A, $B \in M_n$, then $r(A \circ B) \le 2r(A)r(B)$.

Exercise. Deduce (1.7.25) from (1.7.24) using the representation $A = H(A) + S(A)$. Contrast this result with the corresponding inequality for usual matrix multiplication, given in Problem 22 in Section (5.7) of [HJ].

1.7.26   Theorem. If A, $H \in M_n$ and if H is positive definite, then $F'(H \circ A) \subset F'(A)$.

Exercise. Deduce (1.7.26) from (1.7.23).

1.7.27   Example. The example $A = B = \begin{bmatrix} 0 & 2 \\ 0 & 0 \end{bmatrix}$ shows that the constant 2 is the best possible in (1.7.25) and that (1.7.22) cannot be generalized to arbitrary A, $N \in M_n$.

1.7.28   Example. The example $A = N = \begin{bmatrix} 1 & 0 \\ 0 & i \end{bmatrix}$ shows that the "Co" cannot be dropped in (1.7.22), even when both A and N are normal.

Exercise. Show that $F'(A) \subset \cup \{F(H \circ A): H$ is positive definite$\}$.

Problems

1.   Determine all convex sets K such that KS is convex for all convex sets S.

2.   In spite of the failure of (1.7.1a) and example (1.7.4), show that $F(A^2) \subset \text{Co}(F(A)^2)$ for all $A \in M_n$. Conclude that $r(A^2) \leq r(A)^2$ for all $A \in M_n$.

3.   The power inequality for the numerical radius, $r(A^k) \leq r(A)^k$ for all positive integers k [see Problem 23 in Section (1.5)], suggests the conjecture

$$F(A^k) \subset \text{Co}(F(A)^k) \text{ for all } k = 1, 2, \ldots$$

Prove this conjecture and the corresponding numerical radius inequality when A is normal. C.-K. Li and M.-D. Choi have constructed non-normal counterexamples to the conjecture for all $k \geq 2$.

4.   If $A \in M_n(\mathbb{R})$, show that $F(A)$ is symmetric with respect to the real axis, and therefore $F(A^*) = F(A^T) = \overline{F(A)}$.

5.   If $A \in M_n(\mathbb{R})$ is nonsingular, show that $F'(A^{-1}) = F'(A)$.

6.   Under the assumptions of Theorem (1.7.6), show that $\sigma(B^{-1}A) \subset F(A)/F(B)$ also.

7.   If A is normal and B is positive definite, show that the eigenvalues of AB are contained in the convex cone generated by the eigenvalues of A.

8. Suppose that $A \in M_n$ is unitary with eigenvalues $e^{i\theta_1}, \ldots, e^{i\theta_n}$ and suppose, moreover, that $B \in M_n$ is positive semidefinite with eigenvalues $\beta_1 \leq \cdots \leq \beta_n$. Let $\lambda = re^{i\theta}$, $r > 0$, be a given eigenvalue of BA. Show that $\beta_1 \leq r \leq \beta_n$, and, if all $e^{i\theta_j}$ are contained in an arc of the unit circle of length less than or equal to π, show that θ is also contained in this arc.

9. Under the assumptions of (1.7.8), explain why one can prove the apparently better result $\sigma(AB) \subset F(A)F'(B)$. Why isn't this better?

10. Theorem (1.7.8) shows that some angular information about $\sigma(AB)$ may be gained in special circumstances. Note that this might be mated with the value of a matrix norm of A and B to give magnitudinal information also. Give an example.

11. If $A = [a_{ij}] \in M_n$ has positive diagonal entries, define $\hat{A} = [\hat{a}_{ij}]$ by

$$\hat{a}_{ij} \equiv \begin{cases} (|a_{ij}| + |a_{ji}|)/(2a_{ii}) & \text{if } i \neq j \\ \\ 0 & \text{if } i = j \end{cases}$$

Let $\Gamma_t = \{re^{i\theta}: r > 0, -\theta_0 \leq \theta \leq \theta_0\}$, where $\theta_0 = \arcsin(t)$, $0 \leq \theta_0 < \pi/2$ and $0 \leq t < 1$. Show that if $A \in M_n$ has positive diagonal entries and if $\rho(\hat{A}) < 1$, then $F'(A) \subset \Gamma_{\rho(\hat{A})}$.

12. Let $A = \begin{bmatrix} 1 & 1 \\ 0 & 1 \end{bmatrix}$ and $B = \begin{bmatrix} 1 & 0 \\ -1 & 1 \end{bmatrix}$. Compare the angular information about $\sigma(AB)$ given by (1.7.8) with that obtained from direct application of Geršgorin's Theorem (6.1.1,2) in [HJ] to the product AB. Note that $F'(A) = F'(B) = \Gamma_{\frac{1}{2}}$ and $\Gamma_{\frac{1}{2}}\Gamma_{\frac{1}{2}} = \Gamma_{\sqrt{1/2}}$. Note also that $\rho(\hat{A}) = \rho(\hat{B}) = \frac{1}{2}$, so these are cases of equality for the result of the preceding exercise. Either use that exercise or calculate $F'(A)$ and $F'(B)$ directly.

13. Generalize (1.7.10) for sectors of $\mathbb{C}$ other than RHP.

14. Show that $F(A^k) \subset RHP$ for all positive integers k if and only if A is positive definite.

15. Use the ideas of the proof of (1.7.9) to show that the result continues to hold if all matrices are restricted to be real.

16. If $A = H + S$, with H positive definite and S skew-Hermitian, verify that $H^{-\frac{1}{2}}SH^{-\frac{1}{2}}$ is skew-Hermitian and that $H^{-\frac{1}{2}}(H + S)H^{-\frac{1}{2}}$ is normal.

17. Show that for any $A \in M_n$ there is a real number t_0 such that $A + tI$ is congruent to a normal matrix for all $t > t_0$.

18. Let $A_1, ..., A_m \in M_n$. Show that the following statements are equivalent:

(a) $A_1, ..., A_m$ are simultaneously diagonalizable by congruence.

(b) $A_1, ..., A_m$ are simultaneously congruent to commuting normal matrices.

(c) The $2m$ Hermitian matrices $H(A_1)$, $iS(A_1)$, $H(A_2)$, $iS(A_2)$,..., $H(A_m)$, $iS(A_m)$ are simultaneously diagonalizable by congruence.

19. If $A_1, ..., A_{2q} \in M_n$ are Hermitian, show that $A_1, ..., A_{2q}$ are simultaneously diagonalizable by congruence if and only if $A_1 + iA_2$, $A_3 + iA_4$,..., $A_{2q-1} + iA_{2q}$ are simultaneously diagonalizable by congruence.

20. Let a nonsingular $A \in M_n$ be given. Explain why both factors in a polar decomposition $A = PU$ are uniquely determined, where $U \in M_n$ is unitary. Prove that $F'(U) \subset F'(A)$.

21. Let $A \in M_n$ be given with $0 \notin F(A)$ and let $A = PU$ be the polar factorization of A with $U \in M_n$ unitary. Prove that U is a *cramped unitary matrix*, that is, the eigenvalues of U lie on an arc of the unit circle of length less than π. Also show that $\Theta(U) \leq \Theta(A)$, where $\Theta(\cdot)$ is the field angle (1.1.3). Show that a unitary matrix V is cramped if and only if $0 \notin F(V)$.

22. Let $G \subset \mathbb{C}$ be a nonempty open convex set, and let $A \in M_n$ be given. Show that $\sigma(A) \subset G$ if and only if there is a nonsingular matrix $S \in M_n$ such that $F(S^{-1}AS) \subset G$.

23. If $A, B \in M_n$ are both normal with $\sigma(A) = \{\alpha_1, ..., \alpha_n\}$ and $\sigma(B) = \{\beta_1, ..., \beta_n\}$, show that $F(A \circ B)$ is contained in the convex polygon determined by the points $\alpha_i \beta_j$, for $i, j = 1, ..., n$.

24. Let $A \in M_n$ be given. Show that the following are equivalent:

(a) $A^* = S^{-1}AS$ for some $S \in M_n$ with $0 \notin F(S)$.

(b) $A^* = P^{-1}AP$ for some positive definite $P \in M_n$.

(c) A is similar to a Hermitian matrix.

(d) $A = PK$ for $P, K \in M_n$ with P positive definite and K Hermitian.

Compare with (4.1.7) in [HJ].

25. Let $A \in M_n$ be a given normal matrix. Show that there is an $S \in M_n$ with $0 \notin F(S)$ and $A^* = S^{-1}AS$ if and only if A is Hermitian.

26. Let $A \in M_n$ be given. Show that there is a unitary $U \in M_n$ with $0 \notin F(U)$ and $A^* = U^*AU$ if and only if A is Hermitian. Consider $A = \begin{bmatrix} 1 & -1 \\ 1 & -1 \end{bmatrix}$ and $U = \begin{bmatrix} 1 & 0 \\ 0 & -1 \end{bmatrix}$ to show that the identity $A^* = U^*AU$ for a general unitary U does not imply normality of A.

27. (a) If $A \in M_n$ and $0 \in F(A)$, show that $|z| \leq \||A - zI\||_2$ for all $z \in \mathbb{C}$.
(b) Use (a) to show that $\||AB - BA - I\||_2 \geq 1$ for all $A, B \in M_n$; in particular, the identity matrix is not a commutator. See Problem 6 in Section (4.5) for a generalization of this inequality to any matrix norm.
(c) If $A \in M_n$ and there is some $z \in \mathbb{C}$ such that $\||A - zI\||_2 < |z|$, show that $0 \notin F(A)$.

28. Let nonsingular $A, B \in M_n$ be given and let $C \equiv ABA^{-1}B^{-1}$. If A and C are normal, $AC = CA$, and $0 \notin F(B)$, show that $AB = BA$, that is, $C = I$. In particular, use Problem 21 to show that if A and B are unitary, $AC = CA$, and B is a cramped unitary matrix, then $AB = BA$. For an additive commutator analog of this result, see Problem 12 in Section (2.4) of [HJ].

Notes and Further Readings. The results (1.7.6-8) are based upon H. Wielandt, On the Eigenvalues of $A + B$ and AB, *J. Research N.B.S.* 77B (1973), 61-63, which was redrafted from Wielandt's National Bureau of Standards Report #1367, December 27, 1951, by C. R. Johnson. Theorem (1.7.9) is from C. R. Johnson, The Field of Values and Spectra of Positive Definite Multiples, *J. Research N.B.S.* 78B (1974) 197-198, and (1.7.10) was first proved by D. Carlson in A New Criterion for H-Stability of Complex Matrices, *Linear Algebra Appl.* 1 (1968), 59-64. Theorem (1.7.11) is from C. R. DePrima and C. R. Johnson, The Range of $A^{-1}A^*$ in $GL(n,\mathbb{C})$, *Linear Algebra Appl.* 9 (1974), 209-222, and (1.7.12) is from C. R. Johnson, A Lyapunov Theorem for Angular Cones, *J. Research National Bureau Standards* 78B (1974), 7-10. The treatment of simultaneous diagonalization by congruence is a new exposition centered around the field of values, which includes some classical results such as (1.7.17).

1.8   Generalizations of the field of values

There is a rich variety of generalizations of the field of values, some of which have been studied in detail. These generalizations emphasize various algebraic, analytic, or axiomatic aspects of the field of values, making it one of the most generalized concepts in mathematics. With no attempt to be complete, we mention, with occasional comments, several prominent or natural generalizations; there are many others. A natural question to ask about any generalized field of values is whether or not it is always convex: For some generalizations it is, and for some it is not. This gives further insight into the convexity (1.2.2) of the usual field of values, certainly one of its more subtle properties.

The first generalization involves a natural alteration of the inner product used to calculate each point of the field.

1.8.1   *Generalized inner product:* Let $H \in M_n$ be a given positive definite matrix. For any $A \in M_n$, define

$$F_H(A) \equiv \{x^* HAx: x \in \mathbf{C}^n, x^* Hx = 1\}$$

1.8.1a   **Observation.** Since H is positive definite, it can be written as $H = S^* S$ with $S \in M_n$ nonsingular. Then $F_H(A) = F(SAS^{-1})$, so $F_H(A)$ is just the usual field of values of a similarity of A, by a fixed similarity matrix.

Proof: $F_H(A) = \{x^* S^* SAS^{-1} Sx: x \in \mathbf{C}^n, x^* S^* Sx = 1\} = \{y^* SAS^{-1} y: y \in \mathbf{C}^n, y^* y = 1\} = F(SAS^{-1})$.   ☐

Exercise. There are many different matrices $S \in M_n$ such that $S^* S = H$. Why does it not matter which is chosen? *Hint:* If $T^* T = S^* S$, show that ST^{-1} is unitary and apply (1.2.8).

Exercise. Consider the fixed similarity generalization $F^S(A) \equiv F(SAS^{-1})$ for a fixed nonsingular $S \in M_n$. Show that $F^S(\cdot) = F_{S^* S}(\cdot) = F_H(\cdot)$ for $H \equiv S^* S$, so that the fixed similarity and generalized inner product generalizations are the same.

Exercise. Why is $F_H(A)$ always convex?

Exercise. Show that $\sigma(A) \subset F_H(A)$ for any positive definite matrix H. Show further that $\cap \{F_H(A): H \in M_n \text{ is positive definite}\} = Co(\sigma(A))$. *Hint:* Use (2.4.7) in [HJ]. For which matrices A does there exist a positive definite H with $F_H(A) = Co(\sigma(A))$?

Another generalization of the field of values is motivated by the fact that $x^* Ax \in M_1$ when $x \in \mathbf{C}^n$, so the usual field is just the set of determinants of all such matrices for normalized x.

1.8.2 *Determinants of isometric projections:* For any $m \leq n$ and any $A \in M_n$, define

$$F_m(A) \equiv \{\det(X^* A X): X \in M_{n,m}, X^* X = I \in M_m\}$$

Exercise. What are $F_1(\cdot)$ and $F_n(\cdot)$?

Exercise. Construct an example to show that $F_m(A)$ is not always convex.

The so-called k-field of values is another generalization, which is based upon replacement of a single normalized vector x with an orthonormal set. It is more related to the trace than to the determinant.

1.8.3 *The k-field of values:* Let k be a given positive integer. For any $A \in M_n$, define

$$F^k(A) \equiv \{\tfrac{1}{k}(x_1^* A x_1 + \cdots + x_k^* A x_k): X = [x_1 \ldots x_k] \in M_{n,k} \text{ and } X^* X = I\}$$

The set $F^k(A)$ is always convex, but the proof is involved, and this fact is a special case of a further generalization to follow.

Exercise. Show that both $F_m(\cdot)$ and $F^k(\cdot)$ are invariant under unitary similarity of their arguments.

Exercise. What are $F^1(A)$ and $F^n(A)$?

Exercise. If A is normal with eigenvalues $\lambda_1, \ldots, \lambda_n$, what is $F^k(A)$?

Exercise. Show that $F^n(A) \subset F^{n-1}(A) \subset \cdots \subset F^2(A) \subset F^1(A)$.

A further generalization of the k-field of values (1.8.3) is the c-field of values in which the equal positive coefficients $\tfrac{1}{k}, \tfrac{1}{k}, \ldots, \tfrac{1}{k}$ are replaced by arbitrary complex numbers.

1.8.4 *The c-field of values:* Let $c = [c_1, \ldots, c_n]^T \in \mathbb{C}^n$ be a given vector. For any $A \in M_n$, define

$$F^c(A) \equiv \{c_1 x_1^* A x_1 + \cdots + c_n x_n^* A x_n: X = [x_1 \ldots x_n] \in M_n \text{ and } X^* X = I\}$$

Exercise. Show that $F^k(A) = F^c(A)$ for $c = [1,..., 1, 0,..., 0]^T/k \in \mathbb{C}^n$. For what c is $F^c(A) = F(A)$?

Exercise. Is $F^c(\cdot)$ invariant under unitary similarity of its argument?

Exercise. Show that $F^c(A) = \{\text{tr}(CU^*AU): U \in M_n \text{ is unitary}\}$, where $C = \text{diag}(c_1,..., c_n)$.

Unfortunately, the c-field of values $F^c(A)$ is not always convex for every $c \in \mathbb{C}^n$, but it is always convex when c is a *real* vector. More generally, if $c \in \mathbb{C}^n$ is given, $F^c(A)$ is convex for all $A \in M_n$ if and only if the entries of c are collinear as points in the complex plane. If the entries of c are not collinear, there exists a normal $A \in M_n$ such that $F^c(A)$ is not convex. When $n \geq 5$, the ordinary field of values of $A \in M_n$ can be a convex polygon without A being normal, but it is known that A is normal if and only if there is some $c \in \mathbb{R}^n$ with distinct entries such that $F^c(A)$ is a convex polygon.

A set $S \subset \mathbb{C}$ is said to be *star-shaped with respect to a given point* $z_0 \in S$ if for every $z \in S$ the whole line segment $\{\alpha z + (1 - \alpha)z_0: 0 \leq \alpha \leq 1\}$ lies in S; a set $S \subset \mathbb{C}$ is convex if and only if it is star-shaped with respect to every point in S. Although some c-fields of values fail to be convex, it is known that $F^c(A)$ is always star-shaped with respect to the point $\frac{1}{n}(\text{tr } A)(c_1 + \cdots + c_n)$.

Exercise. Construct an example of a complex vector c and a matrix A for which $F^c(A)$ is not convex.

Exercise. Consider the vector set-valued function $^mF(\cdot)$ defined on M_n by

$$^mF(A) = \{[x_1^*Ax_1,..., x_m^*Ax_m]^T: X = [x_1 \ldots x_m] \in M_{n,m} \text{ and } X^*X = I\}$$

Show that $F^c(A)$ is the projection of $^nF(A)$ into $\mathbb{C}$ by the linear functional whose coefficients are the components of the vector c. Show that $^mF(\cdot)$ is invariant under unitary similarity of its argument and that if $y \in {}^mF(A)$, then $Py \in {}^mF(A)$ for any permutation matrix $P \in M_n$.

Naturally associated with the c-field of values is the *c-numerical radius*

$$r_c(A) \equiv \max\{|z|: z \in F^c(A)\}$$

$$= \max\{|\text{tr}(CU^*AU)|: U \in M_n \text{ is unitary}\}$$

where $c = [c_1,..., c_n]^T \in \mathbb{C}^n$ and $C \equiv \mathrm{diag}(c_1,..., c_n)$. When $n \geq 2$, the function $r_c(\cdot)$ is a norm on M_n if and only if $c_1 + \cdots + c_n \neq 0$ and the scalars c_i are not all equal; this condition is clearly met for $c = e_1 = [1, 0,..., 0]^T$, in which case $r_c(\cdot)$ is the usual numerical radius $r(\cdot)$. It is known that if $A, B \in M_n$ are Hermitian, then $r_c(A) = r_c(B)$ for all $c \in \mathbb{R}^n$ if and only if A is unitarily similar to $\pm B$.

A generalization of (1.8.4) is the C-field of values.

1.8.5 *The C-field of values:* Let $C \in M_n$ be given. For any $A \in M_n$, define

$$F^C(A) \equiv \{\mathrm{tr}(CU^*AU): U \in M_n \text{ is unitary}\}$$

Exercise. Show that $F^{V_1^*CV_1}(V_2^*AV_2) = F^C(A)$ if $V_1, V_2 \in M_n$ are unitary.

Exercise. If $C \in M_n$ is normal, show that $F^C(A) = F^c(A)$, where the vector c is the vector of eigenvalues of the matrix C. Thus, the C-field is a generalization of the c-field.

Exercise. Show that $F^A(C) = F^C(A)$ for all $A, C \in M_n$. Deduce that $F^C(A)$ is convex if either C or A is Hermitian, or, more generally, if either A or C is normal and has eigenvalues that are collinear as points in the complex plane.

Known properties of the c-field cover the issue of convexity of C-fields of values for normal C, but otherwise it is not known which pairs of matrices $C, A \in M_n$ produce a convex $F^C(A)$ and for which C the set $F^C(A)$ is convex for all $A \in M_n$.

Associated with the C-field of values are natural generalizations of the numerical radius, spectral norm, and spectral radius:

$$\vertiii{A}_C \equiv \max\{|\mathrm{tr}(CUAV)|: U, V \in M_n \text{ are unitary}\}$$

$$r_C(A) \equiv \max\{|\mathrm{tr}(CUAU^*)|: U \in M_n \text{ is unitary}\}$$

$$\rho_C(A) \equiv \max\left\{\left|\sum_{i=1}^n \lambda_i(A)\lambda_{\pi(i)}(C)\right|: \pi \text{ is a permutation of } 1,..., n\right\}$$

where the eigenvalues of A and C are $\{\lambda_i(A)\}$ and $\{\lambda_i(C)\}$, respectively. For any $A, C \in M_n$, these three quantities satisfy the inequalities

$$\rho_C(A) \le r_C(A) \le \||| A \|||_C \tag{1.8.5a}$$

Exercise. If $C = E_{11} = \text{diag}(1,0,\ldots,0)$, show that $\rho_C(A) = \rho(A)$, $r_C(A) = r(A)$, and $\||| A \|||_C = \||| A \|||_2$.

A generalization of a different sort is a family of objects often called the *Bauer fields of values*. They generalize the fact that the usual field of values is related to the l_2 vector norm $\| \cdot \|_2$. There is one of these generalized fields of values for each vector norm on $\mathbb{C}^n$. If $\| \cdot \|$ is a given vector norm on $\mathbb{C}^n$, let $\| \cdot \|^D$ denote the vector norm that is dual to $\| \cdot \|$. See (5.4.12-17) in [HJ] for information about dual norms that is needed here.

1.8.6 *The Bauer field of values associated with the norm* $\| \cdot \|$ *on* $\mathbb{C}^n$: For any $A \in M_n$, define

$$F_{\| \cdot \|}(A) \equiv \{y^* A x \colon x,\, y \in \mathbb{C}^n \text{ and } \|y\|^D = \|x\| = y^* x = 1\}$$

Notice that $F_{\| \cdot \|}(A)$ is just the range of the sesquilinear form $y^* A x$ over those normalized ordered pairs of vectors $(x,y) \in \mathbb{C}^n \times \mathbb{C}^n$ that are *dual pairs* with respect to the vector norm $\| \cdot \|$ (see (5.4.17) in [HJ]). The set $F_{\| \cdot \|}(A)$ is not always convex, but it always contains the spectrum $\sigma(A)$. There are further norm-related generalizations of the field of values that we do not mention here, but an example of another of these, a very natural one, is the set function $G_F(A)$ defined in (1.5.1).

Exercise. Show that $F_{\| \cdot \|_2}(A) = F(A)$ for every $A \in M_n$.

Exercise. Show that $\sigma(A) \subset F_{\| \cdot \|}(A)$ for every norm $\| \cdot \|$ on $\mathbb{C}^n$.

Exercise. Determine $F_{\| \cdot \|_1} \begin{bmatrix} 1 & 0 \\ -1 & 2 \end{bmatrix}$.

Exercise. Show that $F_{\nu(\cdot)}(A^*) = \overline{F_{\| \cdot \|}(A)}$, where $\nu(\cdot) \equiv \| \cdot \|^D$.

Exercise. If $\|x\|_S \equiv \|Sx\|$ for a nonsingular $S \in M_n$, determine $F_{\| \cdot \|_S}(A)$ in terms of $F_{\| \cdot \|}(\cdot)$.

Exercise. Give an example of an $F_{\| \cdot \|}(A)$ that is not convex.

For $A \in M_n(\mathbb{R})$, it may at first seem unnatural to study $F(A)$, which is defined in terms of vectors in $\mathbb{C}^n$. The motivation for this should be clear, though, from the results presented in this chapter so far. If $A \in M_n(\mathbb{R})$, there is a strictly real object corresponding to $F(A)$ that has been studied.

1.8.7   Definition. *The real field of values:* For any $A \in M_n(\mathbb{R})$, define

$$FR(A) \equiv \{x^T A x : x \in \mathbb{R}^n \text{ and } x^T x = 1\}$$

Notice that $FR(A) = FR([A + A^T]/2)$, so in studying the real field of values it suffices to consider only real symmetric matrices.

Exercise. Show that $FR(A) \subset \mathbb{R}$ and that all real eigenvalues of $A \in M_n(\mathbb{R})$ are contained in $FR(A)$.

Exercise. Show that $FR(A)$ is easy to compute: It is just the real line segment joining the smallest and largest eigenvalues of $H(A) = (A + A^T)/2$. In particular, therefore, $FR(A) = F(H(A))$ is always convex.

Exercise. If $A \in M_n(\mathbb{R})$, show that $FR(A) = F(A)$ if and only if A is symmetric.

Exercise. Which of the basic functional properties (1.2.1-12) does $FR(\cdot)$ share with $F(\cdot)$?

If $x, y \in \mathbb{C}^n$, $y^* y = 1$, and $x^* x = 1$, then $y^* x = 1$ implies that $y = x$ (this is a case of equality in the Cauchy-Schwarz inequality). This suggests another generalization of the field of values in which the Hermitian form $x^* A x$ is replaced by the sesquilinear form $y^* A x$, subject to a relation between y and x.

1.8.8   Definition. *The q-field of values:* Let $q \in [0,1]$ be given. For any $A \in M_n$, define

$$F_q(A) \equiv \{y^* A x : x, y \in \mathbb{C}^n, y^* y = x^* x = 1, \text{ and } y^* x = q\}$$

If $0 \leq q < 1$, $F_q(\cdot)$ is defined only for $n \geq 2$.

Exercise. Show that it is unnecessary to consider $F_q(\cdot)$ for a $q \in \mathbb{C}$ that is not nonnegative, for if $q \in \mathbb{C}$ with $|q| \leq 1$ and $F_q(\cdot)$ is defined analogously ($y^* x = q$), then $F_q(A) = e^{i\theta} F_{|q|}(A)$, where $q = e^{i\theta}|q|$.

Exercise. Let $q = 1$ and show that $F_1(A) = F(A)$ for all A. Thus, $F_1(A)$ is always convex.

Exercise. Let $q = 0$ and show that $F_0(A)$ is always a disc centered at the origin, and hence $F_0(A)$ is always convex. *Hint:* If $z \in F_0(A)$, show that $e^{i\theta}z \in F_0(A)$ for all $\theta \in [0, 2\pi)$. If z_1, z_2 are two given points in $F_0(A)$, show that they are connected by a continuous path lying in $F_0(A)$ (any pair of orthonormal vectors can be rotated continuously into any other pair of orthonormal vectors). It follows that $F_0(A)$ is an annulus centered at the origin. Finally, show that $0 \in F_0(A)$ by considering y^*Ax for an eigenvector x of A and some y that is orthogonal to x.

Exercise. Show that $F_0(A)$ is the set of 1,2 entries of matrices that are unitarily similar to A. The ordinary field of values may be thought of as the set of 1,1 entries of matrices that are unitarily similar to A.

N. K. Tsing has shown that $F_q(A)$ is convex for all $q \in [0, 1]$ and all $A \in M_n$, $n \geq 2$.

Thus far, the classical field of values and all but two of the generalizations of it we have mentioned have been objects that lie intrinsically in two real dimensions $[{}^mF(\cdot) \subset \mathbb{C}^m$ and $FR(A) \subset \mathbb{R}$ for $A \in M_n(\mathbb{R})]$. Another generalization, sometimes called the *shell*, lies in three real dimensions in an attempt to capture more information about the matrix.

1.8.9   Definition. *The Davis-Wielandt shell:* For any $A \in M_n$, define

$$DW(A) \equiv \{[\operatorname{Re} x^*Ax, \operatorname{Im} x^*Ax, x^*(A^*A)x]^T \colon x \in \mathbb{C}^n, x^*x = 1\}$$

Exercise. Let $A \in M_n$ have eigenvalues $\{\lambda_1, ..., \lambda_n\}$. If A is normal, show that $DW(A) = \operatorname{Co}(\{[\operatorname{Re} \lambda_i, \operatorname{Im} \lambda_i, |\lambda_i|^2]^T \colon i = 1, ..., n\})$.

One motivation for the definition of the Davis-Wielandt shell is that the converse of the assertion in the preceding exercise is also true. This is in contrast to the situation for the classical field of values, for which the simple converse to the analogous property (1.2.9) is not valid, as discussed in Section (1.6).

There are several useful multi-dimensional generalizations of the field of values that involve more than one matrix. The first is suggested naturally by thinking of the usual field of values as an object that lies in two real dimensions. For any $A \in M_n$, write $A = A_1 + iA_2$, where $A_1 = H(A) = (A + A^*)/2$ and $A_2 = -iS(A) = -i(A - A^*)/2$ are both Hermitian. Then

$F(A) = \{x^* A x: x \in \mathbf{C}^n,\ x^* x = 1\} = \{x^* A_1 x + i x^* A_2 x: x \in \mathbf{C}^n,\ x^* x = 1\}$, which (since $x^* A_1 x$ and $x^* A_2 x$ are both real) describes the same set in the plane as $\{(x^* A_1 x, x^* A_2 x): x \in \mathbf{C}^n,\ x^* x = 1\}$. Thus, the Toeplitz–Hausdorff theorem (1.2.2) says that the latter set in $\mathbb{R}^2$ is convex for any two Hermitian matrices $A_1,\ A_2 \in M_n$, and we are led to the following generalizations of $F(A)$ and $FR(A)$.

1.8.10   Definition. *The k-dimensional field of k matrices:* Let $k \geq 1$ be a given integer. For any $A_1,..., A_k \in M_n$ define

$$FC_k(A_1,..., A_k) = \{[x^* A_1 x,..., x^* A_k x]^T: x \in \mathbf{C}^n,\ x^* x = 1\} \subset \mathbf{C}^k$$

Similarly, if $A_1,..., A_k \in M_n(\mathbb{R})$, define

$$FR_k(A_1,..., A_k) \equiv \{[x^T A_1 x,..., x^T A_k x]^T: x \in \mathbb{R}^n,\ x^T x = 1\} \subset \mathbb{R}^k$$

Notice that when $k = 1$, $FC_1(A) = F(A)$ and $FR_1(A) = FR(A)$. For $k = 2$, $FC_2([A + A^*]/2,\ -i[A - A^*]/2)$ describes a set in a real two-dimensional subspace of $\mathbf{C}^2$ that is the same as $F(A)$. If the matrices $A_1,..., A_k$ are all Hermitian, then $FC_k(A_1,..., A_k) \subset \mathbb{R}^k \subset \mathbf{C}^k$. In considering FC_k, the case in which all of the k matrices are Hermitian is quite different from the general case in which the matrices are arbitrary, but in studying FR_k, it is convenient to know that we always have $FR_k(A_1,..., A_k) = FR_k([A_1 + A_1^*]/2,..., [A_k + A_k^*]/2)$.

Exercise. For any Hermitian matrices $A_1,\ A_2 \in M_n$, note that $FC_1(A_1) \subset \mathbb{R}^1$ and $FC_2(A_1, A_2) \subset \mathbb{R}^2$ are convex sets. For any real symmetric matrix $A_1 \in M_n(\mathbb{R})$, note that $FR_1(A_1) \subset \mathbb{R}^1$ is convex.

Exercise. For $n = 2$, consider the two real symmetric (Hermitian) matrices

$$A_1 = \begin{bmatrix} -1 & 0 \\ 0 & 1 \end{bmatrix} \text{ and } A_2 = \begin{bmatrix} 0 & 1 \\ 1 & 0 \end{bmatrix}$$

Show that $FR_2(A_1, A_2)$ is *not* convex but that $FC_2(A_1, A_2)$ *is* convex. Compare the latter set with $F(A_1 + iA_2)$. *Hint:* Consider the points in $FR_2(A_1, A_2)$ generated by $x = [1, 0]^T$ and $x = [0, 1]^T$. Is the origin in $FR_2(A_1, A_2)$?

The preceding exercise illustrates two special cases of what is known generally about convexity of the k-dimensional fields.

Convexity of FR_k:

> For $n = 1$, $FR_k(A_1,..., A_k)$ is a single point and is therefore convex for all $A_1,..., A_k \in M_1(\mathbb{R})$ and all $k \geq 1$. For $n = 2$, $FR_1(A)$ is convex for every $A \in M_2(\mathbb{R})$, but for every $k \geq 2$ there are matrices $A_1,..., A_k \in M_2(\mathbb{R})$ such that $FR_k(A_1,..., A_k)$ is not convex. For $n \geq 3$, $FR_1(A_1)$ and $FR_2(A_1, A_2)$ are convex for all $A_1, A_2 \in M_n(\mathbb{R})$, but for every $k \geq 3$ there are matrices $A_1,..., A_k \in M_n(\mathbb{R})$ such that $FR_k(A_1,..., A_k)$ is not convex.

Convexity of FC_k for Hermitian matrices:

> For $n = 1$, $FC_k(A_1,..., A_k)$ is a single point and is therefore convex for all $A_1,..., A_k \in M_1$ and all $k \geq 1$. For $n = 2$, $FC_1(A_1)$ and $FC_2(A_1, A_2)$ are convex for all Hermitian $A_1, A_2 \in M_2$, but for every $k \geq 3$ there are Hermitian matrices $A_1,..., A_k \in M_2$ such that $FC_k(A_1,..., A_k)$ is not convex. For $n \geq 3$, $FC_1(A_1)$, $FC_2(A_1, A_2)$, and $FC_3(A_1, A_2, A_3)$ are convex for all Hermitian $A_1, A_2, A_3 \in M_n$, but for every $k \geq 4$ there are Hermitian matrices $A_1,..., A_k \in M_n$ such that $FR_k(A_1,..., A_k)$ is not convex.

The definition (1.1.1) for the field of values carries over without change to matrices and vectors with quaternion entries, but there are some surprising new developments in this case. Just as a complex number can be written as $z = a_1 + ia_2$ with $a_1, a_2 \in \mathbb{R}$ and $i^2 = -1$, a *quaternion* can be written as $\zeta = a_1 + ia_2 + ja_3 + ka_4$ with $a_1, a_2, a_3, a_4 \in \mathbb{R}$, $i^2 = j^2 = k^2 = -1$, $ij = -ji = k$, $jk = -kj = i$, and $ki = -ik = j$. The *conjugate* of the quaternion $\zeta = a_1 + ia_2 + ja_3 + ka_4$ is $\bar{\zeta} \equiv a_1 - ia_2 - ja_3 - ka_4$; its *absolute value* is $|\zeta| \equiv (a_1^2 + a_2^2 + a_3^2 + a_4^2)^{\frac{1}{2}}$; its *real part* is $\mathrm{Re}\,\zeta \equiv a_1$. The set of quaternions, denoted by $\mathbb{Q}$, is an algebraically closed division ring (noncommutative field) in which the inverse of a nonzero quaternion ζ is given by $\zeta^{-1} \equiv \bar{\zeta}/|\zeta|^2$. The quaternions may be thought of as lying in four real dimensions, and the real and complex fields may be thought of as subfields of $\mathbb{Q}$ in a natural way.

We denote by $M_n(\mathbb{Q})$ the set of n-by-n matrices with quaternion entries and write $\mathbb{Q}^n$ for the set of n-vectors with quaternion entries; for $x \in \mathbb{Q}^n$, x^* denotes the transpose of the entrywise conjugate of x.

1.8.11   Definition. *The quaternion field of values.* For any $A \in M_n(\mathbb{Q})$, define $FQ(A) \equiv \{x^*Ax: \; x \in \mathbb{Q}^n \text{ and } x^*x = 1\}$.

Although the quaternion field of values $FQ(A)$ shares many properties with the complex field of values, it need not be convex even when A is a normal complex matrix. If we set

$$A_1 \equiv \begin{bmatrix} i & 0 & 0 \\ 0 & 1 & 0 \\ 0 & 0 & 1 \end{bmatrix} \text{ and } A_2 \equiv \begin{bmatrix} i & 0 & 0 \\ 0 & i & 0 \\ 0 & 0 & 1 \end{bmatrix}$$

then $FQ(A_1)$ is *not* convex but $FQ(A_2)$ *is* convex; in the classical case, $F(A_1)$ and $F(A_2)$ are identical. It is known that, for a given $A \in M_n(\mathbb{Q})$, $FQ(A)$ is convex if and only if $\{\mathrm{Re}\ \zeta: \; \zeta \in FQ(A)\} = \{\zeta: \; \zeta \in FQ(A) \text{ and } \zeta = \mathrm{Re}\ \zeta\}$, that is, if and only if the projection of $FQ(A)$ onto the real axis is the same as the intersection of $FQ(A)$ with the real axis.

Notes and Further Readings. The generalizations of the field of values that we have mentioned in this section are a selection of only a few of the many possibilities. Several other generalizations of the field of values are mentioned with references in [Hal 67]. Generalization (1.8.1) was studied by W. Givens in Fields of Values of a Matrix, *Proc. Amer. Math. Soc.* 3 (1952), 206-209. Some of the generalizations of the field of values discussed in this section are the objects of current research by workers in the field such as M. Marcus and his students. The convexity results in (1.8.4) are in R. Westwick, A Theorem on Numerical Range, *Linear Multilinear Algebra* 2 (1975), 311-315, and in Y. Poon, Another Proof of a Result of Westwick, *Linear Multilinear Algebra* 9 (1980), 35-37. The converse is in Y.-H. Au-Yeung and N. K. Tsing, A Conjecture of Marcus on the Generalized Numerical Range, *Linear Multilinear Algebra* 14 (1983), 235-239. The fact that the c-field of values is star-shaped is in N.-K. Tsing, On the Shape of the Generalized Numerical Ranges, *Linear Multilinear Algebra* 10 (1981), 173- 182. For a survey of results about the c-field of values and the c-numerical radius, with an extensive bibliography, see C.-K. Li and N.-K. Tsing, Linear Operators that Preserve the c-Numerical Range or Radius of Matrices, *Linear Multilinear Algebra* 23 (1988), 27-46. For a discussion of the generalized spectral norm, generalized numerical radius, and generalized spectral radius introduced in Section (1.8.5), see C.-K. Li, T.-Y. Tam, and N.-K. Tsing, The Generalized Spectral Radius, Numerical Radius and Spectral Norm, *Linear Multilinear Algebra* 16 (1984), 215-237. The convexity of $F_q(A)$ mentioned

in (1.8.8) is shown in N. K. Tsing, The Constrained Bilinear Form and the
C-Numerical Range, *Linear Algebra Appl.* 56 (1984), 195-206. The shell
generalization (1.8.9) was considered independently and in alternate forms
in C. Davis, The Shell of a Hilbert Space Operator, *Acta Sci. Math. (Szeged)*
29 (1968), 69-86 and in H. Wielandt, Inclusion Theorems for Eigenvalues,
U.S. Department of Commerce, National Bureau of Standards, Applied
Mathematics Series 29 (1953), 75-78.   For more information about
$FR_k(A_1,..., A_k)$ see L. Brickman, On the Field of Values of a Matrix, *Proc.
Amer. Math. Soc.* 12 (1961), 61-66.   For a proof that $FC_3(A_1,A_2,A_3)$ is
convex when A_1, A_2, $A_3 \in M_n$ are Hermitian and $n \geq 3$, and for many related
results, see P. Binding, Hermitian Forms and the Fibration of Spheres,
Proc. Amer. Math. Soc. 94 (1985), 581-584.   Also see Y.-H. Au-Yeung and
N. K. Tsing, An Extension of the Hausdorff-Toeplitz Theorem on the
Numerical Range, *Proc. Amer. Math. Soc.* 89 (1983), 215-218.   For references to the literature and proofs of the assertions made about the quaternion
field of values $FQ(A)$, see Y.-H. Au-Yeung, On the Convexity of Numerical
Range in Quaternionic Hilbert Spaces, *Linear Multilinear Algebra* 16 (1984),
93-100. There is also a discussion of the quaternion field of values in Section
II of the 1951 paper by R. Kippenhahn cited at the end of Section (1.6), but
the reader is warned that Kippenhahn's basic Theorem 36 is *false:* the quaternion field of values is *not* always convex.

   There are strong links between the field of values and Lyapunov's
theorem (2.2).   Some further selected readings for Chapter 1 are: C. S.
Ballantine, Numerical Range of a Matrix: Some Effective Criteria, *Linear
Algebra Appl.* 19 (1978), 117-188; C. R. Johnson, Computation of the Field
of Values of a 2-by-2 Matrix, *J. Research National Bureau Standards* 78B
(1974), 105-107; C. R. Johnson, Numerical Location of the Field of Values,
Linear Multilinear Algebra 3 (1975), 9-14; C. R. Johnson, Numerical Ranges
of Principal Submatrices, *Linear Algebra Appl.* 37 (1981), 23-34; F. Murna-
ghan, On the Field of Values of a Square Matrix, *Proc. National Acad. Sci.
U.S.A.* 18 (1932), 246-248; B. Saunders and H. Schneider, A Symmetric
Numerical Range for Matrices, *Numer. Math.* 26 (1976), 99-105; O. Taussky, A Remark Concerning the Similarity of a Finite Matrix A and A^*,
Math. Z. 117 (1970), 189-190; C. Zenger, Minimal Subadditive Inclusion
Domains for the Eigenvalues of Matrices, *Linear Algebra Appl.* 17 (1977),
233-268.

Chapter 2

Stable matrices and inertia

2.0   Motivation

The primary motivation for the study of stable matrices (matrices whose eigenvalues have negative real parts) stems from a desire to understand stability properties of equilibria of systems of differential equations. Questions about such equilibria arise in a variety of forms in virtually every discipline to which mathematics is applied—physical sciences, engineering, economics, biological sciences, etc. In each of these fields it is necessary to study the dynamics of systems whose state changes, according to some rule, over time and, in particular, to ask questions about the long-term behavior of the system. Thus, matrix stability, an initial tool in the study of such questions, is an important topic that has been a major area in the interplay between the applications and evolution of matrix theory.

2.0.1   Stability of equilibria for systems of linear, constant coefficient, ordinary differential equations

Consider the first-order linear constant coefficient system of n ordinary differential equations:

$$\frac{dx}{dt} = A[x(t) - \hat{x}] \tag{2.0.1.1}$$

where $A \in M_n(\mathbb{R})$ and $x(t), \hat{x} \in \mathbb{R}^n$. It is clear that if $x(\hat{t}) = \hat{x}$ at some time $\hat{t}$, $x(t)$ will cease changing at $t = \hat{t}$. Thus, $\hat{x}$ is called an *equilibrium* for this system. If A is nonsingular, $x(t)$ will cease changing *only* when it has reached the equilibrium $\hat{x}$. Central questions about such a system are:

89

(a)   Will $x(t)$ converge to $\hat{x}$ as $t \to \infty$, given an initial point $x(0)$?
(b)   More importantly, will $x(t)$ converge to $\hat{x}$ for *all* choices of the initial data $x(0)$?

If $x(t)$ converges to $\hat{x}$ as $t \to \infty$ for every choice of the initial data $x(0)$, the equilibrium $\hat{x}$ is said to be *globally stable*. It is not difficult to see that the equilibrium is globally stable if and only if each eigenvalue of A has negative real part. A matrix A satisfying this condition is called a *stable matrix*. If we define e^{At} by the power series

$$e^{At} = \sum_{k=0}^{\infty} \frac{1}{k!} A^k t^k$$

then e^{At} is well defined (that is, the series converges) for all t and all A (see Chapter 5 of [HJ]) and $\frac{d}{dt} e^{At} = A e^{At}$. The unique solution $x(t)$ to our differential equation may then be written as

$$x(t) = e^{At}[x(0) - \hat{x}] + \hat{x} \qquad (2.0.1.2)$$

Exercise. Verify that this function $x(t)$ satisfies the differential equation (2.0.1.1).

Exercise. Just as $e^{at} \to 0$ as $t \to \infty$ if and only if the complex scalar a satisfies Re $a < 0$, show that $e^{At} \to 0$ as $t \to \infty$ if and only if each eigenvalue λ of A satisfies Re $\lambda < 0$. Conclude that (2.0.1.2) satisfies $x(t) \to \hat{x}$ as $t \to \infty$ for all choices of the initial data $x(0)$ if and only if A is stable. *Hint:* How are the eigenvalues of e^A related to those of A? If $e^{At} \to 0$ as $t \to \infty$, $t \in \mathbb{R}$, then $(e^A)^k \to 0$ as $k \to \infty$, $k = 1, 2, \dots$. Now use the criterion in Theorem (5.6.12) in [HJ] for a matrix to be *convergent*. Conversely, $||| e^{At} ||| = ||| e^{A[t] + A(t)} ||| \le ||| (e^A)^{[t]} ||| \, ||| e^{A(t)} ||| \le ||| e^{A[t]} ||| \, ||| e^{||| A |||} |||$ for any matrix norm $||| \cdot |||$, where $[t]$ denotes the greatest integer in t and $(t) \equiv t - [t]$ is the fractional part of t. If A is stable, $(e^A)^{[t]} \to 0$ as $t \to \infty$.

2.0.2   Local stability of equilibria for a nonlinear system of ordinary differential equations

Stable matrices are also relevant to the study of more general systems of n not necessarily linear ordinary differential equations

$$\frac{dx}{dt} = f(x(t) - \hat{x})$$
(2.0.2.1)

in which we assume that: (a) each component of $f: \mathbb{R}^n \to \mathbb{R}^n$ has continuous partial derivatives, and (b) $f(x) = 0$ if and only if $x = 0$. Again, $x(t)$ stops changing if and only if it reaches the equilibrium $\hat{x}$ and questions about the stability of this equilibrium arise naturally. Now, however, there are many possible notions of stability, not all of which may be addressed with matrix theory alone. One that may be so addressed is a notion of *local stability*. Suppose we ask whether $x(t)$ converges to $\hat{x}$ for every choice of the initial value $x(0)$ that is *sufficiently close* to $\hat{x}$. That is, will the system converge to equilibrium after arbitrary but small perturbations from equilibrium? The notions of "small" and "sufficiently close" can be made precise, and, by appeal to Taylor's theorem, the system (2.0.2.1) may be replaced by the linear system

$$\frac{dx}{dt} = J_f[x(t) - \hat{x}]$$

in some neighborhood of $\hat{x}$, without altering the qualitative dynamical properties of the original system. Here

$$J_f \equiv \left[\frac{\partial f_i}{\partial x_j}(0) \right] \in M_n(\mathbb{R})$$

is the Jacobian of f evaluated at the origin. Thus, $x(t) \to \hat{x}$ for all initial values $x(0)$ in this neighborhood ($\hat{x}$ is said to be a *locally stable* equilibrium) if all eigenvalues of the Jacobian J_f have negative real parts, that is, if J_f is a stable matrix.

2.1   Definitions and elementary observations

We begin the study of stable matrices by considering the general notion of inertia for matrices in M_n.

2.1.1   Definition. If $A \in M_n$, define:

$i_+(A) \equiv$ the number of eigenvalues of A, counting multiplicities, with positive real part;

$i_-(A)$ ≡   the number of eigenvalues of A, counting multiplicities, with
             negative real part; and

$i_0(A)$ ≡   the number of eigenvalues of A, counting multiplicities, with
             zero real part.

Then, $i_+(A) + i_-(A) + i_0(A) = n$ and the row vector

$$i(A) \equiv [i_+(A),\ i_-(A),\ i_0(A)]$$

is called the *inertia* of A. This generalizes the notion of the inertia of a
Hermitian matrix defined in Section (4.5) of [HJ].

Exercise. Show that $i(-A)^T = \begin{bmatrix} 0 & 1 & 0 \\ 1 & 0 & 0 \\ 0 & 0 & 1 \end{bmatrix} i(A)^T$.

For Hermitian matrices $A, B \in M_n$, Sylvester's law of inertia (Theorem
(4.5.8) in [HJ]) says that B is *congruent to A if and only if $i(A) = i(B)$.

Exercise. For a Hermitian $A \in M_n$, define positive and negative definite and
semidefinite in terms of $i(A)$.

Although the motivation of this chapter is the study of stable matrices,
for reasons purely of mathematical convenience, we shall speak in terms of
matrices all of whose eigenvalues have *positive* real parts and use the term
positive stable to avoid confusion. There is no substantive difference if the
theory is developed this way.

2.1.2   Definition. A matrix $A \in M_n$ is said to be *positive stable* if $i(A) =$
$[n, 0, 0]$, that is, if $i_+(A) = n$.

Exercise. Show that inertia, and therefore stability, is similarity invariant.
Exercise. Show that $A \in M_n$ is positive stable if and only if $-A$ is stable in the
sense arising in the study of differential equations mentioned in Section
(2.0); thus any theorem about positive stable matrices can be translated to
one about stable matrices by appropriate insertion of minus signs.
Exercise. If $A \in M_n$ is positive stable, show that A is nonsingular.

If A is positive stable, then so are many matrices simply related to A.

2.1.3   Observation. If $A \in M_n$ is positive stable, then each of the following is positive stable:

(a) $aA + bI$, $a \geq 0, b \geq 0, a + b > 0$,
(b) A^{-1},
(c) A^*, and
(d) A^T.

Exercise. Prove the four assertions in Observation (2.1.3).

2.1.4   Observation. If $A \in M_n(\mathbb{R})$ is positive stable, then det $A > 0$.

Proof: Since A is real, any complex eigenvalues occur in conjugate pairs whose product contributes positively to the determinant, and, since A is positive stable, any real eigenvalues are positive and hence their product contributes positively to the determinant.   ☐

Exercise. Let $A \in M_n(\mathbb{R})$ be positive stable.   Show that A^k has positive determinant for all integers k.

Exercise. (a) Show by example that the converse of (2.1.4) is false. (b) If A is positive stable, show by example that A^2 need not be.

Exercise. If $A \in M_n$ is positive stable, show that Re tr $A > 0$; if $A \in M_n(\mathbb{R})$, show that tr $A > 0$.

There is, of course, considerable interplay between the *root location problem* of determining conditions under which zeroes of a polynomial lie in some given region, such as the right half of the complex plane, and the problem of determining whether a matrix is positive stable. We will not explore this interplay fully in this chapter, but we note two obvious facts here. All zeroes of the polynomial $p(\cdot)$ lie in the right half-plane if and only if the companion matrix associated with $p(\cdot)$ is positive stable, and a matrix is positive stable if and only if all zeroes of its characteristic polynomial lie in the right half-plane. Thus, conditions for the zeroes of a polynomial to lie in the right half-plane and conditions for positive stability of a matrix may be used interchangeably, where convenient. Also, since the coefficients of the characteristic polynomial of a matrix are just ± the sums of principal minors of a given size, polynomial conditions may often be interpreted in terms of principal minor sums; see (1.2.9-12) in [HJ].

Problems

1. If $A \in M_n$ is nonsingular and has all real eigenvalues, show that A^2 is positive stable.

2. Show that $A \in M_2(\mathbb{R})$ is positive stable if and only if tr $A > 0$ and det $A > 0$.

3. Let $A \in M_n(\mathbb{R})$ be a given real positive stable matrix with characteristic polynomial $p_A(t) = t^n + c_1 t^{n-1} + c_2 t^{n-2} + \cdots + c_{n-1}t + c_n$. Show that the coefficients c_i are all real, nonzero, and strictly alternate in sign: $c_1 < 0$, $c_2 > 0$, $c_3 < 0$,.... Conclude that $E_k(A)$, the sum of all the k-by-k principal minors of the real positive stable matrix A, is positive, $k = 1,..., n$. See (1.2.9-12) in [HJ], where it is shown that $E_k(A)$ is equal to the sum of all the $\binom{n}{k}$ possible products of k eigenvalues of A.

4. Suppose $A \in M_n$ is positive stable. Show that A^k is positive stable for all positive integers k (in fact, all integers) if and only if the eigenvalues of A are real.

5. Recall that a Hermitian matrix has positive leading principal minors if and only if it is positive definite. Show by example that positive leading principal minors are neither necessary nor sufficient for positive stability, even for real matrices.

6. If $A \in M_n$ is partitioned as

$$A = \begin{bmatrix} A_{11} & A_{12} \\ 0 & A_{22} \end{bmatrix}$$

in which $A_{ii} \in M_{n_i}$, $i = 1, 2$ and $n_1 + n_2 = n$, show that $i(A) = i(A_{11}) + i(A_{22})$.

7. Let $A \in M_n(\mathbb{R})$ be given with $n \geq 2$. Show by a diagonal counterexample that the converse to the result of Problem 3 is false; that is, it is possible for A to have all its principal minor sums $E_k(A)$ positive but not be positive stable. What is the smallest value of n for which this can happen? Why? If $E_k(A) > 0$ for $k = 1,..., n$, there is something useful that can be said about the location of $\sigma(A)$: Show that all the eigenvalues of A lie in the open angular wedge $W_n \equiv \{z = re^{i\theta} \in \mathbb{C}: r > 0, |\theta| < \pi - \pi/n\}$. Sketch W_n. What does this say for $n = 2$? What happens for $n = 1$? Conversely, every point in W_n is an eigenvalue of some real n-by-n matrix, all of whose principal minors

are positive (not just the principal minor sums $E_k(A)$); see Theorem (2.5.9).

8. If $A \in M_n$ is Hermitian and is partitioned as

$$A = \begin{bmatrix} B & C \\ C^* & D \end{bmatrix}$$

with B nonsingular, show that $i(A) = i(B) + i(D - C^*B^{-1}C)$, that is, the inertia of a submatrix plus that of its Schur complement (see (0.8.5) in [HJ]).

9. Consider $\begin{bmatrix} -1 & 2 \\ -2 & 3 \end{bmatrix}$ to show that a principal submatrix of a positive stable matrix need not be positive stable.

10. Let $A \in M_n$ be such that $H(A) = \frac{1}{2}(A + A^*)$ is positive definite. Use a field of values argument to show that every principal submatrix of A (including A itself) is positive stable. Use (1.7.7) or (1.7.10) to show that DA is positive stable for every positive diagonal matrix $D \in M_n(\mathbb{R})$ and, more generally, HA is positive stable for every positive definite $H \in M_n$.

2.2   Lyapunov's theorem

In this section we present a basic version of a fundamental theorem about positive stable matrices. It has many interesting generalizations, several of which will be discussed in Section (2.4).

Exercise. If $A \in M_n$ is positive definite, show that A is positive stable.

Exercise. If $A \in M_n$ and if $H(A) \equiv \frac{1}{2}(A + A^*)$ is positive definite, show that A is positive stable but not conversely. *Hint:* Use a field of values argument; there are straightforward direct arguments also. For a counterexample to the converse, try a 2-by-2 positive stable triangular matrix with a large off-diagonal entry.

Positive definite matrices are a special case of positive stable matrices, but Lyapunov's theorem shows that they are intimately related to all positive stable matrices. Just as we might think of Hermitian matrices as a natural matrix analog of real numbers and positive definite Hermitian matrices as an analog of positive real numbers, positive stable matrices are a natural matrix analog of complex numbers with positive real part. This conceptualization is sometimes helpful and sometimes useless. It can be useful in remembering Lyapunov's theorem, for if $g \in \mathbb{R}$ is positive and $a \in \mathbb{C}$, then $2\operatorname{Re} ag = [ag + (ag)^*] > 0$ if and only if $\operatorname{Re} a > 0$.

2.2.1   Theorem (Lyapunov). Let $A \in M_n$ be given. Then A is positive stable if and only if there exists a positive definite $G \in M_n$ such that

$$GA + A^*G = H \qquad (2.2.2)$$

is positive definite. Furthermore, suppose there are Hermitian matrices $G, H \in M_n$ that satisfy (2.2.2), and suppose H is positive definite; then A is positive stable if and only if G is positive definite.

Proof: For any nonsingular $S \in M_n$ we may perform a *congruence of the identity (2.2.2) and obtain $S^* GSS^{-1}AS + S^* A^* S^{*-1} S^* GS = S^* HS$, which may be rewritten as

$$\hat{G}\hat{A} + \hat{A}^*\hat{G} = \hat{H} \qquad (2.2.\hat{2})$$

in which $\hat{G} = S^* GS$, $\hat{A} = S^{-1}AS$, and $\hat{H} = S^* HS$. That is, *congruence applied to (2.2.2) replaces G and H by *congruent Hermitian matrices $\hat{G}$ and $\hat{H}$ and replaces A by a matrix $\hat{A}$ that is similar to A. Recall that *congruence preserves inertia (and therefore positive definiteness) of Hermitian matrices, and, of course, similarity always preserves eigenvalues. Thus, any one of the three matrices G, H, or A in (2.2.2) may be assumed to be in any special form achievable by simultaneous *congruence (for G and H) and similarity (for A).

Now assume that A is positive stable and that $\hat{A}$ is in the modified Jordan canonical form in which any nonzero super-diagonal entries are all equal to $\epsilon > 0$, where ϵ is chosen to be less than the real part of every eigenvalue of A; see Corollary (3.1.13) in [HJ]. Then set $\hat{G} = I \in M_n$ and observe that the resulting matrix $\hat{G}\hat{A} + \hat{A}^*\hat{G} = \hat{A} + \hat{A}^* = 2\mathrm{H}(\hat{A}) = \hat{H} \in M_n$ in (2.2.$\hat{2}$) is Hermitian (even tridiagonal, real, symmetric), has positive diagonal entries (the real parts of the eigenvalues of A), and is strictly diagonally dominant (see Theorem (6.1.10b) in [HJ]). Thus, $\hat{G}$ and $\hat{H}$ are positive definite and the forward implication of the first assertion is proved.

Now suppose there are positive definite matrices G and $H \in M_n$ that satisfy (2.2.2). Then write $GA \equiv B$, so that $B + B^* = H$ is positive definite.

Let $G^{\frac{1}{2}}$ be the unique positive definite square root of G and let $G^{-\frac{1}{2}}$ be its inverse (see Theorem (7.2.6) in [HJ]). Multiplication on both left and right by $G^{-\frac{1}{2}}$ yields $G^{\frac{1}{2}}AG^{-\frac{1}{2}} = G^{-\frac{1}{2}}BG^{-\frac{1}{2}}$, for which $G^{-\frac{1}{2}}BG^{-\frac{1}{2}} + (G^{-\frac{1}{2}}BG^{-\frac{1}{2}})^* = G^{-\frac{1}{2}}HG^{-\frac{1}{2}}$ is positive definite. Since $G^{\frac{1}{2}}AG^{-\frac{1}{2}}$ therefore has positive definite Hermitian part (see the preceding exercise), both it and A (to which it is similar) are positive stable. This verifies the backward implication of both assertions.

To verify the last necessary implication, let $A, G_1 \in M_n$ be such that A is positive stable, G_1 is Hermitian, and $G_1A + A^*G_1 \equiv H_1$ is positive definite. Suppose that G_1 is *not* positive definite. By the forward implication already proved, there is a positive definite $G_2 \in M_n$ such that $G_2A + A^*G_2 \equiv H_2$ is also positive definite. Set $G_1A \equiv B_1$ and $G_2A \equiv B_2$, so that $B_1 = \frac{1}{2}(H_1 + S_1)$ and $B_2 = \frac{1}{2}(H_2 + S_2)$, with $S_1, S_2 \in M_n$ skew-Hermitian. Consider the Hermitian matrix function $G(\alpha) \equiv \alpha G_1 + (1 - \alpha)G_2$ for $\alpha \in [0,1]$. All the eigenvalues of $G(0) = G_2$ are positive, while at least one eigenvalue of $G(1) = G_1$ is not positive. Since the eigenvalues of $G(\alpha)$ are always real and depend continuously on α, there is some $\alpha_0 \in (0,1]$ such that at least one eigenvalue of $G(\alpha_0)$ is zero, that is, $G(\alpha_0)$ is singular. Notice that

$$G(\alpha_0)A = \alpha_0 B_1 + (1 - \alpha_0)B_2$$
$$= \frac{1}{2}\alpha_0 H_1 + \frac{1}{2}(1 - \alpha_0)H_2 + \frac{1}{2}\alpha_0 S_1 + \frac{1}{2}(1 - \alpha_0)S_2$$

has positive definite Hermitian part. This is a contradiction: A matrix with positive definite Hermitian part is nonsingular (its field of values lies in the open right half-plane), while $G(\alpha_0)A$ is singular because $G(\alpha_0)$ is singular. Thus, the original Hermitian matrix G_1 could not fail to be positive definite and the proof is complete. □

The equation (2.2.2) is often referred to as the *Lyapunov equation*, and a matrix G for which $GA + A^*G$ is positive definite is called a *Lyapunov solution* of the Lyapunov equation.

The basic Lyapunov theorem may be refined to indicate the attainable right-hand sides in (2.2.2) and give a more definitive procedure to test for the stability of $A \in M_n$.

Note that (2.2.2) is a system of linear equations for the entries of G. In Chapter 4 the linear matrix equation (2.2.2) will be studied and we shall see (Theorem (4.4.6) and Corollary (4.4.7)) that (2.2.2) has a unique solution G for any given right-hand side H if and only if $\sigma(A^*) \cap \sigma(-A) = \phi$, a condition that is certainly fulfilled if A is positive stable. Thus, as a consequence of general facts about matrix equations, we have the first assertion of the following theorem:

2.2.3   Theorem. Let $A \in M_n$ be positive stable. For each given $H \in M_n$, there is a unique solution G to (2.2.2). If a given right-hand side H is Hermitian, then the corresponding solution G is necessarily Hermitian; if H is positive definite, then G is also positive definite.

Proof: The assertion "H is Hermitian implies that G is Hermitian" follows from the first (uniqueness) assertion, for if G is the solution to $GA + A^*G = H$, then $H = H^* = (GA + A^*G)^* = G^*A + A^*G^*$, and hence $G^* = G$. The assertion "H is positive definite implies that G is positive definite" follows from the last part of (2.2.1).   ▯

Exercise. Let $A \in M_n$ be positive stable. According to (2.2.1) and (2.2.3), the solution G of (2.2.2) is positive definite for each given positive definite H. Show by example, however, that a positive definite G need not produce a positive definite H. *Hint:* Choose a positive stable A such that $H(A) = \frac{1}{2}(A + A^*)$ is not positive definite, showing that $G = I$ qualifies.

Thus, each positive stable A induces, via Lyapunov's theorem, a function $G_A(\cdot) : M_n \to M_n$ given by

$$G_A(H) \equiv \text{the unique solution } G \text{ to } GA + A^*G = H$$

This function is evidently one-to-one, for if $G_A(H_1) = G_A(H_2)$, then $H_1 = G_A(H_1)A + A^*G_A(H_1) = G_A(H_2)A + A^*G_A(H_2) = H_2$. In particular, the function $G_A(\cdot)$ establishes a one-to-one correspondence between the set of positive definite matrices and a (typically proper) subset of itself. When A is diagonalizable, there is a simple explicit formula for the function $G_A(\cdot)$; see Problem 3.

Because of (2.2.1) and (2.2.3), a common form in which Lyapunov's theorem is often seen is the following:

2.2.4   Corollary. Let $A \in M_n$ be given. Then A is positive stable if and only if there is a positive definite matrix G satisfying the equation

$$GA + A^*G = I \qquad\qquad (2.2.5)$$

If A is positive stable, there is precisely one solution G to this equation, and G is positive definite. Conversely, if for a given $A \in M_n$ there exists a positive definite solution G, then A is positive stable and G is the unique solution to the equation.

Exercise. If $A \in M_n$ is positive stable, show that $\sigma(A^*) \cap \sigma(-A) = \phi$.

Exercise. Prove Corollary (2.2.4).

Exercise. Show that $\begin{bmatrix} -1 & -2 \\ 2 & 3 \end{bmatrix}$ is positive stable by solving equation (2.2.5) and checking the solution for positive definiteness.

A definitive test for the positive stability of a given matrix $A \in M_n$ may now be summarized as follows: Choose your favorite positive definite matrix H (perhaps $H = I$), attempt to solve the linear system (2.2.2) for G, and if there is a solution, check G for positive definiteness. The matrix A is stable if and only if a solution G can be found and G passes the test of positive definiteness. This method has the pleasant advantage that the question of positive stability for an arbitrary $A \in M_n$ can be transferred to the question of positive definiteness for a Hermitian matrix. Unfortunately, the cost of this transferal is the solution of an n^2-by-n^2 linear system, although this can be reduced somewhat. Fortunately, a Lyapunov solution sometimes presents itself in a less costly way, but this general procedure can be useful as a tool.

Problems

1.   Show that if G_1 and G_2 are two Lyapunov solutions for A, then $aG_1 + bG_2$ is a Lyapunov solution for A as long as $a, b \geq 0$ and $a + b > 0$. A set with this property is called a (convex) *cone*. Show that the positive definite matrices themselves form a cone and that A is positive stable if and only if its cone of Lyapunov solutions is nonempty and is contained in the cone of positive definite matrices.

2.   (a) Suppose $x(t) = [x_i(t)] \in \mathbb{C}^n$, where each complex-valued function $x_i(t)$ is absolutely integrable over an interval $[a,b] \subset \mathbb{R}$. Show that the matrix $A \equiv \int_a^b x(t)x^*(t)dt \in M_n$ is positive semidefinite.

(b) Let $\lambda_1,..., \lambda_n \in \mathbb{C}$ be given with $\operatorname{Re} \lambda_i > 0$ for $i = 1,..., n$. Show that the matrix $L \equiv [(\bar{\lambda}_i + \lambda_j)^{-1}] \in M_n$ is positive semidefinite and has positive main diagonal entries.

3.   Let a given $A \in M_n$ be diagonalizable and suppose $A = S \Lambda S^{-1}$, $S, \Lambda \in M_n$, S is nonsingular, $\Lambda = \operatorname{diag}(\lambda_1,..., \lambda_n)$ and $\bar{\lambda}_i + \lambda_j \neq 0$ for all $i, j = 1,..., n$. Show that $GA + A^*G = H$ if and only if

$$G = (S^{-1})^*[L(A) \circ (S^* H S)]S^{-1}$$

where $L(A) \equiv [(\bar{\lambda}_i + \lambda_j)^{-1}]$ and $\circ$ denotes the Hadamard product. If A is diagonalizable and positive stable, deduce that the cone of Lyapunov solutions for A is given by

$$\{(S^{-1})^*[L(A) \circ H]S^{-1}: H \in M_n \text{ is positive definite}\}$$

4.   Use Problem 3 to determine the cone of Lyapunov solutions for $A = \begin{bmatrix} 3 & -2 \\ -4 & 3 \end{bmatrix}$.

5.   Use the ideas in Problem 3 to give a different proof of the last implication demonstrated in the proof of Theorem (2.2.1).

6.   If A and B are both positive stable, show that $A + B$ is positive stable if A and B have a common Lyapunov solution. Is it possible for $A + B$ to be positive stable without A and B having a common Lyapunov solution?

7.   Give an example of a matrix $A \in M_2$, with $i_0(A) = 0$, such that (2.2.2) does not have a solution for every Hermitian $H \in M_2$.

8.   If H is Hermitian and there is a unique solution G to (2.2.2), show that G must be Hermitian.

9.   Let $A \in M_n$ be positive stable and let $H \in M_n$ be given. For $t > 0$, let $P(t) \equiv e^{-A^*t}He^{-At}$ and show that

$$G \equiv \int_0^\infty P(t)\, dt$$

exists, satisfies $GA + A^*G = H$, and is positive definite if H is positive definite.

2.3   The Routh-Hurwitz conditions

There is another important stability criterion that focuses on the principal minor sums or characteristic polynomial of $A \in M_n(\mathbb{R})$.

2.3.1   Definition. Let $A \in M_n$. For $k = 1,\dots,n$, $E_k(A)$ is the sum of the $\binom{n}{k}$ principal minors of A of order k, that is, $\pm E_k(A)$ is the coefficient of t^{n-k} in the characteristic polynomial of A. For example, tr $A = E_1(A)$ and det $A = E_n(A)$; see Theorem (1.2.12) in [HJ].

2.3.2   Definition. The *Routh-Hurwitz matrix* $\Omega(A) \in M_n(\mathbb{R})$ associated with $A \in M_n(\mathbb{R})$ is

$$
\Omega(A) = \begin{bmatrix}
E_1(A) & E_3(A) & E_5(A) & \cdots & 0 \\
1 & E_2(A) & E_4(A) & \cdots & 0 \\
0 & E_1(A) & \cdot & & \cdot \\
\cdot & 1 & \cdot & \ddots & \cdot \\
\cdot & 0 & \cdot & \ddots & 0 \\
0 & 0 & 0 & \cdots & E_n(A)
\end{bmatrix}
$$

The diagonal entries of $\Omega(A) = [\omega_{ij}]$ are $\omega_{ii} = E_i(A)$. In the column above ω_{ii} are $\omega_{i-1,i} = E_{i+1}(A)$, $\omega_{i-2,i} = E_{i+2}(A)$, ... up to the first row ω_{1i} or to $E_n(A)$, whichever comes first; all entries above $E_n(A)$ are zero. In the column below ω_{ii} are $\omega_{i+1,i} = E_{i-1}(A)$, $\omega_{i+2,i} = E_{i-2}(A)$, ..., $E_1(A)$, 1, 0, 0, ... down to the last row ω_{ni}.

The Routh-Hurwitz criterion for $A \in M_n(\mathbb{R})$ to be positive stable is that the leading principal minors of $\Omega(A)$ all be positive. Two items are worth noting, however.

(a)   The Routh-Hurwitz stability criterion is a special case of more general and complicated conditions for determining the inertia of $A \in M_n(\mathbb{R})$ in certain circumstances.

(b)   The Routh-Hurwitz criterion may equivalently be stated as a test for "positive stability" of a given polynomial, that is, a test for whether all its zeroes are in the right half-plane.

2.3.3   Theorem (Routh-Hurwitz stability criterion). A matrix $A \in M_n(\mathbb{R})$ is positive stable if and only if the leading principal minors of the Routh-Hurwitz matrix $\Omega(A)$ are positive.

We omit a proof of the Routh-Hurwitz criterion. One modern technique of proof may be described as follows: Write the characteristic polynomial of A in terms of the $E_k(A)$ and construct a matrix of simple form (such as a companion matrix) that has this polynomial as its characteristic polynomial. Then apply Lyapunov's theorem (and perform many algebraic calculations) to obtain the Routh-Hurwitz conditions.

Problems

1. Consider the polynomial $p(t) = t^n + a_{n-1}t^{n-1} + \cdots + a_1 t + a_0$ with real coefficients. Use the Routh-Hurwitz criterion to give necessary and sufficient conditions on the coefficients so that all roots of $p(t)$ have positive real part.

2. Prove (2.3.3) for $n = 2$ and $n = 3$.

3. Consider the matrix

$$A = \begin{bmatrix} x & a & b \\ \alpha & y & c \\ \beta & \gamma & z \end{bmatrix} \in M_3(\mathbb{R})$$

Show that DA is positive stable for every positive diagonal matrix $D \in M_3(\mathbb{R})$ if (a) all principal minors of A are positive, and (b) $xyz > (ac\beta + \alpha\gamma b)/2$.

4. Show that the sufficient condition (b) in Problem 3 is not necessary by considering the matrix

$$A = \begin{bmatrix} 6 & 5 & -1 \\ 1 & 2 & 5 \\ 5 & -3 & 1 \end{bmatrix}$$

Show that $H(A) = \frac{1}{2}(A + A^*)$ is positive definite, so DA is positive stable for all positive diagonal $D \in M_3(\mathbb{R})$; see Problem 10 in Section (2.1).

Further Reading: For a proof of Theorem (2.3.3), see Chapter 3 of [Bar 83].

2.4   Generalizations of Lyapunov's theorem

Lyapunov's fundamental Theorem (2.2.1) may be generalized in a variety of ways. We mention a selection of the many generalizations in this section.

These include:

(a)   Circumstances under which positive stability may be concluded when a solution G of (2.2.2) is positive definite, but the right-hand side H is only positive semidefinite;

(b)   The general inertia result when the right-hand side H of (2.2.2) is positive definite, but the solution G is Hermitian of general inertia; and

(c)   Positive stability conditions involving possibly varying positive definite multiples of A.

For additional generalizations that involve the field of values, see Section (1.7).

Consider the Lyapunov equation

$$GA + A^*G = H, \qquad A, G, H \in M_n \tag{2.4.1}$$

in which we assume that G, and therefore H, is Hermitian. Define

$$S \equiv GA - A^*G \tag{2.4.2}$$

so that S is skew-Hermitian and

$$2GA = H + S \tag{2.4.3}$$

We next give a definitive criterion (2.4.7) for positive stability in the case of a positive definite G for which the matrix H in (2.4.1) is only positive semidefinite. Recall that a positive semidefinite matrix is necessarily Hermitian.

2.4.4   **Definition.** A matrix $A \in M_n$ is said to be *positive semistable* if $i_-(A) = 0$, that is, if every eigenvalue of A has nonnegative real part.

2.4.5   **Lemma.** Let $A \in M_n$. If $GA + A^*G = H$ for some positive definite $G \in M_n$ and some positive semidefinite $H \in M_n$, then A is positive semistable.

Proof: Suppose that $\lambda \in \sigma(A)$ and $Ax = \lambda x$, $x \neq 0$. Then $2GAx = 2\lambda Gx$ and from (2.4.3) we have $(H + S)x = 2\lambda Gx$. Multiplication on the left by x^* gives $x^*(H + S)x = 2\lambda x^* Gx$, and extraction of real parts yields

$$2(\operatorname{Re} \lambda)x^* Gx = x^* Hx \qquad\qquad (2.4.6)$$

Since $x^* Gx > 0$ and $x^* Hx \geq 0$, we conclude that $\operatorname{Re} \lambda \geq 0$.   ▯

If the circumstances of the lemma guarantee positive semistability, then what more is needed for positive stability? This question is answered in the following theorem.

2.4.7   Theorem. Let $A \in M_n$, suppose that $GA + A^* G = H$ for some positive definite $G \in M_n$ and some positive semidefinite $H \in M_n$, and let $S \equiv GA - A^* G$. Then A is positive stable if and only if no eigenvector of $G^{-1}S$ lies in the nullspace of H.

Proof: Suppose x is an eigenvector of $G^{-1}S$ that also lies in the nullspace of H. Then $x \neq 0$ and, for some $\lambda \in \mathbb{C}$, $0 = \lambda Gx - Sx = (\lambda G - S - H)x = (\lambda G - 2GA)x = G(\lambda x - 2Ax)$, so $Ax = \frac{1}{2}\lambda x$. Since $\frac{1}{2}\lambda$ is then an eigenvalue of A and, by (2.4.6), $\operatorname{Re} \lambda = x^* Hx / x^* Gx = 0$ (because x is in the nullspace of H), A is not positive stable. This means that positive stability of A implies the "no eigenvector" condition.

Because of (2.4.5), all eigenvalues of A have nonnegative real part. If A is not positive stable, then there is a $\lambda \in \sigma(A)$ with $\operatorname{Re} \lambda = 0$. From (2.4.6), we conclude that $x^* Hx = 0$ if x is an eigenvector of A associated with λ. Since H is positive semidefinite, this can happen only if $Hx = 0$, that is, x is in the nullspace of H. Combining $(H + S)x = 2\lambda Gx$ (a consequence of x being an eigenvector associated with $\lambda \in \sigma(A)$) and $Hx = 0$, we conclude that $Sx = 2\lambda Gx$, that is, x is an eigenvector of $G^{-1}S$ that lies in the nullspace of H. This means that the "no eigenvector" condition implies positive stability for A.   ▯

Exercise. Show that Theorem (2.4.7) implies the backward implication of the first statement in Theorem (2.2.1).

Exercise. Show that Lemma (2.4.5) is just a generalization to nonstrict inequalities (semidefinite, semistable) of the backward implication in the first statement of Theorem (2.2.1). Prove the lemma by a continuity argument.

Exercise. Unlike the positive stable case, the converse of (2.4.5) does not hold. Give an example of a positive semidefinite $A \in M_n$ for which no positive definite $G \in M_n$ and positive semidefinite $H \in M_n$ exist that satisfy (2.4.1). *Hint:* Try something with nontrivial Jordan block structure associated with a purely imaginary multiple eigenvalue.

2.4.8   Definition. Let $A \in M_n$, $B \in M_{n,m}$. The pair of matrices (A, B) is said to be *controllable* if rank $[B \ \ AB \ \ A^2 B \dots A^{n-1} B] = n$.

The concept of controllability arises in the theory of linear control differential equation systems. Let $A \in M_n(\mathbb{R})$ and $B \in M_{n,m}(\mathbb{R})$ be given. For any given continuous vector function $u(t) \in \mathbb{R}^m$, the initial-value problem

$$\frac{dx}{dt} = Ax(t) + Bu(t), \quad x(0) = x_0$$

has a unique solution $x(t)$ for each given $x_0 \in \mathbb{R}^n$; the vector function $u(\cdot)$ is the *control*. The question of controllability of this system is the following: For each $\hat{x}$, $x_0 \in \mathbb{R}^n$, is there some $\hat{t} < \infty$ and some choice of the control vector $u(\cdot)$ such that the solution satisfies $x(\hat{t}) = \hat{x}$? The answer: There is if and only if the pair (A, B) is controllable in the sense (2.4.8).

2.4.9   Remark. The condition on $G^{-1}S$ in (2.4.7) is known to be equivalent to the statement that the pair (A^*, H) is controllable. Thus, if a positive definite $G \in M_n$ and a positive semidefinite $H \in M_n$ satisfy (2.4.1), then $A \in M_n$ is positive stable if and only if the pair (A^*, H) is controllable.

We next turn to information about the inertia of A that follows if one has a Hermitian (but not necessarily positive definite or positive semidefinite) solution G of the Lyapunov equation (2.4.1) with H positive definite. This is known as the *general inertia theorem*.

2.4.10   Theorem. Let $A \in M_n$ be given. There exists a Hermitian $G \in M_n$ and a positive definite $H \in M_n$ such that $GA + A^*G = H$ if and only if $i_0(A) = 0$. In this event, $i(A) = i(G)$.

Proof: If $i_0(A) = 0$, recall the form (2.2.$\hat{2}$), equivalent to (2.4.1), in which we again take $\hat{A} = [\hat{a}_{ij}]$ to be in modified Jordan canonical form with any nonzero super-diagonal entries equal to $\epsilon > 0$, and choose $\epsilon < \min \{|\operatorname{Re} \lambda| : \lambda \in \sigma(A)\}$. Now choose $\hat{G} = E = \operatorname{diag}(e_1, ..., e_n)$, in which

$$e_i \equiv \begin{cases} 1 \text{ if } \operatorname{Re} \hat{a}_{ii} > 0 \\ -1 \text{ if } \operatorname{Re} \hat{a}_{ii} < 0 \end{cases}, \quad i = 1, ..., n$$

The resulting Hermitian matrix $\hat{H} = E\hat{A} + \hat{A}^* E$ has positive diagonal entries and is strictly diagonally dominant, and therefore it is positive definite. On the other hand, suppose there is a Hermitian G such that $H \equiv GA + A^* G$ is positive definite, and suppose that $i_0(A) \neq 0$. Again appeal to the equivalent form (2.2.$\hat{2}$) and take $\hat{A}$ to be in (usual) Jordan canonical form with a purely imaginary 1,1 entry. Since the 1,1 entry of $\hat{G}$ is real, and all entries below the 1,1 entry in the first column of $\hat{A}$ are zero, a calculation reveals that the 1,1 entry of the matrix $\hat{H}$ resulting in (2.2.$\hat{2}$) is zero. Such an $\hat{H}$ cannot be positive definite, contradicting our assumption that $i_0(A) \neq 0$. We conclude that $i_0(A) = 0$, which completes the proof of the first assertion in the theorem.

If a Hermitian G (necessarily nonsingular) and a positive definite H satisfy (2.4.1), we verify that $i(A) = i(G)$ by induction on the dimension n. For $n = 1$, the assertion is clear: If Re $ga > 0$ and g real, then Re a and g have the same sign. Now suppose the assertion is verified for values of the dimension up to and including $n - 1$ and again appeal to the equivalent form (2.2.$\hat{2}$). We claim that a nonsingular matrix $S \in M_n$ can be chosen so that

$$\hat{G} = S^* GS = \begin{bmatrix} \hat{G}_{11} & 0 \\ 0 & \hat{g}_{22} \end{bmatrix} \text{ and } \hat{A} = S^{-1}AS = \begin{bmatrix} \hat{A}_{11} & \hat{a}_{12} \\ 0 & \hat{a}_{22} \end{bmatrix}$$

in which $\hat{G}_{11} \in M_{n-1}$ is Hermitian and $\hat{A}_{11} \in M_{n-1}$. This may be seen by choosing S_1 so that $S_1^{-1}AS_1$ is in Jordan canonical form and then choosing S_2 of the form $S_2 = \begin{bmatrix} I & x \\ 0 & 1 \end{bmatrix}$, in which $x = -G_{11}^{-1} g_{12}$ and

$$S_1^* GS_1 = \begin{bmatrix} G_{11} & g_{12} \\ g_{12}^* & g_{22} \end{bmatrix}, \ G_{11} \in M_{n-1}, g_{12} \in \mathbb{C}^{n-1}$$

If G_{11} is singular, then the original G may be perturbed slightly, $G \to G + \epsilon I$, with $\epsilon > 0$ sufficiently small so that neither the positive definiteness of H nor the inertia of G is altered—note that G itself must be nonsingular—and so that the resulting G_{11} is nonsingular. Nothing essential about the problem is thus changed and so there is no loss of generality to assume that G_{11} is nonsingular. Then let $S = S_1 S_2$ to produce $\hat{A}$ and $\hat{G}$ in (2.2.$\hat{2}$) that have the claimed form. The induction then proceeds as follows: Since $\hat{G}\hat{A} + \hat{A}^* \hat{G} = \hat{H}$ is positive definite and since principal submatrices of positive definite matrices are positive definite, $\hat{G}_{11}\hat{A}_{11} + \hat{A}_{11}^* \hat{G}_{11} = \hat{H}_{11}$ is positive definite.

Application of the induction hypotheses, together with the observation that the inertia of a block triangular matrix is the sum of the inertias of the blocks, then yields the desired induction step. Note that Re $\hat{g}_{22}\hat{a}_{22}$ = the last diagonal entry of $\hat{H}$ (and H is positive definite), so the real number $\hat{g}_{22}$ has the same sign as Re $\hat{a}_{22}$. This completes the proof of the second assertion and the theorem. ☐

Exercise. Show that Lyapunov's theorem (2.2.1) follows from (2.4.10).

Lyapunov's theorem asks for a single positive definite matrix G such that the Hermitian quadratic form x^*GAx has positive real part for all nonzero $x \in \mathbb{C}^n$. If there is such a G, Lyapunov's theorem guarantees that A is positive stable. This global requirement (one G for all x) can be relaxed considerably to a very local requirement (one G for each x).

2.4.11   Theorem. Let $A \in M_n$ be given. Then A is positive stable if and only if for each nonzero $x \in \mathbb{C}^n$ there exists a positive definite G (which may depend on x) such that Re $x^*GAx > 0$.

Proof: Let $x \in \mathbb{C}^n$ be an eigenvector of A associated with a given $\lambda \in \sigma(A)$ and let $G \in M_n$ be any given positive definite matrix. Then Re $x^*GAx =$ Re $x^*G(\lambda x) =$ Re $x^*Gx\lambda = (x^*Gx)$Re λ has the same sign as Re λ since $x^*Gx > 0$. We conclude that if for each nonzero $x \in \mathbb{C}^n$ there is a positive definite $G \in M_n$ with Re $x^*GAx > 0$, then A must be positive stable. On the other hand, if A is positive stable, apply Lyapunov's Theorem (2.2.1) to produce a G satisfying (2.2.2) with H positive definite. Then, for *any* nonzero $x \in \mathbb{C}^n$, Re $x^*GAx = \frac{1}{2}(x^*GAx + x^*A^*Gx) = \frac{1}{2}x^*Hx > 0$. ☐

2.4.12   Definition. Let $A \in M_n$ and a nonzero vector $x \in \mathbb{C}^n$ be given. Define $L(A,x) \equiv \{G \in M_n : G$ is positive definite and Re $x^*GAx > 0\}$.

Theorem (2.4.11) says that A is positive stable if and only if $L(A,x) \neq \phi$ for all nonzero $x \in \mathbb{C}^n$.

Exercise. Prove the somewhat remarkable fact that, for a given $A \in M_n$, $L(A,x) \neq \phi$ for all nonzero $x \in \mathbb{C}^n$ if and only if $\cap\{L(A,x): 0 \neq x \in \mathbb{C}^n\} \neq \phi$.
Exercise. Prove the forward implication of (2.4.11) by direct argument without using Lyapunov's theorem.

If H, $K \in M_n$ are Hermitian, it is a basic fact that if H is positive definite, then $i(HK) = i(K)$. If H is negative definite, the inertia of HK is also completely determined by that of K. In general, there are multiple possibilities for $i(HK)$ given $i(H)$ and $i(K)$. It is known that $A \in M_n$ is similar to a matrix in $M_n(\mathbb{R})$ if and only if A may be factored as $A = HK$ with H and K Hermitian. These facts are discussed in (4.1.7) in [HJ]. For more completeness in our discussion of general inertia theory, we include without proof a description of the possible values of the inertia of the product HK, given $i(H)$ and $i(K)$, for H, $K \in M_n$ Hermitian and nonsingular, that is, $i_0(H) = i_0(K) = 0$. In this event, of course, $i_-(H) = n - i_+(H)$ and $i_-(K) = n - i_+(K)$, but $i_0(HK)$ may be a nonzero even integer. For simplicity in description, let

$$S(p,q,n) \equiv \{\, |p+q-n|,\ |p+q-n|+2,\ |p+q-n|+4, \ldots, n-|p-q|\,\}$$

for nonnegative integers p, $q \leq n$. Note that $|p+q-n|$ and $n-|p-q|$ have the same parity.

2.4.13   Theorem. If H, $K \in M_n$ are Hermitian and nonsingular with $i_+(H) = p$ and $i_+(K) = q$, then

$$i_+(HK) \in S(p,q,n),\quad i_-(HK) \in S(n-p,q,n)$$

and

$$i_0(HK) \in \{0, 2, 4, \ldots, n-|p+q-n|-|p-q|\}$$

Conversely, if

$$i_1 \in S(p,q,n),\quad i_2 \in S(n-p,q,n)$$

and

$$i_3 \in \{0, 2, 4, \ldots, n-|p+q-n|-|p-q|\}$$

satisfy $i_1 + i_2 + i_3 = n$, then there exist nonsingular Hermitian matrices H, $K \in M_n$ such that

$$i_+(H) = p,\ i_+(K) = q,\ \text{and } i(HK) = (i_1, i_2, i_3)$$

Exercise. Verify (2.4.13) in the special case $n = 2$. *Hint:* Use $\begin{bmatrix} 0 & 1 \\ 1 & 0 \end{bmatrix}$ and $\begin{bmatrix} -1 & 0 \\ 0 & 1 \end{bmatrix}$ to show existence in the "intermediate" cases.

Exercise. Show that $n \in S(p,q,n)$ if and only if $p = q$, and $0 \in S(p,q,n)$ if and only if $p + q = n$.

There are two useful special cases of Theorem (2.4.13) that follow immediately from the preceding characterization of the extremal cases in which $0, n \in S(p,q,n)$.

2.4.14   **Corollary.** Let $H, K \in M_n$ be Hermitian. Then
(a)   If $i_+(HK) = n$, that is, if HK is positive stable, then H and K are nonsingular and $i(H) = i(K)$.
(b)   If H and K are nonsingular and $i_+(HK) = 0$, that is, if $-HK$ is positive stable, then $i_+(H) + i_+(K) = n$ and $i_-(H) + i_-(K) = n$.

Exercise. Verify the two assertions of the preceding corollary.

Exercise. Verify that the following statement is equivalent to (2.4.14a): If $A, B \in M_n$ are such that A is positive stable, B is Hermitian and nonsingular, and AB is Hermitian, then $i(AB) = i(B)$. Consider $A = \begin{bmatrix} -1 & 2 \\ 2 & 3 \end{bmatrix}$, $B_1 = \begin{bmatrix} 1 & 0 \\ 0 & 0 \end{bmatrix}$, and $B_2 = \begin{bmatrix} 0 & 0 \\ 0 & 1 \end{bmatrix}$ to show that the nonsingularity assumption on B may not be omitted. *Hint:* Let $K = B^{-1}$ and $H = AB$.

It is well known that if $A, B \in M_n$ with A positive definite and B Hermitian, then $i(AB) = i(B)$ (see (7.6.3) in [HJ]). The preceding exercise shows that the same conclusion follows from the weaker hypothesis that A is merely positive stable, if we simultaneously strengthen the hypothesis on B to include nonsingularity and assume a special interaction between A and B, namely, that AB is Hermitian. The latter assumption implies that A is similar to a real matrix (see (4.1.7) in [HJ]) and hence, in particular, $\sigma(A)$ must be symmetric with respect to the real axis, the Jordan blocks of A associated with nonreal eigenvalues must occur with a special symmetry, and so forth. Thus, only a rather restricted class of positive stable matrices is covered by this result. As a final application of the general inertia theorem in this section, we now show that we may drop entirely the assumptions that B is nonsingular and AB is Hermitian if we strengthen the assumption on A from positive stability to $\mathrm{H}(A) = \frac{1}{2}(A + A^*)$ being positive definite, that is, $F(A)$ is contained in the right half-plane.

2.4.15   **Theorem.** Let $A, B \in M_n$ with B Hermitian and $A + A^*$ positive definite. Then $i(AB) = i(B)$.

Proof: First assume that B is nonsingular. Since nonsingular congruence preserves positive definiteness, $B^*(A + A^*)B = BAB + BA^*B = B(AB) + (AB)^*B$ is positive definite, and hence $i(AB) = i(B)$ by (2.4.10). If B is singular with $i_0(B) = k$, let $U^*BU = \begin{bmatrix} \Lambda & 0 \\ 0 & 0 \end{bmatrix}$ with $U \in M_n$ unitary and a nonsingular diagonal $\Lambda \in M_{n-k}(\mathbb{R})$. Let

$$U^*AU = \begin{bmatrix} A_{11} & A_{12} \\ A_{21} & A_{22} \end{bmatrix} \text{ and } A_{11} \in M_{n-k}$$

be a partition conformal with that of UBU^*, and observe that the spectrum (and hence the inertia) of AB is the same as that of

$$U^*(AB)U = (U^*AU)(U^*BU) = \begin{bmatrix} A_{11}\Lambda & 0 \\ A_{22}\Lambda & 0 \end{bmatrix}$$

Thus, the spectrum of AB consists of k zeroes together with the spectrum of $A_{11}\Lambda$; in particular, $i_0(AB) = k + i_0(A_{11}\Lambda)$, $i_+(AB) = i_+(A_{11}\Lambda)$, and $i_-(AB) = i_-(A_{11}\Lambda)$. Since $A_{11} + A_{11}^*$ is positive definite (A_{11} is a principal submatrix of a matrix with this property) and Λ is a nonsingular Hermitian matrix, it follows from the first case that $i(A_{11}\Lambda) = i(\Lambda)$. Since $i_0(\Lambda) = 0$, $i_+(\Lambda) = i_+(B)$, and $i_-(\Lambda) = i_-(B)$, it follows that $i(AB) = i(B)$. □

An independent proof of the preceding theorem can be based only on (1.7.8) rather than on the general inertia theorem (2.4.10); see Problem 11.

Problems

1.   Show that $A = \begin{bmatrix} 0 & 1 \\ -1 & 1 \end{bmatrix}$ is positive stable by taking $G = I \in M_2$ in (2.4.1) and applying (2.4.7). Show further that DA is positive stable whenever $D \in M_2$ is a positive diagonal matrix.

2.   Suppose $A = HK$, A, H, $K \in M_n$, with H and K Hermitian and A positive stable. If H is positive definite, use the identity $H^{-\frac{1}{2}}AH^{\frac{1}{2}} = H^{\frac{1}{2}}KH^{\frac{1}{2}}$ to show that K must be positive definite. If H is merely Hermitian, and possibly indefinite, use (2.4.10) to show that $i(H) = i(K)$. This gives a proof of (2.4.14a) that is independent of (2.4.13). Show by example that the converse does not hold.

3.   If $A \in M_n$ and $x \in \mathbb{C}^n$, show that $L(A,x)$ is a convex cone.

4.   Let $A, B \in M_n$ and suppose that $L(A,x) \cap L(B,x) \neq \phi$ for all nonzero $x \in \mathbb{C}^n$. Show that $A + B$ is positive stable. Give an example to show that the converse is not true.

5.   If $A \in M_n$ is nonsingular, show that $L(A^{-1},x) = L(A,A^{-1}x)$, and, if $S \in M_n$ is nonsingular, show that $L(SAS^{-1},x) = S^*L(A,S^{-1}x)S$.

6.   The Lyapunov equation (2.2.2) is the basis for (2.2.1), (2.4.7), (2.4.10) and (2.4.11). Show that there are analogs of all these results based upon the alternate Lyapunov equation $AG + GA^* = H$, with the same assumptions on A, G, and H.

7.   Give another proof of the second statement in (2.4.10), equality of inertias, using the first statement and a continuity argument.

8.   Let $A, G \in M_n$ with G Hermitian and nonsingular, suppose $H \equiv GA + A^*G$ is positive semidefinite, and let $S = GA - A^*G$.
   (a) Show that if no eigenvector of $G^{-1}S$ lies in the nullspace of H, then $i_0(A) = 0$.
   (b) If $i_0(A) = 0$, show that $i(A) = i(G)$. How are these results related to (2.4.7) and (2.4.10)?
   (c) If $i_0(A) = 0$, show that *either* no eigenvector of $G^{-1}S$ lies in the nullspace of H *or* the eigenvectors of $G^{-1}S$ in the nullspace of H are isotropic vectors (1.7.14) for G. Show by example that this second possibility can actually occur.

9.   The *Schwarz matrix*, which often arises in stability analysis, is a tridiagonal matrix of the form

$$B = [b_{ij}] = \begin{bmatrix} 0 & 1 & & & \\ -b_n & 0 & \cdot & \cdot & \\ & \cdot & \cdot & \cdot & \cdot \\ & & \cdot & \cdot & \cdot & \cdot \\ & & & \cdot & 0 & 1 \\ & & & & -b_2 & b_1 \end{bmatrix}$$

for real numbers $b_1, b_2,..., b_n$, that is, $b_{i,i+1} = 1$, $b_{i+1,i} = -b_{n+1-i}$, $b_{n,n} = b_1$, and all other $b_{ij} = 0$. Show that $i_0(B) = 0$ if and only if $b_i \neq 0$, $i = 1,..., n$, and show that B is positive stable if and only if $b_1, b_2,..., b_n > 0$. Can you determine $i(B)$ in general?

10. Let $A = [a_{ij}] \in M_n(\mathbb{R})$ be tridiagonal. Show that if all principal minors

of A are positive then A is positive stable. Show further that "tridiagonal" may be generalized to "no simple circuits of more than two edges in the directed graph of A." Note that this is a circumstance in which the converse of the statement in Problem 3 of Section (2.1) is valid. Compare to the case $n = 2$.

11. Use (1.7.8) to give an independent proof of Theorem (2.4.15).

12. Use the general inertia theorem (2.4.10) to give a short proof that if $A \in M_n$ and $A + A^*$ is positive definite, then DA is positive stable for every positive diagonal $D \in M_n$, and the same is true for any positive definite $H \in M_n$.

Notes and Further Readings. The treatment up to and including (2.4.7) and (2.4.9) is based on D. Carlson, B. Datta, and C. R. Johnson, A Semi-definite Lyapunov Theorem and the Characterization of Tridiagonal D-Stable Matrices, *SIAM J. Alg. Discr. Meth.* 3 (1982), 293-304. Various forms of (2.4.10) were published independently by H. Wielandt, On the Eigenvalues of $A + B$ and AB, *J. Research NBS* 77B (1973), 61-63 (adapted from Wielandt's 1951 NBS Report #1367 by C. R. Johnson), O. Taussky, A Generalization of a Theorem of Lyapunov, *SIAM J. Appl. Math.* 9 (1961), 640-643, and A. Ostrowski and H. Schneider, Some Theorems on the Inertia of General Matrices, *J. Math. Anal. Appl.* 4 (1962), 72-84. Wielandt, though earliest, gave only the equality of inertias portion. Theorem (2.4.11) is from C. R. Johnson, A Local Lyapunov Theorem and the Stability of Sums, *Linear Algebra Appl.* 13 (1976), 37-43. Theorem (2.4.13) may be found in C. R. Johnson, The Inertia of a Product of Two Hermitian Matrices, *J. Math. Anal. Appl.* 57 (1977), 85-90.

2.5 M-matrices, P-matrices, and related topics

In this section we study a very important special class of real positive stable matrices that arise in many areas of application: the M-matrices. They also link the nonnegative matrices (Chapter 8 of [HJ]) with the positive stable matrices. If $X, Y \in M_{m,n}(\mathbb{R})$, we write $X \geq Y$ (respectively, $X > Y$) if all the entries of $X - Y$ are nonnegative (respectively, positive); in particular, $X \geq 0$ means that X has nonnegative entries. For $X \in M_n$, we denote the spectral radius of X by $\rho(X)$.

2.5.1 **Definition.** The set $Z_n \subset M_n(\mathbb{R})$ is defined by

$$Z_n = \{A = [a_{ij}] \in M_n(\mathbb{R}): a_{ij} \leq 0 \text{ if } i \neq j, \, i, j = 1, \dots, n\}$$

The simple sign pattern of the matrices in Z_n has many striking consequences.

Exercise. Show that $A = [a_{ij}] \in Z_n$ if and only if $A \in M_n(\mathbb{R})$ and $A = \alpha I - P$ for some $P \in M_n(\mathbb{R})$ with $P \geq 0$ and some $\alpha \in \mathbb{R}$ (even $\alpha > 0$); one can take any $\alpha \geq \max\{a_{ii}: i = 1, \dots, n\}$.

This representation $A = \alpha I - P$ for a matrix in Z_n is often convenient because it suggests connections with the Perron-Frobenius theory of nonnegative matrices.

2.5.2 **Definition.** A matrix A is called an *M-matrix* if $A \in Z_n$ and A is positive stable.

The following simple facts are very useful in the study of *M*-matrices.

2.5.2.1 **Lemma.** Let $A = [a_{ij}] \in Z_n$ and suppose $A = \alpha I - P$ with $\alpha \in \mathbb{R}$ and $P \geq 0$. Then $\alpha - \rho(P)$ is an eigenvalue of A, every eigenvalue of A lies in the disc $\{z \in \mathbb{C}: |z - \alpha| \leq \rho(P)\}$, and hence every eigenvalue λ of A satisfies $\operatorname{Re} \lambda \geq \alpha - \rho(A)$. In particular, A is an *M*-matrix if and only if $\alpha > \rho(P)$. If A is an *M*-matrix, one may always write $A = \gamma I - P$ with $\gamma = \max\{a_{ii}: i = 1, \dots, n\}$, $P = \gamma I - A \geq 0$; necessarily, $\gamma > \rho(P)$.

Proof: If $A = \alpha I - P$ with $P \geq 0$ and $\alpha \in \mathbb{R}$, notice that every eigenvalue of A is of the form $\alpha - \lambda(P)$, where $\lambda(P)$ is an eigenvalue of P. Thus, every eigenvalue of A lies in a disc in the complex plane with radius $\rho(P)$ and centered at $z = \alpha$. Since $\rho(P)$ is an eigenvalue of P (see Theorem (8.3.1) in [HJ]), $\alpha - \rho(P)$ is a real eigenvalue of A. If A is an *M*-matrix, it is positive stable and hence $\alpha - \rho(P) > 0$. Conversely, if $\alpha > \rho(P)$ then the disc $\{z \in \mathbb{C}: |z - \alpha| \leq \rho(P)\}$ lies in the right half-plane, so A is positive stable. ∎

A very important property of the class of *M*-matrices is that principal submatrices and direct sums of *M*-matrices are again *M*-matrices. Notice that the same is true of the class of positive definite matrices; there are a great many analogies between positive definite matrices and *M*-matrices.

Exercise. If A is an M-matrix, show that any principal submatrix of A is an M-matrix. In particular, conclude that an M-matrix has positive diagonal entries. Show by example that a matrix $A \in Z_n$ with positive diagonal need not be an M-matrix. *Hint:* Use the fact that the spectral radius of a principal submatrix of a nonnegative matrix P is no larger than the spectral radius of P (see Corollary (8.1.20) in [HJ]).

Exercise. Show that a direct sum of M-matrices is an M-matrix. *Hint:* If $A = (\alpha_1 I - P_1) \oplus \cdots \oplus (\alpha_k I - P_k)$ with all $\alpha_i > 0$ and all $P_i \geq 0$, consider $\alpha = \max \alpha_i$ and $P = \alpha I - A$.

It is mathematically intriguing, and important for applications, that there are a great many different (not obviously equivalent) conditions that are necessary and sufficient for a given matrix in Z_n to be an M-matrix. In order to recognize M-matrices in the immense variety of ways in which they arise, it is useful to list several of these. It should be noted that there are many more such conditions and the ones we list are a selection from the more useful ones.

2.5.3   **Theorem.** If $A \in Z_n$, the following statements are equivalent:

2.5.3.1 A is positive stable, that is, A is an M-matrix.

2.5.3.2 $A = \alpha I - P$, $P \geq 0$, $\alpha > \rho(P)$.

2.5.3.3   Every real eigenvalue of A is positive.

2.5.3.4 $A + tI$ is nonsingular for all $t \geq 0$.

2.5.3.5 $A + D$ is nonsingular for every nonnegative diagonal matrix D.

2.5.3.6   All principal minors of A are positive.

2.5.3.7   The sum of all k-by-k principal minors of A is positive for $k = 1,\dots,n$.

2.5.3.8   The leading principal minors of A are positive.

2.5.3.9 $A = LU$, where L is lower triangular and U is upper triangular and all the diagonal entries of each are positive.

2.5.3.10   For each nonzero $x \in \mathbb{R}^n$, there is an index $1 \leq i \leq n$ such that $x_i(Ax)_i > 0$.

2.5.3.11   For each nonzero $x \in \mathbb{R}^n$, there is a positive diagonal matrix D such that $x^T A Dx > 0$.

2.5.3.12   There is a positive vector $x \in \mathbb{R}^n$ with $Ax > 0$.

2.5.3.13   The diagonal entries of A are positive and AD is strictly row diagonally dominant for some positive diagonal matrix D.

2.5.3.14   The diagonal entries of A are positive and $D^{-1}AD$ is strictly row diagonally dominant for some positive diagonal matrix D.

2.5.3.15 The diagonal entries of A are positive and there exist positive diagonal matrices D, E such that DAE is both strictly row diagonally dominant and strictly column diagonally dominant.

2.5.3.16 There is a positive diagonal matrix D such that $DA + A^TD$ is positive definite; that is, there is a positive diagonal Lyapunov solution.

2.5.3.17 A is nonsingular and $A^{-1} \geq 0$.

2.5.3.18 $Ax \geq 0$ implies $x \geq 0$.

We do not give a complete proof of the equivalence of all 18 conditions in (2.5.3), but do present a selected sample of implications. Other implications, as well as an ordering of all 18 that gives an efficient proof of equivalence, should be considered as exercises. Many of the other implications are immediate. A longer list of conditions, together with references to proofs of various implications and of lists of equivalent conditions may be found in [BP1].

Selected implications in (2.5.3). Numbers correspond to final digits of listing in (2.5.3.x).

$1 \Longleftrightarrow 2$: See Lemma 2.5.2.1.

$3 \Rightarrow 2$: Since $A \in Z_n$ may be written $A = \alpha I - P$, $P \geq 0$, and $\alpha - \rho(P) \in \sigma(A)$, (3) means that $\alpha - \rho(P) > 0$ or $\alpha > \rho(P)$.

$2 \Rightarrow 4$: Given (2), $A + tI = (\alpha + t)I - P$ and $\alpha + t > \rho(P)$ for $t \geq 0$. Thus, (2) holds with A replaced by $A + tI$, and since (2) implies (1), $A + tI \in M_n(\mathbb{R})$ is positive stable. Since positive stable real matrices have positive determinant, $A + tI$ is nonsingular if $t \geq 0$.

$2 \Rightarrow 6$: Since (2) is inherited by principal submatrices of A, and because (2) implies (1), it follows that every principal submatrix of A is positive stable. But, since $A \in M_n(\mathbb{R})$, each of these principal submatrices has positive determinant.

$7 \Rightarrow 4$: Expansion of $\det(A + tI)$ as a polynomial in t gives a monic polynomial of degree n whose coefficients are the $E_k(A)$, which are all guaranteed to be positive by (7). Therefore, this polynomial takes on positive values for nonnegative t. Since $A + tI$ has positive determinant for $t \geq 0$, it is nonsingular for $t \geq 0$.

$8 \Rightarrow 9$: This follows from the development of the LU factorization in Section (3.5) of [HJ]. See a following exercise to note that both factors L and U are M-matrices if A is an M-matrix.

$2 \Rightarrow 17$: Since division by a positive number α changes nothing important in (17), we assume $\alpha = 1$. The power series $I + P + P^2 + \cdots$ then converges since $\rho(P) < 1$. Because $P \geq 0$, the limit $(I - P)^{-1} = A^{-1}$ of this series is nonnegative.

$17 \Rightarrow 18$: Suppose that $Ax = y \geq 0$. Then $x = A^{-1}y \geq 0$ because $A^{-1} \geq 0$.

$18 \Rightarrow 2$: Write $A = \alpha I - P$ for some $P \geq 0$ and $\alpha > 0$, and let v be a Perron vector for P, so $0 \neq v \geq 0$ and $Pv = \rho(P)v$. If $\alpha \leq \rho(P)$, then $A(-v) = [\rho(P) - \alpha]v \geq 0$, which contradicts (18), so $\alpha > \rho(P)$.

$2 \Rightarrow 12$: If $P \geq 0$ is irreducible, let $x > 0$ be a Perron-Frobenius right eigenvector of P; if P is reducible, perturb it by placing sufficiently small values $\epsilon > 0$ in the positions in which P has zeroes, and let x be the Perron-Frobenius right eigenvector of the result. Then $Ax = \alpha x - Px$, and Px is either $\rho(P)x$, or as close to it as we like. In either event, Ax is sufficiently close to $[\alpha - \rho(P)]x > 0$ so that $Ax > 0$.

$12 \Rightarrow 13$: Let $D = \text{diag}(x)$ and note that the row sums of AD are just the entries of $ADe = Ax > 0$, where $e = [1, 1, ..., 1]^T \in \mathbb{R}^n$. Since all off-diagonal entries of $A \in Z_n$ are nonpositive, the diagonal entries of AD, and therefore those of A, must then be positive and AD must be strictly row diagonally dominant.

$15 \Rightarrow 16$: If (15) holds, then $DAE + (DAE)^T$ has positive diagonal entries, is strictly diagonally dominant, and is, therefore, positive definite. A congruence by E^{-1} shows that $E^{-1}[DAE + (DAE)^T]E^{-1} = (E^{-1}D)A + A^T(E^{-1}D)$ is positive definite also. This means that $E^{-1}D$ is a positive diagonal Lyapunov solution.

$16 \Rightarrow 6$: Note that condition (16) is inherited by principal submatrices of A and that (16) implies positive stability by Lyapunov's theorem. Thus, every principal submatrix of A is a real positive stable matrix and therefore has positive determinant; hence, (6) holds.

$11 \Rightarrow 1$: This follows immediately from the local version of Lyapunov's theorem (2.4.11).

It should be noted that, for the most part, the conditions of (2.5.3) are *not* equivalent outside of Z_n. The structure imposed by Z_n, mostly because of its relationship with the nonnegative matrices, is remarkable.

Exercise. Show that no two of conditions (2.5.3.1, 6, 7, and 8) are equivalent, in general, in $M_n(\mathbb{R})$.

Exercise. Show that (2.5.3.8 and 9) *are* equivalent as conditions on a matrix $A \in M_n(\mathbb{R})$.

Exercise. Show that (2.5.3.9) can be strengthened and, in fact, an *M*-matrix may be *LU*-factored with L and U both *M*-matrices. If L and U are *M*-matrices, is the product LU an *M*-matrix? *Hint:* Reduce the *M*-matrix A to upper triangular form using only lower triangular type 3 elementary operations and keep careful track.

There are remarkable parallels between the *M*-matrices and the positive definite matrices, for example, the equivalence of (2.5.3.6, 7, and 8). These parallels extend to some of the classical matrix and determinantal inequalities for positive definite matrices discussed in Sections (7.7–8) of [HJ].

2.5.4   Theorem. Let $A, B \in Z_n$ be given and assume that $A = [a_{ij}]$ is an *M*-matrix and $B \geq A$. Then

(a) B is an *M*-matrix,

(b) $A^{-1} \geq B^{-1} \geq 0$, and

(c) $\det B \geq \det A > 0$.

Moreover, A satisfies the determinantal inequalities of

(d)   **Hadamard**: $\det A \leq a_{11} \cdots a_{nn}$;

(e)   **Fischer**: $\det A \leq \det A(\alpha) \det A(\alpha')$ for any $\alpha \subseteq \{1,\dots,n\}$; and

(f)   **Szasz**: $P_{k+1}^{\left[\binom{n-1}{k}\right]^{-1}} \leq P_k^{\left[\binom{n-1}{k-1}\right]^{-1}}$, $k = 1, 2,\dots, n-1$.

In the Szasz inequality, P_k is the product of all *k*-by-*k* principal minors of A.

Proof: Let A and $B = [b_{ij}]$ satisfy the stated conditions. Then (a) follows from (2.5.3.12), for if $x > 0$ and $Ax > 0$, then $Bx \geq Ax > 0$. The inequality (b) now follows from (2.5.3.17) since $A^{-1} - B^{-1} = A^{-1}(B - A)B^{-1}$ is the product of three nonnegative matrices and hence is nonnegative.

To prove (c), proceed by induction on the dimension n. The asserted inequality $\det B \geq \det A$ is trivial for $n = 1$, so suppose it holds for all dimensions $k = 1,\dots, n-1$. Partition A and B as

$$A = \begin{bmatrix} A_{11} & A_{12} \\ A_{21} & a_{nn} \end{bmatrix} \text{ and } B = \begin{bmatrix} B_{11} & B_{12} \\ B_{21} & b_{nn} \end{bmatrix}$$

with $A_{11}, B_{11} \in Z_{n-1}$. The principal submatrices A_{11} and B_{11} are M-matrices and $B_{11} \ge A_{11}$, so det $B_{11} \ge$ det $A_{11} > 0$ by the induction hypothesis. We also have $A^{-1} \ge B^{-1} \ge 0$, det $A > 0$, and det $B > 0$. The n,n entry of the nonnegative matrix $A^{-1} - B^{-1}$ is (det A_{11})/det A - (det B_{11})/det $B \ge 0$, so det $B \ge$ (det B_{11}/det A_{11}) det $A \ge$ det $A > 0$.

The Hadamard inequality (d) follows from the Fischer inequality (e), which follows easily from the determinant inequality (c) as follows: Assume without loss of generality that $\alpha = \{1,..., k\}$, $1 \le k \le n$, and partition

$$A = \begin{bmatrix} A_{11} & A_{12} \\ A_{21} & A_{22} \end{bmatrix}$$

with $A_{11} \in M_k$, $A_{22} \in M_{n-k}$. Define

$$\mathcal{A} = \begin{bmatrix} A_{11} & 0 \\ 0 & A_{22} \end{bmatrix}$$

so that $\mathcal{A} \in Z_n$ and $\mathcal{A} \ge A$. It follows from (c) that det $A \le$ det $\mathcal{A} =$ (det A_{11})(det A_{22}), which is (e).

Szasz's inequality may be deduced with the same technique used in Corollary (7.8.2) in [HJ].   ☐

There are many other M-matrix inequalities that are analogs of classical results for positive definite matrices. See Problem 12(c) for an M-matrix analog of Oppenheim's inequality and see Problem 15 for a result involving Schur complements.

Exercise. Deduce Hadamard's inequality for M-matrices from Fischer's inequality for M-matrices.

Exercise. Prove Szasz's inequality for M-matrices. *Hint:* It is possible to argue directly using Fischer's inequality and (0.8.4) in [HJ], or to use the fact that inverse M-matrices satisfy Hadamard's inequality (Problem 9)—in either event mimicking the proof of Szasz's inequality for positive definite matrices; see (7.8.4) in [HJ].

Within the set Z_n, the M-matrices are natural analogs of the positive definite matrices. The corresponding natural analogs of the positive *semi-*definite matrices are the positive *semi*stable matrices in Z_n. According to Lemma (2.5.2.1), any $A \in Z_n$ can be written as $A = \alpha I - P$ with $\alpha \in \mathbb{R}$ and $P \geq 0$, $\alpha - \rho(P)$ is an eigenvalue of A, and every eigenvalue of A lies in a disc with radius $\rho(P)$ and centered at $z = \alpha$. Thus, if a positive semistable $A \in Z_n$ is represented in this way as $A = \alpha I - P$, we must have $\alpha \geq \rho(P)$ and the only way such a matrix could fail to be positive stable is to have $\alpha = \rho(P)$, that is, A is singular. Thus, a positive semistable matrix $A \in Z_n$ is an *M*-matrix if and only if it is nonsingular. This observation justifies the terminology *singular M-matrix* for an $A \in Z_n$ that is positive semistable but not positive stable. Just as for *M*-matrices, there are a host of equivalent conditions for a matrix $A \in Z_n$ to be a singular *M*-matrix, and many follow easily from the observation that if A is a singular *M*-matrix then $A + \epsilon I$ is an *M*-matrix for all $\epsilon > 0$. For example, the following equivalent conditions for a matrix $A \in Z_n$ are analogs of conditions (2.5.3.x):

A is positive semistable.

$A = \alpha I - P$ with $P \geq 0$ and $\alpha \geq \rho(P)$.

Every real eigenvalue of A is nonnegative.

All principal minors of A are nonnegative.

The sum of all k-by-k principal minors of A is nonnegative for $k = 1, 2, ..., n$.

2.5.5   Definition. A nonsingular matrix $B \in M_n$ is said to be an *inverse M-matrix* if B^{-1} is an *M*-matrix.

It is not surprising that inverse *M*-matrices inherit considerable structure from the *M*-matrices.

Exercise. Show that an inverse *M*-matrix B is a nonnegative matrix that (a) is positive stable; (b) has a positive diagonal Lyapunov solution; and (c) has the property that DB is an inverse *M*-matrix for any positive diagonal matrix D.

Exercise. Show that: (a) The principal minors of an inverse *M*-matrix are positive; (b) Every principal submatrix of an inverse *M*-matrix is an inverse *M*-matrix; and (c) If A is an inverse *M*-matrix and D is a positive diagonal matrix, then $A + D$ is an inverse M-matrix. Verification of these facts requires some effort.

The M-matrices naturally suggest a number of notions (that of a P-matrix, D-stability, etc.) that are also of interest in other contexts.

2.5.6   Definition. A matrix $A \in M_n(\mathbb{R})$ is called a P-*matrix* (respectively, a P_0-*matrix*, or a P_0^+-*matrix*) if all k-by-k principal minors of A are positive (respectively, are nonnegative, or are nonnegative with at least one positive) for $k = 1,..., n$.

Just as for M-matrices, there are many different equivalent conditions for a matrix to be a P-matrix. All of the following are easily shown to be equivalent for a given $A \in M_n(\mathbb{R})$:

2.5.6.1   All the principal minors of A are positive, that is, A is a P-matrix.
2.5.6.2   For each nonzero $x \in \mathbb{R}^n$, some entry of the Hadamard product $x \circ (Ax)$ is positive, that is, for each nonzero $x = [x_i] \in \mathbb{R}^n$ there is some $k \in \{1,..., n\}$ such that $x_k(Ax)_k > 0$.
2.5.6.3   For each nonzero $x \in \mathbb{R}^n$, there is some positive diagonal matrix $D = D(x) \in M_n(\mathbb{R})$ such that $x^T(D(x)A)x > 0$.
2.5.6.4   For each nonzero $x \in \mathbb{R}^n$, there is some nonnegative diagonal matrix $E = E(x) \in M_n(\mathbb{R})$ such that $x^T(E(x)A)x > 0$.
2.5.6.5   Every real eigenvalue of every principal submatrix of A is positive.

Sketch of the implications in (2.5.6). Numbers correspond to final digits of the listing in (2.5.6.x).

$1 \Rightarrow 2$:   Let $0 \neq x \in \mathbb{R}^n$, suppose $x \circ (Ax) \leq 0$, let $J = \{j_1,..., j_k\} \subset \{1,..., n\}$ be the set of indices for which $x_i \neq 0$, $1 \leq k \leq n$, $1 \leq j_1 < \cdots < j_k \leq n$, let $x(J) \equiv [x_{j_1},..., x_{j_k}]^T \in \mathbb{R}^k$, and consider the principal submatrix $A(J) \in M_k$. Notice that $[A(J)x(J)]_i = (Ax)_{j_i}$, $i = 1,..., k$. Then $x(J) \circ [A(J)x(J)] \leq 0$ and there exists a nonnegative diagonal matrix $D \in M_k$ such that $A(J)x(J) = -Dx(J)$; the diagonal entries of D are just the Hadamard quotient of $A(J)x(J)$ by $x(J)$. Then $[A(J) + D]x(J) = 0$, which is impossible since $\det [A(J) + D] \geq \det A(J) > 0$ (see Problem 18) and hence $A(J) + D$ is nonsingular.
$2 \Rightarrow 3$:   Let $0 \neq x \in \mathbb{R}^n$ be given, and let $k \in \{1,..., n\}$ be an index for which

$x_k(Ax)_k > 0$. Then

$$x_k(Ax)_k + \epsilon \sum_{i \neq k} x_i(Ax)_i > 0$$

for some $\epsilon > 0$ and we may take $D(x) \equiv \mathrm{diag}(d_1,..., d_n)$ with $d_k = 1$, $d_i = \epsilon$ if $i \neq k$.

3 ⇒ 4: Trivial.

4 ⇒ 5: Let $J = \{j_1,..., j_k\} \subset \{1,..., n\}$ be given with $1 \leq j_1 < \cdots < j_k \leq n$ and $1 \leq k \leq n$, and suppose $A(J)\xi = \lambda\xi$ with $\lambda \in \mathbb{R}$ and $0 \neq \xi \in \mathbb{R}^k$. Let $x = [x_i] \in \mathbb{R}^n$ be defined by $x_{j_i} \equiv \xi_i$, $i = 1,..., k$ and $x_i = 0$ if $i \notin J$, so $\xi = x(J)$. Let $E = E(x) \in M_n$ be a nonnegative diagonal matrix such that $x^T(EA)x > 0$. Then $0 < x^T(EA)x = (Ex)^T(Ax) = (E(J)x(J))^T(A(J)x(J)) = (E(J)\xi)^T(\lambda\xi) = \lambda(\xi^T E(J)\xi)$, which implies that both λ and $\xi^T E(J)\xi$ are positive.

5 ⇒ 1: Since A is real, its complex nonreal eigenvalues occur in conjugate pairs, whose product is positive. But $\det A$ is the product of the complex nonreal eigenvalues of A and the real ones, which are all positive by (5), so $\det A > 0$. The same argument works for every principal submatrix of A. ☐

Exercise. Show that any real eigenvalues of a P_0^+-matrix are positive. *Hint:* Use the fact that $\det(A + tI) > 0$ if $t > 0$ and A is a P_0^+-matrix; see Problem 18.

2.5.7 Definition. A matrix $A \in M_n$ is called *D-stable* if DA is positive stable for all positive diagonal matrices $D \in M_n$.

Exercise. Note that the class of *P*-matrices is closed under positive diagonal multiplication.

Exercise. Show that every *M*-matrix is a *P*-matrix and is *D*-stable.

Exercise. If $A \in M_n$ is positive definite or, more generally, if $A + A^*$ is positive definite, show that A is *D*-stable.

Exercise. Show that a *D*-stable matrix is nonsingular and that the inverse of a *D*-stable matrix is *D*-stable.

Exercise. A matrix $A \in M_n$ is called *D-semistable* if DA is positive semistable for all positive diagonal matrices $D \in M_n$. Show that a principal submatrix of a D-stable matrix is D-semistable, but is not necessarily D-stable. *Hint:* Suppose A is D-stable and $\alpha \subset \{1,..., n\}$ is an index set. Let E be a diagonal matrix with ones in the diagonal positions indicated by α and a value $\epsilon > 0$ in the other diagonal positions. Now consider DEA, where D is positive diagonal, and show that $\sigma(DEA)$ (approximately) includes $\sigma(D(\alpha)A(\alpha))$.

We next give a basic necessary condition and a basic sufficient condition for D-stability. Unfortunately no effective characterization of D-stability is yet known.

2.5.8   Theorem. Let $A \in M_n(\mathbb{R})$ be given.
  (a)   If A is D-stable, then A is a P_0^+-matrix; and
  (b)   If A has a positive diagonal Lyapunov solution, then A is D-stable.

Proof: Suppose that $A \in M_n(\mathbb{R})$ is D-stable but is not in P_0^+. Consider the principal submatrices $A(\alpha) \in M_k$. If one of them had negative determinant, this would contradict the fact that each must be D-semistable; thus, all k-by-k principal minors are nonnegative. Since $E_k(A)$ is the kth elementary symmetric function of the eigenvalues of A, not all the k-by-k principal minors can be zero because A is real and positive stable. Thus, $A \in P_0^+$ and (a) is proved.

Now suppose there is a positive diagonal matrix $E \in M_n$ such that $EA + A^T E = B$ is positive definite. Let $D \in M_n$ be positive diagonal, consider DA, and notice that $(ED^{-1})(DA) + (DA)^T(ED^{-1}) = EA + A^T E = B$, so the positive definite matrix ED^{-1} is a Lyapunov solution for DA. Thus, DA is positive stable for any positive diagonal matrix D, and A is D-stable. $\square$

Exercise. Give examples to show that neither the converse of (2.5.8a) nor the converse of (2.5.8b) is true.

Exercise. Show that $A \in M_2(\mathbb{R})$ is D-stable if and only if $A \in P_0^+$. Show that $A = [a_{ij}] \in M_3(\mathbb{R})$ is D-stable if (a) $A \in P_0^+$ and (b) $a_{11}a_{22}a_{33} > (a_{12}a_{23}a_{31} + a_{21}a_{32}a_{13})/2$. *Hint:* This may be shown by direct argument or by using the Routh-Hurwitz criterion.

Exercise. Compare (2.5.6.3) and (2.4.11). Why may we not conclude that a P-matrix is positive stable?

Exercise. Show that if $A \in M_2(\mathbb{R})$ is a P-matrix, then A is (positive) stable.

Give an example of a *P*-matrix in $M_3(\mathbb{R})$ that is not positive stable.

Except for $n \leq 2$, there is no subset–superset relationship between the *P*-matrices and the positive stable matrices; not all positive stable matrices are *P*-matrices and not all *P*-matrices are positive stable. However, being an *M*-matrix or a *P*-matrix does have subset–superset implications for eigenvalue location.   There is an open angular wedge anchored at the origin (extending into the left half-plane if $n > 2$), symmetric about the real axis, that is the union of all the eigenvalues of all *P*-matrices in $M_n(\mathbb{R})$. There is a smaller angular wedge anchored at the origin (open if $n > 2$ and equal to the positive real axis if $n = 2$), symmetric about the real axis and strictly contained within the right half-plane, that is the union of all the eigenvalues of all *M*-matrices in $M_n(\mathbb{R})$.

2.5.9   Theorem. Let $A \in M_n(\mathbb{R})$ be given with $n \geq 2$.

(a)   Assume that $E_k(A) > 0$ for all $k = 1,\ldots, n$, where $E_k(A)$ is the sum of all the principal minors of A of order k. In particular, this condition is satisfied if A is a *P*-matrix or a P_0^+-matrix. Then every eigenvalue of A lies in the open angular wedge $W_n \equiv \{z = r e^{i\theta}: |\theta| < \pi - \pi/n, \, r > 0\}$. Moreover, every point in W_n is an eigenvalue of some *n*-by-*n* *P*-matrix.

(b)   Assume that A is an *M*-matrix. Then every eigenvalue of A lies in the open angular wedge $W_n \equiv \{z = r e^{i\theta}: r > 0, |\theta| < \pi/2 - \pi/n\}$ if $n > 2$, and in $W_2 \equiv (0, \infty)$ if $n = 2$. Moreover, every point in W_n is an eigenvalue of some *n*-by-*n* *M*-matrix.

Proofs of the assertions in (a) and (b) are outlined in Problem 7 of Section (2.1) and Problem 20 at the end of this section, respectively. Constructions to show that every point in the respective angular regions is an eigenvalue of a matrix of the given type are given in Problems 23 and 21–22.

2.5.10   Definition. The *comparison matrix* $M(A) = [m_{ij}]$ of a given matrix $A = [a_{ij}] \in M_n$ is defined by

$$m_{ij} \equiv \begin{cases} |a_{ij}| & \text{if } j = i \\ -|a_{ij}| & \text{if } j \neq i \end{cases}$$

A given matrix $A \in M_n$ is called an *H-matrix* if its comparison matrix $M(A)$ is an M-matrix.

Notice that we always have $M(A) \in Z_n$, that $M(A) = A$ if and only if $A \in Z_n$, and that $M(A) = |I \circ A| - (|A| - |I \circ A|)$ is the difference between a nonnegative diagonal matrix and a nonnegative matrix with zero diagonal. Here, $|X| \equiv [|x_{ij}|]$ denotes the *entrywise absolute value* of $X = [x_{ij}] \in M_n$. Furthermore, a given matrix $A \in M_n$ is an M-matrix if and only if $M(A) = A$ and A is an H-matrix. The representation $A = |I \circ A| - (|A| - |I \circ A|) = (I \circ A) - [(I \circ A) - A]$ for an M-matrix is an alternative to the usual representation $A = \alpha I - P$ that can be very useful; see Problem 6(c) and Corollary (5.7.4.1).

Exercise. Show that a given matrix $A \in M_n$ is an H-matrix if and only if there is a diagonal matrix $D \in M_n$ such that AD is strictly row diagonally dominant (see definition below).

Exercise. Show that an H-matrix with real entries is positive stable if and only if it has positive diagonal entries.

Exercise. Show that an H-matrix with positive diagonal entries has a positive diagonal Lyapunov solution and is, therefore, D-stable. *Hint:* Use (2.5.3.15).

Exercise. Show that a strictly diagonally dominant matrix with positive diagonal entries has a positive diagonal Lyapunov solution and is, therefore, D-stable.

Exercise. Show that a nonsingular triangular matrix $T \in M_n$ is an H-matrix.

Exercise. Show that a triangular matrix $T \in M_n$ is D-stable if and only if the diagonal entries have positive real part.

For $A = [a_{ij}] \in M_n$, we say that A is *strictly row diagonally dominant* if

$$|a_{ii}| > \sum_{j \neq i} |a_{ij}| \quad \text{for } i = 1, ..., n$$

and we say that A is *strictly column diagonally dominant* if A^T is strictly row diagonally dominant. A weaker, but also important, concept is the following:

2.5.11   Definition. A matrix $A = [a_{ij}] \in M_n$ is said to be *strictly diagonally dominant of its row entries* (respectively, *of its column entries*) if

$$|a_{ii}| > |a_{ij}| \quad (\text{respectively, } |a_{ji}|)$$

for each $i = 1,\dots, n$ and all $j \neq i$.

Exercise. Note that if $A \in M_n$ is strictly row diagonally dominant, then it must be strictly diagonally dominant of its row entries, but not conversely.

Exercise. Let $A = [a_{ij}] \in M_n(\mathbb{R})$ have positive diagonal entries. Show that A is strictly row diagonally dominant if and only if

$$a_{ii} + \sum_{j \neq i} \pm a_{ij} > 0$$

for all possible 2^{n-1} choices of $\pm$ signs for every $i = 1,\dots, n$.

Exercise. If $A = [a_{ij}] \in M_n(\mathbb{R})$ is strictly row or column diagonally dominant, show that $\det A$ has the same sign as the product $a_{11}a_{22}\cdots a_{nn}$ of its main diagonal entries. *Hint:* Either reduce to the case in which all $a_{ii} > 0$, or let $D \equiv \mathrm{diag}(a_{11},\dots,a_{nn})$ and consider $D + \epsilon(A - D)$ as ϵ decreases from 1 to 0. In either case, Geršgorin's theorem (Theorem (6.1.1) in [HJ]) is the key.

2.5.12   Theorem. If $A \in M_n(\mathbb{R})$ is strictly row diagonally dominant, then A^{-1} is strictly diagonally dominant of its column entries.

Proof: The Levy-Desplanques theorem ensures that A is invertible because 0 cannot be an eigenvalue of a strictly (row or column) diagonally dominant matrix; see (5.6.17) in [HJ]. Let $A = [a_{ij}]$ and denote $A^{-1} = [\alpha_{ij}]$. Since $\alpha_{ij} = (-1)^{i+j}\det A_{ji}/\det A$, where A_{ji} denotes the submatrix of A with row j and column i deleted, it suffices to show that $|\det A_{ii}| > |\det A_{ij}|$ for all $i = 1,\dots, n$ and all $j \neq i$.

Without loss of generality, we take $i = 1$ and $j = 2$ for convenience and, via multiplication of each row of A by ± 1, we may also assume that $a_{ii} > 0$ for $i = 1,\dots, n$. By the preceding exercise, $\det A_{11} > 0$. To complete the proof, it suffices to show that $\det A_{11} \pm \det A_{12} > 0$. But

$$\det A_{11} + \epsilon \det A_{12}$$

$$= \det \begin{bmatrix} a_{22} & a_{23} & \cdots \\ a_{32} & a_{33} & \cdots \\ \vdots & \vdots & \end{bmatrix} + \det \begin{bmatrix} \epsilon a_{21} & a_{23} & \cdots \\ \epsilon a_{31} & a_{33} & \cdots \\ \vdots & \vdots & \end{bmatrix}$$

$$= \det \begin{bmatrix} a_{22} + \epsilon a_{21} & a_{23} & \cdots \\ a_{32} + \epsilon a_{31} & a_{33} & \cdots \\ \vdots & \vdots & \end{bmatrix}$$

By the two preceding exercises, the latter matrix is strictly row diagonally dominant for $\epsilon = \pm 1$, and its determinant is positive because all its diagonal entries are positive. ☐

Exercise. Let $A = [a_{ij}] \in M_n$ be of the form $A = PMD$, in which $P \in M_n(\mathbb{R})$ is a permutation matrix, $D \in M_n$ is diagonal and nonsingular, and $M \in M_n$ is strictly row diagonally dominant. Show that any matrix $B = [b_{ij}] \in M_n$ such that $|b_{ij}| = |a_{ij}|$, $i, j = 1, \ldots, n$, is nonsingular.

One might ask if there is a converse to the observation of the preceding exercise, that is, a kind a converse to the "diagonal dominance implies nonsingularity" version of Geršgorin's theorem. An affirmative answer is provided by a theorem of Camion and Hoffman.

2.5.13   Definition. A matrix $B = [b_{ij}] \in M_n$ is said to be *equimodular* with a given matrix $A = [a_{ij}] \in M_n$ if $|b_{ij}| = |a_{ij}|$ for all $i, j = 1, \ldots, n$.

2.5.14   Theorem. Let $A \in M_n$ be given. Every $B \in M_n$ that is equimodular with A is nonsingular if and only if A may be written as $A = PMD$, where $P \in M_n(\mathbb{R})$ is a permutation matrix, $D \in M_n(\mathbb{R})$ is a positive diagonal matrix, and $M \in M_n$ is strictly row diagonally dominant.

Exercise. Suppose that $A \in M_n(\mathbb{R})$ is a matrix of the type characterized in Theorem (2.5.14). Show that there is a generalized diagonal such that the sign of $\det B$ is determined by the product of the entries along this diagonal for every $B \in M_n(\mathbb{R})$ that is equimodular with A.

Exercise. Show that a matrix of the type characterized in Theorem (2.5.14) is just a permutation multiple of an H-matrix.

Problems

1.   If $A \in M_n$ is an M-matrix and $P \in M_n(\mathbb{R})$ is a permutation matrix, show

that $P^T A P$ is an *M*-matrix.

2. If $A \in M_n$ is an *M*-matrix and $D \in M_n(\mathbb{R})$ is a positive diagonal matrix, show that DA and AD are *M*-matrices.

3. If $A, B \in M_n$ are *M*-matrices, show that AB is an *M*-matrix if and only if $AB \in Z_n$. In particular, if $A, B \in M_2(\mathbb{R})$ are *M*-matrices, show that AB is an *M*-matrix.

4. Consider $A = \begin{bmatrix} .5 & -1 \\ 0 & .5 \end{bmatrix}$ and $B = \begin{bmatrix} .5 & 0 \\ -1 & .5 \end{bmatrix}$ to show that a sum of *M*-matrices need not be an *M*-matrix.

5. If $A, B \in M_n(\mathbb{R})$ are *M*-matrices, show that $\alpha A + (1-\alpha)B$ is an *M*-matrix for all $\alpha \in [0,1]$ if and only if $B^{-1}A$ has no negative real eigenvalues. In particular, if $B^{-1}A$ is an *M*-matrix, show that every convex combination of A and B is an *M*-matrix.

6. (a) Let $A, B \in Z_n$, suppose A is an *M*-matrix, and assume $B \geq A$. Show that $B^{-1}A$ and AB^{-1} are both *M*-matrices, that $B^{-1}A \leq I$ and $AB^{-1} \leq I$, and that $\alpha A + (1-\alpha)B$ is an *M*-matrix for all $\alpha \in [0, 1]$.
(b) Let $A, B \in M_n(\mathbb{R})$ be given *M*-matrices. Show that $B^{-1}A$ is an *M*-matrix if and only if $B^{-1}A \in Z_n$.
(c) Let $A \in M_n(\mathbb{R})$ be a given *M*-matrix and write $A = (I \circ A) - S$ as in the discussion following (2.5.10). Notice that $S \geq 0$, S has zero main diagonal, $(I \circ A) \geq A$, and $(I \circ A) \in Z_n$. Conclude that $(I \circ A)^{-1}A = I - (I \circ A)^{-1}S \equiv I - T$ is an *M*-matrix with $T \geq 0$, so $\rho(T) < 1$. This representation $A = (I \circ A)[I - T]$ (with $T = [t_{ij}] \geq 0$, all $t_{ii} = 0$, and $\rho(T) < 1$) can be very useful; see Corollary (5.7.4.1).

7. Show that (2.5.3.7) implies (2.5.3.3) for general matrices $A \in M_n(\mathbb{R})$. What about the converse?

8. Determine the cases of equality in Hadamard's inequality and Fischer's inequality for *M*-matrices.

9. Show that the determinant inequalities of Hadamard, Fischer, and Szasz hold for inverse *M*-matrices.

10. If $A \in M_n(\mathbb{R})$ is an inverse *M*-matrix, show that there is a positive diagonal matrix D such that AD is strictly diagonally dominant of its row entries.

11. If $A, B \in M_n(\mathbb{R})$ are *M*-matrices and $\circ$ denotes the Hadamard product, show that $A^{-1} \circ B$ is an M-matrix.

12. (a) Let $A, B = [b_{ij}] \in M_n(\mathbb{R})$, assume that A is an M-matrix, and suppose $B \geq 0$ with all $b_{ii} \geq 1$ and $0 \leq b_{ij} \leq 1$ for all $i \neq j$. Use Theorem 2.5.4 to prove that $A \circ B$ is an M-matrix, $A^{-1} \geq (A \circ B)^{-1} \geq 0$, and

$$\det(A \circ B) \geq \det A$$

(b) Use this general determinant inequality and suitable choices of B to prove the Fischer inequality 2.5.4(e) and the Hadamard inequality 2.5.4(d) for M-matrices.

(c) Use the determinant inequality in (a) to prove the following analog of Oppenheim's inequality for positive definite matrices (see Theorem (7.8.6) in [HJ]): If $A, B \in M_n(\mathbb{R})$ are M-matrices with $B^{-1} = [\beta_{ij}]$, then

$$\det(A \circ B^{-1}) \geq (\beta_{11} \cdots \beta_{nn}) \det A \geq \frac{\det A}{\det B}$$

13. If $A = [a_{ij}] \in M_n(\mathbb{R})$ is an M-matrix, show that

$$a_{ii} > \sum_{j \neq i} |a_{ij}|$$

for at least one index i, that is, at least one row is "strictly diagonally dominant."

14. Use the result of Problem 13 to show that if $A = [a_{ij}] \in M_n$ is an H-matrix, there must be at least one index i, $1 \leq i \leq n$, such that

$$|a_{ii}| > \sum_{j \neq i} |a_{ij}|$$

15. Let

$$A = \begin{bmatrix} A_{11} & A_{12} \\ A_{21} & A_{22} \end{bmatrix} \in Z_n$$

be partitioned so that A_{11} and A_{22} are square. Show that A is an M-matrix if and only if both A_{11} and its Schur complement $A_{22} - A_{21}(A_{11})^{-1}A_{12}$ are M-matrices. Compare with Theorem (7.7.6) for positive definite matrices in [HJ].

16. Let $A = [a_{ij}] \in Z_n$ and denote $\tau(A) \equiv \min \{\mathrm{Re}(\lambda): \lambda \in \sigma(A)\}$.

(a) Show that $\tau(A) \in \sigma(A)$; $\tau(A)$ is called the *minimum eigenvalue* of A.

(b) Show that $\tau(A) \geq \min_{1 \leq i \leq n} \sum_{j=1}^{n} a_{ij}$.

(c) If there is a positive vector $x \in \mathbb{R}^n$ such that $Ax \geq \alpha x$, show that $\tau(A) \geq \alpha$.

(d) If A is irreducible, show that the assumption in (c) that x *is positive* may be weakened to x *is nonnegative and nonzero* without affecting the conclusion.

(e) May the assumption of irreducibility in (d) be relaxed? Show by example that it may not be eliminated entirely.

17. Let $A = [a_{ij}] \in M_n(\mathbb{R})$ be given and assume $a_{ii} \neq 0$ for $i = 1,\dots,n$. Define

$$r_i(A) \equiv \sum_{j \neq i} |a_{ij}|/|a_{ii}|, \qquad i = 1,\dots,n$$

Note that A is strictly row diagonally dominant if and only if $r_i(A) < 1$ for $i = 1,\dots,n$.

(a) If A is row diagonally dominant and $A_{ij} \in M_{n-1}(\mathbb{R})$ denotes the submatrix of A obtained by deleting row i and column j, show that $|\det A_{ij}| \leq r_j(A)|\det A_{ii}|$ for all $i \neq j$.

(b) Strengthen (2.5.12) by showing that if A is strictly row diagonally dominant and $A^{-1} \equiv [\alpha_{ij}]$, then $|\alpha_{ii}| \geq |\alpha_{ji}|/r_j(A) > |\alpha_{ji}|$ for all $i = 1,\dots,n$, $j \neq i$.

18. Let $A = [a_{ij}] \in M_n$. Show that

$$\frac{\partial}{\partial a_{ij}} \det A = (-1)^{i+j} \det A_{ij} = \text{the } i,j \text{ cofactor of } A$$

Conclude that if A is a P-matrix and $D \in M_n$ is a nonnegative diagonal matrix, then $\det(P + D) \geq \det P > 0$, and $\det(P + D) > \det P$ if $D \neq 0$. Also conclude that $\det(A + tI) > \det A > 0$ if A is a P_0^+-matrix and $t > 0$.

19. Let $A \in M_n(\mathbb{R})$ be a given *M*-matrix.

(a) Use the representation $A = \alpha I - P$, with $P \geq 0$ and $\alpha > \rho(P)$, to show that $\sigma(A)$ lies in a disc of radius $\rho(P)$ centered at the point α on the positive real axis. Sketch this region and indicate the point corresponding to the Perron root $\rho(P)$.

(b) Let $\rho(A^{-1})$ denote the Perron root of A^{-1} and show that $\tau(A) \equiv$

$1/\rho(A^{-1})$ is a positive real eigenvalue of A with the property that $\text{Re } \lambda \geq \tau(A)$ for all $\lambda \in \sigma(A)$, with equality only for $\lambda \in \tau(A)$. The point $\tau(A)$ is often called the *minimum eigenvalue* of the M-matrix A.

(c) Show that $\det A \geq [\tau(A)]^n$.

(d) Using the representation $A = \alpha I - P$ in (a), show that $\tau(A) = \alpha - \rho(P)$, so this difference is independent of the choice of α and P in the representation.

20. Let $A \in M_n(\mathbb{R})$ be a given M-matrix with $n \geq 2$, and let $\tau(A) \equiv 1/\rho(A^{-1})$ denote the minimum eigenvalue of A, as in Problem 19.

(a) If $\lambda \in \sigma(A)$, show that $\lambda - \tau(A) = r e^{i\theta}$ with $r \geq 0$ and $|\theta| \leq \pi/2 - \pi/n$.

(b) Conclude that $\sigma(A)$ is contained in the open wedge $W_n \equiv \{z = r e^{i\theta}: r > 0, |\theta| < \pi/2 - \pi/n\}$ in the right half-plane if $n > 2$ and in $(0,\infty)$ if $n = 2$, as asserted in Theorem (2.5.9b). Sketch this region.

21. Let $n \geq 2$ and let $Q = [q_{ij}] \in M_n$ denote the cyclic permutation matrix defined by $q_{i,i+1} = 1$ for $i = 1,\dots, n-1$, $q_{n,1} = 1$, and all other entries zero. Let $B = \alpha_1 Q + \alpha_2 Q^2 + \cdots + \alpha_n Q^n$, $\alpha_i \in \mathbb{C}$; a matrix of this form is called a *circulant*. Show that the eigenvalues of the circulant B are $\{\lambda_k = \Sigma_j \alpha_j e^{2\pi i k j/n}: k = 1,\dots, n\}$. If all $\alpha_i \geq 0$ and $\alpha_1 + \cdots + \alpha_n = 1$, then B is a doubly stochastic circulant. Recognize its eigenvalues as points in the closed convex hull of a polygon with n vertices that lies in the wedge L_n defined by (1.5.19), or, alternatively, cite Problem 21 of Section (1.5) to conclude that $\sigma(B) \subset L_n$.

22. Let $n \geq 2$ be given and let λ be a given point (in the right half-plane) such that $\lambda - 1 = r e^{i\theta}$ with $r \geq 0$ and $|\theta| \leq \pi/2 - \pi/n$. Show that λ is an eigenvalue of an n-by-n M-matrix A with minimum eigenvalue $\tau(A) = 1$ (see Problem 19). Conclude that every point in the open wedge $W_n \equiv \{z = r e^{i\theta}: r > 0, |\theta| < \pi/2 - \pi/n\}$ if $n > 2$, or $W_2 \equiv (0,\infty)$ if $n = 2$, is an eigenvalue of an n-by-n M-matrix, as asserted in Theorem (2.5.9b).

23. Let $n \geq 2$ be given and let λ be a given point in the open wedge $W_n = \{z = r e^{i\theta}: r > 0, |\theta| < \pi - \pi/n\}$. Show that λ is an eigenvalue of an n-by-n P-matrix, as asserted in Theorem (2.5.9a).

24. Deduce from Theorem (2.5.9a) that a 2-by-2 P-matrix is positive stable, and give a direct proof as well. Why must there be a 3-by-3 P-matrix that is not positive stable?

25. Let $S \equiv \{\lambda_1,\dots, \lambda_n\} \subset \mathbb{C}$ be given. Show that S is the set of eigenvalues of

an *n*-by-*n* *P*-matrix if and only if the coefficients of the polynomial $p(t) \equiv$ $(t - \lambda_1)(t - \lambda_2)\cdots(t - \lambda_n) = t^n + a_1 t^{n-1} + \cdots + a_{n-1}t + a_n$ are real, nonzero, and strictly alternate in sign, that is, $(-1)^k a_k > 0$ for all $k = 1,...,n$.

26. Let $A \in Z_n$. If A has positive main diagonal entries and is irreducibly diagonally dominant (see (6.2.25-27) in [HJ]), show that A is an *M*-matrix. Is this sufficient condition also necessary?

27. Show that a symmetric *M*-matrix is positive definite and that a symmetric singular *M*-matrix is positive semidefinite.

28. Let $A, B \in M_n$ be *M*-matrices and suppose that $A \geq B$. Using the notation in Problem 19, show that $\tau(A) \geq \tau(B)$.

29. Let $A \in M_n$ be an *M*-matrix and let $D \in M_n$ be a diagonal matrix with positive main diagonal entries $d_1, ..., d_n$. Show that $\tau(DA) \geq \tau(A) \min_{1 \leq i \leq n} d_i$.

30. Provide details for the following alternate proof of the determinant inequality in Theorem 2.5.4(c): Let A and B satisfy the hypotheses of Theorem 2.5.4. Write $B = \beta I - Q$ with $Q \geq 0$ and $\beta > \rho(Q)$. Then $0 \leq Q = \beta I - B \leq \beta I - A \equiv P$, so $A = \beta I - P$ with $P \geq Q \geq 0$ and $\beta > \rho(P)$. If $\lambda_1(P),...,$ $\lambda_n(P)$ are the eigenvalues of P, the eigenvalues of $\beta^{-1}A$ are $1 - \beta^{-1}\lambda_i(P)$, and similarly for the eigenvalues of Q and $\beta^{-1}B$. Compute

$$\log \det(\beta^{-1}A) = \sum_{i=1}^{n} \log(1 - \beta^{-1}\lambda_i(P)) = -\sum_{i=1}^{n}\sum_{k=1}^{\infty} k^{-1}\beta^{-k}(\lambda_i(P))^k$$

$$= -\operatorname{tr}\sum_{k=1}^{\infty} k^{-1}\beta^{-k} P^k \leq -\operatorname{tr}\sum_{k=1}^{\infty} k^{-1}\beta^{-k} Q^k$$

$$= \log \det(\beta^{-1}B)$$

31. The purpose of this problem is to provide a generalization to complex matrices B of the basic results in Theorem (2.5.4(a,b,c)). Let $A, B = [b_{ij}] \in M_n$ be given, assume that A is an *M*-matrix, and suppose that $M(B) \geq A$ (where $M(B)$ is the comparison matrix (2.5.10)), that is, $|b_{ii}| \geq a_{ii}$ for all i and $|b_{ij}| \leq |a_{ij}| = -a_{ij}$ for all $i \neq j$. Show that:

(a) B is an *H*-matrix, that is, $M(B)$ is an *M*-matrix.
(b) B is nonsingular.
(c) $A^{-1} \geq |B^{-1}| \geq 0$.
(d) $|\det B| \geq \det A > 0$.

32. This problem gives an extension of the result on convex combinations of M-matrices in Problems 4 and 5. Let $A, B \in M_n(\mathbb{R})$ be M-matrices with $B \geq A$, so that $C \equiv AB^{-1}$ is an M-matrix, $C \leq I$, and $\alpha A + (1 - \alpha)B$ is an M-matrix for all $\alpha \in [0, 1]$.

(a) Show that $\alpha C + (1 - \alpha)I$ is an M-matrix for all $\alpha \in [0, 1]$.

(b) Verify the identity

$$\alpha A^{-1} + (1 - \alpha)B^{-1} - (\alpha A + (1 - \alpha)B)^{-1}$$
$$= \alpha(1 - \alpha)B^{-1}C^{-1}(I - C)^2(\alpha C + (1 - \alpha)I)^{-1}$$

for all $\alpha \in \mathbb{C}$.

(c) Prove Fan's convexity theorem:

$$(\alpha A + (1 - \alpha)B)^{-1} \leq \alpha A^{-1} + (1 - \alpha)B^{-1} \text{ for all } \alpha \in [0, 1].$$

33. Theorem 2.5.9(a) identifies the precise set $W_n \subset \mathbb{C}$ of all eigenvalues of all n-by-n $P_0^\dagger$-matrices, and part of W_n is in the left half-plane for $n \geq 3$. Show that not all the eigenvalues of a given $P_0^\dagger$-matrix A can have negative real part.

34. Let $A = [a_{ij}]$, $B = [b_{ij}] \in M_n$ be given and suppose B is an M-matrix with $|b_{ij}| \geq |a_{ij}|$ for all $i \neq j$. Using the notation of Problem 19, let $\tau(B) = 1/\rho(B^{-1}) > 0$.

(a) Show that every eigenvalue of A is in the union of discs

$$\bigcup_{i=1}^{n} \{z \in \mathbb{C}: |z - a_{ii}| \leq b_{ii} - \tau(B)\}$$

and compare this result with Fan's theorem (8.2.12) in [HJ].

(b) Suppose, in addition, that $b_{ii} = |\operatorname{Re} a_{ii}|$ for $i = 1,..., n$. If A has exactly k main diagonal entries with negative real parts and exactly $n - k$ main diagonal entries with positive real parts, show that A has exactly k eigenvalues λ such that $\operatorname{Re} \lambda < -\tau(B)$ and $n - k$ eigenvalues λ with $\operatorname{Re} \lambda > \tau(B)$. In particular, $\operatorname{Re} \lambda \neq 0$ for all eigenvalues of A.

35. We have noted that for $n \geq 3$, P-matrices in $M_n(\mathbb{R})$ need not be positive stable. They need not, therefore, be D-stable, although they are closed under positive diagonal multiplication. Show, however, that if $A \in M_n(\mathbb{R})$ is a P-matrix, then *there exists* a positive diagonal matrix $D \in M_n(\mathbb{R})$ such that DA is positive stable. May the hypothesis on A be weakened?

36. Suppose that $A \in M_n$ is a *P*-matrix that also satisfies

$$\det A(\{i_1,\ldots, i_k\},\{j_1,\ldots, j_k\}) \det A(\{j_1,\ldots, j_k\},\{i_1,\ldots, i_k\}) \geq 0$$

whenever $\{i_1,\ldots, i_k\}$ differs from $\{j_1,\ldots, j_k\}$ in at most one index. Show that A satisfies Fischer's inequality, as well as the more general inequality

$$\det A \leq \frac{\det A(\alpha) \ \det \ A(\beta)}{\det \ A(\alpha\cap\beta)} \text{ for } \alpha \cup \beta = \{1,\ldots, n\}$$

Give an example to show that general *P*-matrices need not satisfy Fischer's inequality. Note that many interesting classes of matrices satisfy the hypotheses given on A: positive definite matrices; *M*-matrices; triangular matrices with positive diagonal; and totally positive matrices, for example.

Notes and Further Readings. In addition to [BP1], for further references on, and a discussion of, *M*-matrices, see the surveys Sufficient Conditions for *D*-Stability, *J. Econ. Theory* 9 (1974), 53–62 and Inverse *M*-Matrices, *Linear Algebra Appl.* 47 (1982), 195–216, both by C. R. Johnson. For a proof of Theorem (2.5.14) see P. Camion and A. J. Hoffman, On the Nonsingularity of Complex Matrices, *Pacific J. Math.* 17 (1966), 211–214. For many results about *P*-matrices, *M*-matrices, and related classes, see M. Fiedler and V. Pták, On Matrices with Non-Positive Off-Diagonal Elements and Positive Principal Minors, *Czech. Math. J.* 12 (1962), 382–400 and Chapter 5 of [Fie 86]. For a wealth of inequalities and extensive references to the literature see T. Ando, Inequalities for *M*-matrices, *Linear Multilinear Algebra* 8 (1980), 291–316. More inequalities for Hadamard products of *M*-matrices are in Ando's paper and in Section (5.7). For additional results about the spectra of *M*-matrices and *P*-matrices, see R. B. Kellogg, On Complex Eigenvalues of *M* and *P* Matrices, *Numer. Math.* 19 (1972), 170–175. For example, Kellogg shows that a point $\lambda = re^{i\theta}$ in the right half-plane is an eigenvalue of an *n*-by-*n* *M*-matrix whose graph contains no Hamiltonian circuit $[\mu(A) \leq n-1$ in the notation of Problem 19 of Section (1.5)] if and only if $r > 0$ and $|\theta| < \pi/2 - \pi/(n-1)$.

Chapter 3

Singular value inequalities

3.0   Introduction and historical remarks

Singular values and the singular value decomposition play an important role in high-quality statistical computations and in schemes for data compression based on approximating a given matrix with one of lower rank. They also play a central role in the theory of unitarily invariant norms. Many modern computational algorithms are based on singular value computations because the problem of computing the singular values of a general matrix (like the problem of computing the eigenvalues of a Hermitian matrix) is well-conditioned; for numerous examples see [GV1].

There is a rich abundance of inequalities involving singular values, and a selection from among them is the primary focus of this chapter. But approximations and inequalities were not the original motivation for the study of singular values. Nineteenth-century differential geometers and algebraists wanted to know how to determine whether two real bilinear forms

$$\varphi_A(x,y) = \sum_{i,\,j=1}^{n} a_{ij}x_iy_j \ \text{ and } \ \varphi_B(x,y) = \sum_{i,\,j=1}^{n} b_{ij}x_iy_j, \qquad (3.0.1)$$

$$A = [a_{ij}], \ B = [b_{ij}] \in M_n(\mathbb{R}), \ x = [x_i], \ y = [y_i] \in \mathbb{R}^n$$

were equivalent under independent real orthogonal substitutions, that is, whether there are real orthogonal $Q_1, Q_2 \in M_n(\mathbb{R})$ such that $\varphi_A(x,y) = \varphi_B(Q_1x, Q_2y)$ for all $x, y \in \mathbb{R}^n$. One could approach this problem by finding a canonical form to which any such bilinear form can be reduced by orthogonal substitutions, or by finding a complete set of invariants for a bilinear form under orthogonal substitutions. In 1873, the Italian differential geometer E. Beltrami did both, followed, independently, by the French algebraist C. Jordan in 1874.

Beltrami discovered that for each real $A \in M_n(\mathbb{R})$ there are always real orthogonal Q_1, $Q_2 \in M_n(\mathbb{R})$ such that

$$Q_1^T A Q_2 = \Sigma = \mathrm{diag}(\sigma_1(A), \ldots, \sigma_n(A)) \tag{3.0.2}$$

is a nonnegative diagonal matrix, where $\sigma_1(A)^2 \geq \cdots \geq \sigma_n(A)^2$ are the eigenvalues of AA^T (and also of $A^T A$). Moreover, he found that the (ortho-normal) columns of Q_1 and Q_2 are eigenvectors of AA^T and $A^T A$, respectively. Although Beltrami proposed no terminology for the elements of his canonical form, this is what we now call the *singular value decomposition* for a real square matrix; the *singular values* of A are the numbers $\sigma_1(A) \geq \cdots \geq \sigma_n(A) \geq 0$. The diagonal bilinear form

$$\varphi_{Q_1^T A Q_2}(\xi, \eta) = \sigma_1(A)\xi_1 \eta_1 + \cdots + \sigma_n(A)\xi_n \eta_n \tag{3.0.3}$$

gives a convenient canonical form to which any real bilinear form $\varphi_A(x, y)$ can be reduced by independent orthogonal substitutions, and the eigenvalues of AA^T are a complete set of invariants for this reduction.

Jordan came to the same canonical form as Beltrami, but from a very different point of view. He found that the (necessarily real) eigenvalues of the $2n$-by-$2n$ real symmetric matrix

$$\begin{bmatrix} 0 & A \\ A^T & 0 \end{bmatrix} \tag{3.0.4}$$

are paired by sign, and that its n largest eigenvalues are the desired coeffi-cients $\sigma_1(A), \ldots, \sigma_n(A)$ of the canonical form (3.0.3) (see Problem 2). Jor-dan's proof starts by observing that the largest singular value of A is the maximum of the bilinear form $x^T A y$ subject to the constraints $x^T x = y^T y = 1$, and concludes with an elegant step-by-step deflation in the spirit of our proof of Theorem (3.1.1). The block matrix (3.0.4) has been rediscovered repeatedly by later researchers, and plays a key role in relating eigenvalue results for Hermitian matrices to singular value results for general matrices.

Apparently unaware of the work of Beltrami and Jordan, J. J. Sylvester (1889/90) discovered, and gave yet a third proof for, the factorization (3.0.2) for real square matrices. He termed the coefficients $\sigma_i(A)$ in (3.0.3) the *canonical multipliers* of the bilinear form $\varphi_A(x,y)$. The leading idea for Sylvester's proof was "to regard a finite orthogonal substitution as the product of an infinite number of infinitesimal ones."

In 1902, L. Autonne showed that every nonsingular complex $A \in M_n$ can be written as $A = UP$, where $U \in M_n$ is unitary and $P \in M_n$ is positive definite. Autonne returned to these ideas in 1913/15 and used the fact that A^*A and AA^* are similar [See Problem 1(a) in Section (3.1)] to show that any square complex matrix $A \in M_n$ (singular or nonsingular) can be written as $A = V\Sigma W^*$, where V, $W \in M_n$ are unitary and Σ is a nonnegative diagonal matrix; he gave no name to the diagonal entries of Σ. Autonne recognized that the positive semidefinite factor Σ is determined essentially uniquely by A, but that V and W are not, and he determined the set of all possible unitary factors V, W associated with A [see Theorem (3.1.1′)]. He also recognized that the unitary factors V and W could be chosen to be real orthogonal if A is real, thus obtaining the theorem of Beltrami, Jordan, and Sylvester as a special case; however, Autonne was apparently unaware of their priority for this result. He realized that by writing $A = V\Sigma W^* = (VW^*)(W\Sigma W^*) = (V\Sigma V^*)(VW^*)$ he could generalize his 1902 polar decomposition to the square singular case. Although Autonne did not consider the singular value decomposition of a nonsquare A in his 1915 paper, the general case follows easily from the square case; see Problem 1(b) in Section (3.1).

In his 1915 paper, Autonne also considered special forms that can be achieved for the singular value decomposition of A under various assumptions on A, for example, unitary, normal, real, coninvolutory ($\bar{A} = A^{-1}$), and symmetric. In the latter case, Autonne's discovery that a complex square symmetric A can always be written as $A = U\Sigma U^T$ for some unitary U and nonnegative diagonal Σ gives him priority for a useful (and repeatedly rediscovered) result often attributed in the literature to Schur (1945) or Takagi (1925) (see Section (4.4) of [HJ]). To be precise, one must note that Autonne actually presented a proof for the factorization $A = U\Sigma U^T$ only for nonsingular A, but his comments suggest that he knew that the assumption of nonsingularity was inessential. In any event, the general case follows easily from the nonsingular case; see Problem 5.

The pioneering work on the singular value decomposition in Autonne's 77-page 1915 paper seems to have been completely overlooked by later researchers. In the classic 1933 survey [Mac], Autonne's 1915 paper is referenced, but the singular value decomposition for square complex matrices (MacDuffee's Theorem 41.6) is stated so as to suggest (incorrectly) to the reader that Autonne had proved it only for nonsingular matrices.

In a 1930 paper, Browne cited Autonne's 1913 announcement and used his factorization $A = V\Sigma W^*$ (perhaps for the first time) to prove inequalities for the spectral radius of Hermitian and general square matrices. Browne attached no name to the diagonal entries of Σ, and referred to them merely as "the square roots...of the characteristic roots of AA^*."

In 1931, Wintner and Murnaghan rediscovered the polar decomposition of a nonsingular square complex matrix as well as a special case (nonsingular A) of Autonne's observation that one may always write $A = PU = UQ$ (the same unitary U) for possibly different positive definite P and Q; they also noted that one may choose $P = Q$ if and only if A is normal. Wintner and Murnaghan seemed unaware of any prior work on the polar decomposition.

A complete version of the polar decomposition for a rectangular complex matrix (essentially Theorem (3.1.9) herein) was published in 1935 by Williamson, who credited both Autonne (1902) and Wintner-Murnaghan (1931) for prior solution of the square nonsingular case; Autonne's 1915 solution of the general square case is not cited. Williamson's proof, like Autonne's, starts with the spectral decompositions of the Hermitian matrices AA^* and A^*A. Williamson did not mention the singular value decomposition, so there is no recognition that the general singular value decomposition for a rectangular complex matrix follows immediately from his result.

Finally, in 1939, Eckart and Young gave a clear and complete statement of the singular value decomposition for a rectangular complex matrix [Theorem (3.1.1)], crediting Autonne (1913) and Sylvester (1890) for their prior solutions of the square complex and real cases. The Eckart-Young proof is self-contained and is essentially the same as Williamson's (1930); they do not seem to recognize that the rectangular case follows easily from the square case. Eckart and Young give no indication of being aware of Williamson's result and do not mention the polar decomposition, so there is no recognition that their theorem implies a general polar decomposition for rectangular complex matrices. They give no special name to the nonnegative square roots of "the characteristic values of AA^*," and they view the factorization $A = V\Sigma W^*$ as a generalization of the "principal axis transformation" for Hermitian matrices.

While algebraists were developing the singular value and polar decompositions for finite matrices, there was a parallel and apparently quite independent development of related ideas by researchers in the theory of integral equations. In 1907, E. Schmidt published a general theory of real integral equations, in which he considered both symmetric and nonsymmetric kernels. In his study of the nonsymmetric case, Schmidt introduced a pair of integral equations of the form

$$\varphi(s) \;=\; \lambda \int_a^b K(s,t)\,\psi(t)\,dt$$

and

$$\psi(s) \;=\; \lambda \int_a^b K(t,s)\,\varphi(t)\,dt$$

(3.0.5)

where the functions $\varphi(s)$ and $\psi(s)$ are not identically zero. Schmidt showed that the scalar λ must be real since λ^2 is an eigenvalue of the symmetric (and positive semidefinite) kernel

$$H(s,t) \;=\; \int_a^b K(s,\tau)\,K(t,\tau)\,d\tau$$

If one thinks of $K(s,t)$ as an analog of a matrix A, then $H(s,t)$ is an analog of AA^T. Traditionally, the "eigenvalue" parameter λ in the integral equation literature is the reciprocal of what matrix theorists call an eigenvalue. Recognizing that such scalars λ together with their associated pairs of functions $\varphi(s)$ and $\psi(s)$ are, for many purposes, the natural generalization to the nonsymmetric case of the eigenvalues and eigenfunctions that play a key role in the theory of integral equations with symmetric kernels, Schmidt called λ an "eigenvalue" and the associated pair of functions $\varphi(s)$ and $\psi(s)$ "adjoint eigenfunctions" associated with λ.

Picard (1910) further developed Schmidt's theory of nonsymmetric kernels but, at least in the symmetric case, refers to Schmidt's "eigenvalues" as *singular values* (valeurs singulières). Perhaps in an effort to avoid confusion between Schmidt's two different uses of the term "eigenvalue," later researchers in integral equations seem to have adopted the term "singular value" to refer to the parameter λ in (3.0.5). In a 1937 survey, for example, Smithies refers to "singular values, i.e., E. Schmidt's eigen-values of the nonsymmetric kernel, and not to the eigen-values in the ordinary sense." Smithies also notes that he had been "unable to establish any direct connection between the orders of magnitude of the eigen-values and the singular values when the kernel is not symmetric." Establishing such a connection seems to have remained a theme of Smithies' research for many years, and Smithies' student Chang (1949) succeeded in establishing an indirect connection: Convergence of an infinite series of given powers of the singular values of an integral kernel implies convergence of the infinite series of the same powers of the absolute values of the eigenvalues. Weyl (1949) then showed that there was actually a direct inequality between partial sums of Chang's two series, and the modern theory of singular value inequalities was born.

Although it is not at all concerned with the singular value decomposition or polar decomposition, a seminal 1939 paper of Von Neumann made vital use of facts about singular values in showing that a norm $\| A \|$ on $M_{m,n}$ is unitarily invariant if and only if it is a symmetric gauge function of the square roots of "the proper values of AA^*" [Theorem (3.5.18)]. Despite his familiarity with the integral equation and operator theory literature, Von Neumann never uses the term "singular value," and hence his pages are speckled with square roots applied to the eigenvalues of AA^*. His primary tools are duality and convexity, and since the basic triangle inequality for singular value sums [Corollary (3.4.3)] was not discovered until 1951, Von Neumann's proof is both long and ingenious.

During 1949-50, a remarkable series of papers in the *Proceedings of the National Academy of Sciences (U.S.)* established all of the basic inequalities involving singular values and eigenvalues that are the main focus of this chapter. Let $\lambda_1,\ldots, \lambda_n$ and $\sigma_1,\ldots, \sigma_n$ denote the eigenvalues and singular values of a given square matrix, ordered so that $|\lambda_1| \geq \cdots \geq |\lambda_n|$ and $\sigma_1 \geq \cdots \geq \sigma_n \geq 0$. Apparently motivated by Chang's (1949) analytic results in the theory of integral equations, Weyl (1949) showed that $|\lambda_1 \cdots \lambda_k| \leq \sigma_1 \cdots \sigma_k$ for $k = 1,\ldots, n$ [Theorem (3.3.2)] and deduced that $\varphi(|\lambda_1|) + \cdots + \varphi(|\lambda_k|) \leq \varphi(\sigma_1) + \cdots + \varphi(\sigma_k)$ for $k = 1,\ldots, n$, for any increasing function φ on $[0,\infty)$ such that $\varphi(e^t)$ is convex on $(-\infty,\infty)$; Weyl was particularly interested in $\varphi(s) = s^p$, $p > 0$, since this special case completely explained (and enormously simplified) Chang's results. Ky Fan (1949, 1950) introduced initial versions of variational characterizations of eigenvalue and singular value sums that became basic for much of the later work in this area. Pólya (1950) gave an alternative proof of a key lemma in Weyl's 1949 paper. Pólya's insight [incorporated into Problem 23 of Section (3.3)] led to extensive and fruitful use of properties of doubly stochastic matrices and majorization in the study of singular value inequalities. None of these papers uses the term "singular value;" instead, they speak of the "two kinds of eigenvalues" of a matrix, namely, the eigenvalues of A and those of A^*A.

In a 1950 paper, A. Horn begins by saying, "In this note I wish to present a theorem on the singular values of a product of completely continuous operators on Hilbert space.... The singular values of an operator K are the positive square roots of the eigen-values of K^*K, where K^* is the adjoint of K." He then shows that

$$\sigma_1(AB) \cdots \sigma_k(AB) \leq \sigma_1(A) \cdots \sigma_k(A)\sigma_1(B) \cdots \sigma_k(B)$$

for all $k = 1, 2,....$ [Theorem (3.3.4)] and gives the additive inequalities in Theorem (3.3.14).

In the following year, Ky Fan (1951) extended the work in his 1949-50 papers and obtained the fundamental variational characterization of singular value sums [Theorem (3.4.1)] that enabled him to prove basic inequalities such as Theorem (3.3.16), Corollary (3.4.3), and the celebrated Fan dominance theorem [Corollary (3.5.9(a)(b))]. He also revisited Von Neumann's characterization of all unitarily invariant norms [Theorem (3.5.18)] and showed how it follows easily from his new insights. A novel and fundamental feature of Fan's variational characterizations of singular values is that they are quasilinear functions of A itself, not via A^*A. They are surely the foundation for all of the modern theory of singular value inequalities.

In 1954, A. Horn proved that Weyl's 1949 inequalities are sufficient for the existence of a matrix with prescribed singular values and eigenvalues, and stimulated a long series of other investigations to ascertain whether inequalities originally derived as necessary conditions are sufficiently strong to characterize exactly the properties under study. In this 1954 paper, which, unlike his 1950 paper, is clearly a paper on matrix theory rather than operator theory, Horn uses "singular values" in the context of matrices, a designation that seems to have become the standard terminology for matrix theorists writing in English. In the Russian literature one sees singular values of a matrix or operator referred to as *s-numbers* [GoKr]; this terminology is also used in [Pie].

Problems

1. Use Beltrami's theorem to show that the two real bilinear forms in (3.0.1) are equivalent under independent real orthogonal substitutions if and only if AA^T and BB^T have the same eigenvalues.

2. Let $A \in M_{m,n}$ be given, let $q = \min\{m, n\}$, and let

$$A = \begin{bmatrix} 0 & A \\ A^* & 0 \end{bmatrix} \in M_{m+n}$$

Verify that the characteristic polynomial of A can be evaluated as

$$\det(tI_{m+n} - A) = \det \begin{bmatrix} I_m & t^{-1}A \\ 0 & I_n \end{bmatrix} \det \begin{bmatrix} tI_m & -A \\ -A^* & tI_n \end{bmatrix}$$

$$= \det \begin{bmatrix} tI_m - t^{-1}AA^* & 0 \\ -A^* & tI_n \end{bmatrix} = t^{n-m} \det(t^2 I_m - AA^*)$$

and conclude that the eigenvalues of A are $\pm\sigma_1(A),...,\pm\sigma_q(A)$ and $|m-n|$ additional zeroes. This is a generalization of Jordan's observation about the significance of the matrix (3.0.4).

3.   Suppose $A \in M_{m,n}$ can be written as $A = V\Sigma W^*$, where $V = [v_{ij}] \in M_m$ and $W = [w_{ij}] \in M_n$ are unitary, and $\Sigma = [\sigma_{ij}] \in M_{m,n}$ has $\sigma_{ij} = 0$ for all $i \neq j$, $\sigma_{ii} \geq 0$ for $i = 1,..., \min\{m,n\}$.

   (a) Verify Browne's upper bound on the singular values σ_{ii}:

$$\sigma_{ii} = \Big| \sum_{k=1}^{n} \sum_{j=1}^{m} a_{jk} \bar{v}_{ji} w_{ki} \Big| \leq \sum_{k=1}^{n} \sum_{j=1}^{m} |a_{jk}| \, |v_{ji} w_{ki}| \tag{†}$$

$$\leq \tfrac{1}{2} \sum_{k=1}^{n} \sum_{j=1}^{m} |a_{jk}| \, (|v_{ji}|^2 + |w_{ki}|^2)$$

$$\leq \tfrac{1}{2} \Big[\max_j \sum_{k=1}^{n} |a_{jk}| + \max_k \sum_{j=1}^{m} |a_{jk}| \Big]$$

$$= \tfrac{1}{2} \Big[\, |\!|\!| A |\!|\!|_\infty + |\!|\!| A |\!|\!|_1 \Big]$$

where $|\!|\!| A |\!|\!|_\infty$ and $|\!|\!| A |\!|\!|_1$ denote the maximum absolute row and column sum norms, respectively. Actually, Browne considered only the square case $m = n$ because he was interested in obtaining eigenvalue bounds. He knew that every eigenvalue λ of A satisfies $\min_i \sigma_{ii} \leq |\lambda| \leq \max_i \sigma_{ii}$, and concluded that $\rho(A) \leq \tfrac{1}{2}(|\!|\!| A |\!|\!|_\infty + |\!|\!| A |\!|\!|_1)$. He was apparently unaware of the better bound $\rho(A) \leq \min\{ |||\, A\, |||_{\infty}, |||\, A\, |||_1 \}$.

(b) Use the Cauchy-Schwarz inequality at the point (†) in the preceding argument to give the better bounds

$$\sigma_{ii} \leq \left[\max_k \sum_{j=1}^m |a_{jk}|^{2p} \right]^{\frac{1}{2}} \left[\max_j \sum_{k=1}^n |a_{jk}|^{2(1-p)} \right]^{\frac{1}{2}}$$

for any $p \in [0,1]$. What is this for $p = \frac{1}{2}$, for $p = 0$, and for $p = 1$? Show that Browne's inequality in (a) follows from the case $p = \frac{1}{2}$.

(c) Verify Browne's corollary of the bound in (a): If $A \in M_n$ is Hermitian, then $\rho(A) \leq |||\, A\, |||_{\infty}$. He was apparently unaware that this bound holds for all $A \in M_n$, Hermitian or not.

4. In their 1939 paper, Eckart and Young found a criterion for simultaneous unitary equivalence of two given matrices $A, B \in M_{m,n}$ to real diagonal forms $V^* A W = E \geq 0$, $V^* B W = F$, namely, the two products AB^* and $B^* A$ are both Hermitian.

(a) What does this say when $m = n$ and A and B are both Hermitian?

(b) Verify the necessity and sufficiency of the criterion.

(c) The Eckart-Young criterion can be weakened slightly: AB^* and $B^* A$ are both normal if and only if there exist unitary V and W such that $V^* A W$ and $V^* B W$ are both diagonal (but not necessarily real); see Problem 26 in Section (7.3) of [HJ].

(d) Show that one can write $A = VEW^*$ and $B = VFW^*$ with E and F both nonnegative diagonal if and only if AB^* and $B^* A$ are both positive semidefinite.

5. Let $A \in M_n$ be a given complex symmetric matrix. Assume that one can always write $A = U\Sigma U^T$ when A is nonsingular, where $U, \Sigma \in M_n$, U is unitary, and Σ is nonnegative diagonal. Provide details for the following two arguments to show that one can also factorize a singular A in this way.

(a) Let $u_1, \ldots, u_\nu \in \mathbb{C}^n$ be an orthonormal basis for the nullspace of A, let $U_2 \equiv [u_1 \ \ldots \ u_\nu] \in M_{n,\nu}$, and let $U = [U_1 \ U_2] \in M_n$ be unitary. Show that $U^T A U = A_1 \oplus 0$, where $A_1 \in M_{n-\nu}$ is nonsingular. Now write $A_1 = V_1\Sigma_1 W_1^*$ and obtain the desired factorization for A.

(b) Consider $A_\epsilon = A + \epsilon I = U_\epsilon \Sigma_\epsilon U_\epsilon^T$ for all sufficiently small $\epsilon > 0$ and use the selection principle in Lemma (2.1.8) of [HJ].

Notes and Further Reading. The following classical papers mentioned in this section mark noteworthy milestones in the development of the theory of singular values and the singular value or polar decomposition: E. Beltrami, Sulle Funzioni Bilineari, *Giornale de Matematiche* 11 (1873), 98-106. C. Jordan, Mémoire sur les Formes Bilinéaires, *J. Math. Pures Appl.* (2) 19 (1874), 35-54. J. J. Sylvester, Sur la réduction biorthogonale d'une forme linéo-linéaire à sa forme canonique, *Comptes Rendus Acad. Sci. Paris* 108 (1889), 651-653. J. J. Sylvester, On the Reduction of a Bilinear Quantic of the nth Order to the Form of a Sum of n Products by a Double Orthogonal Substitution, *Messenger of Mathematics* 19 (1890), 42-46. L. Autonne, Sur les Groupes Linéaires, Réels et Orthogonaux, *Bull. Soc. Math. France* 30 (1902), 121-134. E. Schmidt, Zur Theorie der linearen und nichtlinearen Integralgleichungen, *Math. Annalen* 63 (1907), 433-476. É. Picard, Sur un Théorème Général Relatif aux Équations Intégrales de Première Espèce et sur quelques Problèmes de Physique Mathématique, *Rend. Circ. Mat. Palermo* 29 (1910), 79-97. L. Autonne, Sur les Matrices Hypohermitiennes et les Unitaires, *Comptes Rendus Acad. Sci. Paris* 156 (1913), 858-862; this is a short announcement of some of the results in the following detailed paper. L. Autonne, Sur les Matrices Hypohermitiennes et sur les Matrices Unitaires, *Ann. Univ. Lyon, Nouvelle Série I*, Fasc. 38 (1915), 1-77. E. T. Browne, The Characteristic Roots of a Matrix, *Bull. Amer. Math. Soc.* 36 (1930), 705-710. A. Wintner and F. D. Murnaghan, On a Polar Representation of Non-singular Square Matrices, *Proc. National Acad. Sciences (U.S.)* 17 (1931), 676-678. J. Williamson, A Polar Representation of Singular Matrices, *Bull. Amer. Math. Soc.* 41 (1935), 118-123. F. Smithies, Eigenvalues and Singular Values of Integral Equations, *Proc. London Math. Soc.* (2) 43 (1937), 255-279. J. Von Neumann, Some Matrix-Inequalities and Metrization of Matric-Space, *Tomsk Univ. Rev.* 1 (1937), 286-300; reprinted in Vol. 4 of Von Neumann's *Collected Works*, A. H. Taub, ed., Macmillan, N.Y., 1962. C. Eckart and G. Young, A Principal Axis Transformation for Non-Hermitian Matrices, *Bull. Amer. Math. Soc.* 45 (1939), 118-121. S. H. Chang, On the Distribution of the Characteristic Values and Singular Values of Linear Integral Equations, *Trans. Amer. Math. Soc.* 67 (1949), 351-367. H. Weyl, Inequalities Between the Two Kinds of Eigenvalues of a Linear Transformation, *Proc. National Acad. Sciences (U.S)* 35 (1949), 408-411. K. Fan, On a Theorem of Weyl Concerning Eigenvalues of Linear Transformations. I, *Proc. National Acad. Sciences (U.S.)* 35 (1949), 652-655. K. Fan, On a Theorem of Weyl Concerning Eigenvalues of Linear Transformations. II, *Proc. National Acad. Sciences (U.S.)* 36 (1950), 31-35. G. Pólya, Remark on Weyl's Note "Inequalities Between the Two Kinds of

Eigenvalues of a Linear Transformation," *Proc. National Acad. Sciences* 36 (*U.S.*) (1950), 49-51. A. Horn, On the Singular Values of a Product of Completely Continuous Operators, *Proc. National Acad. Sciences* (*U.S.*) 36 (1950), 374-375. K. Fan, Maximum Properties and Inequalities for the Eigenvalues of Completely Continuous Operators, *Proc. National Acad. Sciences* (*U.S.*) 37 (1951), 760-766. A. Horn, On the Eigenvalues of a Matrix with Prescribed Singular Values, *Proc. Amer. Math. Soc.* 5 (1954), 4-7.

3.1   The singular value decomposition

There are many ways to approach a proof of the singular value decomposition of a matrix $A \in M_{m,n}$. A natural route is via the spectral theorem applied to the positive semidefinite matrices AA^* and A^*A (see Problem 1); we present a proof that emphasizes the normlike properties of the singular values.

3.1.1   Theorem. Let $A \in M_{m,n}$ be given, and let $q = \min\{m,n\}$. There is a matrix $\Sigma = [\sigma_{ij}] \in M_{m,n}$ with $\sigma_{ij} = 0$ for all $i \neq j$, and $\sigma_{11} \geq \sigma_{22} \geq \cdots \geq \sigma_{qq} \geq 0$, and there are two unitary matrices $V \in M_m$ and $W \in M_n$ such that $A = V\Sigma W^*$. If $A \in M_{m,n}(\mathbb{R})$, then V and W may be taken to be real orthogonal matrices.

Proof: The Euclidean unit sphere in $\mathbb{C}^n$ is a compact set and the function $f(x) = \|Ax\|_2$ is a continuous real-valued function, so the Weierstrass theorem guarantees that there is some unit vector $w \in \mathbb{C}^n$ such that $\|Aw\|_2 = \max\{\|Ax\|_2 : x \in \mathbb{C}^n, \|x\|_2 = 1\}$. If $\|Aw\|_2 = 0$, then $A = 0$ and the asserted factorization is trivial, with $\Sigma = 0$ and any unitary matrices $V \in M_m$, $W \in M_n$. If $\|Aw\|_2 \neq 0$, set $\sigma_1 \equiv \|Aw\|_2$ and form the unit vector $v \equiv Aw/\sigma_1 \in \mathbb{C}^m$. There are $m-1$ orthonormal vectors $v_2, \ldots, v_m \in \mathbb{C}^m$ so that $V_1 = [v \; v_2 \; \cdots \; v_m] \in M_m$ is unitary and there are $n-1$ orthonormal vectors $w_2, \ldots, w_n \in \mathbb{C}^n$ so that $W_1 = [w \; w_2 \; \cdots \; w_n] \in M_n$ is unitary. Then

$$\bar{A}_1 \equiv V_1^* A W_1 = \begin{bmatrix} v^* \\ v_2^* \\ \vdots \\ v_m^* \end{bmatrix} [Aw \; Aw_2 \; \ldots \; Aw_n]$$

$$= \begin{bmatrix} v_*^* \\ v_2^* \\ \vdots \\ v_m^* \end{bmatrix} [\sigma_1 v \; A w_2 \; \cdots \; A w_n]$$

$$= \begin{bmatrix} \sigma_1 & v^* A w_2 & \cdots & v^* A w_n \\ 0 & & & \\ \vdots & & A_2 & \\ 0 & & & \end{bmatrix}, \quad A_2 \in M_{m-1,n-1}$$

$$= \begin{bmatrix} \sigma_1 & z^* \\ 0 & A_2 \end{bmatrix}, \quad \bar{z} \equiv \begin{bmatrix} v^* A w_2 \\ \vdots \\ v^* A w_n \end{bmatrix} \in \mathbb{C}^{n-1}$$

Now consider the unit vector $\zeta \equiv \begin{bmatrix} \sigma_1 \\ z \end{bmatrix} / (\sigma_1^2 + z^* z)^{\frac{1}{2}} \in \mathbb{C}^n$ and compute

$$\|A(W_1\zeta)\|_2^2 = \|(V_1^* A W_1)\zeta\|_2^2 = \|\bar{A}_1 \zeta\|_2^2$$

$$= [(\sigma_1^2 + z^* z)^2 + \|A_2 z\|_2^2]/(\sigma_1^2 + z^* z)$$

$$\geq \sigma_1^2 + z^* z$$

which is strictly greater than σ_1^2 if $z \neq 0$. Since this would contradict the maximality of $\sigma_1 = \max\{\|Ax\|_2 : x \in \mathbb{C}^n, \|x\|_2 = 1\}$, we conclude that $z = 0$ and $\bar{A}_1 = V_1^* A W_1 = \begin{bmatrix} \sigma_1 & 0 \\ 0 & A_2 \end{bmatrix}$. Now repeat this argument on $A_2 \in M_{m-1,n-1}$ and its successors to obtain the desired unitary matrices V and W as products of the unitary matrices that accomplish the reduction at each step. The matrix $\Sigma = [\sigma_{ij}] \in M_{m,n}$ has $\sigma_{ii} = \sigma_i$ for $i = 1, \ldots, q$. If $m \leq n$, $AA^* = (V\Sigma W^*)(W\Sigma^T V^*) = V\Sigma\Sigma^T V^*$ and $\Sigma\Sigma^T = \operatorname{diag}(\sigma_1^2, \ldots, \sigma_q^2)$, so the nonnegative numbers $\{\sigma_i\}$ are uniquely determined by A; if $m > n$, consider $A^* A$ to arrive at the same conclusion. If A is real, all the calculations required can be performed over the reals, in which case the real unitary matrices V and W are real orthogonal.   ☐

The construction used in the preceding proof shows that if $w \in \mathbb{C}^n$ is a unit vector such that $\|Aw\|_2 = \max\{\|Ax\|_2 : x \in \mathbb{C}^n, \|x\|_2 = 1\}$, then the $n-1$ dimensional subspace of $\mathbb{C}^n$ that is orthogonal to w is mapped by A into a subspace of $\mathbb{C}^n$ that is orthogonal to Aw. It is this geometric fact that permits the successive unitary equivalences that eventually reduce A to Σ.

In the singular value decomposition $A = V\Sigma W^*$ given in Theorem (3.1.1), the quantities $\sigma_i(A) = \sigma_{ii}$, $i = 1, 2, ..., q = \min\{m, n\}$, are the *singular values* of the matrix $A \in M_{m,n}$. We shall usually arrange the singular values of a matrix in decreasing (nonincreasing) order $\sigma_1(A) \geq \sigma_2(A) \geq \cdots \geq \sigma_q(A) \geq 0$. The number of positive singular values of A is evidently equal to the rank of A. The columns of the unitary matrix W are *right singular vectors* of A; the columns of V are *left singular vectors* of A.

If $A \in M_n$ is normal, consider a spectral decomposition $A = U\Lambda U^*$ with a unitary $U \in M_n$, $\Lambda = \mathrm{diag}(\lambda_1, ..., \lambda_n)$, and $|\lambda_1| \geq \cdots \geq |\lambda_n|$. If we write $\Sigma \equiv \mathrm{diag}(|\lambda_1|, ..., |\lambda_n|)$, then $\Lambda = D\Sigma$, where $D = \mathrm{diag}(e^{i\theta_1}, ..., e^{i\theta_n})$ is a diagonal unitary matrix. Thus, $A = U\Lambda U^* = (UD)\Sigma U^* = V\Sigma W^*$, which is a singular value decomposition of A with $V = UD$ and $W = U$. Also, we see that the singular values of a normal matrix are just the absolute values of its eigenvalues. This observation has a converse (see Problem 19) and a generalization to diagonalizable matrices (see Problem 45).

Notice that the singular values of $A = V\Sigma W^*$ are exactly the nonnegative square roots of the q largest eigenvalues of either $A^*A = W\Sigma^T\Sigma W^*$ or $AA^* = V\Sigma\Sigma^T V^*$; the remaining eigenvalues of A^*A and AA^*, if any, are all zero. Consequently, the ordered singular values of A are uniquely determined by A, and the singular values of A and A^* are the same. Furthermore, the singular values of $U_1 A U_2$ are the same as those of A whenever U_1 and U_2 are unitary matrices of appropriate sizes; this expresses the *unitary invariance* (invariance under unitary equivalence) of the set of singular values of a complex matrix.

Unlike the diagonal factor Σ in a singular value decomposition $A = V\Sigma W^*$, the left and right unitary factors V and W are never uniquely determined; the degree of nonuniqueness depends on the multiplicities of the singular values. Let $s_1 > s_2 > \cdots > s_k > 0$ denote the distinct positive singular values of $A \in M_{m,n}$ with respective positive multiplicities $\mu_1, \mu_2, ..., \mu_k$. Then $\mu_1 + \cdots + \mu_k = r = \mathrm{rank}\, A \leq q \equiv \min\{m, n\}$,

$$AA^* = V\Sigma\Sigma^T V^* = V[s_1^2 I_{\mu_1} \oplus \cdots \oplus s_k^2 I_{\mu_k} \oplus 0_{m-r}]V^* \equiv VS_1 V^*$$

and

$$A^*A = W\Sigma^T\Sigma W^* = W[s_1^2 I_{\mu_1} \oplus \cdots \oplus s_k^2 I_{\mu_k} \oplus 0_{n-r}]W^* \equiv WS_2 W^*$$

where $I_{\mu_j} \in M_{\mu_j}$ are identity matrices for $j = 1, ..., k$, $0_{m-r} \in M_{m-r}$, $0_{n-r} \in M_{n-r}$

are zero matrices, and S_1, S_2 are the indicated diagonal matrices. If $\hat{V} \in M_m$ and $\hat{W} \in M_n$ are unitary matrices such that $A = \hat{V}\Sigma\hat{W}^*$, then

$$AA^* = VS_1V^* = \hat{V}S_1\hat{V}^*, \text{ so } S_1(V^*\hat{V}) = (V^*\hat{V})S_1$$

and

$$A^*A = WS_2W^* = \hat{W}S_2\hat{W}^*, \text{ so } S_2(W^*\hat{W}) = (W^*\hat{W})S_2$$

that is, $V^*\hat{V}$ and $W^*\hat{W}$ each commutes with a diagonal matrix with equal diagonal entries grouped together. A simple calculation (see (0.9.1) in [HJ]) reveals the basic fact that $V^*\hat{V}$ and $W^*\hat{W}$ must be block diagonal with diagonal blocks (necessarily unitary) conformal to those of S_1 and S_2, that is,

$$\hat{V} = V[V_1 \oplus \cdots \oplus V_k \oplus \tilde{V}] \text{ and } \hat{W} = W[W_1 \oplus \cdots \oplus W_k \oplus \tilde{W}]$$

for some unitary V_i, $W_i \in M_{\mu_i}$, $i = 1,...,k$, $\tilde{V} \in M_{m-r}$, $\tilde{W} \in M_{n-r}$. But since $\hat{V}\Sigma\hat{W}^* = V\Sigma W^*$, it follows that $V_i = W_i$ for $i = 1,...,k$. Thus, we can characterize the set of all possible left and right unitary factors in a singular value decomposition as follows:

3.1.1′   Theorem. Under the assumptions of Theorem (3.1.1), suppose that the distinct nonzero singular values of A are $s_1 > \cdots > s_k > 0$, with respective multiplicities $\mu_1,..., \mu_k \geq 1$. Let $\mu_1 + \cdots + \mu_k = r$ and let $A = V\Sigma W^*$ be a given singular value decomposition with $\Sigma = \text{diag}(s_1 I_{\mu_1},..., s_k I_{\mu_k}, 0_{q-r}) \in M_{m,n}$. Let $\hat{V} \in M_m$ and $\hat{W} \in M_n$ be given unitary matrices. Then $A = \hat{V}\Sigma\hat{W}^*$ if and only if there are unitary matrices $U_i \in M_{\mu_i}$, $i = 1,...,k$, $\tilde{V} \in M_{m-r}$, and $\tilde{W} \in M_{n-r}$ such that

$$\hat{V} = V[U_1 \oplus \cdots \oplus U_k \oplus \tilde{V}] \text{ and } \hat{W} = W[U_1 \oplus \cdots \oplus U_k \oplus \tilde{W}]$$

One useful special case of the preceding characterization occurs when $m = n$ and $k = r = n$ or $n-1$, that is, A is square and has distinct singular values. In this case, there are diagonal unitary matrices $D_1 = \text{diag}(e^{i\theta_1},..., e^{i\theta_{n-1}}, d_1)$ and $D_2 = \text{diag}(e^{i\theta_1},..., e^{i\theta_{n-1}}, d_2)$ such that $\hat{V} = VD_1$ and $\hat{W} =$

WD_2; if $r = n$ (A is nonsingular), then $d_1 = d_2$ and $D_1 = D_2$.

The construction used in the proof of Theorem (3.1.1) shows that

$\sigma_1(A) = \max \{\|Ax\|_2 : x \in \mathbb{C}^n, \|x\|_2 = 1\}$, so $\sigma_1 = \|Aw_1\|_2$ for some unit vector $w_1 \in \mathbb{C}^n$

$\sigma_2(A) = \max \{\|Ax\|_2 : x \in \mathbb{C}^n, \|x\|_2 = 1, x \perp w_1\}$, so $\sigma_2 = \|Aw_2\|_2$ for some unit vector $w_2 \in \mathbb{C}^n$ such that $w_2 \perp w_1$

$\vdots$

$\sigma_k(A) = \max \{\|Ax\|_2 : x \in \mathbb{C}^n, \|x\|_2 = 1, x \perp w_1, \ldots, w_{k-1}\}$, so $\sigma_k = \|Aw_k\|_2$ for some unit vector $w_k \in \mathbb{C}^n$ such that $w_k \perp w_1, \ldots, w_{k-1}$

$\vdots$

Thus, $\sigma_1(A) = \|\|A\|\|_2$, the spectral norm of A, and each singular value is the norm of A as a mapping restricted to a suitable subspace of $\mathbb{C}^n$. The similarity of this observation to the situation that holds for square Hermitian matrices is not merely superficial, and there is an analog of the Courant-Fischer theorem for singular values.

3.1.2   Theorem. Let $A \in M_{m,n}$ be given, let $\sigma_1(A) \geq \sigma_2(A) \geq \cdots$ be the ordered singular values of A, and let k be a given integer with $1 \leq k \leq \min \{m,n\}$. Then

$$\sigma_k(A) = \min_{\substack{w_1, \ldots, w_{k-1} \in \mathbb{C}^n}} \quad \max_{\substack{x \in \mathbb{C}^n \\ \|x\|_2 = 1 \\ x \perp w_1, \ldots, w_{k-1}}} \|Ax\|_2 \qquad (3.1.2a)$$

$$= \max_{\substack{w_1, \ldots, w_{n-k} \in \mathbb{C}^n}} \quad \min_{\substack{x \in \mathbb{C}^n \\ \|x\|_2 = 1 \\ x \perp w_1, \ldots, w_{n-k}}} \|Ax\|_2 \qquad (3.1.2b)$$

$$= \min_{\substack{S \subset \mathbb{C}^n \\ \dim S = n-k+1}} \quad \max_{\substack{x \in S \\ \|x\|_2 = 1}} \|Ax\|_2 \qquad (3.1.2c)$$

$$= \max_{\substack{S \subset \mathbb{C}^n \\ \dim S = k}} \quad \min_{\substack{x \in S \\ \|x\|_2 = 1}} \|Ax\|_2 \qquad (3.1.2d)$$

where the outer optimizations in $(3.1.2(c,d))$ are over all subspaces S with the indicated dimension.

Proof: Use the "min-max" half of the Courant-Fischer theorem $((4.2.12)$ in [HJ]) to characterize the decreasingly ordered eigenvalues of A^*A:

$$\lambda_k(A^*A) = \min_{\substack{w_1,\ldots,w_{k-1}\in\,\mathbb{C}^n}} \max_{\substack{0\neq x\in\mathbb{C} \\ x\perp w_1,\ldots,w_{k-1}}} \frac{x^*(A^*A)x}{x^*x}$$

$$= \min_{\substack{w_1,\ldots,w_{k-1}\in\,\mathbb{C}^n}} \max_{\substack{x\in\mathbb{C}^n \\ \|x\|_2=1 \\ x\perp w_1,\ldots,w_{k-1}}} \|Ax\|_2^2$$

which proves (3.1.2a) since $\lambda_k(A^*A) = \sigma_k(A)^2$. The same argument with the max-min half of the Courant-Fischer theorem $((4.2.13)$ in [HJ]) proves the characterization (3.1.2b). The alternative formulations $(3.1.2(c,d))$ are equivalent versions of $(3.1.2(a,b))$, in which the specification of x via a stated number of orthogonality constraints is replaced by specification of x via membership in a subspace—the orthogonal complement of the span of the constraints. ☐

The Courant-Fischer theorem implies useful interlacing theorems for eigenvalues of Hermitian matrices, so it is not surprising that interlacing theorems for singular values of complex matrices follow from the preceding variational characterization. The proof of the following result is formally identical to the proof of the classical inclusion principle given for Theorem (4.3.15) in [HJ]. For a different proof, see Theorem (7.3.9) in [HJ].

3.1.3   Corollary. Let $A \in M_{m,n}$ be given, and let A_r denote a submatrix of A obtained by deleting a total of r rows and/or columns from A. Then

$$\sigma_k(A) \geq \sigma_k(A_r) \geq \sigma_{k+r}(A), \quad k = 1,\ldots, \min\{m,n\} \qquad (3.1.4)$$

where for $X \in M_{p,q}$ we set $\sigma_j(X) \equiv 0$ if $j > \min\{p,q\}$.

Proof: It suffices to consider the case $r = 1$, in which any one row or column is deleted, and to show that $\sigma_k(A) \geq \sigma_k(A_1) \geq \sigma_{k+1}(A)$. The general case then follows by repeated application of these inequalities. If A_1 is formed from A by deleting column s, denote by e_s the standard unit basis vector with a 1 in position s. If $x \in \mathbb{C}^n$, denote by $\xi \in \mathbb{C}^{n-1}$ the vector obtained by deleting entry s from x. Now use (3.1.2a) to write

$$
\sigma_k(A) = \min_{\substack{w_1, \ldots, w_{k-1} \in \mathbb{C}^n}} \quad \max_{\substack{x \in \mathbb{C}^n \\ \|x\|_2 = 1 \\ x \perp w_1, \ldots, w_{k-1}}} \quad \|Ax\|_2
$$

$$
\geq \min_{\substack{w_1, \ldots, w_{k-1} \in \mathbb{C}^n}} \quad \max_{\substack{x \in \mathbb{C}^n \\ \|x\|_2 = 1 \\ x \perp w_1, \ldots, w_{k-1}, e_s}} \quad \|Ax\|_2
$$

$$
= \min_{\substack{\omega_1, \ldots, \omega_{k-1} \in \mathbb{C}^{n-1}}} \quad \max_{\substack{\xi \in \mathbb{C}^{n-1} \\ \|\xi\|_2 = 1 \\ \xi \perp \omega_1, \ldots, \omega_{k-1}}} \quad \|A_1 \xi\|_2 = \sigma_k(A_1)
$$

For the second inequality, use (3.1.2b) to write

$$
\sigma_{k+1}(A) = \max_{\substack{w_1, \ldots, w_{n-k-1} \in \mathbb{C}^n}} \quad \min_{\substack{x \in \mathbb{C}^n \\ \|x\|_2 = 1 \\ x \perp w_1, \ldots, w_{n-k-1}}} \quad \|Ax\|_2
$$

$$
\leq \max_{\substack{w_1, \ldots, w_{n-k-1} \in \mathbb{C}^n}} \quad \min_{\substack{x \in \mathbb{C}^n \\ \|x\|_2 = 1 \\ x \perp w_1, \ldots, w_{n-k-1}, e_s}} \quad \|Ax\|_2
$$

$$
= \max_{\substack{\omega_1, \ldots, \omega_{n-k-1} \in \mathbb{C}^{n-1}}} \quad \min_{\substack{\xi \in \mathbb{C}^{n-1} \\ \|\xi\|_2 = 1 \\ \xi \perp \omega_1, \ldots, \omega_{n-k-1}}} \quad \|A_1 \xi\|_2 = \sigma_k(A_1)
$$

If a row of A is deleted, apply the same argument to A^*, which has the same singular values as A.   □

By combining Theorem (3.1.2) with the Courant-Fischer theorem for Hermitian matrices, we can obtain useful inequalities between individual singular values of a matrix and eigenvalues of its Hermitian part.

3.1.5   **Corollary.** Let $A \in M_n$ be given, let $\sigma_1(A) \geq \cdots \geq \sigma_n(A)$ denote its ordered singular values, let $H(A) = \frac{1}{2}(A + A^*)$, and let $\{\lambda_i(H(A))\}$ denote the algebraically decreasingly ordered eigenvalues of $H(A)$, $\lambda_1(H(A)) \geq \cdots \geq \lambda_n(H(A))$. Then

$$\sigma_k(A) \geq \lambda_k(H(A)) \text{ for } k = 1,..., n \qquad (3.1.6a)$$

More generally,

$$\sigma_k(A) \geq \lambda_k(H(UAV)) \text{ for all } k = 1,..., n \text{ and all unitary } U, V \in M_n \qquad (3.1.6b)$$

Proof: For any unit vector $x \in \mathbb{C}^n$, we have

$$x^* H(A) x = \frac{1}{2}(x^* A x + x^* A^* x) = \operatorname{Re} x^* A x \leq |x^* A x|$$

$$\leq \|x\|_2 \|Ax\|_2 = \|Ax\|_2$$

Thus,

$$\lambda_k(H(A)) = \min_{\substack{w_1, \ldots, w_{k-1} \in \mathbb{C}^n}} \max_{\substack{x \in \mathbb{C}^n \\ \|x\|_2 = 1 \\ x \perp w_1, \ldots, w_{k-1}}} x^* H(A) x$$

$$\leq \min_{\substack{w_1, \ldots, w_{k-1} \in \mathbb{C}^n}} \max_{\substack{x \in \mathbb{C}^n \\ \|x\|_2 = 1 \\ x \perp w_1, \ldots, w_{k-1}}} \|Ax\|_2 = \sigma_k(A)$$

The more general assertion (3.1.6b) follows from (3.1.6a) since $\sigma_k(A) = \sigma_k(UAV)$ for every $U, V \in M_n$. $\square$

We now consider several matrix representations associated with the singular value decomposition. One way to phrase the spectral theorem for normal matrices is to say that every n-by-n normal matrix is a complex linear combination of n pairwise orthogonal Hermitian projections. An analogous result, valid for all matrices, follows immediately from the singular value decomposition.

3.1.7   Definition. A matrix $P \in M_{m,n}$ is said to be a *rank r partial isometry* if $\sigma_1(P) = \cdots = \sigma_r(P) = 1$ and $\sigma_{r+1}(P) = \cdots = \sigma_q(P) = 0$, where $q \equiv \min\{m,n\}$. Two partial isometries $P, Q \in M_{m,n}$ (of unspecified rank) are said to be *orthogonal* if $P^* Q = 0$ and $PQ^* = 0$.

The asserted representations in the following theorem are readily verified.

3.1.8   Theorem. Let $A \in M_{m,n}$ have singular value decomposition $A = V \Sigma W^*$ with $V = [v_1 \ \cdots \ v_m] \in M_m$ and $W = [w_1 \ \cdots \ w_n] \in M_n$ unitary, and $\Sigma = [\sigma_{ij}] \in M_{m,n}$ with $\sigma_1 = \sigma_{11} \geq \cdots \geq \sigma_q = \sigma_{qq} \geq 0$ and $q = \min\{m,n\}$. Then

(a) $A = \sigma_1 P_1 + \cdots + \sigma_q P_q$ is a nonnegative linear combination of mutually orthogonal rank one partial isometries, with $P_i = v_i w_i^*$ for $i = 1, \dots, q$.

(b) $A = \mu_1 K_1 + \cdots + \mu_q K_q$ is a nonnegative linear combination of partial isometries with rank $K_i = i$, $i = 1, \dots, q$, such that

(1) $\mu_i = \sigma_i - \sigma_{i+1}$ for $i = 1, \dots, q-1$, $\mu_q = \sigma_q$;
(2) $\mu_i + \cdots + \mu_q = \sigma_i$ for $i = 1, \dots, q$; and
(3) $K_i = VE_i W^*$ for $i = 1, \dots, q$ in which the first i columns of $E_i \in M_{m,n}$ are the respective unit basis vectors $e_1, \dots, e_i$ and the remaining $n - i$ columns are zero.

Another useful representation that follows immediately from the singular value decomposition is the *polar decomposition*.

3.1.9   Theorem. Let $A \in M_{m,n}$ be given.

(a)   If $n \geq m$, then $A = PY$, where $P \in M_m$ is positive semidefinite, $P^2 = AA^*$, and $Y \in M_{m,n}$ has orthonormal rows.
(b)   If $m \geq n$, then $A = XQ$, where $Q \in M_n$ is positive semidefinite, $Q^2 = A^*A$, and $X \in M_{m,n}$ has orthonormal columns.
(c)   If $m = n$, then $A = PU = UQ$, where $U \in M_n$ is unitary, $P, Q \in M_n$ are positive semidefinite, $P^2 = AA^*$, and $Q^2 = A^*A$.

In all cases, the positive semidefinite factors P and Q are uniquely determined by A and their eigenvalues are the same as the singular values of A.

Proof: If $n \geq m$ and $A = V\Sigma W^*$ is a singular value decomposition, write $\Sigma = [S \ 0]$ and $W = [W_1 \ W_2]$, where $S = \text{diag}(\sigma_1(A),..., \sigma_m(A)) \in M_m$ and $W_1 \in M_{n,m}$. Then $A = V[S \ 0][W_1 \ W_2]^* = VSW_1^* = (VSV^*)(VW_1^*)$. Notice that $P \equiv VSV^*$ is positive semidefinite and $Y \equiv VW_1^*$ satisfies $YY^* = VW_1^* W_1 V^* = VIV^* = I$, so Y has orthonormal rows. The assertions in (b) follow from applying (a) to A^*. For (c), notice that $A = V\Sigma W^* = (V\Sigma V^*)(VW^*) = (VW^*)(W\Sigma W^*)$, so we may take $P = V\Sigma V^*$, $Q = W\Sigma W^*$, and $U = VW^*$.   □

Exercise. In the square case (c) of Theorem (3.1.9), use the characterization in Theorem (3.1.1′) to show that all possible unitary factors U (left or right) in the polar decomposition of $A \in M_n$ are of the form $U = V[I \oplus \tilde{U}]W^*$, where the unitary matrix $\tilde{U} \in M_{n-r}$ is arbitrary, $r = \text{rank } A$, and $A = V\Sigma W^*$ is a given singular value decomposition. In particular, conclude that all three factors P, Q, and U are uniquely determined when A is nonsingular.

Problems

1.   Provide the details for the following proof of the singular value decomposition:

(a) Let $A \in M_n$ be given. Then AA^* and A^*A are both normal and have the same eigenvalues, so they are unitarily similar. Let $U \in M_n$ be unitary and such that $A^*A = U(AA^*)U^*$. Then UA is normal, so there is a unitary $X \in M_n$ and a diagonal $\Lambda \in M_n$ such that $UA = X\Lambda X^*$. Write $\Lambda = \Sigma D$, where $\Sigma = |\Lambda|$ is nonnegative and D is a diagonal unitary matrix. Then $A = V\Sigma W^*$ with $V = U^*X$, $W = DX^*$. This is essentially the approach to the singular value decomposition used in L. Autonne's 1915 paper cited in Section (3.0).

(b) If $A \in M_{m,n}$ with $m > n$, let $u_1,..., u_\nu$ be an orthonormal basis for the nullspace of A^*, so $\nu \geq m - n$. Let $U_2 \equiv [u_1 \ ... \ u_{m-n}] \in M_{m,m-n}$ and let $U = [U_1 \ U_2] \in M_m$ be unitary. Then

$$U^*A = \begin{bmatrix} A_1 \\ 0 \end{bmatrix}$$

with $A_1 \in M_n$, so

$$A = U \begin{bmatrix} V\Sigma W^* \\ 0 \end{bmatrix} = U(V \oplus I) \begin{bmatrix} \Sigma \\ 0 \end{bmatrix} W^*$$

If $m < n$, apply this result to A^*.

2.   Let $A \in M_{m,n}$. Explain why the rank of A is exactly the number of its nonzero singular values.

3.   If $A \in M_{m,n}$ and $A = V\Sigma W^*$ is a singular value decomposition, what are singular value decompositions of A^*, A^T, $\bar{A}$ and, if $m = n$ and A is nonsingular, A^{-1}? Conclude that the singular values of A, A^*, A^T, and $\bar{A}$ are all the same, and, if $A \in M_n$ is nonsingular, the singular values of A are the reciprocals of the singular values of A^{-1}.

4.   Let $A = [a_{ij}] \in M_{m,n}$ have a singular value decomposition $A = V\Sigma W^*$ with unitary $V = [v_{ij}] \in M_m$ and $W = [w_{ij}] \in M_n$, and let $q = \min\{m,n\}$.

(a) Show that each $a_{ij} = v_{i1}\bar{w}_{j1}\sigma_1(A) + \cdots + v_{iq}\bar{w}_{jq}\sigma_q(A)$.

(b) Use the representation in (a) to show that

$$\sum_{i=1}^q |a_{ii}| \leq \sum_{k=1}^q \sum_{i=1}^q |v_{ik}w_{ik}| \, \sigma_k(A) \leq \sum_{k=1}^q \sigma_k(A) \tag{3.1.10a}$$

(c) When $m = n$, use (a) and the conditions for equality in the Cauchy-Schwarz inequality to show that

$$\text{Re tr } A \leq \sum_{i=1}^n \sigma_i(A) \tag{3.1.10b}$$

with equality if and only if A is positive semidefinite.

(d) When $m = n$, let $A = U\Delta U^*$ with U unitary and Δ upper triangular, let $\lambda_1(A),...,\lambda_n(A)$ be the main diagonal entries of Δ, and let $D \equiv \text{diag}(e^{i\theta_1},...,e^{i\theta_n})$ with each $\theta_k \in \mathbb{R}$. What are the eigenvalues and singular values of the matrix $AUDU^*$? Use (c) to show that

$$\text{Re tr } A \leq |\text{tr } A| \leq \sum_{i=1}^n |\lambda_i(A)| \leq \sum_{i=1}^n \sigma_i(A) \tag{3.1.10c}$$

See (3.3.35) and (3.3.13a,b) for generalizations of these inequalities.

5.   Recall that a matrix $C \in M_n$ is a *contraction* if $\sigma_1(C) \leq 1$ (and hence $0 \leq$

$\sigma_i(C) \leq 1$ for all $i = 1, 2,..., n$). All matrices in this problem are in M_n.

(a) Show that the unitary matrices are the only rank n (partial) isometries in M_n.

(b) Show that any finite product of contractions is a contraction.

(c) Is a product of partial isometries a partial isometry?

(d) Describe all the rank r partial isometries and contractions in M_n in terms of their singular value decompositions.

(e) Show that $C \in M_n$ is a rank one partial isometry if and only if $C = xy^*$ for some unit vectors $x, y \in \mathbb{C}^n$.

(f) For $1 \leq r < n$, show that every rank r partial isometry in M_n is a convex combination of two distinct unitary matrices in M_n.

(g) Use Theorem (3.1.8) to show that every matrix in M_n is a finite nonnegative linear combination of unitary matrices in M_n.

(h) Use Theorem (3.1.8(b)) to show that a given $A \in M_n$ is a contraction if and only if it is a finite convex combination of unitary matrices in M_n. See Problem 27 for another approach to this result.

6.   Let $A \in M_{m,n}$.   Show that $\sigma_1(A) = \max \{|x^*Ay| : x, y \in \mathbb{C}^n$ are unit vectors$\}$. If $m = n$, show that $\sigma_1(A) = \max \{|\operatorname{tr} AC| : C \in M_n$ is a rank one partial isometry$\}$. See Theorem (3.4.1) for an important generalization of this result.

7.   Let $A \in M_n$ be given and let r be a given integer with $1 \leq r \leq n$. Show that there is a partial isometry $C_r \in M_n$ such that $\sigma_1(A) + \cdots + \sigma_r(A) = \operatorname{tr} AC_r$. What is this when $r = 1$? Compare with Problem 6. See Theorem (3.3.1) for a proof that $\sigma_1 + \cdots + \sigma_r = \max \{|\operatorname{tr} AC_r| : C_r \in M_n$ is a rank r partial isometry$\}$.

8.   Let $A \in M_n$. A vector $x \in \mathbb{C}^n$ such that $Ax = x$ is called a *fixed point* of A; the nonzero fixed points of A are just its eigenvectors corresponding to the eigenvalue $\lambda = 1$.

   (a) If A is a contraction, show that every fixed point of A is a fixed point of A^*.

   (b) Consider $A = \begin{bmatrix} 1 & 1 \\ 0 & 0 \end{bmatrix}$ to show that the assertion in (a) is not generally true if A is not a contraction.

9.   Let $A \in M_n$, and let $U, V \in M_n$ be unitary. Show that the singular values of A and UAV are the same. Are the eigenvalues the same? The eigenvalues of A^2 are the squares of the eigenvalues of A. Is this true for the singular values as well?

10. Let $A \in M_n$.
    (a) Show that $\sigma_1(A) \geq \rho(A) = \max \{|\lambda| : \lambda$ is an eigenvalue of $A\}$.
    (b) Let $x, y \in \mathbb{C}^n$ and let $A \equiv xy^*$. Calculate $\sigma_1(A)$ and $\rho(A)$. What does the general inequality $\sigma_1(A) \geq \rho(A)$ give in this case?

11. Let $A \in M_n$. Show that $\sigma_1(A) \cdots \sigma_n(A) = |\det A| = |\lambda_1(A) \cdots \lambda_n(A)|$.

12. Let $A = \begin{bmatrix} 1 & 1 \\ 0 & 1 \end{bmatrix} \in M_n$. What are the eigenvalues and singular values of A? Verify that $\sigma_1(A) \geq \rho(A)$ and $\sigma_1(A)\sigma_2(A) = |\lambda_1(A)\lambda_2(A)| = |\det A|$.

13. Show that all the singular values of a matrix $A \in M_n$ are equal if and only if A is a scalar multiple of a unitary matrix.

14. Let $1 \leq i \leq n$ be a given integer and consider the function $\sigma_i(\cdot) : M_n \to \mathbb{R}^+$. Show that $\sigma_i(cA) = |c| \, \sigma_i(A)$ for all $A \in M_n$ and all $c \in \mathbb{C}$, and $\sigma_i(A) \geq 0$ for all $A \in M_n$. Is $\sigma_i(\cdot)$ a norm on M_n? Is it a seminorm?

15. Verify the assertions in Theorem (3.1.8).

16. State an analog of (3.1.8b) for normal matrices in M_n.

17. If $A \in M_n$ is positive semidefinite, show that the eigenvalues and singular values of A are the same.

18. Let $A = [a_{ij}] \in M_n$ have eigenvalues $\{\lambda_i(A)\}$ and singular values $\{\sigma_i(A)\}$, and let $A = U\Delta U^* = U(\Lambda + T)U^*$ be a Schur upper triangularization of A, that is, U is unitary, $\Lambda = \text{diag}(\lambda_1(A), \ldots, \lambda_n(A))$, and $T = [t_{ij}]$ is strictly upper triangular. Let $\|A\|_2 = (\text{tr } A^*A)^{\frac{1}{2}}$ denote the Frobenius norm of A.
    (a) Show that

$$\|A\|_2^2 = \sum_{i=1}^{n} \sigma_i(A)^2 = \|\Lambda\|_2^2 + \|T\|_2^2 \geq \sum_{i=1}^{n} |\lambda_i(A)|^2 \quad (3.1.11)$$

with equality if and only if A is normal. Although different Schur upper triangularizations of A can result in different strictly upper triangular parts T in (3.1.11), notice that the quantity

$$\|T\|_2 = \left[\sum_{i=1}^{n} \sigma_i(A)^2 - \sum_{i=1}^{n} |\lambda_i(A)|^2 \right]^{\frac{1}{2}} \quad (3.1.12)$$

has the same value for all of them. This is called the *defect from normality* of A *with respect to the Frobenius norm.*

(b) Let $x_1,..., x_n \geq 0$ be given and let $a_{i,i+1} = \sqrt{x_i}$, $i = 1,..., n-1$, $a_{n1} = \sqrt{x_n}$, and all other $a_{ij} = 0$. Show that the eigenvalues of A are the n nth roots of $(x_1 \cdots x_n)^{\frac{1}{2}}$ and derive the arithmetic-geometric mean inequality $x_1 + \cdots + x_n \geq n(x_1 \cdots x_n)^{1/n}$ from the inequality in (a).

19. Let $A \in M_n$ have singular values $\sigma_1(A) \geq \cdots \geq \sigma_n(A) \geq 0$ and eigenvalues $\{\lambda_i(A)\}$ with $|\lambda_1(A)| \geq \cdots \geq |\lambda_n(A)|$. Show that A is normal if and only if $\sigma_i(A) = |\lambda_i(A)|$ for all $i = 1,..., n$. More generally, if $\sigma_i(A) = |\lambda_i(A)|$ for $i = 1,..., k$, show that A is unitarily similar to $D \oplus B$, where $D = \text{diag}(\lambda_1,..., \lambda_k)$, $B \in M_{n-k}$, and $\sigma_k(A) = |\lambda_k(A)| \geq \sigma_1(B)$.

20. Using the notation of Theorem (3.1.1), let $W = [w_1 \ ... \ w_n]$ and $V = [v_1 \ ... \ v_m]$, where the orthonormal sets $\{v_i\} \subset \mathbb{C}^m$ and $\{w_i\} \subset \mathbb{C}^n$ are left and right singular vectors of A, respectively. Show that $Aw_i = \sigma_i v_i$, $A^* v_i = \sigma_i w_i$, $\|A^* v_i\|_2 = \sigma_i$, and $\|Aw_i\|_2 = \sigma_i$ for $i = 1,...,$ min $\{m,n\}$. These are the matrix analogs of E. Schmidt's integral equations (3.0.5).

21. Provide details for the following proof of the singular value decomposition that relies on the spectral theorem for Hermitian matrices (Theorem (4.1.5) in [HJ]): Let $A \in M_{m,n}$ have rank r. Then $A^* A \in M_n$ is Hermitian and positive semidefinite, so $A^* A = U(\Lambda^2 \oplus 0)U^* = UD(I_r \oplus 0)DU^*$, where $U \in M_n$ is unitary, $\Lambda \in M_r$ is diagonal and positive definite, and $D = \Lambda \oplus I_{n-r} \in M_n$. Then $(AUD^{-1})^*(AUD^{-1}) = I_r \oplus 0$, so $AUD^{-1} = [V_1 \ 0]$, where $V_1 \in M_{m,r}$ has orthonormal columns. If $V = [V_1 \ V_2] \in M_m$ is unitary, then $A = [V_1 \ 0]DU^* = V(\Lambda \oplus 0)U^* = V\Sigma W^*$.

22. Deduce the singular value decomposition (3.1.1) from the polar decomposition (3.1.9).

23. According to Corollary (3.1.3), deleting some rows or columns of a matrix may decrease some of its singular values. However, if the rows or columns deleted are all zero, show that the nonzero singular values are unchanged, and some previously zero singular values are deleted.

24. Let $x \in \mathbb{C}^n$ be given. If one thinks of x as an n-by-1 matrix, that is, $x \in M_{n,1}$, what is its singular value decomposition? What is the singular value?

25. Suppose $A \in M_n$ and $\sigma_1(A) \leq 1$, that is, A is a contraction. If $H(A) \equiv \frac{1}{2}(A + A^*) = I$, show that $A = I$.

26. Let $U \in M_n$ be a given unitary matrix and suppose $U = \frac{1}{2}(A + B)$ for some contractions $A, B \in M_n$. Show that $A = B = U$.

27. Let B_n denote the unit ball of the spectral norm in M_n, that is, $B_n = \{A \in M_n: \sigma_1(A) \leq 1\}$ is the set of contractions in M_n. Like the unit ball of any norm, B_n is a convex set.

    (a) If $A \in B_n$ and if A is an extreme point of B_n, show that A is unitary.

    (b) If $U \in M_n$ is unitary, use Problem 26 to show that U is an extreme point of B_n.

    (c) Conclude that the extreme points of B_n are exactly the unitary matrices in M_n.

    (d) Use (c) to show that $A \in B_n$ if and only if A is a finite convex combination of unitary matrices. See Problem 5 as well as Problem 4 of Section (3.2) for different approaches to this result.

28. Let $A \in M_n$ be given and let $r = \text{rank } A$. Show that A is normal if and only if there is a set $\{x_1,..., x_r\} \subset \mathbb{C}^n$ of r orthonormal vectors such that $|x_i^* A x_i| = \sigma_i(A)$ for $i = 1,..., r$.

29. Let $A \in M_n$, $B \in M_m$. Show that the set of singular values of the direct sum $A \oplus B$ is the union of the sets of singular values of A and B, including multiplicities. Show that $\sigma_1(A \oplus B) = \max \{\sigma_1(A), \sigma_1(B)\}$.

30. Let $A \in M_n$ be given. Use (3.1.4) to show that $\sigma_1(A) \geq$ maximum Euclidean column or row length in A and $\sigma_n(A) \leq$ minimum Euclidean column or row length in A. If A is nonsingular, conclude that $\kappa(A) = \sigma_1(A)/\sigma_n(A)$, the spectral condition number of A, is bounded from below by the ratio of the largest to smallest Euclidean lengths of the set of rows and columns of A. Thus, if a system of linear equations $Ax = b$ is poorly scaled (that is, its ratio of largest to smallest row and column norms is large), then the system must be ill conditioned. This *sufficient* condition for ill conditioning is not *necessary*, however. Give an example of an ill-conditioned A for which all the rows and columns have nearly the same norm.

31. Let $A \in M_n$ be given and have singular values $\sigma_1(A) \geq \cdots \geq \sigma_n(A) \geq 0$. If $H(A) = \frac{1}{2}(A + A^*)$ is positive semidefinite and if $\alpha \geq 0$ is a given scalar, show that $\sigma_i(A + \alpha I)^2 \geq \sigma_i(A)^2 + \alpha^2$ for $i = 1,..., n$. Show by example that this need not be true, and indeed $\sigma_i(A + \alpha I) < \sigma_i(A)$ is possible, if $H(A)$ is not positive semidefinite.

32. Suppose $A \in M_n$ is skew-symmetric, that is, $A = -A^T$. Show that the nonzero singular values of A are of the form $s_1, s_1, s_2, s_2,..., s_k, s_k$, where

$s_1,..., s_k > 0$, that is, A has even rank and its nonzero singular values occur in pairs.

33. The purpose of this problem is to use the singular value decomposition to show that any complex symmetric $A \in M_n$ can be written as $A = U\Sigma U^T$ for some unitary $U \in M_n$ and nonnegative diagonal $\Sigma \in M_n$. The approach suggested is essentially the one used by L. Autonne in his 1915 paper cited in Section (3.0).

(a) Let $D = d_1 I_{n_1} \oplus \cdots \oplus d_r I_{n_r} \in M_n$ be diagonal with $|d_1| > \cdots > |d_r| \geq 0$ and $n_1 + \cdots + n_r = n$, let a unitary $U \in M_n$ be given, and suppose $DU = U^T D$. Show that $U = U_1 \oplus \cdots \oplus U_r$, where each $U_i \in M_{n_i}$ is unitary and $U_i = U_i^T$ if $d_i \neq 0$.

(b) Let $B \in M_n$ be a given diagonalizable matrix. Show that there is a polynomial $p(t)$ of degree at most $n-1$ such that $C \equiv p(B)$ satisfies $C^2 = B$. If, in addition, $B = B^T$, then $C = C^T$. Using the notation and hypothesis of part (a), show that there is a unitary $Z \in M_n$ such that $Z^2 = U$ and $DZ = Z^T D$.

(c) Let $A \in M_n$ be given and let $A = V\Sigma W^*$ be a given singular value decomposition for A. If $A = A^T$, show that $\Sigma(V^T W) = (V^T W)^T \Sigma$. Show that there is a unitary Z such that $V = \overline{W}(Z^T)^2$ and $Z^T \Sigma = \Sigma Z$, and show that $A = (\overline{W}Z^T)\Sigma(\overline{W}Z^T)^T = U\Sigma U^T$, as desired.

34. The interlacing inequalities (3.1.4) have been shown to be necessary constraints on the singular values of a submatrix of given size, relative to the singular values of the overall matrix. They are also known to be sufficient, in the sense that they precisely characterize the possible ranges of singular values for submatrices (of a given size) of a matrix with given singular values. The purpose of this problem is to demonstrate their sufficiency in an interesting special case.

(a) Let $A \in M_n$ have singular values $\sigma_1 \geq \cdots \geq \sigma_n \geq 0$. Denote the singular values of the upper left $(n-1)$-by-$(n-1)$ principal submatrix of A by $s_1 \geq \cdots \geq s_n \geq 0$. Show that

$$\sigma_1 \geq s_1 \geq \sigma_3, \quad \sigma_2 \geq s_2 \geq \sigma_4, ..., \quad \sigma_{n-2} \geq s_{n-2} \geq \sigma_n, \quad \sigma_{n-1} \geq s_{n-1} \geq 0 \quad (3.1.13)$$

(b) Let $\sigma_1 \geq \cdots \geq \sigma_n \geq 0$ be given and let $\Sigma \equiv \mathrm{diag}(\sigma_1,..., \sigma_n)$. Explain why $\{U\Sigma U^T : U \in M_n$ is unitary$\}$ is exactly the set of symmetric matrices in M_n with singular values $\sigma_1,..., \sigma_n$.

(c) Let $\sigma_1 \geq \cdots \geq \sigma_n \geq 0$ and $s_1 \geq \cdots \geq s_{n-1} \geq 0$ satisfy the interlacing inequalities (3.1.13). Show that the set of $n-1$ numbers $\{s_1, -s_2, s_3, -s_4,...\}$ interlaces the set of n numbers $\{\sigma_1, -\sigma_2, \sigma_3, -\sigma_4,...\}$ (in the sense of the hypothesis of Theorem (4.3.10) in [HJ]) when both are put into algebraically decreasing order. Conclude that there is a real symmetric $Q \in M_n(\mathbb{R})$ such that the upper left $(n-1)$-by-$(n-1)$ principal submatrix of $Q \operatorname{diag}(\sigma_1, -\sigma_2, \sigma_3, -\sigma_4,...) Q^T \equiv A$ has eigenvalues $s_1, -s_2, s_3, -s_4,....$ Explain why the singular values of A are $\sigma_1,..., \sigma_n$ and the singular values of the principal submatrix are $s_1,..., s_n$.

(d) Let $A \in M_n$ be a given symmetric matrix with singular values $\sigma_1 \geq \cdots \geq \sigma_n \geq 0$ [for example, $A = \operatorname{diag}(\sigma_1,..., \sigma_n)$], and let $s_1 \geq \cdots \geq s_{n-1} \geq 0$ be given. Explain why there is a unitary $U \in M_n$ such that UAU^T has singular values $\sigma_1,..., \sigma_n$ and its upper left $(n-1)$-by-$(n-1)$ principal submatrix has singular values $s_1,..., s_{n-1}$ if and only if the interlacing inequalities (3.1.13) are satisfied.

35. Let $A \in M_{m,n}$ be given, suppose it has $q = \min\{m,n\}$ distinct singular values, let $A = V\Sigma W^*$ be a singular value decomposition, and let $A = \hat{V}\Sigma\hat{W}^*$ be another given singular value decomposition. If $m \leq n$, show that there is a diagonal unitary $D \in M_m$ such that $\hat{V} = VD$. If $m \geq n$, show that there is a diagonal unitary $D \in M_n$ such that $\hat{W} = WD$. In these two cases, how are the other unitary factors related?

36. Let $A \in M_n$ be given, let $H(A) = \frac{1}{2}(A + A^*)$ denote the Hermitian part of A, and let $S(A) = \frac{1}{2}(A - A^*)$ denote the skew-Hermitian part of A. Order the eigenvalues of $H(A)$ and $S(A)$ so that $\lambda_1(H(A)) \geq \cdots \geq \lambda_n(H(A))$ and $\lambda_1(iS(A)) \geq \cdots \geq \lambda_n(iS(A))$. Use Corollary (3.1.5) to show that

$$\sigma_k(A) \geq \max\{\lambda_k(H(A)), -\lambda_{n-k+1}(H(A)),$$

$$\lambda_k(iS(A)), -\lambda_{n-k+1}(iS(A))\} \qquad (3.1.14)$$

for $k = 1,..., n$.

37. Recall the Loewner partial order on n-by-n Hermitian matrices: $A \succeq B$ if and only if $A - B$ is positive semidefinite. For $A \in M_n$, we write $|A| \equiv (A^*A)^{\frac{1}{2}}$.

(a) If $A \in M_n$ has a singular value decomposition $A = V\Sigma W^*$, use Corollary (3.1.5) to show that $|A| = W\Sigma W^* \succeq 0$. Show that $|UA| = |A|$ for any unitary $U \in M_n$.

(b) If $H(A) = \frac{1}{2}(A + A^*)$ has a spectral decomposition $H(A) = U\Lambda U^*$

with a unitary $U \in M_n$, $\Lambda = \text{diag}(\lambda_1(A),..., \lambda_n(A))$, and $\lambda_1(A) \geq \cdots$
$\geq \lambda_n(A)$, and if $A = V\Sigma W^*$ is a singular value decomposition with $\Sigma = $
$\text{diag}(\sigma_1(A),..., \sigma_n(A))$, use (3.1.6a) to show that $\Sigma \geq \Lambda$, $U\Sigma U^* \geq H(A)$,
and $(UW^*)|A|(UW^*)^* \geq H(A)$, where $Z \equiv UW^*$ is unitary. Conclude
that for each given $X \in M_n$ there is some unitary $Z \in M_n$ (Z depends on
X) such that $Z|X|Z^* \geq H(X)$.
(c) Show that Corollary (3.1.5) is equivalent to the assertion that for
each given $X \in M_n$ there is some unitary $Z \in M_n$ such that $Z|X|Z^* \geq$
$H(X)$.
(d) For each given $X \in M_n$, why is there some unitary $U \in M_n$ such that
$UX \geq 0$?
(e) Let $A, B \in M_n$ be given. Show that there are unitary $U_1, U_2 \in M_n$
such that

$$|A + B| \leq U_1|A|U_1^* + U_2|B|U_2^* \tag{3.1.15}$$

This is often called the *matrix-valued triangle inequality*.
(f) Show by example that $|A + B| \leq |A| + |B|$ is not generally true.

38. Let $A, B \in M_{m,n}$ be given. Show that there exist unitary $V \in M_m$ and
$W \in M_n$ such that $A = VBW$ if and only if A and B have the same singular
values. Thus, the set of singular values of a matrix is a complete set of
invariants for the equivalence relation of unitary equivalence on $M_{m,n}$.

39. Let $A \in M_n$ be given, and let $H(A) = \frac{1}{2}(A + A^*)$ and $S(A) = \frac{1}{2}(A - A^*)$
denote its Hermitian and skew-Hermitian parts, respectively. The singular
values of $H(A)$ and $S(A)$ (which are the absolute values of their eigenvalues),
respectively, are sometimes called the *real singular values* and *imaginary
singular values* of A; in this context, the ordinary singular values are
sometimes called the *absolute singular values* of A. Show that

$$\sum_{i=1}^{n} \sigma_i(A)^2 = \sum_{i=1}^{n} \sigma_i(H(A))^2 + \sum_{i=1}^{n} \sigma_i(S(A))^2$$

40. Let $A \in M_{m,n}$ be given, let $q = \min\{m,n\}$, and let $S = \text{diag}(\sigma_1(A),...,$
$\sigma_q(A))$. Use the singular value decomposition to show that the Hermitian
block matrix

$$\mathcal{A} = \begin{bmatrix} 0 & A \\ A^* & 0 \end{bmatrix} \in M_{m+n}$$

is unitarily similar to the diagonal matrix $D \oplus (-D) \oplus 0_{m+n-2q} \in M_{m+n}$. Thus, the algebraically ordered eigenvalues of A are

$$\sigma_1(A) \geq \cdots \geq \sigma_q(A) \geq \underbrace{0 = \cdots = 0}_{(m+n-2q \text{ times})} \geq -\sigma_q(A) \geq \cdots \geq -\sigma_1(A)$$

This observation, which was the foundation of Jordan's 1874 development of the singular value decomposition, can be useful in converting results about Hermitian matrices into results about singular values, and vice versa; see Sections (7.3-4) of [HJ] for examples.

41. Give an example of $A, B \in M_2$ for which the singular values of AB and BA are different. What about the eigenvalues?

42. Let $A \in M_n$ be given. Use Theorem (3.1.1) to show that:
(a) There exists a unitary $U \in M_n$ such that $A^* = UAU$.
(b) There exists a unitary $U \in M_n$ such that AU is Hermitian (even positive semidefinite).

43. Let $A \in M_n$ be given.
(a) Use Theorem (3.1.1) to show that there is an $\tilde{A} \in M_n$ that is unitarily similar to A and satisfies $\tilde{A}^* \tilde{A} = \Sigma^2 = \text{diag}(\sigma_1(A)^2, \ldots, \sigma_n(A)^2)$. Let $x \in \mathbb{C}^n$ be a unit vector such that $\tilde{A}x = \lambda x$. Compute $\|\tilde{A}\|_2^2 = \|\tilde{A}x\|_2^2$ and show that any eigenvalue $\lambda(A)$ of A satisfies $\sigma_n(A) \leq |\lambda(A)| \leq \sigma_1(A)$.
(b) Use the fact that $\rho(A) \leq \|\| A \|\|_2$ (simply because the spectral norm is a matrix norm) to show that $\sigma_n(A) \leq |\lambda(A)| \leq \sigma_1(A)$ for every eigenvalue $\lambda(A)$ of A.

44. Let $C \in M_n$ be a contraction. Use the singular value decomposition to show that $(I - CC^*)^{\frac{1}{2}} C = C(I - C^*C)^{\frac{1}{2}}$, where each square root is the unique positive semidefinite square root of the indicated matrix. This identity is useful when working with the representation (1.5.21) for the unit ball of the numerical radius norm.

45. Let $A, B \in M_n$ be given and suppose $A = SBS^{-1}$ for some nonsingular $S \in M_n$. Let $\kappa_2(S) \equiv \sigma_1(S)/\sigma_n(S)$ denote the spectral condition number of S. Use Theorem (4.5.9) in [HJ] to show that

$$\sigma_k(B)/\kappa_2(S) \leq \sigma_k(A) \leq \kappa_2(S)\sigma_k(B), \quad k = 1, \ldots, n \qquad (3.1.16)$$

When A is diagonalizable and $A = S\Lambda S^{-1}$ with $\Lambda = \text{diag}(\lambda_1(A), \ldots, \lambda_n(A))$ and $|\lambda_1(A)| \geq \cdots \geq |\lambda_n(A)|$, use these bounds to show that

$$|\lambda_k(A)|/\kappa_2(S) \leq \sigma_k(A) \leq \kappa_2(S)|\lambda_k(A)|, \quad k = 1, \ldots, n \qquad (3.1.17)$$

When A is normal, use (3.1.17) to show that its singular values are the absolute values of its eigenvalues. See Problem 31 in Section (3.3) for another approach to the inequalities (3.1.16-17).

46. Show that $\sigma > 0$ is a singular value of $A \in M_{m,n}$ if and only if the block matrix $\begin{bmatrix} A & -\sigma I \\ -\sigma I & A^* \end{bmatrix} \in M_{m+n}$ is singular.

Notes and Further Readings. The interlacing inequalities between the singular values of a matrix and its submatrices are basic and very useful facts. For the original proof that the interlacing inequalities (3.1.4) describe exactly the set of all possible singular values of a submatrix of given size, given the singular values of the overall matrix, see R. C. Thompson, Principal Submatrices IX: Interlacing Inequalities for Singular Values of Submatrices, *Linear Algebra Appl.* 5 (1972), 1-12. For a different proof, see J. F. Queiró, On the Interlacing Property for Singular Values and Eigenvalues, *Linear Algebra Appl.* 97 (1987), 23-28.

3.2 Weak majorization and doubly substochastic matrices

Because the singular value decomposition $A = V \Sigma W^*$ is a natural generalization of the spectral decomposition $A = U \Lambda U^*$ for a square Hermitian or normal matrix, familiar properties of Hermitian or normal matrices can point the way toward interesting results for general matrices. For example, if $A = [a_{ij}] \in M_n$ is normal and $A = U \Lambda U^*$ with a unitary $U = [u_{ij}] \in M_n$ and diagonal $\Lambda = \text{diag}(\lambda_1, ..., \lambda_n)$, direct computation shows that the vector $a = [a_{ii}] \in \mathbb{C}^n$ of main diagonal entries of A and the vector $\lambda(A) = [\lambda_i(A)] \in \mathbb{C}^n$ of eigenvalues of A are related by the transformation

$$a = S\lambda(A) \tag{3.2.1}$$

in which $S = [\,|u_{ij}|^2\,] = [U \circ \bar{U}] \in M_n$ is doubly stochastic (see Theorem (4.3.33) in [HJ]), that is, S has nonnegative entries and all its row and column sums are one.

Let $A = [a_{ij}] \in M_{m,n}$ have a singular value decomposition $A = V \Sigma W^*$, where $V = [v_{ij}] \in M_m$ and $W = [w_{ij}] \in M_n$ are unitary and $\Sigma = [\sigma_{ij}] \in M_{m,n}$ has $\sigma_{ii} = \sigma_i(A)$ for $i = 1, ..., q \equiv \min\{m, n\}$. A calculation reveals that the vector of diagonal entries $a = [a_{ii}] \in \mathbb{C}^q$ of A and its vector of singular values $\sigma(A) = [\sigma_i(A)] \in \mathbb{C}^q$ are related by the transformation

$$a = Z\sigma(A) \qquad (3.2.2)$$

where $Z = [v_{ij}\bar{w}_{ij}] \in M_q$ is the Hadamard product of the upper-left q-by-q principal submatrices of the unitary matrices V and W. If we let $Q \equiv |Z| = [\,|v_{ij}w_{ij}|\,]$, then we have the entrywise inequalities

$$|a| = [\,|a_i|\,] \le |Z|\,\sigma(A) = Q\,\sigma(A)$$

Notice that the row sums of the nonnegative matrix Q are at most one since

$$\left[\sum_{j=1}^{q} |v_{ij}w_{ij}| \right]^2 \le \sum_{j=1}^{q} |v_{ij}|^2 \sum_{j=1}^{q} |w_{ij}|^2 \le \sum_{j=1}^{m} |v_{ij}|^2 \sum_{j=1}^{n} |w_{ij}|^2 = 1$$

for all $i = 1,\dots, q$; the same argument shows that the column sums of Q are at most one as well.

3.2.3   Definition. A matrix $Q \in M_n(\mathbb{R})$ is said to be *doubly substochastic* if its entries are nonnegative and all its row and column sums are at most one.

The set of doubly substochastic n-by-n matrices is clearly a convex set that contains all doubly stochastic n-by-n matrices. It is useful to know that it is generated in a simple way by the set of doubly stochastic $2n$-by-$2n$ matrices. Let $Q \in M_n$ be doubly substochastic, let $e \equiv [1, \dots, 1]^T \in \mathbb{R}^n$, and let $D_r = \text{diag}(Qe)$ and $D_c = \text{diag}(Q^T e)$ be diagonal matrices containing the row and column sums of Q. The doubly stochastic $2n$-by-$2n$ matrix

$$\begin{bmatrix} Q & I - D_r \\ I - D_c & Q^T \end{bmatrix} \qquad (3.2.4)$$

is a dilation of Q. Conversely, it is evident that any square submatrix of a doubly stochastic matrix is doubly substochastic.

The doubly stochastic matrices are the convex hull of the permutation matrices (Birkhoff's theorem (8.7.1) in [HJ]). There is an analogous characterization of the doubly substochastic matrices.

3.2.5 **Definition.** A matrix $P \in M_n(\mathbb{R})$ is said to be a *partial permutation matrix* if it has at most one nonzero entry in each row and column, and these nonzero entries (if any) are all 1.

It is evident that every partial permutation matrix is doubly substochastic and can be obtained (perhaps in more than one way) by replacing some 1 entries in a permutation matrix by zeroes. Moreover, every square submatrix of a permutation matrix is doubly substochastic.

If $Q \in M_n(\mathbb{R})$ is a given doubly substochastic matrix, construct the $2n$-by-$2n$ doubly stochastic dilation (3.2.4) and use Birkhoff's theorem to express it as a convex combination of $2n$-by-$2n$ permutation matrices. Then Q is the same convex combination of the upper-left n-by-n principal submatrices of these permutation matrices, each of which is doubly substochastic. The conclusion is a doubly substochastic analog of Birkhoff's theorem: Every doubly substochastic matrix is a finite convex combination of partial permutation matrices. Conversely, a finite convex combination of partial permutation matrices is evidently doubly substochastic.

As a final observation, suppose a given doubly substochastic matrix Q is expressed as a convex combination of partial permutation matrices. In each partial permutation matrix summand, replace some zero entries by ones to make it a permutation matrix. The resulting convex combination of permutation matrices is a doubly stochastic matrix, all of whose entries are not less than those of Q. Conversely, a matrix $Q \in M_n(\mathbb{R})$ such that $0 \leq Q \leq S$ for some doubly stochastic $S \in M_n(\mathbb{R})$ is evidently doubly substochastic. We summarize the preceding observations as follows:

3.2.6 **Theorem.** Let $Q \in M_n(\mathbb{R})$ be a given matrix with nonnegative entries. The following are equivalent:

(a) Q is doubly substochastic.
(b) Q has a doubly stochastic dilation, that is, Q is an upper left principal submatrix of a doubly stochastic matrix.
(c) Q is a finite convex combination of partial permutation matrices.
(d) There is a doubly stochastic $S \in M_n(\mathbb{R})$ such that $0 \leq Q \leq S$.

There is an intimate connection between doubly stochastic matrices and (strong) majorization inequalities between the entries of two real n-vectors; see Section (4.3) of [HJ]. If $\lambda = [\lambda_i]$ and $a = [a_i]$ are two given real n-vectors, their entries may be re-indexed in algebraically decreasing order

$$\lambda_{[1]} \geq \cdots \geq \lambda_{[n]} \text{ and } a_{[1]} \geq \cdots \geq a_{[n]}$$

or in algebraically increasing order

$$\lambda_{(1)} \leq \cdots \leq \lambda_{(n)} \text{ and } a_{(1)} \leq \cdots \leq a_{(n)}$$

The inequalities

$$\sum_{i=1}^{k} a_{[i]} \leq \sum_{i=1}^{k} \lambda_{[i]} \quad \text{for } k = 1,\ldots, n \text{ with equality for } k = n \qquad (3.2.7a)$$

are easily seen to be equivalent to the inequalities

$$\sum_{i=1}^{k} a_{(i)} \geq \sum_{i=1}^{k} \lambda_{(i)} \quad \text{for } k = 1,\ldots, n \text{ with equality for } k = n \qquad (3.2.7b)$$

These equivalent families of (strong) majorization inequalities characterize the relationship between the main diagonal entries and eigenvalues of a Hermitian matrix (Theorem (4.3.26) in [HJ]) and are equivalent to the existence of a doubly stochastic $S \in M_n(\mathbb{R})$ such that $a = S\lambda$. It is not surprising that a generalized kind of majorization is intimately connected with doubly substochastic matrices.

3.2.8   Definition. Let $x = [x_i]$, $y = [y_i] \in \mathbb{R}^n$ be given vectors, and denote their algebraically decreasingly ordered entries by $x_{[1]} \geq \cdots \geq x_{[n]}$ and $y_{[1]} \geq \cdots \geq y_{[n]}$. We say that y *weakly majorizes* x if

$$\sum_{i=1}^{k} x_{[i]} \leq \sum_{i=1}^{k} y_{[i]} \quad \text{for } k = 1,\ldots, n \qquad (3.2.9)$$

Notice that in weak majorization there is no requirement that the sums of all the entries of x and y be equal.

3.2.10   Theorem. Let $x = [x_i]$, $y = [y_i] \in \mathbb{R}^n$ be given vectors with nonnegative entries. Then y weakly majorizes x if and only if there is a doubly substochastic $Q \in M_n(\mathbb{R})$ such that $x = Qy$.

Proof: If $Q \in M_n(\mathbb{R})$ is doubly substochastic and $Qy = x$ with $x, y \geq 0$, let $S \in M_n(\mathbb{R})$ be doubly stochastic and such that $0 \leq Q \leq S$. If we adopt the notation of Definition (3.2.8) to denote the algebraically decreasingly ordered entries of a real vector, and if $(Qy)_{i_1} = (Qy)_{[1]}, ..., (Qy)_{i_k} = (Qy)_{[k]}$, then

$$\sum_{i=1}^{k} (Qy)_{[i]} = \sum_{j=1}^{k} (Qy)_{i_j} \leq \sum_{j=1}^{k} (Sy)_{i_j} \leq \sum_{i=1}^{k} (Sy)_{[i]} \leq \sum_{i=1}^{k} y_{[i]}$$

for $k = 1, ..., n$; we invoke Theorem (4.3.33) in [HJ] for the last inequality. Thus, y weakly majorizes $Qy = x$.

Conversely, suppose y weakly majorizes x and $x, y \geq 0$. If $x = 0$, let $Q \equiv 0$. If $x \neq 0$, let ϵ_x and ϵ_y denote the smallest positive entries of x and y, respectively, set $\delta \equiv (y_1 - x_1) + \cdots + (y_n - x_n) \geq 0$, and let m be any positive integer such that $\delta/m \leq \min\{\epsilon_x, \epsilon_y\}$. Let

$$\xi \equiv [x_1, ..., x_n, \delta/m, ..., \delta/m]^T \in \mathbb{R}^{n+m} \text{ and}$$

$$\eta \equiv [y_1, ..., y_n, 0, ..., 0]^T \in \mathbb{R}^{n+m}$$

Then

$$\sum_{i=1}^{k} \xi_{[i]} \leq \sum_{i=1}^{k} \eta_{[i]} \text{ for } k = 1, ..., m + n$$

with equality for $k = m + n$. Thus, there is a strong majorization relationship between ξ and η. By Theorem (4.3.33) in [HJ] there is a doubly stochastic $S \in M_{m+n}(\mathbb{R})$ such that $\xi = S\eta$. If we let Q denote the upper left n-by-n principal submatrix of S, then Q is doubly substochastic and $x = Qy$.   []

3.2.11   Corollary. Let $x = [x_i]$, $y = [y_i] \in \mathbb{R}^n$ be given. Then y weakly majorizes x if and only if there is a doubly stochastic $S \in M_n(\mathbb{R})$ such that the entrywise inequalities $x \leq Sy$ hold.

Proof: If there is a doubly stochastic S such that $x \leq Sy$, then since there is a (strong) majorization relationship between Sy and y, x must be weakly majorized by y. Conversely, suppose x is weakly majorized by y, let $e =$

$[1, ..., 1]^T \in \mathbb{R}^n$, and choose $\kappa \geq 0$ so that $x + \kappa e \geq 0$ and $y + \kappa e \geq 0$. Theorem (3.2.10) guarantees that there is a doubly substochastic $Q \in M_n(\mathbb{R})$ such that $x + \kappa e = Q(y + \kappa e)$. Theorem (3.2.6(d)) ensures that there is a doubly stochastic S such that $Q \leq S$. Then $x + \kappa e = Q(y + \kappa e) \leq Sy + \kappa e$, so $x \leq Sy$. ☐

Problems

1.   The matrix (3.2.4) shows that every n-by-n doubly substochastic matrix Q has a $2n$-by-$2n$ doubly stochastic dilation. Consider $Q = 0$ to show that the dilation sometimes cannot be smaller than $2n$-by-$2n$.

2.   Let $\hat{V}$ and $\hat{W}$ denote any q-by-q submatrices of given unitary matrices $V \in M_m$, $W \in M_n$, $q \leq \min\{m,n\}$. Show that $|\hat{V} \circ \hat{W}|$ (entrywise absolute values) is doubly substochastic.

3.   Let $A = [a_{ij}] \in M_{m,n}$ be given and let $q = \min\{m,n\}$. Let $a = [a_{11}, ..., a_{qq}]^T \in \mathbb{C}^q$ and $\sigma(A) = [\sigma_1(A), ..., \sigma_q(A)]^T \in \mathbb{R}^q$ denote the vectors of main diagonal entries and singular values of A, respectively, with $\sigma_1(A) \geq \cdots \geq \sigma_q(A)$. Let $|a|_{[1]} \geq \cdots \geq |a|_{[q]}$ denote the absolutely decreasingly ordered main diagonal entries of A and let $A = V \Sigma W^*$ be a singular value decomposition of A. Use the discussion at the beginning of this section and Theorem (3.2.10) to verify the weak majorization

$$|a|_{[1]} + \cdots + |a|_{[k]} \leq \sigma_1(A) + \cdots + \sigma_k(A) \text{ for } k = 1,..., q$$

What does this say if $A \in M_n$ is positive semidefinite? For a different proof of these inequalities and some related results, see Problem 21 in Section (3.3).

4.   (a) Consider the partial permutation matrix $\begin{bmatrix} 0 & 1 \\ 0 & 0 \end{bmatrix}$. Notice that it is a convex combination of $\begin{bmatrix} 0 & 1 \\ -1 & 0 \end{bmatrix}$ and $\begin{bmatrix} 0 & 1 \\ 1 & 0 \end{bmatrix}$. A *generalized permutation matrix* is obtained from a permutation matrix by replacing each 1 entry by an entry (real or complex) with absolute value 1. Show that each partial permutation matrix in M_n is a convex combination of two real generalized permutation matrices.
(b) If $G \in M_n$ is a generalized permutation matrix and if $\sigma = [\sigma_1, ..., \sigma_n]^T \in \mathbb{R}^n$ is a nonnegative vector, show that the singular values of $\text{diag}(G\sigma)$ are $\sigma_1,..., \sigma_n$.
(c) Let $\sigma_1 \geq \cdots \geq \sigma_n \geq 0$ and $s_1 \geq \cdots \geq s_n \geq 0$ be given and suppose $s_1 + \cdots$

$+ s_k \leq \sigma_1 + \cdots + \sigma_k$ for $k = 1, \ldots, n$, that is, the values $\{\sigma_i\}$ weakly majorize the values $\{s_i\}$. Show that $\mathrm{diag}(s_1, \ldots, s_n)$ is a finite convex combination of matrices, each of which has the same singular values $\sigma_1, \ldots, \sigma_n$.

(d) Suppose given matrices $A_1, \ldots, A_m \in M_n$ all have the same singular values $\sigma_1 \geq \cdots \geq \sigma_n$. For each $i = 1, \ldots, m$, let $a^{(i)} \in \mathbb{C}^n$ denote the vector of main diagonal entries of A_i, that is, $a^{(i)} \equiv \mathrm{diag}(A_i)$. If $\mu_1, \ldots, \mu_n$ are nonnegative real numbers such that $\mu_1 + \cdots + \mu_n = 1$, show that the entries of the vector $\mu_1 | a^{(1)}| + \cdots + \mu_n | a^{(n)}|$ are weakly majorized by the entries of the vector $\sigma = [\sigma_1, \ldots, \sigma_n]^T$.

(e) Let $\sigma_1 \geq \cdots \geq \sigma_n \geq 0$ and $A \in M_n$ be given. Prove that A is in the convex hull of the set of all n-by-n complex matrices that have the same singular values $\sigma_1, \ldots, \sigma_n$ if and only if the singular values of A are weakly majorized by the given values $\{\sigma_i\}$, that is, $\sigma_1(A) + \cdots + \sigma_k(A) \leq \sigma_1 + \cdots + \sigma_k$ for all $k = 1, \ldots, n$.

(f) Recall that $A \in M_n$ is a *contraction* if $\sigma_1(A) \leq 1$; the set of contractions in M_n is the unit ball of the spectral norm. Use (e) to show that A is a contraction if and only if it is a finite convex combination of unitary matrices. See Problems 4 and 27 of Section (3.1) for different approaches to this result.

5. A given norm $\|\cdot\|$ on $\mathbb{F}^n \equiv \mathbb{R}^n$ or $\mathbb{C}^n$ is *absolute* if $\|x\| = \| \, |x| \, \|$ for all $x \in \mathbb{F}^n$ (entry-wise absolute values); it is *permutation-invariant* if $\|x\| = \|Px\|$ for all $x \in \mathbb{F}^n$ and every permutation matrix $P \in M_n(\mathbb{R})$. It is known (see Theorem (5.5.10) in [HJ]) that $\|\cdot\|$ is absolute if and only if it is *monotone*, that is, $\|x\| \leq \|y\|$ whenever $|x| \leq |y|$. Let $x = [x_i], y = [y_i] \in \mathbb{F}^n$ be given, and suppose $|y|$ weakly majorizes $|x|$. Use Corollary (3.2.11) to show that $\|x\| \leq \|y\|$ for every absolute permutation-invariant norm $\|\cdot\|$ on $\mathbb{R}^n$.

6. Notice that there is an analogy between the dilation (3.2.4) and the dilation discussed in Problem 21 of Section (1.6). There is also a doubly stochastic dilation of a given doubly substochastic matrix that has a property analogous to the power dilation property considered in Problem 24 of Section (1.6). If $Q \in M_n$ is a given doubly substochastic matrix, and if k is a given positive integer, let D_r and D_c be defined as in (3.2.4) and define the $(k+1)$-by-$(k+1)$ block matrix $S = [S_{ij}] \in M_{n(k+1)}$, each $S_{ij} \in M_n$, as follows: $S_{11} = Q$, $S_{12} = I - D_r$, $S_{k+1,1} = I - D_c$, $S_{k+1,2} = Q^T$, $S_{i,i+1} = I$ for $i = 2, \ldots, k$, and all other blocks are zero. Show that $S \in M_{n(k+1)}$ is doubly stochastic and that $S^m = \begin{bmatrix} Q^m & * \\ * & * \end{bmatrix}$ for $m = 1, 2, \ldots, k$. What is this for $k = 1$?

Notes and Further Readings. For a modern survey of the ideas discussed in this section, with numerous references, see T. Ando, Majorization, Doubly Stochastic Matrices, and Comparison of Eigenvalues, *Linear Algebra Appl.* 118 (1989), 163-248. See also [MOl]. Sharp bounds on the size of a doubly stochastic S with the power dilation property discussed in Problem 6 are given in the paper by R. C. Thompson and C.-C. T. Kuo cited in Section (1.6): If $Q = [q_{ij}] \in M_n$ is doubly substochastic, let δ denote the least integer greater than $n - \Sigma_{i,j} q_{ij}$. Then there is a doubly stochastic $S \in M_{n+\mu}$ such that $S^m = \begin{bmatrix} Q^m & * \\ * & * \end{bmatrix}$ for $m = 1, 2, ..., k$ if and only if $\mu \geq k\delta$.

3.3 Basic inequalities for singular values and eigenvalues

It is clear that one cannot prescribe completely independently the eigenvalues and singular values of a matrix $A \in M_n$. For example, if some singular value of A is zero, then A is singular and hence it must have at least one zero eigenvalue. A basic result in this section is a necessary condition (which we show to be sufficient as well in Section (3.6)) indicating the interdependence between the singular values and eigenvalues of a square matrix. We also develop useful inequalities for singular values of products and sums of matrices.

The basic inequalities between singular values and eigenvalues, and the singular value inequalities for products of matrices are both consequences of the following inequality, which follows readily from unitary invariance and the interlacing property for singular values of a submatrix of a given matrix.

3.3.1   Lemma. Let $C \in M_{m,n}$, $V_k \in M_{m,k}$, and $W_k \in M_{n,k}$ be given, where $k \leq \min\{m,n\}$ and V_k, W_k have orthonormal columns. Then

(a) $\sigma_i(V_k^* C W_k) \leq \sigma_i(C)$, $i = 1, ..., k$, and

(b) $|\det V_k^* C W_k| \leq \sigma_1(C) \cdots \sigma_k(C)$.

Proof: Since the respective columns of V_k and W_k can be extended to orthonormal bases of $\mathbb{C}^m$ and $\mathbb{C}^n$, respectively, there are unitary matrices $V \in M_m$ and $W \in M_n$ such that $V = [V_k \ *]$ and $W = [W_k \ *]$. Since $V_k^* C W_k$ is the upper left k-by-k submatrix of $V^* C W$, the interlacing inequalities (3.1.4) and unitary invariance of singular values ensure that $\sigma_i(V_k^* C W_k) \leq \sigma_i(V^* C W) = \sigma_i(C)$, $i = 1, ..., k$, and hence $|\det V_k^* C W_k| = \sigma_1(V_k^* C W_k) \cdots$

$$\sigma_k(V_k^* CW_k) \leq \sigma_1(C) \cdots \sigma_k(C). \qquad\qquad \Box$$

Important necessary conditions relating the singular values and eigenvalues now follow readily in the following result, first proved by H. Weyl in 1949.

3.3.2   Theorem. Let $A \in M_n$ have singular values $\sigma_1(A) \geq \cdots \geq \sigma_n(A) \geq 0$ and eigenvalues $\{\lambda_1(A),..., \lambda_n(A)\} \subset \mathbb{C}$ ordered so that $|\lambda_1(A)| \geq \cdots \geq |\lambda_n(A)|$. Then

$$|\lambda_1(A) \cdots \lambda_k(A)| \leq \sigma_1(A) \cdots \sigma_k(A) \text{ for } k = 1,..., n,$$

$$\text{with equality for } k = n \qquad (3.3.3)$$

Proof: By the Schur triangularization theorem, there is a unitary $U \in M_n$ such that $U^* A U = \Delta$ is upper triangular and diag $\Delta = (\lambda_1,..., \lambda_n)$. Let $U_k \in M_{n,k}$ denote the first k columns of U, and compute

$$U^* A U = [U_k \ *]^* A [U_k \ *] = \begin{bmatrix} U_k^* A U_k & * \\ * & * \end{bmatrix} = \Delta$$

Thus, $U_k^* A U_k = \Delta_k$ is upper triangular since it is the upper left k-by-k principal submatrix of Δ, and diag $\Delta_k = (\lambda_1,..., \lambda_k)$. Now apply the lemma with $C = A$ and $V_k = W_k = U_k$ to conclude that

$$|\lambda_1(A) \cdots \lambda_k(A)| = |\det \Delta_k| = |\det U_k^* A U_k| \leq \sigma_1(A) \cdots \sigma_k(A)$$

When $k = n$, one sees readily from the singular value decomposition that $|\det A| = \sigma_1(A) \cdots \sigma_n(A)$, and, since $\det A = \lambda_1(A) \cdots \lambda_n(A)$, we are done. $\Box$

If the singular values of two matrices are known, what can one say about the singular values of their product? A useful answer is another immediate consequence of Lemma (3.3.1), a theorem about singular values of matrix products first proved by A. Horn in 1950.

3.3.4   Theorem. Let $A \in M_{m,p}$ and $B \in M_{p,n}$ be given, let $q \equiv \min\{n,p,m\}$, and denote the ordered singular values of A, B, and AB by $\sigma_1(A) \geq \cdots \geq \sigma_{\min\{m,p\}}(A) \geq 0$, $\sigma_1(B) \geq \cdots \geq \sigma_{\min\{p,n\}}(B) \geq 0$,   and

$\sigma_1(AB) \geq \cdots \geq \sigma_{min\{m,n\}}(AB) \geq 0$. Then

$$\prod_{i=1}^{k} \sigma_i(AB) \leq \prod_{i=1}^{k} \sigma_i(A)\sigma_i(B), \quad k = 1,\ldots,q \qquad (3.3.5)$$

If $n = p = m$, then equality holds in (3.3.5) for $k = n$.

Proof: Let $AB = V\Sigma W^*$ be a singular value decomposition of the product AB, and let $V_k \in M_{m,k}$ and $W_k \in M_{n,k}$ denote the first k columns of V and W, respectively. Then $V_k^*(AB)W_k = \mathrm{diag}(\sigma_1(AB),\ldots,\sigma_k(AB))$ because it is the upper left k-by-k submatrix of $V^*(AB)W = \Sigma$. Since $p \geq k$, use the polar decomposition (3.1.9(b)) to write the product $BW_k \in M_{p,k}$ as $BW_k = X_k Q$, where $X_k \in M_{p,k}$ has orthonormal columns, $Q \in M_k$ is positive semidefinite, $Q^2 = (BW_k)^*(BW_k) = W_k^* B^* B W_k$, and hence $\det Q^2 = \det W_k^*(B^*B)W_k \leq \sigma_1(B^*B) \cdots \sigma_k(B^*B) = \sigma_1(B)^2 \cdots \sigma_k(B)^2$ by Lemma (3.3.1). Use Lemma (3.3.1) again to compute

$$\sigma_1(AB) \cdots \sigma_k(AB) = |\det V_k^*(AB)W_k| = |\det V_k^* A X_k Q|$$

$$= |\det V_k^* A X_k \det Q| \leq (\sigma_1(A) \cdots \sigma_k(A))(\sigma_1(B) \cdots \sigma_k(B))$$

$$(3.3.6)$$

If $n = p = m$, then $\sigma_1(AB) \cdots \sigma_n(AB) = |\det AB| = |\det A||\det B| = \sigma_1(A) \cdots \sigma_n(A)\sigma_1(B) \cdots \sigma_n(B)$. ☐

The inequalities (3.3.3) and (3.3.5) are a kind of multiplicative majorization, and if A is nonsingular we can take logarithms in (3.3.3) to obtain equivalent ordinary (strong) majorization inequalities

$$\sum_{i=1}^{k} \log |\lambda_i(A)| \leq \sum_{i=1}^{k} \log \sigma_i(A), \quad k = 1,\ldots, n, \text{ with equality for } k = n \quad (3.3.7)$$

Since our next goal is to show that these inequalities may be exponentiated term-by-term to obtain weak majorization inequalities such as

$$\sum_{i=1}^{k} |\lambda_i(A)| \leq \sum_{i=1}^{k} \sigma_i(A), \quad k = 1,\ldots, n$$

it is convenient to establish that there is a wide class of functions $f(t)$ (including $f(t) = e^t$) that preserve systems of inequalities such as (3.3.7); such functions have been termed *Schur-convex* or *isotone*. The next lemma shows that any increasing convex function with appropriate domain preserves weak majorization.

3.3.8   **Lemma.** Let $x_1,..., x_n, y_1,..., y_n$ be $2n$ given real numbers such that $x_1 \geq x_2 \geq \cdots \geq x_n, y_1 \geq y_2 \geq \cdots \geq y_n$, and

$$\sum_{i=1}^{k} x_i \leq \sum_{i=1}^{k} y_i, \quad k=1,..., n \tag{3.3.9a}$$

Let $f(\cdot)$ be a given real-valued function on the interval $[a,b] \equiv [\min\{x_n, y_n\}, y_1]$. If $f(\cdot)$ is increasing and convex on $[a,b]$, then $f(x_1) \geq \cdots \geq f(x_n), f(y_1) \geq \cdots \geq f(y_n)$, and

$$\sum_{i=1}^{k} f(x_i) \leq \sum_{i=1}^{k} f(y_i), \quad k=1,..., n \tag{3.3.9b}$$

Write $x \equiv [x_i], y \equiv [y_i] \in \mathbb{R}^n$. For any $z \equiv [z_i] \in \mathbb{R}^n$ with all $z_i \in [a,b]$, write $f(z) \equiv [f(z_i)] \in \mathbb{R}^n$ and let $f(z)_{[1]} \geq \cdots \geq f(z)_{[n]}$ denote an algebraically decreasingly ordered rearrangement of the entries of $f(z)$. If equality holds for $k = n$ in (3.3.9a) and if $f(\cdot)$ is convex (but not necessarily increasing) on $[a,b] = [y_n, y_1]$, then

$$\sum_{i=1}^{k} f(x)_{[i]} \leq \sum_{i=1}^{k} f(y)_{[i]}, k=1,..., n-1 \text{ and } \sum_{i=1}^{n} f(x_i) \leq \sum_{i=1}^{n} f(y_i) \tag{3.3.9c}$$

Proof: We need to show that $f(y)$ weakly majorizes $f(x)$. Corollary (3.2.11) ensures that there is a doubly stochastic $S \in M_n(\mathbb{R})$ such that $x \leq Sy$, and if equality holds for $k = n$ in (3.3.9a) we may choose S so that $x = Sy$. In the former case, monotonicity of $f(\cdot)$ ensures that $f(x) \leq f(Sy)$, and in the latter case, $f(x) = f(Sy)$. In both cases, $f(Sy)$ weakly majorizes $f(x)$. Thus, it suffices to show that $f(Sy)$ is weakly majorized by $f(y)$. Use Birkhoff's Theorem to write

$$S = \alpha_1 P_1 + \cdots + \alpha_m P_m$$

where all $\alpha_i > 0$, $\alpha_1 + \cdots + \alpha_m = 1$, and each $P_i \in M_n(\mathbb{R})$ is a permutation matrix. Convexity of $f(\cdot)$ ensures that

$$f(Sy) = f(\sum_{i=1}^m \alpha_i P_i y) \leq \sum_{i=1}^m \alpha_i f(P_i y) = \sum_{i=1}^m \alpha_i P_i f(y) = Sf(y)$$

which implies that $f(Sy)$ is weakly majorized by $Sf(y)$. Since there is a strong majorization relationship between $Sf(y)$ and $f(y)$, and since weak majorization is a transitive relation, $f(Sy)$ is weakly majorized by $f(y)$. ☐

3.3.10   Corollary. Let $\alpha_1, ..., \alpha_n, \beta_1, ..., \beta_n$ be $2n$ given nonnegative real numbers such that $\alpha_1 \geq \cdots \geq \alpha_n \geq 0$, $\beta_1 \geq \cdots \geq \beta_n \geq 0$, and

$$\prod_{i=1}^k \alpha_i \leq \prod_{i=1}^k \beta_i, \quad k = 1, ..., n \qquad (3.3.11)$$

Let $f(\cdot)$ be a given real-valued function on the interval $[a,b] \equiv [\min\{\alpha_n, \beta_n\}, \beta_1]$ and define $\varphi(t) \equiv f(e^t)$. Then

(i) We have the weak majorization

$$\sum_{i=1}^k \alpha_i \leq \sum_{i=1}^k \beta_i, \quad k = 1, ..., n \qquad (3.3.12a)$$

(ii) If $\alpha_1 = 0$, set $p \equiv 0$; otherwise, set $p \equiv \max\{i: \alpha_i > 0, i \in \{1, ..., n\}\}$. If $p = 0$, assume $f(\beta_i) \geq f(0)$ for all $i = 1, ..., n$; otherwise, assume that $f(\cdot)$ is increasing on $[a,b]$ and that $\varphi(\cdot)$ is convex on $[\ln \min\{\alpha_p, \beta_p\}, \ln \beta_1]$. Then

$$\sum_{i=1}^k f(\alpha_i) \leq \sum_{i=1}^k f(\beta_i) \text{ for } k = 1, ..., n \qquad (3.3.12b)$$

(iii) If $\alpha_n > 0$ and equality holds for $k = n$ in (3.3.11), assume that $\varphi(\cdot)$ is convex (but not necessarily increasing) on $[\ln \beta_n, \ln \beta_1]$. Then

$$\sum_{i=1}^{n} f(\alpha_i) \leq \sum_{i=1}^{n} f(\beta_i) \qquad (3.3.12c)$$

Proof: The inequalities (3.3.12a) follow from (3.3.12b) with $f(t) = t$, so it suffices to prove (3.3.12b,c). First consider the case $p = n$, in which all α_i and β_i are positive. Then (3.3.11) is equivalent to

$$\sum_{i=1}^{k} \log \alpha_i \leq \sum_{i=1}^{k} \log \beta_i, \qquad k = 1,\dots, n$$

The inequalities (3.3.12b,c) now follow from Lemma (3.3.8) using $\varphi(\cdot)$. Now suppose that $p < n$. The case $p = 0$ is trivial (since our hypotheses ensure that $f(\beta_i) \geq f(0) = f(\alpha_i)$ for all $i = 1,\dots, n$), so assume that $1 \leq p < n$. The validity of (3.3.12b) for $k = 1,\dots, p$ has already been established in the first case considered, so we need to consider only $k = p + 1,\dots, n$. Since monotonicity of $f(\cdot)$ implies that $f(\beta_i) \geq f(0) = f(\alpha_i)$ for all $i = p + 1,\dots, n$, we conclude that

$$\sum_{i=1}^{p+r} f(\beta_i) = \sum_{i=1}^{p} f(\beta_i) + \sum_{i=p+1}^{p+r} f(\beta_i) \geq \sum_{i=1}^{p} f(\alpha_i) + \sum_{i=p+1}^{p+r} f(0)$$

$$= \sum_{i=1}^{p+r} f(\alpha_i) \quad \text{for } r = 1,\dots, n-p$$

$\square$

Although the multiplicative inequalities (3.3.11) imply the additive inequalities (3.3.12), the reverse implication is false; see Problem 3. The following results now follow from Theorem (3.3.2) and Corollary (3.3.10).

3.3.13   Theorem. Let $A \in M_n$ have ordered singular values $\sigma_1(A) \geq \cdots \geq \sigma_n(A) \geq 0$ and eigenvalues $\{\lambda_1(A),\dots, \lambda_n(A)\}$ ordered so that $|\lambda_1(A)| \geq \cdots \geq |\lambda_n(A)|$. Then

(a) $\displaystyle\sum_{i=1}^{k} |\lambda_i(A)| \leq \sum_{i=1}^{k} \sigma_i(A)$ for $k = 1,\ldots, n$

In particular,

(a′) $|\operatorname{tr} A| \leq \displaystyle\sum_{i=1}^{n} \sigma_i(A)$

(b) $\displaystyle\sum_{i=1}^{k} |\lambda_i(A)|^p \leq \sum_{i=1}^{k} \sigma_i(A)^p$ for $k = 1,\ldots, n$ and any $p > 0$

More generally, let $f(\cdot)$ be a given real-valued function on $[0,\infty)$ and define $\varphi(t) \equiv f(e^t)$. If $f(\cdot)$ is increasing on $[0,\infty)$ and $\varphi(\cdot)$ is convex on $(-\infty,\infty)$, then

(c) $\displaystyle\sum_{i=1}^{k} f(|\lambda_i(A)|) \leq \sum_{i=1}^{k} f(\sigma_i(A))$ for $k = 1,\ldots, n$

If $\varphi(\cdot)$ is convex (but not necessarily increasing) on $(-\infty,\infty)$, and if either A is nonsingular or $f(\cdot)$ is continuous on $[0,\infty)$, then

(d) $\displaystyle\sum_{i=1}^{n} f(|\lambda_i(A)|) \leq \sum_{i=1}^{n} f(\sigma_i(A))$

In particular, if A is nonsingular, then

(e) $\displaystyle\sum_{i=1}^{n} |\lambda_i(A)|^p \leq \sum_{i=1}^{n} \sigma_i(A)^p$ for all $p \in \mathbb{R}$

The same reasoning permits us to deduce a host of other inequalities from the inequalities (3.3.5) for the singular values of a product.

3.3.14   Theorem. Let $A \in M_{n,r}$ and $B \in M_{r,m}$ be given, let $q \equiv \min\{n,r,m\}$, and denote the ordered singular values of A, B, and AB by $\sigma_1(A) \geq \cdots \geq \sigma_{\min\{n,r\}} \geq 0$, $\sigma_1(B) \geq \cdots \geq \sigma_{\min\{r,m\}}(B) \geq 0$, and $\sigma_1(AB) \geq \cdots \geq \sigma_{\min\{n,m\}}(AB) \geq 0$. Then

(a) $\displaystyle\sum_{i=1}^{k} \sigma_i(AB) \leq \sum_{i=1}^{k} \sigma_i(A)\sigma_i(B)$ for $k = 1,..., q$

(b) $\displaystyle\sum_{i=1}^{k} [\sigma_i(AB)]^p \leq \sum_{i=1}^{k} [\sigma_i(A)\sigma_i(B)]^p$ for $k = 1,..., q$ and any $p > 0$

More generally, let $f(\cdot)$ be a given real-valued function on $[0,\infty)$ and define $\varphi(t) \equiv f(e^t)$. If $f(\cdot)$ is increasing on $[0,\infty)$ and $\varphi(\cdot)$ is convex on $(-\infty,\infty)$, then

(c) $\displaystyle\sum_{i=1}^{k} f(\sigma_i(AB)) \leq \sum_{i=1}^{k} f(\sigma_i(A)\sigma_i(B))$ for $k = 1,..., q$

If $m = n = r$, if $\varphi(\cdot)$ is convex (but not necessarily increasing) on $(-\infty,\infty)$, and if either A and B are nonsingular or $f(\cdot)$ is continuous on $[0,\infty)$, then

(d) $\displaystyle\sum_{i=1}^{n} f(\sigma_i(AB)) \leq \sum_{i=1}^{n} f(\sigma_i(A)\sigma_i(B))$

In particular, if $m = n = r$ and if A and B are nonsingular, then

(e) $\displaystyle\sum_{i=1}^{n} \sigma_i(AB)^p \leq \sum_{i=1}^{n} \sigma_i(A)^p \, \sigma_i(B)^p$ for all $p \in \mathbb{R}$

For analogs of the inequalities (a) for the Hadamard product, see Theorems (5.5.4) and (5.6.2).

We have so far been concentrating on inequalities involving sums and products of *consecutive* singular values and eigenvalues, as in the Weyl inequalities (3.3.3). There are many other useful inequalities, however, and as an example we consider singular value analogs of the Weyl inequalities for eigenvalues of sums of Hermitian matrices (Theorem (4.3.7) in [HJ]). A preliminary lemma is useful in the proof we give.

3.3.15   Lemma. Let $A \in M_{m,n}$ be given, let $q = \min\{m,n\}$, let A have a singular value decomposition $A = V\Sigma W^*$, and partition $W = [w_1 \; ... \; w_n]$

according to its columns, the right singular vectors of A, whose corresponding singular values are $\sigma_1(A) \geq \sigma_2(A) \geq \cdots$. If $S \equiv \operatorname{Span} \{w_i,..., w_n\}$, then $\max \{\|Ax\|_2 : x \in S, \|x\|_2 = 1\} = \sigma_i(A)$, $i = 1,..., q$.

Proof: If $x = \alpha_i w_i + \cdots + \alpha_n w_n$ and $|\alpha_i|^2 + \cdots + |\alpha_n|^2 = 1$, then $\|Ax\|_2^2 = \|V \Sigma W^* x\|_2^2 = \|\Sigma W^* x\|_2^2 = \sigma_i(A)^2 |\alpha_i|^2 + \cdots + \sigma_q(A)^2 |\alpha_q|^2 \leq \sigma_i(A)^2$. ☐

3.3.16   Theorem. Let $A, B \in M_{m,n}$ be given and let $q = \min \{m,n\}$. The following inequalities hold for the decreasingly ordered singular values of A, B, $A + B$, and AB^*:

(a) $\sigma_{i+j-1}(A + B) \leq \sigma_i(A) + \sigma_j(B)$ (3.3.17)

(b) $\sigma_{i+j-1}(AB^*) \leq \sigma_i(A)\sigma_j(B)$ (3.3.18)

for $1 \leq i, j \leq q$ and $i + j \leq q + 1$. In particular,

(c) $|\sigma_i(A + B) - \sigma_i(A)| \leq \sigma_1(B)$ for $i = 1,..., q$ (3.3.19)

and

(d) $\sigma_i(AB^*) \leq \sigma_i(A)\sigma_1(B)$ for $i = 1,..., q$ (3.3.20)

Proof: Let $A = V \Sigma_A W^*$ and $B = X \Sigma_B Y^*$ be singular value decompositions of A and B with unitary $W = [w_1 \ \cdots \ w_n]$, $Y = [y_1 \ \cdots \ y_n] \in M_n$ and unitary $V = [v_1 \ \cdots \ v_m]$, $X = [x_1 \ \cdots \ x_m] \in M_m$. Let i and j be positive integers with $1 \leq i, j \leq q$ and $i + j \leq q + 1$.

First consider the sum inequalities (3.3.17). Define $S' \equiv \operatorname{Span} \{w_i,..., w_n\}$ and $S'' \equiv \operatorname{Span} \{y_j,..., y_n\}$; notice that $\dim S' = n - i + 1$ and $\dim S'' = n - j + 1$. Then

$$\nu \equiv \dim(S' \cap S'') = \dim S' + \dim S'' - \dim(S' + S'')$$

$$= (n - i + 1) + (n - j + 1) - \dim(S' + S'')$$

$$\geq (n - i + 1) + (n - j + 1) - n = n - (i + j - 1) + 1 \geq 1$$

because of the bounds assumed for i and j. Thus, the subspace $S' \cap S''$ has positive dimension ν, $n - \nu + 1 \leq i + j - 1$, and we can use (3.1.2(c)) and Lemma (3.3.15) to compute

$$\sigma_{i+j-1}(A + B) \leq \sigma_{n-\nu+1}(A + B)$$

$$= \min_{\substack{S \subset \mathbb{C}^n \\ \dim S = \nu}} \max_{\substack{x \in S \\ \|x\|_2 = 1}} \|(A + B)x\|_2$$

$$\leq \max_{\substack{x \in S' \cap S'' \\ \|x\|_2 = 1}} \|(A + B)x\|_2$$

$$\leq \max_{\substack{x \in S' \cap S'' \\ \|x\|_2 = 1}} \|Ax\|_2 + \max_{\substack{x \in S' \cap S'' \\ \|x\|_2 = 1}} \|Bx\|_2$$

$$\leq \max_{\substack{x \in S' \\ \|x\|_2 = 1}} \|Ax\|_2 + \max_{\substack{x \in S'' \\ \|x\|_2 = 1}} \|Bx\|_2 = \sigma_i(A) + \sigma_j(B)$$

Now consider the product inequalities (3.3.18). Use the polar decomposition (3.1.9(c)) to write $AB^* = UQ$, where $U \in M_m$ is unitary and $Q \in M_m$ is positive semidefinite and has the same singular values (which are also its eigenvalues) as AB^*. Let

$$S' = \text{Span}\,\{U^* v_i, ..., U^* v_n\} \text{ and } S'' = \text{Span}\,\{x_j, ..., x_n\}$$

so $\nu \equiv \dim(S' \cap S'') \geq n - (i + j - 1) + 1 \geq 1$, as before. Since $Q = U^* A B^*$, we have $x^* Q x = x^* U^* A B^* x = (A^* U x)^* (B^* x) \leq \|A^* U x\|_2 \|B^* x\|_2$ for any $x \in \mathbb{C}^n$ and hence we can use (3.1.2(c)), the Courant-Fischer theorem (4.2.11) in [HJ], and Lemma (3.3.15) to compute

$$\sigma_{i+j-1}(AB^*) = \sigma_{i+j-1}(Q) \leq \sigma_{n-\nu+1}(Q)$$

$$= \min_{\substack{S \subset \mathbb{C}^n \\ \dim S = \nu}} \max_{\substack{x \in S \\ \|x\|_2 = 1}} x^* Q x$$

$$\leq \max_{\substack{x \in S' \cap S'' \\ \|x\|_2 = 1}} x^* Q x \leq \max_{\substack{x \in S' \cap S'' \\ \|x\|_2 = 1}} \|A^* U x\|_2 \|B^* x\|_2$$

$$\leq \max_{\substack{x \in S' \cap S'' \\ \|x\|_2 = 1}} \|A^* U x\|_2 \max_{\substack{x \in S' \cap S'' \\ \|x\|_2 = 1}} \|B^* x\|_2$$

$$\le \max_{\substack{x \in S' \\ \|x\|_2 = 1}} \|A^* U x\|_2 \max_{\substack{x \in S^* \\ \|x\|_2 = 1}} \|B^* x\|_2 = \sigma_i(A)\,\sigma_j(B)$$

Notice that the increasingly ordered eigenvalues $\lambda_1(Q) \le \cdots \le \lambda_n(Q)$ are related to the decreasingly ordered singular values $\sigma_1(Q) \ge \cdots \ge \sigma_n(Q)$ by $\lambda_i(Q) = \sigma_{n-i+1}(Q)$.

The inequalities $\sigma_i(A + B) \le \sigma_i(A) + \sigma_1(B)$ and $\sigma_i(AB^*) \le \sigma_i(A)\sigma_1(B)$ follow from setting $j = 1$ in (a) and (b). The two–sided bound in the additive case now follows from observing that $\sigma_i(A) = \sigma_i([A + B] - B) \le \sigma_i(A + B) + \sigma_1(-B) = \sigma_i(A + B) + \sigma_1(B)$. ∐

As a final application in this section of the interlacing inequalities for singular values, we consider the following generalization of the familiar and useful fact that $\lim \sigma_1(A^m)^{1/m} = |\lambda_1(A)| = \rho(A)$ as $m \to \infty$ for any $A \in M_n$. This is a theorem of Yamamoto, first proved in 1967.

3.3.21   Theorem. Let $A \in M_n$ be given, and let $\sigma_1(A) \ge \cdots \ge \sigma_n(A)$ and $\{\lambda_1(A),\dots,\lambda_n(A)\}$ denote its singular values and eigenvalues, respectively, with $|\lambda_1(A)| \ge \cdots \ge |\lambda_n(A)|$. Then

$$\lim_{m \to \infty} [\sigma_i(A^m)]^{1/m} = |\lambda_i(A)| \text{ for } i = 1,\dots, n$$

Proof: The case $i = 1$ is a special case of a theorem valid for all prenorms on M_n (see Corollaries (5.7.10) and (5.6.14) in [HJ]). The case $i = n$ is trivial if A is singular, and follows in the nonsingular case by applying the case $i = 1$ to A^{-1}, since $\sigma_1(A^{-1}) = 1/\sigma_n(A)$ and $|\lambda_1(A^{-1})| = |1/\lambda_n(A)|$. Use the Schur triangularization theorem to write $A = U \Delta U^*$ with $U, \Delta \in M_n$, U unitary, and Δ upper triangular with diag $\Delta = (\lambda_1,\dots,\lambda_n)$. For $B = [b_{ij}] \in M_n$, let $B_{[i]}$ denote the i-by-i upper left principal submatrix of B, and let $B_{<i>} \in M_{n-i+1}$ denote the lower right $(n-i+1)$-by-$(n-i+1)$ principal submatrix of B; notice that the entry b_{ii} is in the lower right corner of $B_{[i]}$ and in the upper left corner of $B_{<i>}$.

The upper triangular structure of Δ ensures that $(\Delta^m)_{[i]} = (\Delta_{[i]})^m$ and $(\Delta^m)_{<i>} = (\Delta_{<i>})^m$, as well as $|\lambda_i(\Delta_{[i]})| = |\lambda_i(A)|$ and $|\lambda_1(\Delta_{<i>})| = |\lambda_i(A)|$. Thus, the left half of the interlacing inequalities (3.1.4) gives the lower bound

$$\sigma_i(A^m) = \sigma_i(\Delta^m) \geq \sigma_i((\Delta^m)_{[i]}) = \sigma_i((\Delta_{[i]})^m)$$

Now write $B_m \equiv [\,0 \;\; (\Delta^m)_{<i>}\,] \in M_{n-i+1,n}$ and $T_m \equiv [\,(\Delta^m)_{[i]} \;\; *\,] \in M_{i,n}$, so

$$\Delta^m = \begin{bmatrix} T_m \\ B_m \end{bmatrix}$$

Notice that B_m is obtained from Δ^m by deleting $i-1$ rows, and that the singular values of B_m and $(\Delta^m)_{<i>}$ are the same since $(\Delta^m)_{<i>}$ is obtained from B_m by deleting zero columns. We can now obtain an upper bound by using the right half of (3.1.4) with $r = i-1$ and $k = 1$:

$$\sigma_i(A^m) = \sigma_i(\Delta^m) = \sigma_{1+(i-1)}(\Delta^m) \leq \sigma_1(B_m)$$

$$= \sigma_1((\Delta^m)_{<i>}) = \sigma_1((\Delta_{<i>})^m)$$

Thus, we have the two-sided bounds

$$\sigma_i((\Delta_{[i]})^m)^{1/m} \leq \sigma_i(A^m)^{1/m} \leq \sigma_1((\Delta_{<i>})^m)^{1/m}$$

which yield the desired result since $\sigma_i((\Delta_{[i]})^m)^{1/m} \to |\lambda_i(\Delta_{[i]})| = |\lambda_i(A)|$ and $\sigma_1((\Delta_{<i>})^m)^{1/m} \to |\lambda_1(\Delta_{<i>})| = |\lambda_i(A)|$ as $m \to \infty$.   ☐

Problems

1.   Explain why the case $k = 1$ in (3.3.3) and (3.3.5) are familiar results.

2.   Use (3.3.3) to show that if $A \in M_n$ has rank r, then A has at least $n - r$ zero eigenvalues, $1 \leq r < n$. Could it have more than $n - r$ zero eigenvalues?

3.   Let $\alpha_1 \geq \cdots \geq \alpha_n \geq 0$ and $\beta_1 \geq \cdots \geq \beta_n \geq 0$ be two given ordered sets of nonnegative real numbers.
   (a) For $n = 2$, prove by direct calculation that the multiplicative inequalities $\alpha_1 \leq \beta_1$ and $\alpha_1 \alpha_2 \leq \beta_1 \beta_2$ imply the additive inequalities $\alpha_1 \leq \beta_1$ and $\alpha_1 + \alpha_2 \leq \beta_1 + \beta_2$.
   (b) For $n = 2$, consider $\alpha_1 = 1$, $\alpha_2 = \frac{1}{4}$, $\beta_1 = 5/4$, $\beta_2 = \frac{1}{4}$ and show that the additive inequalities do not imply the multiplicative inequalities. Thus, there is more intrinsic information in the inequalities (3.3.3), (3.3.5), and (3.3.11) than in the inequalities (3.3.13(a)), (3.3.14(a)), and (3.3.12), respectively.

4.   Let A_1, A_2,..., $A_m \in M_n$ for some integer $m \geq 2$. Using the notation of (3.3.14), show that

$$\sum_{i=1}^{k} \sigma_i(A_1 \cdots A_m) \leq \sum_{i=1}^{k} \sigma_i(A_1) \cdots \sigma_i(A_m) \text{ for } k = 1,..., n \quad (3.3.22)$$

5.   Let $A, B \in M_n$.
(a) Why is $\sigma_1(AB) \leq \sigma_1(A)\sigma_1(B)$?
(b) Consider $A = B = \begin{bmatrix} 1 & 1 \\ 0 & 1 \end{bmatrix}$ to show that $\sigma_2(AB) > \sigma_2(A)\sigma_2(B)$ is possible.
(c) Does the example in (b) contradict the known inequality $\sigma_1(AB) + \sigma_2(AB) \leq \sigma_1(A)\sigma_1(B) + \sigma_2(A)\sigma_2(B)$? What are the values of all the terms of this inequality for the example in (b)?

6.   Prove Theorem (3.3.14).

7.   Let $A, B \in M_n$ be given. Show that

$$\prod_{i=1}^{k} (\alpha + t|\lambda_i(A)|) \leq \prod_{i=1}^{k} [\alpha + t\sigma_i(A)] \quad (3.3.23)$$

and

$$\prod_{i=1}^{k} [\alpha + t\sigma_i(AB)] \leq \prod_{i=1}^{k} [\alpha + t\sigma_i(A)\sigma_i(B)] \quad (3.3.24)$$

for all $\alpha, t > 0$ and all $k = 1,..., n$. For $\alpha = 1$, these inequalities are useful in the theory of integral equations.

8.   Provide details for the following proof of Weyl's theorem (3.3.2) that avoids direct use of Lemma (3.3.1): Using the notation of (3.3.2), let $U^* A U = \Delta$ be upper triangular with diag $\Delta = [\lambda_1,..., \lambda_n]^T$, and write $\Delta = [\Delta_1 \ \Delta_2]$ and $\Delta_1 = \begin{bmatrix} \Delta_3 \\ 0 \end{bmatrix}$ with $\Delta_1 \in M_{n,k}$ and $\Delta_3 \in M_k$. Then $\sigma_i(A) = \sigma_i(\Delta) \geq \sigma_i(\Delta_1) = \sigma_i(\Delta_3)$ for $i = 1,..., k$, so $\sigma_1(A) \cdots \sigma_k(A) \geq |\det \Delta_3| = |\lambda_1(A) \cdots \lambda_k(A)|$.

9.   Deduce Weyl's theorem (3.3.2) from the A. Horn singular value product theorem (3.3.4).

10.   Let $A, B \in M_{m,n}$ be given and let $q = \min \{m,n\}$.
(a) Use the inequality (3.3.13(a)) and the product inequalities (3.3.14(a)) to prove *Von Neumann's trace theorem*

$$|\operatorname{tr} A^*B| \le \sum_{i=1}^{q} \sigma_i(A)\sigma_i(B) \tag{3.3.25}$$

(b) Use (a) and the singular value decomposition to show that $\max\{|\operatorname{tr} UA^*VB|: \ U \in M_n \text{ and } V \in M_m \text{ are unitary}\} = \Sigma_i \sigma_i(A)\sigma_i(B)$. Show that there are choices of U and V that yield the maximum and make both $(UA^*)(VB)$ and $(VB)(UA^*)$ positive semidefinite; see Theorem (7.4.10) in [HJ] for an important related result.

11. Use (3.3.17) to explain why the problem of computing singular values of a given matrix is intrinsically well conditioned. Contrast with the analogous situation for eigenvalues; see (6.3.3-4) in [HJ].

12. Prove a two-sided multiplicative bound analogous to the two-sided additive bound in (3.3.19): If $A, B \in M_n$, then $\sigma_1(B)\sigma_i(A) \ge \sigma_i(AB) \ge \sigma_n(B)\sigma_i(A)$ for $i = 1,..., n$. If $B = I + E$, verify the multiplicative perturbation bounds

$$[1 - \sigma_1(E)]\sigma_i(A) \le \sigma_i(A[I+E]) \le [1 + \sigma_1(E)]\sigma_i(A), \ i = 1,..., n \tag{3.3.26}$$

What relative perturbation bounds on $\sigma_i(A[I+E])/\sigma_i(A)$ does this give for $i = 1,...,$ rank A? What does this say for $i >$ rank A?

13. Let $A \in M_{m,n}$ and $X \in M_{n,k}$ be given with $k \le \min\{m,n\}$. Show that $\det(X^*A^*AX) \le [\sigma_1(A) \cdots \sigma_k(A)]^2 \det X^*X$.

14. Let $A \in M_n$. Show that $|\lambda_1(A) \cdots \lambda_k(A)| = \sigma_1(A) \cdots \sigma_k(A)$ for all $k = 1,..., n$ if and only if A is normal.

15. Suppose a given real-valued function $\varphi(u) \equiv \varphi(u_1,..., u_k)$ of k scalar variables has continuous first partial derivatives in the domain $\mathcal{D}_k(L) \equiv \{u = [u_i] \in \mathbb{R}^k: \ L < u_k \le u_{k-1} \le \cdots \le u_2 \le u_1 < \infty\}$, where $L \ge -\infty$ is given. Let $\alpha = [\alpha_i] \in \mathcal{D}_k(L)$ and $\beta = [\beta_i] \in \mathcal{D}_k(L)$ be given, so the real vectors α and β have algebraically decreasingly ordered entries.
   (a) Sketch $\mathcal{D}_k(L)$ for $k = 2$ and finite L; for $L = -\infty$.
   (b) Show that $\mathcal{D}_k(L)$ is convex.
   (c) Explain why $f(t) \equiv \varphi((1-t)\alpha + t\beta)$ is defined and continuously differentiable for $t \in [0,1]$, and

$$\varphi(\beta) - \varphi(\alpha) = \int_0^1 f'(t) \, dt$$

Show that

$$f'(t) = (\beta - \alpha)^T \nabla \varphi = \sum_{i=1}^{k} (\beta_i - \alpha_i) \frac{\partial \varphi}{\partial u_i}$$

$$= (\beta_1 - \alpha_1) \frac{\partial \varphi}{\partial u_1} + \sum_{i=2}^{k-1} \left[\sum_{j=1}^{i-1} (\beta_j - \alpha_j) \right] \left[\frac{\partial \varphi}{\partial u_i} - \frac{\partial \varphi}{\partial u_{i+1}} \right]$$

$$+ \frac{\partial \varphi}{\partial u_k} \sum_{i=1}^{k} (\beta_i - \alpha_i)$$

for all $t \in [0,1]$.

(d) Now suppose that α is weakly majorized by β, that is,

$$\sum_{j=1}^{i} \alpha_j \leq \sum_{j=1}^{i} \beta_j, \quad i = 1,..., k$$

and suppose that $\varphi(\cdot)$ satisfies the inequalities

$$\frac{\partial \varphi}{\partial u_1} \geq \frac{\partial \varphi}{\partial u_2} \geq \cdots \geq \frac{\partial \varphi}{\partial u_k} \geq 0 \text{ at every point in } \mathcal{D}_k(L) \qquad (3.3.27)$$

that is, the vector $\nabla \varphi(u)$ is a point in $\mathcal{D}_k(0)$ for every $u \in \mathcal{D}_k(L)$. Show that $\varphi(\alpha) \leq \varphi(\beta)$, so φ is monotone with respect to weak majorization.

(e) Let $f: \mathbb{R} \to \mathbb{R}$ be a given increasing convex and continuously differentiable function. Show that $\varphi(u) \equiv f(u_1) + \cdots + f(u_k)$ satisfies the inequalities (3.3.27) on $\mathcal{D}_k(-\infty)$. Use a smoothing argument to deduce the result in Lemma (3.3.8), which does not require that $f(\cdot)$ be continuously differentiable.

16. Suppose a given real-valued function $F(u) \equiv F(u_1,..., u_k)$ of k scalar variables has continuous first partial derivatives in the domain $\mathcal{D}_k(0) \equiv \{u = [u_i] \in \mathbb{R}^k: 0 \leq u_k \leq u_{k-1} \leq \cdots \leq u_2 \leq u_1 < \infty\}$, and suppose $u \circ \nabla F(u) \in \mathcal{D}_k(0)$ for every $u \in \mathcal{D}_k(0)$, that is,

$$u_1 \frac{\partial F}{\partial u_1} \geq u_2 \frac{\partial F}{\partial u_2} \geq \cdots \geq u_k \frac{\partial F}{\partial u_k} \geq 0 \text{ at every point } u \in \mathcal{D}_k(0) \quad (3.3.28)$$

(a) Show that the function $\varphi(u) \equiv F(e^{u_1}, ..., e^{u_k})$ satisfies the hypotheses in Problem 15 on $\mathcal{D}_k(-\infty)$ and satisfies the inequalities (3.3.27).

(b) Using the notation of Theorem (3.3.2), show that $F(|\lambda_1(A)|, ..., |\lambda_k(A)|) \leq F(\sigma_1(A), ..., \sigma_k(A))$ for $k = 1, ..., n$.

(c) Using the notation of Theorem (3.3.4), show that $F(\sigma_1(AB), ..., \sigma_k(AB)) \leq F(\sigma_1(A)\sigma_1(B), ..., \sigma_k(A)\sigma_k(B))$ for $k = 1, ..., \min\{m, n, p\}$.

(d) Let the real-valued function $f(t)$ be increasing and continuously differentiable on $[0, \infty)$ and suppose $f(e^t)$ is convex on $(-\infty, \infty)$. Show that $F(u) = F(u_1, ..., u_k) \equiv f(e^{u_1}) + \cdots + f(e^{u_k})$ satisfies the required smoothness conditions and the inequalities (3.3.28). Conclude that the additive weak majorization relations (3.3.13-14(d)) follow from the product inequalities (3.3.2,4) and the general inequalities in (b) and (c).

(e) Let $S_j(u_1, ..., u_k)$ denote the jth elementary symmetric function of the k scalar variables $u_1, ..., u_k$; see (1.2.9) in [HJ]. Show that $S_j(u_1, ..., u_k)$ is a smooth function that satisfies the inequalities (3.3.28).

(f) Using the notation of Theorems (3.3.2,4), conclude that each elementary symmetric function $S_j(u_1, ..., u_k)$ satisfies

$$S_j(|\lambda_1(A)|, ..., |\lambda_k(A)|) \leq S_j(\sigma_1(A), ..., \sigma_k(A))$$

and

$$S_j(\sigma_1(AB), ..., \sigma_k(AB)) \leq S_j(\sigma_1(A)\sigma_1(B), ..., \sigma_k(A)\sigma_k(B))$$

for all appropriate values of k and all $j = 1, ..., k$. What do these inequalities say when $j = 1$?

17. Let $A \in M_n$ be given. Using the notation of Theorem (3.3.2), show that $|\lambda_1(A) \cdots \lambda_k(A)| \leq \sigma_1(A^m)^{1/m} \cdots \sigma_k(A^m)^{1/m}$, $k = 1, ..., n$, $m = 1, 2, ...$. If $f(t)$ is an increasing continuously differentiable real-valued function on $[0, \infty)$, and if $f(e^t)$ is convex on $(-\infty, \infty)$, show that

$$\sum_{i=1}^{k} f(|\lambda_i(A)|) \leq \sum_{i=1}^{k} f(\sigma_i(A^m)^{1/m}), \quad k = 1, ..., n, \ m = 1, 2, ... \quad (3.3.29)$$

which is a generalization of (3.3.13(d)). What happens as $m \to \infty$?

18. Let $A = [a_{ij}]$, $B \in M_{m,n}$ be given, let $q = \min\{m, n\}$, let $[A, B]_F \equiv$ tr B^*A denote the Frobenius inner product on $M_{m,n}$, and let $\|A\|_2 \equiv [A, A]_F^{1/2}$ denote the Frobenius norm (the l_2 norm) on $M_{m,n}$.

(a) Show that

$$\| A \|_2 = \left[\sum_{i=1}^{m} \sum_{j=1}^{n} |a_{ij}|^2 \right]^{\frac{1}{2}} = \left[\sum_{i=1}^{q} \sigma_i(A)^2 \right]^{\frac{1}{2}} \tag{3.3.30}$$

(b) Show that

$$|[A,B]_F| \leq \sigma_1(A)\sigma_1(B) + \cdots + \sigma_q(A)\sigma_q(B) \leq \| A \|_2 \| B \|_2 \tag{3.3.31}$$

which gives an enhancement to the Cauchy-Schwarz inequality for the Frobenius inner product.

(c) Show that

$$\left[\sum_{i=1}^{q} [\sigma_i(A) - \sigma_i(B)]^2 \right]^{\frac{1}{2}} \leq \| A - B \|_2 \tag{3.3.32}$$

which may be thought of as a simultaneous perturbation bound on all of the singular values. How is this related to the perturbation bound (3.3.19) for one singular value? Discuss the analogy with the Hoffman-Wielandt theorem (Theorem (6.3.5) in [HJ]).

19. Ordinary singular values are intimately connected with the Euclidean norm, but many properties of singular values hold for other functions of a matrix connected with arbitrary norms. Let $\| \cdot \|_a$ and $\| \cdot \|_b$ denote two given norms on $\mathbb{C}^m$ and $\mathbb{C}^n$, respectively, and let $q = \min\{m,n\}$. For $A \in M_{m,n}$, define

$$\sigma_k(A; \| \cdot \|_a, \| \cdot \|_b) \equiv \min_{\substack{S \subset \mathbb{C}^n \\ \dim S = n-k+1}} \max_{\substack{x \in S \\ \|x\|_b = 1}} \| Ax \|_a, \quad k = 1, \dots, q$$

where S ranges over all subspaces with the indicated dimension.

(a) Show that $\sigma_k(A; \| \cdot \|_a, \| \cdot \|_b) = 0$ if $k > q$.

(b) Show that $\sigma_1(A; \| \cdot \|_a, \| \cdot \|_b)$ is a norm on $M_{m,n}$.

(c) Show that

$$\sigma_{i+j-1}(A + B; \| \cdot \|_a, \| \cdot \|_b) \leq \sigma_i(A; \| \cdot \|_a, \| \cdot \|_b) + \sigma_j(B; \| \cdot \|_a, \| \cdot \|_b)$$

for all $A, B \in M_{m,n}$, $1 \leq i, j \leq q$, and $i + j \leq q + 1$.

(d) Verify the perturbation bounds

$$| \sigma_i(A + B; \| \cdot \|_a, \| \cdot \|_b) - \sigma_i(A; \| \cdot \|_a, \| \cdot \|_b)| \leq \sigma_1(B; \| \cdot \|_a, \| \cdot \|_b)$$

for all $A, B \in M_{m,n}$ and all $i = 1,..., q$.

20. Let $A \in M_n$ be given.   There are three natural Hermitian matrices associated with A: A^*A, AA^*, and $H(A) \equiv \frac{1}{2}(A + A^*)$.   Weyl's theorem (3.3.2) and A. Horn's theorem (3.6.6) characterize the relationship between the eigenvalues of A and those of A^*A and AA^*.   A theorem of L. Mirsky characterizes the relationship between the eigenvalues of A and $H(A)$.   Let $\{\lambda_i(A)\}$ and $\{\lambda_i(H(A))\}$ denote the eigenvalues of A and $H(A)$, respectively, ordered so that $\operatorname{Re} \lambda_1(A) \geq \cdots \geq \operatorname{Re} \lambda_n(A)$ and $\lambda_1(H(A)) \geq \cdots \geq \lambda_n(H(A))$.

(a) Show that

$$\sum_{i=1}^{k} \operatorname{Re} \lambda_i(A) \leq \sum_{i=1}^{k} \lambda_i(H(A)), \quad k = 1,..., n, \text{ with equality for } k = n \quad (3.3.33)$$

(b) Conversely, if $2n$ given scalars $\{\lambda_i\}_{i=1}^n \subset \mathbb{C}$ and $\{\eta_i\}_{i=1}^n \subset \mathbb{R}$ are ordered so that $\operatorname{Re} \lambda_1 \geq \cdots \geq \operatorname{Re} \lambda_n$ and $\eta_1 \geq \cdots \geq \eta_n$, and if

$$\sum_{i=1}^{k} \operatorname{Re} \lambda_i \leq \sum_{i=1}^{k} \eta_i, \quad k = 1,..., n, \text{ with equality for } k = n$$

show that there is some $A \in M_n$ such that $\{\lambda_i\}$ is the set of eigenvalues of A and $\{\eta_i\}$ is the set of eigenvalues of $H(A)$.

(c) Use (3.1.6a) to show that

$$\sum_{i=1}^{k} \operatorname{Re} \lambda_i(A) \leq \sum_{i=1}^{k} \lambda_i(H(A)) \leq \sum_{i=1}^{k} \sigma_i(A) \quad \text{for } k = 1,..., n \ (3.3.34a)$$

If f is a real-valued convex function on the interval $[\lambda_n(H(A)), \lambda_1(H(A))]$, show that

$$\sum_{i=1}^{n} f(\operatorname{Re} \lambda_i(A)) \leq \sum_{i=1}^{n} f(\lambda_i(H(A))) \quad (3.3.34b)$$

If f is a real-valued increasing convex function on the interval

$[\lambda_n(H(A)), \sigma_1(A)]$, show that

$$\sum_{i=1}^{k} f(\text{Re } \lambda_i(A)) \leq \sum_{i=1}^{k} f(\lambda_i(H(A))) \leq \sum_{i=1}^{k} f(\sigma_i(A)) \quad \text{for } k = 1,..., n(3.3.34c)$$

(d) Let $H \in M_n$ be a given Hermitian matrix, and write $H = \text{Re } H + i\text{Im } H$, where $\text{Re } H = (H + \bar{H})/2$ is real symmetric and $\text{Im } H = (H - \bar{H})/2i$ is real skew-symmetric. Let $\text{Re } H = QDQ^T$, where $Q \in M_n(\mathbb{R})$ is real orthogonal and $D = \text{diag}(\alpha_1,..., \alpha_n)$ is real diagonal. Show that $\alpha_1,..., \alpha_n$ are the main diagonal entries of Q^THQ, and conclude, as in (a), that there is a strong majorization relationship between the eigenvalues of $\text{Re } H$ and the eigenvalues of H.

(e) Let $\lambda_1(\text{Re } H(A)) \geq \cdots \geq \lambda_n(\text{Re } H(A))$ denote the algebraically decreasingly ordered eigenvalues of $\text{Re } H(A) = \frac{1}{2}(H(A) + \overline{H(A)})$. Use (d) to show that $\text{Re } \lambda_i(A)$ can be replaced by $\lambda_i(\text{Re } H(A))$ in (3.3.33) and (3.3.34a-c).

21. Let $A = [a_{ij}] \in M_{m,n}$ be given and let $q = \min\{m,n\}$. Let $a \equiv [a_{11}, ..., a_{qq}]^T \in \mathbb{C}^q$ and $\sigma(A) \equiv [\sigma_1(A), ..., \sigma_q(A)]^T \in \mathbb{R}^q$ denote the vectors of main diagonal entries and singular values of A, respectively, with $\sigma_1(A) \geq \cdots \geq \sigma_q(A)$. Let $|a|_{[1]} \geq \cdots \geq |a|_{[q]}$ denote the absolutely decreasingly ordered main diagonal entries of A.

(a) Use the basic inequalities (3.3.13(a)) or the inequalities (3.1.10a) (which follow directly from the singular value decomposition) and the interlacing inequalities (3.1.4) to show that

$$|a|_{[1]} + \cdots + |a|_{[k]} \leq \sigma_1(A) + \cdots + \sigma_k(A) \quad \text{for } k = 1,..., q \quad (3.3.35)$$

For a different proof of these necessary conditions, see Problem 3 in Section (3.2).

(b) Use Lemma (3.3.8) to deduce the weaker necessary conditions

$$|a|_{[1]}^2 + \cdots + |a|_{[k]}^2 \leq \sigma_1(A)^2 + \cdots + \sigma_k(A)^2 \quad \text{for } k = 1,..., q \quad (3.3.36)$$

Consider $\alpha_1 = 1$, $\alpha_2 = 1/\sqrt{2}$, $s_1 = \sqrt{5}/2$, $s_2 = \frac{1}{2}$ to show that there are values satisfying (3.3.36) that do not satisfy (3.3.35), and hence cannot be the main diagonal entries and singular values of a complex matrix.

(c) Consider $A = \begin{bmatrix} a & b \\ c & d \end{bmatrix} \in M_2$. Show that $[\sigma_1(A) - \sigma_2(A)]^2 = \text{tr } A^*A - 2|\det A| = |a|^2 + |b|^2 + |c|^2 + |d|^2 - 2|ad - bc| \geq (|a| - |d|)^2 + (|b| - |c|)^2 \geq (|a| - |d|)^2$. Using the notation in (a), conclude that

$$|a|_{[1]} - |a|_{[2]} \leq \sigma_1(A) - \sigma_2(A) \text{ for all } A \in M_2 \qquad (3.3.37)$$

(d) Consider $\alpha_1 = s_1 = 1$, $\alpha_2 = \frac{1}{2}$, $s_2 = 3/4$ to show that there are potential values for the main diagonal entries and singular values of a 2-by-2 complex matrix that satisfy the necessary conditions (3.3.35) (and hence satisfy (3.3.36) as well) but do not satisfy the necessary conditions (3.3.37).

(e) Let $\sigma_1 \geq \sigma_2 \geq 0$ and $a_1, a_2 \in \mathbb{C}$ be given with $|a_1| \geq |a_2|$. Suppose $|a_1| + |a_2| \leq \sigma_1 + \sigma_2$ and $|a_1| - |a_2| \leq \sigma_1 - \sigma_2$. Square these inequalities, show that $\pm 2||a_1 a_2| - \sigma_1 \sigma_2| \leq \sigma_1^2 + \sigma_2^2 - (|a_1|^2 + |a_2|^2)$, and conclude that $||a_1 a_2| - \sigma_1 \sigma_2| \leq 2||a_1 a_2| - \sigma_1 \sigma_2| \leq \sigma_1^2 + \sigma_2^2 - (|a_1|^2 + |a_2|^2)$. Sketch the loci of $xy = ||a_1 a_2| - \sigma_1 \sigma_2|$ and $x^2 + y^2 = \sigma_1^2 + \sigma_2^2 - (|a_1|^2 + |a_2|^2)$, $x, y \in \mathbb{R}$, and explain why there exist real ξ, η such that $\xi \eta = |a_1 a_2| - \sigma_1 \sigma_2$ and $\xi^2 + \eta^2 + |a_1|^2 + |a_2|^2 = \sigma_1^2 + \sigma_2^2$. Set

$$B = \begin{bmatrix} |a_1| & \xi \\ \eta & |a_2| \end{bmatrix}$$

show that $\text{tr } B^*B = \sigma_1^2 + \sigma_2^2$ and $\det B = \sigma_1 \sigma_2$, and conclude that B has singular values σ_1 and σ_2. Construct a diagonal unitary D such that $A = DB$ has main diagonal entries a_1, a_2 and singular values σ_1, σ_2.

(f) Explain why there is some $A \in M_2$ with given main diagonal entries $a_1, a_2 \in \mathbb{C}$ and singular values $\sigma_1 \geq \sigma_2 \geq 0$ if and only if the inequalities (3.3.35) (with $n = 2$) and (3.3.37) are satisfied. Of what theorem for general $n = 2, 3, \ldots$ do you think this is a special case?

22. We have shown that (weak) multiplicative majorization inequalities of the form (3.3.5) imply additive weak majorization inequalities of the form (3.3.14). However, not every weak additive majorization comes from a multiplicative one. Show by example that (3.3.5) is false if the ordinary product AB is replaced by the Hadamard product $A \circ B$. Nevertheless, Theorem (5.5.4) says that (3.3.14) does hold when AB is replaced by $A \circ B$.

23. Using the notation of Lemma (3.3.8), provide details for the following

alternative proof that avoids using facts about doubly substochastic matrices and relies only on the basic relationship between strong majorization and doubly stochastic matrices. State carefully the hypotheses and conclusions appropriate to this argument, and notice that one may now have to assume that $f(\cdot)$ is defined, monotone, and convex on an interval that extends to the left of the point $\min\{x_n, y_n\}$. If $\delta_k \equiv \Sigma_{i=1}^k (y_i - x_i) > 0$, choose points x'_{k+1} and y'_{k+1} such that $x_k \geq x'_{k+1}$, $y_k \geq y'_{k+1}$, and $\delta_k = x'_{k+1} - y'_{k+1}$. Let $x'_i = x_i$ and $y'_i = y_i$ for $i = 1, ..., k$ and let $x' \equiv [x'_i]$, $y' \equiv [y'_i] \in \mathbb{R}^{k+1}$. By strong majorization, there is a doubly stochastic $S \in M_{k+1}(\mathbb{R})$ such that $x' = Sy'$, so $f(x_1) + \cdots + f(x_k) + f(x'_{k+1}) \leq f(y_1) + \cdots + f(y_k) + f(y'_{k+1})$. Now use $f(y'_{k+1}) - f(x'_{k+1}) \leq 0$.

24. Based on Lemma (3.3.8), how can the hypotheses of the domain of the function $f(\cdot)$ in Theorems (3.3.13-14) be weakened?

25. What do Theorems (3.3.13-14(d)) say for $f(t) = 1/t$? Can you get the same result from (3.3.13-14(a))? What about $f(t) = t^{\frac{1}{2}}$? If $A, B \in M_n$ are nonsingular, show that

$$\left[\sum_{i=1}^n \sigma_i(AB)^{-\frac{1}{2}} \right]^2 \leq \left[\sum_{i=1}^n \sigma_i(A^{-1}) \right] \left[\sum_{i=1}^n \sigma_i(B^{-1}) \right] \qquad (3.3.38)$$

26. Let $A \in M_n$ be given, and let m be a given positive integer.
(a) Show that

$$\sum_{i=1}^k \sigma_i(A^m) \leq \sum_{i=1}^k \sigma_i(A)^m \quad \text{for } k = 1, ..., n$$

(b) Show that

$$\sum_{i=1}^k \sigma_i(A^m)^p \leq \sum_{i=1}^k \sigma_i(A)^{mp} \quad \text{for } k = 1, ..., n \qquad (3.3.39)$$

for all $p > 0$.
(c) Show that

$$|\operatorname{tr} A^{2m}| \leq \operatorname{tr}[(A^m)^* A^m] \leq \operatorname{tr}[(A^*A)^m] \text{ for all } m = 1, 2, ... \quad (3.3.40)$$

The role of this inequality in statistical mechanics is mentioned in Section (6.5); see Corollary (6.5.32) for some other functions $f(\cdot)$ that satisfy inequalities of the form $f(A^{2m}) \leq f([A^*A]^m)$.

27. Let $A = [a_{ij}] \in M_n$ be a given Hermitian matrix with algebraically decreasingly ordered eigenvalues $\lambda_1(A) \geq \cdots \geq \lambda_n(A)$ and main diagonal entries $a_1 \geq \cdots \geq a_n$. Let $f(t)$ be a given convex function on the interval $[\lambda_n(A), \lambda_1(A)]$.

(a) Show that

$$\sum_{i=1}^{n} f(a_{ii}) \leq \sum_{i=1}^{n} f(\lambda_i(A)) \tag{3.3.41}$$

and, if $f(t)$ is also an increasing function,

$$\sum_{i=1}^{k} f(a_{ii}) \leq \sum_{i=1}^{k} f(\lambda_i(A)) \text{ for } k = 1,\ldots, n \tag{3.3.42}$$

(b) Let $\{x_1,\ldots, x_n\} \subset \mathbb{C}^n$ be a given orthonormal set, and let the values $x_1^*Ax_1,\ldots, x_n^*Ax_n$ be algebraically ordered as $\alpha_1 \geq \cdots \geq \alpha_n$. Show that

$$\sum_{i=1}^{n} f(x_i^*Ax_i) \leq \sum_{i=1}^{n} f(\lambda_i(A)) \tag{3.3.43}$$

and, if $f(\cdot)$ is also increasing,

$$\sum_{i=1}^{k} f(\alpha_i) \leq \sum_{i=1}^{k} f(\lambda_i(A)) \text{ for } k = 1,\ldots, n \tag{3.3.44}$$

28. Let $A \in M_n$ be given. Show that every eigenvalue λ of A satisfies $\sigma_1(A) \geq |\lambda| \geq \sigma_n(A)$.

29. Let $f: \mathbb{R}^+ \to \mathbb{R}$ be a given twice differentiable real-valued function on $(a,b) \subset \mathbb{R}$, and let $\varphi(t) \equiv f(e^t)$. Show that if $f(t)$ is increasing and convex, then $\varphi(t)$ is convex. Conversely, consider $f(t) = 1/t$ and $f(t) = 1/t^{\frac{1}{3}}$ on $(0,\infty)$ to show that $\varphi(t)$ can be convex and f can be either decreasing or concave (but not both at any one point). Comment on the relative strengths of the hypotheses of parts (c) and (d) of Theorems (3.3.13-14).

30. Let $A, B \in M_n$ be given *commuting* matrices. If all eigenvalues are arranged in decreasing absolute value as in Theorem (3.3.21), show that

$$|\lambda_k(AB)| \le |\lambda_{k-j+1}(A)| \, |\lambda_j(B)|, \quad j = 1, ..., k, \quad k = 1, ..., n \; (3.3.45)$$

which implies

$$|\lambda_k(AB)| \le \min \{\rho(A) |\lambda_k(B)|, \rho(B) |\lambda_k(A)|\}, \quad k = 1, ..., n$$
$$(3.3.46a)$$

and

$$\rho(AB) \le \rho(A) \rho(B) \tag{3.3.46b}$$

If, in addition, both A and B are nonsingular, deduce that

$$|\lambda_k(AB)| \ge |\lambda_{k+j-1}(A)| \, |\lambda_{n-j+1}(B)|, \quad j = 1, ..., n-k+1,$$
$$k = 1, ..., n \quad (3.3.47a)$$

and

$$|\lambda_n(AB)| \ge |\lambda_n(A)| \, |\lambda_n(B)| \tag{3.3.47b}$$

Give examples to show that these inequalities need not hold if the hypothesis of commutativity is dropped.

31. Let $A \in M_n$ be given, let $|\lambda_1(A)| \ge \cdots \ge |\lambda_n(A)|$ denote its absolutely decreasingly ordered eigenvalues, and let $\sigma_1(A) \ge \cdots \ge \sigma_n(A)$ denote its decreasingly ordered singular values. If A is normal, we know that $|\lambda_i(A)| = \sigma_i(A)$ for $i = 1, ..., n$, but if A is "almost normal," can we say that $|\lambda_i(A)| \doteq \sigma_i(A)$? The *defect from normality* of A with respect to a given unitarily invariant norm $\|\cdot\|$ is defined in terms of the set of all Schur unitary triangularizations of A:

$$\delta(A; \|\cdot\|) = \inf \{\|T\| \colon A = U(\Lambda + T)U^*, \; U \text{ is unitary}, \; \Lambda \text{ is}$$

diagonal, and T is strictly upper triangular$\}$ (3.3.48)

One might consider A to be "almost normal" if $\delta(A; \|\cdot\|\}$ is small. This measure of nonnormality is perhaps most familiar when $\|\cdot\|$ is the Frobenius norm $\|\cdot\|_2$.
    (a) Show that

$$\delta(A;\|\cdot\|_2)^2 = \sum_{i=1}^{n} [\sigma_i(A)^2 - |\lambda_i(A)|^2] \tag{3.3.49}$$

Conclude that all $|\lambda_i(A)| \doteq \sigma_i(A)$ if and only if A is "almost normal" in the sense that $\delta(A;\|\cdot\|_2)$ is small.

(b) Consider the spectral norm $\|\|\cdot\|\|_2$. Use (3.3.19) and a Schur upper triangularization $A = U(\Lambda + T)U^*$ to show that

$$|\sigma_i(A) - |\lambda_i(A)|| \leq \delta(A;\|\|\cdot\|\|_2) \text{ for } i = 1,...,n \tag{3.3.50}$$

Thus, if A is "almost normal" in the sense that $\delta(A;\|\|\cdot\|\|_2)$ is small, then all $\sigma_i(A) \doteq |\lambda_i(A)|$.

(c) Suppose $A = SBS^{-1}$ for some nonsingular $S \in M_n$. Use (3.3.20) to show that

$$\sigma_i(B)/\kappa_2(S) \leq |\sigma_i(A)| \leq \kappa_2(S)\sigma_i(B) \text{ for } i = 1,...,n \tag{3.3.51}$$

where $\kappa_2(S) \equiv \|\|S\|\|_2 \|\|S^{-1}\|\|_2$ denotes the spectral condition number of S. What does this say when A is normal? See Problem 45 in Section (3.1) for another approach to the inequalities (3.3.51).

Notes and Further Readings. The basic theorems (3.3.2) and (3.3.4) were first published in the classic papers of H. Weyl (1949) and A. Horn (1950) cited at the end of Section (3.0). The proof we give for Theorem (3.3.2) is similar to the proof of the lemma in Section (2.4) of J. P. O. Silberstein, On Eigenvalues and Inverse Singular Values of Compact Linear Operators in Hilbert Space, *Proc. Cambridge Phil. Soc.* 49 (1953), 201-212. Theorem (3.3.4) can be generalized to give inequalities for products of arbitrarily chosen singular values, not just the consecutive ones: If A, $B \in M_n$ and indices $1 \leq i_1 < i_2 < \cdots < i_k \leq n$ are given, then

$$\sigma_{i_1}(AB) \cdots \sigma_{i_k}(AB) \leq \min\{\sigma_{i_1}(A) \cdots \sigma_{i_k}(A)\sigma_1(B) \cdots \sigma_k(B),$$
$$\sigma_1(A) \cdots \sigma_k(A)\sigma_{i_1}(B) \cdots \sigma_{i_k}(B)\} \tag{3.3.52}$$

For a proof and a discussion of related results for arbitrary sums of eigenvalues of sums of Hermitian matrices, see the appendix to [Gan 86], pp. 621-624, or Section 4 of [AmMo]. There are also generalizations of the inequalities (3.3.18): If $A, B \in M_n$ and indices $1 \leq i_1 < \cdots < i_k \leq n$, $1 \leq j_1 < \cdots < j_k \leq n$ are given with $i_k + j_k - k \leq n$, then

$$\prod_{r=1}^{k} \sigma_{i_r + j_r - r}(AB) \leq \prod_{r=1}^{k} \sigma_{i_r}(A) \, \sigma_{j_r}(B) \qquad (3.3.53)$$

For a proof, see R. C. Thompson, On the Singular Values of Matrix Products-II, *Scripta Math.* 29 (1973), 111-114.

The discovery of inequalities of the form (3.3.13b) seems to have been stimulated by results in the theory of integral equations. In the 1949 paper cited at the end of Section (3.0), S.-H. Chang showed with a long analytic argument that if the infinite series $\Sigma_{i=1}^{\infty} \sigma_i(K)^p$ of powers of the singular values of an L^2 kernel K is convergent for some $p > 0$, then the series $\Sigma_{i=1}^{\infty} |\lambda_i(K)|^p$ of the same powers of the absolute eigenvalues of K is also convergent, but Chang had no inequality between these two series. Apparently stimulated by Chang's results, Weyl discovered the basic inequalities in Theorems (3.3.2) and (3.3.13).

The proof given for Yamamoto's theorem (3.3.21) was developed in C. R. Johnson and P. Nylen, Yamamoto's Theorem for Generalized Singular Values, *Linear Algebra Appl.* 128 (1990), 147-158 (where generalizations to other types of singular values are given) and also in R. Mathias, Two Theorems on Singular Values and Eigenvalues, *Amer. Math. Monthly* 97 (1990), 47-50, from which the approach to Weyl's theorem given in Problem 8 is also taken. There is always a version of the right half of the interlacing inequalities (3.1.4) for the generalized singular values associated with two arbitrary norms, as defined in Problem 19; if the two norms have a general monotonicity property, one also gets a version of the left half—see M.-J. Sodupe, Interlacing Inequalities for Generalized Singular Values, *Linear Algebra Appl.* 87 (1987), 85-92. The fact that the weak majorization conditions (3.3.35), together with a general version of (3.3.37), namely, $|a|_{[1]} + \cdots + |a|_{[n-1]} - |a|_{[n]} \leq \sigma_1 + \cdots + \sigma_{n-1} - \sigma_n$, are both necessary and sufficient for $2n$ given scalars $a_1, \ldots, a_n \in \mathbb{C}$ and $\sigma_1 \geq \cdots \geq \sigma_n \geq 0$ to be the main diagonal entries and singular values of some n-by-n complex matrix A is proved in R. C. Thompson, Singular Values, Diagonal Elements, and Convexity, *SIAM J. Appl. Math.* 32 (1977), 39-63. If, in addition, one wants A to be symmetric, then there are some additional conditions that must be

satisfied;   these are described in R. C. Thompson, Singular Values and
Diagonal Elements of Complex Symmetric Matrices, *Linear Algebra Appl.*
26 (1979), 65-106.   For a characterization of the cases in which equality
holds in the inequalities (3.3.3), (3.3.13a), (3.3.33), or (3.3.35) for some
value of k, see C.-K. Li, Matrices with Some Extremal Properties, *Linear
Algebra Appl.* 101 (1988), 255-267.

3.4 Sums of singular values: the Ky Fan k-norms

One way to approach majorization inequalities for the eigenvalues of sums of
Hermitian matrices is via the variational characterization

$$\lambda_1(A) + \cdots + \lambda_k(A)$$

$$= \max\{\operatorname{tr} U_k^* A U_k \colon U_k \in M_{n,k} \text{ and } U_k^* U_k = I\}, \, k = 1,\ldots, n$$

in which the eigenvalues of the Hermitian matrix $A \in M_n$ are arranged in
algebraically decreasing order $\lambda_1 \geq \cdots \geq \lambda_n$ (see Corollary (4.3.18) and
Theorem (4.3.27) in [HJ]).   An analogous characterization of the sum of the
k largest singular values has the pleasant consequence that it makes evident
that this sum is a norm.   The following result is a natural generalization of
Problem 6 in Section (3.1).

3.4.1   Theorem. Let $A \in M_{m,n}$ have singular values $\sigma_1(A) \geq \cdots \geq$
$\sigma_q(A) \geq 0$, where $q = \min\{m,n\}$.   For each $k = 1,\ldots, q$ we have

$$\sum_{i=1}^{k} \sigma_i(A) = \max\{|\operatorname{tr} X^* A Y| \colon X \in M_{m,k}, \, Y \in M_{n,k}, \, X^* X = I = Y^* Y\}$$

$$= \max\{|\operatorname{tr} A C| \colon C \in M_{n,m} \text{ is a rank } k \text{ partial isometry}\}$$

Proof: If $X \in M_{m,k}$ and $Y \in M_{n,k}$ satisfy $X^* X = I = Y^* Y$, notice that
$\operatorname{tr} X^* A Y = \operatorname{tr} A Y X^*$ and $C \equiv Y X^* \in M_{n,m}$ has $C^* C = X Y^* Y X^* = X X^*$.
Since the k largest singular values of $X X^*$ are the same as those of $X^* X = I \in$
M_k, we conclude that $C = Y X^*$ is a rank k partial isometry.   Conversely, if
$C \in M_{n,m}$ is a given rank k partial isometry, then the singular value decomposition of C is

$$C = V\Sigma W^* = [V_k \ *] \begin{bmatrix} I_k & 0 \\ 0 & 0 \end{bmatrix} \begin{bmatrix} W_k^* \\ * \end{bmatrix} = V_k W_k^*$$

where $I_k \in M_k$ is an identity matrix, and $V_k \in M_{n,k}$ and $W_k \in M_{m,k}$ are the first k columns of the unitary matrices $V \in M_n$ and $W \in M_m$, respectively. Thus the two asserted variational formulae are equivalent and it suffices to show that the second equals the indicated singular value sum. But this is an immediate consequence of (3.3.13a) and (3.3.14a). Compute

$$|\mathrm{tr}\ AC| = \left| \sum_{i=1}^{m} \lambda_i(AC) \right| \leq \sum_{i=1}^{m} |\lambda_i(AC)| \leq \sum_{i=1}^{m} \sigma_i(AC)$$

$$\leq \sum_{i=1}^{q} \sigma_i(A)\sigma_i(C) = \sum_{i=1}^{k} \sigma_i(A) \qquad (3.4.2)$$

in which all indicated eigenvalues $\{\lambda_i\}$ and singular values $\{\sigma_i\}$ are arranged in decreasing absolute value. If $A = V\Sigma W^*$ is a singular value decomposition of A, let

$$C_{max} \equiv WP_kV^*, \text{ where } P_k \equiv \begin{bmatrix} I_k & 0 \\ 0 & 0 \end{bmatrix} \in M_{n,m} \text{ and } I_k \in M_k$$

Then $\mathrm{tr}\ AC_{max} = \mathrm{tr}\ V\Sigma W^* WP_kV^* = \mathrm{tr}\ V\Sigma P_kV^* = \mathrm{tr}\ \Sigma P_k = \sigma_1(A) + \cdots + \sigma_k(A)$, so the upper bound in (3.4.2) can be achieved.   ⬜

The preceding result is useful as a quasilinear characterization of the sum of the k largest singular values and it implies majorization relations for sums of singular values of sums of (not necessarily square) matrices.

3.4.3   Corollary. Let $A, B \in M_{m,n}$ have respective ordered singular values $\sigma_1(A) \geq \cdots \geq \sigma_q(A) \geq 0$ and $\sigma_1(B) \geq \cdots \geq \sigma_q(B) \geq 0$, $q \equiv \min\{m,n\}$, and let $\sigma_1(A + B) \geq \cdots \geq \sigma_q(A + B) \geq 0$ be the ordered singular values of $A + B$. Then

$$\sum_{i=1}^{k} \sigma_i(A + B) \leq \sum_{i=1}^{k} \sigma_i(A) + \sum_{i=1}^{k} \sigma_i(B), \quad k = 1, ..., q$$

Proof: Let $P_{n,m;k}$ denote the set of rank k partial isometries in $M_{n,m}$. Use

(3.4.1) and observe that

$$\sum_{i=1}^{k} \sigma_i(A + B)$$

$$= \max \{|\operatorname{tr}(A + B)C| : C \in P_{n,m;k}\}$$

$$= \max \{|\operatorname{tr}(AC + BC)| : C \in P_{n,m;k}\}$$

$$\leq \max \{|\operatorname{tr}(AC)| + |\operatorname{tr}(BC)| : C \in P_{n,m;k}\}$$

$$\leq \max \{|\operatorname{tr}(AC)| : C \in P_{n,m;k}\} + \max \{|\operatorname{tr}(BC)| : C \in P_{n,m;k}\}$$

$$= \sum_{i=1}^{k} \sigma_i(A) + \sum_{i=1}^{k} \sigma_i(B) \qquad\qquad \square$$

Except for the largest singular value, individual singular values need not obey the triangle inequality. For example, with $A = \begin{bmatrix} 1 & 0 \\ 0 & 0 \end{bmatrix}$ and $B = \begin{bmatrix} 0 & 0 \\ 0 & 1 \end{bmatrix}$, it is clear that $\sigma_2(A + B) = \sigma_2(I) = 1 \nleq \sigma_2(A) + \sigma_2(B) = 0$. Nevertheless, the previous result says that the sum of the k largest singular values does obey the triangle inequality. Actually, somewhat more is true.

3.4.4   Corollary. For $A \in M_{m,n}$, let $q \equiv \min \{m,n\}$, and let $N_k(A) \equiv \sigma_1(A) + \cdots + \sigma_k(A)$ denote the sum of the k largest singular values of A. Then

(a) $N_k(\cdot)$ is a norm on $M_{m,n}$ for $k = 1,...,q$.

(b)   When $m = n$, $N_k(\cdot)$ is a matrix norm on M_n for $k = 1,...,n$.

Proof: To prove (a), we must show that $N_k(\cdot)$ is a positive homogeneous function on $M_{m,n}$ that satisfies the triangle inequality (see (5.1.1) in [HJ]). It is clear that $N_k(A) \geq 0$, and since $N_k(A) \geq \sigma_1(A) = \||A\||_2$ is the spectral norm of A it is also clear that $N_k(A) = 0$ if and only if $A = 0$. If $c \in \mathbb{C}$ is a given scalar, then $(cA)^*(cA) = |c|^2 A^* A$, so $\sigma_i(cA) = |c| \sigma_i(A)$ for $i = 1,...,q$. Since (3.4.3) says that $N_k(A + B) \leq N_k(A) + N_k(B)$, we conclude that $N_k(\cdot)$ is a norm on $M_{m,n}$. Now let $m = n$. To show that $N_k(\cdot)$ is a matrix norm on M_n, we must show that $N_k(AB) \leq N_k(A)N_k(B)$ for all $A, B \in M_n$. This follows immediately from (3.3.14a) since

$$N_k(AB) = \sum_{i=1}^{k} \sigma_i(AB) \leq \sum_{i=1}^{k} \sigma_i(A)\sigma_i(B) \leq \sum_{i=1}^{k} \sigma_i(A) \sum_{j=1}^{k} \sigma_j(B)$$

$$= N_k(A)N_k(B) \qquad\qquad \square$$

The function $N_k(A) \equiv \sigma_1(A) + \cdots + \sigma_k(A)$ is often called the *Ky Fan k-norm*. For a different approach to the preceding result, via VonNeumann's theory of unitarily invariant norms, see (7.4.43) and (7.4.54) in [HJ].

The following inequality for singular value sums is very useful in matrix approximation problems; see Problem 18 in Section (3.5).

3.4.5   Theorem. Let $A, B \in M_{m,n}$ be given, and suppose A, B, and $A - B$ have decreasingly ordered singular values $\sigma_1(A) \geq \cdots \geq \sigma_q(A)$, $\sigma_1(B) \geq \cdots \geq \sigma_q(B)$, and $\sigma_1(A-B) \geq \cdots \geq \sigma_q(A-B)$, where $q = \min\{m,n\}$. Define $s_i(A,B) \equiv |\sigma_i(A) - \sigma_i(B)|$, $i = 1,...,q$, and let $s_{[1]}(A,B) \geq \cdots \geq s_{[q]}(A,B)$ denote a decreasingly ordered rearrangement of the values of $s_i(A,B)$. Then

$$\sum_{i=1}^{k} s_{[i]}(A,B) \leq \sum_{i=1}^{k} \sigma_i(A-B) \quad \text{for } k = 1,...,q \qquad (3.4.6)$$

That is, if we denote the vectors of decreasingly ordered singular values of A, B, and $A-B$ by $\sigma(A)$, $\sigma(B)$, and $\sigma(A-B) \in \mathbb{R}^q$, then the entries of $|\sigma(A) - \sigma(B)|$ are weakly majorized by the entries of $\sigma(A-B)$.

Proof: Consider the $(m+n)$-by-$(m+n)$ Hermitian block matrices

$$\tilde{A} \equiv \begin{bmatrix} 0 & A \\ A^* & 0 \end{bmatrix}, \quad \tilde{B} \equiv \begin{bmatrix} 0 & B \\ B^* & 0 \end{bmatrix}$$

and use Jordan's observation (3.0.4) that the eigenvalues of $\tilde{A}$ are $\pm\sigma_1(A),...,$ $\pm\sigma_q(A)$, together with $m+n-2q$ zeroes, and similarly for $\tilde{B}$ (see Theorem (7.3.7) in [HJ]). Denote the vectors of algebraically decreasingly ordered eigenvalues of $\tilde{A}$, $\tilde{B}$, and $\tilde{A} - \tilde{B}$ by $\lambda(\tilde{A})$, $\lambda(\tilde{B})$, and $\lambda(\tilde{A}-\tilde{B}) \in \mathbb{R}^{m+n}$. Notice that the entries of $\lambda(\tilde{A}) - \lambda(\tilde{B})$ are $\pm(\sigma_1(A) - \sigma_1(B)),...,$ $\pm(\sigma_q(A) - \sigma_q(B))$

together with $m + n - 2q$ additional zero entries. Thus, the q algebraically largest entries of $\lambda(\tilde{A}) - \lambda(\tilde{B})$ comprise the set $\{|\sigma_1(A) - \sigma_1(B)|,...,$ $|\sigma_q(A) - \sigma_q(B)|\}$. The q algebraically largest entries of $\lambda(\tilde{A} - \tilde{B})$ comprise the set $\{\sigma_1(\tilde{A} - \tilde{B}),..., \sigma_q(\tilde{A} - \tilde{B})\}$. Since (4.3.27) in [HJ] guarantees that there is a (strong) majorization relationship between all the entries of $\lambda(\tilde{A} - \tilde{B})$ and all the entries of $\lambda(\tilde{A}) - \lambda(\tilde{B})$ (this also follows from Corollary (3.4.3); see (3.4.11b) in Problem 8), there is a weak majorization between the sets of q largest entries of these two vectors: The q algebraically largest entries of $\lambda(\tilde{A}) - \lambda(\tilde{B})$ are weakly majorized by the q algebraically largest entries of $\lambda(\tilde{A} - \tilde{B})$, which is exactly what the inequalities (3.4.6) assert.   []

Problems

1.   Let $A, B \in M_n$ be rank k and rank r Hermitian projections ($A^* = A$ and $A^2 = A$), respectively. Show that AB is a contraction with rank at most min $\{k,r\}$ that need not be Hermitian or a projection (an idempotent).

2.   Consider the function $L_k(A) \equiv |\lambda_1(A)| + \cdots + |\lambda_k(A)|$ on M_n, $k = 1,..., n$; the eigenvalues are arranged in decreasing order $|\lambda_1| \geq \cdots \geq |\lambda_n|$. Is $L_k(\cdot)$ a norm on M_n? Why?

3.   Define the *Ky Fan p-k norm* on $M_{m,n}$ by $N_{k;p}(A) \equiv [\sigma_1(A)^p + \cdots + \sigma_k(A)^p]^{1/p}$, $p \geq 1$, $k = 1,..., \min \{m,n\}$.
   (a) Use Corollary (3.4.3) and Lemma (3.3.8) to show that

$$f(\sigma_1(A + B)) + \cdots + f(\sigma_k(A + B))$$
$$\leq f(\sigma_1(A) + \sigma_1(B)) + \cdots + f(\sigma_k(A) + \sigma_k(B))$$

   for $k = 1,..., \min \{m,n\}$ for every real-valued increasing convex function $f(\cdot)$ on $[0,\infty)$.
   (b) Take $f(t) \equiv t^p$ and use Minkowski's inequality to show that $N_{k;p}(\cdot)$ is a norm on $M_{m,n}$ for all $p \geq 1$ and all $k = 1,..., q = \min \{m,n\}$. When $k = q$, $N_{q;p}(\cdot)$ is often called the *Schatten p-norm*.
   (c) When $m = n$, use Corollary (3.3.14(c)) to show that all the Ky Fan p-k norms are matrix norms on M_n.

4.   Let $A \in M_n$ be given, let $e \equiv [1, ..., 1]^T \in \mathbb{R}^n$, and let $r(X)$ denote the numerical radius norm on M_n.
   (a) Show that the Ky Fan n-norm, often called the *trace norm*, has the

variational characterization

$$N_n(A) = \max \{ |\operatorname{tr} AU| : \ U \in M_n \text{ is unitary} \}$$

$$= \max \{ |e^T (A \circ U)e| : \ U \in M_n \text{ is unitary} \} \qquad (3.4.7)$$

(b) Explain why the point $\operatorname{tr}(AU)$ is in the field of values of nAU and use Problem 23(b) in Section (1.5) to show that $N_n(A) \le 4nr(A)$.

(c) Explain why the point $e^T (A \circ U)e$ is in the field of values of $nA \circ U$ and use Corollary (1.7.24) to show that $N_n(A) \le nr(A)$. Show that this bound is sharp for every $n = 1, 2, \ldots$

5.   Let $A \in M_n$ be given. Use Theorem (3.4.1) to show that

$$\left| \sum_{i=1}^{k} x_i^* UAx_i \right| \le \sigma_1(A) + \cdots + \sigma_k(A), \quad k = 1, \ldots, n \qquad (3.4.8)$$

for any unitary $U \in M_n$ and any orthonormal set $\{x_1, \ldots, x_k\} \subset \mathbb{C}^n$. For each k, show that equality is possible for some choice of U and $\{x_i\}$.

6.   Let $A \in M_n$ be given, let $n_1, \ldots, n_m$ and $\nu_1, \ldots, \nu_m$ be given positive integers with $n_1 + \cdots + n_m = n = \nu_1 + \cdots + \nu_m$, and let $I_{n_i} \in M_{n_i}$ and $I_{\nu_i} \in M_{\nu_i}$ be identity matrices for $i = 1, \ldots, m$. Consider the simple families of mutually orthogonal Hermitian projections

$$\Lambda_1 = I_{n_1} \oplus 0_{n-n_1}, \ \Lambda_2 = 0_{n_1} \oplus I_{n_2} \oplus 0_{n-n_1-n_2}, \ldots, \ \Lambda_m = 0_{n-n_m} \oplus I_{n_m}$$

and

$$D_1 = I_{\nu_1} \oplus 0_{n-\nu_1}, \ D_2 = 0_{\nu_1} \oplus I_{\nu_2} \oplus 0_{n-\nu_1-\nu_2}, \ldots, \ D_m = 0_{n-\nu_m} \oplus I_{\nu_m}$$

and let $\hat{A} \equiv \Lambda_1 A D_1 + \cdots + \Lambda_m A D_m$.

(a) Notice that $\hat{A} = A_1 \oplus \cdots \oplus A_m$, where each $A_i \in M_{n_i, \nu_i}$ is the submatrix of A corresponding to the nonzero rows of Λ_i and nonzero columns of D_i.

(b) Show that the set of singular values of $\hat{A}$ is the union of the sets of singular values of $A_1, \ldots, A_m$, including multiplicities.

(c) If a given singular value $\sigma_i(\hat{A})$ is a singular value of A_k, show that corresponding left and right singular vectors $x^{(i)}$ and $y^{(i)}$ of $\hat{A}$ (that is,

$A x^{(i)} = \sigma_k(A) y^{(i)}$ and $\|x^{(i)}\|_2 = \|y^{(i)}\|_2 = 1$) can be chosen so that all entries of $x^{(i)}$ and $y^{(i)}$ not corresponding to the diagonal 1 entries in Λ_i and D_i, respectively, are zero.

(d) If left and right singular vectors $x^{(1)}, \ldots, x^{(k)}$ and $y^{(1)}, \ldots, y^{(k)}$ of $\hat{A}$ corresponding to the singular values $\sigma_1(\hat{A}) \geq \cdots \geq \sigma_k(\hat{A})$ are chosen as in (c), use Problem 5 to show that

$$\sum_{i=1}^{k} \sigma_i(\hat{A}) = \sum_{i=1}^{k} x_i^* A y_i \leq \sum_{i=1}^{k} \sigma_i(A) \text{ for } k = 1, \ldots, n \qquad (3.4.9)$$

(e) Conclude that the singular values of $\hat{A}$ are weakly majorized by those of A.

7.   Let $P_1, \ldots, P_m \in M_n$ be m given mutually orthogonal Hermitian projections, that is, $P_i = P_i^*$, $P_i^2 = P_i$ for $i = 1, \ldots, m$, $P_i P_j = 0$ for all $i \neq j$, and $\Sigma_i P_i = I$. Let $Q_1, \ldots, Q_m \in M_n$ be m given mutually orthogonal Hermitian projections. The linear operator $A \to \hat{A} \equiv P_1 A Q_1 + \cdots + P_m A Q_m$ on M_n is sometimes called a *pinching* or *diagonal cell operator* because of the form of the special case considered in Problem 6.

(a) Show that

$$\sigma_1(\hat{A}) + \cdots + \sigma_k(\hat{A}) \leq \sigma_1(A) + \cdots + \sigma_k(A) \text{ for } k = 1, \ldots, n \quad (3.4.10)$$

for every $A \in M_n$, so the singular values of $\hat{A}$ are always weakly majorized by those of A; that is, every diagonal cell operator on M_n is norm decreasing with respect to the fundamental Ky Fan k-norms.

(b) As in Problem 3, conclude that $f(\sigma_1(\hat{A})) + \cdots + f(\sigma_k(\hat{A})) \leq f(\sigma_1(A)) + \cdots + f(\sigma_k(A))$ for $k = 1, \ldots, n$ for every real-valued $f(\cdot)$ on $[0, \infty)$ that is increasing and convex.

8.   Let $A, B \in M_n$ be given Hermitian matrices, and let the eigenvalues $\{\lambda_i(A)\}$, $\{\lambda_i(B)\}$, and $\{\lambda_i(A + B)\}$ be arranged in algebraically decreasing order $\lambda_1 \geq \cdots \geq \lambda_n$. Use Corollary (3.4.3) to deduce the strong majorization inequalities

$$\sum_{i=1}^{k} \lambda_i(A + B) \leq \sum_{i=1}^{k} [\lambda_i(A) + \lambda_i(B)] \text{ for } k = 1, \ldots, n$$

with equality for $k = n$ (3.4.11a)

Conclude that

$$\sum_{i=1}^{k} [\lambda_i(A) - \lambda_i(B)] \leq \sum_{i=1}^{k} \lambda_i(A - B) \quad \text{for } k = 1, ..., n \qquad (3.4.11b)$$

9. Corollary (3.4.3), the triangle inequality for the fundamental Ky Fan k-norms, follows easily from the variational characterization (3.4.1), in whose proof we used the A. Horn singular value majorization inequalities (3.3.14a) for the ordinary matrix product to derive (3.4.2). The purpose of this problem is to give a direct proof of Theorem (3.4.1) that does not rely on the inequalities (3.3.14a). The following argument uses nothing about singular values except the singular value decomposition (3.1.1). Let $x = [x_i]$, $y = [y_i] \in \mathbb{R}^n$ be given real nonnegative vectors with $x_1 \geq \cdots \geq x_n \geq 0$ and $y_1 \geq \cdots \geq y_n \geq 0$.

(a) Show that $x_1 y_2 + x_2 y_1 = x_1 y_1 + x_2 y_2 - (x_1 - x_2)(y_1 - y_2) \leq x_1 y_1 + x_2 y_2$.

(b) For any permutation matrix $P \in M_n(\mathbb{R})$, show that $x^T P y \leq x^T y$.

(c) For any doubly stochastic $S \in M_n(\mathbb{R})$, use Birkhoff's theorem (8.7.1) in [HJ] to show that $x^T S y \leq x^T y$.

(d) For any doubly substochastic $Q \in M_n(\mathbb{R})$, use Theorem (3.2.6) to show that $x^T Q y \leq x^T y$.

(e) Using the notation of Theorem (3.4.1) and a singular value decomposition $A = V \Sigma W^*$, show that

$$\max \{|\text{tr}(X^* A Y)| : X \in M_{m,k}, Y \in M_{n,k}, X^* X = I, Y^* Y = I\}$$

$$= \max \{|\text{tr}(X^* \Sigma Y)| : X \in M_{m,k}, Y \in M_{n,k}, X^* X = I, Y^* Y = I\}$$

If $X = [x_{ij}] \in M_{m,k}$ and $Y = [y_{ij}] \in M_{n,k}$ have orthonormal columns, show that $\text{tr}(X^* \Sigma Y) = \sigma(A)^T Z \eta$, where $\eta = [\eta_i] \in \mathbb{R}^q$ has $\eta_1 = \cdots = \eta_k = 1$, $\eta_{k+1} = \cdots = \eta_q = 0$, $\sigma(A) = [\sigma_1(A), ..., \sigma_q(A)]^T \in \mathbb{R}^q$, and $Z \equiv [\bar{x}_{ij} y_{ij}]_{i,j=1}^q$.

(f) Now use Problem 2 in Section (3.2) to conclude that

$$\max \{|\text{tr}(X^* A Y)| : X \in M_{m,k}, Y \in M_{n,k}, X^* X = I, Y^* Y = I\}$$

$$\leq \sigma(A)^T \eta = \sigma_1(A) + \cdots + \sigma_k(A) \qquad (3.4.12)$$

(g) Exhibit X and Y for which equality is achieved in (3.4.12).

Further Reading. The first proof of Theorem (3.4.1) is in the 1951 paper by Ky Fan cited at the end of Section (3.0).

3.5 Singular values and unitarily invariant norms

A norm $\|\cdot\|$ on $M_{m,n}$ is *unitarily invariant* if $\|UAV\| = \|A\|$ for all unitary $U \in M_m$ and $V \in M_n$, and for all $A \in M_{m,n}$. It is clear from the singular value decomposition that if $\|\cdot\|$ is a unitarily invariant norm, then $\|A\| = \|V\Sigma W^*\| = \|\Sigma(A)\|$ is a function only of the singular values of A. The nature and properties of this function are the primary focus of this section.

If $\|\cdot\|$ is a given unitarily invariant norm on $M_{m,n}$, there is a natural sense in which it induces a unitarily invariant norm on $M_{r,s}$ for any r, s with $1 \leq r \leq m$ and $1 \leq s \leq n$: For $A \in M_{r,s}$, define $\|A\| \equiv \|\mathcal{A}\|$, where

$$\mathcal{A} \equiv \begin{bmatrix} A & 0 \\ 0 & 0 \end{bmatrix} \in M_{m,n} \qquad (3.5.0)$$

has been augmented by zero blocks to fill out its size to m-by-n. One checks that the norm on $M_{r,s}$ defined in this way is both well defined and unitarily invariant.

3.5.1   Lemma. Let $\|\cdot\|$ be a given norm on $M_{m,n}$ and let

$$\|A\|^D \equiv \max\{|\operatorname{tr} AC^*|: C \in M_{m,n}, \|C\| = 1\} \qquad (3.5.2)$$

denote its dual norm with respect to the Frobenius inner product. Then $\|\cdot\|$ is unitarily invariant if and only if $\|\cdot\|^D$ is unitarily invariant.

Proof: Suppose $\|\cdot\|$ is unitarily invariant, and let $U \in M_m$, $V \in M_n$ be unitary. Then

$$
\begin{aligned}
\|UAV\|^D &= \max\{|\operatorname{tr}(UAV)C^*|: C \in M_{m,n}, \|C\| = 1\} \\
&= \max\{|\operatorname{tr}(A[U^*CV^*]^*)|: C \in M_{m,n}, \|C\| = 1\} \\
&= \max\{|\operatorname{tr}(AE^*)|: E \in M_{m,n}, \|UEV\| = 1\} \\
&= \max\{|\operatorname{tr}(AE^*)|: E \in M_{m,n}, \|E\| = 1\}
\end{aligned}
$$

$$= \|A\|^D$$

The converse follows from the preceding argument and the *duality theorem* for norms: $(\|A\|^D)^D = \|A\|$ (see Theorem (5.5.14) in [HJ]). ☐

Notice that the duality theorem gives the representation

$$\|A\| = \max\{|\operatorname{tr} AC^*|: C \in M_{m,n}, \|C\|^D = 1\} \qquad (3.5.3)$$

3.5.4   Definition. We write $\mathbb{R}^n_{+\downarrow} \equiv \{x = [x_i] \in \mathbb{R}^n: x_1 \geq \cdots \geq x_n \geq 0\}$. For $X \in M_{m,n}$ and $\alpha = [\alpha_i] \in \mathbb{R}^q_{+\downarrow}$, we write

$$\|X\|_\alpha \equiv \alpha_1 \sigma_1(X) + \cdots + \alpha_q \sigma_q(X)$$

where $q \equiv \min\{m,n\}$.

Because of the ordering convention we have adopted for the vector of singular values $\sigma(X)$, notice that we always have $\sigma(X) \in \mathbb{R}^q_{+\downarrow}$, where $X \in M_{m,n}$ and $q = \min\{m,n\}$.

3.5.5   Theorem. Let m, n be given positive integers and let $q \equiv \min\{m,n\}$.

(a)   For each given nonzero $\alpha \in \mathbb{R}^q_{+\downarrow}$, $\|A\|_\alpha \equiv \alpha_1 \sigma_1(A) + \cdots + \alpha_q \sigma_q(A)$ is a unitarily invariant norm on $M_{m,n}$.

(b)   For each given unitarily invariant norm $\|\cdot\|$ on $M_{m,n}$ there is a compact set $\mathcal{M}(\|\cdot\|) \subset \mathbb{R}^q_{+\downarrow}$ such that

$$\|A\| = \max\{\|A\|_\alpha: \alpha \in \mathcal{M}(\|\cdot\|)\} \qquad (3.5.6)$$

for all $A \in M_{m,n}$. One may take the set $\mathcal{M}(\|\cdot\|)$ to be

$$\mathcal{M}(\|\cdot\|) = \{\sigma(X): \|X\|^D = 1\} \qquad (3.5.7)$$

Proof: Assertion (a) is a simple consequence of summation by parts:

$$\|A\|_\alpha = \sum_{i=1}^q \alpha_i \sigma_i(A) = \sum_{i=1}^{q-1}(\alpha_i - \alpha_{i+1})N_i(A) + \alpha_q N_q(A) \qquad (3.5.8)$$

where $N_k(A) = \sigma_1(A) + \cdots + \sigma_k(A)$ is the Ky Fan k-norm. Since the entries of α are monotone decreasing, (3.5.8) represents $\|A\|_\alpha$ as a nonnegative linear combination of the (unitarily invariant) Ky Fan k-norms, in which at least one coefficient is positive if $\alpha \neq 0$. To prove (b), let $U_1, U_2 \in M_n$ and $V_1, V_2 \in M_m$ be unitary. Then for any $C \in M_{m,n}$ with $\|C\|^D = 1$, unitary invariance and (3.5.3) give

$$\|A\| = \|V_2^* V_1 A U_1 U_2^*\|$$

$$\geq |\mathrm{tr}[(V_2^* V_1 A U_1 U_2^*)C^*]| = |\mathrm{tr}[(V_1 A U_1)(V_2 C U_2)^*]|$$

Now use the singular value decompositions of A and C to select the unitary matrices U_1, V_1, U_2, V_2 so that $V_1 A U_1 = \Sigma(A)$ and $V_2 C U_2 = \Sigma(C)$, from which it follows that

$$\|A\|_{\sigma(C)} = \sum_{i=1}^q \sigma_i(A)\sigma_i(C) \leq \|A\| \quad \text{whenever } C \in M_{m,n} \text{ and } \|C\|^D = 1$$

Using (3.5.3) again, select $C_0 \in M_{m,n}$ such that $\|C_0\|^D = 1$ and

$$\|A\| = \max\{|\mathrm{tr}\, AC^*|: C \in M_{m,n}, \|C\|^D = 1\} = |\mathrm{tr}\, AC_0^*|$$

$$= \left|\sum_{i=1}^m \lambda_i(AC_0^*)\right| \leq \sum_{i=1}^m \sigma_i(AC_0^*) \leq \sum_{i=1}^m \sigma_i(A)\sigma_i(C_0)$$

$$= \|A\|_{\sigma(C_0)} \leq \|A\|$$

Thus, $\|A\| = \|A\|_{\sigma(C_0)}$ and we have shown that

$$\|A\| = \max\{\|A\|_{\sigma(C)}: C \in M_{m,n}, \|C\|^D = 1\}$$

as asserted in (b).   □

There are many useful consequences of the characterization (3.5.6) of unitarily invariant norms.

3.5.9   Corollary. Let $A, B \in M_{m,n}$ be given and let $q = \min\{m,n\}$. The following are equivalent:

(a) $\|A\| \leq \|B\|$ for every unitarily invariant norm $\|\cdot\|$ on $M_{m,n}$.

(b) $N_k(A) \leq N_k(B)$ for $k = 1,..., q$, where $N_k(X) \equiv \sigma_1(X) + \cdots + \sigma_k(X)$ denotes the Ky Fan k-norm.

(c) $\|A\|_\alpha \leq \|B\|_\alpha$ for all $\alpha \in \mathbb{R}^q_{+\downarrow}$.

Proof: Since each Ky Fan k-norm $N_k(\cdot)$ is unitarily invariant, (a) implies (b). The identity (3.5.8) shows that (b) implies (c), and the representation (3.5.6) shows that (c) implies (a).   ⬜

Since the conditions (b) of the preceding corollary are just the assertion that $\sigma(A)$ is weakly majorized by $\sigma(B)$, we see that every unitarily invariant norm is monotone with respect to the partial order on $M_{m,n}$ induced by weak majorization of the vectors of singular values.

3.5.10   Corollary. Let $\|\cdot\|$ be a given unitarily invariant norm, and let $E_{11} \in M_{m,n}$ have the entry 1 in position 1,1 and zeros elsewhere. Then

(a) $\|AB^*\| \leq \sigma_1(A)\|B\|$ for all $A, B \in M_{m,n}$, and

(b) $\|A\| \geq \sigma_1(A)\|E_{11}\|$ for all $A \in M_{m,n}$.

Proof: For each $k = 1,..., q \equiv \min\{m,n\}$ we have

$$N_k(AB^*) = \sum_{i=1}^k \sigma_i(AB^*) \leq \sum_{i=1}^k \sigma_1(A)\sigma_i(B^*)$$

$$= \sigma_1(A)\sum_{i=1}^k \sigma_i(B) = \sigma_1(A)N_k(B) = N_k(\sigma_1(A)B)$$

and

$$N_k(\sigma_1(A)E_{11}) = \sigma_1(A) \leq \sum_{i=1}^k \sigma_i(A) = N_k(A)$$

Then (a) and (b) follow from the equivalence of (3.5.9a,b).   ⬜

The following generalization of Corollary (3.5.9) is a final easy consequence of the characterization (3.5.6) of unitarily invariant norms.

3.5.11   Corollary. Let $f(t_1,..., t_k)$: $\mathbb{R}_+^k \to \mathbb{R}_+$ be a given nonnegative-valued function of k nonnegative real variables that is increasing in each variable separately, that is,

$$f(t_1,..., t_{i-1}, t_i, t_{i+1},..., t_k) \leq f(t_1,..., t_{i-1}, t_i + \epsilon, t_{i+1},..., t_k)$$

for all $\epsilon, t_1,..., t_k \geq 0$ and all $i = 1,..., k$. Let $A, B_1,..., B_k \in M_{m,n}$ be given, and let $q = \min\{m,n\}$. Then $\|A\| \leq f(\|B_1\|,..., \|B_k\|)$ for all unitarily invariant norms $\|\cdot\|$ if and only if $\|A\|_\alpha \leq f(\|B_1\|_\alpha,..., \|B_m\|_\alpha)$ for all $\alpha \in \mathbb{R}_{+\downarrow}^q$.

Proof: The forward implication is immediate. For the reverse implication, let $\alpha_0 \in \mathbb{R}_{+\downarrow}^q$ be given and use the characterization (3.5.6) and the monotonicity of f in each variable to write

$$f(\|B_1\|,..., \|B_k\|) = f(\max \|B_1\|_\alpha,..., \max \|B_k\|_\alpha)$$

$$\geq f(\|B_1\|_{\alpha_0},..., \|B_k\|_{\alpha_0}) \geq \|A\|_{\alpha_0}$$

Since $\alpha_0 \in \mathbb{R}_{+\downarrow}^q$ was arbitrary, we conclude that

$$\|A\| = \max \|A\|_\alpha \leq f(\|B_1\|,..., \|B_k\|)$$

In every case, the indicated maximum is taken over all α in the compact set $\mathcal{N}(\|\cdot\|)$ for which (3.5.6) holds, for example, the set described by (3.5.7).   ☐

We now wish to use Corollary (3.5.11) to derive a general inequality for unitarily invariant norms. To do so, we need the following characterization of positive semidefinite 2-by-2 block matrices.

3.5.12   Lemma. Let $L \in M_m$, $M \in M_n$, and $X \in M_{m,n}$ be given. Then

$$\begin{bmatrix} L & X \\ X^* & M \end{bmatrix} \in M_{m+n} \tag{3.5.13}$$

is positive semidefinite if and only if L and M are positive semidefinite and there is a contraction $C \in M_{m,n}$ (that is, $\sigma_1(C) \le 1$) such that $X = L^{\frac{1}{2}} C M^{\frac{1}{2}}$.

Proof: Suppose L and M are positive definite. By Theorem (7.7.7) in [HJ], the block matrix (3.5.13) is positive semidefinite if and only if

$$1 \ge \rho(X^* L^{-1} X M^{-1}) = \rho(M^{-\frac{1}{2}} X^* L^{-1} X M^{-\frac{1}{2}})$$

$$= \rho[(L^{-\frac{1}{2}} X M^{-\frac{1}{2}})^*(L^{-\frac{1}{2}} X M^{-\frac{1}{2}})] = \sigma_1(L^{-\frac{1}{2}} X M^{-\frac{1}{2}})^2$$

Setting $C \equiv L^{-\frac{1}{2}} X M^{-\frac{1}{2}}$, we have $X = L^{\frac{1}{2}} C M^{\frac{1}{2}}$, as desired. The general case now follows from a limiting argument. $\square$

Now suppose that a given block matrix of the form (3.5.13) is positive semidefinite with $X \in M_{m,n}$ and write $X = L^{\frac{1}{2}} C M^{\frac{1}{2}}$. Then

$$\prod_{i=1}^{k} \sigma_i(X) = \prod_{i=1}^{k} \sigma_i(L^{\frac{1}{2}} C M^{\frac{1}{2}}) \le \prod_{i=1}^{k} \sigma_i(L^{\frac{1}{2}}) \sigma_i(C) \sigma_i(M^{\frac{1}{2}})$$

$$\le \prod_{i=1}^{k} \sigma_i(L)^{\frac{1}{2}} \sigma_i(M)^{\frac{1}{2}} \text{ for } k = 1, \ldots, q = \min\{m, n\}$$

and hence

$$\prod_{i=1}^{k} \sigma_i(X)^p \le \prod_{i=1}^{k} \sigma_i(L)^{p/2} \sigma_i(M)^{p/2} \text{ for } k = 1, \ldots, q$$

for any $p > 0$. Moreover, for any $\alpha = [\alpha_i] \in \mathbb{R}_{+\downarrow}^q$ we also have

$$\prod_{i=1}^{k} \alpha_i \sigma_i(X)^p \le \prod_{i=1}^{k} \alpha_i \sigma_i(L)^{p/2} \sigma_i(M)^{p/2} \text{ for } k = 1, \ldots, q \qquad (3.5.14)$$

Corollary (3.3.10) now gives the inequalities

$$\sum_{i=1}^{q} \alpha_i \sigma_i(X)^p \le \sum_{i=1}^{q} (\alpha_i^{\frac{1}{2}} \sigma_i(L)^{p/2})(\alpha_i^{\frac{1}{2}} \sigma_i(M)^{p/2})$$

$$\leq \left[\sum_{i=1}^{q} \alpha_i \sigma_i(L)^p \right]^{\frac{1}{2}} \left[\sum_{i=1}^{q} \alpha_i \sigma_i(M)^p \right]^{\frac{1}{2}}$$

These inequalities say that

$$\| (X^* X)^{p/2} \|_\alpha \leq [\, \| L^p \|_\alpha \| M^p \|_\alpha \,]^{\frac{1}{2}}$$

for all $p > 0$ and all $\alpha \in \mathbb{R}^q_{+\downarrow}$. If we now apply Corollary (3.5.11) with $f(t_1, t_2) = (t_1 t_2)^{\frac{1}{2}}$, we have proved the following:

3.5.15   Theorem. Let $L \in M_m$, $M \in M_n$, and $X \in M_{m,n}$ be such that the block matrix

$$\begin{bmatrix} L & X \\ X^* & M \end{bmatrix} \in M_{m+n}$$

is positive semidefinite. Then

$$\| (X^* X)^{p/2} \|^2 \leq \| L^p \| \, \| M^p \| \tag{3.5.16}$$

for every $p > 0$ and every unitarily invariant norm $\| \cdot \|$.

Some special cases of this inequality are considered in Problems 7-10. In addition to the quasilinearization (3.5.6), there is another useful characterization of unitarily invariant norms, originally due to Von Neumann.

3.5.17   Definition. A function $g: \mathbb{R}^q \to \mathbb{R}_+$ is said to be a *symmetric gauge function* if

(a) $g(x)$ is a norm on $\mathbb{R}^q$;
(b) $g(x) = g(|x|)$ for all $x \in \mathbb{R}^q$, where $|x| = [|x_i|]$; and
(c) $g(x) = g(Px)$ for all $x \in \mathbb{R}^q$ and every permutation matrix $P \in M_q(\mathbb{R})$.

Thus, a symmetric gauge function on $\mathbb{R}^q$ is an absolute permutation-invariant norm. Since a norm on $\mathbb{R}^q$ is absolute if and only if it is monotone (Theorem (5.5.10) in [HJ]), it follows that a symmetric gauge function is also a monotone norm.

3.5.18   Theorem. If $\| \cdot \|$ is a given unitarily invariant norm on $M_{m,n}$, then there is a symmetric gauge function $g(\cdot)$ on $\mathbb{R}^q$ such that $\| A \| = g(\sigma(A))$ for all $A \in M_{m,n}$. Conversely, if $g(\cdot)$ is a given symmetric gauge function on $\mathbb{R}^q$, then $\| A \| \equiv g(\sigma(A))$ is a unitarily invariant norm on $M_{m,n}$.

Proof: If $\| \cdot \|$ is a given unitarily invariant norm on $M_{m,n}$, for $x \in \mathbb{R}^q$ define $g(x) \equiv \| X \|$, where $X = [x_{ij}] \in M_{m,n}$ has $x_{ii} = x_i$, $i = 1,...,q$, and all other entries are zero. That $g(\cdot)$ is a norm on $\mathbb{R}^q$ follows from the fact that $\| \cdot \|$ is a norm. Absoluteness and permutation-invariance of $g(\cdot)$ on $\mathbb{R}^q$ follow from unitary invariance of $\| \cdot \|$: Consider $X \rightarrow DX$ or $X \rightarrow PXQ$, where D is a diagonal unitary matrix and P, Q are permutation matrices.

Conversely, if $g(\cdot)$ is a given symmetric gauge function on $\mathbb{R}^q$, for $A \in M_{m,n}$ define $\| A \| \equiv g(\sigma(A))$. Positive definiteness and homogeneity of $\| \cdot \|$ follow from the fact that $g(\cdot)$ is a norm, while unitary invariance of $\| \cdot \|$ follows from unitary invariance of singular values. To show that $\| \cdot \|$ satisfies the triangle inequality, let $A, B \in M_{m,n}$ be given and use Corollary (3.4.3) to observe that the entries of the vector $\sigma(A + B)$ are weakly majorized by the entries of $\sigma(A) + \sigma(B)$. Then Corollary (3.2.11) guarantees that there is a doubly stochastic matrix $S \in M_q(\mathbb{R})$ such that $\sigma(A + B) \leq S[\sigma(A) + \sigma(B)]$, and $S = \mu_1 P_1 + \cdots + \mu_N P_N$ can be written as a convex combination of permutation matrices. Now use monotonicity, the triangle inequality, and permutation invariance of $g(\cdot)$ to conclude that

$$\| A + B \| \equiv g(\sigma(A + B)) \leq g(S[\sigma(A) + \sigma(B)])$$

$$\leq g(S\sigma(A)) + g(S\sigma(B))$$

$$\leq \sum_{i=1}^{N} \mu_i[g(P_i\sigma(A)) + g(P_i\sigma(B))]$$

$$= \sum_{i=1}^{N} \mu_i[g(\sigma(A)) + g(\sigma(B))] = g(\sigma(A)) + g(\sigma(B))$$

$$= \| A \| + \| B \| \qquad \qquad \square$$

Problems

1.   Let $\| \cdot \|$ be a given unitarily invariant norm on $M_{m,n}$. For given inte-

gers r, s with $1 \le r \le m$ and $1 \le s \le n$, and for any $A \in M_{r,s}$, define $\|A\| \equiv \|\hat{A}\|$, where $\hat{A} \in M_{m,n}$ is the block matrix (3.5.0). Explain why this function $\|\cdot\|$: $M_{r,s} \to \mathbb{R}$ is well defined, is a norm, and is unitarily invariant.

2.   Let $\|\cdot\|$ be a given unitarily invariant norm on $M_{m,n}$ and let $q = \min\{m,n\}$. Use Theorem (3.5.5) and the identity (3.5.8) to show that

$$\delta \, \sigma_1(A) \equiv \delta \, N_1(A) \le \|A\| \le \delta \, N_q(A) \equiv \delta \, [\sigma_1(A) + \cdots + \sigma_q(A)] \quad (3.5.19)$$

for all $A \in M_{m,n}$, where $\delta \equiv \max \{\sigma_1(C) : C \in M_{m,n} \text{ and } \|C\|^D = 1\}$ is a geometric factor that is independent of A.

3.   Let $\|\cdot\|$ be a given unitarily invariant norm on M_n. Recall that $\|\cdot\|$ is a *matrix norm* if $\|AB\| \le \|A\| \, \|B\|$ for all A, $B \in M_n$. Let $E_{11} \in M_n$ denote the matrix with 1,1 entry 1 and all other entries 0. Show that the following are equivalent:

     (a) $\|\cdot\|$ is a matrix norm on M_n.

     (b) $\|A^m\| \le \|A\|^m$ for all $A \in M_n$ and all $m = 1, 2, \ldots$.

     (c) $\|A^2\| \le \|A\|^2$ for all $A \in M_n$.

     (d) $\|E_{11}\| \ge 1$.

     (e) $\|A\| \ge \sigma_1(A)$ for all $A \in M_n$.

4.   Let $A \in M_{m,n}$ be given and let $\|\cdot\|$ be a given norm on $M_{m,n}$.
     (a) If $\|\cdot\|$ is unitarily invariant, show that $\|BAC\| \le \sigma_1(B)\sigma_1(C)\|A\|$ for all $B \in M_m$ and all $C \in M_n$. In particular, if B and C are contractions, then $\|BAC\| \le \|A\|$.
     (b) A norm $\|\cdot\|$ on $M_{m,n}$ is said to be *symmetric* if

$$\|BAC\| \le \sigma_1(B)\sigma_1(C)\|A\| \text{ for all } B \in M_m \text{ and all } C \in M_n$$

Show that $\|\cdot\|$ is symmetric if and only if it is unitarily invariant.

5.   Use Theorem (3.5.18) to show that the Schatten p-norms $[\sigma_1(A)^p + \cdots + \sigma_q(A)^p]^{1/p}$ and, more generally, the Ky Fan p-k norms $[\sigma_1(A)^p + \cdots + \sigma_k(A)^p]^{1/p}$ are unitarily invariant norms on $M_{m,n}$, $q = \min\{m,n\}$, $1 \le k \le q$, $p \ge 1$.

6.   Let $A \in M_{m,n}$ be given, and let $|A| \equiv (A^*A)^{\frac{1}{2}}$ denote the unique positive semidefinite square root of A^*A, which is the positive semidefinite

factor in the polar decomposition $A = U|A|$. Show that $\|\,A\,\| = \|\,|A|\,\|$ for every unitarily invariant norm $\|\cdot\|$ on $M_{m,n}$.

7. Let $A, B \in M_{m,n}$ be given, and let $\|\cdot\|$ be a given unitarily invariant norm. Apply (3.5.16) to $[A\ B]^*[A\ B]$ and show that

$$\|\,|A^*B\,|^p\,\|^2 \leq \|\,(A^*A)^p\,\|\ \|\,(B^*B)^p\,\| \quad \text{for all } p > 0 \qquad (3.5.20)$$

where $|X|$ is defined in Problem 6. In particular, deduce the inequality

$$\|\,|A^*B\,|^{\frac{1}{2}}\,\|^2 \leq \|\,A\,\|\ \|\,B\,\| \qquad (3.5.21)$$

and a Cauchy-Schwarz inequality for ordinary products and unitarily invariant norms

$$\|\,A^*B\,\|^2 \leq \|\,A^*A\,\|\ \|\,B^*B\,\| \qquad (3.5.22)$$

What does this say when $n = 1$? If a given 2-by-2 block matrix $A \equiv [A_{ij}]_{i,j=1}^2$ is positive semidefinite, with $A_{11} \in M_k$ and $A_{22} \in M_{n-k}$, show that its (unitarily invariant) *norm compression* $A_{\|\cdot\|} \equiv [\,\|A_{ij}\|\,]_{i,j=1}^2 \in M_2$ is positive semidefinite. Show by example that a unitarily invariant norm compression of a positive semidefinite 3-by-3 block matrix need not be positive semidefinite.

8. Let $A, B \in M_{m,n}$ be given and let $\|\cdot\|$ be a given unitarily invariant norm. Apply (3.5.16) to the Hadamard product of $[A^*\ I]^*[A^*\ I]$ and $[I\ B]^*[I\ B]$ and show that

$$\|\,|A \circ B\,|^p\,\|^2 \leq \|\,[(AA^*) \circ I]^p\,\|\ \|\,[(B^*B) \circ I]^p\,\| \quad \text{for all } p > 0 \quad (3.5.23)$$

where $|A|$ is defined as in Problem 6. In particular, deduce a Cauchy-Schwarz inequality for Hadamard products and unitarily invariant norms

$$\|\,A \circ B\,\|^2 \leq \|\,(AA^*) \circ I\,\|\ \|\,(B^*B) \circ I\,\| \leq \|\,AA^*\,\|\ \|\,B^*B\,\| \qquad (3.5.24)$$

What does this say when $n = 1$ and $\|\cdot\|$ is the spectral norm?

9. Let $A, B \in M_n$ be given, and suppose $A = X^*Y$ for some $X, Y \in M_{r,n}$. Consider the positive semidefinite matrices $[X\ Y]^*[X\ Y]$ and

$$\begin{bmatrix} \sigma_1(B)I & B \\ B^* & \sigma_1(B)I \end{bmatrix}$$

and their Hadamard product. Use Theorem (3.5.15) to show that

$$\| A \circ B \| \le \sigma_1(B)c_1(X)c_1(Y) \| I \| \tag{3.5.25}$$

for any unitarily invariant norm $\|\cdot\|$ on M_n, where $c_1(Z)$ denotes the maximum Euclidean column length of $Z \in M_{r,n}$. Take $\|\cdot\|$ to be the spectral norm and deduce Schur's inequality $\sigma_1(A \circ B) \le \sigma_1(A)\sigma_1(B)$. If $A = [a_{ij}]$ is positive semidefinite, take $X = Y = A^{\frac{1}{2}}$ and deduce another inequality of Schur: $\sigma_1(A \circ B) \le \sigma_1(B) \max_i a_{ii}$.

10. If $A \in M_n$ is normal, show that the 2-by-2 block matrix $\begin{bmatrix} |A| & A \\ A^* & |A| \end{bmatrix}$ is positive semidefinite, where $|A|$ is defined in Problem 6. Use Theorem (3.5.15) to show that

$$\| A \circ B \| \le \| \, |A| \circ |B| \, \| \tag{3.5.26}$$

for all normal $A, B \in M_n$ and every unitarily invariant norm $\|\cdot\|$ on M_n. The hypothesis of normality of both A and B is essential here; there are positive definite A and nonnormal B for which this inequality is false for some unitarily invariant norm.

11. Let $\|\cdot\|$ be a given norm on M_n, and define $\nu(A) \equiv \| A^*A \|^{\frac{1}{2}}$.
    (a) If the norm $\|\cdot\|$ satisfies the inequality

$$\|A^*B\|^2 \le \|A^*A\| \, \|B^*B\| \text{ for all } A, B \in M_n \tag{3.5.27}$$

show that $\nu(A)$ is a norm on M_n. In particular, conclude that $\nu(A)$ is a unitarily invariant norm on M_n whenever $\|\cdot\|$ is a unitarily invariant norm on M_n. Explain why the Schatten p-norms $[\sigma_1(A)^p + \cdots + \sigma_n(A)^p]^{1/p}$ are norms of this type for $p \ge 2$.
    (b) Show that the l_1 norm $\|A\|_1 = \Sigma_{i,j}|a_{ij}|$ does *not* satisfy (3.5.27), but that the l_∞ norm $\|A\|_\infty = \max_{i,j} |a_{ij}|$ *does* satisfy (3.5.27); neither of these norms is unitarily invariant.
    (c) Show that the set of norms on M_n that satisfy (3.5.27) is a convex set that is strictly larger than the set of all unitarily invariant norms,

but does not include all norms on M_n.

12. Let $\|\cdot\|$ be a given unitarily invariant norm on M_n. For all $A \in M_n$, show that

$$\|A\| = \min\{\|B^*B\|^{\frac{1}{2}} \|C^*C\|^{\frac{1}{2}}: \; B, C \in M_n \text{ and } A = BC\} \quad (3.5.28)$$

13. For $A \in M_{m,n}$, let $\hat{A}$ be any submatrix of A. Show that $\|\hat{A}\| \leq \|A\|$ for every unitarily invariant norm on $M_{m,n}$, where $\|\hat{A}\|$ is defined as in Problem 1. Give an example of a norm for which this inequality does not always hold.

14. If $A, B \in M_n$ are positive semidefinite and $A \succeq B \succeq 0$, show that $\|A\| \geq \|B\|$ for every unitarily invariant norm $\|\cdot\|$ on M_n.

15. A norm $\nu(\cdot)$ on M_n is said to be *unitary similarity invariant* if $\nu(A) = \nu(UAU^*)$ for all $A \in M_n$ and all unitary $U \in M_n$. Notice that every unitarily invariant norm is unitary similarity invariant.
(a) For any unitarily invariant norm $\|\cdot\|$ on M_n, show that $N(A) \equiv \|A\| + |\mathrm{tr}\, A|$ is a unitary similarity invariant norm on M_n that is not unitarily invariant.
(b) Show that the numerical radius $r(A)$ is a unitary similarity invariant norm on M_n that is not unitarily invariant.

16. Let $\|\cdot\|$ be a given unitary similarity invariant norm on M_n. Show that the function $\nu(A) \equiv \| \, |A| \, \|$ is a unitary similarity invariant function on M_n that is always a prenorm, and is a norm if and only if $\|X\| \leq \|Y\|$ whenever $X, Y \in M_n$ are positive semidefinite and $Y \succeq X \succeq 0$. Here, $|A|$ is defined as in Problem 6.

17. Show that the dual of the Ky Fan k-norm $N_k(A) = \sigma_1(A) + \cdots + \sigma_k(A)$ on $M_{m,n}$ is

$$N_k(A)^D = \max\{N_1(A), N_q(A)/k\} \text{ for } k = 1,\ldots, \; q = \min\{m,n\} \quad (3.5.29)$$

In particular, conclude that the spectral norm $\sigma_1(A) = \|\|A\|\|_2$ and the trace norm $N_q(A) = \sigma_1(A) + \cdots + \sigma_q(A)$ are dual norms.

18. Let $A = [a_{ij}]$, $B = [b_{ij}] \in M_{m,n}$ be given and let $q = \min\{m,n\}$. Define the diagonal matrix $\Sigma(A) = [\sigma_{ij}] \in M_{m,n}$ by $\sigma_{ii} = \sigma_i(A)$, all other $\sigma_{ij} = 0$, where $\sigma_1(A) \geq \cdots \geq \sigma_q(A)$ are the decreasingly ordered singular values of A. Let $\sigma(A) = [\sigma_i(A)]_{i=1}^q \in \mathbb{R}^q$ denote the vector of decreasingly ordered singular values of A; define $\Sigma(B)$ and $\sigma(B)$ similarly.

(a) Show that the singular values of $\Sigma(A) - \Sigma(B)$ are the entries of the vector $|\sigma(A) - \sigma(B)|$ (which are not necessarily decreasingly ordered).
(b) Use Corollary (3.5.9) and Theorem (3.4.5) to show that

$$\|A - B\| \geq \|\Sigma(A) - \Sigma(B)\| \geq |\,\|A\| - \|B\|\,| \qquad (3.5.30)$$

for every unitarily invariant norm $\|\cdot\|$ on $M_{m,n}$.
(c) If $B \in M_{m,n}$ is any matrix with rank $B = r \leq q$, show that

$$\|A - B\| \geq \|\mathrm{diag}(0,\ldots,0,\,\sigma_{r+1}(A),\ldots,\sigma_q(A))\| \qquad (3.5.31)$$

Explain why this has the lower bound $\sigma_q(A)\|E_{11}\|$ if $r < q$. Using the singular value decomposition $A = V\Sigma(A)W^*$, exhibit a matrix B_r for which this inequality is an equality.
(d) Conclude that for a given $A \in M_{m,n}$ and for each given $r = 1,\ldots,$ rank A there is some $B_r \in M_{m,n}$ that is a best rank r approximation to A, simultaneously with respect to every unitarily invariant norm $\|\cdot\|$ on $M_{m,n}$.
(e) Let $m = n$ and show that $\|A - B\| \geq \sigma_n(A)\|E_{11}\|$ for every singular $B \in M_n$, where $E_{11} \in M_n$ has a 1 entry in position 1,1 and zeros elsewhere. Show that this lower bound is sharp; that is, exhibit a singular B for which this inequality is an equality.
(f) Derive the following singular value bounds from (3.5.30):

$$\max_{1 \leq i \leq q} |\sigma_i(A) - \sigma_i(B)| \leq \|\|A - B\|\|_2 \equiv \sigma_1(A - B) \qquad (3.5.32)$$

$$\sum_{i=1}^{q} [\sigma_i(A) - \sigma_i(B)]^2 \leq \|A - B\|_2^2 = \sum_{i,j} |a_{ij} - b_{ij}|^2 \qquad (3.5.33)$$

$$\sum_{i=1}^{q} |\sigma_i(A) - \sigma_i(B)| \leq \|A - B\|_{tr} \equiv \sum_{i=1}^{q} \sigma_i(A - B) \qquad (3.5.34)$$

19. For any given $A \in M_{m,n}$ with singular values $\sigma_1(A) \geq \sigma_2(A) \geq \cdots$ show that

$$\sigma_k(A) = \inf \{\|\|A - B\|\|_2 : B \in M_{m,n} \text{ and rank } B < k\} \qquad (3.5.35)$$

for $k = 2,\ldots,\min\{m,n\}$. This characterization of singular values extends naturally to infinite–dimensional Banach spaces, whereas the notion of eigenvalues of A^*A does not; see $(2.3.1)$ of [Pie].

20. Let $\|\cdot\|$ be a given norm on M_n. We say that $\|\cdot\|$ is *self-adjoint* if $\|A\| = \|A^*\|$ for all $A \in M_n$; that $\|\cdot\|$ is an *induced norm* if there is some norm $\nu(\cdot)$ on $\mathbb{C}^n$ such that $\|A\| = \max \{\nu(Ax): x \in \mathbb{C}^n$ and $\nu(x) = 1\}$ for all $A \in M_n$; and that $\|\cdot\|$ is *derived from an inner product* if there is an inner product $<\cdot,\cdot>$ on $M_n \times M_n$ such that $\|A\| = <A,A>^{\frac{1}{2}}$ for all $A \in M_n$. The spectral norm $\|\|A\|\|_2 = \sigma_1(A)$ is induced by the Euclidean norm on $\mathbb{C}^n$, and the Frobenius norm $\|A\|_2 = (\operatorname{tr} AA^*)^{\frac{1}{2}}$ is derived from the Frobenius inner product; both are unitarily invariant and self-adjoint. According to Theorem $(5.6.36)$ in [HJ], the spectral norm is the only norm on M_n that is both self-adjoint and induced.

(a) Show that $\|\cdot\|$ is both unitarily invariant and induced if and only if $\|\cdot\| = \|\|\cdot\|\|_2$.

(b) Let $g(x)$ be a given symmetric gauge function on $\mathbb{R}^n$. Suppose $g(x)$ is derived from an inner product, so $g(x) = x^T Bx$ for some positive definite $B \in M_n$. Show that $g(e_i) = g(e_j)$ and $g(e_i + e_j) = g(e_i - e_j)$ for all $i \neq j$. Conclude that $B = \beta I$ for some $\beta > 0$ and $g(x) = \beta^{\frac{1}{2}} \|x\|_2$.

(c) Show that $\|\cdot\|$ is both unitarily invariant and derived from an inner product if and only if $\|\cdot\| = \gamma \|\cdot\|_2$ for some $\gamma > 0$.

21. Let $\varphi: M_n \to M_n$ be a given function such that:

(1) $\varphi(A)$ and A have the same singular values for all $A \in M_n$;

(2) $\varphi(\alpha A + \beta B) = \alpha\varphi(A) + \beta\varphi(B)$ for all $A, B \in M_n$ and all $\alpha, \beta \in \mathbb{R}$;

(3) $\varphi(\varphi(A)) = A$ for all $A \in M_n$.

Let $I(\varphi) \equiv \{A \in M_n: \varphi(A) = A\}$ denote the set of matrices that are invariant under φ. Show that:

(a) $\frac{1}{2}(A + \varphi(A)) \in I(\varphi)$ for all $A \in M_n$.

(b) $I(\varphi)$ is a nonempty real linear subspace of M_n.

(c)   For all $B \in I(\varphi)$ and for every unitarily invariant norm $\|\cdot\|$ on M_n, $\|A - \frac{1}{2}(A + \varphi(A))\| = \|\frac{1}{2}(A - B) - \frac{1}{2}\varphi(A - B)\| \leq \|A - B\|$.

Conclude that $\frac{1}{2}(A + \varphi(A))$ is a best approximation to A among all matrices in $I(\varphi)$ with respect to every unitarily invariant norm on M_n. Consider $\varphi(A) = \pm A^*, \pm A^T$, and $\pm \bar{A}$ and verify that the following are best approximations to a given $A \in M_n$ by matrices in the given class with respect to every unitarily invariant norm on M_n: $\frac{1}{2}(A + A^*)$, Hermitian; $\frac{1}{2}(A - A^*)$, skew-Hermitian; $\frac{1}{2}(A + A^T)$, symmetric; $\frac{1}{2}(A - A^T)$, skew-symmetric; $\frac{1}{2}(A + \bar{A})$, real; $\frac{1}{2}(A - \bar{A})$, purely imaginary.

22. Let $A = [A_{ij}]_{i,j=1}^2 \in M_n$ be a given positive semidefinite block matrix with $A_{11} \in M_m$ and $A_{22} \in M_{n-m}$. For a given unitarily invariant norm $\|\cdot\|$ on M_n, define $\|\cdot\|$ on M_k, $k \leq n$, as in Problem 1.

(a) Use Corollary (3.5.11) to show that $\|A\| \leq \|A_{11}\| + \|A_{22}\|$ for every unitarily invariant norm $\|\cdot\|$ on M_n if and only if $N_k(A) \leq N_k(A_{11}) + N_k(A_{22})$ for $k = 1,...,n$.

(b) Prove the inequalities in (a) for the Ky Fan k-norms; conclude that $\|A\| \leq \|A_{11}\| + \|A_{22}\|$ for every unitarily invariant norm $\|\cdot\|$ on M_n.

(c) Show that $\|A_{11} \oplus A_{22}\| \leq \|A\| \leq \|A_{11}\| + \|A_{22}\|$ for every unitarily invariant norm $\|\cdot\|$ on M_n. What does this say for the Frobenius norm? the spectral norm? the trace norm?

(d) If $A = [a_{ij}]_{i,j=1}^n$, show that $\|\mathrm{diag}(a_{11},..., a_{nn})\| \leq \|A\| \leq \|E_{11}\| \, \mathrm{tr}\, A$ for every unitarily invariant norm $\|\cdot\|$ on M_n.

Notes and Further Readings. Results related to (3.5.6) and (3.5.16) are in R. A. Horn and R. Mathias, Cauchy-Schwarz Inequalities Associated with Positive Semidefinite Matrices, *Linear Algebra Appl.* 142 (1990), 63–82. Ky Fan proved the equivalence of Corollary (3.5.9a,b) in 1951; this basic result is often called the *Fan dominance theorem.* The bound (3.5.31) (for the Frobenius norm) was discovered in 1907 by E. Schmidt [see Section 18 of Schmidt's paper cited in Section (3.0)], who wanted to approximate a function of two variables by a sum of products of two univariate functions; it was rediscovered in the context of factorial theories in psychology by C. Eckart and G. Young, The Approximation of One Matrix by Another of Lower Rank, *Psychometrika* 1 (1936), 211-218. The results in Problems 3, 20, and 21 are in C.-K. Li and N.-K. Tsing, On the Unitarily Invariant Norms and Some Related Results, *Linear Multilinear Algebra* 20 (1987), 107-119.

3.6 Sufficiency of Weyl's product inequalities

The eigenvalue–singular value product inequalities in Weyl's theorem (3.3.2) raise the natural question of whether the inequalities (3.3.3) exactly characterize the relationship between the eigenvalues and singular values of a matrix. Alfred Horn proved in 1954 that they do.

The inequalities (3.3.3) are a kind of multiplicative majorization, and if A is nonsingular we can take logarithms to obtain the ordinary majorization inequalities.

$$\sum_{i=1}^{k} \log |\lambda_i(A)| \le \sum_{i=1}^{k} \log \sigma_i(A), \quad k = 1,..., n, \text{ with equality for } k = n \quad (3.6.1)$$

Since majorization inequalities of this type are the exact relationship between the eigenvalues and main diagonal entries of a Hermitian matrix, it is not surprising that this fact is useful in establishing the converse of Theorem (3.3.2). The following lemma of Mirsky is useful in this effort.

3.6.2   Lemma. Let $A \in M_n$ be positive definite and have decreasingly ordered eigenvalues $\lambda_1(A) \ge \cdots \ge \lambda_n(A) > 0$. For $k = 1,..., n$, let A_k denote the upper left k-by-k principal submatrix of A, and let the positive real numbers $r_i(A)$ be defined recursively by

$$r_1(A) \cdots r_k(A) \equiv \det A_k, \quad k = 1,..., n \qquad (3.6.3)$$

Let the decreasingly ordered values of $r_1(A),..., r_n(A)$ be denoted by $\eta_{[1]} \ge \cdots \ge \eta_{[n]}$. Then

$$\eta_{[1]}(A) \cdots \eta_{[k]}(A) \le \lambda_1(A) \cdots \lambda_k(A) \text{ for } k = 1,..., n, \text{ with equality for } k = n \qquad (3.6.4)$$

and A can be represented as $A = \Delta^* \Delta$, where $\Delta = [d_{ij}] \in M_n$ is upper triangular with main diagonal entries $d_{ii} = +\sqrt{r_i(A)}$ and singular values $\sigma_i(\Delta) = +\sqrt{\lambda_i(A)}$ for $i = 1,..., n$, and Δ may be taken to be real if A is real.

Conversely, let $\lambda_1 \ge \cdots \ge \lambda_n > 0$ and $r_1, ..., r_n$ be $2n$ given positive numbers and let $\eta_{[1]} \ge \cdots \ge \eta_{[n]} > 0$ denote the decreasingly ordered values of $r_1,..., r_n$. If

$$\eta_{[1]} \cdots \eta_{[k]} \le \lambda_1 \cdots \lambda_k \text{ for } k = 1,..., n, \text{ with equality for } k = n \quad (3.6.4')$$

then there exists a real upper triangular matrix $\Delta = [d_{ij}] \in M_n(\mathbb{R})$ with main diagonal entries $d_{ii} = +\sqrt{r_i}$ and singular values $\sigma_i(\Delta) = +\sqrt{\lambda_i}$ for $i = 1,..., n$.

The real symmetric positive definite matrix $A \equiv \Delta^T\Delta$ satisfies

$$r_1 \cdots r_k = \det A_k \text{ for } k = 1,\ldots, n \tag{3.6.3'}$$

Proof: For the forward assertion, let $A = \Delta^*\Delta$ be a Cholesky factorization of A with an upper triangular $\Delta = [d_{ij}] \in M_n$ with positive main diagonal entries $\{d_{11},\ldots, d_{nn}\}$, which are the eigenvalues of Δ. Moreover, Δ may be taken to be real if A is real; see Corollary (7.2.9) in [HJ]. Since $\Delta_k^*\Delta_k = A_k$ and $r_1(A) \cdots r_k(A) = \det A_k = (d_{11} \cdots d_{kk})^2$ for $k = 1,\ldots, n$, it is clear that $d_{ii} = r_i(A)^{\frac{1}{2}}$ for $i = 1,\ldots, n$. Moreover, $\sigma_i(\Delta)^2 = \lambda_i(\Delta^*\Delta) = \lambda_i(A)$ for $i = 1,\ldots, n$. Using Weyl's inequalities (3.3.2), we have

$$r_{[1]}(A) \cdots r_{[k]}(A) = \lambda_1(\Delta)^2 \cdots \lambda_k(\Delta)^2$$

$$\leq \sigma_1(\Delta)^2 \cdots \sigma_k(\Delta)^2 = \lambda_1(A) \cdots \lambda_k(A) \text{ for } k = 1,\ldots, n$$

The case $n = 1$ of the converse is trivial, so assume that it holds for matrices of all orders up to and including $n-1$ and proceed by induction. The given inequalities (3.6.4') are equivalent to

$$\sum_{i=1}^{k} \log r_{[i]} \leq \sum_{i=1}^{k} \log \lambda_i, \quad k = 1,\ldots, n, \text{ with equality for } k = n$$

Whenever one has such a majorization, it is always possible to find (see Lemma (4.3.28) in [HJ]) real numbers $\gamma_1,\ldots, \gamma_{n-1}$ satisfying the interlacing inequalities

$$\log \lambda_1 \geq \gamma_1 \geq \log \lambda_2 \geq \gamma_2 \geq \cdots \geq \log \lambda_{n-1} \geq \gamma_{n-1} \geq \log \lambda_n$$

as well as the majorization inequalities

$$\sum_{i=1}^{k} \log r_{[i]} \leq \sum_{i=1}^{k} \gamma_i, \quad k = 1,\ldots, n-1, \text{ with equality for } k = n-1$$

Then

$$\lambda_1 \geq e^{\gamma_1} \geq \lambda_2 \geq e^{\gamma_2} \geq \cdots \geq \lambda_{n-1} \geq e^{\gamma_{n-1}} \geq \lambda_n > 0 \tag{3.6.5}$$

and Lemma (3.3.8) with $f(t) = e^t$ gives

$$\eta_{[1]} \cdots \eta_{[k]} \le e^{\gamma_1} \cdots e^{\gamma_k}, \quad k = 1,..., n-1, \text{ with equality for } k = n-1$$

so by the induction hypothesis there exists a real symmetric positive definite matrix $\hat{A} \in M_{n-1}$ with eigenvalues $e^{\gamma_1},..., e^{\gamma_{n-1}}$ and such that $r_1 \cdots r_k = \det \hat{A}_k$, $k = 1,..., n-1$. Since interlacing inequalities of the type (3.6.5) exactly characterize the possible set of eigenvalues $\{\lambda_1,..., \lambda_n\}$ of a matrix $A \in M_n$ that can be obtained from $\hat{A} \in M_{n-1}$ by bordering it with a single row and column (see Theorem (4.3.10) in [HJ]), we know that there exists a real vector $y \in \mathbb{R}^n$ and a positive real number a such that the eigenvalues of the real symmetric matrix

$$A \equiv \begin{bmatrix} \hat{A} & y \\ y^T & a \end{bmatrix} \in M_n$$

are $\lambda_1,..., \lambda_n$. Since $\lambda_1 \cdots \lambda_n = r_1 \cdots r_n$ is given, we must have $\det A_n = \det A = \lambda_1 \cdots \lambda_n = r_1 \cdots r_n$, as required. The final assertion about the factorization $A = \Delta^T \Delta$ and the properties of Δ have already been established in the proof of the forward assertion.   ∎

3.6.6   Theorem. Let $\sigma_1 \ge \sigma_2 \ge \cdots \ge \sigma_n \ge 0$ be n given nonnegative real numbers, let $\lambda_1,..., \lambda_n \in \mathbb{C}$ be given with $|\lambda_1| \ge |\lambda_2| \ge \cdots \ge |\lambda_n|$, and suppose that

$$\sigma_1 \cdots \sigma_k \ge |\lambda_1 \cdots \lambda_k| \text{ for } k = 1,..., n, \text{ with equality for } k = n \qquad (3.6.7)$$

Then there exists an upper triangular matrix $\Delta \in M_n$ whose main diagonal entries (its eigenvalues) are $\lambda_1,..., \lambda_n$ (in any order) and whose singular values are $\sigma_1,..., \sigma_n$. The matrix Δ may be taken to be real if $\lambda_1,..., \lambda_n$ are real.

Proof: First consider the case in which all the values $\{\sigma_i\}$ and $\{\lambda_i\}$ are positive real numbers. In this event, $\sigma_1^2 \cdots \sigma_k^2 \ge \lambda_1^2 \cdots \lambda_k^2$ for $k = 1,..., n$, with equality for $k = n$. Thus, the converse half of Lemma (3.6.2) guarantees that there is a real upper triangular matrix $\Delta = [d_{ij}] \in M_n(\mathbb{R})$ with main diagonal entries $d_{ii} = +\sqrt{\lambda_i^2} = \lambda_i$ and singular values $\sigma_i(\Delta) = +\sqrt{\sigma_i^2} = \sigma_i$, $i = 1,..., n$.

Now suppose $\{\lambda_i\}$ are all nonnegative and $\lambda_n = 0$, so $\sigma_n = 0$ as well. If $\lambda_2 = \cdots = \lambda_n = 0$, then $\sigma_1 \geq \lambda_1$ and

$$\Delta \equiv \begin{bmatrix} \lambda_1 & \beta & 0 & \cdots & 0 \\ & 0 & \sigma_2 & & \vdots \\ \vdots & & 0 & \ddots & \\ & & & & \sigma_{n-1} \\ 0 & & \cdots & & 0 \end{bmatrix} \in M_n(\mathbb{R}), \quad \beta \equiv (\sigma_1^2 - \lambda_1^2)^{1/2} \geq 0$$

has the desired eigenvalues; it has the desired singular values since

$$\Delta\Delta^T = \mathrm{diag}(\lambda_1^2 + \beta^2 = \sigma_1^2, \sigma_2^2, ..., \sigma_{n-1}^2, 0)$$

Now suppose $\lambda_m > \lambda_{m+1} = \cdots = \lambda_n = 0$ with $2 \leq m < n$. Then $\sigma_{m-1} \geq \sigma_m \geq (\lambda_1 \cdots \lambda_m)/(\sigma_1 \cdots \sigma_{m-1}) \equiv \alpha > 0$ and we know there is a real $B \in M_m$ with eigenvalues $\lambda_1, ..., \lambda_m$ and singular values $\sigma_1, ..., \sigma_{m-1}, \alpha$. Let Q_1 be a real orthogonal matrix such that $Q_1(BB^T)Q_1^T = \mathrm{diag}(\sigma_1^2, ..., \sigma_{m-1}^2, \alpha^2) = CC^T$, where $C = [c_{ij}] \equiv Q_1 B Q_1^T$. Define

$$A = [a_{ij}] \equiv \begin{bmatrix} C & \vdots & \beta \cdots \\ \cdots & \vdots & 0 \\ & & \ddots & \sigma_{m+1} \\ 0 & & & \ddots & \sigma_{n-1} \\ & & & & 0 \end{bmatrix} \in M_n(\mathbb{R}), \quad \beta \equiv (\sigma_m^2 - \alpha^2)^{1/2}$$

That is, $a_{ij} = c_{ij}$ if $1 \leq i, j \leq m$, $a_{m,m+1} = \beta$, $a_{i,i+1} = \sigma_i$ for $i = m+1, ..., n-1$, and all other entries of A are zero. Then the eigenvalues of A are $\lambda_1, ..., \lambda_m$ (the eigenvalues of C) together with $n - m$ zeros, as desired, and

$$AA^T = \mathrm{diag}(\sigma_1^2, ..., \sigma_{m-1}^2, \alpha^2 + \beta^2 = \sigma_m^2, \sigma_{m+1}^2 ..., \sigma_{n-1}^2, 0)$$

so A has the desired singular values. If Q_2 is a real orthogonal matrix such that $\Delta \equiv Q_2 A Q_2^T$ is upper triangular, then Δ is a matrix with the asserted properties, since it has the same eigenvalues and singular values as A.

Finally, suppose nonnegative $\{\sigma_i\}$ and complex $\{\lambda_i\}$ satisfy the inequalities (3.6.7), and let $\lambda_k = e^{i\theta_k}|\lambda_k|$ for $k = 1, ..., n$, $\theta_k \in \mathbb{R}$. Let $A \in M_n(\mathbb{R})$ be upper triangular with eigenvalues $|\lambda_1|, ..., |\lambda_n|$ and singular values $\sigma_1, ..., \sigma_n$, set $D \equiv \mathrm{diag}(e^{i\theta_1}, ..., e^{i\theta_n})$ and take $\Delta \equiv DB$. If $\{\lambda_i\}$ are all real, then D has diagonal entries ± 1 and hence it and Δ are real in this case.

We have now constructed an upper triangular matrix Δ with given singular values and eigenvalues satisfying (3.6.7), in which the main diagonal entries of Δ occur with decreasing moduli. By Schur's triangularization theorem (Theorem (2.3.1) in [HJ]), Δ is unitarily similar to an upper triangular matrix in which the eigenvalues appear on the main diagonal in any prescribed order, and the unitary similarity does not change the singular values.   ☐

Problems

1.   Let $\lambda_1 \geq \cdots \geq \lambda_n > 0$ and $\sigma_1 \geq \cdots \geq \sigma_n > 0$ be $2n$ given positive numbers. Explain why the following three statements are equivalent:

(a) There is some $A \in M_n$ with eigenvalues $\{\lambda_1,..., \lambda_n\}$ and singular values $\{\sigma_1,..., \sigma_n\}$.

(b) $\lambda_1 \cdots \lambda_k \leq \sigma_1 \cdots \sigma_k$ for $k = 1,..., n$, with equality for $k = n$.

(c) There is a positive definite $A \in M_n$ with eigenvalues $\{\sigma_1,..., \sigma_n\}$ and such that $\lambda_1 \cdots \lambda_k = \det A_k$ for $k = 1,..., n$, where A_k denotes the upper left k-by-k principal submatrix of A.

2.   (a) Let

$$A = \begin{bmatrix} r & a \\ 0 & r \end{bmatrix} \text{ and } Q_\varphi = \begin{bmatrix} \cos \varphi & -\sin \varphi \\ \sin \varphi & \cos \varphi \end{bmatrix}$$

where $r > 0$ and $a \in \mathbb{R}$ are given, and let $\theta \in \mathbb{R}$ be given. Show that $\varphi \in \mathbb{R}$ can be chosen so that tr $A Q_\varphi = 2r \cos \theta$, in which case the eigenvalues of $A Q_\varphi$ are $r \cos \theta \pm i r \sin \theta$ and the singular values of $A Q_\varphi$ are the same as those of A.
(b) Using the notation and hypotheses of Theorem (3.6.6), use (a) to show that if any nonreal values λ_i occur only in complex conjugate pairs, then there exists a real upper Hessenberg $\Delta \in M_n(\mathbb{R})$ with eigenvalues $\lambda_1,..., \lambda_n$ and singular values $\sigma_1,..., \sigma_n$. Moreover, Δ may be taken to have the block upper triangular form

$$\Delta = \begin{bmatrix} A_1 & & * \\ & \ddots & \\ 0 & & A_k \end{bmatrix}$$

where each block A_i is a real 1-by-1 matrix or a real 2-by-2 matrix with nonreal complex conjugate eigenvalues.

Notes and Further Readings. Theorem (3.6.6) was first proved in A. Horn, On the Eigenvalues of a Matrix with Prescribed Singular Values, *Proc. Amer. Math. Soc.* 5 (1954), 4-7; the proof we present is adapted from A. Mirsky, Remarks on an Existence Theorem in Matrix Theory Due to A. Horn, *Monatsheft Math.* 63 (1959), 241-243.

3.7 Inclusion intervals for singular values

Since the squares of the singular values of $A \in M_{m,n}$ are essentially the singular values of the Hermitian matrices AA^* and A^*A, inclusion intervals for the singular values of A can be developed by applying Geršgorin regions, norm bounds, and other devices developed in Chapter 6 of [HJ] to AA^* and A^*A. Unfortunately, this approach typically gives bounds in terms of complicated sums of products of pairs of entries of A. Our objective in this section is to develop readily computable upper and lower bounds for individual singular values that are simple functions of the entries of A, rather than their squares and pairwise products.

The upper bounds that we present are motivated by the observation that, for $A = [a_{ij}] \in M_n$, $\sigma_1^2(A) = \rho(A^*A)$, the spectral radius or largest eigenvalue of A^*A. Moreover, the spectral radius of a matrix is bounded from above by the value of any matrix norm (Theorem (5.6.9) in [HJ]), so we have the bound

$$\sigma_1(A) = [\rho(A^*A)]^{\frac{1}{2}} \le [\, |||\, A^*A\, ||| \,]^{\frac{1}{2}} \le [\, |||\, A^*\, ||| \; |||\, A\, ||| \,]^{\frac{1}{2}} \qquad (3.7.1)$$

for any matrix norm $|||\cdot|||$. For the particular choice $|||\cdot||| \equiv |||\cdot|||_1$, the maximum column sum matrix norm (see (5.6.4) in [HJ]), we have $|||\, A^*\, |||_1 = |||\, A\, |||_\infty$, the maximum row sum matrix norm ((5.6.5) in [HJ]), and hence we have the upper bound

$$\sigma_1(A) \le \sqrt{|||\, A\, |||_\infty\, |||\, A\, |||_1} = \left[\, \left[\max_i \sum_{j=1}^n |a_{ij}|\right] \left[\max_j \sum_{i=1}^n |a_{ij}|\right] \,\right]^{\frac{1}{2}} \quad (3.7.2)$$

in terms of the largest absolute row and column sums. If $A \in M_{m,n}$ is not square, application of (3.7.2) to the square matrix $[A\ \ 0] \in M_m$ (if $m > n$) or $\begin{bmatrix} A \\ 0 \end{bmatrix} \in M_n$ (if $m < n$) shows that the bound (3.7.2) is valid for any $A = [a_{ij}] \in M_{m,n}$.

To see how the estimate (3.7.2) can be used to obtain an upper bound on $\sigma_2(A)$, delete from A any row or column whose absolute sum gives the larger of the two quantities $\|\|\| A \|\|\|_1$ or $\|\|\| A \|\|\|_\infty$ and denote the resulting matrix by $A_{(1)}$. Notice that $A_{(1)} \in M_{m-1,n}$ if $\|\|\| A \|\|\|_\infty \geq \|\|\| A \|\|\|_1$ and a row was deleted, or $A_{(1)} \in M_{m,n-1}$ if $\|\|\| A \|\|\|_1 \geq \|\|\| A \|\|\|_\infty$ and a column was deleted. The interlacing property for singular values (Corollary (3.1.3)) ensures that

$$\sigma_1(A) \geq \sigma_1(A_{(1)}) \geq \sigma_2(A) \geq \sigma_2(A_{(1)}) \geq \cdots$$

so

$$\sigma_2(A) \leq \sigma_1(A_{(1)}) \leq [\|\|\| A_{(1)} \|\|\|_1 \; \|\|\| A_{(1)} \|\|\|_\infty]^{\frac{1}{2}} \qquad (3.7.3)$$

Although the bound in (3.7.3) is readily computed, there is another, though weaker, bound that is even easier to compute.

3.7.4   Definition. Let $A = [a_{ij}] \in M_{m,n}$ be given. Arrange the $m + n$ quantities

$$\sum_{j=1}^{n} |a_{ij}|, \quad i = 1,\ldots, m, \quad \text{and} \quad \sum_{i=1}^{m} |a_{ij}|, \quad j = 1,\ldots, n$$

in decreasing order and denote the resulting $m + n$ *ordered absolute line sums* of A by

$$l_1(A) \geq l_2(A) \geq \cdots \geq l_{m+n}(A) \geq 0$$

Observe that $l_1(A) = \max \{\|\|\| A \|\|\|_1, \|\|\| A \|\|\|_\infty\}$ and $l_2(A) \geq \min \{\|\|\| A \|\|\|_1, \|\|\| A \|\|\|_\infty\}$, so

$$\|\|\| A \|\|\|_1 \; \|\|\| A \|\|\|_\infty \leq l_1(A)l_2(A) \qquad (3.7.5)$$

Moreover, after deleting a row or column of A whose absolute sum gives the value $l_1(A)$, we have $l_1(A_{(1)}) \leq l_2(A)$, $l_2(A_{(1)}) \leq l_3(A)$, and, more generally,

$$l_i(A_{(1)}) \leq l_{i+1}(A), \quad i = 1,\ldots, m+n-1, \; A \in M_{m,n} \qquad (3.7.6)$$

Thus, we have the bounds

$$\sigma_2(A) \leq \sigma_1(A_{(1)}) \leq [\|\|\| A_{(1)} \|\|\|_1 \; \|\|\| A_{(1)} \|\|\|_\infty]^{\frac{1}{2}} \leq [l_1(A_{(1)})l_2(A_{(1)})]^{\frac{1}{2}}$$

$$\leq [l_2(A)l_3(A)]^{\frac{1}{2}}$$

The process leading to these bounds can be iterated to give upper bounds for all the singular values. We have already verified the cases $k = 1, 2$ of the following theorem.

3.7.7   Theorem. Let $A \in M_{m,n}$ be given and define $A_{(0)} \equiv A$. For $k = 1, 2, \ldots$ form $A_{(k+1)}$ by deleting from $A_{(k)}$ a row or column corresponding to the largest absolute line sum $l_1(A_{(k)})$. Then

$$\sigma_k(A) \leq \sigma_1(A_{(k-1)}) \leq [\;||| A_{(k-1)} |||_1 \; ||| A_{(k-1)} |||_\infty \;]^{\frac{1}{2}}$$

$$\leq \left[l_1(A_{(k-1)}) l_2(A_{(k-1)}) \right]^{\frac{1}{2}} \leq \left[l_2(A_{(k-2)}) l_3(A_{(k-2)}) \right]^{\frac{1}{2}}$$

$$\leq \cdots$$

$$\leq \left[l_r(A_{(k-r)}) l_{r+1}(A_{(k-r)}) \right]^{\frac{1}{2}} \leq \left[l_k(A) l_{k+1}(A) \right]^{\frac{1}{2}} \qquad (3.7.8)$$

for $k = 1, \ldots, \min\{m,n\}$. The norms $||| X |||_1$ and $||| X |||_\infty$ are the maximum absolute column and row sums of X, respectively.

Proof: Since $A_{(k)}$ is formed from A by deleting a total of k rows or columns, the interlacing property (3.1.4) gives the upper bound $\sigma_{k+1}(A) \leq \sigma_1(A_{(k)})$. The further bounds

$$\sigma_1(A_{(k)}) \leq [\;||| A_{(k)} |||_1 \; ||| A_{(k)} |||_\infty \;]^{\frac{1}{2}} \leq \left[l_1(A_{(k)}) l_2(A_{(k)}) \right]^{\frac{1}{2}}$$

are obtained by applying (3.7.1) and (3.7.5) to $A_{(k)}$. The final chain of bounds involving successive absolute line sums of $A_{(k-1)}, A_{(k-2)}, \ldots, A_{(1)}, A_{(0)} = A$ follow from (3.7.6) and the observation that $(A_{(k)})_{(1)} = A_{(k+1)}$:

$$l_i(A_{(k)}) \leq l_{i+1}(A_{(k-1)}) \leq \cdots \leq l_{i+r}(A_{(k-r)}) \leq \cdots \leq l_{i+k}(A) \qquad \square$$

The weakest of the bounds in (3.7.8) involves the quantities $l_1(A), \ldots,$ $l_{\min\{m,n\}+1}(A)$, which can all be computed directly from the absolute row and column sums of A without any recursion. The better bounds, involving the submatrices $A_{(r)}$ and their two largest absolute line sums, are a little more work to compute but can be surprisingly good.

Exercise. Consider the square matrix

$$A = [a_{ij}] = \begin{bmatrix} 1 & 1 & \cdots & 1 \\ \vdots & & & \\ 1 & & 0 & \end{bmatrix} \in M_n, \, n \geq 3$$

for which $a_{ij} = 1$ if either $i = 1$ or $j = 1$, and $a_{ij} = 0$ otherwise. Verify the entries in the following table, which was obtained by deleting the first row of A to produce $A_{(1)}$ and then deleting the first column of $A_{(1)}$ to produce $A_{(2)}$. How good are the upper bounds that are obtained for $\sigma_2(A)$ and $\sigma_3(A)$?

$k =$	1	2	3	4
$l_k(A)$	n	n	1	1
$l_k(A_{(1)})$	$n-1$	1	1	
$l_k(A_{(2)})$	0	0		
$\lVert A_{(k-1)} \rVert_1$	n	$n-1$		
$\lVert A_{(k-1)} \rVert_\infty$	n	1		
$[l_k(A)\, l_{k+1}(A)]^{\frac{1}{2}}$	n	$n^{\frac{1}{2}}$	1	
$[l_{k-1}(A_{(1)})\, l_k(A_{(1)})]^{\frac{1}{2}}$		$(n-1)^{\frac{1}{2}}$	1	
$[l_{k-2}(A_{(2)})\, l_{k-1}(A_{(2)})]^{\frac{1}{2}}$			0	
$[\lVert A_{(k-1)} \rVert_1 \lVert A_{(k-1)} \rVert_\infty]^{\frac{1}{2}}$	n	$(n-1)^{\frac{1}{2}}$	0	
$\sigma_k(A)$	$(n-\frac{1}{4})^{\frac{1}{2}}+\frac{1}{2}$	$(n-\frac{1}{4})^{\frac{1}{2}}-\frac{1}{2}$	0	

The lower bounds that we present are motivated by recalling from Corollary (3.1.5) that, for $A = [a_{ij}] \in M_n$, the smallest singular value of A has the lower bound $\sigma_n(A) \geq \lambda_{min}(\mathrm{H}(A))$. Since all the eigenvalues of the Hermitian matrix $\mathrm{H}(A) = [h_{ij}] = [\frac{1}{2}(a_{ij} + \bar{a}_{ji})]$ are contained in the union of Geršgorin intervals

$$\bigcup_{k=1}^{n} \left\{ x \in \mathbb{R} : |x - \mathrm{Re}\, a_{kk}| = |x - h_{kk}| \leq \sum_{j \neq k} |h_{kj}| = \frac{1}{2} \sum_{j \neq k} |a_{kj} + \bar{a}_{jk}| \right\}$$

(see Theorem (6.1.1) in [HJ]), we have the simple lower bound

$$\sigma_n(A) \geq \lambda_n(\mathrm{H}(A)) \geq \min_{1 \leq k \leq n} \left\{ \mathrm{Re}\, a_{kk} - \tfrac{1}{2} \sum_{j \neq k} |a_{kj} + \bar{a}_{jk}| \right\} \qquad (3.7.9)$$

in which $\lambda_1 \geq \lambda_2 \geq \cdots \geq \lambda_n$ are the algebraically decreasingly ordered eigenvalues of $\mathrm{H}(A)$. Notice that (3.7.9) is a trivial bound if any main diagonal entry of A has negative real part, an inadequacy that is easy to remedy since the singular values $\sigma_k(A)$ are unitarily invariant functions of A.

3.7.10 Definition. For $A = [a_{ij}] \in M_{m,n}$ define $D_A = \mathrm{diag}(e^{i\theta_1}, \ldots, e^{i\theta_n}) \in M_m$, where $e^{i\theta_k} a_{kk} = |a_{kk}|$ if $a_{kk} \neq 0$ and $\theta_k \equiv 0$ if $a_{kk} = 0$, $k = 1, \ldots, m$.

Since the main diagonal entries of $D_A A$ are nonnegative, (3.7.9) gives the lower bounds

$$\sigma_n(A) = \sigma_n(D_A A) \geq \lambda_n(\mathrm{H}(D_A A))$$

$$\geq \min_{1 \leq k \leq n} \left\{ |a_{kk}| - \tfrac{1}{2} \sum_{j \neq k} |e^{i\theta_k} a_{kj} + e^{-i\theta_j} \bar{a}_{jk}| \right\} \qquad (3.7.11a)$$

$$\geq \min_{1 \leq k \leq n} \left\{ |a_{kk}| - \tfrac{1}{2} \sum_{j \neq k} (|a_{kj}| + |a_{jk}|) \right\}$$

$$= \min_{1 \leq k \leq n} \left\{ |a_{kk}| - \tfrac{1}{2} [R_k'(A) + C_k'(A)] \right\} \qquad (3.7.11b)$$

in which we use the triangle inequality and write

$$R_k'(A) \equiv \sum_{j \neq k} |a_{kj}|, \quad C_k'(A) \equiv \sum_{i \neq k} |a_{ik}|, \quad A = [a_{ij}] \in M_{m,n}$$

as in Definition (1.5.1), for the deleted (that is, the term $|a_{kk}|$ is omitted) absolute kth row and column sums of A. Since there may be some cancellation between $e^{i\theta_k} a_{kj}$ and $e^{-i\theta_j} \bar{a}_{jk}$, the lower bound given by (3.7.11a) is generally better than that given by (3.7.11b), but the latter has the advantage of simplicity since it involves only $2n$ deleted column and row sums of A rather than all the $n(n-1)$ off-diagonal entries of $D_A A$.

In order to extend the bounds in (3.7.11a,b) to all the singular values, it is convenient to introduce notation for the left-hand endpoints of the Geršgorin intervals of $H(D_A A)$.

3.7.12   Definition. Let $A = [a_{ij}] \in M_{m,n}$ be given. Arrange the n *row deficits* of $H(D_A A)$,

$$d_k(H(D_A A)) \equiv |a_{kk}| - \tfrac{1}{2} R'_k(H(D_A A)), \ k = 1,..., n$$

in decreasing order and denote the resulting *ordered row deficits* of A by

$$\alpha_1(A) \geq \alpha_2(A) \geq \cdots \geq \alpha_n(A) = \min_k \{|a_{kk}| - R'_k(H(D_A A))\}$$

Notice that $\alpha_k(A) < 0$ is possible, and the lower bound (3.7.11a) may be restated as

$$\sigma_n(A) \geq \lambda_n(H(D_A A)) \geq \alpha_n(A), \quad A \in M_n \tag{3.7.13}$$

To see how the estimate (3.7.13) can be used to obtain a lower bound on $\sigma_{n-1}(A)$, delete row p *and* column p from A, for p such that the p th row deficit of $H(D_A A)$ has the least possible value $d_p(H(D_A A)) \equiv \alpha_n(A)$; denote the resulting matrix by $A_{[n-1]} \in M_{n-1}$. Then $A_{[n-1]}$ is a principal submatrix of A and $H(D_{A_{[n-1]}} A_{[n-1]})$ is a principal submatrix of $H(D_A A)$, so the interlacing eigenvalues theorem for Hermitian matrices (Theorem (4.3.15) in [HJ]) ensures that

$$\lambda_{n-1}(H(D_A A)) \geq \lambda_{n-1}(H(D_A A)_{[n-1]})$$

(notice that $H(X_{[k]}) = H(X)_{[k]}$). Moreover, the respective ordered row deficits of A are not greater than those of its principal submatrix $A_{[n-1]}$, that is

$$\alpha_k(A_{[n-1]}) \geq \alpha_k(A), \quad k = 1,..., n-1, \quad A \in M_n \tag{3.7.14}$$

and $\sigma_{n-1}(A) = \sigma_{n-1}(D_A A) \geq \lambda_{n-1}(H(D_A A))$ by (3.1.6b). Combining these facts with (3.7.13) applied to $A_{[n-1]} \in M_{n-1}$ gives the bound

$$\sigma_{n-1}(A) \geq \lambda_{n-1}(H(D_A A)) \geq \lambda_{n-1}(H(D_A A)_{[n-1]})$$

$$\geq \alpha_{n-1}(A_{[n-1]}) \geq \alpha_{n-1}(A)$$

The method used to develop this lower bound for $\alpha_{n-1}(A)$ can be iterated to give lower bounds for all of the singular values. We have already verified the cases $k = n$, $n-1$ of the following theorem.

3.7.15   Theorem. Let $A \in M_n$ be given and define $A_{[n]} \equiv A$. For $k = n$, $n-1$, $n-2$,... calculate the least ordered row deficit $\alpha_k(A_{[k]})$ and suppose it is obtained by using row i_k of $\mathrm{H}((D_A A)_{[k]})$. Form $A_{[k-1]}$ by deleting both row i_k and column i_k from $A_{[k]}$. Then

$$\sigma_k(A) \geq \alpha_k(A_{[k]}) \geq \alpha_k(A_{[k+1]}) \geq \cdots \geq \alpha_k(A_{[n-1]})$$

$$\geq \alpha_k(A), \quad k = n, n-1, n-2,..., 2, 1 \qquad (3.7.16)$$

Proof: Since all the concepts needed to prove the general case have already occurred in our discussion of the case $k = n-1$, the following chain of inequalities and citations provides an outline of the proof. A crucial, if simple, point is that $\mathrm{H}((D_A A)_{[k]}) \in M_k$ is a principal submatrix of $\mathrm{H}(D_A A)$ because $A_{[k]} \in M_k$ is a principal submatrix of A. We continue to write the eigenvalues of $\mathrm{H}(\cdot)$ in decreasing order $\lambda_1 \geq \lambda_2 \geq \cdots$.

$$\sigma_k(A) = \sigma_k(D_A A) \qquad \text{[by unitary invariance]}$$

$$\geq \lambda_k(\mathrm{H}(D_A A)) \qquad \text{[by (3.1.6a)]}$$

$$\geq \lambda_k(\mathrm{H}(D_A A)_{[k]}) \qquad \text{[by Hermitian eigenvalue interlacing]}$$

$$\geq \alpha_k(A_{[k]}) \qquad \text{[by (3.7.13)]}$$

$$\geq \alpha_k(A_{[k+1]}) \geq \alpha_k(A_{[k+2]}) \geq \cdots \geq \alpha_k(A) \qquad \text{[by (3.7.14)]}$$

$\square$

Although the preceding theorem has been stated for square complex matrices, it can readily be applied to nonsquare matrices by the familiar device of augmenting the given matrix with a suitable block of zeroes to make it square. We illustrate this observation with an important special case.

3.7.17   Corollary. Let $A = [a_{ij}] \in M_{m,n}$ be given, let $q = \min\{m,n\}$, and let $D_A = \text{diag}(e^{i\theta_1},\ldots,e^{i\theta_q}) \in M_q$ be defined by $e^{i\theta_k} a_{kk} = |a_{kk}| \neq 0$ and $\theta_k = 0$ if $a_{kk} = 0$, $k = 1,\ldots, q$. Then

$$\sigma_q(A) \geq \begin{cases} \displaystyle\min_{1 \leq k \leq n} \left\{ |a_{kk}| - \tfrac{1}{2} \sum_{\substack{j=1 \\ j \neq k}}^{n} |e^{i\theta_k} a_{kj} + e^{-i\theta_j} \bar{a}_{jk}| - \tfrac{1}{2} \sum_{j=n+1}^{m} |a_{jk}| \right\} & \text{if } m \geq n \\[4ex] \displaystyle\min_{1 \leq k \leq m} \left\{ |a_{kk}| - \tfrac{1}{2} \sum_{\substack{j=1 \\ j \neq k}}^{m} |e^{i\theta_k} a_{kj} + e^{-i\theta_j} \bar{a}_{jk}| - \tfrac{1}{2} \sum_{j=m+1}^{n} |a_{kj}| \right\} & \text{if } m \leq n \end{cases}$$

$$\geq \min_{1 \leq k \leq q} \{ |a_{kk}| - \tfrac{1}{2} [R'_k(A) + C'_k(A)] \}$$

Proof: There is no loss of generality if we assume that $m \geq n$, so let $B = [A \ 0] \in M_m$. Then $q = n$ and $\sigma_n(A) = \sigma_n(B) \geq \alpha_n(B)$. Because the lower right $(m-n)$-by-$(m-n)$ submatrix of $H(B)$ is a block of zeroes, the corresponding row deficits

$$d_k(H(D_B B)) = -R'_k(H(D_B B)), \quad k = n+1,\ldots, m$$

are all nonpositive. Now let

$$\tilde{\alpha}_n(B) \equiv \min \{ d_1(H(D_B B)),\ldots, d_n(H(D_B B)) \}$$

If $\tilde{\alpha}_n(B) \geq 0$ then $\sigma_n(B) \geq \alpha_n(B) = \tilde{\alpha}_n(B)$. If $\tilde{\alpha}_n(B) < 0$ then it need not be true that $\alpha_n(B) = \tilde{\alpha}_n(B)$, but the inequality $\sigma_n(B) \geq \tilde{\alpha}_n(B)$ is trivially satisfied. Thus, in either case we have the desired lower bound

$$\sigma_n(A) = \sigma_n(B) \geq \tilde{\alpha}_n(B)$$

$$= \min_{1 \leq k \leq n} \left\{ |a_{kk}| - \tfrac{1}{2} \sum_{\substack{j \neq k \\ j=1}}^{n} |e^{i\theta_k} a_{kj} + e^{-i\theta_j} \bar{a}_{jk}| - \tfrac{1}{2} \sum_{j=n+1}^{m} |a_{jk}| \right\}$$

$$\geq \min_{1 \leq k \leq n} \left\{ |a_{kk}| - \tfrac{1}{2} \sum_{\substack{j \neq k \\ j=1}}^{n} (|a_{jk}| + |a_{kj}|) - \tfrac{1}{2} \sum_{j=n+1}^{m} |a_{jk}| \right\}$$

$$= \min_{1 \leq k \leq n} \left\{ |a_{kk}| - \tfrac{1}{2} [R_k'(A) + C_k'(A)] \right\}$$ ☐

Problems

1. What upper bound is obtained for $\sigma_1(A)$ if the Frobenius norm $\|A\|_2 = [\operatorname{tr} A^*A]^{\frac{1}{2}}$ is used in (3.7.1)?

2. Let $A \in M_n$ be a given nonsingular matrix, and let $\kappa(A) = \sigma_1(A)/\sigma_n(A)$ denote its spectral condition number. Using the notation of definitions (3.7.4,12), show that

$$\frac{\alpha_1(A)}{\sqrt{l_n(A) l_{n+1}(A)}} \leq \kappa(A) \leq \frac{\sqrt{l_1(A) l_2(A)}}{\alpha_n(A)} \tag{3.7.18}$$

Apply this to $A = \begin{bmatrix} 1 & 1 \\ 0 & 1 \end{bmatrix}$, for which $\kappa(A) \approx 2.62$. How good are the bounds?

3. For $A = [a_{ij}] \in M_n$, show that

$$\sigma_n(A) \geq \min_{1 \leq k \leq n} \tfrac{1}{2} \left[[4|a_{kk}|^2 + (R_k' - C_k')^2]^{\frac{1}{2}} - [R_k' + C_k'] \right] \tag{3.7.19}$$

Compare this lower bound with (3.7.11).

Notes and Further Readings. Most of the results in this section are taken from C. R. Johnson, A Geršgorin-Type Lower Bound for the Smallest Singular Value, *Linear Algebra Appl.* 112 (1989), 1-7. The lower bound in (3.7.19) resulted from a discussion with A. Hoffman.

3.8 Singular value weak majorization for bilinear products

Since the basic weak majorization inequalities (3.3.14a) for the ordinary matrix product have an analog for the Hadamard product (see Theorem

(5.5.4)), it is natural to try to characterize the bilinear products for which these inequalities hold. For convenience, we consider square matrices and use $\bullet: M_n \times M_n \to M_n$ to denote a bilinear function, which we interpret as a "product" $(A,B) \to A \bullet B$. Examples of such bilinear products are:

AB (ordinary product)

AB^T

$A^T B$

$A \circ B$ (Hadamard product)

$A \circ B^T$

$A^T \circ B$

$UAVBW$ for given $U, V, W \in M_n$

and any linear combination of these.

If $A = [a_{ij}] \in M_n$ is given and if $E_{ij} \in M_n$ has entry 1 in position i,j and zero entries elsewhere, notice that $\operatorname{tr} AE_{ij} = a_{ji}$. Thus, a given bilinear product $\bullet: M_n \times M_n \to M_n$ is completely determined by the values of $\operatorname{tr}(A \bullet B)C$ for all $A, B, C \in M_n$. Associated with any bilinear product $\bullet$ are two bilinear products $\bullet_L$ and $\bullet_R$ characterized by the adjoint-like identities

$$\operatorname{tr}(A \bullet_L B)C = \operatorname{tr} C(A \bullet_L B) = \operatorname{tr}(C \bullet A)B \text{ for all } A, B, C \in M_n \quad (3.8.1)$$

and

$$\operatorname{tr}(A \bullet_R B)C = \operatorname{tr} A(B \bullet C) \text{ for all } A, B, C \in M_n \quad (3.8.2)$$

The notations $\bullet_L$ and $\bullet_R$ are intended to remind us that the parentheses and position of $\bullet$ are moved one position to the left (respectively, right) in (3.8.1,2).

Exercise. If $\bullet$ is the usual matrix product $A \bullet B = AB$, use the identities (3.8.1,2) to show that $A \bullet_L B = AB$ and $A \bullet_R B = AB$.

Exercise. If $\bullet$ is the Hadamard product $A \bullet B = A \circ B$, show that $A \circ_L B = A^T \circ B$ and $A \circ_R B = A \circ B^T$.

Exercise. If $A \bullet B = UAVBW$ for given $U, V, W \in M_n$, show that $A \bullet_L B = VAWBU$ and $A \bullet_R B = WAUBV$.

3.8.3   Theorem. Let $\bullet: M_n \times M_n \to M_n$ be a given bilinear product. The following are equivalent:

(a)   For all $A, B \in M_n$,

$$\sum_{i=1}^{k} \sigma_i(A \bullet B) \le \sum_{i=1}^{k} \sigma_i(A)\sigma_i(B) \text{ for all } k = 1,\dots, n \quad (3.8.4)$$

(b)   For all $A, B \in M_n$,

$$\sigma_1(A \bullet B) \le \sigma_1(A)\sigma_1(B) \quad (3.8.5a)$$

$$\sum_{i=1}^{n} \sigma_i(A \bullet B) \le \sigma_1(A)\sum_{i=1}^{n} \sigma_i(B) \quad (3.8.5b)$$

$$\sum_{i=1}^{n} \sigma_i(A \bullet B) \le \sigma_1(B)\sum_{i=1}^{n} \sigma_i(A) \quad (3.8.5c)$$

(c)   For all $A, B \in M_n$,

$$\sigma_1(A \bullet B) \le \sigma_1(A)\sigma_1(B) \quad (3.8.6a)$$
$$\sigma_1(A \bullet_L B) \le \sigma_1(A)\sigma_1(B) \quad (3.8.6b)$$
$$\sigma_1(A \bullet_R B) \le \sigma_1(A)\sigma_1(B) \quad (3.8.6c)$$

(d) $|\operatorname{tr} P(Q \bullet R)| \le 1$ for all partial isometries $P, Q, R \in M_n$ such that $\min\{\operatorname{rank} P, \operatorname{rank} Q, \operatorname{rank} R\} = 1$.

(e) $|\operatorname{tr} P(Q \bullet R)| \le \min\{\operatorname{rank} P, \operatorname{rank} Q, \operatorname{rank} R\}$ for all partial isometries $P, Q, R \in M_n$.

Proof: The inequalities in (b) follow easily from (3.8.4) with $k = 1$ and $k = n$. Assuming (b), use Theorem (3.4.1), Theorem (3.3.13a), and (3.3.18) to compute

$$\sigma_1(A \bullet_L B) = \max\{|\operatorname{tr}(A \bullet_L B)C| : C \in M_n, \operatorname{rank} C = 1, \sigma_1(C) \le 1\}$$

$$= \max\{|\operatorname{tr} B(C \bullet A)| : C \in M_n, \operatorname{rank} C = 1, \sigma_1(C) \le 1\}$$

$$\leq \max \left\{ \sum_{i=1}^{n} \sigma_i(B(C \bullet A)): \; C \in M_n, \text{rank } C = 1, \sigma_1(C) \leq 1 \right\}$$

$$\leq \max \left\{ \sigma_1(B) \sum_{i=1}^{n} \sigma_i(C \bullet A): \; C \in M_n, \text{rank } C = 1, \sigma_1(C) \leq 1 \right\}$$

$$\leq \max \left\{ \sigma_1(B)\sigma_1(A) \sum_{i=1}^{n} \sigma_i(C): \; C \in M_n, \text{rank } C = 1, \sigma_1(C) \leq 1 \right\}$$

$$= \sigma_1(A)\sigma_1(B)$$

The assertion (3.8.6c) for $\sigma_1(A \bullet_R B)$ follows in the same way. Now assume (c) and let $P, Q, R \in M_n$ be partial isometries with min $\{\text{rank } P, \text{rank } Q, \text{rank } R\} = 1$. If rank $P = 1$, use (3.3.13a) again and (3.3.14a) to compute

$$|\text{tr } P(Q \bullet R)| \leq \sum_{i=1}^{n} \sigma_i(P(Q \bullet R))$$

$$\leq \sum_{i=1}^{n} \sigma_i(P)\sigma_i(Q \bullet R) = \sigma_1(P)\sigma_1(Q \bullet R)$$

$$= \sigma_1(Q \bullet R) \leq \sigma_1(Q)\sigma_1(R) = 1$$

If rank $Q = 1$ (respectively, rank $R = 1$), one uses (3.8.6b) (respectively, (3.8.6c)) to reach the same conclusion.

Now assume (d) and suppose $r = \text{rank } P \leq \min \{\text{rank } Q, \text{rank } R\}$. Use Theorem (3.1.8a) to write $P = P_1 + \cdots + P_r$ as a sum of mutually orthogonal rank one partial isometries and compute

$$|\text{tr } P(Q \bullet R)| = \left| \text{tr } \sum_{i=1}^{r} P_i(Q \bullet R) \right| = \left| \sum_{i=1}^{r} \text{tr } P_i(Q \bullet R) \right|$$

$$\leq \sum_{i=1}^{r} |\text{tr } P_i(Q \bullet R)| \leq r$$

If Q or R has the minimum rank, the same computation using linearity of $Q \bullet R$ in each factor leads to the desired conclusion.

Finally, assume (e) and use Theorem (3.1.8b) to write

$$A = \sum_{i=1}^{n} \alpha_i Q_i \text{ and } B = \sum_{i=1}^{n} \beta_i R_i$$

as a nonnegative linear combination of partial isometries, in which rank $Q_i =$ rank $R_i = i$ for $i = 1,..., n$ and

$$\sum_{i=j}^{n} \alpha_i = \sigma_j(A) \text{ and } \sum_{i=j}^{n} \beta_i = \sigma_j(B) \text{ for } j = 1,..., n$$

Let P be a given rank k partial isometry and compute

$$|\operatorname{tr} P(A \bullet B)| = \left| \operatorname{tr} P \left[\left[\sum_{i=1}^{n} \alpha_i Q_i \right] \bullet \left[\sum_{j=1}^{n} \beta_j R_j \right] \right] \right|$$

$$= \left| \operatorname{tr} P \left[\sum_{i,j=1}^{n} \alpha_i \beta_j Q_i \bullet R_j \right] \right|$$

$$\leq \sum_{i,j=1}^{n} \alpha_i \beta_j |\operatorname{tr} P(Q_i \bullet R_j)|$$

$$\leq \sum_{i,j=1}^{n} \alpha_i \beta_j \min \{k,i,j\}$$

$$= \sum_{l=1}^{k} \left[\sum_{i=l}^{n} \alpha_i \right] \left[\sum_{j=l}^{n} \beta_j \right]$$

$$= \sum_{l=1}^{k} \sigma_l(A) \sigma_l(B)$$

The desired inequality now follows from Theorem (3.4.1).   □

The criteria (3.8.6a–c) permit one to verify the basic weak majorization inequalities (3.8.4) for a given bilinear product • by checking simple submultiplicativity of the three products •, $•_L$, and $•_R$ for the spectral norm. Since $• = •_L = •_R$ for the ordinary product $A • B = AB$, the criteria (3.8.6a–c) reduce to the single criterion $\sigma_1(AB) \leq \sigma_1(A)\, \sigma_1(B)$, which is just the fact that the spectral norm is a matrix norm.

For the Hadamard product ∘, we have $A \circ_L B = A^T \circ B$ and $A \circ_R B = A \circ B^T$. Since $\sigma_1(X^T) = \sigma_1(X)$ for all $X \in M_n$, verifying the criteria (3.8.6a–c) for the Hadamard product reduces to checking that $\sigma_1(A \circ B) \leq \sigma_1(A)\, \sigma_1(B)$, a fact that was proved by Schur in 1911 (see Problem 31 in Section (4.2), Theorem (5.5.1), and Problem 7 in Section (5.5) for four different proofs of this basic inequality).

There are interesting and natural bilinear products that do *not* satisfy the basic weak majorization inequalities (3.8.4); see Problem 3.

Problems

1. Show that $(•_L)_L = •_R$, $(•_R)_R = •_L$, and $(•_R)_L = (•_L)_R = •$ for any bilinear product •: $M_n \times M_n \to M_n$.

2. Show that any bilinear product •: $M_n \times M_n \to M_n$ may be represented as

$$A • B = \sum_{k=1}^{N} U_k A V_k B W_k \quad \text{for all } A, B \in M_n \tag{3.8.7}$$

where U_k, V_k, and W_k are determined by •, and $N \leq n^6$.

3. Let $n = pq$ and consider the block matrices $A = [A_{ij}]_{i,j=1}^{p}$, $B = [B_{ij}]_{i,j=1}^{p} \in M_n$, where each A_{ij}, $B_{ij} \in M_q$. Consider the two bilinear products $\square_1$ and $\square_2$ defined on $M_n \times M_n$ by

$$A \,\square_1\, B = [A_{ij}B_{ij}]_{i,j=1}^{p} \tag{3.8.8}$$

and

$$A \,\square_2\, B = \Big[\, \sum_{k=1}^{q} A_{ik}B_{kj} \,\Big]_{i,j=1}^{p} \tag{3.8.9}$$

(a) When $p = 1$, show that $\square_1$ is the usual matrix product and $\square_2$ is the Hadamard product, while if $q = 1$ then $\square_1$ is the Hadamard product and $\square_2$ is the usual matrix product. Thus, for a nonprime n, we have two natural discrete families of bilinear products on $M_n \times M_n$ that include the usual and Hadamard products.

(b) Use the defining identities (3.8.1,2) to show that

$$A\,(\square_1)_L\,B = A^{bt}\,\square_1\,B \text{ and } A\,(\square_1)_R\,B = A\,\square_1\,B^{bt} \qquad (3.8.10a)$$

and

$$A\,(\square_2)_L\,B = (A^{bt})^T\,\square_2\,B \text{ and } A\,(\square_2)_R\,B = A\,\square_2\,(B^{bt})^T \qquad (3.8.10b)$$

where $A^{bt} \equiv [A_{ji}]_{i,j=1}^p$ denotes the *block transpose*; notice that $(A^{bt})^T = [A_{ij}^T]_{i,j=1}^p$, that is, each block in $(A^{bt})^T$ is the ordinary transpose of the corresponding block in A.

(c) Consider $p = q = 2$, $n = 4$,

$$A = \begin{bmatrix} E_{21} & E_{11} \\ E_{22} & E_{12} \end{bmatrix} \text{ and } B = \begin{bmatrix} E_{12} & E_{22} \\ E_{11} & E_{21} \end{bmatrix}$$

and verify that $\square_1$ does not satisfy the inequalities (3.8.6b,c).

(d) Consider $p = q = 2$, $n = 4$,

$$A = \begin{bmatrix} E_{12} & E_{22} \\ E_{11} & E_{21} \end{bmatrix} \text{ and } B = \begin{bmatrix} E_{21} & E_{11} \\ E_{22} & E_{12} \end{bmatrix}$$

and verify that $\square_2$ does not satisfy the inequalities (3.8.6b,c).

(e) Use the method in Problem 7 of Section (5.5) to show that $\square_1$ and $\square_2$ satisfy the inequality (3.8.6a).

Thus, the bilinear products $\square_1$ and $\square_2$ are always submultiplicative with respect to the spectral norm but do not always satisfy the basic weak majorization inequalities (3.8.4).

Notes and Further Readings. The results in this section are taken from R. A. Horn, R. Mathias, and Y. Nakamura, Inequalities for Unitarily Invariant Norms and Bilinear Matrix Products, *Linear Multilinear Algebra* 30 (1991), 303-314, which contains additional results about the bilinear products $\square_1$ and $\square_2$. For example, if we let • denote either $\square_1$ or $\square_2$, there is a common generalization of the inequalities (3.5.22) and (3.5.24) for the ordinary and Hadamard products:

$$\| A \bullet B \|^2 \leq \| A^* A \| \, \| B^* B \| \qquad\qquad (3.8.11)$$

for all unitarily invariant norms $\| \cdot \|$ on M_n. Moreover, there is also a generalization of the submultiplicativity criterion in Problem 3 of Section (3.5): For a given unitarily invariant norm $\| \cdot \|$, $\| A \bullet B \| \leq \| A \| \, \| B \|$ for all $A, B \in M_n$ if and only if $\| A \| \geq \sigma_1(A)$ for all $A \in M_n$. Thus, it is possible for a bilinear product to share many important properties with the ordinary (and Hadamard) product without satisfying the weak majorization inequalities (3.8.4). See also C. R. Johnson and P. Nylen, Converse to a Singular Value Inequality Common to the Hadamard and Conventional Product, *Linear Multilinear Algebra* 27 (1990), 167–187.

Chapter 4

Matrix equations and the Kronecker product

4.0 Motivation

4.0.1 Lyapunov's equation

In Section (2.2), the equation

$$XA + A^*X = H \qquad (4.0.1.1)$$

arose in connection with the study of matrix stability. The matrix $A \in M_n$ is given and the right-hand side $H \in M_n$ is Hermitian. The unknown matrix $X \in M_n$ is to be determined if possible. We are interested in the solvability of (4.0.1.1) in terms of properties of A and/or H.

It is important to note that the left-hand side of the matrix equation (4.0.1.1) is linear in the unknown X. This linearity may be thought of in two ways: If we define the mapping $L: M_n \to M_n$ by $L(X) \equiv XA + A^*X$, then $L(\alpha X + \beta Y) = \alpha L(X) + \beta L(Y)$; but also, the n^2 entries of X, thought of as a vector in $\mathbf{C}^{n^2}$, are transformed linearly to another n^2-vector by the transformation L. This latter observation leads to a mechanism for solving (4.0.1.1) using routine algorithms for systems of linear equations. For example, if $n = 2$, (4.0.1.1) may be rewritten as

$$
\begin{array}{llll}
(a_{11}+\bar{a}_{11})x_{11} + & \bar{a}_{21}x_{21} + & a_{21}x_{12} & = h_{11} \\
\bar{a}_{12}x_{11} + & (a_{11}+\bar{a}_{22})x_{21} + & & a_{21}x_{22} = h_{21} \\
a_{12}x_{11} + & & (\bar{a}_{11}+a_{22})x_{12} + & \bar{a}_{21}x_{22} = h_{12} \\
& a_{12}x_{21} + & \bar{a}_{12}x_{12} + & (a_{22}+\bar{a}_{22})x_{22} = h_{22}
\end{array}
$$

where $X = \begin{bmatrix} x_{11} & x_{12} \\ x_{21} & x_{22} \end{bmatrix}$, $A = \begin{bmatrix} a_{11} & a_{12} \\ a_{21} & a_{22} \end{bmatrix}$, and $H = \begin{bmatrix} h_{11} & h_{12} \\ h_{21} & h_{22} \end{bmatrix}$.

239

in matrix notation, this becomes

$$
\begin{bmatrix}
a_{11}+\bar{a}_{11} & \bar{a}_{21} & a_{21} & 0 \\
\bar{a}_{12} & a_{11}+\bar{a}_{22} & 0 & a_{21} \\
a_{12} & 0 & \bar{a}_{11}+a_{22} & \bar{a}_{21} \\
0 & a_{12} & \bar{a}_{12} & a_{22}+\bar{a}_{22}
\end{bmatrix}
\begin{bmatrix}
x_{11} \\ x_{21} \\ x_{12} \\ x_{22}
\end{bmatrix}
=
\begin{bmatrix}
h_{11} \\ h_{21} \\ h_{12} \\ h_{22}
\end{bmatrix}
\qquad (4.0.1.2)
$$

Notice that the matrices X (unknown) and H (known) become vectors in this formulation of the equation (4.0.1.1); we have adopted the convention of stacking the first column on top of the second, the second on top of the third, etc., to produce the vector form. All entries of the coefficient matrix arise from entries of A.

The process of creating equations like (4.0.1.2) from (4.0.1.1) is straightforward but tedious, and is usually not very informative. We would like a more efficient way to represent the coefficient matrix and the linear system for problems of this sort—a way that allows us to deduce properties of the linear system (4.0.1.2) directly from properties of A and H. Such a method is a goal of this chapter.

4.0.2   Commutativity with a given matrix

Given a matrix $A \in M_n$, we may wish to know all matrices $X \in M_n$ that commute with A, that is, all matrices X such that

(a) $AX = XA$

The commutativity condition (a) is equivalent to the linear matrix equation

(b) $AX - XA = 0$ $\qquad\qquad\qquad\qquad\qquad\qquad$ (4.0.2.1)

In the 2-by-2 case, this linear system may be written again in the extended form

$$
\begin{bmatrix}
0 & a_{12} & -a_{21} & 0 \\
a_{21} & a_{22}-a_{11} & 0 & -a_{21} \\
-a_{12} & 0 & a_{11}-a_{22} & a_{12} \\
0 & -a_{12} & a_{21} & 0
\end{bmatrix}
\begin{bmatrix}
x_{11} \\ x_{21} \\ x_{12} \\ x_{22}
\end{bmatrix}
=
\begin{bmatrix}
0 \\ 0 \\ 0 \\ 0
\end{bmatrix}
$$

We are interested in the nullspace of the coefficient matrix. Since the sum of the entries in the first and last rows of the coefficient matrix is 0 (reflecting the fact that scalar multiples of the identity are in the nullspace), the nullspace has dimension at least one.

Often, some sort of pre-processing of a matrix equation, prior to presenting it in this extended form, is useful at a conceptual or analytical (if not computational) level. For example, if $S \in M_n$ is nonsingular, an equation that is equivalent (has the same solutions) to (b) is

(c) $\quad S^{-1}ASS^{-1}XS - S^{-1}XSS^{-1}AS = 0$

which, with $B = S^{-1}AS$ and $Y = S^{-1}XS$, may be written as

(d) $\quad BY - YB = 0$

This is the same form as (b). Appropriate choice of the similarity S, for example, one that puts A into upper triangular form, Jordan canonical form, or some other special form, may make the equation easier to analyze. It is clear that solutions X to (b) may be recovered from solutions Y to (d).

Although the manipulative techniques developed here are extremely useful for revealing the theory of matrix equations, they need not lead to efficient numerical techniques for solving these equations. For efficiency, it may be important to take account of the special structure, sparsity, and symmetries of the resulting linear systems, which are typically of size n^2 if the original matrices are of order n.

4.1   Matrix equations

In an equation involving several matrices, one or more is usually designated as "unknown" and we wish to determine whether or not the equation is solvable, that is, if there are values of the unknown matrices that result in the desired equality, and, if it is solvable, what the set of solutions is. We shall follow the convention of designating unknown matrices with upper case letters from the end of the alphabet. Often interest focuses upon the extent to which these questions can be answered in terms of properties of the known matrices in the equation.

We shall be interested primarily in *linear* (in the unknown matrices) matrix equations. Some examples of frequently occurring types are

(a) $\quad AX = B,$

(b) $AX + XB = C,$
(c) $AXB = C,$
(d) $A_1 XB_1 + A_2 XB_2 + \cdots + A_k XB_k = C,$ and
(e) $AX + YB = C.$

The two examples discussed in Section (4.0), namely, Lyapunov's equation (4.0.1.1) and the commutativity equation (4.0.2.1) are special cases of (b). Equation (a) is a special case of (c) which, in turn, is a special case of (d), while (b) is a special case of both (d) and (e). The form (b) arises in some derivations of the Jordan canonical form, as well as in other questions involving similarity.

Some examples of *nonlinear* matrix equations are the *square root*

(f) $X^2 - A = 0$

which is discussed in Chapter 6, and the *matrix Riccati equation*

(g) $X^T AX + B^T X + X^T B + C = 0, \qquad A, B, C, X \in M_n(\mathbb{R})$

which arises in mathematical control theory. When discussing the matrix Riccati equation it is usually assumed that A and C are symmetric and one is interested in symmetric solutions X.

Exercise. Show that (g) may be written in partitioned form as

$$[X^T \ \ I] \begin{bmatrix} A & B \\ B^T & C \end{bmatrix} \begin{bmatrix} X \\ I \end{bmatrix} = 0$$

Exercise. Show that the matrix equation $AX + X^T B = 0$ is linear in the entries of X. What about $AX + X^* B = 0$?

4.2   The Kronecker product

A notion that is useful in the study of matrix equations and other applications, and is of interest in its own right, is the *Kronecker product, direct product,* or *tensor product* of matrices. This product is defined for two matrices of arbitrary sizes over any ring, though we shall be interested in matrices over a field $\mathbb{F} = \mathbb{R}$ or $\mathbb{C}$.

4.2.1   Definition. The *Kronecker product* of $A = [a_{ij}] \in M_{m,n}(\mathsf{F})$ and $B = [b_{ij}] \in M_{p,q}(\mathsf{F})$ is denoted by $A \circledast B$ and is defined to be the block matrix

$$A \circledast B \equiv \begin{bmatrix} a_{11}B & \cdots & a_{1n}B \\ \vdots & \ddots & \vdots \\ a_{m1}B & \cdots & a_{mn}B \end{bmatrix} \in M_{mp,nq}(\mathsf{F})$$

Notice that $A \circledast B \neq B \circledast A$ in general.

Exercise. If $x, y \in \mathsf{F}^n$, show that $xy^T = x \circledast y^T$. What about xy^* if $x, y \in \mathbb{C}^n$?

4.2.2   Definition. Let $A \in M_{m,n}(\mathsf{F})$. The kth *Kronecker power* $A^{\circledast k}$ is defined inductively for all positive integers k by $A^{\circledast 1} \equiv A$ and

$$A^{\circledast k} \equiv A \circledast A^{\circledast (k-1)}, \qquad k = 2, 3, \dots$$

Exercise. If $A \in M_{m,n}(\mathsf{F})$, what size matrix is $A^{\circledast k}$?

Exercise. Show that the Kronecker product of $I \in M_m$ and $I \in M_n$ is $I \in M_{mn}$ and that $A \circledast 0 = 0 \circledast A = 0$ always.

Some very basic properties of the Kronecker product include:

4.2.3 $(\alpha A) \circledast B = A \circledast (\alpha B)$ for all $\alpha \in \mathsf{F}$, $A \in M_{m,n}(\mathsf{F})$, $B \in M_{p,q}(\mathsf{F})$

4.2.4 $(A \circledast B)^T = A^T \circledast B^T$ for $A \in M_{m,n}(\mathsf{F})$, $B \in M_{p,q}(\mathsf{F})$

4.2.5 $(A \circledast B)^* = A^* \circledast B^*$ for $A \in M_{m,n}$, $B \in M_{p,q}$

4.2.6 $(A \circledast B) \circledast C = A \circledast (B \circledast C)$ for $A \in M_{m,n}(\mathsf{F})$, $B \in M_{p,q}(\mathsf{F})$, $C \in M_{r,s}(\mathsf{F})$

4.2.7 $(A + B) \circledast C = A \circledast C + B \circledast C$ for $A, B \in M_{m,n}(\mathsf{F})$ and $C \in M_{p,q}(\mathsf{F})$

4.2.8 $A \circledast (B + C) = A \circledast B + A \circledast C$ for $A \in M_{m,n}(\mathsf{F})$ and $B, C \in M_{p,q}(\mathsf{F})$

Exercise. Verify (4.2.3-8).

Exercise. Show that if $A, B \in M_n$ are Hermitian, then $A \circledast B$ is Hermitian.

Exercise. Show that $A \bullet B = 0$ if and only if $A = 0$ or $B = 0$.

The set $M_{m,n}(\mathbf{F})$ is itself a vector space over the field $\mathbf{F}$. In the study of matrix equations, it is often convenient to consider members of $M_{m,n}(\mathbf{F})$ as vectors by ordering their entries in a conventional way. We adopt the common convention of stacking columns, left to right.

4.2.9   Definition. With each matrix $A = [a_{ij}] \in M_{m,n}(\mathbf{F})$, we associate the vector vec $A \in \mathbf{F}^{mn}$ defined by

$$\text{vec } A \equiv [a_{11}, ..., a_{m1}, a_{12}, ..., a_{m2}, ..., a_{1n}, ..., a_{mn}]^T$$

Exercise. Show that if $A \in M_{m,n}(\mathbf{F})$ and $B \in M_{n,m}(\mathbf{F})$, then tr $AB =$ vec$(A^T)^T$vec B. If $A, B \in M_{m,n}$, what is a similar representation for tr A^*B? Conclude that the function $<B,A> \equiv$ tr A^*B is an inner product on $M_{m,n}$. What is this when $n = 1$?

A simple, but fundamental, property of the Kronecker product is the following *mixed-product property*, which involves both the ordinary matrix product and the Kronecker product.

4.2.10   Lemma. Let $A \in M_{m,n}(\mathbf{F})$, $B \in M_{p,q}(\mathbf{F})$, $C \in M_{n,k}(\mathbf{F})$, and $D \in M_{q,r}(\mathbf{F})$. Then $(A \bullet B)(C \bullet D) = AC \bullet BD$.

Proof: The proof is an exercise in partitioned multiplication. Let $A = [a_{ih}]$ and $C = [c_{hj}]$. Partitioning according to the sizes of B and D, $A \bullet B = [a_{ih}B]$ and $C \bullet D = [c_{hj}D]$. The i,j block of $(A \bullet B)(C \bullet D)$ is

$$\sum_{h=1}^{n} a_{ih} B c_{hj} D = \left[\sum_{h=1}^{n} a_{ih} c_{hj} \right] BD$$

But this is the i,j entry of AC times the block BD, which is the i,j block of $AC \bullet BD$. $\square$

4.2.11   Corollary. If $A \in M_m(\mathbf{F})$ and $B \in M_n(\mathbf{F})$ are nonsingular, then so is $A \bullet B$, and $(A \bullet B)^{-1} = A^{-1} \bullet B^{-1}$.

Proof: Consider the calculation $(A \bullet B)(A^{-1} \bullet B^{-1}) = AA^{-1} \bullet BB^{-1} = I \bullet I =$

I, in which the various identity matrices are of the appropriate dimensions. □

We can now determine the eigenvalues of the Kronecker product of two square complex matrices.

4.2.12   Theorem. Let $A \in M_n$ and $B \in M_m$. If $\lambda \in \sigma(A)$ and $x \in \mathbb{C}^n$ is a corresponding eigenvector of A, and if $\mu \in \sigma(B)$ and $y \in \mathbb{C}^m$ is a corresponding eigenvector of B, then $\lambda\mu \in \sigma(A \bullet B)$ and $x \bullet y \in \mathbb{C}^{nm}$ is a corresponding eigenvector of $A \bullet B$. Every eigenvalue of $A \bullet B$ arises as such a product of eigenvalues of A and B. If $\sigma(A) = \{\lambda_1,...,\lambda_n\}$ and $\sigma(B) = \{\mu_1,...,\mu_m\}$, then $\sigma(A \bullet B) = \{\lambda_i\mu_j: i = 1,..., n, j = 1,..., m\}$ (including algebraic multiplicities in all three cases). In particular, $\sigma(A \bullet B) = \sigma(B \bullet A)$.

Proof: Suppose $Ax = \lambda x$ and $By = \mu y$ with $x, y \neq 0$. Use (4.2.10) to compute $(A \bullet B)(x \bullet y) = Ax \bullet By = \lambda x \bullet \mu y = \lambda\mu(x \bullet y)$.   Schur's unitary triangularization theorem (Theorem (2.3.1) in [HJ]) ensures that there are unitary matrices $U \in M_n$ and $V \in M_m$ such that $U^*AU = \Delta_A$ and $V^*BV = \Delta_B$ are upper triangular. Then

$$(U \bullet V)^*(A \bullet B)(U \bullet V) = (U^*AU) \bullet (V^*BV) = \Delta_A \bullet \Delta_B$$

is upper triangular and is similar to $A \bullet B$. The eigenvalues of A, B, and $A \bullet B$ are exactly the main diagonal entries of Δ_A, Δ_B, and $\Delta_A \bullet \Delta_B$, respectively, and the main diagonal of $\Delta_A \bullet \Delta_B$ consists of the n^2 pairwise products of the entries on the main diagonals of Δ_A and Δ_B.   □

Exercise. Let $\hat{\lambda}_1,...,\hat{\lambda}_p$ be the distinct eigenvalues of $A \in M_n$, with respective algebraic multiplicities $s_1,..., s_p$, and let $\hat{\mu}_1,...,\hat{\mu}_q$ be the distinct eigenvalues of $B \in M_n$ with respective algebraic multiplicities $t_1,..., t_q$. Show that the algebraic multiplicity of τ in $\sigma(A \bullet B)$ is exactly

$$\sum_{\hat{\lambda}_i\hat{\mu}_j=\tau} s_i t_j$$

that is, we consider all the ways (possibly none) that τ arises as a product, one factor from $\sigma(A)$ and the other from $\sigma(B)$.

4.2.13   Corollary. If $A \in M_n$ and $B \in M_m$ are positive (semi)definite

Hermitian matrices, then $A \circledast B$ is also positive (semi)definite Hermitian.

Proof: The Kronecker product $A \circledast B$ is Hermitian if A and B are Hermitian by (4.2.5), and its eigenvalues are all positive by Theorem (4.2.12) if the eigenvalues of A and B are positive. A Hermitian matrix with positive eigenvalues is positive definite. The argument is similar if the eigenvalues of A and B are nonnegative. $\square$

Recall that the *singular values* of $A \in M_{m,n}$ are the square roots of the min $\{m, n\}$ largest eigenvalues (counting multiplicities) of A^*A, usually denoted

$$\sigma_1(A) \geq \sigma_2(A) \geq \cdots \geq \sigma_q(A) \geq 0, \qquad q = \min \{m, n\} \qquad (4.2.14)$$

and that the *singular value decomposition* of A is $A = V\Sigma W^*$, where $V \in M_m$ and $W \in M_n$ are unitary and $\Sigma = [\sigma_{ij}] \in M_{m,n}$ has $\sigma_{ii} = \sigma_i(A)$, $i = 1,...,$ min $\{m,n\}$, and all other entries zero; see Chapter 3 and Sections (7.3,4) of [HJ]. The rank of A is evidently the number of its nonzero singular values. The following result about the singular values of a Kronecker product follows immediately from the mixed-product property (4.2.10).

4.2.15   Theorem. Let $A \in M_{m,n}$ and $B \in M_{p,q}$ have singular value decompositions $A = V_1\Sigma_1 W_1^*$ and $B = V_2\Sigma_2 W_2^*$, and let rank $A = r_1$ and rank $B = r_2$. Then $A \circledast B = (V_1 \circledast V_2)(\Sigma_1 \circledast \Sigma_2)(W_1 \circledast W_2)^*$. The nonzero singular values of $A \circledast B$ are the $r_1 r_2$ positive numbers $\{\sigma_i(A)\sigma_j(B): 1 \leq i \leq r_1, 1 \leq j \leq r_2\}$ (including multiplicities). Zero is a singular value of $A \circledast B$ with multiplicity min $\{mp, nq\} - r_1 r_2$. In particular, the singular values of $A \circledast B$ are the same as those of $B \circledast A$, and rank $(A \circledast B) = \text{rank} \, (B \circledast A) = r_1 r_2$.

Exercise. If $\{x_1,..., x_n\} \subset \mathbb{C}^m$ and $\{y_1,..., y_q\} \subset \mathbb{C}^p$ are linearly independent sets, show that $\{x_i \circledast y_j: i = 1,..., n, j = 1,..., q\}$ is a linearly independent set in $\mathbb{C}^{mp}$. *Hint:* One of many approaches is to let $A = [x_1 \ \cdots \ x_n] \in M_{m,n}$ and $B = [y_1 \ \cdots \ y_q] \in M_{p,q}$ and compute rank$(A \circledast B)$.

In our discussion of the Kronecker product so far, we have encountered several instances in which some property of $A \circledast B$ is the same as that of $B \circledast A$, for example, the eigenvalues when both A and B are square, the rank and, more generally, the singular values. In addition, when $A \circledast B$ is square, the same phenomenon is observed for the determinant (Problem 13). If A

and B are both square, we find that $\operatorname{tr}(A \bullet B) = \operatorname{tr}(B \bullet A)$ (Problem 12) and that $A \bullet B$ is normal if and only if $B \bullet A$ is normal (Problems 9-11). There is a simple, but important, fact underlying these observations: Regardless of the sizes of A and B, $A \bullet B$ is always permutation equivalent to $B \bullet A$, that is, $B \bullet A = P(A \bullet B)Q$ for some permutation matrices P and Q. Moreover, when A and B are both square, one may take $P = Q^T$, so the permutation equivalence between $A \bullet B$ and $B \bullet A$ is actually a permutation similarity; see Corollary (4.3.10).

We complete this section by giving some results that relate the Kronecker product and the field of values $F(\cdot)$. Recall that the product of two sets S_1, $S_2 \subseteq \mathbb{C}$ is defined as $S_1 S_2 \equiv \{s_1 s_2 : s_1 \in S_1, s_2 \in S_2\}$.

4.2.16   Theorem. Let $A \in M_m$, $B \in M_n$. Then

(a) $F(A \bullet B) \supseteq \operatorname{Co}(F(A)F(B)) \supseteq F(A)F(B)$.

(b) If A is normal, then $F(A \bullet B) = \operatorname{Co}(F(A)F(B))$.

(c) If $e^{i\theta}A$ is positive semidefinite for some $\theta \in [0,2\pi)$, then $F(A \bullet B) = F(A)F(B)$.

Proof: First we show that $F(A)F(B) \subseteq F(A \bullet B)$, a fact that does not depend on any assumptions about A or B. Suppose $x \in \mathbb{C}^m$, $y \in \mathbb{C}^n$, and $x^*x = 1 = y^*y$, and let $(x^*Ax)(y^*By)$ be a given point in the set $F(A)F(B)$. Then $(x \bullet y)^*(x \bullet y) = (x^*x)(y^*y) = 1$ and

$$(x^*Ax)(y^*By) = (x \bullet y)^*(A \bullet B)(x \bullet y) \in F(A \bullet B)$$

Thus $F(A)F(B) \subseteq F(A \bullet B)$ and hence $\operatorname{Co}(F(A)F(B)) \subseteq \operatorname{Co}(F(A \bullet B)) = F(A \bullet B)$ since the field of values of every matrix is a convex set. This proves the assertions in (a).

Now assume that A is normal. We need to show that $F(A \bullet B) \subseteq \operatorname{Co}(F(A)F(B))$. Let $U \in M_m$ be unitary and such that $U^*AU = D = \operatorname{diag}(\alpha_1,\ldots, \alpha_m)$ is diagonal. Then $F(A) = \operatorname{Co}(\{\alpha_1,\ldots, \alpha_m\})$. Since $(U \bullet I)^*(A \bullet B)(U \bullet I) = D \bullet B$ and $U \bullet I$ is unitary (see Problem 2), we have $F(A \bullet B) = F(D \bullet B)$ by the unitary invariance property (1.2.8); $F(A) = F(D)$ for the same reason. Let $x \in \mathbb{C}^{nm}$ be a given unit vector. We must show that $x^*(D \bullet B)x \in \operatorname{Co}(F(D)F(B))$. Partition x as

$$x = [x_1^T, ..., x_m^T]^T, \quad x_i \in \mathbb{C}^n, \quad i = 1, ..., m$$

Then

$$\sum_{i=1}^{m} x_i^* x_i = x^* x = 1$$

Assume for the moment that all $x_i \neq 0$. Then

$$x^*(D \otimes B)x = \sum_{i=1}^{m} \alpha_i(x_i^* B x_i) = \sum_{i=1}^{m} x_i^* x_i \left[\alpha_i \left(\frac{x_i^* B x_i}{x_i^* x_i} \right) \right]$$

which is a convex combination of the terms

$$\alpha_i \left[\frac{x_i^* B x_i}{x_i^* x_i} \right], \quad i = 1, ..., m$$

each of which is a point in $F(A)F(B)$ since $(x_i^* B x_i / x_i^* x_i) \in F(B)$. This means that $x^*(D \otimes B)x \in \text{Co}(F(A)F(B))$. The case in which some $x_i = 0$ now follows from a limiting argument since $F(A)$, $F(B)$, and hence $\text{Co}(F(A)F(B))$ are all compact. This proves (b).

Finally, observe that $F(e^{i\theta}A \otimes B) = e^{i\theta}F(A \otimes B)$ and $F(e^{i\theta}A)F(B) = e^{i\theta}F(A)F(B)$, so (c) follows from (b) if we show that $F(A)F(B)$ is convex when A is positive semidefinite. If A is positive definite and its eigenvalues $\{\alpha_i\}$ are arranged in increasing order $0 < \alpha_1 \leq ... \leq \alpha_n$, then $F(A)$ is the interval $[\alpha_1, \alpha_n]$. If $x_1, x_2 \in F(A)F(B)$, then $x_i = a_i b_i$ for $a_i \in [\alpha_1, \alpha_n]$ and $b_i \in F(B)$ for $i = 1, 2$. If $\theta_1, \theta_2 \geq 0$ and $\theta_1 + \theta_2 = 1$, then since $F(B)$ is convex we have $\theta_1 x_1 + \theta_2 x_2 = \theta_1 a_1 b_1 + \theta_2 a_2 b_2 = t(t_1 b_1 + t_2 b_2) \in tF(B) \subseteq F(A)F(B)$, where $0 < \alpha_1 \leq t \equiv \theta_1 a_1 + \theta_2 a_2 \leq \alpha_n$ and $t_i = \theta_i a_i / t$ for $i = 1, 2$. This shows that $F(A)F(B)$ is convex when A is positive definite. Convexity of this product when A is positive semidefinite and singular follows from a limiting argument since both $F(A)$ and $F(B)$ are closed.   ☐

The following result about Hadamard products is an immediate consequence of Theorem (4.2.16). It supplies a proof of Theorem (1.7.22) that was deferred in Section (1.7).

4.2.17   Corollary. If $A, N \in M_n$ and if N is normal, then $F(N \circ A) \subset$

$Co(F(N)F(A))$.

Proof: Since N is normal and $N \circ A$ is a principal submatrix of $N \otimes A$, we have $Co(F(N)F(A)) = F(N \otimes A) \supset F(N \circ A)$ by (4.2.16b) and (1.2.11). ☐

Problems

1. Let $A \in M_n$ and $B \in M_m$ be given. Show that $\det(A \otimes B) = (\det A)^m (\det B)^n = \det(B \otimes A)$. Thus, $A \otimes B$ is nonsingular if and only if both A and B are nonsingular.

2. Let $U \in M_n$ and $V \in M_m$ be given. Show that $U \otimes V \in M_{nm}$ is unitary if and only if there is some $r > 0$ such that rU and $r^{-1}V$ are both unitary.

3. If $A \in M_n$ is similar to $B \in M_n$ via S, and $C \in M_m$ is similar to $E \in M_m$ via T, show that $A \otimes C$ is similar to $B \otimes E$ via $S \otimes T$.

4. If $A \in M_{m,n}$ and $B \in M_{n,m}$, show that $(A \otimes B) \in M_{mn}$ has zero as an eigenvalue with algebraic multiplicity at least $|n - m| \min\{m, n\}$.

5. Let $A, B \in M_{m,n}$ be given. Show that $A \otimes B = B \otimes A$ if and only if either $A = cB$ or $B = cA$ for some $c \in \mathbb{C}$. Consider $A \in \mathbb{C}^2$, $B \in \mathbb{C}^3$ to show that the situation is very different when A and B do not have the same dimensions. What can you say in this case?

6. Let $A \in M_{m,n}$ and $B \in M_{p,q}$ be given.
    (a) Provide details for a proof of Theorem (4.2.15).
    (b) Let A have rank r_1 and set $\nu_1 \equiv \min\{m,n\} - r_1$; let B have rank r_2 and set $\nu_2 \equiv \min\{p,q\} - r_2$. Show that $A \otimes B$ always has exactly $r_1 r_2$ nonzero singular values. Show that $A \otimes B$ has exactly $\nu_1 \nu_2 + \nu_1 r_2 + \nu_2 r_1$ zero singular values when A and B have the same shape (that is, both $m \geq n$ and $p \geq q$ or both $m \leq n$ and $p \leq q$). What happens when A and B do not have the same shape?
    (c) Use Theorem (4.2.12) and the fact that the eigenvalues of $(A \otimes B)(A \otimes B)^*$ are the squares of the singular values of $A \otimes B$ to prove Theorem (4.2.15).
    (d) Let $\hat{\sigma}_1 > \cdots > \hat{\sigma}_k \geq 0$ be the distinct singular values of A with respective multiplicities $s_1, \ldots, s_k$, so $s_1 + \cdots + s_k = \min\{m,n\}$. Let $\hat{\tau}_1 > \cdots > \hat{\tau}_l \geq 0$ be the distinct singular values of B with respective multiplicities $t_1, \ldots, t_l$, so $t_1 + \cdots + t_k = \min\{p,q\}$. If A and B have the same shape, show that the multiplicity of ω as a singular value of $A \otimes B$

is

$$\sum_{\hat{\sigma}_i \hat{\tau}_j = \omega} s_i t_j$$

Note that this formula includes a count of the zero singular values. What happens if A and B do not have the same shape?

(e) State and verify an assertion about forming singular vectors of $A \otimes B$ from singular vectors of A and B analogous to the result about eigenvectors given in Theorem (4.2.12).

7.   Analyze the polar decomposition of $A \otimes B$ in terms of polar decompositions of $A \in M_{m,n}$ and $B \in M_{p,q}$.

8.   Show that "similar" may be replaced by "congruent" in Problem 3.

9.   If $A \in M_n$ and $B \in M_m$ are normal, show that $A \otimes B$ is normal.

10.   Let $A \in M_n$ and $B \in M_m$. Consider the converse to Problem 9. If $B = 0$, show that $A \otimes B$ can be normal without both A and B being normal. If neither A nor B is zero, however, show that $A \otimes B$ is normal if and only if both A and B are normal.

11.   Let $A \in M_n$ and $B \in M_m$. Use Problem 10 or the defect from normality criterion in Problem 26 of Section (5.6) of [HJ] to show that $A \otimes B$ is normal if and only if $B \otimes A$ is normal.

12.   Let $A \in M_{m,n}$ and $B \in M_{p,q}$ be given with $mp = nq$. Then $A \otimes B$ and $B \otimes A$ are square, so their traces are defined. Show by example that $\operatorname{tr}(A \otimes B)$ and $\operatorname{tr}(B \otimes A)$ need not be equal under these conditions. If $m = n$ and $p = q$, however, show by direct calculation as well as by applying Theorem (4.2.12) that $\operatorname{tr}(A \otimes B) = (\operatorname{tr} A)(\operatorname{tr} B) = \operatorname{tr}(B \otimes A)$.

13.   Let $A \in M_{m,n}$ and $B \in M_{p,q}$ be given with $mp = nq$. (Then $A \otimes B$ and $B \otimes A$ are square, so their determinants are defined.) Generalize part of the result in Problem 1 to show that $\det(A \otimes B) = \det(B \otimes A)$ under these conditions, but note that $A \otimes B$ can be nonsingular only if both A and B are square.

14.   Show that the mixed-product property (4.2.10) generalizes in two different ways:

(a) $(A_1 \bullet A_2 \bullet \cdots \bullet A_k)(B_1 \bullet B_2 \bullet \cdots \bullet B_k) =$
$$A_1 B_1 \bullet A_2 B_2 \bullet \cdots \bullet A_k B_k, \text{ and}$$

(b) $(A_1 \bullet B_1)(A_2 \bullet B_2) \cdots (A_k \bullet B_k) =$
$$(A_1 A_2 \cdots A_k) \bullet (B_1 B_2 \cdots B_k).$$

for matrices of appropriate sizes.

15. If $A \in M_n$ is positive definite, show that $A^{\bullet k}$ is positive definite for all $k = 1, 2, \dots$.

16. Let $A \in M_n$ and define the matrix $\Pi(A) \in M_{n!}$ in the following way. There are $n!$ distinct permutation matrices in M_n; let $P_1, P_2, \dots, P_{n!}$ be a fixed enumeration of them. Then let the i, j entry of $\Pi(A)$ be the product of the main diagonal entries of $P_i^T A P_j$. Notice that every main diagonal entry of $\Pi(A)$ is the product of the main diagonal entries of A, and the other entries of $\Pi(A)$ are the products of entries along other diagonals of A. A "diagonal" of A is a set of entries $a_{1,\tau(1)}, \dots, a_{n,\tau(n)}$, where $\tau(\cdot)$ is any one of the $n!$ distinct permutations of the integers $1, \dots, n$. Show that if A is Hermitian, then $\Pi(A)$ is also Hermitian, and if A is positive definite, then $\Pi(A)$ is also positive definite. Show that both det A and per A (see (0.3.2) in [HJ]) are always eigenvalues of $\Pi(A)$. Write out $\Pi(A)$ for a general $A \in M_3$.

17. Let $A \in M_n$ and $B \in M_m$, and let $\|\cdot\|_2$ denote the usual Euclidean norm. Show that $\|x \bullet y\|_2 = \|x\|_2 \|y\|_2$ and $\|(A \bullet B)(x \bullet y)\|_2 = \|Ax\|_2 \|By\|_2$ for all $x \in \mathbb{C}^n$ and all $y \in \mathbb{C}^m$. How is the spectral norm of $A \bullet B$ related to the spectral norms of A and B?

18. Show that if $A \in M_n$, $B \in M_m$, and $C \in M_p$, then

$$(A \bullet B) \bullet C = (A \bullet C) \bullet (B \bullet C)$$

but it need not be correct that

$$A \bullet (B \bullet C) = (A \bullet B) \bullet (A \bullet C)$$

see Corollary (4.3.16).

19. Let $p(s,t) = \sum_{i,j} c_{ij} s^i t^j$ be a polynomial in two scalar variables s and t, and define

$$p_\bullet(A,B) \equiv \sum_{i,j} c_{ij} A^i \bullet B^j$$

for $A \in M_{n_1}$ and $B \in M_{n_2}$, where $A^0 \equiv I_{n_1} \in M_{n_1}$ and $B^0 \equiv I_{n_2} \in M_{n_2}$. Show that $\sigma(p_\bullet(A,B)) = \{p(\lambda_k, \mu_l)\}$, where $\sigma(A) = \{\lambda_1, ..., \lambda_{n_1}\}$ and $\sigma(B) = \{\mu_1, ..., \mu_{n_2}\}$. What about multiplicities? Consider $p(s,t) = st$ and deduce the eigenvalue result in Theorem (4.2.12). Consider $p(s,t) = s + t = t^0 s^1 + t^1 s^0$ and find the eigenvalues of the *Kronecker sum* $(I_{n_2} \otimes A) + (B \otimes I_{n_1})$.

20. Recall that $A, B \in M_{m,n}$ are said to be *equivalent* if there exist nonsingular matrices $X \in M_m$ and $Y \in M_n$ such that $B = XAY$. Show that if $A_1, B_1 \in M_{m,n}$ are equivalent and $A_2, B_2 \in M_{p,q}$ are equivalent, then $A_1 \otimes A_2$ and $B_1 \otimes B_2$ are equivalent.

21. If $A \in M_{m,n}$ and $B \in M_{p,q}$, show in at least two ways that rank $(A \otimes B)$ = (rank A)(rank B). Use this to show, in the setting of Problem 20, that if A_1 and B_1 are equivalent and $A_1 \otimes A_2$ and $B_1 \otimes B_2$ are equivalent, then A_2 and B_2 are equivalent.

22. Let $A \in M_{m,n}$ and $B \in M_{n,k}$ be given. Show that vec$(AB) =$ $(I \otimes A)$vec B.

23. (a) When is $A \otimes B = I$?
(b) Let $U \in M_{m,n}$ and $V \in M_{p,q}$ be given with $mp = nq$. Show that $U \otimes V$ is unitary if and only if $m = n$, $p = q$, $U = cU_1$, $V = dV_1$, U_1 and V_1 are unitary, and $|cd| = 1$. Compare with Problem 2.

24. Let $A, B, C, D \in M_n$ be given Hermitian matrices. If $A - B$, $C - D$, B, and C are positive semidefinite, show that $A \otimes C - B \otimes D$ is positive semidefinite.

25. Let $A \in M_{m,n}$, $B \in M_{p,q}$, $X \in M_{q,n}$, and $Y \in M_{p,m}$. Show that

$$(\text{vec } Y)^T [A \otimes B](\text{vec } X) = \text{tr }(A^T Y^T BX).$$

26. Let $A \in M_{m,n}$, $B \in M_{p,q}$ be given and let $I_p \in M_p$, $I_n \in M_n$, $I_q \in M_q$, $I_m \in M_m$ be identity matrices. Show that $(A \otimes I_p)(I_n \otimes B) = A \otimes B =$ $(I_m \otimes B)(A \otimes I_q)$. Conclude that if $m = n$ and $p = q$, that is, both A and B are square, then $A \otimes B = (A \otimes I_p)(I_m \otimes B) = (I_m \otimes B)(A \otimes I_p)$, so $A \otimes I_p$ commutes with $I_m \otimes B$ in this case.

27. Let $X \in M_{m,n}$ have singular values $\sigma_1(X) \geq \cdots \geq \sigma_q(X) \geq 0$, where $q = $ min $\{m,n\}$, and recall that the *Schatten p-norms* of X are the unitarily invariant matrix norms $N_p(X) \equiv [\sigma_1(X)^p + \cdots + \sigma_q(X)^p]^{1/p}$, $1 \leq p \leq \infty$ (see Section (7.4) of [HJ]). For $p = 2$, this is the Frobenius norm; for $p = 1$, this

is the trace norm; the limiting case $p \to \infty$ gives the spectral norm. Use Theorem (4.2.15) to show that $N_p(A \otimes B) = N_p(A)N_p(B)$ for all $A \in M_{m,n}$, all $B \in M_{r,s}$, and all $p \in [1,\infty]$.

28. Let $X \in M_{m,n}$. Let $\||X\||_1$ and $\||X\||_\infty$ denote the maximum absolute column sum and maximum absolute row sum matrix norms, respectively. Let $\|X\|_p$ denote the entrywise l_p norm of X, $1 \leq p \leq \infty$; for $p = 2$ this is the Frobenius norm; when $m = n$, this is a matrix norm for $p = 1$, 2. Let $\|\cdot\|$ denote any of these norms and show that $\|A \otimes B\| = \|A\| \, \|B\|$ for all $A \in M_{m,n}$, $B \in M_{p,q}$.

29. Give an example of a norm $\|\cdot\|$ on matrices such that $\|A \otimes B\| < \|A\| \, \|B\|$ for some A, $B \in M_2$. Give an example of another norm on matrices such that $\|A \otimes B\| > \|A\| \, \|B\|$ for some A, $B \in M_2$.

30. Let $A \in M_n$, $B \in M_m$ be given.
   (a) Show that $(A \otimes I)^k = A^k \otimes I$ and $(I \otimes B)^k = I \otimes B^k$, $k = 1, 2, \ldots$.
   (b) For any polynomial $p(t)$, show that $p(A \otimes I) = p(A) \otimes I$ and $p(I \otimes B) = I \otimes p(B)$.
   (c) Use the power series definition of the matrix exponential $e^X = I + X + X^2/2! + \cdots$ to show that $e^{A \otimes I} = e^A \otimes I$ and $e^{I \otimes B} = I \otimes e^B$.
   (d) It is a fact that if $C, D \in M_n$ commute, then $e^{C+D} = e^C e^D$; see Theorem (6.2.38). Use this fact and Problem 26 to show that $e^{A \otimes I + I \otimes B} = e^A \otimes e^B$.

31. Let A, $B \in M_{m,n}$ be given. Use Theorem (4.2.15), Corollary (3.1.3), and the fact that the Hadamard product $A \circ B$ is a submatrix of the Kronecker product $A \otimes B$ to show that $\||A \circ B\||_2 \leq \||A\||_2 \, \||B\||_2$. See Theorem (5.5.1) and Problem 7 in Section (5.5) for three additional proofs of this basic inequality.

Notes and Further Readings. For a classical summary of properties of the Kronecker product, see Sections VII-VIII of [Mac]. For a survey of Kronecker product results in signal processing with an emphasis on applications to fast Fourier transforms, see P. A. Regalia and Sanjit K. Mitra, Kronecker Products, Unitary Matrices and Signal Processing Applications, *SIAM Review* 31 (1989), 586-613. The basic mixed-product property in Lemma (4.2.10) seems to have appeared first in C. Stéphanos, Sur une Extension du Calcul des Substitutions Linéaires, *J. Math. pures Appl.* (5) 6 (1900), 73-128. Theorem (4.2.16) is in C. R. Johnson, Hadamard Products of Matrices,

Linear Multilinear Algebra 1 (1974), 295-307. Some of the topics of this and subsequent sections are covered in [Grah]; however, the reader should be aware that there are many errors therein.

Some historians of mathematics have questioned the association of the ⊛ product in definition (4.2.1) with Kronecker's name on the grounds that there is no known evidence in the literature for Kronecker's priority in its discovery or use. Indeed, Sir Thomas Muir's authoritative history [Mui] calls det $(A \circledast B)$ the *Zehfuss determinant* of A and B because the determinant identity in Problem 1 appears first in an 1858 paper of Johann Georg Zehfuss (Über eine gewisse Determinante, *Zeit. für Math. und Physik* 3 (1858), 298-301). Following Muir's lead, a few later authors have called $A \circledast B$ the *Zehfuss matrix* of A and B, for example, D. E. Rutherford, On the Condition that Two Zehfuss Matrices be Equal, *Bull. Amer. Math. Soc.* 39 (1933), 801-808, and A. C. Aitken, The Normal Form of Compound and Induced Matrices, *Proc. Lond. Math. Soc.* 38 (2) (1935), 354-376. However, a series of influential texts at and after the turn of the century permanently associated Kronecker's name with the ⊛ product, and this terminology is nearly universal today. For further discussion (including claims by others to independent discovery of the determinant result in Problem 1) and numerous references, see H. V. Henderson, F. Pukelsheim, and S. R. Searle, On the History of the Kronecker Product, *Linear Multilinear Algebra* 14 (1983), 113-120.

4.3   Linear matrix equations and Kronecker products

The Kronecker product can be used to give a convenient representation for many linear matrix transformations and linear matrix equations. A key observation is the following:

4.3.1   Lemma. Let $A \in M_{m,n}(\mathbb{F})$, $B \in M_{p,q}(\mathbb{F})$, and $C \in M_{m,q}(\mathbb{F})$ be given and let $X \in M_{n,p}(\mathbb{F})$ be unknown. The matrix equation

(a) $AXB = C$

is equivalent to the system of qm equations in np unknowns given by

(b) $(B^T \circledast A) \operatorname{vec} X = \operatorname{vec} C$

that is, $\text{vec}(AXB) = (B^T \circledast A)\text{vec } X$.

Proof: For a given matrix Q, let Q_k denote the kth column of Q. Let $B = [b_{ij}]$. Then

$$(AXB)_k = A(XB)_k = AXB_k$$

$$= A\left[\sum_{i=1}^{p} b_{ik}X_i\right] = [b_{1k}A \quad b_{2k}A \quad \cdots \quad b_{pk}A] \text{ vec } X$$

$$= (B_k^T \circledast A) \text{ vec } X$$

Thus,

$$\text{vec}(AXB) = \begin{bmatrix} B_1^T \circledast A \\ \vdots \\ B_q^T \circledast A \end{bmatrix} \text{vec } X$$

But this product is just $B^T \circledast A$ since the transpose of a column of B is a row of B^T. Thus, $\text{vec } C = \text{vec}(AXB) = (B^T \circledast A) \text{ vec } X$, which completes the proof. []

We may now rewrite each of the linear matrix equations (a) - (e) of Section (4.1) using the Kronecker product and the $\text{vec}(\cdot)$ notation:

(a) $AX = B$ becomes
(a′) $(I \circledast A) \text{ vec } X = \text{vec } B$;

(b) $AX + XB = C$ becomes
(b′) $[(I \circledast A) + (B^T \circledast I)] \text{ vec } X = \text{vec } C$;

(c)   is covered by Lemma (4.3.1);

(d) $A_1XB_1 + \cdots + A_kXB_k = C$ becomes
(d′) $[B_1^T \circledast A_1 + \cdots + B_k^T \circledast A_k] \text{ vec } X = \text{vec } C$; and

(e) $AX + YB = C$ becomes
(e′) $(I \circledast A) \text{ vec } X + (B^T \circledast I) \text{ vec } Y = \text{vec } C$.

256   Matrix equations and the Kronecker product

Exercise. Verify that (a′), (b′), (d′), and (e′) are equivalent to (a), (b), (d), and (e), respectively.

As mentioned in Section (4.0.2), pre-processing of matrix equations can often be very useful. It is difficult to indicate the variety of transformations that could be used, but a general principle is that it is usually desirable to maintain the form of the equation. An example is the transformation of A and B in (b) by independent similarities. Begin with

$$AX + XB = C$$

Multiplication on the left by a nonsingular matrix S and on the right by a nonsingular matrix T produces

$$SAXT + SXBT = SCT$$

and insertion of the identity, appropriately factored, gives

$$(SAS^{-1})SXT + SXT(T^{-1}BT) = SCT$$

which may be rewritten as

$$A'X' + X'B' = C'$$

This is an equivalent equation of the same form as the original one, but now A and B are replaced by independent similarity transformations of themselves. A transformation of this type may well make analysis of such an equation simpler, and solutions to the original version can be recovered easily from the transformed one.

There are, of course, Kronecker versions of the two matrix equations discussed in Section (4.0). Lyapunov's equation

$$XA + A^*X = H$$

may be written as

$$[(A^T \otimes I) + (I \otimes A^*)]\,\text{vec}\,X = \text{vec}\,H$$

and the commutativity equation

$$AX - XA = 0$$

may be written as

$$[(I \otimes A) - (A^T \otimes I)] \operatorname{vec} X = 0$$

The left-hand side of each of the preceding linear matrix equations (a) - (e) gives a particular linear transformation from matrices to matrices. There are, of course, many other interesting linear transformations of matrices, and Kronecker products and the vec($\cdot$) function often offer a convenient means to study them.

Because the mapping vec: $M_{m,n} \to \mathbb{C}^{mn}$ given by $X \to \operatorname{vec} X$ is an isomorphism, and because any linear transformation $T: \mathbb{C}^r \to \mathbb{C}^s$ corresponds to a unique matrix $K \in M_{s,r}$ ($T(x) = Kx$ for all $x \in \mathbb{C}^r$), we have the following simple, but very useful, general principle.

4.3.2   Lemma. Let $T: M_{m,n} \to M_{p,q}$ be a given linear transformation. There exists a unique matrix $K(T) \in M_{pq,mn}$ such that $\operatorname{vec}[T(X)] = K(T) \operatorname{vec} X$ for all $X \in M_{m,n}$.

Sometimes one is given a class of linear transformations $T: M_{m,n} \to M_{p,q}$ with particular properties and the problem is to show that the corresponding coefficient matrices $K(T)$ have a special form, often some kind of Kronecker product form. The classical problems of characterizing the linear transformations from M_n to M_n that preserve the determinant or spectrum are examples of problems of this type, and they have elegant solutions; see Section (4.6). Another example is the problem of characterizing the linear derivations on M_n.

4.3.3   Definition. A linear transformation $T: M_n \to M_n$ is a *derivation* if $T(XY) = T(X)Y + XT(Y)$ for all $X, Y \in M_n$.

Exercise. Let $C \in M_n$ be given and let $T(X) \equiv CX - XC$ for $X \in M_n$. Show that T is a linear transformation and a derivation.

It may not be obvious that the preceding exercise gives the *only* examples of linear derivations on M_n, but that assertion is the interesting half of the following theorem.

4.3.4   Theorem. Let $T: M_n \to M_n$ be a given linear transformation. Then T is a derivation if and only if there is some $C \in M_n$ such that $T(X) = CX - XC$ for all $X \in M_n$.

Proof: The sufficiency of the representation $T(X) = CX - XC$ has been established in the preceding exercise; we must prove its necessity. Consider the defining identity for a derivation

$$T(XY) = T(X)Y + XT(Y)$$

valid for all $X, Y \in M_n$, which is equivalent to

$$\mathrm{vec}[T(XY)] = \mathrm{vec}[T(X)Y] + \mathrm{vec}[XT(Y)]$$

Let $K(T) \in M_{n^2}$ be the unique coefficient matrix corresponding to T according to Lemma (4.3.2) and note that $\mathrm{vec}(AB) = (I \bullet A)\,\mathrm{vec}\,B$ for all $A, B \in M_n$. Then

$$\mathrm{vec}[T(XY)] = K(T)\,\mathrm{vec}\,XY = K(T)(I \bullet X)\,\mathrm{vec}\,Y$$

$$\mathrm{vec}[T(X)Y] = [I \bullet T(X)]\,\mathrm{vec}\,Y$$

$$\mathrm{vec}[XT(Y)] = (I \bullet X)\,\mathrm{vec}[T(Y)] = (I \bullet X)K(T)\,\mathrm{vec}\,Y$$

Thus,

$$K(T)(I \bullet X)\,\mathrm{vec}\,Y = [I \bullet T(X)]\,\mathrm{vec}\,Y + (I \bullet X)K(T)\,\mathrm{vec}\,Y$$

for all $X, Y \in M_n$, which implies that

$$K(T)(I \bullet X) = I \bullet T(X) + (I \bullet X)K(T)$$

or

$$K(T)(I \bullet X) - (I \bullet X)K(T) = I \bullet T(X) \qquad (4.3.5)$$

for all $X \in M_n$. Write $K(T)$ in partitioned form as $K(T) = [K_{ij}]$, where each $K_{ij} \in M_n$. Then (4.3.5) says that

$$T(X) = K_{ii}X - XK_{ii}, \qquad i = 1, \ldots, n$$

$$(4.3.6)$$

$$K_{ij}X - XK_{ij} = 0, \qquad i, j = 1, ..., n, \ i \neq j$$

In particular, choose $i = 1$ in the first identities to get

$$T(X) = K_{11}X - XK_{11} \tag{4.3.7}$$

for all $X \in M_n$, a representation for T of the desired form. $\qquad\square$

The mapping $T: M_{m,n} \rightarrow M_{n,m}$ given by $T(X) = X^T$ is a simple, but important, linear transformation, whose coefficient matrix $K(T)$ given in Lemma (4.3.2) is readily characterized and has several useful properties. Since vec X and vec X^T contain the same entries in different positions, it is clear that the linear mapping vec $X \rightarrow$ vec X^T must be given by a permutation.

4.3.8 **Theorem.** Let m, n be given positive integers. There is a unique matrix $P(m,n) \in M_{mn}$ such that

$$\text{vec } X^T = P(m,n) \text{ vec } X \quad \text{for all } X \in M_{m,n} \tag{4.3.9a}$$

This matrix $P(m,n)$ depends only on the dimensions m and n and is given by

$$P(m,n) = \sum_{i=1}^{m} \sum_{j=1}^{n} E_{ij} \otimes E_{ij}^T = [E_{ij}^T]_{\substack{i=1,...,m \\ j=1,...,n}} \tag{4.3.9b}$$

where each $E_{ij} \in M_{m,n}$ has entry 1 in position i,j and all other entries are zero. Moreover, $P(m,n)$ is a permutation matrix and $P(m,n) = P(n,m)^T = P(n,m)^{-1}$.

Proof: Let $X = [x_{ij}] \in M_{m,n}$, let $E_{ij} \in M_{m,n}$ be the unit matrices described in the statement of the theorem, and notice that $E_{ij}^T X E_{ij}^T = x_{ij} E_{ij}^T$ for all $i = 1, ..., m$ and $j = 1, ..., n$. Thus,

$$X^T = \sum_{i=1}^{m} \sum_{j=1}^{n} x_{ij} E_{ij}^T = \sum_{i=1}^{m} \sum_{j=1}^{n} E_{ij}^T X E_{ij}^T$$

Now use the identity for the vec of a threefold matrix product given in Lemma (4.3.1) to write

$$\text{vec } X^T = \sum_{i=1}^m \sum_{j=1}^n \text{vec}(E_{ij}^T X E_{ij}^T) = \sum_{i=1}^m \sum_{j=1}^n (E_{ij} \circledast E_{ij}^T) \text{ vec } X$$

which verifies (4.3.9). Since $(X^T)^T = X$ and $X^T \in M_{n,m}$, we have vec $X = P(n,m)$ vec $X^T = P(n,m) P(m,n)$ vec X, so $P(n,m) = P(m,n)^{-1}$. Finally, let $\mathcal{E}_{ij}$ denote the unit matrices in $M_{n,m}$, notice that $\mathcal{E}_{ij} = E_{ji}^T$, and compute

$$P(n,m) = \sum_{i=1}^n \sum_{j=1}^m \mathcal{E}_{ij} \circledast \mathcal{E}_{ij}^T = \sum_{i=1}^n \sum_{j=1}^m E_{ji}^T \circledast E_{ji}$$

$$= \sum_{i=1}^m \sum_{j=1}^n (E_{ij} \circledast E_{ij}^T)^T = P(m,n)^T$$

Since $P(m,n)$ is a matrix of zeroes and ones for which $P(m,n)^{-1} = P(m,n)^T$, it is a permutation matrix. □

4.3.10   Corollary. Let positive integers m, n, p, and q be given and let $P(p,m) \in M_{pm}$ and $P(n,q) \in M_{nq}$ denote the permutation matrices defined by (4.3.9). Then

$$B \circledast A = P(m,p)^T (A \circledast B) P(n,q) \tag{4.3.11}$$

for all $A \in M_{m,n}$ and $B \in M_{p,q}$, so $B \circledast A$ is always permutation equivalent to $A \circledast B$. When $m = n$ and $p = q$,

$$B \circledast A = P(n,p)^T (A \circledast B) P(n,p) \tag{4.3.12}$$

for all $A \in M_n$ and $B \in M_p$, so $B \circledast A$ is permutation similar to $A \circledast B$ when both A and B are square, and the permutation similarity depends only on the dimensions m and n. More generally, let $A_1, \ldots, A_k \in M_{m,n}$ and $B_1, \ldots, B_k \in M_{p,q}$ be given. Then

$$B_1 \circledast A_1 + \cdots + B_k \circledast A_k$$
$$= P(m,p)^T [A_1 \circledast B_1 + \cdots + A_k \circledast B_k] P(n,q) \tag{4.3.13}$$

Proof: Consider the linear mapping $T: M_{n,q} \to M_{m,p}$ given by $T(X) \equiv AXB^T = Y$, so $Y^T = BX^T A^T$. Then Lemma (4.3.1) gives

$$\text{vec } Y = \text{vec}(AXB^T) = (B \circledast A)\text{vec } X \qquad (4.3.14)$$

and

$$\text{vec } Y^T = \text{vec}(BX^TA^T) = (A \circledast B)\text{vec } X^T$$

for all $X \in M_{n,q}$. Use Theorem (4.3.8) to write vec $Y^T = P(m,p)$ vec Y and vec $X^T = P(n,q)$ vec X so that

$$P(m,p) \text{ vec } Y = (A \circledast B) P(n,q) \text{ vec } X$$

which is equivalent to

$$\text{vec } Y = P(m,p)^T (A \circledast B) P(n,q) \text{ vec } X \qquad (4.3.15)$$

Comparison of (4.3.15) with (4.3.14) verifies (4.3.11).   ☐

Now that we know how $A \circledast B$ and $B \circledast A$ are related, the underlying reason for some of the special relationships developed in Section (4.2) between them is now clarified; see Problems 12-13. When A and B are square, the fact that $A \circledast B$ is similar to $B \circledast A$ means that both must have the same Jordan canonical form. If $A = SJS^{-1}$ and $B = TKT^{-1}$, where J and K are the respective Jordan canonical forms of A and B, then

$$A \circledast B = (SJS^{-1}) \circledast (TKT^{-1}) = (S \circledast T)(J \circledast K)(S \circledast T)^{-1}$$

so the Jordan canonical form of $A \circledast B$ is the same as that of $J \circledast K$. Since J and K are direct sums of Jordan blocks, we need to consider how their Kronecker product depends on their direct summands.

4.3.16   Corollary. Let $A_1,..., A_r$ and $B_1,..., B_s$ be given square complex matrices. Then $(A_1 \circledast \cdots \circledast A_r) \circledast (B_1 \circledast \cdots \circledast B_s)$ is permutation similar to the direct sum of $A_i \circledast B_j$, $i = 1,..., r$, $j = 1,..., s$.

Proof: There is no loss of generality to consider $r = s = 2$, as the general case follows by induction. A direct computation shows that

$$(A_1 \circledast A_2) \circledast (B_1 \circledast B_2) = [A_1 \circledast (B_1 \circledast B_2)] \circledast [A_2 \circledast (B_1 \circledast B_2)]$$

which is permutation similar to

$$[(B_1 \oplus B_2) \otimes A_1] \oplus [(B_1 \oplus B_2) \otimes A_2]$$

$$= (B_1 \otimes A_1) \oplus (B_2 \otimes A_1) \oplus (B_1 \otimes A_2) \oplus (B_2 \otimes A_2)$$

which is, finally, permutation similar to

$$(A_1 \otimes B_1) \oplus (A_1 \otimes B_2) \oplus (A_2 \otimes B_1) \oplus (A_2 \otimes B_2) \qquad \qquad \square$$

If the Jordan blocks in the Jordan canonical form of $A \in M_n$ are $J_1,...,$ J_p and those of $B \in M_m$ are $K_1,..., K_q$, we see that the Jordan canonical form of $A \otimes B$ is the direct sum of the Jordan blocks in the Jordan canonical forms of $J_i \otimes K_j$, $i = 1,..., p$, $j = 1,..., q$. Each pair of Jordan blocks, one associated with an eigenvalue of A and one associated with an eigenvalue of B, contributes independently to the Jordan structure of $A \otimes B$.

Thus, to determine the Jordan structure of $A \otimes B$, it suffices to determine the Jordan block structure associated with the one eigenvalue of the Kronecker product of two given Jordan blocks. The answer hinges upon the four zero/nonzero possibilities for the eigenvalues of the two blocks. The three possibilities involving at least one zero eigenvalue follow easily from the mixed-product property (see Problem 14), but no short proof for the nonsingular case seems to be known. We state the result but do not include the lengthy proof that seems to be required; however, some cases, once stated, may be verified in a straightforward way.

4.3.17   Theorem. Suppose that in the Jordan canonical form of $A \in M_n$ there is a Jordan block of size p associated with an eigenvalue $\lambda \in \sigma(A)$ and that in the Jordan canonical form of $B \in M_m$ there is a Jordan block of size q associated with an eigenvalue $\mu \in \sigma(B)$. Then, independent of any other eigenvalues or Jordan structure of A and B, the contribution of this pair of blocks to the Jordan canonical form of $A \otimes B$ is as follows:

(a)   If $\lambda \neq 0$ and $\mu \neq 0$, then associated with the eigenvalue $\lambda\mu$ there is one Jordan block of size $p + q - (2k-1)$ for each $k = 1, 2,...,$ min $\{p, q\}$.

(b)   If $\lambda \neq 0$ and $\mu = 0$, then associated with the eigenvalue 0 there are p Jordan blocks of size q.

(c)   If $\lambda = 0$ and $\mu \neq 0$, then associated with the eigenvalue 0 there are q Jordan blocks of size p.

(d)   If $\lambda = 0$ and $\mu = 0$, then associated with the eigenvalue 0 there are two Jordan blocks of each size $k = 1, 2,..., \min\{p, q\} - 1$ (these blocks are absent if $\min\{p,q\} = 1$), and $|p - q| + 1$ blocks of size $\min\{p, q\}$.

Exercise. In the context of Theorem (4.3.17), check in each case to see that the sum of the sizes of all the blocks described is pq. Why should this be?

Exercise. Show with and without Theorem (4.3.17) that the Kronecker product of two diagonalizable square matrices is diagonalizable.

Exercise. Suppose that $A \in M_n$ is nilpotent. Show that the largest–dimension Jordan block in the Jordan canonical form of $A \otimes B$ is no larger than the largest–dimension Jordan block in the Jordan canonical form of A.

Exercise. Could

$$\begin{bmatrix} 0 & 1 & 0 & 0 \\ 0 & 0 & 1 & 0 \\ 0 & 0 & 0 & 1 \\ 0 & 0 & 0 & 0 \end{bmatrix}$$

be the Jordan canonical form of the Kronecker product of two matrices in M_2? How arbitrary may the Jordan canonical form of a Kronecker product of two nonsingular matrices be?

Exercise. Show that statements (b) and (c) of the theorem are equivalent to the following: If exactly one of λ, μ is zero, the Jordan canonical form of $J_p(\lambda) \otimes J_q(\mu)$ consists of $\max\{p \operatorname{sgn}|\lambda|, q \operatorname{sgn}|\mu|\}$ blocks, each of size $\max\{q \operatorname{sgn}|\lambda|, p \operatorname{sgn}|\mu|\}$. Here, $\operatorname{sgn} z \equiv z/|z|$ if $z \neq 0$ and $\operatorname{sgn} 0 \equiv 0$.

Problems

1.   Show that the equation (e) may be written in partitioned form as

$$[I \otimes A \quad B^T \otimes I] \begin{bmatrix} \operatorname{vec}(X) \\ \operatorname{vec}(Y) \end{bmatrix} = \operatorname{vec} C$$

2.   Let $A, B, C \in M_n$. Show that the equation $AXB = C$ has a unique solution $X \in M_n$ for every given C if and only if both A and B are nonsingu-

lar. If either A or B is singular, show that there is a solution X if and only if rank $B^T \bullet A = $ rank $[B^T \bullet A \quad$ vec $C]$.

3.   Let $A \in M_{m,n}$, $B, X \in M_{n,m}$, and $C \in M_m$. Show that the equation $AX + X^TB = C$ can be rewritten as $[(I \bullet A) + (B^T \bullet I)F]$vec $X = [(I \bullet A) + F(I \bullet B^T)]$vec $X = $ vec C, where $F = P(m,m)$ is the permutation matrix given by (4.3.9). What is F when $A, B \in M_2$?

4.   Determine an n^2-by-n^4 matrix Q of zeros and ones such that vec$(AB) = Q$ vec$(A \bullet B)$ for all $A, B \in M_n$.

5.   Let $S \in M_n$ be a given nonsingular matrix. Show that the *similarity map* $T: M_n \to M_n$ given by $T(X) \equiv S^{-1}XS$ is a linear transformation on M_n. What are the eigenvalues of the linear transformation T? If the fixed matrix S is diagonalizable, what are the corresponding eigenvectors of T?

6.   Can the transpose map be represented in the general form (4.3.1a)? That is, are there fixed matrices $A, B \in M_{n,m}$ such that $X^T = AXB$ for all $X \in M_{m,n}$?

7.   For $A \in M_n$ with eigenvalues $\{\lambda_1,\ldots, \lambda_n\}$, define *spread(A)* $\equiv$ max $\{|\lambda_i - \lambda_j|: 1 \leq i, j \leq n\}$. Consider the linear transformation ad$_A: M_n \to M_n$ defined by ad$_A(X) \equiv AX - XA$. Show that $\sigma($ad$_A) = \{\lambda_i - \lambda_j;\ i, j = 1,\ldots, n\}$, and that $\rho($ad$_A) = $ spread(A).

8.   Given the linear transformation T, explain how to determine the coefficient matrix $K(T)$ in Lemma (4.3.2).

9.   Let $T: M_n \to M_n$ be a given linear derivation. First use the definition (4.3.3), then also use the representation in (4.3.4) to show that:

(a) $T(I) = 0$,

(b) $T(A^{-1}) = -A^{-1}T(A)A^{-1}$ for all nonsingular $A \in M_n$, and

(c) $T(ABC) = T(A)BC + AT(B)C + ABT(C)$ for all $A, B, C \in M_n$.

Also show that tr $T(A) = 0$ for all $A \in M_n$.

10.   Let $T: M_n \to M_n$ be a given linear derivation, let $K(T) \in M_{n^2}$ be such that vec$[T(X)] = K(T)$ vec X for all $X \in M_n$, and let $K(T) = \lfloor K_{ij} \rfloor$ with all $K_{ij} \in M_n$. Show that $K(T) = (I \bullet K_{11}) - (K_{11}^T \bullet I)$. Calculate all K_{ij} in terms of K_{11} and show that the identities (4.3.6) follow from (4.3.7).

11.   What are the linear derivations on M_1?

12. Let $A \in M_{m,n}$ and $B \in M_{p,q}$ be given. Use Corollary (4.3.10) to show that $A \otimes B$ and $B \otimes A$ have the same singular values.

13. Let $A \in M_n$ and $B \in M_m$ be given square matrices. Use Corollary (4.3.10) to show that [counting multiplicities in (e) and (f)]

(a) $\det(A \otimes B) = \det(B \otimes A)$.

(b) $\mathrm{tr}(A \otimes B) = \mathrm{tr}(B \otimes A)$.

(c)   If $A \otimes B$ is normal, then so is $B \otimes A$.

(d)   If $A \otimes B$ is unitary, then so is $B \otimes A$.

(e)   The eigenvalues of $A \otimes B$ are the same as those of $B \otimes A$.

(f)   The singular values of $A \otimes B$ are the same as those of $B \otimes A$.

14. Use the mixed-product property (4.2.10) to prove the assertions in (b), (c), and (d) of Theorem (4.3.17). Unfortunately, it does not seem to be possible to prove (a) so easily.

15. Let $A \in M_m$ and $B \in M_n$ be given and suppose neither A nor B is the zero matrix. Use Theorem (4.3.17) to show that $A \otimes B$ is diagonalizable if and only if both A and B are diagonalizable.

16. Let $A \in M_{m,n}$ and $B \in M_{n,p}$ be given. Show that

$$\mathrm{vec}(AB) = (I \otimes A)\,\mathrm{vec}\,B = (B^T \otimes A)\,\mathrm{vec}\,I = (B^T \otimes I)\,\mathrm{vec}\,A$$

In each case, what size is I?

17. When $m = n$, show that the basic permutation matrix $P(n,n)$ defined by (4.3.9) is symmetric and has eigenvalues ± 1 with respective multiplicities $\frac{1}{2}n(n \pm 1)$. Conclude that $\mathrm{tr}\,P(n,n) = n$ and $\det P(n,n) = (-1)^{n(n-1)/2}$. In general, it is known that $\det P(m,n) = (-1)^{m(m-1)n(n-1)/4}$ and $\mathrm{tr}\,P(m,n) = 1 + \gcd(m-1, n-1)$, where $\gcd(r,s)$ denotes the greatest common divisor of the integers r and s.

18. For vectors $x \in \mathbf{C}^m$ and $y \in \mathbf{C}^n$, show that $x \otimes y^T = xy^T = y^T \otimes x$. What does (4.3.11) give in this case? Use (4.3.9a) to show that $P(m,1) = I \in M_m$ and $P(1,n) = I \in M_n$.

19. There are natural generalizations of the product-reversing identity (4.3.11) to Kronecker products of three or more matrices. For $A \in M_{m,n}$,

$B \in M_{p,q}$, and $C \in M_{s,t}$, show that

$$C \bullet A \bullet B = P(mp,s)^T [A \bullet B \bullet C] \, P(nq,t)$$

$$= P(p,ms)^T [B \bullet C \bullet A] \, P(q,nt)$$

$$= P(p,ms)^T \, P(m,ps)^T [A \bullet B \bullet C] \, P(n,qt) \, P(q,nt)$$

$$= P(p,ms)^T \, P(m,ps)^T \, P(s,mp)^T [C \bullet A \bullet B] \, P(t,nq) \, P(n,qt) \, P(q,nt)$$

20. Use the identities in Problem 19 to deduce that

$$P(s,mp) = P(ms,p) \, P(ps,m)$$

and that

$$P(s,mp) \, P(m,ps) \, P(p,ms) = I \in M_{mps}$$

Now show that $\{P(ms,p), P(ps,m), P(pm,s)\}$ is a commuting family.

21. Let $A = a \in M_{m,1}$, $B = b^T \in M_{1,q}$, and $C \in M_{s,t}$ be given. Use Problems 18 and 19 to show that

$$b^T \bullet C \bullet a = [C \bullet (ab^T)] \, P(t,q)$$

Deduce that

$$e_i^T \bullet I_n \bullet e_i = (I_n \bullet E_{ii}) \, P(n,m)$$

where $e_i \in \mathbb{C}^m$ denotes the ith standard unit basis vector, $E_{ii} = e_i e_i^T$, and $I_n \in M_n$ is the identity matrix. Conclude that

$$P(m,n) = \sum_{i=1}^{m} e_i \bullet I_n \bullet e_i^T, \qquad e_i \in \mathbb{C}^m \qquad (4.3.18a)$$

Also show that

$$P(m,n) = \sum_{j=1}^{n} e_j^T \bullet I_m \bullet e_j, \qquad e_j \in \mathbb{C}^n \qquad (4.3.18b)$$

where $I_m \in M_m$ is the identity matrix and $e_j \in \mathbb{C}^n$ denotes the jth standard unit basis vector. These explicit representations for $P(m,n)$ are alternatives to (4.3.9b).

22. Let $A = [a_{ij}]$, $B = [b_{ij}] \in M_{m,n}$ be given. Show that

$$\operatorname{tr}[P(m,n)\,(A^T \bullet B)] = \operatorname{tr}(A^T B) = \sum_{i,j} a_{ij} b_{ij}$$

where $P(m,n)$ is the permutation matrix defined by (4.3.9).

23. Let $A \in M_{m,n}$ be given and let $P(m,n)$ be the permutation matrix defined by (4.3.9). Let $B \equiv P(m,n)\,(A^* \bullet A)$. Show that:

    (a) B is Hermitian;

    (b) $\operatorname{tr} B = \operatorname{tr} A^* A = \Sigma_i\, \sigma_i(A)^2$ is the sum of the squares of the singular values of A; and

    (c) $B^2 = A A^* \bullet A^* A$.

Notes and Further Readings. The basic identity

$$\operatorname{vec}(ABC) = (C^T \bullet A)\,\operatorname{vec} B$$

in Lemma (4.3.1) seems to have appeared first in W. E. Roth, On Direct Product Matrices, *Bull. Amer. Math. Soc.* 40 (1934), 461–468; it has been rediscovered many times since. The characterization of a linear derivation as a commutator with a fixed matrix given in Theorem (4.3.4) has an extension to the infinite-dimensional case; see J. Dixmeier, *Les Algebres d'Operateurs dans l'Espace Hilbertien*, 2nd ed., Gauthier-Villars, Paris, 1969, pp. 310–312. The Jordan canonical form of a Kronecker product (4.3.17) has an interesting history going back to its discovery and first attempts at proofs by Aitken and Roth in 1934. See R. Brualdi, Combinatorial Verification of the Elementary Divisors of Tensor Products, *Linear Algebra Appl.* 71 (1985), 31–47, for a modern proof, historical discussion, and references. One finds many different names in the literature for the permutation matrix $P(m,n)$ with the basic properties described in Theorem (4.3.8) and Corollary (4.3.10), for example, *vec-permutation matrix, elementary operator, permutation matrix, shuffle matrix, permuted identity matrix, universal flip matrix,* and *commutation matrix.* For a discussion of the history of this matrix, many identities, and 63 references to the literature, see H. V. Henderson and S. R. Searle, The Vec-Permutation Matrix, the Vec Operator and Kronecker Products: A Review, *Linear Multilinear Algebra* 9 (1981), 271–288.

4.4   Kronecker sums and the equation $AX + XB = C$

We illustrate the application of the Kronecker product to matrix equations with an analysis of the equation

$$AX + XB = C \tag{4.4.1}$$

in which $A \in M_n$, $B \in M_m$, and $X, C \in M_{n,m}$. This equation includes several important special cases, such as Lyapunov's equation and the commutativity equation. In Kronecker form, (4.4.1) may be written as

$$[(I_m \otimes A) + (B^T \otimes I_n)]\, \text{vec}\, X = \text{vec}\, C \tag{4.4.2}$$

The Kronecker form (4.4.2) of (4.4.1), as well as the Kronecker forms of other matrix equations discussed in the preceding section, indicate that matrices of the special form

$$(I \otimes A) + (B^T \otimes I) \tag{4.4.3}$$

arise naturally in the study of matrix equations and therefore merit special study.

4.4.4   Definition. Let $A \in M_n$ and $B \in M_m$. The mn-by-mn matrix

$$(I_m \otimes A) + (B \otimes I_n)$$

is called the *Kronecker sum* of A and B; the first identity matrix is in M_m and the second identity matrix is in M_n. Thus, $I_m \otimes A$, $B \otimes I_n$, and their sum are in M_{mn}. Whenever the dimension of an identity matrix is clear from the context, we often omit explicit indication of it.

Just as the Kronecker product of A and B has as its eigenvalues all possible pairwise products of the eigenvalues of A and B, the Kronecker sum of A and B has as its eigenvalues all possible pairwise sums of the eigenvalues of A and B.

4.4.5   Theorem. Let $A \in M_n$ and $B \in M_m$ be given. If $\lambda \in \sigma(A)$ and $x \in \mathbb{C}^n$ is a corresponding eigenvector of A, and if $\mu \in \sigma(B)$ and $y \in \mathbb{C}^m$ is a corresponding eigenvector of B,   then $\lambda + \mu$ is an eigenvalue of the

Kronecker sum $(I_m \bullet A) + (B \bullet I_n)$, and $y \bullet x \in \mathbb{C}^{nm}$ is a corresponding eigenvector. Every eigenvalue of the Kronecker sum arises as such a sum of eigenvalues of A and B, and $I_m \bullet A$ commutes with $B \bullet I_n$. If $\sigma(A) = \{\lambda_1,...,\lambda_n\}$ and $\sigma(B) = \{\mu_1,...,\mu_m\}$, then $\sigma((I_m \bullet A) + (B \bullet I_n)) = \{\lambda_i + \mu_j : i = 1,..., n, j = 1,..., m\}$ (including algebraic multiplicities in all three cases). In particular, $\sigma((I_m \bullet A) + (B \bullet I_n)) = \sigma((I_n \bullet B) + (A \bullet I_m))$.

Proof: Use the mixed-product property (4.2.10) to calculate

$$(I_m \bullet A)(B \bullet I_n) = B \bullet A = (B \bullet I_n)(I_m \bullet A)$$

Thus, $I_m \bullet A$ and $B \bullet I_n$ commute. Let $U \in M_n$ and $V \in M_m$ be unitary matrices (guaranteed by Schur's triangularization theorem (2.3.1) in [HJ]) such that $U^*AU = \Delta_A$ and $V^*BV = \Delta_B$ are upper triangular. Then $W = V \bullet U \in M_{mn}$ is unitary and a calculation reveals that the matrices

$$W^*(I_m \bullet A)W = I_m \bullet \Delta_A = \begin{bmatrix} \Delta_A & & & 0 \\ & \Delta_A & & \\ & & \cdot & \\ & & & \cdot \\ 0 & & & \Delta_A \end{bmatrix}$$

(there are m copies of $\Delta_A \in M_n$) and

$$W^*(B \bullet I_n)W = \Delta_B \bullet I_n = \begin{bmatrix} \mu_1 I_n & & * \\ & \cdot & \\ & & \cdot \\ 0 & & \mu_m I_n \end{bmatrix}$$

are upper triangular. Then

$$W^*[(I_m \bullet A) + (B \bullet I_n)]W = (I_m \bullet \Delta_A) + (\Delta_B \bullet I_n)$$

is upper triangular with the eigenvalues of the Kronecker sum on its diagonal. Inspection of the diagonal of $(I_m \bullet \Delta_A) + (\Delta_B \bullet I_n)$ shows that each diagonal entry of Δ_B is paired with all diagonal entries of Δ_A, thus verifying the assertion about the eigenvalues of the Kronecker sum (including multiplicities).   ☐

Exercise. Verify the remaining assertions in Theorem (4.4.5). State and verify a formula for the multiplicity of τ as an eigenvalue of the Kronecker sum $(I_m \otimes A) + (B \otimes I_n)$ analogous to the formula related to the Kronecker product in the exercise following Theorem (4.2.12)

Theorem (4.4.5) allows us to analyze the coefficient matrix of equation (4.4.2), which is just the Kronecker sum of A and B^T.

4.4.6   Theorem. Let $A \in M_n$ and $B \in M_m$. The equation $AX + XB = C$ has a unique solution $X \in M_{n,m}$ for each $C \in M_{n,m}$ if and only if $\sigma(A) \cap \sigma(-B) = \phi$.

Proof: Since the eigenvalues of B^T are the same as those of B, the coefficient matrix of (4.4.2) has, in view of (4.4.5), a zero eigenvalue if and only if $\sigma(A) \cap \sigma(-B) \neq \phi$. Thus, (4.4.1) is nonsingular in X if and only if $\sigma(A) \cap \sigma(-B) = \phi$. □

Exercise. Let $A \in M_n$ and $B \in M_m$. Show that $AX - XB = 0$ has a nonzero solution $X \in M_{n,m}$ if and only if $\sigma(A) \cap \sigma(B) \neq \phi$.

We may also apply (4.4.6) to Lyapunov's equation to obtain the solvability conditions referred to in Section (2.2).

4.4.7   Corollary. Let $A \in M_n$. The equation $XA + A^*X = C$ has a unique solution $X \in M_n$ for each $C \in M_n$ if and only if $\sigma(A) \cap \overline{\sigma(-A)} = \phi$, that is, if and only if $\lambda \in \sigma(A)$ implies $-\bar{\lambda} \notin \sigma(A)$.

Proof: In (4.4.6), replace A with A^* and B with A to obtain the necessary and sufficient condition $\sigma(A^*) \cap \sigma(-A) = \phi$. Since $\sigma(A^*) = \overline{\sigma(A)}$ and $\sigma(-A) = -\sigma(A)$, we are finished. □

Exercise. Verify that the coefficient matrix for the linear equation $XA + A^*X = C$ in Kronecker form is

$$(I \otimes A^*) + (A^T \otimes I) \tag{4.4.8}$$

and that it is

$$(I \otimes A) + (\bar{A} \otimes I) \tag{4.4.9}$$

if the equation is written in the alternate form $AX + XA^* = C$. Show that the spectrum of the coefficient matrix is the same in either event.

Exercise. Suppose the equation $XA + A^*X$ has a unique solution for every right-hand side C. Show that the solution X is Hermitian whenever C is Hermitian. *Hint:* Calculate the adjoint of the equation.

We mention for completeness the special case of (4.4.7) that arises in stable matrix theory. If $\sigma(A)$ lies in the open right half-plane, then clearly $\sigma(A) \cap \overline{\sigma(-A)} = \phi$.

4.4.10   Corollary. Suppose that $A \in M_n$ is positive stable. Then the equation $XA + A^*X = C$ has a unique solution $X \in M_n$ for each $C \in M_n$. Moreover, if C is Hermitian, then so is X.

Proof: We need only observe that $X^*A + A^*X = C^*$, so if C is Hermitian we must have $X^* = X$ by uniqueness.   ☐

Thus far, we have concentrated on the situation in which A and B are such that the matrix equation $AX + XB = C$ is uniquely solvable for every right-hand side C. However, there are many interesting situations in which this is not the case. For the commutativity equation (4.0.2.1), for example, $m = n$, $B = -A$, and the condition $\sigma(A) \cap \sigma(-B) = \phi$ fails completely. Thus, there are some matrices C for which $AX - XA = C$ has no solution X (either for a given A, or for any A), and some matrices C for which there are many solutions. A classical theorem of Shoda (4.5.2) reveals an easy way to tell whether a given C can be written as $C = AX - XA$ for some A and some X; the question of whether $C = AX - XA$ for some X for a given A is more subtle.

If the equation (4.4.1) is singular, which happens if and only if $\sigma(A) \cap \sigma(-B) \neq \phi$, what is the dimension of the nullspace of the linear transformation $X \rightarrow AX + XB$? Consider first the case in which both A and B are Jordan blocks (see (3.1.1) in [HJ]).

4.4.11   Lemma. Let $J_r(0) \in M_r$ and $J_s(0) \in M_s$ be singular Jordan blocks. Then $X \in M_{r,s}$ is a solution to $J_r(0)X - XJ_s(0) = 0$ if and only if

$$X = [0 \ \ Y], \ Y \in M_r, 0 \in M_{r,s-r} \text{ if } r \leq s, \text{ or}$$

$$(4.4.12)$$

$$X = \begin{bmatrix} Y \\ 0 \end{bmatrix}, \quad Y \in M_s, \ 0 \in M_{r-s,s} \ \text{if } r \geq s$$

where

$$Y \equiv \begin{bmatrix} a_0 & a_1 & a_2 & \cdot & \cdot & \cdot \\ & a_0 & a_1 & \cdot & \cdot & \cdot \\ & & a_0 & a_1 & \cdot & \\ & & & \cdot & \cdot & \cdot \\ 0 & & & & a_0 & a_1 \\ & & & & & a_0 \end{bmatrix} = [y_{ij}]$$

is, in either case, an arbitrary upper triangular Toeplitz matrix with $y_{ij} = a_{i-j}$. The dimension of the nullspace of the linear transformation

$$X \to J_r(0)X - XJ_s(0)$$

is $\min\{r,s\}$.

Proof: Let $X = [x_{ij}]$, so that

$$[J_r(0)X]_{ij} = x_{i+1,j}$$

$$[XJ_s(0)]_{ij} = x_{i,j-1}$$

for $i = 1, 2, ..., r$ and $j = 1, 2, ..., s$

where we define $x_{r+1,j} \equiv 0 \equiv x_{i,0}$. Then $[J_r(0)X - XJ_s(0)]_{ij} = 0$ for all $i = 1, 2, ..., r$ and $j = 1, 2, ..., s$ if and only if

$$x_{i+1,j+1} = x_{i,j} \text{ for } i = 1, ..., r \text{ and } j = 0, 1, 2, ..., s-1$$

where $x_{i,0} \equiv 0 \equiv x_{r+1,j}$. The resulting matrix X has the form (4.4.12) and it depends linearly on $\min\{r,s\}$ free parameters. $\qquad\square$

Let $\sigma(A) = \{\lambda_i\}$ and $\sigma(B) = \{\mu_i\}$. Since equation (4.4.1) is equivalent to $SCR = SAXR + SXBR = (SAS^{-1})(SXR) - (SXR)(R^{-1}(-B)R)$ for any nonsingular $S \in M_n$ and $R \in M_m$, we can choose S and R so that

$$SAS^{-1} = \begin{bmatrix} J_{n_1}(\lambda_1) & & 0 \\ & \cdot \cdot \cdot & \\ 0 & & J_{n_p}(\lambda_p) \end{bmatrix} = J_A$$

$$R(-B)R^{-1} = \begin{bmatrix} J_{m_1}(-\mu_1) & & 0 \\ & \ddots & \\ 0 & & J_{m_q}(-\mu_q) \end{bmatrix} = J_{(-B)}$$

are both in Jordan canonical form. It suffices to determine the dimension of the nullspace of the linear transformation $X \to J_A X - X J_{(-B)}$. Write the unknown matrix X in partitioned form as

$$X = \begin{bmatrix} X_{11} & \cdots & X_{1q} \\ \vdots & \ddots & \vdots \\ X_{p1} & \cdots & X_{pq} \end{bmatrix}, \quad X_{ij} \in M_{n_i, m_j}, \quad i = 1, 2, \dots, p, \quad j = 1, 2, \dots, q$$

The equation $J_A X - X J_{(-B)} = 0$ is equivalent to the set of pq equations

$$J_{n_i}(\lambda_i) X_{ij} - X_{ij} J_{m_j}(-\mu_j) = 0 \text{ for } i = 1, 2, \dots, p, \; j = 1, 2, \dots, q \quad (4.4.13)$$

each of which, by (4.4.6), has only the trivial solution $X_{ij} = 0$ if $\lambda_i + \mu_j \neq 0$. If $\lambda_i + \mu_j = 0$, however, and if we use the identity $J_k(\lambda) = \lambda I + J_k(0)$, equation (4.4.13) becomes

$$J_{n_i}(0) X_{ij} - X_{ij} J_{m_j}(0) = 0$$

which, by Lemma (4.4.11), has a solution space of dimension min $\{n_i, m_j\}$. This argument proves the following:

4.4.14   Theorem. Let $A \in M_n$ and $B \in M_m$, and let

$$A = S \begin{bmatrix} J_{n_1}(\lambda_1) & & 0 \\ & \ddots & \\ 0 & & J_{n_p}(\lambda_p) \end{bmatrix} S^{-1}, \; n_1 + n_2 + \cdots + n_p = n$$

$$B = R \begin{bmatrix} J_{m_1}(\mu_1) & & 0 \\ & \ddots & \\ 0 & & J_{m_q}(\mu_q) \end{bmatrix} R^{-1}, \; m_1 + m_2 + \cdots + m_q = m$$

be the respective Jordan canonical forms of A and B. The dimension of the nullspace of the linear transformation $L: M_{n,m} \to M_{n,m}$ given by $L: X \to AX + XB$ is

$$\sum_{i=1}^{p} \sum_{j=1}^{q} \nu_{ij}$$

where $\nu_{ij} = 0$ if $\lambda_i \neq -\mu_j$ and $\nu_{ij} = \min\{n_i, m_j\}$ if $\lambda_i = -\mu_j$.

We have already noted that the case $m = n$ and $B = -A$ is of special interest, since (4.4.1) is always singular in this event.

4.4.15   Corollary. Let $A \in M_n$ be a given matrix. The set of matrices in M_n that commute with A is a subspace of M_n with dimension at least n; the dimension is equal to n if and only if A is nonderogatory.

Proof: Using the notation of (4.4.14), let $B = -A$, and then we have $p = q$, $n_i = m_i$, and $\mu_i = -\lambda_i$ for $i = 1, 2, ..., p$. The dimension of the nullspace of $L: X \to AX - XA$ is

$$\sum_{i,j=1}^{p} \nu_{ij} \geq \sum_{i=1}^{p} \nu_{ii} = \sum_{j=1}^{p} n_i = n$$

with equality if and only if $\lambda_i \neq \lambda_j$ for all $i \neq j$, which happens if and only if each eigenvalue of A is associated with exactly one Jordan block. This happens if and only if each eigenvalue of A has geometric multiplicity one, which is equivalent to A being nonderogatory (see (3.2.4.1) in [HJ]).   ☐

Given $A \in M_n$, it is clear that any matrix of the form $p(A)$ commutes with A, where $p(t)$ is any polynomial. On the other hand, consideration of $A = I$ makes it clear that for some A, there are matrices that commute with A but are not polynomials in A.

Exercise. Consider $A = \begin{bmatrix} 0 & 1 \\ 0 & 0 \end{bmatrix}$. Show by direct calculation that $B = \begin{bmatrix} a & b \\ c & d \end{bmatrix}$ commutes with A if and only if $c = 0$ and $a = d$, that is, if and only if $B = p(A)$ with $p(t) = a + bt$.

When must *every* matrix that commutes with A be a polynomial in A? To answer this question, it is convenient to introduce two subspaces of M_n related to a given matrix A.

4.4.16   Definition. Let $A \in M_n$ be a given matrix. The *centralizer* of A is the set

$$C(A) \equiv \{B \in M_n: AB = BA\}$$

of all matrices that commute with A. The set of all polynomials in A is the set

$$P(A) \equiv \{p(A): p(t) \text{ is a polynomial}\}$$

There are some immediate facts about, and a relationship between, the sets $C(A)$ and $P(A)$.

4.4.17 **Theorem.** Let $A \in M_n$ be a given matrix and let $q_A(t)$ denote the minimal polynomial of A. Then

    (a) $P(A)$ and $C(A)$ are both subspaces of M_n,

    (b) $P(A) \subseteq C(A)$,

    (c)   degree $q_A(t) = \dim P(A) \leq n$, and

    (d) $\dim C(A) \geq n$ with equality if and only if A is nonderogatory.

Proof: The assertions in (a) are easily verified, and (b) is immediate since any polynomial in A commutes with A. The dimension of $P(A)$ is the cardinality of a maximal independent set in $\{I, A, A^2, ...\}$, which is exactly the degree of a monic polynomial of least degree that annihilates A, so the degree of the minimal polynomial of A, which is not greater than n by the Cayley-Hamilton theorem, is the dimension of $P(A)$. The assertions in (d) are the content of (4.4.15).   □

We always have $P(A) \subseteq C(A)$, but if $P(A) = C(A)$ then both subspaces must have the same dimension, which, by (4.4.17(c) and (d)) happens if and only if $\dim P(A) = n = \dim C(A)$. But $\dim P(A) = n$ if and only if the degree of the minimal polynomial $q_A(t)$ is n. This is the case if and only if the minimal and characteristic polynomials of A are the same, which can happen if and only if A is nonderogatory. This observation, together with (4.4.17(d)), improves (3.2.4.2) in [HJ], which is the "only if" part of the following:

4.4.18 **Corollary.** A matrix $A \in M_n$ is nonderogatory if and only if every matrix that commutes with A is a polynomial in A.

If $A \in M_n$ is derogatory (not nonderogatory), then how can one detect if a given matrix B is a polynomial in A? Commutativity with A, and hence with $p(A)$, is necessary but not sufficient. Commutativity with the larger set $C(A)$ is sufficient.

4.4.19   Theorem. Let $A \in M_n$ be a given matrix. A matrix $B \in M_n$ is a polynomial in A if and only if B commutes with every matrix that commutes with A.

Proof: The forward implication is immediate, so we must establish the backward implication. The set of matrices that commute with every matrix in $C(A)$ is a subspace of $C(A)$ that contains $P(A)$, so in order to show that this subspace is precisely $P(A)$, it will suffice to show that its dimension is equal to the degree of the minimal polynomial of A. As in the proof of Theorem (4.4.14), it suffices to assume that the given matrix A is in Jordan canonical form

$$ A = \begin{bmatrix} J_{n_1}(\lambda_1) & & 0 \\ & \ddots & \\ 0 & & J_{n_p}(\lambda_n) \end{bmatrix} \equiv \begin{bmatrix} A_1 & & 0 \\ & A_2 \ddots & \\ 0 & & A_k \end{bmatrix} $$

where $k \leq p$, the Jordan blocks with the same eigenvectors are placed contiguously, and each $A_i \in M_{r_i}$ is composed of the direct sum of all the Jordan blocks of A that have the same eigenvalue. If $X \in C(A)$, write

$$ X = \begin{bmatrix} X_{11} & \cdots & X_{1k} \\ \vdots & \ddots & \vdots \\ X_{k1} & \cdots & X_{kk} \end{bmatrix}, \; X_{ij} \in M_{r_i, r_j} \tag{4.4.20} $$

and observe that $AX = XA$ implies that $A_i X_{ij} = X_{ij} A_j$, for all $i, j = 1, 2, \ldots, k$, which has only the trivial solution if $i \neq j$ by (4.4.6). Therefore, every matrix in $C(A)$ has the block diagonal form

$$ X = \begin{bmatrix} X_{11} & & 0 \\ & \ddots & \\ 0 & & X_{kk} \end{bmatrix}, \; X_{ii} \in M_{r_i} \tag{4.4.21} $$

Two matrices of this form commute if and only if their respective diagonal blocks commute. Thus, to prove the theorem it suffices to assume that $k = 1$, $r_1 = n$, and

$$
A = \begin{bmatrix} J_{n_1}(\lambda) & & 0 \\ & \ddots & \\ 0 & & J_{n_p}(\lambda) \end{bmatrix}
$$

with $n_1 + n_2 + \cdots + n_p = n$ and $n_1 \geq n_2 \geq \cdots \geq n_p \geq 1$, and to prove that the subspace of $C(A)$ consisting of those matrices that commute with every matrix in $C(A)$ has dimension $n_1 = \max\{n_1, n_2, ..., n_p\}$, since this is the degree of the minimal polynomial of A.

Notice that $J_k(\lambda)J_k(\mu) = [\lambda I + J_k(0)][\mu I + J_k(0)] = \lambda\mu I + \lambda J_k(0) + \mu J_k(0) + [J_k(0)]^2 = J_k(\mu)J_k(\lambda)$ for any $\lambda, \mu \in \mathbb{C}$, so the matrix

$$
Z \equiv \begin{bmatrix} J_{n_1}(\lambda+1) & & & 0 \\ & J_{n_2}(\lambda+2) & & \\ & & \ddots & \\ 0 & & & J_{n_p}(\lambda+p) \end{bmatrix} \in M_n
$$

whose blocks have distinct eigenvalues, commutes with A. Now let

$$
Y = \begin{bmatrix} Y_{11} & \cdots & Y_{1p} \\ \vdots & \ddots & \vdots \\ Y_{p1} & \cdots & Y_{pp} \end{bmatrix} \in M_n, \quad Y_{ij} \in M_{n_i, n_j}
$$

be a matrix that commutes with every matrix that commutes with A. Since Y must commute with Z, the argument we have already used to reduce (4.4.20) to the form (4.4.21) shows that Y must have the block diagonal form

$$
Y = \begin{bmatrix} Y_{11} & & 0 \\ & \ddots & \\ 0 & & Y_{pp} \end{bmatrix}, \quad Y_{ii} \in M_{n_i}
$$

Moreover, the identity $Y_{ii}J_{n_i}(\lambda+i) = J_{n_i}(\lambda+i)Y_{ii}$ is equivalent to $Y_{ii}J_{n_i}(0) = J_{n_i}(0)Y_{ii}$, so Lemma (4.4.11) says that each Y_{ii} is an upper triangular Toeplitz matrix that depends linearly on n_i parameters. The p matrices $Y_{11}, ..., Y_{pp}$ are not independent, however. Define the p-by-p block matrix

$$W_1 \equiv \begin{bmatrix} 0 & W_{12} & & 0 \\ & 0 & \ddots & \\ & & & \ddots \\ 0 & & & 0 \end{bmatrix} \in M_n$$

where the partitioning is conformal to that of A and all the blocks are zero except for the 1,2 block $W_{12} \in M_{n_1, n_2}$, which has the form

$$W_{12} = \begin{bmatrix} I \\ 0 \end{bmatrix}, \quad I \in M_{n_2}, \quad 0 \in M_{n_1-n_2, n_2}$$

This is possible because we have arranged that $n_1 \geq n_2 \geq \cdots \geq n_p \geq 1$. Because $W_{12} J_{n_2}(\lambda) = J_{n_1}(\lambda) W_{12}$, W_1 commutes with A and hence W_1 must also commute with Y. The identity $W_1 Y = Y W_1$ implies that $W_{12} Y_{22} = Y_{11} W_{12}$, which shows that Y_{22} is determined (in a linear way) by Y_{11}. The precise form of the dependence of Y_{22} on Y_{11} is exhibited by the block identity

$$W_{12} Y_{22} = \begin{bmatrix} I \\ 0 \end{bmatrix} Y_{22} = \begin{bmatrix} Y_{22} \\ 0 \end{bmatrix} = Y_{11} W_{12} = \begin{bmatrix} Y'_{11} & F \\ 0 & G \end{bmatrix} \begin{bmatrix} I \\ 0 \end{bmatrix} = \begin{bmatrix} Y'_{11} \\ 0 \end{bmatrix}$$

where we have partitioned Y_{11} so that $Y'_{11} \in M_{n_2}$ is the upper left n_2-by-n_2 principal submatrix of Y_{11}. Thus, $Y_{22} = Y'_{11}$, so Y_{22} is completely determined by Y_{11}.

One can continue in this way for $i = 2, 3, \ldots, p - 1$ to determine $W_i \in M_n$ to be a p-by-p block matrix with a nonzero block only in the $i, i+1$ position, and this block has the form

$$\begin{bmatrix} I \\ 0 \end{bmatrix} \in M_{n_i, n_{i+1}}, \quad I \in M_{n_{i+1}}, \quad 0 \in M_{n_i-n_{i+1}, n_{i+1}}$$

The same argument shows that each W_i commutes with A, and each $Y_{i+1,i+1} = Y'_{ii}$ is the upper left n_{i+1}-by-n_{i+1} principal submatrix of Y_{ii}.

Thus, all the blocks Y_{ii} are determined by Y_{11} (in fact, each Y_{ii} is the upper left n_i-by-n_i principal submatrix of Y_{11}), and Y_{11} depends linearly on n_1 parameters. The dimension of the subspace of matrices Y that commute with every matrix in $C(A)$ is therefore n_1.   □

If $A \in M_m$ and $B \in M_n$ do not have disjoint spectra, then the equation $AX - XB = C$ is singular and has a solution $X \in M_{m,n}$ only for certain right-hand sides C. How might one characterize the relationship of C to A and B that permits this equation to have a solution? For any $X \in M_{m,n}$ consider the identity

$$\begin{bmatrix} I & X \\ 0 & I \end{bmatrix}\begin{bmatrix} A & C \\ 0 & B \end{bmatrix}\begin{bmatrix} I & -X \\ 0 & I \end{bmatrix} = \begin{bmatrix} A & -AX+XB+C \\ 0 & B \end{bmatrix}$$

which, since

$$\begin{bmatrix} I & X \\ 0 & I \end{bmatrix}^{-1} = \begin{bmatrix} I & -X \\ 0 & I \end{bmatrix}$$

says that

$$\begin{bmatrix} A & C \\ 0 & B \end{bmatrix} \text{ is similar to } \begin{bmatrix} A & -AX+XB+C \\ 0 & B \end{bmatrix}$$

Thus, if there is a solution X to the equation $AX - XB = C$, then

$$\begin{bmatrix} A & C \\ 0 & B \end{bmatrix} \text{ is similar to } \begin{bmatrix} A & 0 \\ 0 & B \end{bmatrix}$$

The converse holds as well.

4.4.22   Theorem. Let $A \in M_m$, $B \in M_n$, and $C \in M_{m,n}$ be given. There is some $X \in M_{m,n}$ such that $AX - XB = C$ if and only if

$$\begin{bmatrix} A & C \\ 0 & B \end{bmatrix} \text{ is similar to } \begin{bmatrix} A & 0 \\ 0 & B \end{bmatrix}$$

Proof: The necessity of the similarity condition has already been shown, so let us assume that there is a nonsingular $S \in M_{m+n}$ such that

$$S\begin{bmatrix} A & 0 \\ 0 & B \end{bmatrix}S^{-1} = \begin{bmatrix} A & C \\ 0 & B \end{bmatrix}$$

Define two linear transformations $T_i: M_{m+n} \to M_{m+n}$, $i = 1, 2$, by

$$T_1(X) \equiv \begin{bmatrix} A & 0 \\ 0 & B \end{bmatrix}X - X\begin{bmatrix} A & 0 \\ 0 & B \end{bmatrix} \tag{4.4.23}$$

$$T_2(X) \equiv \begin{bmatrix} A & C \\ 0 & B \end{bmatrix}X - X\begin{bmatrix} A & 0 \\ 0 & B \end{bmatrix} \tag{4.4.24}$$

By the similarity hypothesis, $T_2(X) = ST_1(S^{-1}X)$ for all $X \in M_{m+n}$, so $T_2(X)$ $= 0$ if and only if $T_1(S^{-1}X) = 0$. In particular, the nullspaces of T_1 and T_2 must have the same dimension.

Now write

$$X = \begin{bmatrix} X_{11} & X_{12} \\ X_{21} & X_{22} \end{bmatrix}$$

with $X_{11} \in M_m$, $X_{12} \in M_{m,n}$, $X_{21} \in M_{n,m}$, $X_{22} \in M_n$, and calculate

$$T_1(X) = \begin{bmatrix} AX_{11}-X_{11}A & AX_{12}-X_{12}B \\ BX_{21}-X_{21}A & BX_{22}-X_{22}B \end{bmatrix}$$

$$T_2(X) = \begin{bmatrix} AX_{11}-X_{11}A+CX_{21} & AX_{12}-X_{12}B+CX_{22} \\ BX_{21}-X_{21}A & BX_{22}-X_{22}B \end{bmatrix}$$

If the nullspace of T_2 contains a matrix of the form $\begin{bmatrix} X_{11} & X_{12} \\ 0 & -I \end{bmatrix}$ then we are done, for $X = X_{12}$ would then satisfy $AX - XB = C$, as desired. In order to show that there is a matrix of this special form in the nullspace of T_2, consider the two linear transformations

$$\mathcal{T}_i : \text{nullspace } T_i \to M_{n,m+n}, \quad i = 1, 2$$

given by the projections

$$\mathcal{T}_i : \begin{bmatrix} X_{11} & X_{12} \\ X_{21} & X_{22} \end{bmatrix} \to [X_{21} \ X_{22}], \quad i = 1, 2$$

Notice that $X \in$ nullspace $\mathcal{T}_1$ if and only if $X_{21} = 0$, $X_{22} = 0$, $AX_{11} - X_{11}A = 0$, and $AX_{12} - X_{12}B = 0$, and that exactly the same four conditions characterize the nullspace of $\mathcal{T}_2$, so the nullspaces of $\mathcal{T}_1$ and $\mathcal{T}_2$ are identical. Moreover, notice that

$$\text{range } \mathcal{T}_1 = \{[X_{21} \ X_{22}] : BX_{21} - X_{21}A = 0 \text{ and } BX_{22} - X_{22}B = 0\}$$

and

$$\text{range } \mathcal{T}_2 = \{[X_{21} \ X_{22}] : BX_{21} - X_{21}A = 0, \ BX_{22} - X_{22}B = 0, \text{ and there} \\ \text{exist } X_{11}, \ X_{12} \text{ such that } CX_{21} = X_{11}A - AX_{11} \\ \text{and } CX_{22} = X_{12}B - AX_{12}\}$$

so range $T_2 \subseteq$ range T_1. The basic identity relating the dimensions of the nullspace, range, and domain of a linear mapping ensures that

$$\dim \text{nullspace } T_i + \dim \text{range } T_i = \dim \text{nullspace } T_i, \quad i = 1, 2$$

(see (0.2.3) in [HJ]). Since we have shown that nullspace $T_1 =$ nullspace T_2 and dim nullspace $T_1 = \dim$ nullspace T_2, it follows that dim range $T_1 = \dim$ range T_2, and hence the inclusion of the range of T_2 in that of T_1 is actually an equality: range $T_2 =$ range T_1. Finally, notice that $\begin{bmatrix} 0 & 0 \\ 0 & -I \end{bmatrix} \in$ nullspace T_1, so $[0 \ -I] \in$ range $T_1 =$ range T_2, from which it follows that there is some $X \in$ nullspace T_2 of the form

$$\begin{bmatrix} X_{11} & X_{12} \\ 0 & -I \end{bmatrix}$$

which is what we needed to show. ▯

The proof we have given for Theorem (4.4.22), while completely elementary, is, like all other known proofs of this result, nonconstructive. Given that $\begin{bmatrix} A & 0 \\ 0 & B \end{bmatrix}$ and $\begin{bmatrix} A & C \\ 0 & B \end{bmatrix}$ are similar via a given nonsingular $S \in M_{m+n}$, for many purposes it would be desirable to have an algorithm to construct a matrix X (we know it exists) such that $\begin{bmatrix} I & X \\ 0 & I \end{bmatrix}$ accomplishes this similarity.

A similar argument can be used to prove a second theorem that is related to the more general linear matrix equation $AX - YB = C$. Since this equation has two independent matrix unknowns X and Y, it may be expected to have a solution under weaker conditions on the given matrices A, B, and C than the more restrictive equation $AX - XB = C$. This expectation is realized—one merely replaces *similarity* in Theorem (4.4.22) with the weaker notion of *equivalence* and obtains a general characterization of the relationship among A, B, and C that permits the equation $AX - YB = C$ to have a solution. Recall that matrices $A, B \in M_{m,n}$ are *equivalent* if there exist nonsingular matrices $P \in M_m$ and $Q \in M_n$ such that $B = PAQ$. Two matrices of the same size are equivalent if and only if they have the same rank (see Problem 6 in Section (3.5) of [HJ]).

4.4.25   Theorem. Let $A \in M_{m,n}$, $B \in M_{p,q}$, and $C \in M_{m,q}$ be given. There are matrices $X \in M_{m,q}$ and $Y \in M_{n,p}$ such that $AX - YB = C$ if and only if

$$\text{rank} \begin{bmatrix} A & C \\ 0 & B \end{bmatrix} = \text{rank} \begin{bmatrix} A & 0 \\ 0 & B \end{bmatrix}$$

Proof: For any $X \in M_{n,q}$ and any $Y \in M_{m,p}$ we have

$$\begin{bmatrix} I & -Y \\ 0 & I \end{bmatrix} \begin{bmatrix} A & 0 \\ 0 & B \end{bmatrix} \begin{bmatrix} I & X \\ 0 & I \end{bmatrix} = \begin{bmatrix} A & AX-YB \\ 0 & B \end{bmatrix}$$

so if the equation $AX - YB = C$ has a solution X, Y, the asserted equivalence is clear. Conversely, define two linear transformations $S_i : M_{n+q} \times M_{m+p} \rightarrow M_{m+p,n+q}$, $i = 1, 2$, by

$$S_1(X,Y) \equiv \begin{bmatrix} A & 0 \\ 0 & B \end{bmatrix} X - Y \begin{bmatrix} A & 0 \\ 0 & B \end{bmatrix} \tag{4.4.26}$$

$$S_2(X,Y) \equiv \begin{bmatrix} A & C \\ 0 & B \end{bmatrix} X - Y \begin{bmatrix} A & 0 \\ 0 & B \end{bmatrix} \tag{4.4.27}$$

where $X \in M_{n+q}$ and $Y \in M_{m+p}$. If there are nonsingular matrices $P \in M_{m+p}$ and $Q \in M_{n+q}$ such that

$$P \begin{bmatrix} A & 0 \\ 0 & B \end{bmatrix} Q = \begin{bmatrix} A & C \\ 0 & B \end{bmatrix}$$

then $S_2(X,Y) = PS_1(QX,P^{-1}Y)$ for all X, Y, and hence $S_2(X,Y) = 0$ if and only if $S_1(QX,P^{-1}Y) = 0$. In particular, the nullspaces of S_1 and S_2 must have the same dimension.

Now write

$$X = \begin{bmatrix} X_{11} & X_{12} \\ X_{21} & X_{22} \end{bmatrix} \in M_{n+q} \text{ and } Y = \begin{bmatrix} Y_{11} & Y_{12} \\ Y_{21} & Y_{22} \end{bmatrix} \in M_{m+p}$$

with $X_{11} \in M_n$, $X_{22} \in M_q$, $Y_{11} \in M_m$, $Y_{22} \in M_p$, and calculate

$$S_1(X,Y) = \begin{bmatrix} AX_{11}-Y_{11}A & AX_{12}-Y_{12}B \\ BX_{21}-Y_{21}A & BX_{22}-Y_{22}B \end{bmatrix}$$

$$S_2(X,Y) = \begin{bmatrix} AX_{11}-Y_{11}A+CX_{21} & AX_{12}-Y_{12}B+CX_{22} \\ BX_{21}-Y_{21}A & BX_{22}-Y_{22}B \end{bmatrix}$$

If the nullspace of S_2 contains a point (X,Y) with

$$X = \begin{bmatrix} X_{11} & X_{12} \\ X_{21} & -I \end{bmatrix} \text{ and } Y = \begin{bmatrix} Y_{11} & Y_{12} \\ Y_{21} & -I \end{bmatrix}$$

then we are done, for then $X = X_{12}$ and $Y = Y_{12}$ would satisfy $AX - YB = C$, as desired. Consider the two linear transformations

$$S_i: \text{nullspace } S_i \rightarrow M_{q,n} \times M_q \times M_{p,m} \times M_p, \quad i = 1, 2$$

given by the projections

$$S_i \left(\begin{bmatrix} X_{11} & X_{12} \\ X_{21} & X_{22} \end{bmatrix}, \begin{bmatrix} Y_{11} & Y_{12} \\ Y_{21} & Y_{22} \end{bmatrix} \right) \rightarrow (X_{21}, X_{22}, Y_{21}, Y_{22}), \quad i = 1, 2$$

The nullspaces of S_1 and S_2 are exactly the same since $S_1(X,Y) = S_2(X,Y)$ for all X, Y such that $X_{21} = 0$ and $X_{22} = 0$. Furthermore, range $S_2 \subseteq$ range S_1 since the points in the ranges of both S_1 and S_2 must all satisfy the relations $BX_{21} - Y_{21}A = 0$ and $BX_{22} - Y_{22}B = 0$, while the points in the range of S_2 must satisfy the additional relations $CX_{21} = Y_{11}A - AX_{11}$ and $CX_{22} = Y_{12}B - AX_{12}$. From the general identities for rank and nullity

$$\text{dim nullspace } S_i + \text{dim range } S_i = \text{dim nullspace } S_i, \quad i = 1, 2$$

we conclude that dim range $S_1 =$ dim range S_2 and hence range $S_1 =$ range S_2. Since $\left(\begin{bmatrix} 0 & 0 \\ 0 & -I \end{bmatrix}, \begin{bmatrix} 0 & 0 \\ 0 & -I \end{bmatrix} \right)$ is in the nullspace of S_1, the point $(0, -I, 0, -I)$ is in the common range of S_1 and S_2, so there is some point (X,Y) in the nullspace of S_2 with X and Y of the form

$$X = \begin{bmatrix} X_{11} & X_{12} \\ X_{21} & -I \end{bmatrix} \text{ and } Y = \begin{bmatrix} Y_{11} & Y_{12} \\ Y_{21} & -I \end{bmatrix}$$

which is what we needed to show. ☐

The preceding result may be thought of as a generalization of the familiar fact that the square system of linear equations $Ax = c$ has a solution $x \in \mathbb{C}^n$ for a given $c \in \mathbb{C}^n$ if and only if $A \in M_n$ and $[A \ c] \in M_{n,n+1}$ have the same rank.

Problems

1.   Consider $A \equiv I \in M_n$ to explain in a concrete case why a matrix B that commutes with A need not be a polynomial in A, but if B commutes with every matrix that commutes with A, then B must be a polynomial in A. What do polynomials in A look like in this case?

2.   Let $A, B \in M_n$ be given matrices. Show that the equation $AX - XB = 0$ has a nonsingular solution $X \in M_n$ if and only if A and B are similar.

3.   If $A \in M_n$ is nonsingular, then the matrix equation $AX = B$ has the unique solution $X = A^{-1}B$. Analyze the solvability of $AX = B$ in general, given $A \in M_{m,n}$, $B \in M_{m,p}$. Notice that this is a generalization of the solvability theory for the linear system $Ax = b$. You may wish to utilize some results from that theory and phrase your answer in analogous terms.

4.   If $A \in M_n$ is given, show that there is a nonsingular matrix X such that $AX = XA^T$. Interpret this result.

5.   The proof given for (4.4.5) relies on the fact that $I \otimes A$ and $B \otimes I$ can be simultaneously upper triangularized, but this is not the only way to prove the result. Provide details for the following argument that the additive property (4.4.5) of the Kronecker sum follows directly from the multiplicative property (4.2.12) of the Kronecker product: Note that the eigenvalues of the matrix $(I_n + \epsilon A) \otimes (I_m + \epsilon B) = (I_n \otimes I_m) + \epsilon(I_n \otimes B) + \epsilon(A \otimes I_m) + \epsilon^2(A \otimes B)$ are $\{(1 + \epsilon\lambda_i)(1 + \epsilon\mu_j)\}$, and hence the eigenvalues of the matrix $[(I_n + \epsilon A) \otimes (I_m + \epsilon B) - I_n \otimes I_m]/\epsilon$ are $\lambda_i + \mu_j + \epsilon\lambda_i\mu_j$ for all $\epsilon \neq 0$. Now let $\epsilon \to 0$.

6.   According to (4.4.5), $\sigma(F + G) = \sigma(F) + \sigma(G)$ if F and G are of the special form $F = I \otimes A$ and $G = B \otimes I$, which commute. Recall that if F and G are commuting square matrices of the same size, regardless of form, $\sigma(F + G) \subseteq \sigma(F) + \sigma(G)$. Show by example that the containment is not, in general, an equality.

7.   Show that the necessary and sufficient condition in (4.4.7) for solvability of Lyapunov's equation is that if the set of eigenvalues of A is reflected across the imaginary axis in the complex plane, then no eigenvalue of $-A$ is in this reflected set. What does this say about an eigenvalue that is zero or purely imaginary?

8.   Let $A \in M_n$ be a given matrix, and let $\lambda_1, ..., \lambda_n$ be the eigenvalues of A.

Show that the matrix equation $AX - XA = \lambda X$ has a nontrivial $(X \neq 0)$ solution $X \in M_n$ if and only if $\lambda = \lambda_i - \lambda_j$ for some i, j.

9. Let $A \in M_n$ be a given matrix with p distinct eigenvalues $\{\lambda_1, \lambda_2, \ldots, \lambda_p\}$, and suppose that the Jordan form of A with respect to each eigenvalue λ_i is a direct sum of the form $J_{n_{i,1}}(\lambda_i) \oplus \cdots \oplus J_{n_{i,k_i}}(\lambda_i)$, arranged so that $n_{i,1} \geq n_{i,2} \geq \cdots \geq n_{i,k_i} \geq 1$. Show that $C(A)$, the set of matrices $X \in M_n$ that commute with A, is a subspace of M_n of dimension

$$\nu = \sum_{r=1}^{p} \sum_{i,j=1}^{k_r} \min\{n_{r,i}, n_{r,j}\} = \sum_{r=1}^{p} \sum_{i,j=1}^{k_r} n_{r,max\{i,j\}}$$

$$= \sum_{r=1}^{p} \left[n_{r,1} + 3n_{r,2} + 5n_{r,3} + \cdots + (2k_r - 1)\, n_{r,k_r} \right]$$

Notice that the *values* of the distinct eigenvalues of A play no role in this formula, and the dimension of $C(A)$ is determined only by the *sizes* of the Jordan blocks of A associated with the distinct eigenvalues of A.

10. Show that $A \in M_n$ is normal if and only if $C(A) = C(A^*)$.

11. Discuss the statement and proof of Theorem (4.4.22) in the special case in which $A \in M_n$ and $B = \lambda \in \mathbb{C}$ are given, $C = y \in \mathbb{C}^n$ is arbitrary, and $X = x \in \mathbb{C}^n$ is unknown. What does this have to do with eigenvalues of A?

12. Let A, B, C be as in Theorem (4.4.22). Examine the proof of the theorem and explain why $\begin{bmatrix} A & C \\ 0 & B \end{bmatrix}$ is similar to $\begin{bmatrix} A & 0 \\ 0 & B \end{bmatrix}$ if and only if the respective nullspaces of the two operators defined by (4.4.23-24) have the same dimensions.

13. Let $A \in M_n$ have eigenvalues λ_i with respective algebraic multiplicities m_i, $i = 1, \ldots, k$, and suppose $\lambda_i \neq \lambda_j$ if $i \neq j$. Use Theorems (4.4.22) and (4.4.6) (don't just quote the Jordan canonical form theorem) to show that A is similar to a block upper triangular matrix $T_1 \oplus \cdots \oplus T_k$, where each $T_i \in M_{m_i}$ is upper triangular and has all diagonal entries equal to λ_i.

14. Let $A \in M_{m,n}$ and $C \in M_{m,q}$ be given. Use Theorem (4.4.25) to show that the system of linear equations $AX = C$ has a solution $X \in M_{n,q}$ if and only if A and $[A \ C]$ have the same rank. Comment on the special case $m = n$, $q = 1$.

15. Let A, B, C be as in Theorem (4.4.25). Examine the proof of the theorem and explain why $\begin{bmatrix} A & C \\ 0 & B \end{bmatrix}$ has the same rank as $\begin{bmatrix} A & 0 \\ 0 & B \end{bmatrix}$ if and only if the respective nullspaces of the two linear transformations defined by (4.4.26-27) have the same dimension.

16. Two matrices $F, G \in M_m$ are said to be *consimilar* if there is a nonsingular $S \in M_n$ such that $F = SG\bar{S}^{-1}$ (see Section (4.6) of [HJ]). Let $A \in M_m$, $B \in M_n$, and $C \in M_{m,n}$ be given. If there is some $X \in M_{m,n}$ such that $A\bar{X} - XB = C$, show that $\begin{bmatrix} A & C \\ 0 & B \end{bmatrix}$ is consimilar to $\begin{bmatrix} A & 0 \\ 0 & B \end{bmatrix}$. The proof of Theorem (4.4.22) may be modified to prove the converse.

17. Let $A \in M_n$ be a given nonderogatory matrix. Provide details for the following Kronecker product proof that $C(A) = P(A)$, which is the easy half of (4.4.18); for another direct proof see (3.2.4.2) in [HJ]. We are interested in an upper bound on the dimension of the nullspace of the linear transformation $T(X) = AX - XA$, which is at least n since the independent set $\{I, A, A^2, ..., A^{n-1}\}$ is in the nullspace, and which is the same as the dimension of the nullspace of the Kronecker sum $I \otimes A - A^T \otimes I$. Since A is similar to the companion matrix of its characteristic polynomial

$$C = \begin{bmatrix} 0 & & & 0 & -a_0 \\ 1 & \ddots & & & \vdots \\ & \ddots & \ddots & & \vdots \\ & & 1 & 0 & -a_{n-2} \\ 0 & & & 1 & -a_{n-1} \end{bmatrix}$$

it suffices to consider $I \otimes C - C^T \otimes I$, which, by explicit calculation, is easily seen to have rank at least $n(n-1)$ (focus on $-I$ blocks in the $n-1$ upper-diagonal positions). Thus, the dimension of the nullspace of $I \otimes C - C^T \otimes I$ is at most $n^2 - n(n-1)$.

18. Let $A \in M_m$, $B \in M_n$ be given.
(a) Make a list of the norms considered in Problems 27-28 of Section (4.2) that satisfy the condition $\|I\| = 1$ for the identity matrix $I \in M_p$ for all $p = 1, 2, ...$
(b) Explain why $\|A \otimes I_n + I_m \otimes B\| \leq \|A\| + \|B\|$ for all of the norms listed in (a), and show that strict inequality is possible in each case for some choice of A and B.
(c) Show that $\|A \otimes I_n + I_n \otimes A\| = 2\|A\|$ when $\|\cdot\|$ is the maximum absolute row sum or maximum absolute column sum matrix norm.

(d) Show that the identity in (c) does not always hold when $\|\cdot\|$ is the spectral norm, but it does hold if A is normal.

19.   Use the fact that the Frobenius norm $\|\cdot\|_2$ is generated by the Frobenius inner product, $\|X\|_2 = [\operatorname{tr} X^* X]^{\frac{1}{2}}$ for any $X \in M_{m,n}$, to show that $\|A \bullet I_n + I_m \bullet B\|_2^2 = n\|A\|_2^2 + m\|B\|_2^2 + 2\operatorname{Re}(\operatorname{tr} A)(\operatorname{tr} \overline{B})$ for all $A \in M_m$, $B \in M_n$.

20.   How many square roots can a matrix with a positive spectrum have, if each square root is required to have a positive spectrum? For a different approach to answering this question, see Problem 37 in Section (6.4).

(a) Let $B \in M_n$ be given. Note that $C(B) \subseteq C(B^2)$ and give an example to show that this containment can be strict. If all the eigenvalues of B are positive, use the result in Problem 9 to show that $C(B) = C(B^2)$ and that, more generally, $C(B) = C(B^k)$ for all $k = 1, 2, \ldots$.

(b) Suppose all the eigenvalues of $A \in M_n$ are positive. Use the result in (a) to show that there is at most one $B \in M_n$ with a positive spectrum such that $B^2 = A$, that is, A has at most one square root, all of whose eigenvalues are positive. Since there is always a B with only positive eigenvalues such that $B^2 = A$ and B is a polynomial in A [see Example (6.2.14)], it follows that a complex matrix with a positive spectrum has a unique square root with a positive spectrum. Compare this with Theorem (7.2.6) in [HJ], which concerns positive semidefinite matrices.

(c) Explain why "square root" may be replaced by "kth root" in (b) for each positive integer k.

Notes and Further Readings. The original statements and proofs of Theorems (4.4.22) and (4.4.25) are in W. Roth, The Equations $AX - YB = C$ and $AX - XB = C$ in Matrices, *Proc. Amer. Math. Soc.* 3 (1952), 392-396. The proofs presented for these two theorems are adapted from H. Flanders and H. K. Wimmer, On the Matrix Equation $AX - XB = C$ and $AX - YB = C$, *SIAM J. Appl. Math.* 32 (1977), 707-710. The consimilarity analog of Roth's theorem noted in Problem 16 is in J. H. Bevis, F. J. Hall, and R. E. Hartwig, Consimilarity and the Matrix Equation $A\overline{X} - XB = C$ in [UhGr]; this paper also contains consimilarity analogs of many other results in this section, including a conspectral analog of Theorem (4.4.6) for the equation $A\overline{X} - XB = C$.

4.5   Additive and multiplicative commutators and linear preservers

4.5.1   Definition. Let $A, B \in M_n$. A matrix of the form $AB - BA$ is called an *additive commutator*. If A and B are nonsingular, a matrix of the form $ABA^{-1}B^{-1}$ is called a *multiplicative commutator*.

Exercise. Show that two given matrices $A, B \in M_n$ commute if and only if their additive commutator is 0. If they are nonsingular, show that they commute if and only if their multiplicative commutator is I.

Exercise. Show that an additive commutator has trace 0 and a multiplicative commutator has determinant 1.

Exercise. Show that the property of being either an additive or multiplicative commutator is a similarity invariant.

Which matrices are realizable as additive or multiplicative commutators? We address the additive case first.

4.5.2   Theorem. A matrix $C \in M_n$ may be written as $C = XY - YX$ for some $X, Y \in M_n$ if and only if tr $C = 0$.

Proof: Since tr $XY =$ tr YX, we need only show that if tr $C = 0$, then the (nonlinear) matrix equation $XY - YX = C$ can be solved for $X, Y \in M_n$. To do so, we may replace C by any matrix similar to C because the property of being a commutator is similarity invariant. Recall from (1.3.4) that any matrix $A \in M_n$ is unitarily similar to a matrix whose diagonal entries are all equal (and hence are all equal to $\frac{1}{n}$tr A). Thus, we may assume without loss of generality that $C = [c_{ij}]$ with $c_{11} = c_{22} = \cdots = c_{nn} = 0$. Now choose $X = \operatorname{diag}(x_1, x_2, \ldots, x_n)$, in which $x_1, \ldots, x_n$ are pairwise distinct, but otherwise arbitrary, complex numbers. With X fixed and C given, $XY - YX = C$ becomes a linear matrix equation in Y that has a simple solution. If $Y = [y_{ij}]$, then $XY - YX = [(x_i - x_j)y_{ij}]$. Thus,

$$
y_{ij} = \begin{cases} \dfrac{c_{ij}}{x_i - x_j} & \text{if } i \neq j \\[6pt] \text{arbitrary} & \text{if } i = j \end{cases}
$$

gives a solution since $x_i \neq x_j$ when $i \neq j$, and $c_{ii} = 0$ for $i = 1, \ldots, n$; there may be many other solutions [see Corollary (4.4.15)].   □

Now consider the problem of characterizing multiplicative commutators. As in the case of additive commutators, we shall find that every matrix in M_n satisfying the obvious necessary condition is a multiplicative commutator, but first we make some helpful observations of a broader nature. Recall that a *scalar matrix* is a scalar multiple of the identity matrix.

4.5.3 Lemma. Suppose that $A \in M_n$ is not a scalar matrix, and let $\alpha \in \mathbb{C}$ be given. Then there is a matrix similar to A that has α in its 1,1 position and has at least one nonzero entry below the diagonal in its first column.

Exercise. Verify (4.5.3). *Hint:* First do the case $n = 2$ by direct calculation, and then show that verification of the general case may be based upon the case $n = 2$.

4.5.4 Theorem. Let $A \in M_n$ be given and suppose rank $A = k \leq n$; if $k = n$, assume that A is not a scalar matrix. Let $b_1, b_2,..., b_n$ and $c_1, c_2,..., c_n$ be given complex numbers, exactly $n - k$ of which are zero; if $k = n$, assume that $b_1 b_2 \cdots b_n c_1 c_2 \cdots c_n = \det A$. Then there is a matrix $B \in M_n$ with eigenvalues $b_1, b_2,..., b_n$ and a matrix $C \in M_n$ with eigenvalues $c_1, c_2,..., c_n$ such that $A = BC$.

Proof: If $k = 0$, then $A = 0$ and $c_1,..., c_n$ may be ordered so that $b_i c_{j_i} = 0$, $i = 1,..., n$. The choice $B = \mathrm{diag}(b_1,...,b_n)$ and $C = \mathrm{diag}(c_{j_1},...,c_{j_n})$ then verifies the assertion in this event.

We now assume that $k \geq 1$ and may therefore assume without loss of generality that $b_1 c_1 \neq 0$ in order to complete the proof via induction on n. The case $n = 1$ is clear and the case $n = 2$ is a straightforward calculation. As both the hypothesis and conclusion of the theorem are invariant under simultaneous similarity, we may assume that A is in any form achievable via similarity. Because $A = [a_{ij}]$ is not a scalar matrix, using Lemma (4.5.3) we may suppose that $a_{11} = b_{11} c_{11}$ and that $[a_{21},..., a_{n1}]^T \neq 0$.

Now, assume that $n \geq 3$ and that B and C in the desired factorization $A = BC$ have the form

$$B = \begin{bmatrix} b_1 & 0 \\ B_{21} & B_{22} \end{bmatrix} \text{ and } C = \begin{bmatrix} c_1 & C_{12} \\ 0 & C_{22} \end{bmatrix}$$

and partition A conformally as

$$A = \begin{bmatrix} b_1 c_1 & A_{12} \\ A_{21} & A_{22} \end{bmatrix}$$

Here, A_{12}^T, A_{21}, B_{21}, $C_{12}^T \in \mathbb{C}^{n-1}$ and A_{22}, B_{22}, $C_{22} \in M_{n-1}$. In order for there to be a solution of this form, we must have $C_{12} = A_{12}/b_1$, $B_{21} = A_{21}/c_1$, and $A_{22} = B_{22}C_{22} + B_{21}C_{12}$. It follows that B_{22} and C_{22} must satisfy $B_{22}C_{22} = A_{22} - A_{21}A_{12}/b_1 c_1$. If $A_{22} - A_{21}A_{12}/b_1 c_1 \in M_{n-1}$ is not a nonzero scalar matrix, the induction hypothesis may be applied because

$$\det(A_{22} - A_{21}A_{12}/b_1 c_1) = (\det A)/b_1 c_1$$

and

$$\mathrm{rank}(A_{22} - A_{21}A_{12}/b_1 c_1) = k - 1$$

(see the following exercise). This would ensure matrices B_{22} and C_{22} with respective spectra $\{b_2,...,b_n\}$ and $\{c_2,...,c_n\}$, which would complete the proof.

It thus suffices to show via similarity that, for $n \geq 3$, it may be assumed that $A_{22} - A_{21}A_{12}/b_1 c_1$ is not a nonzero scalar matrix. In the rank deficient case $k < n$, this is guaranteed by the fact that $\mathrm{rank}(A_{22} - A_{21}A_{12}/b_1 c_1) = k - 1$. Thus, assume that A is nonsingular, $n \geq 3$, and $A_{22} - A_{21}A_{12}/b_1 c_1 = \alpha I$ for $0 \neq \alpha \in \mathbb{C}$. Since rank $A > 2$, the range of A_{22} contains vectors that are not multiples of A_{21}. Thus we may choose $w \in \mathbb{C}^{n-1}$ such that $w^T A_{21} = 0$ and $w^T A_{22} \neq 0$, and let

$$S \equiv \begin{bmatrix} 1 & w^T \\ 0 & I \end{bmatrix} \in M_n$$

Then

$$S^{-1}AS = \begin{bmatrix} b_1 c_1 & A_{12} + b_1 c_1 w^T - w^T A_{22} \\ A_{21} & A_{22} + A_{21} w^T \end{bmatrix}$$

Now compute

$$(A_{22} + A_{21}w^T) - A_{21}(A_{12} + b_1 c_1 w^T - w^T A_{22})/b_1 c_1$$

$$= A_{22} - A_{21}A_{12}/b_1 c_1 + A_{21}w^T A_{22}/b_1 c_1$$

$$= \alpha I + A_{21}w^T A_{22}/b_1 c_1$$

Since $A_{21} \neq 0$ and $w^T A_{22} \neq 0$, this matrix is a rank 1 perturbation of a scalar matrix and is, therefore, not a scalar matrix. $\square$

Exercise. Let x, $y \in \mathbb{C}^n$ and $B \in M_n$ be given, and set $A \equiv \begin{bmatrix} 1 & z^T \\ y & B \end{bmatrix} \in M_{n+1}$. Use elementary column operations (or do a right multiplication by $\begin{bmatrix} 1 & -z^T \\ 0 & I \end{bmatrix}$) to show that

$$\det A = \det \begin{bmatrix} 1 & 0 \\ y & B - yz^T \end{bmatrix} = \det (B - yz^T)$$

and explain why rank $A =$ rank $(B - yz^T) + 1$. How is this result used in the inductive step of the preceding proof?

Exercise. Show that the factors B and C guaranteed in (4.5.4) may be simultaneously upper and lower triangularized, respectively, by similarity.

The multiplicative analog of Theorem (4.5.2) now follows readily; it is also originally due to K. Shoda.

4.5.5   Theorem. A matrix $A \in M_n$ may be written as $A = XYX^{-1}Y^{-1}$ for some nonsingular X, $Y \in M_n$ if and only if $\det A = 1$.

Proof. The necessity of the determinant condition is immediate. Conversely, suppose that $\det A = 1$. If A is not a scalar matrix, let $b_1, \dots, b_n$ be distinct nonzero scalars and let $c_i = b_i^{-1}$ for $i = 1, \dots, n$. Apply Theorem (4.5.4) to write $A = XZ$, with $\sigma(X) = \{b_1, \dots, b_n\}$ and $\sigma(Z) = \{c_1, \dots, c_n\}$. But Z is then similar to X^{-1}, so that $Z = YX^{-1}Y^{-1}$ for some nonsingular $Y \in M_n$. It follows that $A = XZ = XYX^{-1}Y^{-1}$, as was to be shown. If $A = \alpha I$ for some $\alpha \in \mathbb{C}$, then $\alpha^n = 1$ and we may take $X = \text{diag}(\alpha, \alpha^2, \dots, \alpha^n)$ and $Z = \text{diag}(1, \alpha, \alpha^2, \dots, \alpha^{n-1})$. Again, Z is similar to X^{-1} and the proof is completed as before. $\square$

Exercise. If $A \in M_n$ has determinant 1 and is not a scalar matrix, show that $A = XYX^{-1}Y^{-1}$ in which the respective determinants of X and Y may be arbitrarily prescribed (but nonzero).

Exercise. Use (4.5.2) and (4.5.5) to note that (a) the sum of two additive commutators is an additive commutator, and (b) the product of two multiplicative commutators is a multiplicative commutator. It follows that the set of additive commutators forms an additive subgroup of M_n, and the set of multiplicative commutators forms a multiplicative subgroup of $GL(n,\mathbb{C})$, the multiplicative group of nonsingular n-by-n complex matrices. It is difficult to verify these closure properties without (4.5.2) and (4.5.5). In a general group G, the set of all commutators in G (with respect to the group operation of G) is *not* always a subgroup of G.

Over the years, there has been much work done to characterize the linear transformations (on the vector space of matrices) that preserve a given feature of a matrix, for example, the determinant, eigenvalues, rank, positive definiteness, similarity class, congruence class, idempotence, M-matrices, etc. Two classical examples of such results are the characterizations of the determinant and spectrum preservers first obtained in 1897 by Frobenius, who showed that a linear transformation $T: M_n \to M_n$ such that $\det T(X) = c \det X$ for some constant $c \ne 0$ for every $X \in M_n$ must either be of the form $T(X) = AXB$ or of the form $T(X) = AX^TB$ for all $X \in M_n$, where $A, B \in M_n$ are fixed matrices such that $\det AB = c$. If, on the other hand, $T(X)$ always has the same spectrum as X, Frobenius showed that either $T(X) = SXS^{-1}$ or $T(X) = SX^TS^{-1}$ for all X, for some fixed nonsingular S.

The connection between Frobenius' results and Kronecker products is apparent from (4.3.1). According to Lemma (4.3.2), each linear mapping $T: M_n \to M_n$ can be represented by a matrix in M_{n^2}, that is, there is a fixed matrix $K(T) = [K_{ij}] \in M_{n^2}$, each $K_{ij} \in M_n$, such that $\text{vec } T(X) = K(T) \text{ vec } X$ for every $X \in M_n$. By (4.3.1), the mapping $T(\cdot)$ has the form $T(X) = AXB$ if and only if the matrix $K(T)$ has the form $K(T) = B^T \otimes A$, that is, all the blocks K_{ij} have the form $K_{ij} = b_{ji}A$, where $B = [b_{ij}]$. Thus, to prove Frobenius' results, or similar theorems about other types of linear preservers, one must show that certain assumptions about the linear mapping T imply that the matrix $K(T)$ has a special Kronecker product structure. A sample of some of the known results about linear preservers is given in the following theorems.

4.5.6   Theorem. Let $T: M_n \to M_n$ be a given linear transformation. The following are equivalent:

(a)   There exists a nonzero constant $c \in \mathbb{C}$ such that $\det T(X) = c \det X$ for all $X \in M_n$;

(b)   There exists a nonzero constant $c \in \mathbb{C}$ such that $\det T(H) = c \det H$ for every Hermitian $H \in M_n$;

(c)   rank $T(X) = $ rank X for all $X \in M_n$;

(d)   rank $T(X) = 1$ for all $X \in M_n$ with rank $X = 1$;

(e)   rank $T(X) = n$ for all $X \in M_n$ with rank $X = n$, that is, $T(X)$ is nonsingular whenever X is nonsingular; and

(f)   There exist nonsingular matrices $A, B \in M_n$ such that *either* $T(X) = AXB$ for all $X \in M_n$, *or* $T(X) = AX^T B$ for all $X \in M_n$.

4.5.7   Theorem: Let $T: M_n \to M_n$ be a given linear transformation. The following are equivalent:

(a) $\sigma(T(X)) = \sigma(X)$ for every $X \in M_n$, that is, X and $T(X)$ have the same characteristic polynomials;

(b) $\sigma(T(H)) = \sigma(H)$ for every Hermitian $H \in M_n$, that is, H and $T(H)$ have the same characteristic polynomials;

(c) $\det T(X) = \det X$ and tr $T(X) = $ tr X for all $X \in M_n$; and

(d)   There exists a nonsingular $S \in M_n$ such that *either* $T(X) = SXS^{-1}$ for all $X \in M_n$ *or* $T(X) = SX^T S^{-1}$ for all $X \in M_n$.

Problems

1.   Give an alternate proof of (4.5.2) by first using the same reduction to a matrix C with zero diagonal entries and letting $X = \operatorname{diag}(1, 2, \ldots, n)$. Then write the resulting linear system in Y in Kronecker form and analyze its coefficient matrix. Recall that a (singular) linear system is solvable if and only if the right-hand side (vec C in this case) is orthogonal to the nullspace of the adjoint of the coefficient matrix.

2.   If $A \in M_n$ and $B \in M_m$, show that $A \bullet B$ is an additive commutator if and only if either A or B is an additive commutator.

3.   Use (4.5.2) to show that the identity matrix (and, more generally, any nonzero nonnegative diagonal matrix) cannot be an additive commutator.

4.   Consider the real vector space V of all polynomials with real coefficients and consider the linear transformations $T_1, T_2: V \to V$ defined by $T_1(p(t)) \equiv$

$(d/dt)p(t)$ and $T_2(p(t)) \equiv tp(t)$. Show that $T_1(T_2(p(t))) - T_2(T_1(p(t))) = p(t)$ for all $p(t) \in V$, that is, $T_1 T_2 - T_2 T_1 = I$. Does this contradict Problem 3?

5.   Let $A, B \in M_n$ and let $\|\|\cdot\|\|$ be a matrix norm on M_n. Provide the details for the following proof that $AB - BA \neq I$: If $AB - BA = I$, then $2A = A(AB - BA) + (AB - BA)A = A^2 B - BA^2$ and $A^{k+1} B - BA^{k+1} = (k + 1)A^k$ for $k = 1, 2, \ldots$. Thus,

$$(k + 1) \|\| A^k \|\| \leq 2 \|\| A^{k+1} \|\| \, \|\| B \|\| \leq 2 \|\| A^k \|\| \, \|\| A \|\| \, \|\| B \|\|$$

for all $k = 1, 2, \ldots$ and hence $\|\| A^m \|\| = 0$ if $m + 1 > 2 \|\| A \|\| \, \|\| B \|\|$. But then $A^m = 0 \Rightarrow A^{m-1} = 0 \Rightarrow \cdots \Rightarrow A = 0$. The virtue of this proof (due to H. Wielandt) is that it does not use the trace function; it extends verbatim to the infinite-dimensional case to show that $AB - BA \neq I$ for bounded linear operators A, B on a Banach space or, more generally, for elements A, B of a Banach algebra. These ideas arise naturally in quantum mechanics; the position operator X and the momentum operator P satisfy the commutation rule $XP - PX = (ih/2\pi)I$, where h is Planck's constant. Wielandt's result gave a simple solution to a long-open question in quantum mechanics and showed that operators satisfying such a commutation rule cannot be bounded.

6.   The identity matrix cannot be a commutator, but how close to the identity can a commutator be? If $\|\|\cdot\|\|$ is a given matrix norm on M_n, show that $\|\| I - C \|\| \geq 1$ for any additive commutator $C \in M_n$, that is, there are no additive commutators in the $\|\|\cdot\|\|$-open ball of radius 1 around I.

7.   If $C \in M_n$ and tr $C = 0$, show that $XY - YX = C$ can be solved with the spectrum of one of the factors arbitrary, except that repeats are not allowed. Actually, more can be said: The spectra of both factors can be chosen arbitrarily, except that repeats are not, in general, allowed in one of the factors, but this is more difficult to prove.

8.   Let $X, Y \in M_n$ be given solutions to the commutator equation $XY - YX = C$ for a given $C \in M_n$. Show that X may be replaced by $X + Z$ if and only if $Z \in M_n$ commutes with Y.

9.   Let $C = \begin{bmatrix} 0 & 1 \\ 0 & 0 \end{bmatrix}$ so that tr $C = 0$. Show that there is no pair $X, Y \in M_2$ such that X has an eigenvalue α with multiplicity 2, Y has an eigenvalue β with multiplicity 2, and $C = XY - YX$.

10. (a) Show that $A \in M_n$ is a product of four positive definite (Hermitian) matrices if and only if det $A > 0$ and $A \neq \alpha I$ with $\alpha \in \mathbb{C}$ not positive.
(b) Show that $A \in M_n$ is a product of five positive definite (Hermitian) matrices if and only if det $A > 0$.

11. Recall that $A \in M_n$ is an *involution* if $A^2 = I$. Show that if $A \in M_n$ and det $A = \pm 1$, then A can be written as a product of at most four involutions.

12. A matrix $A \in M_n$ is said to be be *unipotent* if $A - I$ is nilpotent. Show that if $A \in M_n$ and det $A = 1$, then A can always be written as a product of three unipotent matrices; only two factors are necessary if A is not a scalar matrix.

13. Let $C \in M_n$ have tr $C = 0$ and consider the possible solutions X, Y to the equation $XY - YX = C$. Show that one may always choose X to be Hermitian. If C is real and skew-symmetric, show that one may choose X and Y to be real and symmetric. If C is real and symmetric, show that one may choose X and Y to be real with X symmetric and Y skew-symmetric.

14. Provide details for A. Wintner's proof that given matrices $A, B \in M_n$ cannot have $AB - BA = I$: If $AB = BA + I$, $\lambda \in \sigma(BA)$, and $(BA)x = \lambda x$ for some nonzero $x \in \mathbb{C}^n$, then $(AB)x = (\lambda + 1)x$, so $\lambda + 1 \in \sigma(AB)$; induction gives a contradiction. This argument, like the one in Problem 5, avoids use of the trace function and can be extended to the infinite-dimensional case to show that the commutator of two bounded linear operators on a Banach space cannot be a nonzero scalar multiple of the identity operator.

15. Let $A, B \in M_n$ be given nonsingular matrices. When is it possible to write $A = X_1 X_2 X_3$ and $B = X_{\tau(1)} X_{\tau(2)} X_{\tau(3)}$ for some nonsingular $X_1, X_2, X_3 \in M_n$, where τ is some permutation of 1, 2, 3? Theorem (4.5.5) plays a key role in answering this question.
(a) Show that $A = CD$ and $B = DC$ for some nonsingular $C, D \in M_n$ if and only if A is similar to B. Conclude that $A = X_1 X_2 X_3$ and $B = X_2 X_3 X_1$ or $B = X_3 X_1 X_2$ if and only if A is similar to B.
(b) Consider the case $A = X_1 X_2 X_3$ and $B = X_2 X_1 X_3$. Show that the following are equivalent ($\alpha \Rightarrow \beta \Rightarrow \gamma \Rightarrow \alpha$):

(α)   det $A =$ det B.

(β) $AB^{-1} = XYX^{-1}Y^{-1}$ for some nonsingular $X, Y \in M_n$.

(γ) $A = X_1 X_2 X_3$ and $B = X_2 X_1 X_3$.

(c) Now consider the case $A = X_1 X_2 X_3$ and $B = X_1 X_3 X_2$. Modify the argument in (b) to show that these factorizations are possible if and only if det $A = $ det B.

(d) Finally, consider the case $A = X_1 X_2 X_3$ and $B = X_3 X_2 X_1$. Show that the following are equivalent $(\alpha \Rightarrow \cdots \Rightarrow \epsilon \Rightarrow \alpha)$:

(α) det $A = $ det B.

(β) $AB^{-1} = XYX^{-1}Y^{-1}$ for some nonsingular $X, Y \in M_n$.

(γ) $A = CDBC^{-1}D^{-1}$ for some nonsingular $C, D \in M_n$.

(δ) $C^{-1}AD = DBC^{-1}$ for some nonsingular $C, D \in M_n$.

(ϵ) $A = X_1 X_2 X_3$ and $B = X_3 X_2 X_1$.

(e) To answer the question posed at the beginning of the problem, show that the asserted simultaneous three-term factorizations of A and B are possible if and only if: det $A = $ det B when τ is not a cyclic permutation of $1, 2, 3$; A is similar to B when τ is a cyclic permutation of $1, 2, 3$ that is not the identity; $A = B$ when τ is the identity permutation.

Further Readings. Shoda's theorems (4.5.2,5) are in K. Shoda, Einige Sätze über Matrizen, *Japan J. Math.* 13 (1936), 361-365. Shoda showed that (4.5.2) holds for matrices over any field with characteristic zero. The case of (4.5.4) for nonsingular A was given in A. Sourour, A Factorization Theorem for Matrices, *Linear Multilinear Algebra* 19 (1986), 141-147, which also considers matrices over general fields, not just over ℂ. Applications of (4.5.4) to obtain previously known results such as Problems 10-12 are also in Sourour's paper, as are references to the original sources for these results. For example, the results on products of positive definite matrices in Problem 10 are originally due to C. Ballantine, Products of Positive Definite Matrices III, *J. Algebra* 10 (1968), 174-182 (for real matrices) and Products of Positive Definite Matrices IV, *Linear Algebra Appl.* 3 (1970), 79-114 (for complex matrices); these papers also contain characterizations of products of two (easy: see Problem 9 in Section (7.6) of [HJ]) and three (apparently much harder than the other cases) positive definite matrices. Being a multiplicative commutator is actually equivalent to having determinant equal to 1 for matrices over very general fields, there being only a few exceptions—see R. C. Thompson, Commutators of Matrices With Prescribed Determinant, *Canad. J. Math.* 20 (1968), 203-221. Since the Schur triangularization is not available over general fields, other tools, such as companion matrices and the rational form (see Chapter 3 of [HJ]), are often used in the general algebraic setting. Shoda's original version of Theorem (4.5.5) is in his 1936 paper cited previously. Shoda's exact result was that if $\mathbb{F}$ is a field with infinitely many elements, and if $A \in M_n(\mathbb{F})$ has all its eigenvalues in $\mathbb{F}$, then $\det A = 1$ if and only if there exist nonsingular $C, D \in M_n(\mathbb{F})$ such that $A = CDC^{-1}D^{-1}$; moreover, D may be chosen to have n distinct eigenvalues. Shoda also showed that if $A \in M_n(\mathbb{R})$, then $\det A = 1$ if and only if A is the product of two multiplicative commutators in $M_n(\mathbb{R})$. The first results on linear preservers seem to be due to F. G. Frobenius (Über die Darstellung der endlichen Gruppen durch lineare Substitutionen, *Sitzungsberichte der König. Preuss. Akad. der Wissenschaften zu Berlin* (1897), 944-1015), who showed the equivalence of (a) and (f) of (4.5.6) as well as the equivalence of (a) and (d) of (4.5.7). I. Schur generalized Frobenius' determinant preserver result to linear transformations $T: M_{m,n} \to M_{m,n}$ in Einige Bemerkungen zur Determinantentheorie, *Sitzungsberichte der Preuss. Akad. der Wissenschaften, Phys.-Math. Kl.* (1925), 454-463. For a modern discussion of some of the results cited in (4.5.6-7) and references to papers where other results are proved, see M. Marcus and B. N. Moyls, Transformations on Tensor Product Spaces, *Pacific Math. J.* 9 (1959), 1215-1221. The elegant norm argument in Problem 5 is in H. Wielandt, Über die Unbeschränktheit der Operatoren der Quantenmechanik, *Math. Annalen* 121 (1949), 21. For generalizations and variations on the results in Theorem (4.5.5) and Problem 15, see Ky Fan, Some Remarks on Commutators of Matrices, *Arch. Math.* 5 (1954), 102-107. Fan shows that: A unitary U can be written as $U = VWV^{-1}W^{-1}$ with unitary V and W if and only if $\det U = 1$, and unitary matrices A, B can be written as $A = X_1X_2X_3$ and $B = X_3X_2X_1$ with unitary X_1, X_2, and X_3 if and only if $\det A = \det B$; a normal A with $\det A = 1$ can always be written as $A = XYX^{-1}Y^{-1}$ with normal X, Y; a Hermitian A with $\det A = 1$ can be written as $A = XYX^{-1}Y^{-1}$ with Hermitian X, Y if and only if A and A^{-1} have the same eigenvalues.

Chapter 5

The Hadamard product

5.0 Introduction

In this chapter, we study a matrix product that is much simpler than the conventional product, but is much less widely understood. It arises naturally in a variety of ways and enjoys considerable rich structure.

5.0.1 **Definition.** The *Hadamard product* of $A = [a_{ij}] \in M_{m,n}$ and $B = [b_{ij}] \in M_{m,n}$ is defined by $A \circ B \equiv [a_{ij}b_{ij}] \in M_{m,n}$.

The conformability condition for the Hadamard product is simply that both matrices have the same dimensions, and the multiplication is carried out entry by entry, as in conventional matrix addition. Simply imagine the superposition of one array upon the other with a single scalar multiplication performed in each cell.

5.0.2 **Example.** If

$$A = \begin{bmatrix} 2 & 0 & -7i \\ -3 & \pi & 4 \end{bmatrix} \text{ and } B = \begin{bmatrix} -1 & 14 & -1 \\ 0 & 3 & 2 \end{bmatrix}$$

then

$$A \circ B = \begin{bmatrix} -2 & 0 & 7i \\ 0 & 3\pi & 8 \end{bmatrix}$$

The Hadamard product is sometimes called the *entrywise product*, for obvious reasons, or the *Schur product*, because of some early and basic results about the product obtained by Issai Schur. The most basic of these is the closure of the cone of positive semidefinite matrices under the Hadamard

298

product (5.2.1).

The Hadamard product arises in a wide variety of ways. Several examples, such as trigonometric moments of convolutions of periodic functions, products of integral equation kernels (especially the relationship with Mercer's theorem), the weak minimum principle in partial differential equations, and characteristic functions in probability theory (Bochner's theorem), are discussed in Section (7.5) of [HJ]. In combinatorial theory, the Hadamard product arises in the study of association schemes. In operator theory, the Hadamard product for infinite matrices has been studied for the alternate analytical structure it induces on the set of linear operators and because of interest in properties of the linear transformation on matrices given by the Hadamard product with a fixed matrix. Several additional illustrations of how the Hadamard product arises are given in the following subsections.

5.0.3   The matrix $A \circ (A^{-1})^T$

5.0.3a   The relationship between diagonal entries and eigenvalues of a diagonalizable matrix

Suppose $B = [b_{ij}] \in M_n$ is a diagonalizable matrix with eigenvalues $\lambda_1, \lambda_2, ..., \lambda_n$. Thus, there is a nonsingular matrix $A \in M_n$ such that

$$B = A \left[\mathrm{diag}(\lambda_1, \lambda_2, ..., \lambda_n)\right] A^{-1}$$

A calculation then reveals that

$$\begin{bmatrix} b_{11} \\ b_{22} \\ \vdots \\ b_{nn} \end{bmatrix} = \left[A \circ (A^{-1})^T \right] \begin{bmatrix} \lambda_1 \\ \lambda_2 \\ \vdots \\ \lambda_n \end{bmatrix}$$

Thus, the vector of eigenvalues of B is transformed to the vector of its diagonal entries by the coefficient matrix $A \circ (A^{-1})^T$.

In the event that A is unitary (so B is normal), $A \circ (A^{-1})^T = A \circ \overline{A}$ is doubly stochastic (entrywise nonnegative with all row and column sums one). If, in addition, the eigenvalues $\lambda_1, \lambda_2, ..., \lambda_n$ are real (so B is Hermitian), it follows that the diagonal entries majorize the eigenvalues. See (4.3.24) and what follows in [HJ] for more details.

5.0.3b   The relative gain array of chemical engineering process control

For nonsingular $A \in M_n(\mathbb{R})$, the matrix $A \circ (A^{-1})^T$ is used extensively in one approach to the engineering design of chemical plants. There, the *gain matrix* A describes the relation between the inputs and outputs in a chemical process, and the matrix $A \circ (A^{-1})^T$ is called the *relative gain array*. The designer inspects the relative gain array with the goal of determining a pairing of inputs and outputs to be used in facility design. The book [McA] gives an exposition of the motivation for such an approach and includes several case studies of its implementation.

5.0.4   Hadamard products and the Lyapunov equation $GA + A^*G = H$

Let $A \in M_n$ be given with spectrum $\sigma(A) = \{\lambda_1, ..., \lambda_n\}$. According to Theorem (4.4.6), the Lyapunov equation

$$GA + A^*G = H$$

has a unique solution $G = G_A(H)$ for every Hermitian right-hand side H if and only if $\overline{\lambda}_i + \lambda_j \neq 0$ for all $i, j = 1, ..., n$. Let us make this assumption about the spectrum of A; in particular, this condition is clearly satisfied if we make the stronger assumption that A is positive stable. If, in addition, A is diagonalizable with $A = S\Lambda S^{-1}$ and $\Lambda = \text{diag}(\lambda_1, ..., \lambda_n)$, there is a simple formula for the function $G_A(H)$ that involves the Hadamard product. Write

$$GA + A^*G = GS\Lambda S^{-1} + (S^{-1})^* \overline{\Lambda} S^* G = H$$

which is equivalent to

$$(S^*GS)\Lambda + \overline{\Lambda}(S^*GS) = S^* HS$$

For any matrix $B = [b_{ij}] \in M_n$, $B\Lambda + \overline{\Lambda}B = [b_{ij}\lambda_j] + [\overline{\lambda}_i b_{ij}] = [(\overline{\lambda}_i + \lambda_j)b_{ij}]$, which is a Hadamard product. If we invoke the assumption that $\overline{\lambda}_i + \lambda_j \neq 0$ and define $L(A) \equiv [(\overline{\lambda}_i + \lambda_j)^{-1}]$, we have $S^*GS = L(A) \circ (S^*HS)$, which yields the explicit formula

$$G_A(H) = (S^{-1})^*[L(A) \circ (S^* H S)]S^{-1}, \qquad A = S\Lambda S^{-1}$$

for the solution to the Lyapunov equation $GA + A^*G = H$ when A is diagonalizable and $\overline{\lambda}_i + \lambda_j \neq 0$ for all $\lambda_i, \lambda_j \in \sigma(A)$.

5.0.5   Covariance matrices and Hadamard products

Let $X = [X_1, \ldots, X_n]^T$ and $Y = [Y_1, \ldots, Y_n]^T$ be real or complex fourth-order vector random variables with zero means, that is, $E(X) = [E(X_i)] = 0$ and $E(Y) = 0$, where $E(\cdot)$ denotes the expectation operator. The *covariance matrix* of X has the general form

$$\text{Cov}(X) \equiv E([X - E(X)][X - E(X)]^*)$$

which takes the special form $\text{Cov}(X) = E(XX^*) = [E(X_i\overline{X}_j)]$ in our case since $E(X) = 0$. If X and Y are independent, the second-order vector random variable $Z \equiv X \circ Y$ has zero mean and has a covariance matrix that is obtained from the covariance matrices of X and Y as a Hadamard product:

$$\text{Cov}(Z) = E(ZZ^*) = [E(X_i Y_i \overline{X}_j \overline{Y}_j)] = [E(X_i \overline{X}_j Y_i \overline{Y}_j)]$$

$$= [E(X_i\overline{X}_j)E(Y_i\overline{Y}_j)] = \text{Cov}(X) \circ \text{Cov}(Y)$$

One pleasant property of a covariance matrix is that it is easily shown to be positive semidefinite:

$$\xi^* \text{Cov}(X)\xi = \sum_{i,j=1}^{n} E(X_i\overline{X}_j)\overline{\xi}_i\xi_i = \sum_{i,j=1}^{n} E(\overline{\xi}_i X_i \xi_j \overline{X}_j)$$

$$= E\left[\sum_{i,j=1}^{n} \overline{\xi}_i X_i \xi_j \overline{X}_j\right] = E\left[\left|\sum_{i=1}^{n} \overline{\xi}_i X_i\right|^2\right] \geq 0$$

for any $\xi = [\xi_i] \in \mathbb{C}^n$. In this calculation we have used the facts that the expectation operator is linear and positive. Thus, since the Hadamard product of two covariance matrices is itself a covariance matrix, we may immediately infer that the Hadamard product of two covariance matrices is positive semidefinite. Since every positive semidefinite matrix is the covariance matrix of a Gaussian random vector, these observations give a probabilistic proof of the Schur product theorem.

Problems

1.   (a) What is the identity element of $M_{m,n}$ under the Hadamard product?
(b) Which elements of $M_{m,n}$ have inverses under the Hadamard product?

2.   A matrix $A \in M_n$ is called *essentially triangular* if A is similar via a permutation matrix to an upper triangular matrix. Show that the relative gain array $A \circ (A^{-1})^T$ associated with a nonsingular essentially triangular matrix $A \in M_n$ is I, the n-by-n conventional identity. Give an example of a 4-by-4 nonsingular matrix that is *not* essentially triangular, but whose relative gain array is I, and show that 4 is the minimum dimension in which this can occur.

3.   If $A \in M_n$ is nonsingular, show that all the row and column sums of $A \circ (A^{-1})^T$ are equal to 1. Use this to show that the trace of a diagonalizable matrix is equal to the sum of its eigenvalues; extend this result to all matrices by a continuity argument.

4.   Suppose that $A \in M_n$ is diagonalizable, with $A = S \Lambda S^{-1}$ for some nonsingular $S \in M_n$ and $\Lambda = \text{diag}(\lambda_1,..., \lambda_n)$. If all the entries of $S \circ (S^{-1})^T$ are nonnegative, show that the minimum of the real part of the eigenvalues of A is less than or equal to the minimum of the real part of its diagonal entries. Under these circumstances, conclude that A cannot be positive stable if it has any diagonal entries in the left half-plane. In particular, show that a normal matrix cannot be positive stable if it has any diagonal entries in the left half-plane.

5.   We have observed in (5.0.3a) that the diagonal entries of a diagonalizable matrix are related to its eigenvalues by a matrix that is a Hadamard product. Show that the diagonal entries of any matrix $A \in M_n$ (no diagonalizability assumption needed) are related to its singular values by a matrix that is a Hadamard product. In particular, if $A = V \Sigma W^*$ is a singular value decomposition of $A = [a_{ij}]$ with $\Sigma = \text{diag}(\sigma_1,..., \sigma_n)$, show that $a = [V \circ \overline{W}]\sigma$, where $a \equiv [a_{11}, ..., a_{nn}]^T$ and $\sigma \equiv [\sigma_1, ..., \sigma_n]^T$.

Notes and Further Readings. What we now know as the Schur product theorem and related results are in J. (Issai) Schur, Bemerkungen zur Theorie der beschränkten Bilinearformen mit unendlich vielen Veränderlichen, *J. Reine Angew. Math.* 140 (1911), 1-28. In this paper, Schur states and proves the quantitative form (5.3.4) of the entrywise product theorem, and he introduces it as a result that "despite its simplicity seems not to be known." He also proves the inequality $||| A \circ B |||_2 \leq ||| A |||_2 \, ||| B |||_2$; see Theorem (5.5.1). Since Schur seems to have made the first systematic study of the algebraic properties of the entrywise product, there is ample justification to call it the *Schur product*, and some writers have done so.

The term *Hadamard product* seems to be more common, however, perhaps because this appellation was used—exactly once and apparently for the first time in print—in the 1948 first edition of [Hal 58]. This may have been done because of a well-known paper of Hadamard (Théorème sur les Séries Entières, *Acta. Math.* 22 (1899), 55-63), in which he studied two Maclaurin series $f(z) = \Sigma_n a_n z^n$ and $g(z) = \Sigma_n b_n z^n$ with positive radii of convergence and their composition $h(z) \equiv \Sigma_n a_n b_n z^n$ obtained as the coefficientwise product. Hadamard showed that $h(\cdot)$ can be obtained from $f(\cdot)$ and $g(\cdot)$ by an integral convolution, and he proved that any singularity z_1 of $h(\cdot)$ must be of the form $z_1 = z_2 z_3$, where z_2 is a singularity of $f(\cdot)$ and z_3 is a singularity of $g(\cdot)$. Even though Hadamard did not study entrywise products of matrices in this paper, the enduring influence of the results in it as well as his mathematical eminence seems to have linked firmly—at least for analysts—the name of Hadamard with term-by-term products of all kinds.

As is often the case in the history of science, both Schur and Hadamard were anticipated in part by even earlier work. In 1894, Th. Moutard (Notes sur les Équations Derivées Partielles, *J. de L'Ecole Polytechnique* 64 (1894), 55-69) showed that if two real symmetric n-by-n matrices $A = [a_{ij}]$ and $B = [b_{ij}]$ are positive definite, then the sum of the entries of their entrywise product $\Sigma_{ij} a_{ij} b_{ij}$ is positive. Moutard used this result to obtain a uniqueness theorem for boundary-value problems for elliptic partial differential equations, but he did not consider the quadratic form $x^*(A \circ B)x$ at all. Also interested in partial differential equations, L. Fejér devoted the entire first section of a 1918 paper (Über die Eindeutigkeit der Lösung der Linearen Partiellen Differentialgleichung zweiter Ordnung, *Math. Zeit.* 1 (1918), 70-79) to a review of Moutard's result and to a simple proof that it implies what he termed the "remarkable theorem of J. Schur."

The connection between Gaussian vector random variables and positive semidefinite matrices mentioned in (5.0.5) is discussed in H. Cramér, *Mathematical Methods in Statistics*, Princeton University Press, Princeton, 1946. For a survey of basic results on, and applications of, Hadamard products as well as a comprehensive bibliography, see R. A. Horn, The Hadamard Product, in [Joh], pp. 87–169.

5.1   Some basic observations

We record here several elementary algebraic facts that are very useful in analyzing the Hadamard product. Of course, the Hadamard product of complex matrices, unlike the usual matrix product, is commutative.

Our first observation relates the Hadamard product to the Kronecker product by identifying $A \circ B$ as a submatrix of $A \otimes B$. Recall [HJ, Section (0.7)] that for index sets $\alpha \subseteq \{1, 2,..., m\}$ and $\beta \subseteq \{1, 2,..., n\}$, the submatrix of $C \in M_{m,n}$ lying in the rows indicated by α and the columns indicated by β is denoted by $C(\alpha,\beta)$, and, when $m = n$ and $\beta = \alpha$, we abbreviate $C(\alpha,\alpha)$ by $C(\alpha)$.

5.1.1   Lemma. If $A, B \in M_{m,n}$, then

$$A \circ B = (A \otimes B)(\alpha,\beta)$$

in which $\alpha = \{1, m + 2, 2m + 3,..., m^2\}$ and $\beta = \{1, n + 2, 2n + 3, ..., n^2\}$. In particular, if $m = n$, $A \circ B$ is a principal submatrix of $A \otimes B$.

Proof: Inspect the positions indicated for the submatrix.   ⬚

The next observation indicates that conventional multiplication by a diagonal matrix and Hadamard multiplication "commute."

5.1.2   Lemma. If $A, B \in M_{m,n}$ and if $D \in M_m$ and $E \in M_n$ are diagonal, then

$$D(A \circ B)E = (DAE) \circ B = (DA) \circ (BE)$$
$$= (AE) \circ (DB) = A \circ (DBE)$$

Exercise. Verify (5.1.2).

The special case of (5.1.2) in which $B = J$, the matrix each of whose entries is 1, shows that diagonal equivalence is the same as Hadamard multiplication by a rank one matrix with no zero entries.

For a given $x \in \mathbb{C}^n$, $\mathrm{diag}(x) \in M_n$ denotes the diagonal matrix whose main diagonal entries are the entries of x. It is convenient to write $D_x \equiv \mathrm{diag}(x)$, and we note that $D_x e = x$, where $e \equiv [1, 1,..., 1]^T \in \mathbb{R}^n$.

The next two observations indicate further relations between Hadamard and conventional matrix multiplication.

5.1.3   Lemma. Let $A, B \in M_{m,n}$ and let $x \in \mathbb{C}^n$. Then the ith diagonal entry of the matrix $AD_x B^T$ coincides with the ith entry of the vector $(A \circ B)x$, $i = 1,..., m$.

Proof: If $A = [a_{ij}]$, $B = [b_{ij}]$, and $x = [x_j]$, then

$$(AD_x B^T)_{ii} = \sum_{j=1}^{n} a_{ij} x_j b_{ij} = \sum_{j=1}^{n} a_{ij} b_{ij} x_j = [(A \circ B)x]_i$$

for $i = 1, 2,..., m$. $\square$

Exercise. Show that the ith row sum of $A \circ B$ is the ith diagonal entry of AB^T under the assumptions in (5.1.3).

Exercise. Use (5.1.2) and the preceding exercise to verify (5.1.3).

Exercise. Use (5.1.3) to verify the assertion about the relationship between main diagonal entries and eigenvalues stated in (5.0.3a).

The next observation involves a mixed triple product.

5.1.4   Lemma. Let $A, B, C \in M_{m,n}$. The ith diagonal entry of the matrix

$$(A \circ B)C^T$$

coincides with the ith diagonal entry of the matrix

$$(A \circ C)B^T$$

for $i = 1,..., m$.

Exercise. Verify (5.1.4) by calculating the ith diagonal entry of each side, as in the proof of (5.1.3).

Exercise. Deduce (5.1.3) from (5.1.4). *Hint:* Let each row of the matrix C in (5.1.4) coincide with the vector x of (5.1.3) and calculate.

Exercise. Show that $\mathrm{tr}[(A \circ B)C^T] = \mathrm{tr}[(A \circ C)B^T]$ under the assumptions in (5.1.4).

Our next observation is that the sesquilinear form generated by a Hadamard product may be written as a trace.

5.1.5   Lemma. For $A, B \in M_{m,n}$, $y \in \mathbb{C}^m$, and $x \in \mathbb{C}^n$, we have

$$y^*(A \circ B)x = \mathrm{tr}(D_y^* A D_x B^T)$$

Proof: We have

$$y^*(A \circ B)x = e^T D_y^*(A \circ B)x = e^T[(D_y^* A \circ B)x] = \mathrm{tr}(D_y^* A D_x B^T)$$

The second equality follows from Lemma (5.1.2), and the third follows from Lemma (5.1.3) with $D_y^* A$ in place of A. $\square$

Exercise. Show that the function $\langle \cdot, \cdot \rangle: M_{m,n} \times M_{m,n} \to \mathbb{C}$ defined by $\langle A, B \rangle \equiv \mathrm{tr}(AB^*)$ is an inner product (usually called the *Frobenius inner product*), and that $y^*(A \circ B)x = \langle D_y^* A, \overline{BD_x} \rangle$.

Exercise. Show that $x^*(A \circ B)x = \mathrm{tr}(D_x^* A D_x B^T) = \langle D_x^* A D_x, \overline{B} \rangle$.

Our next observation does not deal directly with Hadamard products, but it is useful in calculations with inequalities involving the Hadamard product.

5.1.6   Lemma. Suppose that $A \in M_n$ is positive definite and let $B \equiv A^{-1}$. For any conformal partitioning

$$A = \begin{bmatrix} A_{11} & A_{12} \\ A_{12}^* & A_{22} \end{bmatrix} \text{ and } B = \begin{bmatrix} B_{11} & B_{12} \\ B_{12}^* & B_{22} \end{bmatrix}$$

in which $A_{11}, B_{11} \in M_k$, the matrix

$$A - \begin{bmatrix} B_{11}^{-1} & 0 \\ 0 & 0 \end{bmatrix} = \begin{bmatrix} A_{11} - B_{11}^{-1} & A_{12} \\ A_{12}^* & A_{22} \end{bmatrix}$$

is positive semidefinite and has rank $n - k$.

Proof: Since B_{11}^{-1} may be written as a Schur complement $B_{11}^{-1} = A_{11} - A_{12} A_{22}^{-1} A_{12}^*$ [HJ, (0.7.3)], we have

$$A - \begin{bmatrix} B_{11}^{-1} & 0 \\ 0 & 0 \end{bmatrix} = \begin{bmatrix} A_{12} A_{22}^{-1} A_{12}^* & A_{12} \\ A_{12}^* & A_{22} \end{bmatrix} = \begin{bmatrix} A_{12}(A_{22})^{-\frac{1}{2}} \\ (A_{22})^{\frac{1}{2}} \end{bmatrix} \begin{bmatrix} A_{12}(A_{22})^{-\frac{1}{2}} \\ (A_{22})^{\frac{1}{2}} \end{bmatrix}^*$$

and the assertion follows. Here $(A_{22})^{\frac{1}{2}}$ denotes the unique positive definite Hermitian square root of A_{22} and $(A_{22})^{-\frac{1}{2}}$ denotes its inverse.   ∏

Our final observation is that the rank function is Hadamard submultiplicative.

5.1.7   Theorem. Let $A, B \in M_{m,n}$. Then

$$\text{rank}(A \circ B) \le (\text{rank } A)(\text{rank } B)$$

Proof: Any matrix of rank r may be written as a sum of r rank one matrices, each of which is an outer product of two vectors [HJ, Section (1.4), Problems 1 and 2]. Thus, if rank $A = r_1$ and rank $B = r_2$, we have

$$A = \sum_{i=1}^{r_1} x_i y_i^* \text{ and } B = \sum_{j=1}^{r_2} u_j v_j^*$$

in which $x_i, u_j \in \mathbb{C}^m$ and $y_i, v_j \in \mathbb{C}^n$, $i = 1,\ldots, r_1$ and $j = 1,\ldots, r_2$. Then

$$A \circ B = \sum_{i=1}^{r_1} \sum_{j=1}^{r_2} (x_i \circ u_j)(y_i \circ v_j)^*$$

which shows that $A \circ B$ is a sum of at most $r_1 r_2$ rank one matrices. Thus, $\text{rank}(A \circ B) \le r_1 r_2 = (\text{rank } A)(\text{rank } B)$.   ∏

Exercise. Use (5.1.1), the fact that rank is multiplicative with respect to the Kronecker product, and the fact that rank is nonincreasing with respect to extraction of submatrices to give an alternate proof of (5.1.7).

Problems

1. Which of the following classes is closed under the Hadamard product, for matrices of a given size: Hermitian matrices, symmetric matrices, skew-Hermitian matrices, unitary matrices, normal matrices, entrywise nonnegative matrices, nonsingular matrices, singular matrices?

2. In what class is the Hadamard product of a Hermitian and a skew-Hermitian matrix? Of a skew-Hermitian and a skew-Hermitian?

3. Let $\rho(A)$ denote the Perron root (spectral radius) of an entrywise nonnegative matrix $A \in M_n(\mathbb{R})$. Use (5.1.1), properties of the Kronecker product (Chapter 4), and properties of nonnegative matrices [HJ, Chapter 8] to show that $\rho(A \circ B) \leq \rho(A)\rho(B)$ whenever $A, B \in M_n(\mathbb{R})$ have nonnegative entries.

4. Give an example to show that the spectral radius is not Hadamard submultiplicative on M_n, that is, exhibit $A, B \in M_n$ such that $\rho(A \circ B) > \rho(A)\rho(B)$.

5. Use Lemma (5.1.5) and the Cauchy-Schwarz inequality for the Frobenius inner product to show that $||| A \circ B |||_2 \leq ||| A |||_2 ||| B |||_2$ for all $A, B \in M_n$, that is, the spectral norm is Hadamard submultiplicative.

5.2   The Schur product theorem

The fact that the class of positive (semi)definite Hermitian matrices of a given size is closed under Hadamard multiplication is fundamental for several reasons. It is a useful fact by itself (see Section (7.5) of [HJ]), and it is a very useful lemma in proving other results about the Hadamard product. It was perhaps the first significant fact published about the Hadamard product, many results in the field may be viewed as generalizations or analogs of it, and it is a good example to illustrate that, from an algebraic point of view, the Hadamard product may be a more "natural" product, in some circumstances, than the ordinary matrix product—the class of positive definite matrices is closed under Hadamard multiplication but *not* under ordinary matrix multiplication.

The basic fact in its broadest statement is given in the following theorem. Note that the Hadamard product of two Hermitian matrices is Hermitian, and a positive semidefinite matrix is necessarily Hermitian.

5.2.1   Theorem. If $A, B \in M_n$ are positive semidefinite, then so is $A \circ B$. If, in addition, B is positive definite and A has no diagonal entry equal to 0, then $A \circ B$ is positive definite. In particular, if both A and B are positive definite, then so is $A \circ B$.

Proof: Consider the second assertion and use (5.1.5) to write

$$x^*(A \circ B)x = \operatorname{tr}(D_x^* A D_x B^T) \qquad (5.2.2)$$

for any given $x \in \mathbb{C}^n$. Recall (Theorem (7.6.3) in [HJ]) that the inertia of the product of a Hermitian matrix and a positive definite matrix is the same as that of the Hermitian matrix itself. For any $x \in \mathbb{C}^n$, the matrix $D_x^* A D_x$ is positive semidefinite; if $x \neq 0$ and no diagonal entry of A is zero, then $D_x^* A D_x \neq 0$ and hence it has at least one positive eigenvalue. If B is positive definite, then so is B^T. Thus, all the eigenvalues of $(D_x^* A D_x)B^T$ are nonnegative and at least one is positive, from which it follows that $\operatorname{tr}(D_x^* A D_x B^T) > 0$. From (5.2.2) we conclude that $A \circ B$ is positive definite. The first assertion now follows by applying the second to $A_\epsilon \equiv A + \epsilon I$ and $B_\epsilon \equiv B + \epsilon I$, $\epsilon > 0$, and letting $\epsilon \to 0$. The last assertion follows from the observation that if $A = [a_{ij}]$ is positive definite, then $e_i^* A e_i = a_{ii} > 0$ for all $i = 1,..., n$.   □

Exercise. Prove the first and third assertions of Theorem (5.2.1) using Lemma (5.1.1), the fact that eigenvalues multiply in the Kronecker product [Theorem (4.2.12)], and the fact that a principal submatrix of a positive semidefinite matrix is positive semidefinite ((7.1.2) in [HJ]). Is it possible to verify the second assertion in this way?

The proof we have given for Theorem (5.2.1) indicates another way in which facts about the Hadamard product and the ordinary product are linked. Section (7.5) of [HJ] contains further discussion of the ideas surrounding this fundamental theorem.

Problems

1.   Let $A \in M_n$ be positive semidefinite and let $\alpha \subsetneq \{1, 2, ..., n\}$ be a given

index set. Show that $A(\alpha)$ is positive definite if and only if $x^*Ax = 0$ and $x \neq 0$ imply there is an $i \notin \alpha$ such that $x_i \neq 0$.

2.   If $A, B, C \in M_n$ are Hermitian with $A \succeq B$ and C positive semidefinite, use (5.2.1) to show that $A \circ C \succeq B \circ C$; see Section (7.7) of [HJ] for a discussion of the Loewner partial order $\succeq$.

3.   If $A, B, C, D \in M_n$ are Hermitian with $A \succeq B \succeq 0$ and $C \succeq D \succeq 0$, show that $A \circ C \succeq B \circ D$.

4.   Let positive semidefinite $A, B \in M_n$ have nonzero eigenvalues $\lambda_1, \ldots, \lambda_{r_1}$ and $\mu_1, \ldots, \mu_{r_2}$, respectively (including multiplicities). Let $\{x_1, \ldots, x_{r_1}\}$ and $\{y_1, \ldots, y_{r_2}\}$ be corresponding independent sets of eigenvectors of A and B, respectively, that is, $Ax_i = \lambda_i x_i$ and $By_j = \mu_j y_j$. Show that $A \circ B$ is positive definite if and only if $\{x_i \circ y_j: i = 1, \ldots, r_1, j = 1, \ldots, r_2\}$ spans $\mathbb{C}^n$.

5.   Give another proof of (5.2.1) using the result of Problem 4.

6.   Show that the Schur product theorem (5.2.1) is equivalent to the Moutard-Fejér theorem: Let $A = [a_{ij}] \in M_n$ be given. Then A is positive definite if and only if

$$\sum_{i,\, j=1}^{n} a_{ij} b_{ij} > 0 \qquad\qquad (5.2.3)$$

for every nonzero positive semidefinite $B = [b_{ij}] \in M_n$.

7.   Let $A \in M_n$ be given. Show that $\mathrm{Re}\, \mathrm{tr}(AB) > 0$ for all nonzero positive semidefinite $B \in M_n$ if and only if $H(A) \equiv \frac{1}{2}(A + A^*)$ is positive definite.

8.   Give an example of $A, B \in M_n$ such that A is positive semidefinite, B is positive definite, and $A \circ B$ is positive semidefinite but not positive definite.

9.   Provide details for the following alternative argument that the identity (5.2.2) implies the full statement of Theorem (5.2.1): Let $C \equiv B^{\frac{1}{2}T} D_z^* A^{\frac{1}{2}}$, so $x^*(A \circ B)x = \mathrm{tr}(CC^*) \geq 0$. If equality holds, then $C = 0$, so $D_z^* A^{\frac{1}{2}} = 0$ (B is nonsingular). The squared Euclidean row lengths of $A^{\frac{1}{2}}$ are the diagonal entries of A, which are all positive, so no row of $A^{\frac{1}{2}}$ is zero and hence $D_z = 0$.

10. Let $A, B \in M_n$ be positive definite with $A = [a_{ij}]$. Show that $A \circ B = AB$ if and only if both A and B are diagonal. Use Hadamard's inequality

$$\det A \leq a_{11} \cdots a_{nn} \qquad\qquad (5.2.4)$$

5.2   The Schur product theorem

in which equality holds if and only if A is diagonal, and Oppenheim's inequality (Section (7.8) of [HJ])

$$\det(A \circ B) \geq (a_{11} \cdots a_{nn})\det B \tag{5.2.5}$$

11. Using the notation of Problem 10, show that Oppenheim's inequality (5.2.5) implies Hadamard's inequality (5.2.4).

12. Let $A, B \in M_n$ be positive semidefinite. Use Oppenheim's inequality and Hadamard's inequality to show that

$$\det(A \circ B) \geq (\det A)(\det B) \tag{5.2.6}$$

13. Let $A \in M_n$ be normal, and write $A = U \Lambda U^*$ with $\Lambda = \mathrm{diag}(\lambda_1, \ldots, \lambda_n)$ and a unitary $U = [u_{ij}] \in M_n$. Show that $|A| \equiv (A^*A)^{\frac{1}{2}}$, defined in Problem 6 of Section (3.5), may be written as $|A| \equiv U\,\mathrm{diag}(|\lambda_1|, \ldots, |\lambda_n|)\, U^*$. Notice that $|A| = (A^2)^{\frac{1}{2}}$ if A is Hermitian, and $|A| = A$ if A is positive semidefinite. If $A, B \in M_n$ are both Hermitian, show that $|A| \circ |B| \geq A \circ B$, that is, $|A| \circ |B| - (A \circ B)$ is positive semidefinite. Deduce that $\lambda_i(|A| \circ |B|) \geq \lambda_i(A \circ B)$ for $i = 1, \ldots, n$, where $\lambda_1 \geq \cdots \geq \lambda_n$ are the eigenvalues. See Problem 10 in Section (3.5) for related results.

14. Use Problem 13 to show that if $A, B \in M_n$ are Hermitian, then $||| A \circ B |||_2 \leq ||| \, |A| \circ |B| \, |||_2$.

15. Let $A, B \in M_n$ be given positive semidefinite matrices.
   (a) Show that $AB = 0$ if and only if $\mathrm{tr}\, AB = 0$. Conclude that the zero matrix is the only positive semidefinite matrix that is orthogonal to a positive definite matrix in the Frobenius inner product on M_n.
   (b) Let $x \in \mathbb{C}^n$ be given. Show that the following are equivalent:

   (1) $D_x^* A D_x B^T = 0$,

   (2) $\mathrm{tr}(D_x^* A D_x B^T) = 0$,

   (3) $x^*(A \circ B)x = 0$, and

   (4) $(A \circ B)x = 0$.

   (c) Use the implication (b)(4) $\Rightarrow$ (b)(1) to give yet another proof that $A \circ B$ is nonsingular if B is positive definite and A has positive main diagonal entries.

5.3   Generalizations of the Schur product theorem

Theorem (5.2.1) establishes the closure of the classes of positive definite and positive semidefinite matrices under the Hadamard product. These are qualitative results, and in this section we are primarily interested in various quantitative refinements of them.

Definition. When $X \in M_n$ is a matrix with real eigenvalues, $\lambda_{min}(X)$ and $\lambda_{max}(X)$ denote its algebraically smallest and largest eigenvalues, respectively.

A relatively weak quantitative statement follows directly from recognizing that the Hadamard product of two positive semidefinite matrices $A, B \in M_n$ is a principal submatrix of the Kronecker product $A \otimes B$ (Lemma (5.1.1)). Since the eigenvalues of $A \otimes B$ are just the pairwise products of the eigenvalues of A and B, its smallest and largest eigenvalues are $\lambda_{min}(A)\lambda_{min}(B)$ and $\lambda_{max}(A)\lambda_{max}(B)$. But the eigenvalues of any principal submatrix of a Hermitian matrix lie between its smallest and largest eigenvalues (Theorem (4.3.15) in [HJ]), so we have the upper and lower bounds

$$\lambda_{min}(A \circ B) \geq \lambda_{min}(A)\lambda_{min}(B)$$

and

$$\lambda_{max}(A \circ B) \leq \lambda_{max}(A)\lambda_{max}(B)$$

whenever $A, B \in M_n$ are positive semidefinite.

These estimates can be useful, but for many purposes they are not sharp enough. For example, if A is positive definite and $B = J$, the matrix each of whose entries is 1 (whose eigenvalues are n and 0), the preceding estimates yield the trivial lower bound $\lambda_{min}(A) = \lambda_{min}(A \circ J) \geq \lambda_{min}(A)\lambda_{min}(J) = \lambda_{min}(A) \, 0 = 0$. Also, if A is positive definite and $B = A^{-1}$, these estimates yield $\lambda_{min}(A \circ A^{-1}) \geq \lambda_{min}(A)\lambda_{min}(A^{-1}) = \lambda_{min}(A)/\lambda_{max}(A)$. Here again, the lower bound obtained is not very good; we shall see that a much better lower bound, namely, 1, is available.

We first present a pair of inequalities illustrating that the transpose often occurs naturally, as in Lemma (5.1.5), and that it may sometimes be removed. Note that if $B \in M_n(\mathbb{C})$ is Hermitian, $B^T = \overline{B} \neq B$ if B has any nonreal entries.

5.3.1   **Theorem.** Let $A, B \in M_n$ be given positive semidefinite Hermitian matrices. Then

(a) $\lambda_{min}(A \circ B) \geq \lambda_{min}(AB^T)$

and

(b) $\lambda_{min}(A \circ B) \geq \lambda_{min}(AB)$

In order to prove Theorem (5.3.1) we use the following lemma, which is an analog of the familiar fact that the sum of squares of the absolute values of the eigenvalues of a square matrix is dominated by the square of its Frobenius norm (see Chapters 2 and 5 of [HJ]). Let $\| A \|_2 \equiv [\mathrm{tr}(A^*A)]^{\frac{1}{2}}$, the Frobenius norm of $A \in M_{m,n}$.

5.3.2   **Lemma.** Let $C, E \in M_n$ be given with E symmetric $(E = E^T)$ and C nonsingular and normal. Then $\| C^{-1}EC^T \|_2 \geq \| E \|_2$.

Proof: Let $C = U^*DU$ be a unitary diagonalization of C. Then

$$\| C^{-1}EC^T \|_2 = \| U^*D^{-1}UEU^TD\overline{U} \|_2 = \| D^{-1}UEU^TD \|_2$$

$$\geq \| UEU^T \|_2 = \| E \|_2$$

The second and last equalities follow from unitary invariance of the Frobenius norm (Chapter 5 of [HJ]), and the inequality is due to the fact that UEU^T is symmetric and $|t|^{-1} + |t| \geq 2$ for all $0 \neq t \in \mathbb{C}$.   ⬜

We may now prove Theorem (5.3.1). It suffices to consider only positive definite A, B, as the asserted inequality is trivial if either is singular. Let $\|\|\cdot\|\|_2 \equiv (\lambda_{max}(A^*A))^{\frac{1}{2}}$ denote the *spectral norm* on M_n. We first prove (5.3.1b) as follows:

$$\lambda_{min}(A \circ B) = \min_{\| x \|_2 = 1} x^*(A \circ B)x = \min_{\| D_x \|_2 = 1} \mathrm{tr}(D_x^*AD_xB^T)$$

$$= \min_{\| D_x \|_2 = 1} \mathrm{tr}(B^{\frac{1}{2}T}D_x^*A^{\frac{1}{2}}A^{\frac{1}{2}}D_xB^{\frac{1}{2}T})$$

$$= \min_{\|D_z\|_2=1} \mathrm{tr}([A^{\frac{1}{2}}D_zB^{\frac{1}{2}T}]^*[A^{\frac{1}{2}}D_zB^{\frac{1}{2}T}])$$

$$= \min_{\|D_z\|_2=1} \|A^{\frac{1}{2}}D_zB^{\frac{1}{2}T}\|_2^2 = \min_{\|D_z\|_2=1} \|A^{\frac{1}{2}}B^{\frac{1}{2}}B^{-\frac{1}{2}}D_zB^{\frac{1}{2}T}\|_2^2$$

$$\geq \||(A^{\frac{1}{2}}B^{\frac{1}{2}})^{-1}\||_2^{-2} \min_{\|D_z\|_2=1} \|B^{-\frac{1}{2}}D_zB^{\frac{1}{2}T}\|_2^2$$

$$\geq \||(B^{-\frac{1}{2}}A^{-\frac{1}{2}})\||_2^{-2} \cdot 1$$

$$= \left[\lambda_{max}([B^{-\frac{1}{2}}A^{-\frac{1}{2}}]^*[B^{-\frac{1}{2}}A^{-\frac{1}{2}}])\right]^{-1} = \left[\lambda_{max}(A^{-\frac{1}{2}}B^{-1}A^{-\frac{1}{2}})\right]^{-1}$$

$$= \left[\lambda_{max}(A^{\frac{1}{2}}[A^{-\frac{1}{2}}B^{-1}A^{-\frac{1}{2}}]A^{-\frac{1}{2}})\right]^{-1} = [\lambda_{max}(B^{-1}A^{-1})]^{-1}$$

$$= \lambda_{min}(AB)$$

The second inequality follows from (5.3.2), while the first uses the fact that $\|RS\|_2 \leq \||R\||_2\|S\|_2$ for any $R, S \in M_n$. The statement (5.3.1a) is proved similarly by inserting the factor $B^{\frac{1}{2}T}B^{-\frac{1}{2}T}$ instead:

$$\lambda_{min}(A \circ B) = \min_{\|D_z\|_2=1} \|A^{\frac{1}{2}}B^{\frac{1}{2}T}B^{-\frac{1}{2}T}D_zB^{\frac{1}{2}T}\|_2^2$$

$$\geq \||(A^{\frac{1}{2}}B^{\frac{1}{2}T})^{-1}\||_2^{-2} \min_{\|D_z\|_2=1} \|B^{-\frac{1}{2}T}D_zB^{\frac{1}{2}T}\|_2^2$$

$$\geq \||(A^{\frac{1}{2}}B^{\frac{1}{2}T})^{-1}\||_2^{-2} \cdot 1 = [\lambda_{max}(B^{-1}TA^{-1})]^{-1}$$

$$= \lambda_{min}(AB^T)$$

In this case, the second inequality is due to the familiar fact that the sum of squares of the absolute values of the eigenvalues of any matrix is no greater than the square of its Frobenius norm. In all cases the superscript $\frac{1}{2}$ denotes the unique positive definite square root. □

For a more algebraic proof of Theorem (5.3.1), see Problem 1 in Section (5.4).

Exercise. Show that Theorem (5.3.1) implies the weaker inequality

$$\lambda_{min}(A \circ B) \geq \lambda_{min}(A)\lambda_{min}(B)$$

Hint: $\lambda_{min}(AB) = \lambda_{min}(A^{\frac{1}{2}}BA^{\frac{1}{2}}) = \min\{x^*A^{\frac{1}{2}}BA^{\frac{1}{2}}x: x^*x = 1\} \geq$
$\min\{\lambda_{min}(B)\|A^{\frac{1}{2}}x\|_2^2: x^*x = 1\} \geq \lambda_{min}(B)\lambda_{min}(A).$

Exercise. Show by example that λ_{min} may not be replaced by λ_{max} in the statements in Theorem (5.3.1), whether or not the direction of the inequality is reversed. It is an open question whether there is any analog of Theorem (5.3.1) for λ_{max}, as there is for the inequality of the preceding exercise.

Exercise. Show by example that the inequalities in Theorem (5.3.1) are not valid if A is assumed to be only Hermitian. What about the weaker inequality $\lambda_{min}(A \circ B) \geq \lambda_{min}(A)\lambda_{min}(B)$?

5.3.3   Example. If A or $B \in M_n(\mathbb{R})$ the statements (5.3.1a,b), of course, coincide. In general, when both A and B have complex entries, however, the spectra of AB and AB^T can differ and (5.3.1a,b) can provide different estimates. For example, if

$$A = \begin{bmatrix} 2 & i \\ -i & 1 \end{bmatrix} \text{ and } B = \begin{bmatrix} 3 & -2i \\ 2i & 2 \end{bmatrix}$$

then $\sigma(AB) = \{2 \pm \sqrt{2}\}$, $\sigma(AB^T) = \{6 \pm \sqrt{34}\}$, and $\sigma(A \circ B) = \{4 \pm \sqrt{8}\}$. In this instance, we have

$$\lambda_{min}(A \circ B) = 4 - \sqrt{8} > \lambda_{min}(AB) = 2 - \sqrt{2} > \lambda_{min}(AB^T) = 6 - \sqrt{34}$$

$$> \lambda_{min}(A)\lambda_{min}(B) = \left[\frac{3-\sqrt{5}}{2}\right]\left[\frac{5-\sqrt{17}}{2}\right]$$

Remark: If $X \in M_n$ has all real eigenvalues, order them algebraically and denote them by

$$\lambda_{min}(X) = \lambda_1(X) \leq \lambda_2(X) \leq \cdots \leq \lambda_n(X) = \lambda_{max}(X)$$

For positive definite $A, B \in M_n$, Theorem (5.3.1b) states that

$$\lambda_1(AB) \leq \lambda_1(A \circ B)$$

and Oppenheim's inequality $(\det A)(\det B) \leq \det(A \circ B)$ (Theorem (7.8.6) in [HJ]) implies that

$$\prod_{i=1}^{n} \lambda_i(AB) \leq \prod_{i=1}^{n} \lambda_i(A \circ B)$$

These two facts suggest a family of inequalities for the increasingly ordered eigenvalues of AB and $A \circ B$ for positive definite $A, B \in M_n$

$$\prod_{i=1}^{k} \lambda_i(AB) \leq \prod_{i=1}^{k} \lambda_i(A \circ B), \qquad k = 1,\dots, n \qquad (5.3.3a)$$

that has been proved by G. Visick and by T. Ando. The related inequalities

$$\prod_{i=1}^{k} \lambda_i(AB^T) \leq \prod_{i=1}^{k} \lambda_i(A \circ B), \quad k = 1,\dots, n \qquad (5.3.3b)$$

generalizing Theorem (5.3.1a) have also been proved by Ando and by Visick.

We now give a generalization of Theorem (5.2.1) that emphasizes the special role of the diagonal entries of one of the factors. We give two proofs: The first illustrates that a useful quantitative result follows directly from the qualitative statement in Theorem (5.2.1), while the second uses properties of quadratic forms that are vital to obtaining sharper statements about eigenvalues of a Hadamard product.

5.3.4   Theorem. Let $A, B \in M_n$ be positive semidefinite. Any eigenvalue $\lambda(A \circ B)$ of $A \circ B$ satisfies

$$\lambda_{min}(A)\lambda_{min}(B) \leq [\min_{1 \leq i \leq n} a_{ii}] \lambda_{min}(B)$$

$$\leq \lambda(A \circ B)$$

$$\leq [\max_{1 \leq i \leq n} a_{ii}] \lambda_{max}(B) \leq \lambda_{max}(A)\lambda_{max}(B)$$

Proof #1: Since $B - \lambda_{min}(B)I$ and A are positive semidefinite, so is $A \circ (B - \lambda_{min}(B)I)$. Let $x = [x_i]$ be a unit eigenvector of $A \circ B$ corresponding to an eigenvalue $\lambda(A \circ B)$. Then

$$0 \leq x^*[A \circ (B - \lambda_{min}(B)I)]x = x^*(A \circ B)x - \lambda_{min}(B)x^*(A \circ I)x$$

$$= \lambda(A \circ B) - \lambda_{min}(B) \sum_{i=1}^{n} |x_i|^2 a_{ii}$$

$$\leq \lambda(A \circ B) - \lambda_{min}(B) \min_i a_{ii}$$

which gives the asserted lower bound. The calculation for the upper bound is similar.

Proof #2: Let $x = [x_i] \in \mathbb{C}^n$ be a unit eigenvector of $A \circ B$ corresponding to an eigenvalue $\lambda(A \circ B)$. Since A is positive semidefinite, there is some $C \in M_n$ such that $A = CC^*$ ($C = A^{\frac{1}{2}}$, for example). Partition C according to its columns as $C = [c_1 \ c_2 \ ... \ c_n]$ so that

$$A = \sum_{k=1}^{n} c_k c_k^*$$

Also, let $C = [c_{ik}]$, so c_{ik} denotes the ith component of c_k. Then

$$\lambda(A \circ B) = x^*(A \circ B)x = \mathrm{tr}(D_x^* A D_x B^T)$$

$$= \sum_{k=1}^{n} \mathrm{tr}(D_x^* c_k c_k^* D_x B^T)$$

$$= \sum_{k=1}^{n} (c_k^* D_x) B^T (D_x^* c_k) = \sum_{k=1}^{n} (\bar{x} \circ c_k)^* B^T (\bar{x} \circ c_k)$$

Thus,

$$\lambda(A \circ B) \geq \sum_{k=1}^{n} \lambda_{min}(B^T) \|\bar{x} \circ c_k\|_2^2$$

$$= \lambda_{min}(B) \sum_{k=1}^{n} \sum_{i=1}^{n} |\bar{x}_i c_{ik}|^2$$

$$= \lambda_{min}(B) \sum_{i=1}^{n} |x_i|^2 \sum_{k=1}^{n} |c_{ik}|^2$$

$$= \lambda_{min}(B) \sum_{i=1}^{n} |x_i|^2 a_{ii} \geq \lambda_{min}(B) \min_{1 \leq i \leq n} a_{ii}$$

The calculation for the upper bound is similar. The outer inequalities now follow from the fact that $\lambda_{min}(A) \leq a_{ii} \leq \lambda_{max}(A)$. $\square$

Exercise. Deduce Theorem (5.2.1) from Theorem (5.3.4), and note that all parts of Theorem (5.2.1) follow directly from Theorem (5.3.4).

Exercise. Theorems (5.3.1) and (5.3.4) ensure that each of $\lambda_{min}(AB)$, $\lambda_{min}(AB^T)$, and $\lambda_{min}(A)\lambda_{min}(B)$ is a lower bound for $\lambda_{min}(A \circ B)$ when $A, B \in M_n$ are positive semidefinite. Show by examples that each of these three lower bounds can be strictly greater than the other two.

The facts about the field of values of the Hadamard product discussed at the end of Section (1.7) generalize Theorem (5.2.1), as do the main results of this section. In particular, note the assertion in Corollary (1.7.23).

We now wish to present a generalization of the qualitative assertions in Theorem (5.2.1) to a situation in which the matrices are presented in partitioned form. The circumstances discussed in the following theorem occur naturally in the theory of monotone and convex matrix functions discussed in Section (6.6). The key observation about the hypotheses of Theorem (5.2.1) is that a positive semidefinite matrix B has no diagonal entry equal to zero if and only if $B = C^2$ for some Hermitian matrix C with no zero columns, that is, each column of C has (full) rank one.

5.3.5   Lemma. Let $n, p, n_1, ..., n_p$ be given positive integers with $n_1 + \cdots + n_p = n$. Let $B, C \in M_n$ be given Hermitian matrices. Suppose that:

(a) $B = C^2$, and

(b) B and C are partitioned conformally as $B = [B_{ij}]_{i,j=1}^{p}$ and $C = [C_{ij}]_{i,j=1}^{p}$ with $B_{ij}, C_{ij} \in M_{n_i, n_j}$ for $i, j = 1, ..., p$.

For a given integer i with $1 \leq i \leq p$, B_{ii} is positive definite if and only if $C_i \equiv [C_{ki}]_{k=1}^{p}$, the ith block column of C, has full rank n_i.

Proof: For any $z \in \mathbb{C}^{n_i}$, we have

$$z^* B_{ii} z = \sum_{k=1}^{p} z^* (C_{ki})^* C_{ki} z = \sum_{k=1}^{p} \| C_{ki} z \|_2^2 \geq 0$$

with equality if and only if $C_{ki} z = 0$ for all $k = 1, ..., p$. Thus, $z^* B_{ii} z = 0$ if and only if $C_i z = 0$, so the positive semidefinite block B_{ii} is singular if and only if the block column C_i is rank deficient.   □

5.3.6   Theorem. Let $n, p, n_1, ..., n_p$ be given positive integers with $n_1 + \cdots + n_p = n$, and let $A, B \in M_n$ be given positive semidefinite matrices. Suppose that:

(a) A and B are partitioned conformally as $A = [A_{ij}]_{i,j=1}^{p}$ and $B = [B_{ij}]_{i,j=1}^{p}$ with $A_{ij}, B_{ij} \in M_{n_i, n_j}$ for $i, j = 1, ..., p$;

(b) $A_{ij} = \alpha_{ij} J_{n_i, n_j}$, $\alpha_{ij} \in \mathbb{C}$, $i, j = 1, ..., p$, where $J_{r,s}$ denotes an r-by-s matrix all of whose entries are 1;

(c) $A \equiv [\alpha_{ij}]_{i,j=1}^{p} \in M_p$ is positive definite; and

(d) B_{ii} is positive definite for $i = 1, ..., p$.

Then $A \circ B$ is positive definite.

Proof: Since $B \succeq 0$, there is a Hermitian matrix C such that $B = C^2$. Partition C conformally to A and B as $C = [C_{ij}]_{i,j=1}^{p}$ and let $C_k \equiv [C_{ik}]_{i=1}^{p}$ denote the kth block column of C. Since each diagonal block B_{ii} is positive definite, Lemma (5.3.5) ensures that each block column C_i has full rank n_i, $i = 1, ...,$ p. Partition any given $z \in \mathbb{C}^n$ as $z = [z_i]_{i=1}^{p}$, with $z_i \in \mathbb{C}^{n_i}$ for $i = 1, ..., p$. Then

$$z^* (A \circ B) z = z^* (A \circ C^2) z = \sum_{i,j=1}^{p} \alpha_{ij} z_i^* (C_i)^* C_j z_j$$

$$= \sum_{k=1}^{p} \sum_{i,j=1}^{p} \alpha_{ij}\, x_i^*(C_{ki})^* C_{kj} x_j$$

$$= \sum_{k=1}^{p} [(C_k)^* \,\square_1\, x]^* A\,[(C_k)^* \,\square_1\, x] \geq 0$$

where $(C_k)^* \,\square_1\, x \equiv [(C_{ik})^* x_i]_{i=1}^{p}$, as in Problem 3 of Section (3.8). Suppose $x^*(A \circ B)x = 0$. Since $A \succ 0$, it follows that $(C_k)^* \,\square_1\, x = 0$ for all $k = 1,\ldots, p$. Thus, for each $i = 1,\ldots, p$ we have $(C_{ik})^* x_i = C_{ki} x_i = 0$ for all $k = 1,\ldots, p$, that is, $C_i x_i = 0$. Since each C_i has full column rank, this means that all $x_i = 0$. We conclude that $x = 0$ and $A \circ B \succ 0$.   □

5.3.7   Corollary. Let $n, p, n_1,\ldots, n_p$ be given positive integers with $n_1 + \cdots + n_p = n$. Let $A, B, C \in M_n$ be given Hermitian matrices, and suppose that:

(a) $A, B,$ and C are partitioned conformally as $A = [A_{ij}]_{i,j=1}^{p}$, $B = [B_{ij}]_{i,j=1}^{p}$, and $C = [C_{ij}]_{i,j=1}^{p}$ with $A_{ij}, B_{ij}, C_{ij} \in M_{n_i, n_j}$ for $i, j = 1,\ldots, p$;

(b) $A_{ij} = \alpha_{ij} J_{n_i, n_j}$ and $B_{ij} = \beta_{ij} J_{n_i, n_j}$ with $\alpha_{ij}, \beta_{ij} \in \mathbb{C}$ for $i, j = 1,\ldots, p$, where $J_{r,s}$ denotes an r-by-s matrix all of whose entries are 1;

(c) $A \equiv [\alpha_{ij}]_{i,j=1}^{p}$ is positive definite;

(d) $\beta_{ij} \neq 0$ for $i, j = 1,\ldots, p$; and

(e) Each block column $C_j \equiv [C_{ij}]_{i=1}^{p} \in M_{n, n_j}$ has full rank n_j, $j = 1,\ldots, p$.

Then $A \circ (B \circ C)^2$ is positive definite.

Proof: For each $j = 1,\ldots, p$, let $B_j \equiv [B_{ij}]_{i=1}^{p}$ denote the jth block column of B. Because of the special block structure of B and the fact that all the entries of B are nonzero, for each $j = 1,\ldots, p$ the rows of $B_j \circ C_j$ have the same span as the rows of C_j. Thus, $\mathrm{rank}(B_j \circ C_j) = \mathrm{rank}\, C_j = n_j$ for $j = 1,\ldots, p$. Lemma (5.3.5) now ensures that all of the diagonal blocks of

$(B \circ C)^2$ are positive definite, so the asserted result follows from Theorem (5.3.6). ∎

Problems

1. Deduce (5.2.1) from (1.7.23).

2. Show that $\| RS \|_2 \leq \| R \|_2 \| S \|_2$ for all $R \in M_n$ and all $S \in M_{n,k}$.

3. Explain how the result of Problem 2 is used to verify the first inequality in the proof of (5.3.1b).

4. Do the inequalities in (5.3.4) remain valid if B is assumed to be only Hermitian?

5. Let $B \in M_n$ be positive semidefinite. Show that

$$\min \{\lambda_{min}(A \circ B):\ A \succ 0 \text{ and } \lambda_{min}(A) = 1\} = \min_{1 \leq i \leq n} b_{ii}$$

and

$$\max \{\lambda_{max}(A^{-1} \circ B):\ A \succ 0 \text{ and } \lambda_{min}(A) = 1\} = \max_{1 \leq i \leq n} b_{ii}$$

6. Let $A \in M_n$ be positive definite. Show that the matrix $A \circ \overline{A}$ is nonnegative and positive definite and that $[\lambda_{min}(A)]^2 \leq \lambda_{min}(A \circ \overline{A}) \leq \lambda_{max}(A \circ \overline{A}) \leq [\lambda_{max}(A)]^2$. Something interesting happens when the last inequality is an equality; see Theorem (5.5.16).

7. What does Theorem (5.3.6) say in the special cases $p = 1$ and $p = n$? In what sense is Theorem (5.3.6) a generalization of Theorem (5.2.1)?

Notes and Further Reading. For a proof of (5.3.3b) see R. Bapat and V. Sunder, On Majorization and Schur Products, *Linear Algebra Appl.* 72 (1985), 107-117. Statements (5.3.1b) and (5.3.2), as well as example (5.3.3), may be found in C. R. Johnson and L. Elsner, The Relationship between Hadamard and Conventional Multiplication for Positive Definite Matrices, *Linear Algebra Appl.* 92 (1987), 231-240. Statement (5.3.1a) is from A Note on the Hadamard Product of Matrices, *Linear Algebra Appl.* 49 (1983), 233-235, by M. Fiedler. As noted in (5.0), (5.3.4) was first proved by I. Schur. For proofs of (5.3.3a,b) see T. Ando, Majorization Relations for Hadamard Products and G. Visick, A Weak Majorization Involving the Matrices $A \circ B$ and AB, both to appear in *Linear Algebra Appl.*

5.4   The matrices $A \circ (A^{-1})^T$ and $A \circ A^{-1}$

As noted in (5.0.3), the special Hadamard products $A \circ (A^{-1})^T$ and $A \circ A^{-1}$ arise naturally; not surprisingly, they also have special mathematical structure.

5.4.1   Definition. If $A \in M_n$ is nonsingular, we denote

$$\Phi(A) \equiv A \circ A^{-1}$$

and

$$\Phi_T(A) \equiv A \circ (A^{-1})^T$$

We record some simple algebraic facts about Φ and Φ_T in the following lemma.

5.4.2   Lemma. For any nonsingular $A \in M_n$,

(a)   All the row and column sums of $\Phi_T(A)$ are equal to 1;

(b)   For any nonsingular diagonal matrices $D, E \in M_n$,

(i) $\Phi_T(DAE) = \Phi_T(A)$, and

(ii) $\Phi(DAE) = (D^{-1}E)^{-1}\Phi(A)(D^{-1}E)$;

(c)   For any permutation matrices $P, Q \in M_n$, $\Phi_T(PAQ) = P\Phi_T(A)Q$;

(d) $\Phi_T(P) = P$ for any permutation matrix P; and

(e) $\Phi(A^{-1}) = \Phi(A)$, $\Phi(A^T) = \Phi(A)^T$, and $\Phi_T(A^{-1}) = \Phi_T(A^T)$.

Exercise. Verify each of the assertions in (5.4.2). *Hint:* Use (5.1.2) for (5.4.2b).

Exercise. If $0 \neq \alpha \in \mathbb{C}$, show that $\Phi(\alpha A) = \Phi(A)$ and $\Phi_T(\alpha A) = \Phi_T(A)$ for all nonsingular $A \in M_n$.

Exercise. Show that $1 \in \sigma(\Phi_T(A))$ for every nonsingular $A \in M_n$.

A key fact about the functions Φ and Φ_T is the following:

5.4.3   Theorem. Let $A \in M_n$ be positive definite. Then

(a) $\lambda_{min}(\Phi(A)) \geq 1$

and

(b) $\lambda_{min}(\Phi_T(A)) = 1$

Proof: Part (a) follows immediately from (5.3.1b), while part (b) follows from (5.3.1a) together with the fact that $1 \in \sigma(\Phi_T(A))$ because of (5.4.2a). ▯

Of course, as with (5.3.1), the two matrices $\Phi(A)$ and $\Phi_T(A)$ coincide if the Hermitian matrix A is real, that is, if A is real symmetric and positive definite. From (5.4.3) it follows that in the positive semidefinite partial order

$$\Phi(A) \succeq I \qquad\qquad\qquad (5.4.4a)$$

and

$$\Phi_T(A) \succeq I \qquad\qquad\qquad (5.4.4b)$$

for every positive definite $A \in M_n$. Since $1 \in \sigma(\Phi_T(A))$, the inequality (5.4.4b) is never strict.

5.4.5   Example. The inequalities (5.4.3a) and (5.4.4a) can be strict for $A \in M_3$. Consider the positive definite matrix

$$A = \begin{bmatrix} 3 & 1-i & -i \\ 1+i & 2 & 1 \\ i & 1 & 1 \end{bmatrix}$$

Then

$$A \circ A^{-1} - I = \begin{bmatrix} 2 & -1+i & 1-i \\ -1-i & 3 & -2-i \\ 1+i & -2+i & 3 \end{bmatrix}$$

is positive definite (its smallest eigenvalue is .4746), so $\Phi(A) \succ I$.

Exercise. If $A \in M_2$ is positive definite, show that the inequality (5.4.3a) is always an equality.

As is often the case with matrix inequalities, the inequalities (5.4.4) may be refined by insertion of intermediate terms. The following result

The Hadamard product

proves a generalization of the inequalities (5.4.4a,b) with a straightforward argument that does not rely on the inequalities (5.4.3).

5.4.6   Theorem. Let a given positive definite matrix $A \in M_n$ and its inverse $B \equiv A^{-1}$ be partitioned conformally as

$$A = \begin{bmatrix} A_{11} & A_{12} \\ A_{12}^* & A_{22} \end{bmatrix} \text{ and } B = \begin{bmatrix} B_{11} & B_{12} \\ B_{12}^* & B_{22} \end{bmatrix}$$

with A_{11} square. Then

(a) $\Phi(A) \succeq \begin{bmatrix} \Phi(B_{11}) & 0 \\ 0 & \Phi(A_{22}) \end{bmatrix} \succeq I$

and

$\Phi(A) \succeq \begin{bmatrix} \Phi(A_{11}) & 0 \\ 0 & \Phi(B_{22}) \end{bmatrix} \succeq I$

and

(b) $\Phi_T(A) \succeq \begin{bmatrix} \Phi_T(B_{11}^{-1}) & 0 \\ 0 & \Phi_T(A_{22}) \end{bmatrix} \succeq I$

and

$\Phi_T(A) \succeq \begin{bmatrix} \Phi_T(A_{11}) & 0 \\ 0 & \Phi_T(B_{22}^{-1}) \end{bmatrix} \succeq I$

Proof: By (5.1.6),

$$A_1 \equiv A - \begin{bmatrix} B_{11}^{-1} & 0 \\ 0 & 0 \end{bmatrix} \text{ and } B_1 \equiv B - \begin{bmatrix} 0 & 0 \\ 0 & A_{22}^{-1} \end{bmatrix}$$

are positive semidefinite, and $A_1 \circ B_1$ is positive semidefinite by (5.2.1). This means that

$$A \circ B \succeq A \circ \begin{bmatrix} 0 & 0 \\ 0 & A_{22}^{-1} \end{bmatrix} + B \circ \begin{bmatrix} B_{11}^{-1} & 0 \\ 0 & 0 \end{bmatrix} = \begin{bmatrix} \Phi(B_{11}) & 0 \\ 0 & \Phi(A_{22}) \end{bmatrix}$$

which is the first inequality involving the first intermediate matrix in (a). The second inequality then follows from repeated application of the first until each block is 1-by-1, giving I. The inequalities involving the second intermediate matrix are verified similarly by using

$$A_2 \equiv A - \begin{bmatrix} 0 & 0 \\ 0 & B_{22}^{-1} \end{bmatrix} \text{ and } B_2 \equiv B - \begin{bmatrix} A_{11}^{-1} & 0 \\ 0 & 0 \end{bmatrix}$$

in place of A_1 and B_1. The inequalities (b) are demonstrated similarly by using B_1^T in place of B_1 and B_2^T in place of B_2.   ☐

Many variants on (5.4.6a,b) may be obtained by using other principal submatrices resulting from index sets that are not necessarily contiguous.

Exercise. Write down the inequalities analogous to (5.4.6b) that result from using A_1^T and A_2^T (in the notation of the proof).

5.4.7   Example. Another natural conjecture for a refinement of (5.4.4a) is

$$\Phi(A) \succeq \begin{bmatrix} \Phi(A_{11}) & 0 \\ 0 & \Phi(A_{22}) \end{bmatrix} \succeq I$$

using the notation of (5.4.6). Unfortunately, this one is not valid, as shown by the counterexample

$$A = \begin{bmatrix} 3 & 6 & 4 & 5 \\ 6 & 15 & 9 & 11 \\ 4 & 9 & 6 & 7 \\ 5 & 11 & 7 & 9 \end{bmatrix} \in M_4$$

which is positive definite. For the partition in which $A_{11} = \begin{bmatrix} 3 & 6 \\ 6 & 15 \end{bmatrix}$ and $A_{22} = \begin{bmatrix} 6 & 7 \\ 7 & 9 \end{bmatrix}$, we have

$$\Phi(A) - \begin{bmatrix} \Phi(A_{11}) & 0 \\ 0 & \Phi(A_{22}) \end{bmatrix} = \begin{bmatrix} 13 & 10 & -8 & -15 \\ 10 & 10 & -9 & -11 \\ -8 & -9 & 7.2 & 9.8 \\ -15 & -11 & 9.8 & 16.2 \end{bmatrix}$$

which is not positive semidefinite; the principal submatrix $\begin{bmatrix} 10 & -9 \\ -9 & 7.2 \end{bmatrix}$ has negative determinant. For inequalities of the type mentioned in (5.4.6) to be correct, one block must be chosen from each of A and A^{-1}.

Exercise. Show that the conjectured inequalities in (5.4.7) *are* correct for $n = 2$ and 3, so 4 is the smallest value of n for which a counterexample can be found.

Exercise. Show that successive refinement of (5.4.6a) can yield strings of inequalities such as

$$\Phi(A) \succeq \begin{bmatrix} 1 & 0 \\ 0 & \Phi(A(\{2,...,n\})) \end{bmatrix} \succeq \begin{bmatrix} I_2 & 0 \\ 0 & \Phi(A(\{3,...,n\})) \end{bmatrix} \succeq \cdots$$

$$\succeq \begin{bmatrix} I_k & 0 \\ 0 & \Phi(A(\{k+1,...,n\})) \end{bmatrix} \succeq \cdots \succeq I$$

Exercise. State and verify the corresponding string of inequalities based upon (5.4.6b).

Since $I = AA^{-1}$ for any nonsingular $A \in M_n$, (5.4.4a) has the striking interpretation

$$A \circ A^{-1} \succeq AA^{-1} \tag{5.4.8}$$

for any positive definite Hermitian $A \in M_n$, that is, the Hadamard product dominates the ordinary product of A and A^{-1} in the positive semidefinite partial ordering. One may ask how general this phenomenon is; that is, for which positive definite pairs A, B is

$$A \circ B \succeq AB$$

always correct? We have in mind pairs A, B in which $B = f(A)$ is determined in a definite way by A, with $f(A) = A^{-1}$ as the model example. It turns out that the inverse function is essentially the only example of this phenomenon.

5.4.9   Definition. A function $f(\cdot)$ that maps the class of positive definite n-by-n matrices into M_n is said to be an *ordinary function* if there are n functions $f_i: \mathbb{R}^n_+ \to \mathbb{R}_+$, $i = 1,..., n$, such that for any unitary diagonalization

$$A = U^* \text{diag}(\lambda_1,..., \lambda_n) U, \quad U \in M_n \text{ unitary, all } \lambda_i \in \mathbb{R}_+ = (0, \infty)$$

we have

$$f(A) = U^* \mathrm{diag}(f_1(\lambda_1,\ldots,\lambda_n),\ldots,f_n(\lambda_1,\ldots,\lambda_n))U$$

This definition ensures that $f(A)$ commutes with A, which is necessary for $Af(A)$ to be Hermitian, and that $f(A)$ actually "depends" upon A. The assumption that f_i maps $\mathbb{R}^n_+$ to $\mathbb{R}_+$ ensures that $f(A)$ is positive definite whenever A is positive definite, so an ordinary function maps the cone of positive definite matrices into itself. Within these constraints, which are natural for our question, the definition of an ordinary function is quite broad.

Exercise. Show that polynomials with positive coefficients, inversion, and exponentiation are all examples of ordinary functions, and that in each of these cases each function f_i depends upon λ_i only and all the functions f_i are the same. Show that the classical adjoint $f(A) = (\det A)A^{-1}$ is an ordinary function in which not all of the functions f_i are the same, and each f_i depends not only on λ_i.

Exercise. Show that the set of ordinary functions is convex.

A converse for (5.4.4a) is the following:

5.4.10   Theorem. Let $f(\cdot)$ be a given ordinary function on positive definite matrices in M_n. Then $A \circ f(A) \succeq Af(A)$ for all positive definite $A \in M_n$ if and only if $f(A) = \alpha(A)A^{-1}$ for each positive definite $A \in M_n$, where $\alpha(\cdot)$ is a positive real-valued function on the n-by-n positive definite matrices.

Proof: To prove (5.4.10) we need only establish the necessity of the asserted property of $f(\cdot)$. Because the asserted inequality is positively homogeneous, the sufficiency follows from (5.4.4a). Let $f_1, f_2,\ldots, f_n: \mathbb{R}^n_+ \to \mathbb{R}_+$ be the functions that induce the ordinary function $f(\cdot)$, and let $\lambda_1, \lambda_2,\ldots, \lambda_n > 0$ be given positive scalars. We assert that

$$\lambda_1 f_1(\lambda_1,\ldots,\lambda_n) = \lambda_2 f_2(\lambda_1,\ldots,\lambda_n) = \cdots = \lambda_n f_n(\lambda_1,\ldots,\lambda_n)$$

Then, since these are the eigenvalues of $Af(A)$, it follows that $Af(A) = \lambda_i f_i(\lambda_1,\ldots,\lambda_n)I$ for every $i = 1,\ldots, n$. Thus, $f(A) = \alpha(A)A^{-1}$ with $\alpha(A) = \lambda_1 f_1(\lambda_1,\ldots,\lambda_n)$ a positive scalar function, as asserted. Our strategy is to show that $\lambda_k f_k(\lambda_1,\ldots,\lambda_n) = \lambda_j f_j(\lambda_1,\ldots,\lambda_n)$ for every pair of distinct indices

k, j. Without loss of generality (the same proof works for any pair) we show this for $k = 1$, $j = 2$. Let

$$U_1 = \frac{1}{\sqrt{2}}\begin{bmatrix} 1 & 1 \\ 1 & -1 \end{bmatrix}, \quad U_2 = \frac{1}{\sqrt{2}}\begin{bmatrix} -1 & 1 \\ 1 & 1 \end{bmatrix}, \text{ and } A_i = U_i^*\begin{bmatrix} \lambda_1 & 0 \\ 0 & \lambda_2 \end{bmatrix}U_i$$

for $i = 1, 2$. We consider $\hat{A}_i \equiv A_i \bullet \operatorname{diag}(\lambda_3,\ldots, \lambda_n)$, $i = 1, 2$. Since

$$(\hat{A}_i \circ f(\hat{A}_i)) - \hat{A}_i f(\hat{A}_i) = [(A_i \circ f(A_i)) - A_i f(A_i)] \bullet 0_{n-2}$$

where $0_{n-2} \in M_{n-2}$ is the zero matrix, it suffices to consider the implications of the inequalities

$$A_i \circ f(A_i) \succeq A_i f(A_i), \quad i = 1, 2$$

for 2-by-2 matrices. A calculation reveals that

$$A_1 = \frac{1}{2}\begin{bmatrix} \lambda_1 + \lambda_2 & \lambda_1 - \lambda_2 \\ \lambda_1 - \lambda_2 & \lambda_1 + \lambda_2 \end{bmatrix}, \quad f(A_1) = \frac{1}{2}\begin{bmatrix} f_1(\lambda) + f_2(\lambda) & f_1(\lambda) - f_2(\lambda) \\ f_1(\lambda) - f_2(\lambda) & f_1(\lambda) + f_2(\lambda) \end{bmatrix}$$

and

$$A_1 f(A_1) = \frac{1}{2}\begin{bmatrix} \lambda_1 f_1(\lambda) + \lambda_2 f_2(\lambda) & \lambda_1 f_1(\lambda) - \lambda_2 f_2(\lambda) \\ \lambda_1 f_1(\lambda) - \lambda_2 f_2(\lambda) & \lambda_1 f_1(\lambda) + \lambda_2 f_2(\lambda) \end{bmatrix}$$

in which we have used $f_i(\lambda)$ to denote $f_i(\lambda_1,\ldots,\lambda_n)$, $i = 1, 2$. It follows that $A_1 \circ f(A_1) - A_1 f(A_1) =$.

$$\frac{1}{4}\begin{bmatrix} \lambda_2 f_1(\lambda) + \lambda_1 f_2(\lambda) - [\lambda_1 f_1(\lambda) + \lambda_2 f_2(\lambda)] & 3\lambda_2 f_2(\lambda) - \lambda_1 f_1(\lambda) - \lambda_1 f_2(\lambda) - \lambda_2 f_1(\lambda) \\ 3\lambda_2 f_2(\lambda) - \lambda_1 f_1(\lambda) - \lambda_1 f_2(\lambda) - \lambda_2 f_1(\lambda) & \lambda_2 f_1(\lambda) + \lambda_1 f_2(\lambda) - [\lambda_1 f_1(\lambda) + \lambda_2 f_2(\lambda)] \end{bmatrix}$$

Since $A_1 \circ f(A_1) - A_1 f(A_1)$ is positive semidefinite, it is necessary that $\lambda_2 f_2(\lambda) \geq \lambda_1 f_1(\lambda)$ because $[1 \ \ 1][A_1 \circ f(A_1) - A_1 f(A_1)]\begin{bmatrix} 1 \\ 1 \end{bmatrix}$ must be nonnegative. Parallel calculations involving

$$A_2 = \frac{1}{2}\begin{bmatrix} \lambda_1 + \lambda_2 & \lambda_2 - \lambda_1 \\ \lambda_2 - \lambda_1 & \lambda_1 + \lambda_2 \end{bmatrix}$$

reverse the roles of the first and second variables and functions and lead to the second necessary condition $\lambda_1 f_1(\lambda) \geq \lambda_2 f_2(\lambda)$. We conclude that $\lambda_1 f_1(\lambda)$

$= \lambda_2 f_2(\lambda)$ and that $\lambda_k f_k(\lambda) = \lambda_j f_j(\lambda)$ in general, verifying our assertion.   ☐

If we consider Φ or Φ_T as a mapping from the nonsingular matrices in M_n into M_n, it is natural to ask what iteration of Φ or Φ_T can produce. We denote the kth iterate of Φ applied to A by $\Phi^{(k)}(A) \equiv \Phi(\Phi^{(k-1)}(A))$, with $\Phi^{(0)}(A) \equiv A$, and similarly for $\Phi_T^{(k)}$. Here, there is an implicit assumption that successive iterates are nonsingular, so that Φ or Φ_T may be applied repeatedly.

Exercise. Give examples of nonsingular $A, B \in M_n$ such that $\Phi_T(A)$ and $\Phi(B)$ are singular.

Exercise. If $A \in M_n$ is positive definite, show that $\Phi^{(k)}(A)$ and $\Phi_T^{(k)}(A)$ are nonsingular for all $k = 1, 2, \dots.$

For which $A \in M_n$ do $\Phi^{(k)}(A)$ or $\Phi_T^{(k)}(A)$ converge as $k \to \infty$ and to what? In general, this is an open question, but for positive definite matrices and for H-matrices (2.5.10), precise results are known. We state them without proofs.

5.4.11   Theorem. If $A \in M_n$ is positive definite, then

(a) $\lim\limits_{k \to \infty} \Phi^{(k)}(A) = I$

and

(b) $\lim\limits_{k \to \infty} \Phi_T^{(k)}(A) = I$

It is worth comparing Theorem (5.4.11) with the matrix inequalities (5.4.4a,b), which makes it clear that the convergence is from above and that the limits are as "small" as possible.

Exercise. If $A \in M_n$ is an M-matrix, show that all diagonal entries of $\Phi_T(A)$ are at least 1, and that $\Phi_T(A) - I$ is a singular M-matrix.

5.4.12   Theorem. If $A \in M_n$ is an H-matrix, then

$$\lim\limits_{k \to \infty} \Phi^{(k)}(A) = I$$

In (5.0.3) we discussed how Φ_T maps the eigenvalues of a diagonalizable matrix into its main diagonal entries. The most general result in this regard, whose proof we also omit, indicates that the trace condition is the only constraint, with an obvious exception.

5.4.13   Theorem. Let $A \in M_n$ and n scalars $b_{11},..., b_{nn} \in \mathbb{C}$ be given, and suppose A is not a scalar matrix, that is, $A \neq \alpha I$ for any $\alpha \in \mathbb{C}$. There is some $B \in M_n$ that is similar to A and has diagonal entries $b_{11},..., b_{nn}$ if and only if

$$b_{11} + b_{22} + \cdots + b_{nn} = \text{tr } A$$

Theorem (5.4.13) remains valid if the data and solutions are restricted to be real, or if they come from any field of characteristic zero or characteristic at least n.

Problems

1.   We have deduced the matrix inequalities (5.4.4) from the inequalities (5.3.1) via (5.4.3). However, the proof of (5.4.6) gives an independent, perhaps more elementary, and certainly more algebraic, proof of the inequalities (5.4.4). Provide the details of and justification for the steps in the following proof of (5.3.1) from (5.4.4). First, $\lambda_{min}(AB)I = \lambda_{min}(B^{\frac{1}{2}}AB^{\frac{1}{2}})I \preceq B^{\frac{1}{2}}AB^{\frac{1}{2}}$, which implies $B^{-\frac{1}{2}}\lambda_{min}(AB)B^{-\frac{1}{2}} \preceq A$, or $\lambda_{min}(AB)B^{-1} \preceq A$. Then, $A \circ B \succeq \lambda_{min}(AB)(B^{-1} \circ B) \succeq \lambda_{min}(AB)I$, from which (5.3.1b) follows. Give the details for an analogous proof of (5.3.1a).

2.   Use the same ideas as in Problem 1 to show that

$$A \circ B \preceq \lambda_{max}(AB)B \circ B^{-1}$$

and

$$A \circ B \preceq \lambda_{max}(AB^T)B \circ (B^{-1})^T$$

and thus that

$$\lambda_{max}(A \circ B) \leq \begin{cases} \lambda_{max}(AB)\lambda_{max}(B \circ B^{-1}) \\ \lambda_{max}(AB^T)\lambda_{max}(B \circ (B^{-1})^T) \end{cases}$$

for any two positive definite matrices $A, B \in M_n$. Why must the factors $\lambda_{max}(B \circ B^{-1})$ and $\lambda_{max}(B \circ (B^{-1})^T)$ appear here, while there are no similar factors in (5.3.1)? Note that the roles of A and B may be interchanged and that A need only be positive semidefinite.

3.   Let $x = [x_i]$, $z = [z_i] \in \mathbb{C}^n$ be two given vectors with not all x_i equal. Use (5.4.13) to show that there is a nonsingular $A \in M_n$ such that $z = \Phi_T(A)x$ if and only if $x_1 + \cdots + x_n = z_1 + \cdots + z_n$. Show that the condition that not all x_i are equal cannot be dropped.

4.   The *condition number* of a nonsingular matrix $A \in M_n$ relative to a given matrix norm $\|\|\cdot\|\|$ is $\kappa(A) \equiv \|\|A\|\| \, \|\|A^{-1}\|\|$ (see Section (5.8) of [HJ]). Because upper bounds for errors in the solution of linear systems associated with A are proportional to $\kappa(A)$, diagonal scaling (that is, replacement of A by DAE, where D and E are nonsingular diagonal matrices) is often used in an attempt to make the condition number of the scaled system as small (that is, as close to 1) as possible. It is therefore of interest to know how much one could hope to improve the condition number of a given nonsingular A by diagonal scaling. Show that if one uses the spectral norm and its associated condition number, then

$$\inf \{\kappa(DAE): D, E \in M_n \text{ are nonsingular}\} \geq \|\|\Phi_T(A)\|\|_2$$

Give examples to show that equality may, but need not always, occur in this bound.

5.   Show that the function $f(A) \equiv A \circ \overline{A}$ on M_n maps the cone of positive definite matrices into itself but is not an ordinary function.

Notes and Further Readings. For proofs of Theorems (5.4.11-13) and related results, see C. R. Johnson and H. M. Shapiro, Mathematical Aspects of the Relative Gain Array $A \circ A^{-T}$, *SIAM J. Algebraic and Discrete Methods* 7 (1986), 627-644. The counterexample in (5.4.7) and the refinements of (5.4.4a) found in Theorem (5.4.6) are in R. Bapat and M. K. Kwong, A Generalization of $A \circ A^{-1} \geq I$, *Linear Algebra Appl.* 93 (1987), 107-112. Theorem (5.4.10) and Example (5.4.5) are from C. R. Johnson and L. Elsner, The Relationship between Hadamard and Conventional Multiplication for Positive Definite Matrices, *Linear Algebra Appl.* 92 (1987), 231-240. As referenced there also, (5.4.4b) is due to M. Fiedler and (5.4.4a) to C. R. Johnson.

5.5   Inequalities for Hadamard products of general matrices: an overview

Thus far we have devoted most of our attention to Hadamard products on special classes of matrices, principally the positive semidefinite Hermitian matrices. In this section, we consider inequalities for general matrices in M_n and $M_{m,n}$, although we include some special inequalities for Hermitian or positive semidefinite matrices. We survey many classical and recent inequalities for the Hadamard product, several of which are linked as special cases of a new basic family of inequalities (5.6.2). The next section is devoted to this master family of inequalities and how it implies some of the inequalities in this section. For completeness when this section is used for reference, we have indicated after statements of some of the results that they follow from the master inequality (5.6.2) or one of its consequences. For continuity of exposition, the reader may wish to defer consideration of proofs of these results until the end of Section (5.6), where they are all collected.

Recall that the singular values $\sigma_1(A) \geq \sigma_2(A) \geq \cdots \geq \sigma_p(A) \geq 0$ of $A \in M_{m,n}$ are the square roots of the $p = \min\{m,n\}$ possibly nonzero eigenvalues of A^*A; see Sections (7.3) and (7.4) of [HJ]. For convenience, we adopt the convention that $\sigma_i(A) \equiv 0$ if $i > p$. Recall also that the nonzero eigenvalues of A^*A and AA^* coincide. The largest singular value $\sigma_1(\cdot)$ is a unitarily invariant function of A ($\sigma_1(UAV) = \sigma_1(A)$ for all unitary U, $V \in M_n$) that is actually a matrix norm on M_n, denoted $\||\cdot\||_2$ and called the *spectral norm* (see Chapters 5 and 7 of [HJ]).

The earliest and one of the simplest inequalities involving Hadamard products and singular values is due to Schur. It says that the spectral norm is submultiplicative with respect to the Hadamard product (as well as with respect to the ordinary matrix product).

5.5.1   Theorem. For every $A, B \in M_{m,n}$,

$$\sigma_1(A \circ B) \leq \sigma_1(A)\sigma_1(B)$$

Exercise. Show that it suffices to verify Theorem (5.5.1) for the case $m = n$.

Exercise. Verify Theorem (5.5.1) using Lemma (5.1.1), Theorem (4.2.15), and Theorem (7.3.9) in [HJ]. See Problem 7 for alternative proofs.

Exercise. Let $A, B \in M_n$ be positive semidefinite. Show that the bound

$$\lambda_{max}(A \circ B) \leq \lambda_{max}(A)\lambda_{max}(B)$$

mentioned at the beginning of Section (5.3) is a special case of Theorem (5.5.1).

The bound in Theorem (5.5.1) may be improved in a variety of ways, and many Hadamard product inequalities are generalizations of it.

5.5.2   Definition. For $A = [a_{ij}] \in M_{m,n}$, denote the decreasingly ordered Euclidean row lengths of A by

$$r_1(A) \geq r_2(A) \geq \cdots \geq r_m(A)$$

and the decreasingly ordered Euclidean column lengths by

$$c_1(A) \geq c_2(A) \geq \cdots \geq c_n(A)$$

That is, $r_k(A)$ is the kth largest value of $\left[\sum_{j=1}^{n} |a_{ij}|^2 \right]^{\frac{1}{2}}$, $i = 1,..., m$, and similarly for $c_k(A)$.

Exercise. For $A \in M_{m,n}$, show that $\{r_1(A)^2,..., r_m(A)^2\}$ is the set of diagonal entries of AA^* and that $\{c_1(A)^2,..., c_n(A)^2\}$ is the set of diagonal entries of A^*A.

Exercise. For $A \in M_{m,n}$ show that $r_1(A) \leq \sigma_1(A)$ and $c_1(A) \leq \sigma_1(A)$. *Hint:* Use the preceding exercise and the fact that the largest diagonal entry of a positive semidefinite matrix is no larger than its largest eigenvalue.

One significant refinement of Theorem (5.5.1) is the following:

5.5.3   Theorem. For every $A, B \in M_{m,n}$ we have

$$\sigma_1(A \circ B) \leq r_1(A)c_1(B) \leq \begin{Bmatrix} r_1(A)\sigma_1(B) \\ \sigma_1(A)c_1(B) \end{Bmatrix} \leq \sigma_1(A)\sigma_1(B)$$

Proof: In view of the preceding exercise, it suffices to verify only the first inequality. Since $\sigma_1(A \circ B) = \max |x^*(A \circ B)y|$ over all unit vectors x, y, we estimate $|x^*(A \circ B)y|$ for $x \in \mathbb{C}^m$, $y \in \mathbb{C}^n$ as follows:

$$|x^*(A \circ B)y| = \left| \sum_{i,j} \bar{x}_i a_{ij} b_{ij} y_j \right|$$

$$\leq \left[\sum_{i,j} |x_i a_{ij}|^2 \right]^{\frac{1}{2}} \left[\sum_{i,j} |b_{ij} y_j|^2 \right]^{\frac{1}{2}}$$

$$= \left[\sum_i |x_i|^2 \sum_j |a_{ij}|^2 \right]^{\frac{1}{2}} \left[\sum_j |y_j|^2 \sum_i |b_{ij}|^2 \right]^{\frac{1}{2}}$$

$$= \left[\sum_i |x_i|^2 \, r_i(A)^2 \right]^{\frac{1}{2}} \left[\sum_j |y_j|^2 \, c_j(B)^2 \right]^{\frac{1}{2}}$$

$$\leq \left[\sum_i |x_i|^2 \, r_1(A)^2 \right]^{\frac{1}{2}} \left[\sum_j |y_j|^2 \, c_1(B)^2 \right]^{\frac{1}{2}}$$

$$= r_1(A) c_1(B) \|x\|_2 \|y\|_2$$

and the asserted inequality follows. The first inequality is the Cauchy-Schwarz inequality. □

Exercise. Use the commutativity of the Hadamard product to show that $\sigma_1(A \circ B) \leq c_1(A) r_1(B)$ under the same circumstances as in (5.5.3).

Exercise. Analogs of (5.5.1) or (5.5.3) for single indices greater than 1 are, in general, false. Show by example that $\sigma_2(A \circ B) \leq \sigma_2(A)\sigma_2(B)$ need not hold for $A, B \in M_{m,n}$. *Hint:* Consider $B = J$, the matrix with all entries +1, or any other rank one matrix.

There are useful generalizations of the inequality (5.5.1) that involve the sums of singular values that define the Ky Fan k-norms.

5.5.4   Theorem. For every $A, B \in M_{m,n}$ we have

$$\sum_{i=1}^{k} \sigma_i(A \circ B) \leq \sum_{i=1}^{k} \sigma_i(A)\sigma_i(B), \quad k = 1, \dots, n$$

This follows from Theorem (5.5.19); see the discussion following the proof of Lemma (5.6.17).

The sequence of inequalities in Theorem (5.5.4) includes Theorem (5.5.1) as the case $k = 1$ and an inequality reminiscent of Cauchy-Schwarz in the case $k = n$. The inequalities (5.5.4) are weak majorization inequalities (see Chapter 7 of [HJ] and Chapter 3); other weak majorization inequalities will appear later.

One use of the inequalities (5.5.4) is to characterize those unitarily invariant norms on matrices that, like the spectral norm, are Hadamard submultiplicative. Recall that a norm $N(\cdot)$ on $M_{m,n}$ is *unitarily invariant* (see Section (3.5)) if $N(UAV) = N(A)$ for all unitary $U \in M_m$, $V \in M_n$. An important theorem of von Neumann characterizes the unitarily invariant norms as those functions on $M_{m,n}$ that are symmetric gauge functions of the singular values (see Theorem (3.5.18)).

5.5.5   Definition. A norm $N(\cdot)$ on $M_{m,n}$ is *Hadamard submultiplicative* if $N(A \circ B) \leq N(A)N(B)$ for all $A, B \in M_{m,n}$.

The inequality (5.5.1) says that the spectral norm is Hadamard submultiplicative. Note that for $N(\cdot)$ to be a *norm* on M_n (as opposed to a *matrix norm*), the submultiplicativity property $N(AB) \leq N(A)N(B)$ need not be satisfied for all $A, B \in M_n$. If it is satisfied, the norm is called a *matrix norm*. Hadamard submultiplicativity is simply the analogous property for the Hadamard product, and it is not immediately clear that the two types of submultiplicativity should be related. However, remarkably, they are equivalent for unitarily invariant norms. This is another example of the surprisingly close link between the Hadamard and ordinary products.

5.5.6   Definition. A norm $N(\cdot)$ on M_n is *spectrally dominant* if $N(A) \geq \rho(A)$, the spectral radius of A, for every $A \in M_n$. A norm $N(\cdot)$ on $M_{m,n}$ is *singular value dominant* if $N(A) \geq \sigma_1(A)$ for all $A \in M_{m,n}$.

If a norm on M_n is either (ordinary) submultiplicative or singular value dominant, it is spectrally dominant, but the converse implication is not correct in general.

5.5.7   Theorem. Let $N(\cdot)$ be a unitarily invariant norm on M_n. The following are equivalent:

(a) $N(\cdot)$ is singular value dominant.

(b) $N(\cdot)$ is submultiplicative, that is, $N(\cdot)$ is a matrix norm.

(c) $N(\cdot)$ is Hadamard submultiplicative.

(d) $N(\cdot)$ is spectrally dominant.

Proof: The proof consists of the following sequence of implications: (a) $\Rightarrow$ (b) $\Rightarrow$ (d) $\Rightarrow$ (a) $\Rightarrow$ (c) $\Rightarrow$ (a). Let $g(\cdot)$ be the symmetric gauge function associated with the unitarily invariant norm $N(\cdot)$, according to von Neumann's theorem ((7.4.24) in [HJ]), so $g(x)$ is an absolute (and hence monotone) vector norm on $\mathbb{R}^n$ that is a permutation-invariant function of the entries of x and $N(A) = g([\sigma_1(A), \sigma_2(A), ..., \sigma_n(A)]^T)$ for all $A \in M_n$.

(a) $\Rightarrow$ **(b):**
$$
\begin{aligned}
N(AB) &= g([\sigma_1(AB), \sigma_2(AB), ..., \sigma_n(AB)]^T) \\
&\leq g(\sigma_1(A)[\sigma_1(B), \sigma_2(B), ..., \sigma_n(B)]^T) \\
&= \sigma_1(A)g([\sigma_1(B), \sigma_2(B), ..., \sigma_n(B)]^T) \\
&= \sigma_1(A)N(B) \leq N(A)N(B)
\end{aligned}
$$

The first inequality uses the Weyl inequalities (3.3.20) and the monotonicity of the symmetric gauge function $g(\cdot)$ [see (3.5.17)]. The second equality uses the homogeneity of $g(\cdot)$, and the last inequality uses the hypothesis (a).

(b) $\Rightarrow$ **(d):** This implication is well known; see Theorem (5.6.9) in [HJ].

(d) $\Rightarrow$ **(a):** If $A = V \Sigma W^*$ is a singular value decomposition of A with unitary V, $W \in M_n$ and $\Sigma = \mathrm{diag}(\sigma_1(A), ..., \sigma_n(A))$, then $N(A) = N(V \Sigma W^*) = N(\Sigma) \geq \rho(\Sigma) = \sigma_1(A)$. Here we have used the unitary invariance of $N(\cdot)$ and the hypothesis (d) applied to Σ.

(a) $\Rightarrow$ **(c):** Recall that $G(A) \leq G(B)$ for every unitarily invariant norm $G(\cdot)$ on M_n if and only if $N_k(A) \leq N_k(B)$ for $k = 1, ..., n$, where $N_k(C) \equiv \sigma_1(C) + \cdots + \sigma_k(C)$ is the sum of the k largest singular values of C [Corollary (3.5.9)]. Using Theorem (5.5.4) we have

$$
N_k(A \circ B) = \sum_{i=1}^{k} \sigma_i(A \circ B) \leq \sum_{i=1}^{k} \sigma_i(A)\sigma_i(B) \leq \sum_{i=1}^{k} \sigma_1(A)\sigma_i(B)
$$

$$= N_k(\sigma_1(A)B), \quad k = 1, 2, ..., n$$

It follows that $N(A \circ B) \leq N(\sigma_1(A)B)$ because $N(\cdot)$ is unitarily invariant. But $N(\sigma_1(A)B) = \sigma_1(A)N(B) \leq N(A)N(B)$ because of (a), which yields (c).

(c) $\Rightarrow$ (a): Let $E_{11} = \text{diag}(1,0,0,..., 0) \in M_n$ and again let $\Sigma = \text{diag}(\sigma_1(A),..., \sigma_n(A))$. We then have $\sigma_1(A)N(E_{11}) = N(\Sigma \circ E_{11}) \leq N(\Sigma)N(E_{11}) = N(A)N(E_{11})$. The first equality uses the homogeneity of $N(\cdot)$; the inequality uses the hypothesis (c), and the last equality uses the singular value decomposition and the unitary invariance of $N(\cdot)$. Because of the positivity of $N(\cdot)$, we may divide by $N(E_{11})$ to complete the proof of this implication and of the theorem. $\square$

The spectral norm is the most obvious example of a norm of the type characterized in Theorem (5.5.7), so (5.5.7) may be viewed as a generalization of (5.5.1). Of course, (5.5.7a) makes it clear that the spectral norm is "minimal" among Hadamard submultiplicative, unitarily invariant norms on M_n, but the theorem also makes it clear that many other familiar norms are examples.

Exercise. Show in two ways that the Ky Fan k-norms are Hadamard submultiplicative: Use Theorem (5.5.7) and then use only Theorem (5.5.4).

Exercise. Show that the Schatten p-norms (the unitarily invariant norms on M_n associated with the l_p norms as the symmetric gauge functions, $p \geq 1$) are both Hadamard and ordinary submultiplicative. Note that the Schatten 2-norm is just the Frobenius norm (the Euclidean norm on matrices).

It is an important classical fact that there is a strong majorization between the diagonal entries of a Hermitian matrix and its eigenvalues (Theorem (4.3.26) in [HJ]). In particular, if $B = [b_{ij}] \in M_n$ is Hermitian, with decreasingly ordered eigenvalues

$$\lambda_1(B) \geq \lambda_2(B) \geq \cdots \geq \lambda_n(B)$$

and decreasingly ordered diagonal entries

$$d_1(B) \geq d_2(B) \geq \cdots \geq d_n(B)$$

we have

$$\sum_{i=1}^{k} d_i(B) \le \sum_{i=1}^{k} \lambda_i(B), \quad k = 1, \dots, n \qquad (5.5.8)$$

with equality for $k = n$. Since the diagonal entries of B are just the eigenvalues of $I \circ B$, these inequalities may be written suggestively as

$$\sum_{i=1}^{k} \lambda_i(I \circ B) \le \sum_{i=1}^{k} \lambda_i(B), \quad k = 1, \dots, n \qquad (5.5.9)$$

This classical majorization theorem may be generalized very nicely by replacing the identity matrix by any correlation matrix.

5.5.10   Definition. A positive semidefinite Hermitian matrix is a *correlation matrix* if each of its diagonal entries is equal to 1.

Exercise. Show that each positive definite Hermitian matrix in M_n is diagonally congruent to a unique correlation matrix.

5.5.11   Theorem. Let $A, B \in M_n$ be given Hermitian matrices and assume that A is a correlation matrix. Arrange the eigenvalues of $A \circ B$ and B in decreasing order $\lambda_1 \ge \cdots \ge \lambda_n$. Then

$$\sum_{i=1}^{k} \lambda_i(A \circ B) \le \sum_{i=1}^{k} \lambda_i(B), \quad k = 1, \dots, n$$

Exercise. Show that to prove Theorem (5.5.11) for a Hermitian B, it suffices to prove it when B is positive definite. *Hint:* Translate B by a scalar multiple of I and see what happens to the inequalities. Show that

$$k\lambda_{min}(B) \le \sum_{i=1}^{k} \lambda_i(A \circ B) \le \sum_{i=1}^{k} \lambda_i(B), \quad k = 1, \dots, n$$

and explain why this shows that Hadamard multiplication by a correlation matrix tends to squeeze together the eigenvalues of a Hermitian matrix.

Theorem (5.5.11) is a special case of the following more general result.

5.5.12   Theorem. Let $A, B \in M_n$ be given positive semidefinite Hermitian matrices. Arrange the eigenvalues of $A \circ B$ and B and the main diagonal entries $d_i(A)$ of A in decreasing order $\lambda_1 \geq \cdots \geq \lambda_n$ and $d_1(A) \geq \cdots \geq d_n(A)$. Then

$$\sum_{i=1}^{k} \lambda_i(A \circ B) \leq \sum_{i=1}^{k} d_i(A)\lambda_i(B), \quad k = 1,..., n$$

This follows directly from the special case of Theorem (5.5.19) in which A is positive semidefinite, since $\lambda_i(B) = \sigma_i(B)$ when B is positive semidefinite. It also follows from Theorem (5.6.2) with $X = Y = A^{\frac{1}{2}}$; see the discussion following the proof of Lemma (5.6.17).

Exercise. Deduce Theorem (5.5.11) from Theorem (5.5.12). *Hint:* Use the exercise preceding Theorem (5.5.12).

For a fixed matrix $A \in M_{m,n}$, Hadamard multiplication by A induces a linear transformation $\mathcal{H}_A : M_{m,n} \to M_{m,n}$ defined by

$$\mathcal{H}_A(B) \equiv A \circ B \tag{5.5.13}$$

Operator theorists have been interested in the Hadamard multiplier transformation $\mathcal{H}_A$ and in its induced norm for various choices of the norm on the underlying vector space $M_{m,n}$, especially the spectral norm.

5.5.14   Definition. If $G(\cdot)$ is a given norm on $M_{m,n}$, the norm $G^h(\cdot)$ induced by G via the Hadamard product is defined by

$$G^h(A) \equiv \max \{ G(\mathcal{H}_A(B)) : G(B) = 1 \}$$
$$= \max \{ G(A \circ B) : G(B) = 1 \}$$

Exercise. Show that for the spectral norm we have $\|\| A \|\|_2^h \leq \|\| A \|\|_2$ for all $A \in M_{m,n}$. *Hint:* Use (5.5.1).

Exercise. Show that $\|\| A \|\|_2^h \leq c_1(A)$ and $\|\| A \|\|_2^h \leq r_1(A)$ for all $A \in M_{m,n}$.

Some of the remaining results of this section as well as (5.5.3) were originally motivated by an interest in better understanding the norm $\||\cdot\||_2^h$. One of these is a Cauchy-Schwarz type inequality for the Hadamard product. Recall that $\overline{A}$ is just the entrywise complex conjugate of $A \in M_{m,n}$, so $A \circ \overline{A}$ has nonnegative entries. The following bound is, like Theorem (5.5.3), a refinement of Theorem (5.5.1).

5.5.15   Theorem. For any $A, B \in M_{m,n}$, we have

$$\||A \circ B\||_2 \leq \||A \circ \overline{A}\||_2^{\frac{1}{2}} \||B \circ \overline{B}\||_2^{\frac{1}{2}} \leq \||A\||_2 \||B\||_2$$

Proof: The second inequality follows from Theorem (5.5.1), so we need only prove the first. Since the spectral norm $\||\cdot\||_2$ is induced by the Euclidean norm $\|\cdot\|_2$ on $\mathbb{C}^n$, , let $x = [x_j] \in \mathbb{C}^n$ be such that $\|x\|_2 = 1$ and $\||A \circ B\||_2 = \|(A \circ B)x\|_2$. Also let $A = [a_{ij}]$ and $B = [b_{ij}]$. Then

$$\|(A \circ B)x\|_2^2 = \sum_i \left| \sum_j a_{ij} b_{ij} x_j \right|^2$$

$$\leq \sum_i \left[\sum_j |a_{ij}| \, |x_j|^{\frac{1}{2}} \, |b_{ij}| \, |x_j|^{\frac{1}{2}} \right]^2$$

$$\leq \sum_i \left[\sum_j |a_{ij}|^2 |x_j| \right] \left[\sum_k |b_{ik}|^2 |x_k| \right]$$

$$\leq \left[\sum_i \left[\sum_j |a_{ij}|^2 |x_j| \right]^2 \right]^{\frac{1}{2}} \left[\sum_i \left[\sum_j |b_{ij}|^2 |x_j| \right]^2 \right]^{\frac{1}{2}}$$

$$= \|(A \circ \overline{A})|x|\|_2 \, \|(B \circ \overline{B})|x|\|_2$$

$$\leq \||A \circ \overline{A}\||_2 \|x\|_2 \||B \circ \overline{B}\||_2 \|x\|_2$$

$$= \||A \circ \overline{A}\||_2 \||B \circ \overline{B}\||_2$$

The second and third inequalities are applications of the Cauchy-Schwarz inequality; the last inequality is because $\|\cdot\|_2$ is compatible with $\||\cdot\||_2$; and the last equality is because x is a unit vector. The choice of x yields the asserted inequality upon taking square roots.   □

Exercise. Use Theorem (5.5.15) to show that $\||A \circ A\||_2 \leq \||A \circ \overline{A}\||_2$ for all

$A \in M_{m,n}$. More generally, use facts about nonnegative matrices ((8.1.9) and Theorem (8.1.18) in [HJ]) or the definition of the spectral norm as an induced norm to show that $||| B |||_2 \leq ||| \, |B| \, |||_2$ for all $B \in M_{m,n}$. *Hint:* $\sigma_1(B)^2 = \rho(B^*B) \leq \rho(|B^*B|) \leq \rho(|B|^T|B|)$; alternatively, $\| Bx \|_2 = \| \, |B| \, |x| \, \|_2 \leq \| \, |B| \, |x| \, \|_2$.

Exercise. For $A \in M_{m,n}$, show that $||| A |||_2^h = ||| A |||_2$ if and only if $||| A \circ \overline{A} |||_2 = ||| A |||_2^h$.

Exercise. If $U \in M_n$ is unitary, show that $||| U |||_2^h = ||| U |||_2 = 1$. *Hint:* $||| U \circ \overline{U} |||_2 \leq 1$ by Theorem (5.5.1), and $||| U \circ \overline{U} |||_2 \geq \rho(U \circ \overline{U}) = 1$ since $U \circ \overline{U} \geq 0$ and all its row sums are 1.

Exercise. Suppose that $U \in M_p$ is unitary and that $C \in M_{m-p,n-p}$ is a contraction ($||| C |||_2 \leq 1$). Show that the block matrix

$$A = \begin{bmatrix} U & 0 \\ 0 & C \end{bmatrix} \in M_{m,n}$$

satisfies $||| A |||_2^h = ||| A |||_2$.

For completeness, we mention without proof a structural characterization of the matrices A such that $||| A |||_2^h = ||| A |||_2$.

5.5.16 Theorem. Let $A \in M_{m,n}$ be given. The following three statements are equivalent:

(a) $||| A |||_2^h = ||| A |||_2$.

(b) $||| A \circ \overline{A} |||_2 = ||| A |||_2^h$.

(c) There exist a nonnegative number α, permutation matrices $P \in M_m$ and $Q \in M_n$, a unitary matrix $U \in M_p$, $p \leq m, n$, and a contraction $C \in M_{m-p,n-q}$ such that

$$A = \alpha P \begin{bmatrix} U & 0 \\ 0 & C \end{bmatrix} Q$$

Yet another useful set of numbers associated with $A \in M_{m,n}$ is the following:

5.5.17   Definition. For $A \in M_{m,n}$, denote the decreasingly ordered main diagonal entries of $P(A) \equiv (AA^*)^{\frac{1}{2}}$ by

$$p_1(A) \geq p_2(A) \geq \cdots \geq p_m(A) \geq 0$$

and the decreasingly ordered main diagonal entries of $Q(A) \equiv (A^*A)^{\frac{1}{2}}$ by

$$q_1(A) \geq q_2(A) \geq \cdots \geq q_n(A) \geq 0$$

Exercise. For any $A \in M_{m,n}$, show that

$$p_i(A) \leq r_i(A), \quad i = 1, \ldots, m$$

and

$$q_j(A) \leq c_j(A), \quad j = 1, \ldots, n$$

Another estimate of $||| A |||_2^h$ may be obtained from the following fact.

5.5.18   Theorem. For every $A, B \in M_{m,n}$, we have

$$\sigma_1(A \circ B) \leq [p_1(A)q_1(A)]^{\frac{1}{2}} \sigma_1(B)$$

In particular, if $m = n$ and $A = [a_{ij}]$ is positive semidefinite, then

$$\sigma_1(A \circ B) \leq [\max_i a_{ii}] \, \sigma_1(B)$$

Exercise. Show that $||| A |||_2^h \leq [p_1(A)q_1(A)]^{\frac{1}{2}}$ and that Theorem (5.5.18) implies Theorem (5.5.1).

Exercise. Verify that the assertion in Theorem (5.5.18) about the special case in which A is positive semidefinite follows from the general inequality. *Hint:* Show that $p_1(A) = q_1(A) = \max_i a_{ii}$ in this case. Notice that this generalizes the upper bound in Theorem (5.3.4) since no assumption is made about B.

Theorem (5.5.18), which suggests a link between Theorems (5.5.4) and (5.5.12), may be generalized in a variety of ways. First, it may be extended to a sequence of weak majorization inequalities just as (5.5.1) is extended to (5.5.4).

5.5.19   Theorem. For every $A, B \in M_{m,n}$, we have

$$\sum_{i=1}^{k} \sigma_i(A \circ B) \leq \sum_{i=1}^{k} [p_i(A)q_i(A)]^{\frac{1}{2}} \, \sigma_i(B), \quad k = 1, 2, ..., \min\{m, n\}$$

In particular, if $m = n$ and $A = [a_{ij}]$ is positive semidefinite, then

$$\sum_{i=1}^{k} \sigma_i(A \circ B) \leq \sum_{i=1}^{k} d_i(A) \, \sigma_i(B), \quad k = 1, 2, ..., n$$

where $d_1(A) \geq \cdots \geq d_n(A)$ denote the decreasingly ordered main diagonal entries of A.

This is a special case of Theorem (5.5.25).

Exercise. Show by examples that neither the inequality $\sigma_1(A \circ B) \leq p_1(A)\sigma_1(B)$ nor the inequality $\sigma_1(A \circ B) \leq q_1(A)\sigma_1(B)$ holds for general $A, B \in M_{m,n}$.

Exercise. Show that Theorem (5.5.19) implies Theorem (5.5.12), and therefore (5.5.11) as well.

Exercise. Verify that the assertion in Theorem (5.5.19) about the special case in which A is positive semidefinite follows from the general inequality. *Hint:* Show that $p_i(A) = q_i(A) = d_i(A)$ in this case. Notice that this generalizes Theorem (5.5.12) since no assumption is made about B.

Before generalizing Theorem (5.5.19), we mention some inequalities of similar form involving the Euclidean column lengths c_i and row lengths r_i defined in (5.5.2). The first of these extends part of Theorem (5.5.3) just as Theorem (5.5.4) extends Theorem (5.5.1). Even though statements such as (compare Theorem (5.5.19))

$$\sum_{i=1}^{k} \sigma_i(A \circ B) \leq \sum_{i=1}^{k} p_i(A)\sigma_i(B), \quad k = 1, ..., m$$

or

$$\sum_{i=1}^{k} \sigma_i(A \circ B) \leq \sum_{i=1}^{k} q_i(A)\sigma_i(B), \quad k = 1, ..., n$$

are *not* generally valid, corresponding weaker statements involving c_i or r_i are valid.

5.5.20   **Theorem.** For every $A, B \in M_{m,n}$, we have

$$\text{(a)} \quad \sum_{i=1}^{k} \sigma_i(A \circ B) \leq \sum_{i=1}^{k} c_i(A) \sigma_i(B), \quad k = 1, \dots, n$$

and

$$\text{(b)} \quad \sum_{i=1}^{k} \sigma_i(A \circ B) \leq \sum_{i=1}^{k} r_i(A) \sigma_i(B), \quad k = 1, \dots, m$$

These inequalities follow from Theorem (5.6.2) by choosing $X = I$ and $Y = A$, or $X = A^*$ and $Y = I$; see the discussion following the proof of Lemma (5.6.17).

The following inequalities also involve the quantities c_i and r_i.

5.5.21   **Theorem.** For every $A, B \in M_{m,n}$, we have

$$\sum_{i=1}^{k} \sigma_i(A \circ B) \leq \sum_{i=1}^{k} [c_i(A) r_i(A)]^{\frac{1}{2}} \sigma_i(B), \quad k = 1, 2, \dots, \min\{m,n\}$$

Exercise. Use the arithmetic-geometric mean inequality to show that Theorem (5.5.21) follows from Theorem (5.5.19).

Remark: It is natural to ask whether inequalities of the form

$$\sum_{i=1}^{k} \sigma_i(A \circ B) \leq \sum_{i=1}^{k} c_i(A)^\alpha r_i(A)^{1-\alpha} \sigma_i(B) \qquad (5.5.22)$$

are valid for $0 \leq \alpha \leq 1$ and $k = 1, \dots, \min\{m,n\}$. The only values of α for which these inequalities have been proved are $\alpha = 0, \frac{1}{2}$, and 1; there are no known counterexamples for other values of α.

We complete this section by describing a class of generalizations of Theorem (5.5.19). In order to do so we introduce some notation. For $A \in$

$M_{m,n}$, let $A = V\Sigma W^*$ be a singular value decomposition of A with $V \in M_m$ and $W \in M_n$ unitary and

$$\Sigma = \begin{bmatrix} \sigma_1(A) & & & 0 & \vdots \\ & \sigma_2(A) & & & \vdots & 0 \\ 0 & & \ddots & & \vdots \\ & & & \sigma_m(A) & \vdots \end{bmatrix} \in M_{m,n}$$

Here we assume for simplicity and without any loss of generality that $m \leq n$. Let $\alpha \equiv [\alpha_1,..., \alpha_m]^T \in \mathbb{R}^m$ and define

$$P(A)^\alpha \equiv V\Sigma^\alpha V^* \equiv V\operatorname{diag}(\sigma_1(A)^{\alpha_1},..., \sigma_m(A)^{\alpha_m})\, V^*$$

and

$$Q(A)^\alpha \equiv W\Sigma^\alpha W^* \equiv W\operatorname{diag}(\sigma_1(A)^{\alpha_1},..., \sigma_m(A)^{\alpha_m}, 0,..., 0)\, W^*$$

in which we adopt the convention that $0^\beta \equiv 0$ for all $\beta \in \mathbb{R}$. Now, let $p_i(A,\alpha)$ be the ith largest entry on the diagonal of $P(A)^{2\alpha}$ and let $q_i(A,\alpha)$ be the ith largest entry on the diagonal of $Q(A)^{2\alpha}$. Then

$$p_1(A,\alpha) \geq p_2(A,\alpha) \geq \cdots \geq p_m(A,\alpha) \geq 0$$

and

$$q_1(A,\alpha) \geq q_2(A,\alpha) \geq \cdots \geq q_n(A,\alpha) \geq 0 \tag{5.5.23}$$

As usual, denote the m-vector of ones by $e \in \mathbb{R}^m$.

Exercise. Verify that $p_i(A, \tfrac{1}{2}e) = p_i(A)$, $i = 1,..., m$, and $q_i(A, \tfrac{1}{2}e) = q_i(A)$, $i = 1,..., n$, where $\{p_i\}$ and $\{q_i\}$ are defined in (5.5.17).

We shall again make use of geometric means of the quantities p_i and q_i and define

$$t_i(A,\alpha) \equiv [p_i(A,\alpha)q_i(A,e-\alpha)]^{\frac{1}{2}}, \quad i = 1,..., m \tag{5.5.24}$$

We can now state the promised generalization of Theorem (5.5.19).

5.5.25 **Theorem.** For every $A, B \in M_{m,n}$, $m \leq n$, and any $\alpha \in \mathbb{R}^m$, we have

$$\sum_{i=1}^k \sigma_i(A \circ B) \leq \sum_{i=1}^k t_i(A,\alpha)\sigma_i(B), \quad k = 1,..., m$$

This follows from Theorem (5.6.2) by choosing $X = \Sigma^\alpha V^*$ and $Y = \Sigma^{e-\alpha} W^*$, where $A = V\Sigma W^*$ is a singular value decomposition; see the discussion following the proof of Lemma (5.6.17).

Exercise. Show that Theorem (5.5.19) is a special case of Theorem (5.5.25) in which $\alpha = \frac{1}{2}e$.

Problems

1.   Prove the classical majorization inequalities (5.5.8) using the observations of (5.0.3a) and the fact that the relative gain array $\Phi(U)$ of a unitary matrix U is doubly stochastic (nonnegative with row and column sums equal to one).

2.   Show that Theorem (5.5.4) follows from Theorem (5.5.19).

3.   Use Theorem (5.5.7) to show that the set of unitarily invariant matrix norms on M_n is convex.

4.   Show by example that the set of matrix norms on M_n is not convex for $n \geq 2$. See Problem 9 in Section (5.6) of [HJ].

5.   Theorem (5.5.7) contains twelve implications that are valid under the assumption of unitary invariance of the norm $N(\cdot)$ but may or may not be true without this assumption. Show that the implications (a) $\Rightarrow$ (d) and (b) $\Rightarrow$ (d) are valid without the assumption of unitary invariance, but that none of the remaining ten implications are.

6.   For $A = [a_{ij}] \in M_n$, let $\|A\|_\infty \equiv \max\{|a_{ij}|: 1 \leq i, j \leq n\}$. Show that a fifth necessary and sufficient condition

(e) $N(A) \geq \|A\|_\infty$ for all $A \in M_n$

may be added to the list in Theorem (5.5.7).

7.   Provide details for the following alternative proofs of Theorem (5.5.1) when $m = n$. See Problem 31 in Section (4.2) for another proof.
(a) Let $\|\|\cdot\|\|_1$ denote the maximum absolute column sum matrix norm on M_n. Then $\|\|C\|\|_2^2 = \rho(C^*C) \leq \|\|C^*C\|\|_1 \leq \|\|C^*\|\|_1 \|\|C\|\|_1$ for any $C \in M_n$, so $\|\|A \circ B\|\|_2^2 \leq \|\|A^* \circ B^*\|\|_1 \|\|A \circ B\|\|_1$. But

$$\||| A \circ B \|||_1 = \max_i \sum_k | a_{ki} b_{ki} | \le \max_i \left[\left[\sum_k | a_{ki} |^2 \right] \left[\sum_k | b_{ki} |^2 \right] \right]^{\frac{1}{2}}$$

$$\le c_1(A) c_1(B) \le \||| A \|||_2 \||| B \|||_2$$

and consequently $\||| A^* \circ B^* \|||_2 \le r_1(A) r_1(B)$. This is Schur's original proof of Theorem (5.5.1).

(b) Partition $A = [a_1 \ \ldots \ a_n]$, $B = [b_1 \ \ldots \ b_n]$, and the identity matrix $I = [e_1 \ \ldots \ e_n]$ according to their columns. Set $\mathcal{A} \equiv [\mathrm{diag}(a_1) \ \ldots \ \mathrm{diag}(a_n)] \in M_{n,n^2}$ and $\mathcal{B} \equiv [e_1 \otimes b_1 \ \ldots \ e_n \otimes b_n] \in M_{n^2,n}$. Verify that $A \circ B = \mathcal{A}\mathcal{B}$ (ordinary product), so ordinary submultiplicativity of the spectral norm gives $\||| A \circ B \|||_2 = \||| \mathcal{A}\mathcal{B} \|||_2 \le \||| \mathcal{A} \|||_2 \||| \mathcal{B} \|||_2$. Compute the diagonal matrices $\mathcal{B}^*\mathcal{B}$ and $\mathcal{A}\mathcal{A}^*$ (for the latter it is convenient to write $A^T = [\alpha_1 \ \ldots \ \alpha_n]$ so that $\alpha_1^T, \ldots, \alpha_n^T$ are the rows of A) and show that $\||| \mathcal{B} \|||_2 = c_1(B)$ and $\||| \mathcal{A} \|||_2 = r_1(A)$. Deduce Theorem (5.5.1) as in (a).

8. A nonnegative square matrix A is *doubly stochastic* if all its row and column sums are 1. A square matrix of the form $B = U \circ \overline{U}$ for some unitary U is said to be *orthostochastic*.

(a) Show that every orthostochastic matrix is doubly stochastic.

(b) Use the following example to show that not every doubly stochastic matrix is orthostochastic:

$$A = \begin{bmatrix} \frac{1}{2} & 0 & \frac{1}{2} \\ \frac{1}{2} & \frac{1}{2} & 0 \\ 0 & \frac{1}{2} & \frac{1}{2} \end{bmatrix}$$

(c) Use Theorem (5.5.1) to show that every orthostochastic matrix B has $\||| B \|||_2 \le 1$ and conclude that $\||| B \|||_2 = 1$, so every orthostochastic matrix is a *radial* matrix.

(d) More generally, show that *every doubly stochastic matrix is a radial matrix*; see Problems 8, 24, and 27 in Section (1.5).

9. This problem extends the discussion of the solution of the Lyapunov equation given in Section (5.0.4). Let $A \in M_n$ be a given diagonalizable positive stable matrix with $A = S\Lambda S^{-1}$ and eigenvalues $\lambda_1, \ldots, \lambda_n \in RHP$.

(a) Consider

$$\int\limits_0^\infty \left| \sum_{k=1}^n x_k\, e^{-\lambda_k t} \right|^2 dt$$

to show that the *Cauchy matrix* $L(A) \equiv [1/(\bar\lambda_i + \lambda_j)]$ is positive semidefinite.

(b) Use Theorem (5.5.18) to verify the following bound for the solution operator $G_A(\cdot)$ for the Lyapunov equation $GA + A^*G = H$

$$\| G_A(H) \|_2 \le \frac{\kappa(S)^2}{2\,\min\limits_i \mathrm{Re}\ \lambda_i(A)} \| H \|_2$$

Here, $\kappa(S) \equiv \| S \|_2 \| S^{-1} \|_2$ is the spectral condition number of S. What does this give when A is normal?

Notes and Further Readings. The original proofs of Theorems (5.5.4) and (5.5.7) are in R. A. Horn and C. R. Johnson, Hadamard and Conventional Submultiplicativity for Unitarily Invariant Norms on Matrices, *Linear Multilinear Algebra* 20 (1987), 91-106, which also contains a set of examples to show which pairwise implications among the four conditions in Theorem (5.5.7) fail for norms that are not unitarily invariant. A proof of Theorem (5.5.16) and related results (including the original proof of Theorem (5.5.15)) is in S. C. Ong, On the Schur Multiplier Norm of Matrices, *Linear Algebra Appl.* 56 (1984), 45-55. Theorems (5.5.11,12) were first proved by R. Bapat and V. Sunder, On Majorization and Schur Products, *Linear Algebra Appl.* 72 (1985), 107-117. The motivating special case (5.5.18) of Theorem (5.5.19) is itself a special case of an inequality given by I. Schur in his seminal 1911 paper cited at the end of Section (5.0) (see Schur's Satz VI and the second footnote on his p. 13); Theorem (5.5.18) was also noted independently by M. Walter in On the Norm of a Schur Product, *Linear Algebra Appl.* 79 (1986), 209-213. For some extensions and applications of the key idea in Problem 7b, see C. R. Johnson and P. Nylen, Largest Singular Value Submultiplicativity, *SIAM J. Matrix Analysis Appl.* 12 (1991), 1-6.

5.6   Singular values of a Hadamard product: a fundamental inequality

Several of the results discussed in the preceding section are families of inequalities of the form

$$\sum_{i=1}^{k} \sigma_i(A \circ B) \leq \sum_{i=1}^{k} f_i(A)\sigma_i(B), \quad k = 1,..., \min\{m,n\} \qquad (5.6.1)$$

in which the factors $f_i(A)$ form a decreasing sequence that somehow depends upon A. The inequalities (5.5.4, 11, 12, 19, 20a, 20b, 21, and 25) are all of this form. Our goal in this section is to present and prove a general family of inequalities that includes each of the inequalities in the list just mentioned, thus providing proofs that, in some cases, were omitted in Section (5.5). In addition, the method of proof will be used to produce some additional inequalities that are not of the type (5.6.1). One of these is a family of multiplicative weak majorization inequalities.

Our principal result links the conventional product with the Hadamard product in inequalities involving the Euclidean column lengths defined in (5.5.2).

5.6.2   Theorem. Let $A, B \in M_{m,n}$ be given and let $A = X^* Y$ be a given factorization of A in which $X \in M_{r,m}$ and $Y \in M_{r,n}$. Then

$$\sum_{i=1}^{k} \sigma_i(A \circ B) \leq \sum_{i=1}^{k} c_i(X)c_i(Y)\sigma_i(B), \quad k = 1,..., \min\{m,n\}$$

We first note that the general (nonsquare) case of the theorem follows from the square case $m = n$ by augmenting nonsquare matrices to square ones with rectangular blocks of zeroes. Thus, without loss of generality, we may assume for the proof that $m = n$. For other potential uses, however, we state and verify some of the supporting facts for the proof in the general (nonsquare) case. Recall that a matrix $C \in M_{m,n}$ is a *contraction* if $\sigma_1(C) \leq 1$, and that the set of contractions in $M_{m,n}$ (the unit ball in $M_{m,n}$ with respect to the spectral norm) is a compact set. We prepare for the proof of (5.6.2) by developing several facts we shall need in addition to those already available. The first is the same as Lemma (3.5.12); we record it here for convenience.

5.6.3 **Lemma.** Let $L \in M_m$ and $M \in M_n$ be given positive semidefinite matrices, and let $A \in M_{m,n}$. The block matrix

$$\begin{bmatrix} L & A \\ A^* & M \end{bmatrix} \in M_{m+n} \tag{5.6.4}$$

is positive semidefinite if and only if there is a contraction $C \in M_{m,n}$ such that $A = L^{\frac{1}{2}} C M^{\frac{1}{2}}$.

For a given $A \in M_{m,n}$ there are several useful choices of L and M that make the block matrix (5.6.4) positive semidefinite. For example, if $A = X^* Y$ for some $X \in M_{r,m}$ and $Y \in M_{r,n}$, then

$$\begin{bmatrix} X^* X & A \\ A^* & Y^* Y \end{bmatrix} = \begin{bmatrix} X^* X & X^* Y \\ Y^* X & Y^* Y \end{bmatrix} = [X \ Y]^*[X \ Y] \tag{5.6.5}$$

is obviously positive semidefinite. Lemma (5.6.3) also implies that

$$\begin{bmatrix} \sigma_1(A)I & A \\ A^* & \sigma_1(A)I \end{bmatrix} \tag{5.6.6}$$

is positive semidefinite, since $A/\sigma_1(A)$ is always a contraction.

We may now prove a fact weaker than (5.6.2), which will be the first major step in the proof of (5.6.2).

5.6.7 **Lemma.** Let $A, B \in M_n$ be given. Then

$$\sum_{i=1}^{k} \sigma_i(A \circ B) \leq \sum_{i=1}^{k} c_i(X)c_i(Y) \, \sigma_1(B), \quad k = 1, \ldots, n \tag{5.6.8}$$

for any $X, Y \in M_{r,n}$ such that $A = X^* Y$.

Proof: Since the block matrices

$$\begin{bmatrix} X^* X & A \\ A^* & Y^* Y \end{bmatrix} \quad \text{and} \quad \begin{bmatrix} \sigma_1(B)I & B \\ B^* & \sigma_1(B)I \end{bmatrix}$$

are positive semidefinite, so is their Hadamard product

$$\begin{bmatrix} X^*X & A \\ A^* & Y^*Y \end{bmatrix} \circ \begin{bmatrix} \sigma_1(B)I & B \\ B^* & \sigma_1(B)I \end{bmatrix}$$

$$= \begin{bmatrix} \sigma_1(B)I \circ (X^*X) & A \circ B \\ (A \circ B)^* & \sigma_1(B)I \circ (Y^*Y) \end{bmatrix}$$

by Theorem (5.2.1). By Lemma (5.6.3) there is a contraction $C \in M_n$ such that

$$A \circ B = \sigma_1(B)[I \circ (X^*X)]^{\frac{1}{2}} C [I \circ (Y^*Y)]^{\frac{1}{2}}$$

Since the singular values of $[I \circ (X^*X)]^{\frac{1}{2}}$ are the square roots of the main diagonal entries of X^*X, which are the Euclidean lengths of the columns of X (and similarly for the Y^*Y term), and all singular values of C are at most 1, we conclude from Theorem (3.3.4) that

$$\prod_{i=1}^{k} \sigma_i(A \circ B) \le \prod_{i=1}^{k} c_i(X)c_i(Y)\sigma_1(B), \qquad k = 1,...,n \qquad (5.6.9)$$

and, according to Corollary (3.3.10), (5.6.9) implies (5.6.8).   ☐

To proceed to boost the inequality (5.6.8) to (5.6.2) we use the following special case of Lemma (5.1.4):

$$\text{tr}[(A \circ B)C] = \text{tr}[(A \circ C^T)B^T] \qquad (5.6.10)$$

for any $A, B, C \in M_n$.

5.6.11   Lemma. Let $A, K_r, K_s \in M_n$ be given, and assume that $\sigma_1(K_r) = \cdots = \sigma_r(K_r) = 1$, $\sigma_1(K_s) = \cdots = \sigma_s(K_s) = 1$, and $\sigma_{r+1}(K_r) = \cdots = \sigma_n(K_r) = \sigma_{s+1}(K_s) = \cdots = \sigma_n(K_s) = 0$. Then

$$|\text{tr}[(A \circ K_r)K_s]| \le \sum_{i=1}^{\min\{r,s\}} c_i(X)c_i(Y) \qquad (5.6.12)$$

for any $X, Y \in M_{r,n}$ such that $X^*Y = A$.

Proof: Using (5.6.10), we may assume without loss of generality that $s \leq r$. Let $\lambda_1,..., \lambda_n$ be the eigenvalues of $(A \circ K_r)K_s$ and compute

$$|\mathrm{tr}[(A \circ K_r)K_s]| = \left|\sum_{i=1}^{n}\lambda_i\right| \leq \sum_{i=1}^{n}|\lambda_i|$$

$$\leq \sum_{i=1}^{n}\sigma_i[(A \circ K_r)K_s] \leq \sum_{i=1}^{n}\sigma_i(A \circ K_r)\sigma_i(K_s)$$

$$= \sum_{i=1}^{s}\sigma_i(A \circ K_r) \leq \sum_{i=1}^{s}c_i(X)c_i(Y)\sigma_1(K_r)$$

$$= \sum_{i=1}^{s}c_i(X)c_i(Y)$$

We have used (3.3.13a) for the second inequality, Theorem (3.3.14a) for the third inequality, and Lemma (5.6.7) for the last inequality.   ☐

5.6.13   Definition. A matrix that has r singular values $+1$ and all other singular values zero (for example, K_r in Lemma (5.6.11)) is called a *rank r partial isometry*.

Theorem (3.1.8) gives two useful representations of any matrix as a nonnegative linear combination of partial isometries. For convenience, we state the portion of Theorem (3.1.8) that we need here.

5.6.14   Lemma. Let $B \in M_n$ have ordered singular values $\sigma_1(B) \geq \cdots \geq \sigma_n(B) \geq 0$. Then B may be written as a sum

$$B = \sum_{j=1}^{n}\beta_j K_j \tag{5.6.15}$$

in which each K_j is a rank j partial isometry, $\beta_j \geq 0$, $j = 1,..., n$, and

$$\sum_{j=k}^{n}\beta_j = \sigma_k(B), \quad k = 1,..., n \tag{5.6.16}$$

A final lemma now allows us to complete a proof of Theorem (5.6.2).

5.6.17   Lemma. For any $C \in M_n$ and any positive integer k with $1 \leq k \leq n$, there is a rank k partial isometry $C_k \in M_n$ such that

$$\sum_{i=1}^{k} \sigma_i(C) = \text{tr}(C\, C_k)$$

Proof: If $C = V \text{diag}(\sigma_1(C),\dots, \sigma_n(C))\, W^*$ is a singular value decomposition of C, take $C_k \equiv W(E_{11} + \cdots + E_{kk}) V^*$. ∐

Let $A, B \in M_n$ and $1 \leq k \leq n$ be given, let $A = X^* Y$ for some $X, Y \in M_{r,n}$, and let C_k be a rank k partial isometry with the property for $A \circ B$ that is guaranteed by Lemma (5.6.17). Then

$$\sum_{i=1}^{k} \sigma_i(A \circ B) = \text{tr}[(A \circ B)C_k]$$

$$= \text{tr}\left[\left[A \circ \sum_{j=1}^{n} \beta_j K_j \right] C_k \right] \qquad \text{[by (5.6.14)]}$$

$$= \sum_{j=1}^{n} \beta_j \, \text{tr}[(A \circ K_j)C_k] \leq \sum_{j=1}^{n} \beta_j |\, \text{tr}[(A \circ K_j)C_k]|$$

$$\leq \sum_{j=1}^{n} \beta_j \left[\sum_{i=1}^{\min\{j,k\}} c_i(X)c_i(Y) \right] \qquad \text{[by (5.6.11)]}$$

$$= \sum_{i=1}^{k} c_i(X)c_i(Y) \left[\sum_{j=i}^{n} \beta_j \right] = \sum_{i=1}^{k} c_i(X)c_i(Y)\sigma_i(B) \qquad \text{[by (5.6.16)]}$$

This completes our proof of Theorem (5.6.2). ∐

We next indicate how the weak majorization inequalities of the form (5.6.1) from Section (5.5) may be deduced from Theorem (5.6.2). Each case results from a particular factorization of $A = X^* Y$.

To show that Theorem (5.5.12) follows from Theorem (5.6.2), note that $\lambda_i(B) = \sigma_i(B)$, $i = 1,..., n$, since B is positive semidefinite, and factor A as $A = A^{\frac{1}{2}}A^{\frac{1}{2}}$. Of course, Theorem (5.5.11) follows from Theorem (5.5.12).

To show that Theorem (5.5.25) follows from Theorem (5.6.2), factor A as $A = (\Sigma^{\alpha} V^*)^*(\Sigma^{e-\alpha} W^*)$, where $A = V\Sigma W^*$ is a singular value decomposition of A. Of course, Theorem (5.5.19) is the special case of Theorem (5.5.25) resulting from the factorization of A as $A = (\Sigma^{\frac{1}{2}} V^*)^*(\Sigma^{\frac{1}{2}} W^*)$, that is, $\alpha = \frac{1}{2}e$.

The inequality (5.5.4) may now be derived from Theorem (5.5.19) with a little more argument. By (5.5.8) we have

$$\sum_{i=1}^{k} p_i(A) \leq \sum_{i=1}^{k} \sigma_i(A) \text{ and } \sum_{i=1}^{k} q_i(A) \leq \sum_{i=1}^{k} \sigma_i(A), \ k = 1,..., n$$

Use the arithmetic-geometric mean inequality to conclude that

$$\sum_{i=1}^{k} [p_i(A)q_i(A)]^{\frac{1}{2}} \leq \frac{1}{2}\sum_{i=1}^{k} [p_i(A) + q_i(A)] \leq \sum_{i=1}^{k} \sigma_i(A)$$

for $k = 1,..., n$. Thus, the singular values $\sigma_i(A)$ weakly majorize the numbers $[p_i(A)q_i(A)]^{\frac{1}{2}}$, $i = 1,..., n$.

Exercise. Use summation by parts to verify the following *replacement principle* for majorization inequalities.

5.6.18   Lemma. Let $\alpha_1 \geq \cdots \geq \alpha_n$, $\beta_1 \geq \cdots \geq \beta_n$, and $\gamma_1 \geq \cdots \geq \gamma_n$ be given decreasingly ordered real numbers, and assume that all $\gamma_i \geq 0$. If

$$\sum_{i=1}^{k} \alpha_i \leq \sum_{i=1}^{k} \beta_i, \ k = 1,..., n$$

then

$$\sum_{i=1}^{k} \alpha_i \gamma_i \leq \sum_{i=1}^{k} \beta_i \gamma_i, \ k = 1,..., n$$

Using the replacement principle, we may take $\gamma_i \equiv \sigma_i(B)$ and replace the numbers $\alpha_i \equiv [p_i(A)q_i(A)]^{\frac{1}{2}}$ on the right-hand side of the inequality in Theorem (5.5.19) by the numbers $\beta_i \equiv \sigma_i(A)$ to obtain Theorem (5.5.4).

Theorem (5.5.20a) is obtained from Theorem (5.6.2) by factoring $A = IA$, that is, $X = I$, $Y = A$; and (5.5.20b) is obtained by factoring $A = AI$, that is, $X = A^*$, $Y = I$.

Using ideas of the proof of Theorem (5.6.2) we may prove further inequalities of a different type for $\sigma_i(A \circ B)$. The idea is to apply Lemma (5.6.3) to the Hadamard product of other choices of positive semidefinite block matrices. Let $A_i \in M_{m,n}$ and suppose that

$$A_i = X_i^* Y_i, \quad X_i \in M_{r,m}, \quad Y \in M_{r,n} \tag{5.6.19}$$

is a factorization of A_i, $i = 1,..., p$. Define the block matrices

$$A_i \equiv \begin{bmatrix} X_i^* X_i & A_i \\ A_i^* & Y_i^* Y_i \end{bmatrix}, \quad i = 1,..., p \tag{5.6.20}$$

each of which is positive semidefinite by (5.6.5).

Using Theorem (5.2.1) repeatedly, we find that the p-fold Hadamard product $A_1 \circ A_2 \circ \cdots \circ A_p$ is positive semidefinite, from which it follows that

$$\prod_{i=1}^{k} \sigma_i(A_1 \circ \cdots \circ A_p)$$

$$\leq \prod_{i=1}^{k} \sigma_i([X_1^* X_1] \circ \cdots \circ [X_p^* X_p])^{\frac{1}{2}} \sigma_i([Y_1^* Y_1] \circ \cdots \circ [Y_p^* Y_p])^{\frac{1}{2}} \tag{5.6.21}$$

Exercise. Verify (5.6.21).

5.6.22   Theorem. For any $A, B \in M_{m,n}$ we have

$$\prod_{i=1}^{k} \sigma_i(A \circ B) \leq \prod_{i=1}^{k} c_i(A) r_i(B), \quad k = 1,..., \min\{m,n\}$$

Proof: Use the case $p = 2$ of (5.6.21) and factor A as $A = IA$ and B as $B = BI$.   □

356   The Hadamard product

Once again, via Corollary (3.3.10) the multiplicative inequalities in Theorem (5.6.22) imply the following additive ones.

5.6.23   Theorem. For any $A, B \in M_{m,n}$ we have

$$\sum_{i=1}^{k} \sigma_i(A \circ B) \leq \sum_{i=1}^{k} c_i(A) r_i(B), \quad k = 1,\ldots, \min\{m,n\}$$

Problem

1.   For $A, B \in M_{m,n}$, show that

$$\prod_{i=1}^{k} \sigma_i(A \circ B) \leq \prod_{i=1}^{k} \sigma_i((A^*A) \circ (B^*B))^{\frac{1}{2}}, \quad k = 1,\ldots, \min\{m,n\}$$

Notes and Further Readings. The class of inequalities (5.6.2) and the derivation of prior inequalities from it (as well as more elaborate inequalities) is from The Singular Values of a Hadamard Product: A Basic Inequality, *Linear Multilinear Algebra* 21 (1987), 345-365, by T. Ando, R. A. Horn, and C. R. Johnson. Some additional references related to ideas of Sections (5.5) and (5.6) include: [Pau]; K. Okubo, Hölder-Type Norm Inequalities for Schur Products of Matrices, *Linear Algebra Appl.* 91 (1987), 13-28;   F. Zhang, Another Proof of a Singular Value Inequality Concerning Hadamard Products of Matrices, *Linear Multilinear Algebra* 22 (1988), 307-311.

5.7   Hadamard products involving nonnegative matrices and M-matrices

In this chapter we have thus far considered Hermitian (usually positive semidefinite) matrices and general matrices. We now turn our attention to special facts about Hadamard products involving the closely related classes of nonnegative matrices (see Chapter 8 of [HJ]), M-matrices, and H-matrices (see Section (2.5)). There is a rich variety of these; some resemble facts about positive definite matrices, and some are notably different.

We shall continue to use notation employed previously. The spectral radius of $A \in M_n$ is denoted by $\rho(A)$; if all the entries of A are nonnegative

$(A \geq 0)$, the Perron-Frobenius theorem guarantees that $\rho(A) \in \sigma(A)$. If $A \in Z_n$ (A is real and all the off-diagonal entries of A are nonnegative), and if we denote min $\{\text{Re}(\lambda): \lambda \in \sigma(A)\}$ by $\tau(A)$ (see Lemma (2.5.2.1) and Problem 19 in Section 2.5), then we know that $\tau(A) \in \sigma(A)$ and $\tau(A)$ is the "minimum" eigenvalue of A—the eigenvalue of A that lies furthest to the left in the complex plane.

Exercise. If $A \in M_n$ is an *M*-matrix, recall that $A \in Z_n$ and $A^{-1} \geq 0$. Show that $\tau(A) = [\rho(A^{-1})]^{-1}$; if A is written as $A = \alpha I - P$, $\alpha > 0$, $P \geq 0$, then $\tau(A) = \alpha - \rho(P)$.

For convenience, we extend the definition of the function $\tau(\cdot)$ to general matrices via the comparison matrix $M(A)$ defined in (2.5.10). Recall that $A = M(A)$ whenever A is an *M*-matrix.

5.7.0   Definition. For any $A \in M_n$, $\tau(A) \equiv \tau(M(A))$.

Notice that $A \in M_n$ is an *H*-matrix if and only if $\tau(A) > 0$. In the spirit of the Hadamard product, we shall want to discuss component-wise real powers of nonnegative matrices; if $A = [a_{ij}] \geq 0$ and $\alpha \in \mathbb{R}$, we denote the matrix $[a_{ij}^\alpha]$ by $A^{(\alpha)}$. As in prior sections, we use the convention $0^0 \equiv 0$ to ensure continuity in α for $\alpha \geq 0$.

We first discuss some basic closure results and elementary observations. For completeness we define a variant of the Hadamard product.

5.7.1   Definition. Let $A = [a_{ij}]$, $B = [b_{ij}] \in M_{m,n}$. The *Fan product* of A and B is denoted by $A \star B \equiv C = [c_{ij}] \in M_{m,n}$ and is defined by

$$c_{ij} = \begin{cases} -a_{ij} b_{ij} & \text{if } i \neq j \\ a_{ii} b_{ii} & \text{if } i = j \end{cases}$$

5.7.2   Observation. (a) If $A, B \in M_n$ are *M*-matrices, then so is $A \star B$. (b) If $A, B \in M_n$ are *H*-matrices, then so is $A \star B$, and $A \circ B$ is nonsingular. Thus, the classes of *M*-matrices and *H*-matrices are both closed under the Fan product $\star$.

Proof: (a) Let $A, B \in Z_n$ be given *M*-matrices. Clearly, $A \star B \in Z_n$, so verification of any of the equivalent conditions in Theorem (2.5.3) is sufficient to conclude that $A \star B$ is an M-matrix. Taking (2.5.3.13), let D and E be positive diagonal matrices such that both AD and BE are strictly row diagonally dominant. One checks that $(AD) \star (BE)$ is also strictly row diagonally dominant (if $a \geq \Sigma_i \, \alpha_i$ and $b \geq \Sigma_j \, \beta_j$, then $ab \geq (\Sigma_i \, \alpha_i)(\Sigma_j \, \beta_j) \geq \Sigma_i \, \alpha_i \beta_i$). Since $(AD) \star (BE) = (A \star B)(DE)$, another invocation of (2.5.3.13) gives the desired conclusion that the Fan product $A \star B$ is an M-matrix.

(b) If A and B are H-matrices, then their comparison matrices $M(A)$ and $M(B)$ are M-matrices, so $M(A \star B) = M(A) \star M(B)$ is an M-matrix by (a), and hence $A \star B$ is an H-matrix. We saw in (a) that there is a positive diagonal matrix F such that $M(A \star B)F$ is strictly row diagonally dominant, and this same F also makes $(A \circ B)F$ strictly row diagonally dominant. Thus, $(A \circ B)F$, and hence $A \circ B$ itself, is nonsingular by the Levy-Desplanques theorem (Theorem (6.1.10) in [HJ]). □

5.7.3   Observation. The class of nonnegative matrices is closed under the Hadamard product: If A, $B \in M_n$, $A \geq 0$, and $B \geq 0$, then $A \circ B \geq 0$.

A simple estimate for $\rho(A \circ B)$ is the following:

5.7.4   Observation. If $A, B \in M_n$, $A \geq 0$, and $B \geq 0$, then $\rho(A \circ B) \leq \rho(A)\rho(B)$.

Proof: We have $\rho(A \circledast B) = \rho(A)\rho(B)$ by Theorem (4.2.12). But since $A \circledast B \geq 0$ and $A \circ B$ is a principal submatrix of $A \circledast B$ by Lemma (5.1.1), $\rho(A \circ B) \leq \rho(A \circledast B) = \rho(A)\rho(B)$ because of the monotonicity of the Perron root with respect to extraction of principal submatrices [HJ, Corollary (8.1.20)]. □

Exercise. Show that the inequality in (5.7.4) can be very weak by taking $B = J$, the matrix of all ones, but also give an example to show that equality can occur.

Exercise. Show by example that the analog of (5.7.4) in which the Hadamard product is replaced by the ordinary product is not valid in general.

Using the estimate in (5.7.4), we can significantly sharpen the information on the Fan product of two M-matrices given in (5.7.2).

5.7.4.1 Corollary. Let A, $B \in M_n(\mathbb{R})$ be given M-matrices. Then $A^{-1} \circ B^{-1} \geq (A \star B)^{-1}$, and hence $\tau(A \star B) \geq \tau(A)\tau(B)$.

Proof: Define $R \equiv (I \circ A) - A$ and $S \equiv I - (I \circ A)^{-1}A$, so $A = (I \circ A) - R = (I \circ A)[I - S]$, $S = [s_{ij}] \geq 0$, and all $s_{ii} = 0$. Then $(I \circ A)^{-1}A = I - S \in Z_n$ is an M-matrix by the criterion in (2.5.3.12), so $\rho(S) < 1$ by Lemma (2.5.2.1). In the same way, write $B = (I \circ B)[I - T]$, where $T = [t_{ij}] \geq 0$, all $t_{ii} = 0$, and $\rho(T) < 1$. Then $A \star B = (I \circ A)(I \circ B)[I - (S \circ T)]$ and $\rho(S \circ T) \leq \rho(S)\rho(T) < 1$, so

$$(A \star B)^{-1} = (I - S \circ T)^{-1}(I \circ B)^{-1}(I \circ A)^{-1}$$

$$= \left[\sum_{k=0}^{\infty} (S \circ T)^k \right] (I \circ B)^{-1}(I \circ A)^{-1}$$

$$\leq \left[\left[\sum_{k=0}^{\infty} S^k \right] \circ \left[\sum_{k=0}^{\infty} T^k \right] \right] (I \circ B)^{-1}(I \circ A)^{-1}$$

$$= [(I - S)^{-1} \circ (I - T)^{-1}](I \circ B)^{-1}(I \circ A)^{-1}$$

$$= [(I \circ A)(I - S)]^{-1} \circ [(I \circ B)(I - T)]^{-1}$$

$$= A^{-1} \circ B^{-1}$$

Finally, $\rho((A \star B)^{-1}) \leq \rho(A^{-1} \circ B^{-1}) \leq \rho(A^{-1})\rho(B^{-1})$ by (5.7.4) again, so $\tau(A \star B) = \rho((A \star B)^{-1})^{-1} \geq \rho(A^{-1})^{-1}\rho(B^{-1})^{-1} = \tau(A)\tau(B)$, as asserted. ◻

One of the purposes of this section is to give some results much stronger than (5.7.4), again by analyzing the Hadamard product directly, rather than arguing via the Kronecker product.

Another closure result is more subtle than (5.7.2) or (5.7.3), but is more precisely analogous to the closure of the positive definite Hermitian matrices under the Hadamard product (5.2.1). In fact, the statements are formally identical when one realizes that the class of inverse positive definite matrices is just exactly the positive definite matrices themselves.

5.7.5 Theorem. If A, $B \in M_n$ are M-matrices, then so is $A \circ B^{-1}$.

Proof: Whether or not a given matrix is an M-matrix is unaltered by left and/or right multiplication by a positive diagonal matrix [see the exercise following (2.5.7)]. As $A \in Z_n$ and has positive diagonal and $B^{-1} \geq 0$, we have that $A \circ B^{-1} \in Z_n$ and has positive diagonal. Since a diagonally dominant matrix with positive diagonal in Z_n is an M-matrix, in order to verify the assertion it suffices, using (5.1.2), to replace A by $\hat{A} = AD$ and B with $\hat{B} = EB$, D and E positive diagonal matrices, so that the resulting matrix $\hat{A} \circ \hat{B}^{-1}$ is (row) diagonally dominant. Use (2.5.3.13) to choose D so that $\hat{A}$ is row diagonally dominant and to choose E so that $\hat{B}$ is column diagonally dominant (by applying (2.5.3.13) to B^T). Then, by (2.5.12), again arguing on the transpose, $\hat{B}^{-1}$ is diagonally dominant of its row entries. But it is an immediate calculation to see that the Hadamard product of a *row* diagonally dominant matrix and a matrix diagonally dominant of its *row* entries is *row* diagonally dominant, which completes the proof.   ☐

Exercise. Show that if $A \in M_n$ is an M-matrix, then $\Phi(A) = A \circ A^{-1}$ and $\Phi_T(A) = A \circ (A^{-1})^T$ are both M-matrices.

Exercise. If $A \in M_n$ is an M-matrix, show that $\tau(\Phi_T(A)) = 1$.

Later in this section, we shall study the minimum eigenvalue of $A \circ A^{-1}$ for an M-matrix $A \in M_n$.

5.7.6   Example. In view of the closure results (5.7.2) and (5.7.5) involving M-matrices, it is worth noting that the set of inverse M-matrices is *not* closed under Hadamard multiplication. Let

$$A^{-1} = \begin{bmatrix} 6 & -1.8 & -1 & -1 & -1 & -1.0001 \\ -1 & 6 & -2 & -1 & -1 & -1 \\ -1 & -1 & 6 & -2 & -1 & -1 \\ -1 & -1 & -1 & 6 & -2 & -1 \\ -1 & -1 & -1 & -1 & 6 & -2 \\ -2 & -1 & -1 & -1 & -1 & 6 \end{bmatrix}$$

and

$$B^{-1} = \begin{bmatrix} 6 & -3 & -3 & 0 & 0 & -0.0001 \\ -4 & 6 & -1 & 0 & 0 & 0 \\ -1 & -5 & 6 & 0 & 0 & 0 \\ 0 & 0 & 0 & 8 & -2 & -6 \\ 0 & 0 & 0 & -3 & 8 & -5 \\ 0 & 0 & 0 & -7 & -4 & 8 \end{bmatrix}$$

so A and B are inverse M-matrices. However, $A \circ B$ is not an inverse M-matrix, as the 3,5 and 3,6 entries of $(A \circ B)^{-1}$ are both positive.

Exercise. Show that in M_2 and M_3 the set of inverse *M*-matrices *is* closed under the Hadamard product.

We next turn to inequalities for Hadamard products of Hadamard powers of nonnegative matrices. If $A = [a_{ij}] \in M_{m,n}(\mathbb{R})$ has nonnegative entries and $\alpha \geq 0$, we write $A^{(\alpha)} \equiv [a_{ij}^\alpha]$ for the αth Hadamard power of A. As we saw in Section (5.3), simple inequalities may often be strengthened. In this case our primary result may be viewed as a considerable strengthening of (5.7.4).

5.7.7   Theorem. Let $A_i \in M_n$, $A_i \geq 0$, $i = 1,...,k$, and suppose that $\alpha_i \geq 0$, $i = 1,..., k$, satisfy $\alpha_1 + \cdots + \alpha_k \geq 1$. Then

$$\rho(A_1^{(\alpha_1)} \circ \cdots \circ A_k^{(\alpha_k)}) \leq \rho(A_1)^{\alpha_1} \cdots \rho(A_k)^{\alpha_k}$$

Our proof of (5.7.7) will be based upon two special cases involving $k = 1$ and $k = 2$.

5.7.8   Lemma. If $A \in M_n$, $A \geq 0$, and $\alpha \geq 1$, then

$$\rho(A^{(\alpha)}) \leq \rho(A)^\alpha$$

Proof: It suffices to assume that A is irreducible. In this event a positive diagonal matrix D may be chosen so that $D^{-1}AD$ has all row sums equal to $\rho(A)$. Since $\rho([D^{-1}AD]^{(\alpha)}) = \rho((D^\alpha)^{-1}A^{(\alpha)}D^\alpha) = \rho(A^{(\alpha)})$ and the desired inequality is homogeneous of degree 1 in A, we may assume without loss of generality that A is row stochastic. But then, for $\alpha \geq 1$, $A^{(\alpha)} \leq A$, so $\rho(A^{(\alpha)}) \leq \rho(A) = 1 = \rho(A)^\alpha$, as was to be shown.   ▯

5.7.9   Lemma. Suppose that $A, B \in M_n(\mathbb{R})$ and $\alpha \in \mathbb{R}$ satisfy $A \geq 0$, $B \geq 0$, and $0 \leq \alpha \leq 1$. Then

$$\rho(A^{(\alpha)} \circ B^{(1-\alpha)}) \leq \rho(A)^\alpha \rho(B)^{1-\alpha}$$

Proof: The inequality is clearly valid if $\rho(A) = 0$ or $\rho(B) = 0$ or if $\alpha = 0$ or $\alpha = 1$. Via a standard continuity argument we may assume that A and B are irreducible, and, then, as in the proof of (5.7.8) we may suppose that both $A = [a_{ij}]$ and $B = [b_{ij}]$ are row stochastic. Under these assumptions we must now show that $C \equiv A^{(\alpha)} \circ B^{(1-\alpha)}$ satisfies $\rho(C) \le 1$. Hölder's inequality for the row sums of C (see Appendix B of [HJ]) gives

$$\sum_{j=1}^{n} c_{kj} = \sum_{j=1}^{n} a_{kj}^{\alpha} b_{kj}^{1-\alpha} \le \left[\sum_{j=1}^{n} a_{kj}\right]^{\alpha} \left[\sum_{j=1}^{n} b_{kj}\right]^{1-\alpha} = 1^{\alpha} 1^{1-\alpha} = 1$$

so the maximum row sum matrix norm of C, and hence its spectral radius, is at most 1.   ☐

We may now prove the general assertion in (5.7.7) by induction. Because of (5.7.8) it suffices to consider the case in which $\alpha_1 + \cdots + \alpha_k = 1$. Note that we may assume $\alpha_k < 1$, and define $B \ge 0$ by

$$B^{(1-\alpha_k)} \equiv A_1^{(\alpha_1)} \circ \cdots \circ A_{k-1}^{(\alpha_{k-1})}$$

If we define $\beta_j \equiv \alpha_j/(1 - \alpha_k)$, $j = 1,..., k-1$, then $\beta_1 + \cdots + \beta_{k-1} = 1$ and

$$B = A_1^{(\beta_1)} \circ \cdots \circ A_{k-1}^{(\beta_{k-1})}$$

Now,

$$\rho\left[A_1^{(\alpha_1)} \circ \cdots \circ A_k^{(\alpha_k)}\right] = \rho\left[B^{(1-\alpha_k)} \circ A_k^{(\alpha_k)}\right]$$

$$\le \rho(B)^{1-\alpha_k} \rho(A_k)^{\alpha_k}$$

$$\le \left[\rho(A_1)^{\beta_1} \cdots \rho(A_{k-1})^{\beta_{k-1}}\right]^{1-\alpha_k} \rho(A_k)^{\alpha_k}$$

$$= \rho(A_1)^{\alpha_1} \cdots \rho(A_k)^{\alpha_k}$$

as was to be shown. The first inequality follows from Lemma (5.7.9), and the second is based upon the natural induction hypothesis.   ☐

5.7.10   **Remark.** If each A_i, $i = 1,..., k$, is irreducible and $\alpha_1 + \cdots + \alpha_k = 1$, equality occurs in the inequality of Theorem (5.7.7) if and only if there exist positive diagonal matrices $D_i \in M_n$ and positive scalars γ_i, $i = 2,..., n$, such that

$$\gamma_i A_i = D_i^{-1} A_1 D_i, \; i = 2,\ldots, k$$

Exercise. Show that Observation (5.7.4) is a special case of Theorem (5.7.7).

Exercise. Use a field-of-values argument (Chapter 1) to show that if $A \in M_n$ is nonnegative, then $\rho(A) \leq \rho(\tfrac{1}{2}[A + A^T])$.

Thus, forming the *entrywise arithmetic mean* of $A \geq 0$ and A^T cannot decrease the spectral radius. The dual fact that forming the *entrywise geometric mean* cannot increase the spectral radius follows from Theorem (5.7.7).

5.7.11 **Corollary.** If $A \in M_n$ and $A \geq 0$, then

$$\rho[A^{(\tfrac{1}{2})} \circ (A^T)^{(\tfrac{1}{2})}] \leq \rho(A)$$

Proof: According to (5.7.7),

$$\rho[A^{(\tfrac{1}{2})} \circ (A^T)^{(\tfrac{1}{2})}] \leq \rho(A)^{\tfrac{1}{2}} \rho(A^T)^{\tfrac{1}{2}}$$

But, since $\rho(A^T) = \rho(A)$, the right-hand side may be replaced by $\rho(A)$. ☐

The case of equality in the inequality in Corollary (5.7.11) may be analyzed directly in the irreducible case using (5.7.10). It is interesting that it involves the notion of diagonal symmetrizability.

Exercise. Show that equality occurs in the inequality in Corollary (5.7.11) if A is symmetric or if A is similar to a symmetric matrix via a diagonal similarity.

5.7.12 **Theorem.** For $A \in M_n$, $A \geq 0$, and A irreducible, the following are equivalent:

(a) Equality occurs in the inequality in Corollary (5.7.11).

(b) A^T is similar to A by a diagonal matrix.

(c) A is similar to a symmetric matrix via a diagonal matrix.

Proof: Since A^T is irreducible when A is, the equivalence of (a) and (b) follows from (5.7.10). The equivalence of (b) and (c) follows from the following sequence of equivalent statements:

$$D^{-1} A^T D = A$$
$$\Leftrightarrow \quad A^T D = DA$$
$$\Leftrightarrow \quad DA \text{ is symmetric}$$
$$\Leftrightarrow \quad D^{-\frac{1}{2}}(DA)D^{-\frac{1}{2}} = D^{\frac{1}{2}} A D^{-\frac{1}{2}} \text{ is symmetric.}$$

Here, as usual, $D^{\frac{1}{2}}$ is the positive diagonal matrix whose diagonal entries are the square roots of those of D, and $D^{-\frac{1}{2}}$ is its inverse. $\qquad \Box$

For general nonnegative $A, B \in M_n$ and $0 < \alpha < 1$, $\rho[\alpha A + (1 - \alpha)B]$ may be larger than, equal to, or smaller than $\alpha\rho(A) + (1 - \alpha)\rho(B)$, so the spectral radius is *not* a convex function of its nonnegative matrix argument.

Exercise. Verify the previous statement with examples to exhibit each eventuality.

However, it is an important fact that the spectral radius, considered as a function of the *diagonal entries* of a nonnegative matrix $A \in M_n$, is a convex function from $\mathbb{R}^n_+$ to $\mathbb{R}_+$. This may be proved using (5.7.7) and the technique of linearization.

5.7.13   Corollary. Let $A = [a_{ij}]$, $B = [b_{ij}]$, $D_1, D_2 \in M_n$ be given nonnegative matrices, and suppose that D_1 and D_2 are diagonal. Define $C(A,B,\alpha) = [c_{ij}] \in M_n$ by

$$c_{ij} = \begin{cases} (a_{ij})^\alpha (b_{ij})^{1-\alpha} & \text{if } i \neq j \\ \alpha a_{ii} + (1 - \alpha)b_{ii} & \text{if } i = j \end{cases} \quad \text{for } \alpha \in [0,1]$$

The following inequalities then hold for all $\alpha \in [0,1]$:

(a) $\rho(C(A,B,\alpha)) \le \alpha\rho(A) + (1 - \alpha)\rho(B)$

(b) $\rho(\alpha[A + D_1] + (1 - \alpha)[A + D_2]) = \rho(A + \alpha D_1 + (1 - \alpha)D_2)$

$$\le \alpha\rho(A + D_1) + (1 - \alpha)\rho(A + D_2)$$

Proof: By continuity, there is no loss of generality to assume that A and B are positive matrices. For small $\epsilon > 0$, a computation reveals that

$$(I + \epsilon A)^{(\alpha)} \circ (I + \epsilon B)^{(1-\alpha)} = I + \epsilon C(A,B,\alpha) + \cdots$$

For any positive $X \in M_n$ and small $\epsilon > 0$, Theorem (6.3.12) in [HJ] gives the perturbation formula

$$\rho(I + \epsilon X) = 1 + \epsilon \rho(X) + \cdots \tag{5.7.14}$$

Now use the preceding formulae to calculate

$$\rho((I + \epsilon A)^{(\alpha)} \circ (I + \epsilon B)^{(1-\alpha)}) = \rho(I + \epsilon C(A,B,\alpha) + \cdots)$$

$$= 1 + \epsilon \rho(C(A,B,\alpha)) + \cdots$$

and

$$\rho((I + \epsilon A)^{(\alpha)})\rho((I + \epsilon B)^{(1-\alpha)})$$

$$= (1 + \epsilon \alpha \rho(A) + \cdots)(1 + \epsilon[1 - \alpha]\rho(B) + \cdots)$$

$$= 1 + \epsilon[\alpha \rho(A) + (1 - \alpha)\rho(B)] + \cdots$$

The inequality in Lemma (5.7.9) now gives assertion (a). The second assertion follows immediately from the first if we replace A and B in (a) by $A + D_1$ and $A + D_2$, respectively.   □

Exercise. If Theorem (5.7.7) for arbitrary k (instead of just $k = 2$) were used in the proof of Corollary (5.7.13), what would be the resulting analog of (5.7.13a)?

We complete this discussion of facts associated with Theorem (5.7.7) by mentioning results for *H*-matrices dual to Theorem (5.7.7) and (5.7.10).

5.7.15   Theorem. If $A_1,...,A_k \in M_n$ are *H*-matrices and $\alpha_1,...,\alpha_k \geq 0$ satisfy $\sum\limits_{i=1}^{k} \alpha_i \geq 1$, then

$$\tau(A_1^{(\alpha_1)} \circ \cdots \circ A_k^{(\alpha_k)}) \geq \tau(A_1)^{\alpha_1} \cdots \tau(A_k)^{\alpha_k}$$

Here, $A^{(\alpha)}$ is again defined entrywise and *any* scalar definition of a^α such that $|a^\alpha| = |a|^\alpha$ is allowed.

5.7.16   Remark. If each A_i, $i = 1,..., k$, is irreducible and $\Sigma_i \alpha_i = 1$, equality occurs in the inequality in Theorem (5.7.15) if and only if there exist positive diagonal matrices $D_i \in M_n$ and positive scalars γ_i, $i = 2,..., n$, such that

$$\gamma_i |A_i| = D_i^{-1} |A_1| D_i, \qquad i = 2,..., k$$

As usual, the vertical bars denote entrywise absolute value.

Exercise. Show that the result about $\tau(\cdot)$ in Corollary (5.7.4.1) is a special case of Theorem (5.7.15).

Exercise. Write out the inequality in Theorem (5.7.15) entrywise for the important case in which $A_1,..., A_k$ are M-matrices.

We next define several scalar quantities naturally associated with a nonnegative matrix. Though apparently quite different, they turn out to be identical. For a nonnegative matrix $A \in M_n$, the quantity $\rho(A^{(t)})^{1/t}$ is nonincreasing in t for $t \geq 1$ because of Lemma (5.7.8) and is, of course, bounded below by 0.

5.7.17   Definition. For a given nonnegative $A \in M_n$, define

$$h_1(A) \equiv \lim_{t \to \infty} \rho(A^{(t)})^{1/t}$$

Because of the preceding comment, this limit exists for any square nonnegative matrix A.

Exercise. For any nonnegative $A \in M_n$, verify that $\rho(A^{(t)})^{1/t}$ is nonincreasing in t for $t \geq 1$ and explain why $h_1(A) = \lim_{m \to \infty} \rho(A^{(m)})^{1/m}$, where $m \in \{1, 2, 3,...\}$.

5.7.18   Definition. For a given nonnegative $A \in M_n$, define

$$h_2(A) \equiv \sup \left\{ \frac{\rho(A \circ B)}{\rho(B)} : B \geq 0 \text{ and } \rho(B) > 0 \right\}$$

$$= \sup \{ \rho(A \circ B) : B \geq 0 \text{ and } \rho(B) \leq 1 \}$$

These two suprema exist and are equal because of Observation (5.7.4) and homogeneity of the spectral radius function.

Exercise. Verify that $h_2(\cdot)$ is well-defined and $\rho(A) \geq h_2(A) \geq \rho(A)/n$.

5.7.19   Definition. Recall that the norm $\|\cdot\|_\infty$ denotes the maximum absolute value of the entries of its argument [HJ, Section (5.6)]. For a nonnegative matrix $A \in M_n$, define

$$h_3(A) \equiv \inf \{ \|DAD^{-1}\|_\infty : D \in M_n \text{ is a positive diagonal matrix} \}$$

5.7.20   Definition. The *maximum (simple) cycle geometric mean* for a nonnegative matrix $A \in M_n$ is

$$h_4(A) \equiv \max \left[\prod_{j=1}^{k} a_{i_j i_{j+1}} \right]^{1/k}$$

in which $k + 1$ is identified with 1 and the maximum is taken over all sequences of distinct indices $i_1, \ldots, i_k \leq n$ and over all $k = 1, \ldots, n$.

Exercise. If, in the definition of $h_4(\cdot)$, we omit the distinctness requirement and allow arbitrary positive integers k, giving an alternate quantity h_4', show that

$$h_4'(A) = h_4(A)$$

for all nonnegative $A \in M_n$.

5.7.21   Theorem. For each nonnegative matrix $A \in M_n$,

$$h_1(A) = h_2(A) = h_3(A) = h_4(A)$$

Proof: We establish the assertion by verifying the following four inequalities for a given nonnegative $A \in M_n$:

$$h_1(A) \le h_2(A) \le h_3(A) \le h_4(A) \le h_1(A)$$

$h_1 \le h_2$: Since $h_2(A) = 0$ if and only if $\rho(A) = 0$, in which case $h_1(A) = 0$ as well, we assume $\rho(A) > 0$ and $h_2(A) > 0$. For any nonnegative $B \in M_n$, we have $\rho(A \circ B) \le \rho(A)\rho(B) = 0$ if $\rho(B) = 0$, and $\rho(A \circ B) = \rho(A \circ (B/\rho(B))\rho(B)$ if $\rho(B) > 0$, so in either case we have $\rho(A \circ B) \le h_2(A)\rho(B)$. For any nonnegative integer m we have

$$\rho(A^{(m)}) = \rho(A \circ A^{(m-1)}) \le h_2(A)\rho(A^{(m-1)}) \le \cdots \le h_2(A)^m[\rho(A)/h_2(A)]$$

and hence

$$h_1(A) = \lim_{m \to \infty} \rho(A^{(m)})^{1/m} \le h_2(A) \lim_{m \to \infty} [\rho(A)/h_2(A)]^{1/m} = h_2(A)$$

$h_2 \le h_3$: Because of the entrywise monotonicity of the Perron root [HJ, Theorem (8.1.18)], we have for any positive diagonal matrix $D \in M_n$ and for any nonnegative $B \in M_n$

$$\rho(A \circ B) = \rho(D[A \circ B]D^{-1}) = \rho([D^{-1}AD] \circ B) \le \|D^{-1}AD\|_\infty \rho(B)$$

and hence $h_2 \le h_3$.

$h_3 \le h_4$: It suffices to consider irreducible A, so that $\rho(A) > 0$. We construct a positive diagonal matrix $D = \text{diag}(d_1, \ldots, d_n)$ such that $\|DAD^{-1}\|_\infty \le h_4(A)$. [In fact, the largest entry of DAD^{-1} will be equal to $h_4(A)$]. Since the assertion is homogeneous in the entries of A, we may assume $h_4(A) = 1$. Let $D \equiv \text{diag}(d_1, \ldots, d_n)$, and set $d_1 \equiv 1$. Now, for each given j, $2 \le j \le n$, consider all sequences of integers (with each integer between 1 and n) $1 = i_1, i_2, \ldots, i_{k-1}, i_k = j$ and define

$$d_j \equiv \max a_{i_1 i_2} a_{i_2 i_3} \cdots a_{i_{k-1} i_k}$$

This maximum is well defined and positive since: (1) only the finite number of sequences without repetition need be considered because $h_4(A) = 1$, and

(2) A is irreducible, so that at least one sequence yields a positive product. Then

$$\frac{d_i a_{ij}}{d_j} \leq 1$$

since the numerator is a sequence product from 1 to j (via i), or 0, while the denominator is the *maximum* sequence product from 1 to j. Since $h_4(A) = 1$, we have found a D with the desired properties.

$h_4 \leq h_1$: Choose a sequence of indices $i_1, i_2, ..., i_k$ for which the maximum in the definition of $h_4(A)$ is attained. Define A_0 by replacing each entry of A other than $a_{i_1 i_2}, a_{i_2 i_3}, ..., a_{i_k i_1}$ with 0. Then $A_0 \leq A$, and

$$h_4(A) = \rho(A_0) = \rho(A_0^{(t)})^{1/t} \leq \rho(A^{(t)})^{1/t}$$

The inequality follows from the monotonicity of the Perron root. Taking limits yields the asserted inequality, which completes the proof.   □

Exercise. For nonnegative $A \in M_n$, define

$$h_5(A) = \limsup \left[\prod_{j=1}^{k} a_{i_j i_{j+1}} \right]^{1/k}$$

in which $i_1, i_2, ..., i_{k+1} \leq n$ are indices and k runs through positive integers. These are *path geometric means*. Show that $h_5(A) = h_1(A)$ for each nonnegative $A \in M_n$.

Exercise. For nonnegative $A, B \in M_n$, show that $h_1(A \circ B) \leq h_1(A)h_1(B)$.

Exercise. For nonnegative $A \in M_n$, show that $h_1(A^{(t)}) = h_1(A)^t$.

Exercise. For nonnegative $A \in M_n$, show that $h_1(A) \leq \rho(A) \leq n h_1(A)$.

Exercise. Construct examples to show that neither $h_1(AB) \leq h_1(A)h_1(B)$ nor $h_1(A + B) \leq h_1(A) + h_1(B)$ holds for all nonnegative $A, B \in M_n$.

According to Theorem (5.7.5), if $A \in M_n$ is an M-matrix, then so is $\Phi(A) = A \circ A^{-1}$. In particular,

$$\tau(A \circ A^{-1}) > 0$$

This observation raises a natural question: Is there a *positive* lower bound for $\tau(A \circ A^{-1})$ for all M-matrices A of a given size? We answer this question affirmatively in Corollary (5.7.32). Moreover, we might also ask if $\tau(A \circ A^{-1})$ can be uniformly bounded above for all M-matrices A. The latter question is related to the notion of diagonal symmetrizability, and an answer is given in the following theorem.

5.7.22   Theorem. Let $A \in M_n$ be an M-matrix. Then

$$\tau(A \circ A^{-1}) \leq 1$$

Furthermore, if A is irreducible, then $\tau(A \circ A^{-1}) = 1$ if and only if there is a positive diagonal matrix D such that AD is symmetric.

Our proof of Theorem (5.7.22) will be based upon an interesting trace inequality for the product $A^{-1}A^T$ when $A \in M_n$ is an M-matrix.

Exercise. If $A \in M_n$ and $0 \notin F(A)$ (the field of values—Chapter 1), show that $|\text{tr}(A^{-1}A^*)| \leq n$. *Hint:* Use Theorem (1.7.11).

Exercise. For general $A \in M_n(\mathbb{R})$, show by example that $\text{tr}(A^{-1}A^T)$ may be arbitrarily large.

In general, $0 \in F(A)$ is possible when A is an M-matrix. Even though some positive diagonal multiple DA of a given M-matrix $A \in M_n$ satisfies $0 \notin F(DA)$ [see (2.5.3.16)], $\text{tr}(A^{-1}A^T)$ is not obviously related to $\text{tr}((DA)^{-1}(DA)^T)$. Thus, the following trace inequality does not seem to follow from the general fact about matrices satisfying $0 \notin F(A)$ noted in the first of the preceding two exercises.

5.7.23   Theorem. Let $A \in M_n$ be an M-matrix. Then

$$\text{tr}(A^{-1}A^T) \leq n$$

with equality if and only if A is symmetric.

Before discussing Theorem (5.7.23), we first show that Theorem

(5.7.22) may be proved using Theorem (5.7.23).

Proof of (5.7.22): For reducible A, the matrices A and A^{-1}, and thus $A \circ A^{-1}$, have irreducible components determined by the same index sets. Since $\tau(A \circ A^{-1})$ is attained for one of these, it suffices to consider only irreducible A to verify the general inequality in Theorem (5.7.22). Since $A \circ A^{-1}$ is then irreducible, we may suppose that there is a positive eigenvector x of $A \circ A^{-1}$ corresponding to $\tau(A \circ A^{-1})$. Let the diagonal matrix D be defined by $x = De$, where the vector $e \in \mathbb{R}^n$ has all entries 1. Then $(A \circ A^{-1})x = \tau(A \circ A^{-1})x$ implies that $[(AD) \circ (AD)^{-1}]e = [D^{-1}(A \circ A^{-1})D]e = \tau(A \circ A^{-1})e$ and, therefore, that

$$e^T[(AD) \circ (AD)^{-1}]e = n\tau(A \circ A^{-1})$$

or, by Lemma (5.1.5), that

$$\mathrm{tr}[(AD)^{-1}(AD)^T] = n\tau(A \circ A^{-1})$$

Since AD is an M-matrix, Theorem (5.7.23) implies that $n\tau(A \circ A^{-1}) \leq n$, which is the inequality in Theorem (5.7.22). If A is actually irreducible, Theorem (5.7.23) further implies that $\tau(A \circ A^{-1}) = 1$ if and only if AD is symmetric, which completes the proof. □

Exercise. (a) Show by example that for a general (possibly reducible) M-matrix $A \in M_n$, $\tau(A \circ A^{-1}) = 1$ may occur without A being symmetrizable by a positive diagonal multiple. (b) Show that for a general M-matrix A, $\tau(A \circ A^{-1}) = 1$ if and only if A has an irreducible component that may be symmetrized by a positive diagonal matrix.

Since $\mathrm{tr}(A^{-1}A^T) = e^T(A \circ A^{-1})e$ by Lemma (5.1.5), the inequality in Theorem (5.7.23) is equivalent to

$$e^T(A \circ A^{-1})e \leq n \tag{5.7.24}$$

for any M-matrix $A \in M_n$. Like many matrix inequalities, the inequality (5.7.24) may be interpolated via insertion of finer inequalities. The basic inequality we exhibit is the following:

5.7.25   Theorem. Let $A \in M_n$ be an M-matrix. Then

$$e^T(A \circ A^{-1})e \leq 1 + e^T(A(\{i\}') \circ A(\{i\}')^{-1})e$$

in which all entries of the vector e (of whatever size) are 1 and $\{i\}' \equiv \{j: j \in \{1,...,n\}, j \neq i\}$.

Successive applications of Theorem (5.7.25) give the inequalities

$$e^T(A \circ A^{-1})e \leq k + e^T(A(\{i_1,...,i_k\}') \circ A(\{i_1,...,i_k\}')^{-1})e \qquad (5.7.26)$$

for any k indices $1 \leq i_1 < i_2 < \cdots < i_k \leq n$ and any M-matrix $A \in M_n$, in particular

$$e^T(A \circ A^{-1})e \leq 1 + e^T(A(\{1\}') \circ A(\{1\}')^{-1})e$$
$$\leq 2 + e^T(A(\{1,2\}') \circ A(\{1,2,\}')^{-1})e$$
$$\cdot$$
$$\cdot$$
$$\cdot$$
$$\leq n-1 + e^T(A(\{n\}) \circ A(\{n\})^{-1})e = n \qquad (5.7.27)$$

Each of these implies (5.7.24) and thus Theorem (5.7.23).

Thus, we wish to prove Theorem (5.7.25). To do this, we employ a fact about nonnegative matrices.

5.7.28   Lemma. Let $P = [p_{ij}] \in M_n$ be an irreducible nonnegative matrix with right Perron eigenvector $u > 0$ and left Perron eigenvector $v^T > 0$. Then

$$u^T P v \geq v^T P u = \rho(P)v^T u$$

Proof: By the weighted arithmetic-geometric mean inequality,

$$\prod_{\substack{i,j=1 \\ p_{ij}>0}}^{n} \left[\frac{u_i v_j/u^T P v}{v_i u_j/v^T P u} \right]^{\frac{p_{ij} v_i u_j}{v^T P u}}$$

$$\leq \sum_{\substack{i,j=1 \\ p_{ij}>0}}^{n} \left[\frac{u_i v_j / u^T P v}{v_i u_j / v^T P u} \right] \frac{p_{ij} v_i u_j}{v^T P u} = 1$$

Thus,

$$\frac{u^T P v}{v^T P u} \geq \prod_{i,j=1}^{n} \left(\frac{u_i}{v_i} \right)^{\dfrac{p_{ij} v_i u_j}{v^T P u}} \prod_{i,j=1}^{n} \left(\frac{v_j}{u_j} \right)^{\dfrac{p_{ij} v_i u_j}{v^T P u}}$$

$$= \prod_{i=1}^{n} \left(\frac{u_i}{v_i} \right)^{\dfrac{\sum_{j=1}^{n} v_i p_{ij} u_j}{v^T P u}} \prod_{j=1}^{n} \left(\frac{v_j}{u_j} \right)^{\dfrac{\sum_{i=1}^{n} v_i p_{ij} u_j}{v^T P u}}$$

$$= \prod_{i=1}^{n} \left(\frac{u_i}{v_i} \right)^{\dfrac{v_i u_i}{v^T u}} \prod_{j=1}^{n} \left(\frac{v_j}{u_j} \right)^{\dfrac{u_j v_j}{v^T u}} = 1 \qquad\qquad \Box$$

Proof of (5.7.25): We may assume without loss of generality that $i = 1$. Set $A(\{1\}') = A_1$. Then the desired inequality is equivalent to

$$e^T \left[A \circ A^{-1} - \begin{bmatrix} 1 & 0 \\ 0 & A_1 \circ A_1^{-1} \end{bmatrix} \right] e \leq 0$$

The matrix $B \equiv A - \dfrac{\det A}{\det A_1} E_{11}$ is a singular M-matrix, and a partitioned matrix calculation reveals that

$$A \circ A^{-1} - \begin{bmatrix} 1 & 0 \\ 0 & A_1 \circ A_1^{-1} \end{bmatrix} = \frac{1}{\det A} (B \circ \operatorname{adj} B)$$

We make two observations about such a singular M-matrix B. First, if $u, v \geq 0$ are nonzero vectors in $\mathbb{R}^n$ such that $Bu = 0$ and $v^T B = 0$, then $u^T B v \leq 0$. To see this we may assume without loss of generality that B is irreducible and that u and v are positive. Then $B = \rho I - P$ with $P \geq 0$ irreducible and

$\rho = \rho(P)$ and we calculate $u^T B v = \rho u^T v - u^T P v \leq \rho u^T v - v^T P u = \rho u^T v - \rho v^T u$ $= 0$. The inequality is an application of Lemma (5.7.28), as u and v are right and left Perron eigenvectors of P.

Second, our B has rank $n-1$, and hence adj B is a rank one matrix of the form cuv^T, $c > 0$, in which $u \geq 0$ is a right Perron eigenvector of B and $v \geq 0$ is a left Perron eigenvector of B.

We may now complete the proof with the following calculation:

$$e^T \left[A \circ A^{-1} - \begin{bmatrix} 1 & 0 \\ 0 & A_1 \circ A_1^{-1} \end{bmatrix} \right] e$$

$$= \frac{1}{\det A} e^T (B \circ \operatorname{adj} B) e = \frac{1}{\det A} e^T (B \circ [cuv^T]) e$$

$$= \frac{c}{\det A} u^T B v \leq 0 \qquad\qquad \square$$

Theorem (5.7.22) settles the question of an upper bound for $\tau(A \circ A^{-1})$ when A is an M-matrix; we next address the question of a positive lower bound.

5.7.29   Example. Define $A \in M_n$, $n \geq 2$, by $A \equiv I - tC$, where

$$C \equiv \begin{bmatrix} 0 & 1 & 0 & & & 0 \\ 0 & 0 & 1 & 0 & \ldots & 0 \\ \vdots & & & \ddots & & 0 \\ 0 & & & & \ddots & 1 \\ 1 & 0 & & & \ldots & 0 \end{bmatrix}$$

Then A is an M-matrix for $0 \leq t < 1$, and a calculation reveals that

$$A \circ A^{-1} = \frac{1}{1-t^n} [I - t^2 C]$$

Thus,

$$\tau(A \circ A^{-1}) = \frac{1-t^2}{1-t^n} = \frac{1+t}{1+t+t^2+\cdots+t^{n-1}}$$

Since t may be arbitrarily close to 1, this shows that no general inequality better than

$$\tau(A \circ A^{-1}) > \tfrac{2}{n} \tag{5.7.30}$$

is possible for *M*-matrices $A \in M_n$, $n \geq 2$. The case $n = 2$ is special and is treated in the following exercise. It is known that (5.7.30) is valid for all $n \geq 2$; we prove in Corollary (5.7.32) that the lower bound $\tfrac{1}{n}$ is valid.

Exercise. Show that for any *M*-matrix $A \in M_2$, $\tau(A \circ A^{-1}) = 1$.

In pursuit of a lower bound for $\tau(A \circ A^{-1})$, we also exhibit lower bounds for $\tau(A \circ B^{-1})$ for independent *M*-matrices $A, B \in M_n$. These bounds, though tight for their level of generality, do not seem strong enough to yield, upon specialization, the lower bound of $\tfrac{2}{n}$ for $\tau(A \circ A^{-1})$. Note, however, that each of these generalizes Theorem (5.7.5) and also that the first of these is an analog for *M*-matrices of the lower bound in Theorem (5.3.4).

5.7.31   Theorem. Let $A, B \in M_n$ be *M*-matrices and let $B^{-1} \equiv [\beta_{ij}]$. Then

$$\tau(A \circ B^{-1}) \geq \tau(A) \min_{1 \leq i \leq n} \beta_{ii}$$

Proof: Use (2.5.3.14) to choose D so that $D^{-1}BD$ is row diagonally dominant, and note that the diagonal entries of $(D^{-1}BD)^{-1}$ are the same as those of B^{-1}. Now $\tau(A \circ B^{-1}) = \tau(D^{-1}(A \circ B^{-1})D) = \tau(A \circ (D^{-1}BD)^{-1})$. Since $(D^{-1}BD)^{-1}$ is diagonally dominant of its column entries by Theorem (2.5.12), $A \circ (D^{-1}BD)^{-1} \geq A \operatorname{diag}(\beta_{11}, \ldots, \beta_{nn})$, and by the result of Problem 28 in Section (2.5), $\tau(A \circ B^{-1}) \geq \tau(A \operatorname{diag}(\beta_{11}, \ldots, \beta_{nn}))$. But, by the result of Problem 29 in Section (2.5), $\tau(A \operatorname{diag}(\beta_{11}, \ldots, \beta_{nn})) \geq \tau(A) \min \beta_{ii}$, which completes the proof. ☐

Theorem (5.7.31) may now be used to bound $\tau(A \circ A^{-1})$ when $A \in M_n$ is an *M*-matrix.

5.7.32   Corollary. For each *M*-matrix $A \in M_n$, we have

$$\tau(A \circ A^{-1}) > \tfrac{1}{n}$$

Proof: Since the diagonal entries of A^{-1} are positive, we may choose a positive diagonal matrix D so that each diagonal entry of $(DA)^{-1}$ is 1. Since DA is an M-matrix, each eigenvalue of $(DA)^{-1}$ has positive real part. We then have

$$\tau(A \circ A^{-1}) = \tau(D(A \circ A^{-1})D^{-1}) = \tau((DA) \circ (DA)^{-1})$$

$$\geq \tau(DA) = \frac{1}{\rho[(DA)^{-1}]} > \frac{1}{\operatorname{tr}[(DA)^{-1}]} = \tfrac{1}{n}$$

The first inequality is an application of Theorem (5.7.31) under the assumption about DA, while the second is due to the fact that $\operatorname{tr}[(DA)^{-1}] > \rho[(DA)^{-1}]$, the excess being the sum of the remaining eigenvalues, each of which has positive real part.   ▯

Exercise. Show by example that equality may be attained in the inequality of Theorem (5.7.31), but that the inequality is strict if A and B are irreducible. Can the conditions under which the inequality is strict be further relaxed?

We complete this discussion of $\tau(A \circ A^{-1})$ with another inequality for $\tau(A \circ B^{-1})$.

5.7.33   Theorem. Let $A, B \in M_n$ be M-matrices with B irreducible. Suppose that $u > 0$ and $v > 0$ are, respectively, right and left Perron eigenvectors of B. Define $w \equiv u \circ v$. Then

$$\tau(A \circ B^{-1}) \geq \frac{\tau(A)}{\tau(B)} \frac{\min \{w_1, \ldots, w_n\}}{w_1 + \cdots + w_n}$$

Proof: Let $B^{-1} \equiv [\beta_{ij}]$. The strategy is simply to estimate β_{ii}, and thus $\min \beta_{ii}$, and then apply Theorem (5.7.31).

Let $D = \operatorname{diag}(v)$. Since DB is a column diagonally dominant M-matrix, it follows from Theorem (2.5.12) that $(DB)^{-1}$ is diagonally dominant of its row entries. This means that for each pair i, j

$$\beta_{ij} \leq \beta_{ii} \frac{v_j}{v_i}$$

From $B^{-1}u = \tau(B)^{-1}u$, we have for each $i = 1,\ldots, n$ that

$$\tau(B)^{-1}u_i = \sum_{j=1}^{n} \beta_{ij}u_j \leq \sum_{j=1}^{n} \beta_{ii}\frac{v_j u_j}{v_i}$$

It follows that

$$\frac{w_i}{\tau(B)\,e^T w} = \frac{u_i v_i}{\tau(B)\,u^T v} \leq \beta_{ii}$$

Since the denominator in this lower bound for β_{ii} does not depend on i, substitution into the inequality of (5.7.31) yields the asserted inequality. $\Box$

Exercise. What can one say if the matrix B in (5.7.33) is reducible? Under what circumstances is the inequality strict?

Exercise. Sinkhorn's Theorem says that for each positive matrix $C \in M_n$ there exist positive diagonal matrices $D_1, D_2 \in M_n$ such that $D_1 C D_2$ is doubly stochastic. Use this fact to show that the irreducible case of Corollary (5.7.32) may be deduced from (5.7.33), and then show that the general case of the corollary follows from the irreducible case.

Problems

1.   Let $B \in M_n$ be an M-matrix with inverse $B^{-1} = [\beta_{ij}]$. Use Theorem (5.7.31) to show that

$$\min \{\tau(A \circ B^{-1}): A \in M_n \text{ is an } M\text{-matrix with } \tau(A) = 1\} = \min_{1 \leq i \leq n} \beta_{ii}$$

Compare this result with Problem 5 in Section (5.3).

2.   For the inequality in Theorem (5.7.7), (5.7.10) gives a characterization of the case of equality when the A_i are irreducible, $i = 1,\ldots, k$, and $\alpha_1 + \cdots + \alpha_k = 1$. Give a complete description of the case of equality (for general nonnegative A_i and/or $\alpha_1 + \cdots + \alpha_k \geq 1$).

3.   For the inequality in Theorem (5.7.15), a characterization of the case of equality is given when the A_i are irreducible, $i = 1,\ldots, k$, and $\alpha_1 + \cdots + \alpha_k = 1$ in (5.7.16). Give a complete description of the case of equality (for general H-matrices A_i and/or $\alpha_1 + \cdots + \alpha_k = 1$).

4. For real numbers $\alpha_1, \alpha_2, ..., \alpha_k \geq 0$ and given nonnegative matrices $A_1, ..., A_k \in M_n$, define $A(\alpha) \equiv A_1^{(\alpha_1)} \circ A_2^{(\alpha_2)} \circ \cdots \circ A_k^{(\alpha_k)}$. Suppose that $\rho(A(\alpha))$ is never 0. Show that the function $f: \mathbb{R}_+^k \to \mathbb{R}$ defined by $f(\alpha_1, ..., \alpha_k) \equiv \log \rho(A(\alpha))$ is convex.

5. Prove the analog of Oppenheim's inequality (see Theorem (7.8.6) and Problems 5 and 6 in Section (7.8) of [HJ]) for an M-matrix and an inverse M-matrix. Let $A, B \in M_n$ be M-matrices with $B^{-1} \equiv [\beta_{ij}]$. Show that

$$\det(A \circ B^{-1}) \geq \beta_{11}\beta_{22}\cdots\beta_{nn} \det A \geq \frac{\det A}{\det B}$$

6. Show that equality occurs in the inequality of Lemma (5.7.28) if and only if v is a multiple of u and use this to verify the statement about equality in (5.7.23,24).

7. Let $P = [p_{ij}]$, $Q = [q_{ij}] \in M_n$ be nonzero with $P \geq 0$, $Q \geq 0$, and let $u, v, x, y \in \mathbb{R}^n$ be positive vectors. Define $w \equiv y \circ (Px)/y^T Px$ and $z^T \equiv x^T \circ (y^T P)/y^T Px$. Show that

$$\frac{v^T Qu}{y^T Px} \geq \prod_{j=1}^{n} \left(\frac{u_j}{x_j}\right)^{z_j} \prod_{i=1}^{n} \left(\frac{v_i}{y_i}\right)^{w_i} \prod_{\substack{i,j=1 \\ p_{ij} \neq 0}}^{n} \left(\frac{q_{ij}}{p_{ij}}\right)^{\frac{p_{ij}y_i x_j}{y^T Px}} \quad (5.7.34)$$

8. To what extent may (5.7.34) be generalized to the situation in which the vectors u, v, x, y are nonnegative?

9. Let $0 \neq P = [p_{ij}] \in M_n$ satisfy $P \geq 0$ and suppose $u, v, x, y \in \mathbb{R}^n$ are positive vectors. Deduce that

$$\frac{v^T Pu}{u^T Px} \geq \prod_{j=1}^{n} \left(\frac{u_j}{x_j}\right)^{z_j} \prod_{i=1}^{n} \left(\frac{v_i}{y_i}\right)^{w_i}$$

in which w, z are as defined in Problem 7. Deduce Lemma (5.7.28) from this.

10. Let $P = [p_{ij}] \in M_n$ be an irreducible nonnegative matrix. Suppose that $x > 0$ is a right and $y > 0$ is a left Perron eigenvector of P, normalized so that $y^T x = 1$. Then, for any nonnegative matrix $Q = [q_{ij}] \in M_n$, and any positive vectors $u, v \in \mathbb{R}^n$, show that

$$v^T Q u \geq \rho(P) \prod_{i=1}^{n} \left[\frac{u_i v_i}{x_i y_i} \right]^{x_i y_i} \prod_{\substack{i,\,j=1 \\ p_{ij} > 0}}^{n} \left[\frac{q_{ij}}{p_{ij}} \right]^{\frac{p_{ij} y_i x_j}{\rho(A)}}$$

11. Let $P = [p_{ij}] \in M_n$ be an irreducible nonnegative matrix with right Perron eigenvector $x > 0$ and left Perron eigenvector $y > 0$ normalized so that $y^T x = 1$. Let u and v be any positive vectors such that $u \circ v = x \circ y$, and let $Q = [q_{ij}] \in M_n$ be any nonnegative matrix. Deduce that

$$v^T Q u \geq \rho(P) \prod_{\substack{i,\,j=1 \\ p_{ij} > 0}}^{n} \left[\frac{q_{ij}}{p_{ij}} \right]^{\frac{p_{ij} y_i x_j}{\rho(P)}}$$

In particular, conclude that

$$v^T P u \geq \rho(P)$$

and use this to give another proof of Lemma (5.7.28).

12. Show by example that no constant multiple of $\tau(A)\tau(B)^{-1}$ is a valid *upper* bound for $\tau(A \circ B^{-1})$ for all *M*-matrices $A, B \in M_n$.

13. Define a partial order on nonnegative matrices as follows: Let $A = [a_{ij}]$, $B = [b_{ij}] \in M_n$ be nonnegative. We say $A \geq_c B$ if each cycle product from A is greater than or equal to the corresponding one from B, that is, $a_{i_1 i_2} a_{i_2 i_3} \cdots a_{i_k i_1} \geq b_{i_1 i_2} b_{i_2 i_3} \cdots b_{i_k i_1}$ for every sequence of k integers, $1 \leq i_1, i_2, \ldots, i_k \leq n$. Show that $A \geq_c B$ if and only if there exists a positive diagonal matrix D such that $D^{-1}AD \geq B$.

14. For an irreducible nonnegative matrix $P \in M_n$, show that the following statements are equivalent:

(a)   There exists a nonsingular diagonal matrix D such that $D^{-1}PD$ is symmetric.

(b)   There exists a positive diagonal matrix E such that PE is symmetric.

(c)   There exists a positive diagonal matrix F such that FP is symmetric.

(d)   There exist positive diagonal matrices G and H such that GPH is symmetric.

(e)   There exists a positive diagonal matrix K such that $K^{-1}PK = P^T$.

(f)   For each sequence $i_1, i_2, ..., i_k \in \{1, 2, ..., n\}$, we have

$$p_{i_1 i_2} p_{i_2 i_3} \cdots p_{i_k i_1} = p_{i_1 i_k} p_{i_k i_{k-1}} \cdots p_{i_2 i_1}.$$

(g) $\rho(P^{(\frac{1}{2})} \circ P^{T(\frac{1}{2})}) = \rho(P)$.

(h) $\tau(B \circ B^{-1}) = 1$ for $B = \alpha I - P$ and $\alpha > \rho(P)$.

(i)   If $x, y > 0$ are right and left Perron eigenvectors of P, and $D \equiv [\mathrm{diag}(x)\,\mathrm{diag}(y)^{-1}]^{\frac{1}{2}}$, then $D^{-1}PD$ is symmetric.

(j) $\rho(P^{(\alpha)} \circ P^{T(\alpha)}) = \rho(P)$ for some $0 < \alpha < 1$.

15. Let $A, B \in M_n(\mathbb{R})$ be given M-matrices.
     (a) Prove the following lower bound for the matrix inequality between the Fan product and the Hadamard product in Corollary (5.7.4):

$$A^{-1} \circ B^{-1} \geq (A \star B)^{-1} \geq |(A \circ B)^{-1}|$$

     (b) Prove that $|\det(A \circ B)| \geq \det(A \star B) \geq \tau(A \star B)^n \geq \tau(A)^n \tau(B)^n$.

Notes and Further Readings. The closure result (5.7.5) may be found in A Hadamard Product Involving *M*-Matrices, *Linear Multilinear Algebra* 4 (1977), 261-264 by C. R. Johnson. The first discussion of the Fan product is in K. Fan, Inequalities for *M*-Matrices, *Proc. Koninkl. Nederl. Akademie van Wetenschappen*, Series A, 67 (1964), 602-610. Many inequalities for the Fan and Hadamard products as well as numerous references to the literature are in T. Ando, Inequalities for *M*-Matrices, *Linear Multilinear Algebra* 8 (1980), 291-316. Example (5.7.6) is in Closure Properties of Certain Positivity Classes of Matrices under Various Algebraic Operations, *Linear Algebra Appl.* 97 (1987), 243-247 by C. R. Johnson and was found in collaboration with T. Markham and M. Neumann. The material from (5.7.7) through (5.7.21) is primarily from Symmetric Matrices Associated with a Nonnegative Matrix, by C. R. Johnson and J. Dias daSilva in *Circuits, Systems, and Signal Processing* 9 (1990), 171-180, and The Perron Root of a Weighted Geometric Mean of Nonnegative Matrices, *Linear Multilinear Algebra* 24 (1988), 1-13, by L. Elsner, C. R. Johnson, and J. Dias daSilva.

The latter contains references to earlier results, such as (5.7.13) and some of
the relations in (5.7.21). The results from (5.7.22) to the end of the section
are partly based upon (1) A Trace Inequality for M-matrices and the Symmetrizability of a Real Matrix by a Positive Diagonal Matrix, *Linear
Algebra Appl.* 71 (1985), 81-94, by M. Fiedler, C. R. Johnson, T. Markham,
and M. Neumann; (2) M. Fiedler and T. Markham, An Inequality for the
Hadamard Product of an M-matrix and an Inverse M-matrix, *Linear
Algebra Appl.* 101 (1988), 1-8; (3) some joint work of R. Bapat and C. R.
Johnson; and (4) several observations of R. Bapat (e.g., Problem 7). The
authors would particularly like to thank R. Bapat for several pleasant
conversations leading to useful insights into the Hadamard product. For a
comprehensive discussion of Sinkhorn's theorem (mentioned in the last
exercise of the section), see R. A. Brualdi, S. V. Parter, and H. Schneider,
The Diagonal Equivalence of a Nonnegative Matrix to a Stochastic Matrix,
J. Math. Anal. 16 (1966), 31-50. For a proof of the lower bound (5.7.30), see
L. Ching and C. Ji-Cheng, On a Bound for the Hadamard Product of an
M-Matrix and its Inverse, *Linear Algebra Appl.* 144 (1991), 171-178.

Chapter 6

Matrices and functions

6.0 Introduction

If A is a matrix and $f(\cdot)$ is a function, what could be meant by $f(A)$? There are many interesting examples of functions $f\colon M_{m,n} \to M_{p,q}$, depending on the values of m, n, p, and q and the nature of the functional relationship. Familiar examples of such functions are tr A and det A (if $m = n$), vec A, $A \to AA^*$ $(p = q = m)$, $A \to A^*A$ $(p = q = n)$, $A \to$ the kth compound of A $(p = \binom{m}{k}, \ q = \binom{n}{k})$, $A \to A \otimes B$ $(B \in M_{r,s}, \ p = mr, \ q = ns)$, $A \to f(A)$ $(f(\cdot)$ a polynomial, $p = q = m = n)$, $A \to A \circ A$ $(p = m, \ q = n)$.

If $m = n = 1$, we have $f\colon M_{1,1} \to M_{p,q}$, which is a matrix-valued function of a single complex or real variable. One can study differentiation, integration, power series, etc., in this context as a straightforward generalization of scalar-valued functions. We consider some examples of this type in Sections (6.5-6).

Most of this book has been devoted to the study of square matrices, and so we shall be particularly interested in the case $f\colon M_n \to M_n$. This gives a rich class of natural functions because it includes polynomial functions of a matrix (Section (6.1)). Analytic functions with a power series expansion are natural generalizations of polynomials and are discussed in Section (6.2), where we also examine a natural definition of $f(A)$ when $f(\cdot)$ cannot be represented by a power series.

The point of view taken in Sections (6.1) and (6.2) is that $f(A)$ should be defined so as to generalize the natural notion of a polynomial function of a matrix, but there are other useful notions of $f(A)$ that are fundamentally different. For example, if one thinks of a matrix $A \in M_n$ as arising from the discretization of a continuous integral kernel $K(x,y)\colon [a,b] \times [a,b] \to \mathbb{C}$, $A = [a_{ij}] \equiv [K(x_i, x_j)]$, then one is led to consider (pointwise) powers of the kernel

$K^2(x,y)$, $K^3(x,y)$,..., and their discretizations, which are the Hadamard powers $A \circ A$, $A \circ A \circ A$, etc.   More general functions of the kernel, $f(K(x,y))$, yield Hadamard functions $[f(K(x_i,x_j))] = [f(a_{ij})]$ of the matrix, which we study in Section (6.3).

In Section (6.4) we consider some classes of nonlinear matrix equations of the form $f(X) = A$ that can be solved using canonical forms, matrix factorizations, and other tools developed in preceding chapters; we have already considered linear matrix equations in Chapter 4.   Finding a square root $(X^2 = A)$ or a logarithm $(e^X = A)$ of a matrix are important instances in which one wants to solve a nonlinear matrix equation.

A matrix $A(t) = [a_{ij}(t)]$ whose entries are functions of a scalar parameter t is said to be a continuous, integrable, or differentiable function of t if all its entries are, respectively, continuous, integrable, or differentiable functions of t.   In Section (6.5) we show how elementary calculus with matrix functions leads to some remarkable matrix identities and inequalities.

Because of noncommutativity, the straightforward expression for the derivative of a polynomial function of a matrix $A(t)$, though easily calculated, has a complicated form.   In Section (6.6) we derive a general formula for the derivative of $f(A(t))$, where $f(\cdot)$ is a sufficiently differentiable function with suitable domain and $A(t)$ is a differentiably parametrized family of matrices.   When $A(t)$ is diagonalizable, this formula has a simple form involving the Hadamard product, and we use it to characterize matrix functions that are analogs of ordinary monotone and convex real-valued functions.

6.1   Polynomial matrix functions and interpolation

In the context of matrix functions, there are several natural notions of *polynomial* and associated notions of *polynomial function of a matrix*.

If $p(t) = a_m t^m + a_{m-1} t^{m-1} + \cdots + a_1 t + a_0$ is an ordinary scalar-valued polynomial with given coefficients a_0, a_1,..., $a_m \in \mathbb{C}$, then for any $A \in M_n$ one can define

$$p(A) \equiv a_m A^m + a_{m-1} A^{m-1} + \cdots + a_1 A + a_0 I \qquad (6.1.1)$$

The minimal and characteristic polynomials are polynomial functions of a matrix of this kind.

On the other hand, if one has a matrix-valued polynomial

$$P(t) \equiv A_m t^m + A_{m-1} t^{m-1} + \cdots + A_1 t + A_0 \tag{6.1.2}$$

with given matrix coefficients $A_0, A_1, ..., A_m \in M_n$, then for any $A \in M_n$ there are two equally natural definitions of $P(A)$:

$$P_l(A) \equiv A_m A^m + A_{m-1} A^{m-1} + \cdots + A_1 A + A_0 \tag{6.1.3a}$$

or

$$P_r(A) \equiv A^m A_m + A^{m-1} A_{m-1} + \cdots + A A_1 + A_0 \tag{6.1.3b}$$

Unless A commutes with all the coefficients $A_0, A_1, ..., A_m$, $P_l(A)$ and $P_r(A)$ may be different. Polynomials with matrix coefficients arise naturally when one is dealing with matrices whose entries are themselves polynomials. For example, the matrix adj($tI - A$), which arises in the proof of the Cayley-Hamilton theorem for matrices whose entries come from a commutative ring (see Problem 3 in Section (2.4) of [HJ]), is of the form (6.1.2).

Whether one is dealing with polynomials that have scalar or matrix coefficients, the evaluation of $p(A)$, $P_l(A)$, or $P_r(A)$ can often be simplified, especially if the degree m of the polynomial is large. Because the powers of a given matrix $A \in M_n$ greater than $n-1$ can always be expressed as a linear combination of lower powers (this is one consequence of the Cayley-Hamilton theorem), polynomial functions of the form (6.1.1) and (6.1.3a,b) can always be reformulated so as to involve at most $A, A^2, ..., A^{n-1}$; the reformulation depends on the matrix A, however.

If $g(t) = t^k + b_{k-1} t^{k-1} + \cdots + b_1 t + b_0$ is any monic polynomial of degree $k \le m$, the Euclidean algorithm (polynomial long division) can be used to write

$$p(t) = g(t)q(t) + r(t)$$

where $q(t)$ (the quotient) and $r(t)$ (the remainder) are uniquely determined polynomials, the degree of $q(t)$ is $m - k$, and the degree of $r(t)$ is at most $k-1$. If the polynomial $g(t)$ is chosen so that $g(A) = 0$ (and hence $g(\cdot)$ depends upon A), then $p(A) = g(A)q(A) + r(A) = r(A)$. One can always choose $g(t)$ to be the characteristic polynomial of A, so the degree of $r(t)$ never needs to be greater than $n-1$, but if the objective is to obtain a polynomial $r(t)$ of smallest degree to evaluate, then the best one can do is to choose $g(t)$ to be the minimal polynomial of A. We emphasize that since

$g(t)$ has to be chosen to annihilate A, the remainder $r(t)$ depends on A. If $m \geq n$, there is no single polynomial $r(t)$ of degree less than n such that $p(A) = r(A)$ for all $A \in M_n$.

Thus, to evaluate a polynomial function of the form (6.1.1) for $A \in M_n$, one has to deal (at least, in principle) only with polynomials of degree at most $n-1$.

If $A \in M_n$ is diagonalizable and a diagonalization $A = S\Lambda S^{-1}$ with $\Lambda = \mathrm{diag}(\lambda_1, \lambda_2, ..., \lambda_n)$ is known, then (6.1.1) assumes the simple form

$$p(A) = p(S\Lambda S^{-1}) = Sp(\Lambda)S^{-1} = S \begin{bmatrix} p(\lambda_1) & & 0 \\ & \ddots & \\ 0 & & p(\lambda_n) \end{bmatrix} S^{-1} \quad (6.1.4)$$

In deriving this formula, we have used the important property of scalar polynomial functions $p(t)$ that $p(SAS^{-1}) = Sp(A)S^{-1}$; this property is shared by scalar analytic functions but not by other functions such as matrix-valued polynomials.

If A is not necessarily diagonalizable and has *Jordan canonical form* $A = SJS^{-1}$ with

$$J = \begin{bmatrix} J_{n_1}(\lambda_1) & & 0 \\ & \ddots & \\ 0 & & J_{n_s}(\lambda_s) \end{bmatrix} \quad (6.1.5)$$

as in (3.1.12) in [HJ], then

$$p(A) = p(SJS^{-1}) = Sp(J)S^{-1} = S \begin{bmatrix} p(J_{n_1}(\lambda_1)) & & 0 \\ & \ddots & \\ 0 & & p(J_{n_s}(\lambda_s)) \end{bmatrix} S^{-1} \quad (6.1.6)$$

Moreover, because of the special structure of a k-by-k Jordan block $J_k(\lambda)$, there is a simple general formula for $p(J_k(\lambda))$. Write $J_k(\lambda) = \lambda I + N$, where $N \equiv J_k(0)$. Then

$$J_k(\lambda)^j = (\lambda I + N)^j = \sum_{i=0}^{j} \binom{j}{i} \lambda^{j-i} N^i$$

and all terms with $i \geq k$ are zero because $N^k = 0$. This gives

$$p(J_k(\lambda)) = \sum_{j=0}^{m} a_j J_k(\lambda)^j = \sum_{j=0}^{m} \sum_{i=0}^{j} a_j \begin{bmatrix} j \\ i \end{bmatrix} \lambda^{j-i} N^i$$

$$= \sum_{i=0}^{m} \left\{ \sum_{j=i}^{m} \begin{bmatrix} j \\ i \end{bmatrix} a_j \lambda^{j-i} \right\} N^i = \sum_{i=0}^{m} \frac{1}{i!} \left\{ \sum_{j=i}^{m} \frac{j!}{(j-i)!} a_j \lambda^{j-i} \right\} N^i$$

or

$$p(J_k(\lambda)) = \sum_{i=0}^{m} \frac{1}{i!} p^{(i)}(\lambda) N^i$$

$$= \sum_{i=0}^{\mu} \frac{1}{i!} p^{(i)}(\lambda) N^i, \qquad \mu \equiv \min \{m, k-1\} \quad (6.1.7)$$

$$= \begin{bmatrix} p(\lambda) & p'(\lambda) & \tfrac{1}{2}p''(\lambda) & \cdot & \cdot & \cdot & \cdot & \cdot & \frac{1}{(k-1)!} p^{(k-1)}(\lambda) \\ 0 & p(\lambda) & p'(\lambda) & \cdot & & & & & \vdots \\ 0 & 0 & p(\lambda) & \cdot & & \cdot & & & \vdots \\ \cdot & \cdot & & & \cdot & & \cdot & & \vdots \\ \cdot & \cdot & & & & \cdot & & & \tfrac{1}{2}p''(\lambda) \\ \cdot & \cdot & & & & & \cdot & & p'(\lambda) \\ 0 & 0 & & & & & & & p(\lambda) \end{bmatrix} \quad (6.1.8)$$

in which all entries in the ith superdiagonal are $p^{(i)}(\lambda)/i!$, the normalized ith derivative. Only derivatives up to order $k-1$ are required.

An important, and perhaps somewhat surprising, consequence of the formulae (6.1.6) and (6.1.8) is that, for any polynomial $p(t)$, the value of $p(A)$ is determined solely by the values of $p(t)$ and certain of its derivatives at the eigenvalues of A.

6.1.9   Theorem. Let $A \in M_n$ be given and let $p(t)$ and $r(t)$ be given polynomials.

(a)   The value for $p(A)$ given by the polynomial function of a matrix (6.1.1) is the same as the value given by the formulae (6.1.6-8).

(b)   Let A have minimal polynomial $q_A(t) = (t-\lambda_1)^{r_1} \cdots (t-\lambda_\mu)^{r_\mu}$ with $\lambda_1, \ldots, \lambda_\mu$ distinct.   Then $r(A) = p(A)$ if and only if $r^{(u)}(\lambda_i) = p^{(u)}(\lambda_i)$ for $u = 0, 1, \ldots, r_i - 1$ and $i = 1, \ldots, \mu$.

Any polynomial $r(t)$ that satisfies the conditions in (6.1.9(b)) is said to *interpolate* $p(t)$ *and its derivatives at the roots of* $q_A(t) = 0$. If $g(t) = (t - t_1)^{s_1} \cdots (t - t_\nu)^{s_\nu}$ is any polynomial that annihilates A (for example, $g(t)$ might be the characteristic polynomial of A), then the minimal polynomial $q_A(t)$ divides $g(t)$. If $r(t)$ is a polynomial that interpolates $p(t)$ and its derivatives at the roots of $g(t) = 0$, that is, $r^{(u)}(t_i) = p^{(u)}(t_i)$ for $u = 0, 1, \ldots, s_i - 1$ and $i = 1, \ldots, \nu$, then $r(t)$ interpolates $p(t)$ and its derivatives on the roots of $q_A(t) = 0$ (and satisfies some additional conditions as well), so $r(A) = p(A)$. This observation suggests that interpolation may play a useful role in evaluating $p(A)$, an idea to which we return at the end of this section.

If the Jordan canonical form (6.1.5) of A is known and if the derivatives $p'(t)$, $p''(t)$, etc., of $p(t)$ have been calculated, then (6.1.6) is easily written down by inspection using (6.1.8). The ith block contains $p(\lambda_i)$ on the main diagonal, $p'(\lambda_i)$ on the first superdiagonal, $\frac{1}{2}p''(\lambda_i)$ on the second superdiagonal, and so on until the single entry $p^{(n_i - 1)}(\lambda_i)/(n_i - 1)!$ appears in the upper right corner of the n_i-by-n_i diagonal block corresponding to $J_{n_i}(\lambda_i)$. Notice that one needs to know derivatives of $p(t)$ only up to one less than the size of the largest block in the Jordan canonical form of A. This method of calculating $p(A)$ is the basis of a method to calculate nonpolynomial functions of a matrix, which we discuss in the next section.

So far, we have been concentrating on how to define and compute a polynomial function of a matrix, but we also need to know about analytic properties such as continuity and differentiability. Because of noncommutativity of M_n, some care may be required to deduce these properties from the corresponding properties of scalar polynomials. The following theorem takes care of continuity; we discuss differentiability in Sections (6.5-6).

6.1.10   **Theorem.** Let $\|\|\cdot\|\|$ be a given matrix norm on M_n.

(a)   Let $p(t) = a_m t^m + \cdots + a_1 t + a_0$ be a given scalar-valued polynomial, and let $p_{abs}(t) \equiv |a_m| t^m + \cdots + |a_1| t + |a_0|$. If $p(A)$ is defined by (6.1.1), then

$$\|\| p(A + E) - p(A) \|\| \le p_{abs}(\|\| A \|\| + \|\| E \|\|) - p_{abs}(\|\| A \|\|)$$

$$\le \|\| E \|\| \, p'_{abs}(\|\| A \|\| + \|\| E \|\|) \qquad (6.1.11)$$

for all $A, E \in M_n$.

(b) Let $P_l(A)$ and $P_r(A)$ be polynomials with matrix coefficients given by (6.1.3a-b), and let $P_{abs}(t) \equiv \| A_m \| t^m + \cdots + \| A_1 \| t + \| A_0 \|$. Then

$$\left. \begin{array}{l} \| P_l(A + E) - P_l(A) \| \\[2ex] \| P_r(A + E) - P_r(A) \| \end{array} \right\} \leq P_{abs}(\| A \| + \| E \|) - P_{abs}(\| A \|)$$

$$\leq \| E \| P'_{abs}(\| A \| + \| E \|) \quad (6.1.12)$$

In particular, $p(A)$, $P_l(A)$, and $P_r(A)$ are all continuous functions on M_n that are Lipschitz continuous on each compact subset K of M_n, that is, for each such K there is some positive $L = L(K)$ such that $\| f(B) - f(A) \| \leq L \| B - A \|$ for all $A, B \in K$.

Proof: The basic fact common to all three cases is the noncommutative binomial expansion for $(A + E)^k$, which is easily illustrated for $k = 3$:

$$(A + E)^3 = A^3 + (A^2 E + AEA + EA^2) + (AE^2 + EAE + E^2 A) + E^3$$

Now subtract A^3, take the norm, and use the triangle inequality and submultiplicativity:

$$\| (A + E)^3 - A^3 \| \leq 3 \| A \|^2 \| E \| + 3 \| A \| \| E \|^2 + \| E \|^3$$

$$= (\| A \| + \| E \|)^3 - \| A \|^3$$

A simple induction shows that

$$\| (A + E)^k - A^k \| \leq (\| A \| + \| E \|)^k - \| A \|^k \quad (6.1.13)$$

for all $k = 1, 2, \ldots$, and hence

$$\| p(A + E) - p(A) \| \leq \sum_{k=0}^{m} |a_k| \left[(\| A \| + \| E \|)^k - \| A \|^k \right]$$

$$= p_{abs}(\| A \| + \| E \|) - p_{abs}(\| A \|)$$

$$= \| E \| \, p'_{abs}(\xi)$$

$$\leq \| E \| \, p'_{abs}(\| A \| + \| E \|)$$

as asserted, where ξ is some point in the interval $[\| A \|, \| A \| + \| E \|]$. The mean value theorem has been employed for the final equality and inequality, where we note that $p'_{abs}(t)$ is an increasing function since it is a polynomial with nonnegative coefficients. Continuity of $p(A)$ now follows from ordinary continuity of the scalar polynomial $p_{abs}(t)$. The assertion about Lipschitz continuity on a compact subset $K \subset M_n$ follows from taking $A, B \in K$, $E \equiv B - A$, and noting that

$$\| f(A + [B - A]) - f(A) \| \leq \| B - A \| \, p'_{abs}(\| A \| + \| B - A \|)$$

$$\leq L \| B - A \|$$

where $L \equiv \max \{ p'_{abs}(\| A \| + \| B - A \|) : A, B \in K \}$. A similar argument proves the assertions in (b). ☐

For the final topic in this section, we return to the problem of determining, for a given $A \in M_n$ and a given scalar-valued polynomial $p(t)$ of degree $\geq n$, a polynomial $r(t)$ of degree $\leq n-1$ such that $p(A) = r(A)$. We have already observed that *any* polynomial $r(t)$ that interpolates $p(t)$ and certain of its derivatives on the spectrum of A will have this property. Specifically, if $g(t)$ is a given polynomial that annihilates A (and hence is divisible by the minimal polynomial of A), and if $g(t)$ is given in factored form as

$$g(t) = \prod_{i=1}^{\mu} (t - \lambda_i)^{s_i}, \quad \lambda_1, \ldots, \lambda_\mu \in \mathbb{C} \text{ distinct}, \, s_1 + \cdots + s_\mu = k, \text{ all } s_i \geq 1$$

then we require a polynomial $r(t)$ such that $r^{(u)}(\lambda_i) = p^{(u)}(\lambda_i)$ for $u = 0, 1, \ldots, s_i - 1$ and $i = 1, \ldots, \mu$. Is there a polynomial of degree $\leq k-1$ that satisfies these conditions? If so, how can it be constructed and how does it depend on the data?

What we have here is a special case of an interpolation problem that arises again in subsequent sections, so it is worthwhile to examine it carefully and record some important facts.

6.1.14   The Lagrange-Hermite interpolation problem. Let $\lambda_1, \ldots, \lambda_\mu$ be μ given distinct real or complex numbers. For each $i = 1, \ldots, \mu$ let s_i be a given positive integer and let $s_1 + \cdots + s_\mu = k$. Let $g(t) \equiv (t - \lambda_1)^{s_1} \cdots (t - \lambda_\mu)^{s_\mu}$. Let $f(\cdot)$ be a given real- or complex-valued function that is defined and has derivatives up to order $s_i - 1$ at λ_i, $i = 1, \ldots, \mu$. The *Lagrange-Hermite interpolation problem* is to find a polynomial $r(t)$ of degree $\leq k - 1$ such that $r^{(u)}(\lambda_i) = f^{(u)}(\lambda_i)$ for $u = 0, 1, \ldots, s_i - 1$ and $i = 1, \ldots, \mu$. Such a polynomial $r(t)$ is said to *interpolate $f(t)$ and its derivatives at the roots of $g(t) = 0$.*

In the context in which the interpolation problem arises in this section, $f(t)$ is itself a polynomial, but this plays no essential role in constructing a solution and we shall later wish to consider the interpolation problem (6.1.14) for nonpolynomial functions.

Does the Lagrange-Hermite interpolation problem have a solution? It is easy to see from general principles that it must. Let $r(t) = \alpha_{k-1} t^{k-1} + \cdots + \alpha_1 t + \alpha_0$ be a polynomial of degree $\leq k - 1$ with unknown coefficients α_0, $\alpha_1, \ldots, \alpha_{k-1}$. Use the notation in (6.1.14) and notice that the homogeneous interpolation problem

$$r^{(u)}(\lambda_i) = 0 \text{ for } u = 0, 1, \ldots, s_i - 1 \text{ and } i = 1, \ldots, \mu \qquad (6.1.14a)$$

is a system of k homogeneous linear equations for the k unknowns $\alpha_0, \alpha_1, \ldots, \alpha_{k-1}$. If it has a nontrivial solution, these equations assert that the polynomial $r(t) = \alpha_{k-1} t^{k-1} + \cdots + \alpha_1 t + \alpha_0$ of degree $\leq k - 1$ is not identically zero and has *at least* k zeroes, counting multiplicities. But this contradicts the assertion of the fundamental theorem of algebra that $r(t)$ can have *at most* $k - 1$ zeroes and forces us to conclude that the homogeneous interpolation problem (6.1.14a) has only the obvious solution $r(t) \equiv 0$, so $\alpha_0 = \alpha_1 = \cdots = \alpha_{k-1} = 0$. The standard alternative for systems of linear equations (Section (0.5) of [HJ]) now guarantees that the nonhomogeneous system

$$r^{(u)}(\lambda_i) = f^{(u)}(\lambda_i) \text{ for } u = 0, 1, \ldots, s_i - 1 \text{ and } i = 1, \ldots, \mu$$

of linear equations for the unknown coefficients $\alpha_0, \alpha_1, \ldots, \alpha_{k-1}$ has a unique solution for each given set of values $f^{(u)}(\lambda_i)$ to be interpolated. Thus, *the Lagrange-Hermite interpolation problem always has a solution, and it is unique.* It is convenient to have available several representations for the interpolating polynomial that solves the problem (6.1.14).

If all $s_i = 1$, then $\mu = k$ and the distinct interpolation points λ_i all have multiplicity one. *Lagrange's formula* for the solution in this case is

$$r(t) = \sum_{i=1}^{k} f(\lambda_i) \prod_{\substack{j=1 \\ j \neq i}}^{k} \frac{t - \lambda_j}{\lambda_i - \lambda_j} = \sum_{i=1}^{k} f(\lambda_i) \frac{\psi(t)}{\psi'(\lambda_i)(t - \lambda_i)}$$

$$= \sum_{i=1}^{k} \varphi_i(\lambda_i) \frac{\psi(t)}{(t - \lambda_i)} \tag{6.1.15}$$

where $\psi(t) \equiv (t - \lambda_1) \cdots (t - \lambda_k)$ and $\varphi_i(t) \equiv f(t)(t - \lambda_i)/\psi(t)$. Notice that the polynomial $r(t)$ is invariant under permutations of the given points $\lambda_1, \ldots, \lambda_k$.

If some $s_i \geq 1$, the *Lagrange-Hermite formula* for the solution is

$$r(t) = \sum_{i=1}^{\mu} \left\{ \left[\sum_{u=0}^{s_i - 1} \frac{1}{u!} \frac{d^u}{dt^u} \frac{f(t)}{\prod_{\substack{v=1 \\ v \neq i}}^{\mu} (t - \lambda_v)^{s_v}} \bigg|_{t = \lambda_i} (t - \lambda_i)^u \right] \left[\prod_{\substack{j=1 \\ j \neq i}}^{\mu} (t - \lambda_j)^{s_j} \right] \right\}$$

$$= \sum_{i=1}^{\mu} \left\{ \left[\sum_{u=0}^{s_i - 1} \frac{1}{u!} \varphi_i^{(u)}(\lambda_i)(t - \lambda_i)^u \right] \frac{\psi(t)}{(t - \lambda_i)^{s_i}} \right\} \tag{6.1.16}$$

where $\psi(t) \equiv (t - \lambda_1)^{s_1} \cdots (t - \lambda_\mu)^{s_\mu}$ and $\varphi_i(t) \equiv f(t)(t - \lambda_i)^{s_i}/\psi(t)$. Notice that this reduces to the Lagrange formula (6.1.15) if all $s_i = 1$.

There is a third representation for the interpolation polynomial $r(t)$ that includes both (6.1.15-16) as special cases and makes it unnecessary to distinguish between the cases of distinct and coalescent interpolation points. It utilizes *divided differences* of the function $f(\cdot)$, which are defined as follows: Let $t_1, \ldots, t_k$ be k given real or complex numbers, which we will ultimately permit to have repeated values but which for the time being we assume are *distinct*. Set

$$\Delta^0 f(t_1) \equiv f(t_1) \text{ and } \Delta f(t_1, t_2) \equiv \Delta^1 f(t_1, t_2) \equiv \frac{f(t_1) - f(t_2)}{t_1 - t_2} \tag{6.1.17a}$$

and define inductively

$$\Delta^i f(t_1,\ldots, t_i, t_{i+1}) \equiv \frac{\Delta^{i-1} f(t_1,\ldots, t_{i-1}, t_i) - \Delta^{i-1} f(t_1,\ldots, t_{i-1}, t_{i+1})}{t_i - t_{i+1}},$$

$$i = 1,\ldots, k-1 \qquad (6.1.17b)$$

Then *Newton's formula*

$$r(t) = f(t_1) + \Delta f(t_1,t_2)(t-t_1) + \Delta^2 f(t_1,t_2,t_3)(t-t_1)(t-t_2) + \cdots$$

$$+ \Delta^{k-1} f(t_1,\ldots, t_k)(t-t_1) \cdots (t-t_{k-1}) \qquad (6.1.18)$$

produces a polynomial $r(t)$ of degree $\leq k-1$. Notice that $r(t_1) = f(t_1)$, and $r(t_2) = f(t_1) + \Delta f(t_1,t_2)(t_2-t_1) = f(t_1) - [f(t_1)-f(t_2)] = f(t_2)$. As shown in Problem 18, $r(t_i) = f(t_i)$ for *all* $i = 1,\ldots, k$, so Newton's formula (6.1.18) and Lagrange's formula (6.1.15) both represent exactly the same interpolating polynomial $r(t)$ when the points $t_1,\ldots, t_k$ are distinct and $t_i = \lambda_i$, $i = 1,\ldots, k$. In particular, the coefficient of t^{k-1} in both formulae must be the same, so we have the identity

$$\Delta^{k-1} f(t_1,\ldots, t_k) = \sum_{i=1}^{k} f(t_i) \prod_{\substack{j=1 \\ j \neq i}}^{k} (t_i - t_j)^{-1} \qquad (6.1.19)$$

Thus, each of the divided differences defined by (6.1.17) is a symmetric and continuous function of its arguments, so long as the points $t_1,\ldots, t_k$ are distinct and $f(\cdot)$ is continuous.

In order to relate the Newton formula (6.1.18) to the Lagrange–Hermite formula (6.1.16), we must consider how to define the divided differences (6.1.17) when some of their arguments are allowed to coalesce. Since the divided differences are continuous functions of their arguments when the arguments are *distinct*, it is clear that there is at most one way to define them so that they are continuous for *all* values of their arguments. We now show how to do so under two broad sets of hypotheses on $f(\cdot)$ and its domain.

First suppose that $f(\cdot)$ is analytic in a simply connected open set $D \subset \mathbb{C}$ that contains $t_1,\ldots, t_k$ as interior points. If $\Gamma \subset D$ is a simple rectifiable curve that encloses the points $t_1, ..., t_k$, the Cauchy integral theorem gives

$$\frac{1}{2\pi i}\oint_\Gamma \frac{f(t)\ dt}{(t-t_1)\ \cdots\ (t-t_k)} = \sum_{i=1}^k f(t_i) \prod_{\substack{j=1 \\ j \neq i}}^k (t_i - t_j)^{-1} \tag{6.1.20}$$

when the points $t_1, ..., t_k$ are distinct; by (6.1.19), this is also equal to the divided difference $\Delta^{k-1}f(t_1, ..., t_k)$. Since the left-hand side of (6.1.20) is a continuous (even analytic) function of $t_1, ..., t_k$, it provides a representation for the divided difference for arbitrary points $t_1, ..., t_k \in D$, distinct or not:

$$\Delta^{k-1}f(t_1, ..., t_k) = \frac{1}{2\pi i}\oint_\Gamma \frac{f(t)\ dt}{(t-t_1)\ \cdots\ (t-t_k)} \tag{6.1.21}$$

Permutation-invariance of this expression with respect to the k variables $t_1, ..., t_k$ is evident, as is its value when all the points $t_1, ..., t_k$ are the same:

$$\Delta^{k-1}f(\lambda, ..., \lambda) = \frac{1}{2\pi i}\oint_\Gamma \frac{f(t)\ dt}{(t-\lambda)^k} = \frac{1}{(k-1)!}f^{(k-1)}(\lambda) \tag{6.1.22}$$

Sometimes one needs to interpolate a function $f(\cdot)$ that is not necessarily analytic, but is defined on a real interval and is (sufficiently often) differentiable there as a function of a single real variable. There is a representation for the divided differences of $f(\cdot)$ that exhibits their continuity in this case as well. Let $f(\cdot)$ be a real- or complex-valued function that is $(k-1)$-times continuously differentiable on a convex set $D \subset \mathbb{C}$ that contains $t_1, ..., t_k$ in its relative interior; if all the points $t_1, ..., t_k$ are collinear (in particular, if they are all real), D may be a line segment containing them. Consider the $(k-1)$-fold iterated integral

$$\phi_{k-1}(t_1, ..., t_k) \equiv$$

$$\int_0^1 \cdots \int_0^{\tau_{k-3}} \int_0^{\tau_{k-2}} f^{(k-1)}[t_1 + (t_2 - t_1)\tau_1 + \cdots + (t_k - t_{k-1})\tau_{k-1}]\ d\tau_{k-1} \cdots d\tau_1$$

$$\tag{6.1.23}$$

which is well defined because the argument of the integrand is a convex combination of $t_1,..., t_k$. If the points $t_1,..., t_k$ are distinct, one integration gives the identity

$$\phi_{k-1}(t_1,..., t_k) = \frac{\phi_{k-2}(t_1,..., t_{k-2}, t_k) - \phi_{k-2}(t_1,..., t_{k-1})}{t_k - t_{k-1}}$$

which can be recursively computed for $k - 3, k - 4,...,$ and one checks that

$$\phi_1(t_1, t_2) = \frac{f(t_1) - f(t_2)}{t_1 - t_2}$$

Comparison of these identities with the definition (6.1.17) of the divided differences shows that $\phi_{k-1}(t_1,..., t_k) = \Delta^{k-1} f(t_1,..., t_k)$ when the points $t_1,..., t_k$ are distinct. Since the right-hand side of (6.1.23) is a continuous function of $t_1,..., t_k \in D$, it provides a representation for the divided differences for arbitrary $t_1,..., t_k \in D$, distinct or not:

$$\Delta^{k-1} f(t_1,..., t_k) =$$

$$\int_0^1 \cdots \int_0^{\tau_{k-3}} \int_0^{\tau_{k-2}} f^{(k-1)}[t_1 + (t_2 - t_1)\tau_1 + \cdots + (t_k - t_{k-1})\tau_{k-1}] \, d\tau_{k-1} \cdots d\tau_1$$

$$(6.1.24)$$

If this iterated integral is thought of as a multiple integral over the convex hull of $t_1,..., t_k$, it is evident that its value is independent of the ordering of the points $t_1,..., t_k$, and hence the same is true of the divided difference $\Delta^{k-1} f(t_1,..., t_k)$. Moreover, if $t_1 = \cdots = t_k = \lambda$, we have

$$\Delta^{k-1} f(\lambda,..., \lambda) = \int_0^1 \cdots \int_0^{\tau_{k-3}} \int_0^{\tau_{k-2}} f^{(k-1)}(\lambda) \, d\tau_{k-1} \cdots d\tau_1$$

$$= \frac{1}{(k-1)!} f^{(k-1)}(\lambda) \qquad (6.1.25)$$

We now summarize what we have learned about polynomial interpolation.

6.1.26   Theorem. Let k be a given positive integer, let $f(\cdot)$ be a given scalar-valued function, and assume that either

(a) $f(\cdot)$ is analytic on a simply-connected open set $D \subset \mathbb{C}$; set $\mathcal{D} \equiv D$,

or

(b) $f(\cdot)$ is $(k-1)$-times continuously differentiable on a convex set $D \subset \mathbb{C}$; let $\mathcal{D}$ denote the relative interior of D.

In either case, for $t_1, \dots, t_k \in \mathcal{D}$ let the polynomial $r(t)$ be defined by the Newton formula

$$r(t) \equiv f(t_1) + \Delta f(t_1, t_2)(t - t_1) + \Delta^2 f(t_1, t_2, t_3)(t - t_1)(t - t_2) + \cdots$$

$$+ \Delta^{k-1} f(t_1, \dots, t_k)(t - t_1) \cdots (t - t_{k-1}) \qquad (6.1.27a)$$

where the divided differences are defined by (6.1.21) in case (a) and by (6.1.24) in case (b). Write

$$r(t) = \alpha_{k-1}(t_1, \dots, t_k) t^{k-1} + \cdots + \alpha_1(t_1, \dots, t_k) t + \alpha_0(t_1, \dots, t_k) \qquad (6.1.27b)$$

Then

(1)   All the divided differences $\Delta^{i-1} f(t_1, \dots, t_i)$ and the coefficients $\alpha_i(t_1, \dots, t_k)$, $i = 0, 1, \dots, k-1$, are continuous permutation-invariant functions of $t_1, \dots, t_k \in \mathcal{D}$.

(2)   For a given choice of the points $t_1, \dots, t_k$, let $\lambda_1, \dots, \lambda_\mu$ denote their distinct values, where λ_i has multiplicity s_i, $i = 1, \dots, \mu$, and $s_1 + \cdots + s_\mu = k$. Then $r^{(u)}(\lambda_i) = f^{(u)}(\lambda_i)$ for $u = 0, 1, \dots, s_i - 1$ and $i = 1, \dots, \mu$, so $r(t)$ is the solution of the Lagrange-Hermite interpolation problem (6.1.14).

Proof: Continuity of the coefficients of $r(t)$ follows from expanding the Newton formula (6.1.27a), collecting terms, and using the fact that the divided differences are continuous under either of the two hypotheses on $f(\cdot)$ and its domain. We already know that the divided differences are permutation-invariant functions of their arguments.

   Although the formulae (6.1.27a,b) for $r(t)$ seem to depend on the ordering of the points $t_1, \dots, t_k$, we now argue that, in fact, it does not. This assertion is evidently true when the points $t_1, \dots, t_k$ are distinct, for we know in both cases (a) and (b) that $r(t)$ is exactly the Lagrange interpolating polynomial (6.1.15), which is invariant under permutations of $t_1, ..., t_k$. By continuity of the coefficients of $r(t)$, invariance of $r(t)$ under permutations of $t_1, ..., t_k$ must be preserved even if some or all of these points coalesce.

Since we may arrange the points $t_1, ..., t_k$ in any convenient order without changing $r(t)$, first let $t_1 = \cdots = t_{s_1} = \lambda_1$ and use (6.1.22) and (6.1.25) to write

$$r(t) = f(\lambda_1) + f'(\lambda_1)(t - \lambda_1) + \cdots$$
$$+ \frac{1}{(s_1 - 1)!} f^{(s_1-1)}(t - \lambda_1)^{s_1-1} + R(t)(t - \lambda_1)^{s_1}$$

where $R(t)$ is a polynomial. This representation of $r(t)$ makes it clear that $r^{(u)}(\lambda_1) = f^{(u)}(\lambda_1)$ for $u = 0, 1, ..., s_1 - 1$. Each of the distinct values of λ_i may now be listed first in this same way to show that $r(t)$ interpolates $f(t)$ and its derivatives to the prescribed order at all of the distinct points $\lambda_1, ..., \lambda_\mu$. $\square$

We shall make frequent use of the preceding theorem when $k = n$ and the interpolation points $t_1, ..., t_n$ are the eigenvalues $\lambda_1, ..., \lambda_n$ of a matrix $A \in M_n$. Since the eigenvalues of A depend continuously on A (see Appendix D of [HJ]), the coefficients $\alpha_i(\lambda_1, ..., \lambda_n)$ of the interpolating polynomial given by (6.1.27b) also depend continuously on A. These coefficients are permutation-invariant functions of $\lambda_1, ..., \lambda_n$, and it is convenient to think of them as functions of A itself.

6.1.28   Corollary. Let n be a given positive integer, let $f(\cdot)$ be a given scalar-valued function, and let $A \in M_n$ have eigenvalues $\lambda_1, ..., \lambda_n$ (including algebraic multiplicities) and characteristic polynomial $p_A(t)$. Assume that either

(a) $f(\cdot)$ is analytic on a simply connected open set $D \subset \mathbb{C}$; let $\mathcal{D} \equiv \{A \in M_n : \lambda_1, ..., \lambda_n \in D\}$,

or

(b) $f(\cdot)$ is $(n-1)$-times continuously differentiable on a convex set $D \subset \mathbb{C}$; let $\mathcal{D} \equiv \{A \in M_n : \lambda_1, ..., \lambda_n$ are in the relative interior of $D\}$.

In either case, for $A \in \mathcal{D}$, define the *Newton polynomial*

$$r_A(t) \equiv f(\lambda_1) + \Delta f(\lambda_1, \lambda_2)(t - \lambda_1) + \Delta^2 f(\lambda_1, \lambda_2, \lambda_3)(t - \lambda_1)(t - \lambda_2) + \cdots$$

$$+ \Delta^{n-1} f(\lambda_1, \ldots, \lambda_n)(t - \lambda_1) \cdots (t - \lambda_{n-1}) \qquad (6.1.29a)$$

where the divided differences are given by (6.1.21) in case (a) and by (6.1.24) in case (b), and write

$$r_A(t) = \alpha_{n-1}(A) t^{n-1} + \cdots + \alpha_1(A) t + \alpha_0(A) \qquad (6.1.29b)$$

Then:

(1) Each coefficient $\alpha_i(A)$, $i = 0, 1, \ldots, n-1$, is a continuous function of $A \in \mathcal{D}$.

(2) For each $A \in \mathcal{D}$, $r_A(t)$ interpolates $f(t)$ and its derivatives at the roots of $p_A(t) = 0$, that is, $r_A^{(u)}(\lambda_i) = f^{(u)}(\lambda_i)$ for $u = 0, 1, \ldots, s_i - 1$ and $i = 1, \ldots, \mu$, where s_i is the algebraic multiplicity of the eigenvalue λ_i.

(3) Let $A_\epsilon, B_\epsilon \in M_n$ be given for all ϵ in a deleted neighborhood of zero, suppose $A_\epsilon \to A$ and $B_\epsilon \to B$ as $\epsilon \to 0$, and suppose $A, A_\epsilon \in \mathcal{D}$. Then $r_{A_\epsilon}(B_\epsilon) \to r_A(B)$; in particular, $r_{A_\epsilon}(A_\epsilon) \to r_A(A)$.

Proof: The first two assertions are immediate consequences of the preceding theorem. For the third, write

$$r_{A_\epsilon}(B_\epsilon) = [r_{A_\epsilon}(B_\epsilon) - r_{A_\epsilon}(B)] + [r_{A_\epsilon}(B) - r_A(B)] + r_A(B)$$

$$= \sum_{k=0}^{n-1} \alpha_k(A_\epsilon)(B_\epsilon^k - B^k) + \sum_{k=0}^{n-1} [\alpha_k(A_\epsilon) - \alpha_k(A)] B^k + r_A(B)$$

Then (3) follows from noting that, for each $k = 0, 1, \ldots, n-1$, $\alpha_k(A_\epsilon)$ is bounded as $\epsilon \to 0$, and $B_\epsilon^k \to B^k$ and $\alpha_k(A_\epsilon) \to \alpha_k(A)$ as $\epsilon \to 0$. $\quad\Box$

Problems

1. Let $A \in M_n$ be given. If A is not diagonalizable, explain why there could not be a polynomial $g(t)$ with simple roots such that $g(A) = 0$. If $g(t)$ is a

polynomial with some multiple roots and $g(A) = 0$, must A be nondiagonalizable?

2.   Show that if $P(t)$ is a matrix-valued polynomial of the form (6.1.2), then $P(t) \equiv 0$ for all $t \in \mathbb{R}$ if and only if all coefficients $A_k = 0$.

3.   Use Problem 2 to show that if $P(t)$ and $Q(t)$ are matrix-valued polynomials of the form (6.1.2) with possibly different degrees and coefficients, then $P(t) \equiv Q(t)$ for all $t \in \mathbb{R}$ if and only if the respective degrees and coefficients are identical.

4.   The Euclidean algorithm can be used to simplify the evaluation of matrix-valued polynomials of a matrix of the forms (6.1.3a,b). Let $P(t) = A_m t^m + A_{m-1} t^{m-1} + \cdots + A_1 t + A_0$, where $A_m, \ldots, A_0 \in M_n$ are given and t is a scalar variable. Let $g(t) = t^k + b_{k-1} t^{k-1} + \cdots b_1 t + b_0$ be a given monic polynomial of degree $k \leq m$. Use the Euclidean algorithm to show that $P(t) = g(tI)Q(t) + R(t) = Q(t)g(tI) + R(t)$, where $Q(t)$ (the quotient) and $R(t)$ (the remainder) are matrix-valued polynomials that are uniquely determined by the requirement that the degree of $R(t)$ (the highest power of t that occurs) is at most $k-1$. Show that $P_l(X) = Q_l(X)g(X) + R_l(X)$ and $P_r(X) = g(X)Q_r(X) + R_r(X)$ for all $X \in M_n$, where $P_l(X)$ and $P_r(X)$ are computed from $P(t)$ by the formulae (6.1.3a,b); $Q_l(X)$ and $Q_r(X)$ (respectively, $R_l(X)$ and $R_r(X)$) are computed in the same ways from $Q(t)$ (respectively, from $R(t)$). For a given $A \in M_n$, now take $g(t)$ to be the minimal or characteristic polynomial of A, or any other polynomial that annihilates A, and conclude that $P_l(A) = R_l(A)$ and $P_r(A) = R_r(A)$. Since degree $R(t) \leq k-1 \leq m-1$, evaluating $P_r(A)$ or $P_l(A)$ in this way may require much less work than straightforward computation of $P_l(A)$ or $P_r(A)$ if $m >> k$.

5.   Consider a matrix-valued polynomial of the form

$$P(t) = A_m t^m B_m + A_{m-1} t^{m-1} B_{m-1} + \cdots + A_1 t B_1 + A_0 B_0$$

where $A_0, \ldots, A_m \in M_{p,n}$ and $B_0, \ldots, B_m \in M_{n,q}$ are given. For $A \in M_n$, define

$$\hat{P}(A) \equiv A_m A^m B_m + A_{m-1} A^{m-1} B_{m-1} + \cdots + A_1 A B_1 + A_0 B_0$$

and note that this includes both (6.1.3a,b) as special cases. If $g(t)$ is a given monic polynomial of degree $k \leq m$ that annihilates a given $A \in M_n$, investigate whether it is possible to apply the Euclidean algorithm to each term of

$P(t)$ and get $\hat{P}(A) = R(A)$, where $R(t)$ is a matrix-valued polynomial of the same form as $P(t)$, but of degree at most $k-1$. What is fundamentally different from the situation in Problem 4, where this kind of reduction is possible?

6.   Provide details for a proof of Theorem (6.1.10b).

7.   Verify the identity

$$X^k - Y^k = \sum_{m=0}^{k-1} X^m (X - Y) Y^{k-1-m} \qquad (6.1.30)$$

for any $X, Y \in M_n$. Use it to derive an alternative to the bound (6.1.13):

$$||| (A + E)^k - A^k ||| \le k ||| E ||| [\max \{ ||| A + E |||, ||| A ||| \}]^{k-1}$$

$$\le k ||| E ||| [||| A ||| + ||| E |||]^{k-1} \qquad (6.1.31)$$

Show that this is a poorer bound than (6.1.13) for all A, all $k > 1$, and all nonzero E.

8.   Use the definition of the derivative to show that $\frac{d}{dt} p(At) = A p'(At)$ for any scalar-valued polynomial $p(t)$. What are the corresponding formulae for $\frac{d}{dt} P_l(At)$ and $\frac{d}{dt} P_r(At)$ if $P_l(At)$ and $P_r(At)$ are given by (6.1.3a-b)?

9.   Consider the polynomial matrix function $P(t) = At^2 + Bt + C$, where $A, B, C \in M_n$ are all positive definite. Show that if $\det P(\lambda) = 0$, then $\mathrm{Re}\,\lambda < 0$.

10.   Show that the formula (6.1.6) reduces to the form (6.1.4) if A is diagonalizable.

11.   Let a given $A \in M_n$ have at most two distinct eigenvalues λ_1, λ_2, suppose the minimal polynomial of A has degree at most two, and let $p(t)$ be a given polynomial. Use the Newton formulae (6.1.27a,b) with $k = 2$ to compute explicitly the polynomial $r(t)$ that interpolates $p(t)$ and its derivatives at the points $t_1 = \lambda_1$ and $t_2 = \lambda_2$. Verify in this case that the coefficients of $r(t)$ are continuous functions of λ_1, λ_2 that are invariant under interchange of λ_1 and λ_2. Show that

$$p(A) = \begin{cases} \dfrac{p(\lambda_1) - p(\lambda_2)}{\lambda_1 - \lambda_2} A + \dfrac{\lambda_1 p(\lambda_2) - \lambda_2 p(\lambda_1)}{\lambda_1 - \lambda_2} I & \text{if } \lambda_1 \neq \lambda_2 \\[2mm] p'(\lambda_1)A + [p(\lambda_1) - \lambda_1 p'(\lambda_1)]I & \text{if } \lambda_1 = \lambda_2 \end{cases}$$

12. Provide details for the following constructive proof that the Lagrange-Hermite interpolation problem (6.1.14) always has a unique solution: Recall from (0.9.11) in [HJ] that the k-by-k *Vandermonde matrix* in k variables $t_1, \ldots, t_k$ is

$$V = \begin{bmatrix} 1 & t_1 & t_1^2 & \cdots & t_1^{k-1} \\ 1 & t_2 & t_2^2 & \cdots & t_2^{k-1} \\ \vdots & \vdots & \vdots & & \vdots \\ 1 & t_k & t_k^2 & \cdots & t_k^{k-1} \end{bmatrix} \tag{6.1.32}$$

and that

$$\det V = \prod_{\substack{i,j=1 \\ i>j}}^{k} (t_i - t_j) \tag{6.1.33}$$

Let $r(t) = \alpha_{k-1} t^{k-1} + \cdots + \alpha_1 t + \alpha_0$ and formulate the interpolation problem (6.1.14) as a system of k linear equations for the unknown coefficients $\alpha_0, \ldots, \alpha_{k-1}$. Show that the matrix of the resulting linear system can be obtained from the Vandermonde matrix: For $i = 1, \ldots, s_1$, differentiate row i of V $i-1$ times with respect to t_i and then set $t_1 = \cdots = t_{s_1} = \lambda_1$; for $i = 1, \ldots, s_2$, differentiate row $s_1 + i$ of V $i-1$ times with respect to t_{s_1+i} and then set $t_{s_1+1} = \cdots = t_{s_1+s_2} = \lambda_2$; proceed similarly with $\lambda_3, \ldots, \lambda_\mu$. Show that the determinant of the system is obtained from the formula (6.1.33) by performing the same sequence of differentiations and substitutions, and that the result is

$$\pm \prod_{i=1}^{\mu} \prod_{l=0}^{s_i-1} l! \prod_{i>j} (\lambda_i - \lambda_j)^{s_i s_j} \tag{6.1.34}$$

13. Show that the Lagrange-Hermite interpolation formula (6.1.16) reduces to the Lagrange interpolation formula (6.1.15) if each interpolation point λ_i

has multiplicity $s_i = 1$, $i = 1,...,\mu$.

14. Let $A \in M_n$ be given, let $p(t)$ be a given polynomial, and let $g(t) = (t-\lambda_1)^{s_1} \cdots (t-\lambda_\mu)^{s_\mu}$ be a given monic polynomial that annihilates A, where $\lambda_1,..., \lambda_\mu$ are distinct and all $s_i \geq 1$. Not all of the roots of $g(t) = 0$ need be eigenvalues of A, but of course they all are if $g(t)$ is the minimal or characteristic polynomial of A.

(a) If A is diagonalizable, use (6.1.15) to verify *Sylvester's formula*

$$p(A) = \sum_{i=1}^{\mu} p(\lambda_i) A_i \tag{6.1.35}$$

where each

$$A_i \equiv \prod_{\substack{j=1 \\ j \neq i}}^{\mu} \frac{1}{\lambda_i - \lambda_j} (A - \lambda_j I) \tag{6.1.36}$$

Notice that $A_i = 0$ if λ_i is not an eigenvalue of A, and that $\{A_i: \lambda_i \in \sigma(A), i = 1,...,\mu\}$ is determined solely by A, independently of the annihilating polynomial $g(t)$.

(b) If A is not necessarily diagonalizable, use (6.1.16) to verify *Buchheim's formula*

$$p(A) = \sum_{i=1}^{\mu} \left[\sum_{u=0}^{s_i-1} \frac{1}{u!} \varphi_i^{(u)}(\lambda_i) (A - \lambda_i I)^u \right] \prod_{\substack{j=1 \\ j \neq i}}^{\mu} (A - \lambda_j I)^{s_j} \tag{6.1.37}$$

where $\varphi_i(t) \equiv p(t) (t-\lambda_i)^{s_i}/g(t)$.   Show that Buchheim's formula reduces to Sylvester's formula (6.1.35) when all $s_i = 1$.

(c) Use a partial fractions expansion to define polynomials $h_i(t)$ by

$$\frac{1}{g(t)} = \sum_{i=1}^{\mu} \frac{h_i(t)}{(t - \lambda_i)^{s_i}} \tag{6.1.38a}$$

and use them to define the polynomials

$$q_i(t) \equiv \frac{h_i(t)g(t)}{(t-\lambda_i)^{s_i}} = h_i(t) \prod_{j \neq i} (t-\lambda_j)^{s_j}, \quad i = 1,\dots,\mu \quad (6.1.38b)$$

Show that $q_1(t) + \cdots + q_\mu(t) \equiv 1$ for all t. Use these identities to show that

$$q_i^{(u)}(\lambda_p) = 0 \quad \text{for each } p \neq i \text{ and } u = 0, 1,\dots, s_p - 1$$

and that

$$q_i(\lambda_i) = 1 \quad \text{and} \quad q_i^{(u)}(\lambda_i) = 0 \quad \text{for } u = 1,\dots, s_i - 1 \text{ and } i = 1,\dots,\mu$$

Now define the Taylor polynomials

$$w_i(t) \equiv \sum_{k=0}^{s_i-1} \frac{1}{k!} p^{(k)}(\lambda_i)(t-\lambda_i)^k, \quad i = 1,\dots,\mu$$

and show that $w_i^{(u)}(\lambda_i) = p^{(u)}(\lambda_i)$ for $u = 0, 1,\dots, s_i - 1$ and $i = 1,\dots,\mu$. Finally, let

$$r(t) \equiv \sum_{i=1}^{\mu} q_i(t) w_i(t)$$

and show that $r^{(u)}(\lambda_i) = p_i^{(u)}(\lambda_i)$ for $u = 0, 1,\dots, s_i - 1$ and $i = 1,\dots,\mu$. Explain why $p(A) = r(A)$, which is *Schwerdtfeger's formula*

$$p(A) = \sum_{i=1}^{\mu} A_i \sum_{k=0}^{s_i-1} \frac{1}{k!} p^{(k)}(\lambda_i)(A - \lambda_i I)^k \quad (6.1.39)$$

where each $A_i \equiv q_i(A)$.

(d) Continue with the same notation as in (c). Write the Jordan canonical form of A as $A = S[J(\mu_1) \oplus \cdots \oplus J(\mu_m)]S^{-1}$, where $\mu_1,\dots,\mu_m$ are the distinct eigenvalues of A, and each matrix $J(\mu_i)$ is the direct sum of all the Jordan blocks of A that have eigenvalue μ_i. Notice that $\{\mu_1,\dots,\mu_m\} \subset \{\lambda_1,\dots,\lambda_\mu\}$, but not every λ_i need be an eigenvalue of A if $g(t)$ is not the minimal or characteristic polynomial of A. Use (6.1.6-8) and the interpolation properties of the polynomials $q_i(t)$ at the points λ_j to show that

$$A_i \equiv q_i(A) = S\left[J(\mu_1;\lambda_i) \oplus \cdots \oplus J(\mu_m;\lambda_i)\right]S^{-1} \in M_n, \quad i = 1,\dots, \mu \quad (6.1.40)$$

where each matrix $J(\mu_i;\lambda_p)$ is the same size as $J(\mu_i)$, $J(\mu_i;\lambda_p) = I$ if $\lambda_p = \mu_i$, and $J(\mu_i;\lambda_p) = 0$ if $\lambda_p \ne \mu_i$. In particular, $J(\mu_i;\lambda_p) = 0$ for all $i = 1,\dots, m$ if λ_p is not an eigenvalue of A, so $A_i = 0$ in this case. Conclude that $\{A_i: \lambda_i \in \sigma(A), \ i = 1,\dots, \mu\}$, the set of *Frobenius covariants of A*, is determined solely by A independently of the choice of the annihilating polynomial $g(t)$; these are exactly the matrices A_i that are obtained when $g(t)$ is the minimal polynomial of A. Show that:

(d1) $A_1 + \cdots + A_\mu = I$;   (6.1.40.1)

(d2) $A_i A_j = 0$ if $i \ne j$;   (6.1.40.2)

(d3) $A_i^k = A_i$ for all $k = 1, 2,\dots$;   (6.1.40.3)

(d4) Each A_i is a polynomial in A, so each A_i commutes with
     A;   (6.1.40.4)

(d5) $A_i(A - \lambda_i I)^k = 0$ for all $k \ge r_i$, where r_i is the multiplic-
     ity of λ_i as a zero of the minimal polynomial of A, and
     we define $r_i \equiv 0$ if $\lambda_i \notin \sigma(A)$;   (6.1.40.5)

(d6) The Jordan canonical form of AA_i is $J(\lambda_i) \oplus 0$ if $\lambda_i \in \sigma(A)$
     and is 0 otherwise;   (6.1.40.6)

(d7) $A^k = A_1(AA_1)^k + \cdots + A_\mu(AA_\mu)^k$ for all $k = 0, 1,\dots$;   (6.1.40.7)

(d8) $q(A) = A_1 q(AA_1) + \cdots + A_\mu q(AA_\mu)$ for *any* poly-
     nomial $q(t)$.   (6.1.40.8)

(d9) Each A_j can be written as

$$A_j = \frac{1}{2\pi i} \oint_{\Gamma_j} (tI - A)^{-1}\, dt$$

where Γ_j is any simple closed rectifiable curve enclosing
a domain D_j such that $\lambda_j \in D_j$, but no other eigenvalues
of A are contained in the closure of D_j.   (6.1.40.9)

These properties play an important role in Section (6.6). Explain why

$$p(A) = \sum_{i=1}^{\mu} A_i p(AA_i) = \sum_{i=1}^{\mu} A_i \sum_{k=0}^{s_i-1} \frac{1}{k!} p^{(k)}(\lambda_i)(AA_i - \lambda_i A_i)^k$$

$$= \sum_{i=1}^{\mu} A_i \sum_{k=0}^{s_i-1} \frac{1}{k!} p^{(k)}(\lambda_i) A_i (A - \lambda_i I)^k$$

$$= \sum_{i=1}^{\mu} q_i(A) \sum_{k=0}^{s_i-1} \frac{1}{k!} p^{(k)}(\lambda_i)(A - \lambda_i I)^k$$

$$= \sum_{\substack{i=1,\ldots,\mu \\ \lambda_i \in \sigma(A)}} A_i \sum_{k=0}^{r_i-1} \frac{1}{k!} p^{(k)}(\lambda_i)(A - \lambda_i I)^k \qquad (6.1.41)$$

which gives another derivation, and a refinement, of Schwerdtfeger's formula. The last outer sum is over only those indices i such that λ_i is an eigenvalue of A; for such an index, r_i is the multiplicity of λ_i as a zero of the minimum polynomial of A. Notice that the formula (6.1.41) is exactly the same as (6.1.39) when $g(t)$ is the minimal polynomial of A. Explain how this formula illustrates the principle expressed in Theorem (6.1.9(b)). Also explain how (6.1.41) may be viewed as a way of gluing together the formulae (6.1.7). Show that Schwerdtfeger's formulae (6.1.39,41) reduce to Sylvester's formula (6.1.35) when all $s_i = 1$, and that the matrices A_i defined in (6.1.36) are the Frobenius covariants of A (as defined in general by (6.1.40)) when A is diagonalizable.

15. Schwerdtfeger's formulae (6.1.39, 41) have straightforward generalizations to polynomials in two or more *commuting* matrix variables. For example, let $p(s,t)$ be a given polynomial in two scalar variables

$$p(s,t) = \sum_{i=0}^{m_1} \sum_{j=0}^{m_2} \alpha_{ij} s^i t^j$$

and let $A, B \in M_n$ be given commuting matrices, so

$$p(A,B) = \sum_{i=0}^{m_1} \sum_{j=0}^{m_2} \alpha_{ij} A^i B^j$$

Let $\{\lambda_1,..., \lambda_\mu\}$ and $\{\eta_1,..., \eta_\nu\}$ be the distinct eigenvalues of A and B, respectively, let $g(t) = (t-\lambda_1)^{\alpha_1} \cdots (t-\lambda_\mu)^{\alpha_\mu}$ and $h(t) = (t-\eta_1)^{\beta_1} \cdots (t-\eta_\nu)^{\beta_\nu}$ be the minimal polynomials of A and B, respectively, and let $\{A_i\}_{i=1}^\mu$ and $\{B_j\}_{j=1}^\nu$ denote the Frobenius covariants of A and B, respectively, computed using (6.1.40). Show that

$$p(A,B) = \sum_{i=0}^{m_1} \sum_{j=0}^{m_2} A_i B_j \sum_{k=0}^{\alpha_i-1} \sum_{l=0}^{\beta_j-1} \frac{1}{k!l!} p_{k,l}(\lambda_i,\eta_j)(A-\lambda_i I)^k (B-\eta_j I)^l$$

where

$$p_{k,l}(\lambda_i,\eta_j) = \frac{\partial^{k+l}}{\partial s^k \partial t^l} p(s,t)\Big|_{s=\lambda_i, t=\eta_j}$$

What is the corresponding generalization of (6.1.39)?

16. (a) Show that the Lagrange polynomial given by (6.1.15) has the asserted interpolation property.
(b) Show that the Lagrange–Hermite polynomial given by (6.1.16) has the asserted interpolation property.

17. Use (6.1.16) to show that if $A \in M_n$ is any matrix with minimal polynomial $q_A(t) = (t-1)^2$, then $A^{100} = 100A - 99I$. Verify that the formula in Problem 11 gives the same result.

18. Let $t_1,..., t_k$ be distinct and let $1 \le i \le k-1$ be a given integer. Use the definition (6.1.17) of the divided differences to show that

$$\Delta^{i-1} f(t_1,..., t_{i-1}, t_{i+1}) = \Delta^{i-1} f(t_1,..., t_i) + \Delta^i f(t_1,..., t_{i+1})(t_{i+1} - t_i)$$

and, if $i \ge 2$,

$$\Delta^{i-2} f(t_1,..., t_{i-2}, t_{i+1}) = \Delta^{i-2} f(t_1,..., t_{i-1}) + \Delta^{i-1} f(t_1,..., t_i)(t_{i+1} - t_{i-1})$$
$$+ \Delta^i f(t_1,..., t_{i+1})(t_{i+1} - t_i)(t_{i+1} - t_{i-1})$$

Proceed by induction to show that

$$f(t_{i+1}) = f(t_1) + \Delta f(t_1, t_2)(t_{i+1} - t_1) + \cdots$$
$$+ \Delta^i f(t_1, \ldots, t_{i+1})(t_{i+1} - t_1) \cdots (t_{i+1} - t_i)$$

Conclude that the polynomial $r(t)$ given by the Newton formula (6.1.18) has $r(t_i) = f(t_i)$ for $i = 1, \ldots, k$ when the points $t_1, \ldots, t_k$ are distinct.

19. If some of the points $t_1, \ldots, t_k$ in (6.1.21) coalesce, let $\lambda_1, \ldots, \lambda_\mu$ denote their distinct values, with respective multiplicities $s_1, \ldots, s_\mu$, $s_1 + \cdots + s_\mu = k$. Use the Cauchy integral formula to show that, under the hypotheses for (6.1.21),

$$\Delta^{k-1} f(t_1, \ldots, t_k) = \frac{1}{2\pi i} \oint_\Gamma \frac{f(t)\, dt}{(t - \lambda_1)^{s_1} \cdots (t - \lambda_\mu)^{s_\mu}}$$

$$= \sum_{i=1}^{\mu} \frac{1}{(s_i - 1)!} \frac{d^{s_i - 1}}{dt^{s_i - 1}} \left[\frac{f(t)(t - \lambda_i)}{\psi(t)} \right] \Bigg|_{t = \lambda_i}$$

(6.1.42)

where $\psi(t) \equiv (t - \lambda_1)^{s_1} \cdots (t - \lambda_\mu)^{s_\mu}$. If $t_1 = \cdots = t_k = \lambda$, deduce that

$$\Delta^{k-1} f(\lambda, \ldots, \lambda) = \frac{1}{(k-1)!} f^{(k-1)}(\lambda)$$

20. Under the hypotheses for (6.1.24), show that

$$|\Delta^{k-1} f(t_1, \ldots, t_k)| \le \frac{1}{(k-1)!} \max \{ |f^{(k-1)}(t)| :$$

$$t \text{ is in the convex hull of } t_1, \ldots, t_k \} \quad (6.1.43)$$

If D is a real interval and $f(\cdot)$ is real-valued, show that

$$\Delta^{k-1} f(t_1, \ldots, t_k) = \frac{1}{(k-1)!} f^{(k-1)}(\tau) \text{ for some } \tau \in [\min_i t_i, \max_i t_i]$$

(6.1.44)

In particular, if $[\alpha,\beta] \subset D$ is a given compact interval, conclude that $\Delta^{k-1} f(t_1,\ldots, t_k)$ is uniformly bounded for all $t_1,\ldots, t_k \in [\alpha,\beta]$.

21. Under the hypotheses for (6.1.21), show that

$$|\Delta^{k-1} f(t_1,\ldots, t_k)| \le \frac{|\Gamma| \ \max\{|f(t)|: \ t \in \Gamma\}}{2\pi \, [\min \ \{| t \ - \ t_i|: \ t \in \Gamma, \ i = 1,\ldots, k\}]^k} \qquad (6.1.45)$$

where $|\Gamma|$ denotes the arclength of Γ. In particular, if $K \subset D$ is a given compact set, conclude that $\Delta^{k-1} f(t_1,\ldots, t_k)$ is uniformly bounded for all $t_1,\ldots, t_k \in K$.

22. Let $f : D \to \mathbb{R}$ be a given real-valued function on a real interval D. One says that $f(\cdot)$ is *convex* on D if

$$(1 - t)f(a) + tf(b) - f((1 - t)a + tb) \ge 0 \qquad (6.1.46)$$

for all $a, b \in D$ and all $t \in [0,1]$. Suppose $f(\cdot)$ is twice continuously differentiable on D. Use (6.1.44) to show that $f(\cdot)$ is convex on D if and only if $f''(\xi) \ge 0$ for all $\xi \in D$.

Further Readings. For more details, and for alternate forms of the interpolation formulae, see Chapters V and VIII of [Gan 59, Gan 86]. See Chapter I of [Gel] for a careful treatment of divided differences and representations of the solution of the Lagrange–Hermite interpolation problem.

6.2   Nonpolynomial matrix functions

If $f(t)$ is a continuous scalar-valued function of a real or complex variable that is not necessarily a polynomial, how should one define the matrix-valued function $f(A)$ for $A \in M_n$? It seems natural to require that $f(A)$ be a continuous function of A and that, just as in (6.1.4),

$$f(A) = S \begin{bmatrix} f(\lambda_1) & 0 \\ & \ddots & \\ 0 & & f(\lambda_n) \end{bmatrix} S^{-1} \qquad (6.2.1)$$

whenever A is diagonalizable and $A = S\Lambda S^{-1}$, $\Lambda = \mathrm{diag}(\lambda_1,\ldots, \lambda_n)$.

Does (6.2.1) give a well-defined value to $f(A)$, or does it depend on the ordering chosen for the eigenvalues of A or the choice of the diagonalizing similarity S? Suppose $t_1,..., t_k$ are the distinct values of the eigenvalues $\lambda_1,..., \lambda_n$ and let $A = TMT^{-1}$ be any diagonalization of A with $M = \text{diag}(\mu_1,..., \mu_n)$. Then $\mu_1,..., \mu_n$ is merely a permutation of $\lambda_1,..., \lambda_n$ and Lagrange's formula (6.1.15) may be used to construct a polynomial $r(t)$ of degree $\leq k-1$ such that $r(t_i) = f(t_i)$ for $i = 1,..., k$. Then

$$f(A) = S \operatorname{diag}(f(\lambda_1),..., f(\lambda_n)) S^{-1} = S \operatorname{diag}(r(\lambda_1),..., r(\lambda_n)) S^{-1}$$

$$= S r(\Lambda) S^{-1} = r(S\Lambda S^{-1}) = r(A) = r(TMT^{-1})$$

$$= T r(M) T^{-1} = T \operatorname{diag}(r(\mu_1),..., r(\mu_n)) T^{-1}$$

$$= T \operatorname{diag}(f(\mu_1),..., f(\mu_n)) T^{-1}$$

so the value of $f(A)$ given by (6.2.1) is independent of the diagonalization used to represent A.

Since the diagonalizable matrices are dense in M_n, there is at most one continuous way to define $f(A)$ on M_n that agrees with (6.2.1) when A is diagonalizable, and we shall show that there is actually a way to do so if $f(t)$ is smooth enough. It is perhaps surprising that the requirement that $f(A)$ be a continuous function of $A \in M_n$ implies that $f(t)$ must be a differentiable function of t when $n \geq 2$.

Suppose that the scalar-valued function $f(t)$ is defined in a neighborhood of a given point $\lambda \in \mathbb{C}$, and consider the diagonalizable matrices

$$A_\epsilon \equiv \begin{bmatrix} \lambda & 1 \\ 0 & \lambda + \epsilon \end{bmatrix}, \quad \epsilon \neq 0$$

which have $A_\epsilon = S_\epsilon \Lambda_\epsilon S_\epsilon^{-1}$ with

$$\Lambda_\epsilon = \begin{bmatrix} \lambda & 0 \\ 0 & \lambda + \epsilon \end{bmatrix}, \quad S_\epsilon = \begin{bmatrix} 1 & 1 \\ 0 & \epsilon \end{bmatrix}, \text{ and } S_\epsilon^{-1} = \frac{1}{\epsilon}\begin{bmatrix} \epsilon & -1 \\ 0 & 1 \end{bmatrix}, \quad \epsilon \neq 0$$

Then $f(A_\epsilon)$ is defined by (6.2.1) for all sufficiently small nonzero ϵ and

$$f(A_\epsilon) = S_\epsilon \begin{bmatrix} f(\lambda) & 0 \\ 0 & f(\lambda + \epsilon) \end{bmatrix} S_\epsilon^{-1} = \begin{bmatrix} f(\lambda) & [f(\lambda + \epsilon) - f(\lambda)]/\epsilon \\ 0 & f(\lambda + \epsilon) \end{bmatrix} \quad (6.2.2)$$

Since $A_\epsilon \to J_2(\lambda) = \begin{bmatrix} \lambda & 1 \\ 0 & \lambda \end{bmatrix}$ as $\epsilon \to 0$, if $f(A)$ is continuous at $A = J_2(\lambda)$ it is necessary that $f(t)$ be both continuous and differentiable at $t = \lambda$ and the value of $f(J_2(\lambda))$ must be

$$f(J_2(\lambda)) = \lim_{\epsilon \to 0} f(A_\epsilon) = \begin{bmatrix} f(\lambda) & f'(\lambda) \\ 0 & f(\lambda) \end{bmatrix}$$

Thus, for $n \geq 2$, a given open set $D \subset \mathbb{C}$, and each compact set $K \subset D$, in order for the matrix-valued function $f(A)$ defined by (6.2.1) to be a continuous function of A on the closure of the set of diagonalizable matrices in M_n whose spectra lie in K, it is necessary that the scalar-valued function $f(t)$ be analytic in D. If $n \geq 2$ and $f(A)$ is a continuous function of A on all of M_n, then $f(t)$ must be an entire analytic function.

Now suppose that $f(t)$ is analytic in a (small) disc with positive radius R centered at $t = \lambda \in \mathbb{C}$, and consider the diagonalizable matrices

$$A_\epsilon \equiv \begin{bmatrix} \lambda & 1 & & & 0 \\ & \lambda+\epsilon & 1 & & \\ & & \ddots & \ddots & \\ & & & & 1 \\ 0 & & & & \lambda+(k-1)\epsilon \end{bmatrix} \in M_k, \; \epsilon \neq 0$$

which have $A_\epsilon = S_\epsilon \Lambda_\epsilon S_\epsilon^{-1} = S_\epsilon \operatorname{diag}(\lambda, \lambda+\epsilon,..., \lambda+(k-1)\epsilon) S_\epsilon^{-1}$. For all sufficiently small $\epsilon \neq 0$, let $r_{A_\epsilon}(t)$ and $r_{J_k(\lambda)}(t)$ denote the interpolating polynomials given by Newton's formula (6.1.29) with $n = k$, so

$$r_{A_\epsilon}(\lambda + [j-1]\epsilon) = f(\lambda + [j-1]\epsilon) \text{ for } j = 1,..., k$$

and

$$r_{J_k(\lambda)}(t) = f(\lambda) + f'(\lambda)(t-\lambda) + \cdots + f^{(k-1)}(\lambda)(t-\lambda)^{k-1}/(k-1)!$$

Then (6.2.1) gives

$$f(A_\epsilon) = S_\epsilon \operatorname{diag}(f(\lambda), f(\lambda+\epsilon),..., f(\lambda+(k-1)\epsilon)) S_\epsilon^{-1}$$

$$= S_\epsilon \operatorname{diag}(r_{A_\epsilon}(\lambda), r_{A_\epsilon}(\lambda+\epsilon),..., r_{A_\epsilon}(\lambda+(k-1)\epsilon)) S_\epsilon^{-1}$$

$$= r_{A_\epsilon}(A_\epsilon)$$

Since $A_\epsilon \to J_k(\lambda)$ as $\epsilon \to 0$, Corollary (6.1.28(3)) guarantees that $r_{A_\epsilon}(A_\epsilon) \to r_{J_k(\lambda)}(J_k(\lambda))$, so continuity of $f(A)$ at $A = J_k(\lambda)$ requires that

$$f(J_k(\lambda)) = r_{J_k(\lambda)}(J_k(\lambda))$$

$$= \begin{bmatrix} f(\lambda) & f'(\lambda) & \tfrac{1}{2}f''(\lambda) & \cdot & \cdot & \cdot & \cdot & \cdot & \frac{1}{(k-1)!}f^{(k-1)}(\lambda) \\ 0 & f(\lambda) & f'(\lambda) & \cdot & & & & & \vdots \\ 0 & 0 & f(\lambda) & \cdot & & \cdot & & & \vdots \\ \cdot & \cdot & & & \cdot & & \cdot & & \\ \cdot & \cdot & & & & \cdot & & & \tfrac{1}{2}f''(\lambda) \\ \cdot & \cdot & & & & & \cdot & & f'(\lambda) \\ 0 & 0 & & & & & & & f(\lambda) \end{bmatrix}$$

$$(6.2.3)$$

Comparison of this necessary condition with (6.1.6,8) suggests the following general definition for $f(A)$.

6.2.4   Definition. Let $A \in M_n$ be a given matrix with minimal polynomial $q_A(t) = (t - \lambda_1)^{r_1} \cdots (t - \lambda_\mu)^{r_\mu}$, where $\lambda_1, ..., \lambda_\mu$ are distinct and all $r_i \geq 1$, and let A have Jordan canonical form $A = SJS^{-1}$ with

$$J = \begin{bmatrix} J_{n_1}(\lambda_{\mu_1}) & & 0 \\ & \ddots & \\ 0 & & J_{n_s}(\lambda_{\mu_s}) \end{bmatrix} \qquad (6.2.5)$$

in which $1 \leq \mu_i \leq \mu$ for $i = 1, ..., s$, each diagonal block $J_k(\lambda)$ is a k-by-k Jordan block with eigenvalue λ, and $n_1 + \cdots + n_s = n$. Let $f(t)$ be a given scalar-valued function of a real or complex variable t such that each λ_i is in the domain D of $f(\cdot)$, each λ_i with $r_i > 1$ is in the interior of D, and $f(\cdot)$ is $(r_i - 1)$-times differentiable at each λ_i for which $r_i > 1$. Then

$$f(A) \equiv Sf(J)S^{-1} \equiv S \begin{bmatrix} f(J_{n_1}(\lambda_{\mu_1})) & & 0 \\ & \ddots & \\ 0 & & f(J_{n_s}(\lambda_{\mu_s})) \end{bmatrix} S^{-1} \qquad (6.2.6)$$

where each diagonal block is defined by

$$f(J_k(\lambda)) \equiv \begin{bmatrix} f(\lambda) & f'(\lambda) & \tfrac{1}{2}f''(\lambda) & \cdot & \cdot & \cdot & \cdot & \cdot & \frac{1}{(k-1)!}f^{(k-1)}(\lambda) \\ 0 & f(\lambda) & f'(\lambda) & \cdot & & & & & \vdots \\ 0 & 0 & f(\lambda) & \cdot & \cdot & & & & \vdots \\ \cdot & \cdot & & \cdot & \cdot & \cdot & & & \vdots \\ \cdot & \cdot & & & \cdot & \cdot & & & \tfrac{1}{2}f''(\lambda) \\ \cdot & \cdot & & & & \cdot & & & f'(\lambda) \\ 0 & & 0 & & & & \cdot & & f(\lambda) \end{bmatrix} \qquad (6.2.7)$$

This matrix-valued function $f(A)$ is the *primary matrix function* associated with the scalar-valued *stem function* $f(t)$.

Notice that the definition of the primary matrix function $f(A)$, for a given $A \in M_n$, involves only *local* properties of the stem function $f(t)$ at or near the spectrum of A. For example, if $A \in M_n$ has eigenvalues of various multiplicities at the points 1 and 3, consider the stem function

$$f(t) \equiv \begin{cases} t^2 & \text{if } |t-1| < \tfrac{1}{4} \\ e^t & \text{if } |t-3| < \tfrac{1}{4} \\ t^3 & \text{if } |t-5| < \tfrac{1}{4} \end{cases}$$

Then the primary matrix function $f(A)$ is defined by (6.2.4), and the value of $f(A)$ is independent of the definition of $f(t)$ near $t = 5$; it would be unchanged if $f(t)$ were redefined to be $\cos t$ there. Notice that $f(X)$ is defined for all $X \in M_n$ whose eigenvalues are contained in the union of disjoint open discs of radius $\tfrac{1}{4}$ around 1, 3, and 5. Because of the uniqueness of analytic continuation, there is no way to extend $f(t)$ to an analytic function on a connected open set containing the points 1, 3, and 5. However, if we restrict our attention to the real line, there are infinitely many different ways to extend $f(t)$ to an infinitely differentiable (but not real-analytic) function on the real interval $(0,6)$; each such extension gives a well-defined value for $f(X)$ for any $X \in M_n$ such that $\sigma(X) \subset (0,6)$, and each has the same value at $X = A$. The fact that the definition of $f(A)$ as a primary matrix function permits enormous arbitrariness in the behavior of $f(t)$ away from the spectrum of A indicates that eventually we shall have to make some assumptions to ensure the global coherence of $f(t)$ if we hope to have $f(X)$ be a continuous function of $X \in M_n$. Initially, however, we shall study only the static properties of the primary matrix function $f(A)$ for a single given $A \in M_n$.

Theorem (6.1.9(a)) shows that the primary matrix function $f(A)$ has the same value as (6.1.1) for any $A \in M_n$ if $f(t)$ is a polynomial, and it has the same value as (6.2.1) if $A \in M_n$ is diagonalizable and $f(t)$ is not necessarily a polynomial. The following theorem shows that it also has the same value in an important intermediate case between these two extremes.

6.2.8   Theorem. Let $f(t)$ be a scalar-valued analytic function with a power series representation $f(t) = a_0 + a_1 t + a_2 t^2 + \cdots$ that has radius of convergence $R > 0$. If $A \in M_n$ and $\rho(A) < R$, the matrix power series $f(A) \equiv a_0 I + a_1 A + a_2 A^2 + \cdots$ converges with respect to every norm on M_n and its sum is equal to the primary matrix function $f(A)$ associated with the stem function $f(t)$.

Proof: Lemma (5.6.10) in [HJ] guarantees that there is a matrix norm $\|\|\cdot\|\|$ on M_n such that $\|\| A \|\| < R$. It then follows from Theorem (5.6.15) in [HJ] that the infinite series $f(A) \equiv a_0 I + a_1 A + a_2 A^2 + \cdots$ is convergent with respect to every norm on M_n. In particular, the scalar infinite series for each entry of $f(A)$ is absolutely convergent. If A has Jordan canonical form given by (6.2.5), substitution of $A = SJS^{-1}$ into the power series for $f(t)$ shows that $f(A)$ satisfies (6.2.6). Finally, the calculations leading to (6.1.7) show (with $m = \infty$ and $N \equiv J_k(0)$) that

$$f(J_k(\lambda)) = \sum_{i=0}^{m} \frac{1}{i!} f^{(i)}(\lambda) N^i = \sum_{i=0}^{k-1} \frac{1}{i!} f^{(i)}(\lambda) N^i$$

since $N^k = 0$. This is exactly (6.2.7).   □

We now develop some of the fundamental properties of a primary matrix function.

6.2.9   Theorem. Let $A \in M_n$ be given and have minimal polynomial $q_A(t) = (t - \lambda_1)^{r_1} \cdots (t - \lambda_\mu)^{r_\mu}$, where $\lambda_1, ..., \lambda_\mu$ are distinct and all $r_i \geq 1$. Let $f(t)$ and $g(t)$ be given scalar-valued functions whose domains include the points $\lambda_1, ..., \lambda_\mu$. For each λ_i with $r_i > 1$, assume that λ_i is in the interior of the domains of $f(t)$ and $g(t)$ and that each function is $(r_i - 1)$-times differentiable at λ_i. Let $f(A)$ and $g(A)$ be the primary matrix functions associated with the stem functions $f(t)$ and $g(t)$. Then:

(a)   There is a polynomial $r(t)$ of degree $\leq n-1$ such that $f(A) =$

$r(A)$; $r(t)$ may be taken to be any polynomial that interpolates $f(t)$ and its derivatives at the roots of $q_A(t) = 0$.

(b)   The primary matrix function $f(A)$ is well defined, that is, the value of $f(A)$ is independent of the particular Jordan canonical form used to represent A.

(c) $f(TAT^{-1}) = Tf(A)T^{-1}$ for any nonsingular $T \in M_n$.

(d) $f(A)$ commutes with any matrix that commutes with A.

(e) $g(A) = f(A)$ if and only if $g^{(u)}(\lambda_i) = f^{(u)}(\lambda_i)$ for $u = 0, 1,..., r_i - 1$ and $i = 1,..., \mu$.

(f)   If $A = A_1 \oplus \cdots \oplus A_k$, then $f(A) = f(A_1) \oplus \cdots \oplus f(A_k)$.

(g)   If A has Jordan canonical form $J_{n_1}(\lambda_{\mu_1}) \oplus \cdots \oplus J_{n_s}(\lambda_{\mu_s})$ with $1 \leq \mu_i \leq \mu$ for every $i = 1,..., s$ and $n_1 + \cdots + n_s = n$, then the Jordan canonical form of $f(A)$ is the same as that of $f(J_{n_1}(\lambda_{\mu_1})) \oplus \cdots \oplus f(J_{n_s}(\lambda_{\mu_s}))$. In particular, the eigenvalues of $f(A)$ are $f(\lambda_{\mu_1})$ (n_1 times),..., $f(\lambda_{\mu_s})$ (n_s times).

Proof: Consider the polynomial $r(t)$ given by the Lagrange–Hermite formula (6.1.16) with $s_i \equiv r_i$, $i = 1,..., \mu$. Then $r^{(u)}(\lambda_i) = f^{(u)}(\lambda_i)$ for $u = 0, 1,..., r_i - 1$ and $i = 1,..., \mu$, so a comparison of (6.1.6,8) with (6.2.6–7) shows that $f(A) = r(A)$.

To prove (b), suppose $A = SJS^{-1} = TJ'T^{-1}$, where J and J' are Jordan matrices. Then $J' = PJP^T$ for some (block) permutation matrix P, and hence (6.2.6–7) gives $f(J') = Pf(J)P^T$. Let $r(t)$ be a polynomial such that $r(J) = f(J)$, as guaranteed in (a). Then

$$Tf(J')T^{-1} = TPf(J)P^TT^{-1} = (TP)r(J)(TP)^{-1}$$
$$= r((TP)J(TP)^{-1}) = r(TPJP^TT^{-1})$$
$$= r(TJ'T^{-1}) = r(A) = r(SJS^{-1})$$
$$= Sr(J)S^{-1} = Sf(J)S^{-1}$$

so the value of the primary matrix function $f(A)$ is independent of the Jordan canonical form used to represent A.

For the assertion (c), the key observation is that the Jordan canonical form of A is not changed by a similarity of A. If $A = SJS^{-1}$, then $TAT^{-1} = (TS)J(TS)^{-1}$, so

$$f(TAT^{-1}) \equiv (TS)f(J)(TS)^{-1} = T[Sf(J)S^{-1}]T^{-1} \equiv Tf(A)T^{-1}$$

Assertion (d) follows from (a), since a matrix that commutes with A must commute with any polynomial in A. Assertion (e) follows immediately from Definition (6.2.4).

The final assertions (f) and (g) follow immediately from Definition (6.2.4) if the similarity that reduces the direct sum to Jordan canonical form is chosen to be the direct sum of the similarities that reduce each direct summand to Jordan canonical form. $\Box$

The fact that a primary matrix function $f(A)$ is a polynomial in A is very important, but it is also important to understand that this polynomial depends on A. There need not be a *single* polynomial $r(t)$ such that $r(X) = f(X)$ for all $X \in M_n$ for which $f(X)$ is defined.

Example. Let $A \in M_n$ have minimal polynomial $q_A(t) = (t-1)^2(t-2)$, and suppose one wishes to compute $\cos A$. We know that $\cos A = r(A)$ for any polynomial $r(t)$ such that $r(1) = \cos 1$, $r'(1) = \cos' t|_{t=1} = -\sin 1$, and $r(2) = \cos 2$. The Lagrange-Hermite interpolation formula (6.1.16) gives a polynomial of degree two that meets these conditions:

$$r(t) = (t-2)\left[\frac{\cos t}{(t-2)}\Big|_{t=1} + \frac{d}{dt}\left(\frac{\cos t}{t-2}\right)\Big|_{t=1}(t-1)\right] + (t-1)^2\left[\frac{\cos t}{(t-1)^2}\Big|_{t=2}\right]$$

$$= (t-2)\left[\frac{\cos 1}{(1-2)} + \frac{(-\sin t)(t-2) - \cos t}{(t-2)^2}\Big|_{t=1}(t-1)\right] + (t-1)^2\left[\frac{\cos 2}{(2-1)^2}\right]$$

$$= (t-2)\left[-\cos 1 + (\sin 1 - \cos 1)(t-1)\right] + (t-1)^2(\cos 2)$$

This is an explicit polynomial of degree 2 such that $r(A) = \cos A$ for any matrix $A \in M_n$ of any size whose minimal polynomial is $(t-1)^2(t-2)$; indeed, this statement remains true if the minimal polynomial of A is a *divisor* of $(t-1)^2(t-2)$. Notice that in calculating the polynomial $r(t)$ from the interpolation formula, the only information about the function $f(t)$ that is used is its values and derivatives at the eigenvalues λ_i. Its behavior elsewhere is irrelevant, so any two functions with these same values and derivatives at λ_i will yield the same polynomial $r(t)$.

In the preceding example, and in Theorem (6.2.9), the minimal polynomial of A was used to construct a polynomial $r(t)$ such that $r(A) = f(A)$, but any monic annihilating polynomial will serve the same purpose if $f(t)$ is smooth enough. If $h(t)$ is a monic polynomial such that $h(A) = 0$, then the minimal polynomial $q_A(t)$ divides $h(t)$, so every root of $q_A(t) = 0$ is a root of $h(t) = 0$ and the multiplicity of each given root of $q_A(t) = 0$ is less than or equal to its multiplicity as a root of $h(t) = 0$. Thus, if $f(t)$ has sufficiently many derivatives at the eigenvalues of A and (6.1.16) or (6.1.18) is used to calculate a polynomial $r(t)$ that interpolates $f(t)$ and its derivatives at the roots of $h(t) = 0$, then $r(t)$ and its derivatives interpolate $f(t)$ and its derivatives up to orders high enough at the eigenvalues of A to ensure that $r(A) = f(A)$; the higher-order and extraneous interpolations forced on $r(t)$ by $h(t)$ have no effect on the value of $r(A)$. For example, if one has finitely many matrices $A_1, \ldots, A_m$, one can construct a single polynomial $r(t)$ such $f(A_i) = r(A_i)$ for all $i = 1, \ldots, m$ (if $f(t)$ is smooth enough) by letting $h(t)$ be the least common multiple of the minimal or characteristic polynomials of $A_1, \ldots, A_m$. In particular, the polynomial $r(t)$ that interpolates $f(t)$ and its derivatives at the roots of the characteristic polynomial of A also has $f(A) = r(A)$.

The fundamental properties of a primary matrix function given in Theorem (6.2.9) can be used to show that many identities in the algebra of stem functions carry over directly to matrix algebra identities for the associated primary matrix functions. If $r_1(t)$ and $r_2(t)$ are polynomials and the polynomials $r_3(t) \equiv r_1(t) r_2(t)$ and $r_4(t) \equiv r_1(r_2(t))$ are their product and composition, respectively, then the matrix $r_3(A)$ can be written as a product of two matrix factors, $r_3(A) = r_1(A) r_2(A)$, and $r_4(A)$ can be expressed as a polynomial in the matrix $r_2(A)$, $r_4(A) = r_1(r_2(A))$. It would be convenient if these properties held for primary matrix functions that are not necessarily polynomials, and they do.

6.2.10   Corollary. Let $A \in M_n$, and suppose $f(t)$ and $g(t)$ satisfy the hypotheses of Theorem (6.2.9). Then

(a)   If $f(t) = c$ is a constant function, then $f(A) = cI$.

(b)   If $f(t) = t$, then $f(A) = A$.

(c)   If $h(t) \equiv f(t) + g(t)$, then $h(A) = f(A) + g(A)$.

(d)   If $h(t) \equiv f(t) g(t)$, then $h(A) = f(A) g(A) = g(A) f(A)$.

(e)   Suppose $g(\lambda) \neq 0$ for every $\lambda \in \sigma(A)$ and let $h(t) \equiv f(t)/g(t)$. Then $g(A)$ is nonsingular and $h(A) = f(A)[g(A)]^{-1} = [g(A)]^{-1}f(A)$.

Proof: Assertions (a), (b), and (c) follow immediately from Definition (6.2.4). To prove (d), invoke Theorem (6.2.9(a)) to construct polynomials $r_1(t)$ and $r_2(t)$ such that $f(A) = r_1(A)$ and $g(A) = r_2(A)$, and let $r(t) \equiv r_1(t)r_2(t)$. Theorem (6.2.9(e)) guarantees that $r_1(t)$ and $r_2(t)$ interpolate $f(t)$ and $g(t)$ and their derivatives, respectively, at the zeroes of the characteristic polynomial of A. Direct computation using the product rule shows that $h(t)$ has derivatives up to order $r_i - 1$ at each eigenvalue λ_i of A, and $h^{(u)}(\lambda_i) = r^{(u)}(\lambda_i)$ for $u = 0, 1,..., r_i - 1$ and $i = 1,..., \mu$. Thus, $h(A)$ is defined and Theorem (6.2.9(e)) ensures that

$$h(A) = r(A) = r_1(A)r_2(A) = f(A)g(A)$$

The commutativity assertion follows from the fact that both $f(A)$ and $g(A)$ are polynomials in A.

Now let $g(t)$ satisfy the hypotheses of assertion (e) and let $\gamma(t) \equiv 1/g(t)$. Then $\gamma(t)$ has the same number of derivatives as $g(t)$ at the eigenvalues of A, so the primary matrix function $\gamma(A)$ is defined. Since $1 = g(t)\gamma(t)$, $I = g(A)\gamma(A)$ and hence $\gamma(A) = [g(A)]^{-1}$. Thus, (e) follows from (d). The final commutativity assertion follows as in (d).   □

6.2.11   Corollary. Let $A \in M_n$ and $g(t)$ satisfy the hypotheses of Theorem (6.2.9). Let $f(t)$ be a scalar-valued function of a real or complex variable t, and assume that each point $g(\lambda_i)$ is in the domain D' of $f(\cdot)$, each point $g(\lambda_i)$ with $r_i > 1$ is in the interior of D', and $f(\cdot)$ is $(r_i - 1)$-times differentiable at each point $g(\lambda_i)$ such that $r_i > 1$. Let $h(t) \equiv f(g(t))$. Then the primary matrix function $h(A)$ is defined and $h(A) = f(g(A))$, where the right-hand side of this identity is the composition of the two primary matrix functions $A \to g(A)$ and $g(A) \to f(g(A))$.

Proof: Let A have Jordan canonical form given by (6.2.5), so $g(A)$ is given by (6.2.6). By Theorem (6.2.9(g)), the Jordan canonical form of $g(A)$ is the direct sum of the Jordan canonical forms of $g(J_{n_i}(\lambda_{\mu_i}))$, $i = 1,..., s$. But each Jordan block in the Jordan canonical form of $g(J_{n_i}(\lambda_{\mu_i}))$ has size at most n_i, which is no larger than the exponent corresponding to λ_{μ_i} in the minimal polynomial of A, $i = 1,..., s$. Thus, the domain and differentiability assumptions on $g(t)$ and $f(t)$ are adequate to ensure that the primary matrix functions $g(A)$ and $f(g(A))$ are both defined by (6.2.4). Similarly, repeated application of the chain rule shows that the function $h(t)$ meets the domain and differentiability conditions necessary to define the primary matrix function $h(A)$ by (6.2.4). Use the Lagrange-Hermite formula (6.1.16) to construct polynomials $r_1(t)$ and $r_2(t)$ such that $r_1^{(u)}(g(\lambda_i)) = f^{(u)}(g(\lambda_i))$ and $r_2^{(u)}(t_i) = g^{(u)}(t_i)$ for $u = 0, 1,..., r_i - 1$ and $i = 1,..., \mu$. Then $r_1(g(A)) = f(g(A))$, $r_2(A) = g(A)$, and $r_1(r_2(A)) = r_1(g(A)) = f(g(A))$ by Theorem (6.2.9(e)). Now consider the polynomial $r_3(t) \equiv r_1(r_2(t))$, which, since it is a composition of polynomials, satisfies $r_3(A) = r_1(r_2(A))$. Repeated use of the chain rule shows that $r_3^{(u)}(\lambda_i) = h^{(u)}(\lambda_i)$ for $u = 0, 1,..., r_i - 1$ and $i = 1,..., \mu$, so $h(A) = r_3(A)$ by Theorem (6.2.9(e)) again. We conclude that $h(A) = r_3(A) = r_1(r_2(A)) = f(g(A))$. ☐

A word of caution: A natural application of the preceding Corollary is to inverse functions $f(t)$ and $g(t)$, that is, $f(g(t)) = t$. If $g(t)$ is globally one-to-one on its domain, then its inverse $f(t)$ is a well-defined function on the range of $g(t)$ and the Corollary can be applied straightforwardly if $g(t)$ is smooth enough; an example of this situation is $g(t) \equiv e^t$ on $D' = \mathbb{R}$ and $f(t) \equiv \log t$ (real-valued) on $D = (0,\infty)$. However, if $g(t)$ is not globally one-to-one, then its natural "inverse" is not a *function* (a "multivalued function" is not a *function*) as required by the hypotheses of Corollary (6.2.11). For example, $g(t) \equiv e^t$ on $D' = \mathbb{C}$ omits zero and takes every other value in $\mathbb{C}$ infinitely often, so its "inverse" $f(t) = \log t$ is not a "function" on $\mathbb{C} - \{0\}$. This problem may be resolved by restricting $g(t)$ to a *principal domain* on which it is one-to-one, but then some care must be used to interpret the conclusions of the Corollary.

Despite the warning just expressed about incautious use of inverse functions, the local character of a primary matrix function permits use of a *local* inverse function to obtain a useful sufficient condition for existence of a solution to the matrix equation $f(X) = A$.

6.2.12 Corollary. Let a given $A \in M_n$ have minimal polynomial $q_A(t) = (t - \lambda_1)^{r_1} \cdots (t - \lambda_\mu)^{r_\mu}$, where $\lambda_1,..., \lambda_\mu$ are distinct and all $r_i \geq 1$. Let $f(t)$ be a given scalar-valued function of a real or complex variable. Suppose there are points $t_1,..., t_\mu$ in the domain of $f(\cdot)$ such that $f(t_i) = \lambda_i$. For each point t_i such that $r_i > 1$, suppose *either* that $f(t)$ is an analytic function of a complex variable t in a (two-dimensional complex) neighborhood of t_i, *or* that $f(t)$ is a real-valued function of a real variable t in a (one-dimensional real) neighborhood of t_i and that $f(t)$ is (r_i-1)-times differentiable at t_i. If $f'(t_i) \neq 0$ for each i for which $r_i > 1$, $i = 1,..., \mu$, then there exists an $X_0 \in M_n$ such that $f(X_0) = A$. Moreover, there is a scalar-valued stem function $g(s)$ such that $X_0 = g(A)$ is a primary matrix function, and hence X_0 is a polynomial in A.

Proof: If all $r_i = 1$ there is nothing to prove, so let i be an index for which $r_i > 1$. Under the stated hypotheses, the inverse function theorem guarantees that there is an open neighborhood N_i of λ_i on which there is a function $g_i(s)$ for which $f(g_i(s)) \equiv s$, and $g_i(s)$ is (r_i-1)-times differentiable at $s = \lambda_i$. Set $N_i \equiv \{\lambda_i\}$ for those i for which $r_i = 1$. There is no loss of generality to assume that $N_1,..., N_\mu$ are disjoint. Let $D' \equiv \cup_{i=1}^\mu N_i$. For $s \in D'$, define $g(s) \equiv g_i(s)$ if $s \in N_i$, $i = 1,..., \mu$. Then $f(g(s)) = s$ for all $s \in D'$, the primary matrix function $g(A)$ is defined, and Corollary (6.2.11) ensures that $f(g(A)) = A$. ☐

The *sufficient* conditions in the preceding corollary are not *necessary* for the existence of a solution to $f(X) = A$. For example, consider

$$X_0 = \begin{bmatrix} 0 & 0 & 1 & 0 \\ 0 & 0 & 0 & 1 \\ 0 & 1 & 0 & 0 \\ 0 & 0 & 0 & 0 \end{bmatrix} \text{ and } A = \begin{bmatrix} 0 & 1 & 0 & 0 \\ 0 & 0 & 0 & 0 \\ 0 & 0 & 0 & 1 \\ 0 & 0 & 0 & 0 \end{bmatrix} = J_2(0) \oplus J_2(0)$$

Note that $X_0 = PJ_4(0)P^T$ for a suitable permutation matrix P, and that $(X_0)^2 = A$. The function $f(t) = t^2$ does not satisfy the basic condition of the inverse function theorem: $f'(0) \neq 0$. Nevertheless, $f(X_0) = A$. Moreover, X_0 is not a primary matrix function of A since no Jordan block of a primary matrix function of A can be larger than the largest Jordan block of A.

There are a great many identities for matrices and functions that follow from the three preceding corollaries.

6.2.13   Example. Let $A \in M_n$ be nonsingular and let $f(t) \equiv 1/t$ have domain $D = \mathbb{C} - \{0\}$. Then $f(t)$ is an analytic function on D and every eigenvalue of A is in the interior of D, so $f(A)$ may be evaluated as a primary matrix function; we have seen in (6.2.10(e)) that $f(A) = A^{-1}$, the ordinary matrix inverse. Since $f^{(u)}(t) = (-1)^u u!/t^{u+1}$, (6.2.7) shows that the inverse of a nonsingular Jordan block $J_k(\lambda)$ is the upper triangular Toeplitz matrix

$f(J_k(\lambda))$ with first row $1/\lambda, -1/\lambda^2, 1/\lambda^3, \ldots, (-1)^{k+1}/\lambda^k$. Thus, if the Jordan canonical form of a nonsingular matrix is known, its inverse can be written down by inspection.

6.2.14   Example. Let $A \in M_n$ be nonsingular and consider $f(t) = t^2$ with domain $D = \mathbb{C}$. Since $f'(t) \neq 0$ for all $t \neq 0$, Corollary (6.2.12) guarantees that there is some $X_0 \in M_n$ such that $X_0^2 = A$. Moreover, X_0 may be taken to be a primary matrix function of A, $X_0 = g(A)$, and $g(s) = \sqrt{s}$ at or in a disjoint neighborhood of each eigenvalue of A, where either choice of the square root may be taken in each neighborhood and at each isolated λ_i. Since $g(A)$ is a polynomial in A, we conclude that *every nonsingular matrix $A \in M_n$ has a square root (a solution to $X^2 = A$) that is a polynomial in A*. Although no coherence in the choice of square roots in the different neighborhoods is necessary, it is possible. For example, consider the slit plane $D_\theta \equiv \mathbb{C} - \mathbb{Z}_\theta$, where $\mathbb{Z}_\theta = \{r e^{i\theta}: r \geq 0\}$ is a closed ray from 0 to ∞, choose θ so that no eigenvalue of A lies on the ray $\mathbb{Z}_\theta$, and let $g(t) \equiv \sqrt{t}$ on D_θ, where a fixed choice is made for the branch of the square root (either will do). Then $g(t)$ is an analytic function on D_θ, $f(t)$ is analytic on all of $\mathbb{C} \supset g(D_\theta)$, and $h(t) \equiv f(g(t)) = t$ on D_θ. Thus, for all $Y \in M_n$ such that $\sigma(Y) \subset D_\theta$, the primary matrix function $g(Y)$ given by Definition (6.2.4) is a "square root" of Y, that is, $g(Y)^2 = Y$. If a nonsingular matrix A is symmetric or normal, for example, then its square root given by $g(A)$ will also be symmetric or normal since these properties of A are preserved by any polynomial in A. Not all possible square roots of a nonsingular matrix arise in this way, however. For example, $\begin{bmatrix} 1 & 0 \\ 0 & -1 \end{bmatrix}$ is a square root of the identity matrix I that cannot be a polynomial in I and hence cannot be a primary function of I; see Section (6.4) for further discussion of this point.

6.2.15   Example. Let $A \in M_n$ be nonsingular, and consider $f(t) = e^t$. Since $f'(t) \neq 0$ for every $t \in \mathbb{C}$, Corollary (6.2.12) guarantees that there exists some $X_0 \in M_n$ such that $e^{X_0} = A$, X_0 may be taken to be a primary matrix function of A, $g(A) = X_0$, and $g(s) = \log s$ at or in a disjoint neighborhood of every eigenvalue of A, where any choice of argument for the logarithm may be taken in each neighborhood and at each isolated point λ_i. We conclude, as in the preceding example, that *every nonsingular matrix $A \in M_n$ has a logarithm (a solution to $e^X = A$) that is a polynomial in A*. As in the case of the square root, it is possible to make a coherent choice of the logarithm function. For example, consider the slit plane D_θ defined in Example

(6.2.14), where θ is chosen so that no eigenvalue of A lies on the ray $\mathcal{R}_\theta$. Let $g(t) \equiv \log t$ on D_θ, where a fixed choice is made for the branch of the logarithm on D_θ (any one will do). Then $g(t)$ is an analytic function on D_θ, $f(t)$ is analytic on all of $\mathbb{C} \supset g(D_\theta)$, and $h(t) \equiv f(g(t)) = t$ on D_θ. Thus, for all $Y \in M_n$ such that $\sigma(Y) \subset D_\theta$, the primary matrix function $g(Y) = \log Y$ given by (6.2.4) is a "logarithm" of Y, that is, $e^{\log Y} = Y$. Since $f(t) = e^t$ is an entire analytic function, its associated primary matrix function may be evaluated either with (6.2.4) or by using the power series for e^t, which converges for all matrix arguments:

$$\sum_{k=0}^{\infty} \frac{1}{k!} (\log A)^k = A$$

for every nonsingular $A \in M_n$, where $\log A$ is the primary matrix function defined by (6.2.4) using any choice of argument for the log function at or near the spectrum of A.

6.2.16   Example. Let $g(t) = e^t$, so $g(A)$ is defined for all $A \in M_n$. If the domain of $g(t)$ is restricted to a doubly infinite horizontal strip

$$\mathcal{D}_y \equiv \{z \in \mathbb{C} \colon y < \operatorname{Im} z < y + 2\pi\}$$

for any given $y \in \mathbb{R}$, then $g(t)$ is one-to-one on $\mathcal{D}_y$ and its range $g(\mathcal{D}_y) = D_y$ is the slit plane considered in the preceding example. If $f(t) \equiv \log t$ on D_y, where the branch of the logarithm is chosen so that $f(D_y) = \mathcal{D}_y$ [that is, $f(t)$ is continuous on D_y and $f(e^{i(y+\pi)}) = i(y + \pi)$], then the hypotheses of Corollary (6.2.11) are met. We conclude that $f(g(A)) = \log e^A = A$ and $g(f(g(A))) = e^{\log e^A} = e^A = g(A)$ for all $A \in M_n$ such that $\sigma(A) \subset \mathcal{D}_y$. Thus, every matrix A whose eigenvalues are not spread apart too far vertically in the complex plane (that is, the spectrum of A is confined to an open horizontal strip with height 2π) is a primary matrix function logarithm of a nonsingular matrix $B = e^A$, and is therefore a polynomial in its argument. However, not all possible "logarithms" of a nonsingular matrix B are achieved in this way. For example, $C = \begin{bmatrix} 0 & 0 \\ 0 & 2\pi i \end{bmatrix}$ is a "logarithm" of the identity I since $e^C = I$, but C cannot be a primary matrix function of I since its two eigenvalues are not equal.

6.2.17   Example. Let $f(t) = t^k$ for a given positive integer k, let $D' = \mathbb{C} - \{0\}$ denote the punctured complex plane, and let $g(t) = e^t$. Then $g(t)$ is analytic on $D = \mathbb{C}$, $g(D) = D'$, and $h(t) \equiv f(g(t)) = e^{kt}$. It follows from Corollary (6.2.11) that $e^{kA} = h(A) = f(g(A)) = (e^A)^k$ for any $A \in M_n$ and $k = \pm 1, \pm 2, \ldots$ Choosing $k = -1$ gives $e^{-A} = (e^A)^{-1}$.

6.2.18   Example. The functions $\cos t \equiv (e^{it} + e^{-it})/2$ and $\sin t \equiv (e^{it} - e^{-it})/2i$ are entire analytic functions, as are $\sin^2 t$ and $\cos^2 t$, and $\sin^2 t + \cos^2 t = 1$ for all $t \in \mathbb{C}$. The matrix functions $\sin A$ and $\cos A$ may be evaluated either from their power series or as primary matrix functions. It follows from Corollary (6.2.10) that $\cos^2 A + \sin^2 A = I$ for every $A \in M_n$.

6.2.19   Example. In the preceding examples, matrix identities have been derived from function identities, but the reverse is also possible. Let $f(t)$ and $g(t)$ be scalar-valued functions of a real or complex variable t that are k-times differentiable at a point $t = \lambda$ in the interior of their domains, let $h(t) \equiv f(t)g(t)$, and let $A \equiv J_{k+1}(\lambda)$. Then Corollary (6.2.10(d)) gives $h(A) = f(A)g(A)$, in which each term is to be evaluated using (6.2.7). The $1, k+1$ entry of $h(A)$ is $h^{(k)}(\lambda)/k!$. Identifying this with the $1, k+1$ entry of the product $f(A)g(A)$ gives *Leibniz's rule* for the kth derivative of a product, $k = 1, 2, \ldots$:

$$\frac{d^k}{dt^k}[f(t)g(t)] = \sum_{m=0}^{k} \begin{bmatrix} k \\ m \end{bmatrix} f^{(m)}(t)\, g^{(k-m)}(t), \quad k = 1, 2, \ldots \quad (6.2.20)$$

6.2.21   Example. Another, but more complicated, function identity follows from Corollary (6.2.11). Let $f(t)$ and $g(t)$ be scalar-valued functions of a real or complex variable. Suppose λ and $g(\lambda)$ are interior points of the domains of $g(\cdot)$ and $f(\cdot)$, respectively, and assume that each function is k-times differentiable there. Let $h(t) \equiv f(g(t))$ and let $A = J_{k+1}(\lambda)$. Then Corollary (6.2.11) gives $h(A) = f(g(A))$, and the $1, k+1$ entry of $h(A)$ is $h^{(k)}(t)/k!$. To evaluate the corresponding entry in $f(g(A))$, consider the polynomials

$$r_1(t) \equiv \sum_{p=0}^{k} \frac{1}{p!} g^{(p)}(\lambda)(t-\lambda)^p$$

and

$$r_2(t) \equiv \sum_{m=0}^{k} \frac{1}{m!} f^{(m)}(t)\Big|_{t=g(\lambda)} (t - g(\lambda))^m$$

for which $g(A) = r_1(A)$ and

$$f(g(A)) = r_2(g(A)) = r_2(r_1(A))$$

$$= \sum_{m=0}^{k} \frac{1}{m!} f^{(m)}(t)\Big|_{t=g(\lambda)} \left[\sum_{p=1}^{k} \frac{1}{p!} g^{(p)}(\lambda) N^p \right]^m \quad (6.2.22)$$

where $N \equiv J_{k+1}(\lambda) - \lambda I = J_{k+1}(0)$. The multinomial formula permits us to expand the mth power terms and gives

$$\left[\sum_{p=1}^{k} \frac{1}{p!} g^{(p)}(\lambda) N^p \right]^m$$

$$= \sum_{\substack{\alpha_1, \ldots, \alpha_k \ge 0 \\ \alpha_1 + \cdots + \alpha_k = m}} \frac{m!}{\alpha_1! \cdots \alpha_k!} \left[\frac{g^{(1)}(\lambda)}{1!} N \right]^{\alpha_1} \cdots \left[\frac{g^{(k)}(\lambda)}{k!} N^k \right]^{\alpha_k}$$

$$= \sum_{\substack{\alpha_1, \ldots, \alpha_k \ge 0 \\ \alpha_1 + \cdots + \alpha_k = m}} \frac{m!}{\alpha_1! \cdots \alpha_k!} \left[\frac{g^{(1)}(\lambda)}{1!} \right]^{\alpha_1} \cdots \left[\frac{g^{(k)}(\lambda)}{k!} \right]^{\alpha_k} N^{\alpha_1 + 2\alpha_2 + \cdots + k\alpha_k}$$

In this sum, the coefficient of N^k, which is the only term that contributes to the $1, k+1$ entry of (6.2.22), is

$$\sum_{I(m,k)} \frac{m!}{\alpha_1! \cdots \alpha_k!} \prod_{u=1}^{k} \left[\frac{g^{(u)}(\lambda)}{u!} \right]^{\alpha_u} \quad (6.2.23)$$

where $I(m,k)$ is the set of all nonnegative integers $\alpha_1, \ldots, \alpha_k$ such that $\alpha_1 + \cdots + \alpha_k = m$ and $\alpha_1 + 2\alpha_2 + \cdots + k\alpha_k = k$. Substituting (6.2.23) into (6.2.22) gives *Faa di Bruno's rule* for the kth derivative of a composite function, $k = 1, 2, \ldots$:

$$\frac{d^k}{dt^k}[f(g(t))] = \sum_{m=1}^{k} f^{(m)}(t)\Big|_{t=g(t)} \sum_{I(m,k)} \frac{m!}{\alpha_1! \cdots \alpha_k!} \prod_{u=1}^{k} \left[\frac{g^{(u)}(t)}{u!}\right]^{\alpha_u} \quad (6.2.24)$$

For the study of inhomogeneous matrix equations and other purposes, it is important to know which Jordan canonical forms can be achieved by a primary matrix function $f(A)$ and which are excluded. If the Jordan canonical form of A is $J_{n_1}(\lambda_1) \oplus \cdots \oplus J_{n_s}(\lambda_s)$, the Jordan canonical form of $f(A)$ is the direct sum of the Jordan canonical forms of $f(J_{n_i}(\lambda_i))$, $i = 1,..., s$. The Jordan canonical form of $f(J_k(\lambda))$ may have one or more blocks, each of which has eigenvalue $f(\lambda_i)$, but of course none of these blocks can have size greater than k. The primary matrix function $f(\cdot)$ either preserves the size k of the block $J_k(\lambda)$, or it splits it into two or more smaller blocks.

For example, consider $f(t) = t^2$, acting on M_2. Every complex number is in the range of $f(\cdot)$, so all 1-by-1 Jordan blocks can be achieved in the range of $f(\cdot)$ on M_2. The only way a Jordan canonical form of $f(A)$ can have a 2-by-2 block is for the Jordan canonical form of A to be a 2-by-2 block $J_2(\lambda) = \begin{bmatrix} \lambda & 1 \\ 0 & \lambda \end{bmatrix}$. But

$$f(J_2(\lambda)) = J_2(\lambda)^2 = \begin{bmatrix} \lambda^2 & 2\lambda \\ 0 & \lambda^2 \end{bmatrix}$$

and

$$f(J_2(\lambda)) - f(\lambda)I = \begin{bmatrix} 0 & 2\lambda \\ 0 & 0 \end{bmatrix}$$

so for all $\lambda \neq 0$ the Jordan canonical form of $f(J_2(\lambda))$ is $J_2(f(\lambda)) = J_2(\lambda^2)$; all nonsingular 2-by-2 Jordan blocks can be achieved in the range of $f(\cdot)$ on M_2. For $\lambda = 0$, however, $f(J_2(0)) = \begin{bmatrix} 0 & 0 \\ 0 & 0 \end{bmatrix} = [0] \oplus [0]$, so in this case alone, $f(\cdot)$ splits the Jordan canonical form into smaller blocks; notice that $f'(0) = 0$. Thus, we have a complete list of all the Jordan canonical forms that can be achieved in the range of $f(\cdot)$ on M_2; notice that the nilpotent block $J_2(0) = \begin{bmatrix} 0 & 1 \\ 0 & 0 \end{bmatrix}$ is not on the list.

Using (6.2.7), we can predict easily when the Jordan canonical form of $f(J_k(\lambda))$ has size k and when it splits into two or more smaller blocks. The general fact that is relevant here is that if λ is an eigenvalue of $B \in M_m$, then $m - \text{rank}(B - \lambda I)$ equals the total number of all Jordan blocks of B of all sizes with eigenvalue λ. If $f'(\lambda) \neq 0$, the special band structure of $f(J_k(\lambda))$ shows that $f(J_k(\lambda)) - f(\lambda)I$ has rank $k-1$, so the Jordan canonical form of $f(J_k(\lambda))$ must be $J_k(f(\lambda))$ in this case. On the other hand, if $f'(\lambda) = 0$ and either $k =$

2 or $f'(\lambda) \ne 0$, then $f(J_k(\lambda)) - f(\lambda)I$ has rank $k - 2$, so $f(\cdot)$ splits the Jordan block into two smaller blocks, the sum of whose sizes must, of course, be k. Observation of how the first nonzero diagonal of the nilpotent matrix

$$f(J_k(\lambda)) - f(\lambda)I$$

propagates when it is squared, cubed, and so forth, shows that this matrix has index of nilpotence $k/2$ if k is even and $(k+1)/2$ if k is odd (it may be helpful to consider $k = 4$ and 5 as model cases). Since the index of nilpotence equals the size of the largest Jordan block, and the sum of the sizes of the two blocks must be k, we come to the following conclusion when $f'(\lambda) = 0$ and either $k = 2$ or $f'(\lambda) \ne 0$: The Jordan canonical form of $f(J_k(\lambda))$ is $J_{k/2}(\lambda) \oplus J_{k/2}(\lambda)$ if k is even ($k \equiv 0 \bmod 2$) and is $J_{(k+1)/2}(\lambda) \oplus J_{(k-1)/2}(\lambda)$ if k is odd ($k \equiv 1 \bmod 2$).

A similar argument shows that if $f'(\lambda) = f'(\lambda) = 0$ and either $k = 3$ or $f^{(3)}(\lambda) \ne 0$, then the Jordan canonical form of $f(J_k(\lambda))$ has exactly three blocks, and an analysis of the ranks of powers of $f(J_k(\lambda)) - f(\lambda)I$ gives their sizes for each of the three values of $k \bmod 3$; see Problem 21.

The general principle here is both simple and useful, and it provides algebraic insight into the existence result in Corollary (6.2.12) that was proved with a theorem from analysis; see Problem 43.

6.2.25   Theorem. Let $J_k(\lambda)$ be a k-by-k Jordan block with eigenvalue λ, suppose $f(t)$ is $(k-1)$-times differentiable at $t = \lambda$, and let $f(A)$ be the primary matrix function defined by (6.2.4). If $f'(\lambda) \ne 0$, the Jordan canonical form of $f(J_k(\lambda))$ is the single block $J_k(f(\lambda))$. Let p be a given integer between 1 and k and let $k = pq + r$, with $0 \le r < p$, that is, $k \equiv r \bmod p$. If $f'(\lambda) = f'(\lambda) = \cdots = f^{(p-1)}(\lambda) = 0$ and either $p = k$ or $f^{(p)}(\lambda) \ne 0$, the Jordan canonical form of $f(J_k(\lambda))$ splits into exactly p blocks, each of which has eigenvalue $f(\lambda)$; there are $p - r$ blocks $J_q(f(\lambda))$ and r blocks $J_{q+1}(f(\lambda))$.

Proof: Only the last assertion requires verification, and it is trivial for $p = k$, so assume $p \le k - 1$. For convenience, note that the sizes and numbers of blocks with eigenvalue $f(\lambda)$ are the same as the sizes and numbers of blocks in the Jordan canonical form of $[J_k(1)]^p$, a k-by-k matrix with ones on the p th superdiagonal and zeroes elsewhere. Explicit calculation shows that the superdiagonal of ones in $[J_k(1)^p]^m = J_k(1)^{pm}$ is on the mp th superdiagonal for $m = 1, \ldots, q$, so it moves to the right p positions for each successive power

$m = 1,..., q;$ $J_k(\lambda)^{pq}$ has ones down the pqth superdiagonal; and $[J_k(1)]^{p(q+1)} = 0.$ Thus, if we set $\Delta_m \equiv \text{rank } J_k(1)^{pm} - \text{rank } J_k(1)^{p(m+1)}$ with $\Delta_0 \equiv p,$ then $\Delta_m = p$ for $m = 0, 1,..., q-1,$ while $\Delta_q = k - pq = r.$ This sequence of rank differences is the same as that observed for powers of the direct sum of $p - r$ copies of $J_q(1)$ with r copies of $J_{q+1}(1),$ which verifies the asserted Jordan canonical form. $\square$

Although we have *motivated* the definition of a primary matrix function by a continuity argument, we have so far been examining only *static* properties of a primary matrix function—the value and properties of $f(A)$ for a single matrix $A,$ which depend only on local properties of the stem function $f(t).$ We now show that, under appropriate hypotheses on the stem function $f(t),$ the primary matrix function $f(A)$ is continuous on a suitably defined domain in $M_n.$

The natural domains on which a primary matrix function $f(A)$ is continuous are described in the hypotheses of Corollary (6.1.28). Since the most common cases are when the domain of the stem function $f(t)$ is an open set in the complex plane or an open real interval, it is convenient to make the following definition.

6.2.26 **Definition.** Let n be a given positive integer and let $D \subset \mathbb{C}$ be given. If $n = 1,$ define $A_n(D) \equiv \{f(t): f(t)$ is a continuous scalar-valued function on $D\}.$ If $n \geq 2,$ suppose either that

(a) $D \subset \mathbb{C}$ is a simply connected open set, in which case define $A_n(D) \equiv \{f(t): f(t)$ is a scalar-valued analytic function on $D\};$

or

(b) $D = (a,b) \subset \mathbb{R}$ is an open real interval, in which case define $A_n(D) \equiv \{f(t): f(t)$ is an $(n-1)$-times continuously differentiable scalar-valued function on $D\}.$

In all of the preceding cases, define $\mathcal{D}_n(D) \equiv \{A \in M_n: \sigma(A) \subset D\}.$

In all cases of the definition, $A_n(D)$ is an algebra of functions that includes all polynomials and all rational functions that have no poles in $D.$ Every $f(t) \in A_n(D)$ is a stem function whose associated primary matrix function $f(A)$ is defined for all $A \in \mathcal{D}_n(D).$

6.2.27 **Theorem.** Let $n,$ $p,$ and q be given positive integers and let $D \subset \mathbb{C}$ be given. If $n \geq 2,$ let D be either a simply connected open subset of $\mathbb{C}$ or an open real interval. Then

(1)   For every $f(t) \in A_n(D)$, the primary matrix function $f(A)$ is continuous on $\mathcal{D}_n(D)$.

(2)   If $f: \mathcal{D}_n(D) \rightarrow M_{p,q}$ and $g: \mathcal{D}_n(D) \rightarrow M_{p,q}$ are given continuous functions on $\mathcal{D}_n(D)$, then $f(A) = g(A)$ for all $A \in \mathcal{D}_n(D)$ if and only if $f(A) = g(A)$ for all diagonalizable $A \in \mathcal{D}_n(D)$.

Proof: The case $n = 1$ is trivial, so assume $n \geq 2$. Assertion (1) follows immediately from Corollary (6.1.28(3)) if we note that $f(A) = r_A(A)$ for all $A \in \mathcal{D}_n(D)$, where the polynomial $r_A(t)$ is given by the Newton formula (6.1.29) and interpolates $f(t)$ and its derivatives at the zeroes of the characteristic polynomial of A. The second assertion is an immediate consequence of continuity of $f(\cdot)$ and $g(\cdot)$ and the fact that the diagonalizable matrices are dense in $\mathcal{D}_n(D)$; $f(\cdot)$ and $g(\cdot)$ need not be primary matrix functions. ▯

The preceding theorem not only assures us of the continuity of a primary matrix function, but also provides a uniqueness property that can be an efficient way to verify certain kinds of matrix identities without onerous computations. For example, if D is a simply connected open subset of $\mathbb{C}$ or is an open real interval, and if $f(t)$, $g(t) \in A_n(D)$, then $h(t) \equiv f(t)g(t) \in A_n(D)$. The primary matrix functions $f(A)$, $g(A)$, and $h(A)$ are all continuous functions of A on $\mathcal{D}_n(D)$ by (6.2.27(1)), as is the product $\eta(A) \equiv f(A)g(A)$ of primary matrix functions. But $\eta(\Lambda) = h(\Lambda)$ for all diagonal $\Lambda \in \mathcal{D}_n(D)$ because of the scalar identity $h(\lambda) = f(\lambda)g(\lambda)$, and hence for any diagonalizable $A = S\Lambda S^{-1} \in \mathcal{D}_n(D)$ we have

$$h(A) = h(S\Lambda S^{-1}) = Sh(\Lambda)S^{-1} = Sf(\Lambda)g(\Lambda)S^{-1}$$

$$= Sf(\Lambda)S^{-1}Sg(\Lambda)S^{-1} = f(S\Lambda S^{-1})g(S\Lambda S^{-1})$$

$$= f(A)g(A)$$

Thus, (6.2.27(2)) guarantees that $\eta(A) = h(A)$ for all $A \in \mathcal{D}_n(D)$.

The same kind of argument permits easy verification of explicit functional identities. For example, $f(t) = \sin 2t$ and $g(t) = 2(\sin t)(\cos t)$ are both analytic functions on $D = \mathbb{C}$, so the primary matrix functions $f(A)$ and $g(A)$ are defined and continuous for all $A \in \mathcal{D}_n(D)$ by (6.2.27(1)). Since $f(t) = g(t)$ for all $t \in \mathbb{C}$, it follows that $f(A) = g(A)$ for all diagonal, and hence for all diagonalizable, $A \in M_n$. Thus, $f(A) = g(A)$ and

$$\sin 2A = 2(\sin A)(\cos A) = 2(\cos A)(\sin A)$$

for all $A \in M_n$ by (6.2.27(2)).

Theorem (6.2.27) can also be used to prove an important integral representation for the primary matrix function $f(A)$ when the stem function $f(t)$ is an analytic function; see Problem 33 for another proof.

6.2.28   Theorem. Let n be a given positive integer, let $f(t)$ be a given complex-valued analytic function on a simply connected open set $D \subset \mathbb{C}$, and let $f(A)$ be the primary matrix function associated with the stem function $f(t)$. Then

$$f(A) = \frac{1}{2\pi i} \oint_{\Gamma} f(t)\, (tI - A)^{-1} dt \qquad (6.2.29)$$

for each $A \in \mathcal{D}_n(D)$, where $\Gamma \subset D$ is any simple closed rectifiable curve that strictly encloses all of the eigenvalues of A.

Proof: Let $A \in \mathcal{D}_n(D)$ be given, let $\Gamma \subset D$ be a given simple closed rectifiable curve that strictly encloses all the eigenvalues of A, and let $\Phi(A)$ denote the value of the right-hand side of (6.2.29). The n^2 entrywise line integrals that comprise $\Phi(A)$ are defined since $(tI - A)^{-1}$ is a continuous function of $t \in \Gamma$ (no eigenvalues of A lie on Γ). Moreover, A is contained in an open neighborhood $M(A) \subset \mathcal{D}_n(D)$ whose closure $M(A)^{cl}$ (a compact set) has the property that for every $X \in M(A)^{cl}$, all eigenvalues of X lie strictly inside Γ. Since Γ is a compact set, the Cartesian product $\Gamma \times M(A)^{cl}$ is also compact. Thus, the continuous function $(tI - X)^{-1}$ on the compact set $\Gamma \times M(A)^{cl}$ is uniformly bounded there. The identity

$$(tI - A)^{-1} - (tI - X)^{-1} = (tI - X)^{-1}[A - X](tI - A)^{-1}$$

then shows that $(tI - X)^{-1} \to (tI - A)^{-1}$ uniformly in $t \in \Gamma$ as $X \to A$, and hence $\Phi(A)$ is a continuous function on $D_n(D)$. If $A \in \mathcal{D}_n(D)$ is diagonalizable and $A = S\Lambda S^{-1}$ with $\Lambda = \operatorname{diag}(\lambda_1, ..., \lambda_n)$, the Cauchy integral theorem gives

$$\Phi(A) = \frac{1}{2\pi i} \oint_{\Gamma} f(t)\, (tI - A)^{-1} dt = \frac{1}{2\pi i} \oint_{\Gamma} f(t)\, (tI - S\Lambda S^{-1})^{-1} dt$$

$$= S \operatorname{diag}\left[\frac{1}{2\pi i}\oint_\Gamma f(t)\,(t-\lambda_1)^{-1}dt,\ldots,\frac{1}{2\pi i}\oint_\Gamma f(t)\,(t-\lambda_n)^{-1}dt\right]S^{-1}$$

$$= S \operatorname{diag}(f(\lambda_1),\ldots,f(\lambda_n))\,S^{-1}$$

$$= f(A)$$

We conclude from Theorem (6.2.27) that $\Phi(A) = f(A)$ for all $A \in \mathcal{D}_n(D)$.　　$\square$

Exercise. With the preceding arguments as a model, use Theorem (6.2.27) to verify all of the identities in Corollaries (6.2.10-11) without computations. In each case, argue that each side of the desired identity is a continuous function on $\mathcal{D}_n(D)$ and that the identity holds for all diagonalizable matrices.

Although we know that a sufficiently smooth scalar stem function gives a continuous primary matrix function, it may fail to be Lipschitz continuous; for an example, see Problem 37. There is one important case, however, in which Lipschitz continuity is assured. The following theorem also gives an independent proof of the fact that an analytic function of a matrix is continuous on $\mathcal{D}_n(D)$ if the function is represented by a single power series whose disk of convergence includes D.

6.2.30   **Theorem.** Let $\|\|\cdot\|\|$ be a given matrix norm on M_n, let

$$f(t) \equiv \sum_{k=0}^{\infty} a_k t^k$$

have radius of convergence $R > 0$, and define

$$f_{abs}(t) \equiv \sum_{k=0}^{\infty} |a_k|\, t^k$$

Let $A, E \in M_n$ be given with $\|\|A\|\| + \|\|E\|\| < R$. Then

$$\|\|f(A+E)-f(A)\|\| \le f_{abs}(\|\|A\|\| + \|\|E\|\|) - f_{abs}(\|\|A\|\|)$$

$$\le \|\|E\|\|\,f'_{abs}(\|\|A\|\| + \|\|E\|\|) \tag{6.2.31}$$

where $f(A)$ is evaluated either as a primary matrix function according to (6.2.4), or, equivalently, as a power series $f(A) = a_0 I + a_1 A + a_2 A^2 + \cdots$, and similarly for $f(A + E)$. In particular, $f(A)$ is a continuous function on $\{A \in M_n : \rho(A) < R\}$ that is Lipschitz continuous on each compact subset K of $\{A \in M_n : |\!|\!| A |\!|\!| < R/2\}$, that is, for each such K there is some positive $L = L(K)$ such that $|\!|\!| f(B) - f(A) |\!|\!| \leq L |\!|\!| B - A |\!|\!|$ for all $A, B \in K$. One may take the Lipschitz constant to be $L = \max \{f'_{abs}(|\!|\!| A |\!|\!| + |\!|\!| B - A |\!|\!|) : A, B \in K\}$.

Proof: Use (6.1.13) to compute

$$|\!|\!| f(A + E) - f(A) |\!|\!| = \left|\!\left|\!\left| \sum_{k=0}^{\infty} a_k [(A + E)^k - A^k] \right|\!\right|\!\right|$$

$$\leq \sum_{k=0}^{\infty} |a_k| \, |\!|\!| [(A + E)^k - A^k] |\!|\!|$$

$$\leq \sum_{k=0}^{\infty} |a_k| \, [(|\!|\!| A |\!|\!| + |\!|\!| E |\!|\!|)^k - |\!|\!| A |\!|\!|^k]$$

$$= f_{abs}(|\!|\!| A |\!|\!| + |\!|\!| E |\!|\!|) - f_{abs}(|\!|\!| A |\!|\!|)$$

$$= |\!|\!| E |\!|\!| \, f'_{abs}(\xi) \leq |\!|\!| E |\!|\!| \, f'_{abs}(|\!|\!| A |\!|\!| + |\!|\!| E |\!|\!|)$$

where $|\!|\!| A |\!|\!| \leq \xi \leq |\!|\!| A |\!|\!| + |\!|\!| E |\!|\!|$ and we use the mean value theorem and the fact that $f'_{abs}(t)$ is monotone increasing. The continuity assertion follows from the fact that if $\rho(A) < R$, then there is *some* matrix norm $|\!|\!| \cdot |\!|\!|$ on M_n for which $|\!|\!| A |\!|\!| < R$ (Lemma (5.6.10) in [HJ]), and $f_{abs}(|\!|\!| A |\!|\!| + |\!|\!| E |\!|\!|) \to f_{abs}(|\!|\!| A |\!|\!|)$ as $E \to 0$ since $f_{abs}(t)$ is an analytic, and hence continuous, function of $t \in [0, R]$. The final assertion about Lipschitz continuity is proved in the same way as the corresponding assertion in Theorem (6.1.10). $\qquad \Box$

The exponential function $f(t) = e^t$ gives a particularly simple application of this bound since, for it, $R = \infty$, $f_{abs}(\cdot) = f(\cdot)$ because all its Taylor series coefficients are positive, and $f(|\!|\!| A |\!|\!| + |\!|\!| E |\!|\!|) = f(|\!|\!| A |\!|\!|) f(|\!|\!| E |\!|\!|)$ from the functional equation for the exponential.

6.2.32   Corollary. For any $A, E \in M_n$, and for any matrix norm $|||\cdot|||$ on M_n,

$$|||\, e^{A+E} - e^A \,||| \leq [\exp(|||\, E \,|||) - 1] \exp(|||\, A \,|||)$$

$$\leq |||\, E \,||| \exp(|||\, E \,|||) \exp(|||\, A \,|||)$$

In particular, the function $f : A \to e^A$ is continuous on M_n and is Lipschitz continuous on each compact subset of M_n.

If $f(t)$ is represented by a power series $f(t) = a_0 + a_1 t + a_2 t^2 + \cdots$ that has radius of convergence $R > 0$, then $f(t)$ is an analytic function on $D_R = \{t \in \mathbb{C} : |t| < R\}$. Theorem (6.2.30) shows directly that $f(A) \equiv a_0 I + a_1 A + a_2 A^2 + \cdots$ (computed via the power series) is continuous on $\mathcal{D}_n(D_R)$. We also know that the primary matrix function $f(A)$ [computed via (6.2.4) from the scalar-valued stem function $f(t)$] is continuous on $\mathcal{D}_n(D_R)$. Since these two ways to evaluate $f(A)$ agree on all diagonal, and hence on all diagonalizable, matrices in $\mathcal{D}_n(D_R)$, it follows from (6.2.27(2)) that they must agree on all of $\mathcal{D}_n(D_R)$. This is an independent and essentially computation-free proof of Theorem (6.2.8).

There are many instances in which a primary matrix function provides a natural continuation of a matrix function defined as a power series, beyond its natural circle of convergence.

6.2.33   Example. The scalar power series

$$\Phi(t) \equiv -\sum_{k=1}^{\infty} \frac{1}{k} t^k$$

has radius of convergence $R = 1$, and in the unit disk $UD = \{t \in \mathbb{C} : |t| < 1\}$ it represents the principal branch of the function $\log(1 - t)$, that is, $\Phi(t)$ is continuous in UD, $e^{\Phi(t)} = 1 - t$ for all $t \in UD$, and $\Phi(0) = \log 1 = 0$. The infinite series for the matrix function

$$\Phi(A) \equiv -\sum_{k=1}^{\infty} \frac{1}{k} A^k$$

is convergent for all $A \in \mathcal{D}_n(UD)$ and is a continuous function of $A \in \mathcal{D}_n(UD)$, though it is not obvious that $e^{\Phi(A)} = A$ there. Theorem (6.2.27(2)) guarantees that the primary matrix function $\log(I-A)$ (using the principal branch of the logarithm) has the same value as $\Phi(A)$ when $A \in \mathcal{D}_n(UD)$, but it is defined and continuous on a much larger domain than $\mathcal{D}_n(D)$: all $A \in M_n$ whose spectrum lies in the complex plane with a slit from 1 to ∞ along any half-line that intersects the unit circle only at $t = 1$. Throughout this larger domain, we know that $e^{\log(I-A)} = I-A$, which implies, in particular, that $e^{\Phi(A)} = A$ for all $A \in \mathcal{D}_n(UD)$. Thus, the primary matrix function $\log(I-A)$ provides a natural continuous extension of the power series $\Phi(A)$ to a much larger domain.

So far, continuity is the only smoothness property of a primary matrix function that has concerned us. For applications to differential equations, however, for a fixed $A \in M_n$ one wants to be able to differentiate $f(tA)$ with respect to the scalar parameter t, and it is easy to do so. The problem of differentiating the matrix function $f(A(t))$ for a general differentiably parameterized family $A(t) \in M_n$ is more complicated, and we defer discussion of this general case to Section (6.6). The following is a simple but useful special case of Corollary (6.6.19).

6.2.34   Theorem. Let a given $A \in M_n$ have minimal polynomial $q_A(t) = (t - \lambda_1)^{r_1} \cdots (t - \lambda_\mu)^{r_\mu}$, where $\lambda_1, \ldots, \lambda_\mu$ are distinct and all $r_i \geq 1$. Let t_0 be a given nonzero scalar, and suppose $f(t)$ is a given scalar-valued function of a real or complex variable t such that each point $t_0 \lambda_i$ is in the interior of the domain of $f(t)$. Assume that $f(t)$ is $(r_i - 1)$-times differentiable in an open neighborhood of each point $t_0 \lambda_i$ and is r_i-times differentiable at each $t_0 \lambda_i$, $i = 1, \ldots, \mu$. Then

(a) The primary matrix function $f(tA)$ associated with the stem function $f(t)$ is defined for all t in an open neighborhood of $t = t_0$.

(b) The primary matrix function $f'(t_0 A)$ associated with the stem function $f'(t)$ is defined.

(c) $f(tA)$ is a differentiable function of t at $t = t_0$, and

$$\frac{d}{dt} f(tA)\Big|_{t=t_0} = A f'(t_0 A) = f'(t_0 A) A \qquad (6.2.35)$$

In particular, for $f(t) = e^t$ we have

$$\frac{d}{dt}e^{tA} = Ae^{tA} = e^{tA}A \text{ for all } A \in M_n \qquad (6.2.36)$$

Proof: The differentiability and domain assumptions are sufficient to define the primary matrix functions $f(tA)$ (for t in a neighborhood of t_0) and $f'(t_0A)$. Let A have Jordan canonical form $A = SJS^{-1}$, where J has the form (6.2.5). Then

$$f(tA) = f(S[tJ]S^{-1}) = Sf(tJ)S^{-1}$$
$$= S[f(tJ_{n_1}(\lambda_1)) \oplus \cdots \oplus f(tJ_{n_s}(\lambda_s))]S^{-1}$$

so it suffices to prove (6.2.35) in the special case of a single Jordan block.

Let $J_k(\lambda)$ be one of the Jordan blocks in the Jordan canonical form of A, so $f(t)$ is at least k-times differentiable at $t = t_0\lambda$ and is at least $(k-1)$-times differentiable in a neighborhood of $t_0\lambda$. For $t \neq 0$, let $S_t \equiv \text{diag}(1, 1/t, ..., 1/t^{k-1}) \in M_k$, and observe that $tJ_k(\lambda) = S_t J_k(t\lambda)S_t^{-1}$. Then $f(tJ_k(\lambda)) = S_t f(J_k(t\lambda))S_t^{-1}$, which, by (6.2.7), is an upper triangular Toeplitz matrix whose $1, j$ entry is $t^{j-1}f^{(j-1)}(t\lambda)/(j-1)!$, $j = 1, ..., k$; the latter expression for $f(tJ_k(\lambda))$ is also valid for $t = 0$, when it gives the diagonal matrix $f(0)I$. Substitution of $f'(\cdot)$ for $f(\cdot)$ now shows that $f'(t_0J_k(\lambda))$ is an upper triangular Toeplitz matrix whose $1, j$ entry is $t_0^{j-1}f^{(j)}(t_0\lambda)/(j-1)!$, $j = 1, ..., k$. Explicit differentiation now shows that $\frac{d}{dt}f(tJ_k(\lambda))\big|_{t=t_0}$ is an upper triangular Toeplitz matrix whose $1, 1$ entry is $\lambda f'(t_0\lambda)$ and whose $1, j$ entry is $t_0^{j-2}f^{(j-1)}(t_0\lambda)/(j-2)! + \lambda t_0^{j-1}f^{(j)}(t_0\lambda)/(j-1)!$, $j = 2, ..., k$. Since these are also the entries in the first row of the upper triangular Toeplitz matrix $f'(t_0J_k(\lambda))J_k(\lambda)$, we have $\frac{d}{dt}f(tJ_k(\lambda))\big|_{t=t_0} = f'(t_0J_k(\lambda))J_k(\lambda)$, as asserted. The commutativity assertion $f'(t_0A)A = Af'(t_0A)$ follows from the fact that the primary matrix function $f'(t_0A)$ is a polynomial in t_0A. []

The important formula (6.2.36) shows that a solution to the first-order vector initial-value problem $\frac{d}{dt}x(t) = Ax(t)$, $x(0) \in \mathbb{C}^n$ given, is $x(t) = e^{tA}x(0)$ for any $A \in M_n$ and any given initial value $x(0)$. The general theory of differential equations guarantees that this problem has a *unique* solution. Exploitation of this fact leads to a simple method to calculate e^{tA} for any $A \in M_n$ that avoids both power series and the Jordan canonical form; see Theorem (6.5.35).

We have been interested mainly in the primary matrix function $f(A)$

for all $A \in M_n$ whose spectrum lies in a suitably defined domain, and we have found that to ensure continuity of $f(A)$ it is necessary to assume that $f(t)$ has continuous derivatives up to order $n-1$. But if we restrict our attention just to *normal* matrices, and especially to the important special case of *Hermitian* matrices, there is no longer any need to assume differentiability of the stem function $f(t)$ in order to obtain continuity of the primary matrix function $f(A)$.

6.2.37   Theorem. Let $D \subset \mathbb{C}$ be a given set and let $\mathcal{N}_n(D) \equiv \{A \in M_n: A$ is normal and $\sigma(A) \subset D\}$. If $f(t)$ is a continuous scalar-valued function on D, then the primary matrix function

$$f(A) = U \begin{bmatrix} f(\lambda_1) & 0 \\ 0 & \ddots \\ & & f(\lambda_n) \end{bmatrix} U^*$$

is continuous on $\mathcal{N}_n(D)$, where $A = U\Lambda U^*$, $\Lambda = \mathrm{diag}(\lambda_1,\ldots,\lambda_n)$, and $U \in M_n$ is unitary.

Proof: Let $A \in M_n$ be given and let $A = U\Lambda U^*$ with U unitary and $\Lambda = \mathrm{diag}(\lambda_1,\ldots,\lambda_n)$. Let $p(t)$ be a polynomial such that $p(\lambda_i) = f(\lambda_i)$ for $i = 1,\ldots, n$, so $f(A) = p(A)$. Let $|\|\cdot\||_2$ denote the spectral norm on M_n, let $\epsilon > 0$ be given, and choose a $\delta > 0$ small enough so that

(a) $|\| p(A) - p(B) \||_2 \leq \epsilon/3$ whenever $B \in M_n$ is such that $|\| A - B \||_2 \leq \delta$ (use Theorem (6.1.10)),

(b) $|p(t) - p(\lambda_i)| \leq \epsilon/3$ whenever $|t - \lambda_i| \leq \delta$, $i = 1,\ldots, n$, and

(c) $|f(t) - f(\lambda_i)| \leq \epsilon/3$ whenever $t \in D$ and $|t - \lambda_i| \leq \delta$, $i = 1,\ldots, n$.

Let $B \in \mathcal{N}_n(D)$ satisfy $|\| A - B \||_2 \leq \delta$. By the Hoffman-Wielandt theorem (Theorem (6.3.5) in [HJ]), there is always a unitary $V \in M_n$ such that $B = VMV^*$, $M = \mathrm{diag}(\mu_1,\ldots,\mu_n)$, and $|\mu_i - \lambda_i| \leq \delta$ for all $i = 1,\ldots, n$. Using the unitary invariance of the spectral norm, we have

$$|\| f(A) - f(B) \||_2 = |\| p(A) - p(B) + p(B) - f(B) \||_2$$

$$\leq |\| p(A) - p(B) \||_2 + |\| p(B) - f(B) \||_2$$

$$= |\| p(A) - p(B) \||_2 + |\| V[p(M) - f(M)]V^* \||_2$$

$$= \||| \, p(A) - p(B) \, \||_2 + \||| \, p(M) - p(\Lambda) + p(\Lambda) - f(M) \, \||_2$$

$$\leq \||| \, p(A) - p(B) \, \||_2 + \||| \, p(M) - p(\Lambda) \, \||_2 + \||| \, p(\Lambda) - f(M) \, \||_2$$

$$= \||| \, p(A) - p(B) \, \||_2 + \max_{1 \leq i \leq n} \, |p(\mu_i) - p(\lambda_i)| + \max_{1 \leq i \leq n} \, |f(\lambda_i) - f(\mu_i)|$$

$$\leq \epsilon/3 + \epsilon/3 + \epsilon/3 = \epsilon \qquad\qquad\qquad \Box$$

We have seen that many familiar functional identities for functions of a single variable have natural analogs for primary matrix functions, and that they can be discovered or verified with tools such as Theorem (6.2.27) and Corollaries (6.2.10-11). Unfortunately, many functional identities involving two or more independent variables do not translate directly into identities for primary matrix functions if the independent matrix variables do not commute. To illustrate this point, consider the exponential function $f(t) = e^t$, an entire analytic function whose power series converges for all $t \in \mathbb{C}$, and whose primary matrix function e^A is therefore defined and continuous for all $A \in M_n$.

Sometimes it is convenient to compute e^A as a primary matrix function rather than as a power series. For example, this point of view makes it clear immediately that if $\lambda_1, \ldots, \lambda_n$ are the eigenvalues of A, then $e^{\lambda_1}, \ldots, e^{\lambda_n}$ are the eigenvalues of e^A, which is therefore nonsingular for all $A \in M_n$. Also, for

$$A = \begin{bmatrix} 0 & 1 \\ 0 & 0 \end{bmatrix} \quad \text{and} \quad B = \begin{bmatrix} 0 & 0 \\ 1 & 0 \end{bmatrix}$$

we can read off at once from (6.2.7) that

$$e^A = \begin{bmatrix} 1 & 1 \\ 0 & 1 \end{bmatrix} \quad \text{and} \quad e^B = \begin{bmatrix} 1 & 0 \\ 1 & 1 \end{bmatrix}$$

To evaluate e^{A+B}, however, one can avoid diagonalizing the idempotent matrix $A + B$ by using the power series to compute

$$e^{A+B} = e^{B+A} = \sum_{k=0}^{\infty} \frac{1}{k!} (A+B)^k = \sum_{k=0}^{\infty} \left[\frac{1}{(2k)!} I + \frac{1}{(2k+1)!} (A+B) \right]$$

$$= \cosh(1)\, I + \sinh(1)\, (A + B) = \begin{bmatrix} \dfrac{e + e^{-1}}{2} & \dfrac{e - e^{-1}}{2} \\[2ex] \dfrac{e - e^{-1}}{2} & \dfrac{e + e^{-1}}{2} \end{bmatrix}$$

Now compute

$$e^A \cdot e^B = \begin{bmatrix} 2 & 1 \\ 1 & 1 \end{bmatrix} \quad \text{and} \quad e^B \cdot e^A = \begin{bmatrix} 1 & 1 \\ 1 & 2 \end{bmatrix}$$

These calculations show that e^{A+B}, $e^A \cdot e^B$, and $e^B \cdot e^A$ can all be different, and warn us that familiar functional identities for scalar functions may not carry over to functions of matrices.

6.2.38   Theorem. Let A, $B \in M_n$ be given. If A and B commute, then $e^{A+B} = e^A \cdot e^B = e^B \cdot e^A$. In particular, $e^{-A} = (e^A)^{-1}$ and $e^{mA} = (e^A)^m$ for any $A \in M_n$ and any integer $m = \pm 1, \pm 2, \dots$.

Proof: Use the power series for e^t to compute

$$e^{A+B} = \sum_{k=0}^{\infty} \frac{1}{k!} (A + B)^k = \sum_{k=0}^{\infty} \frac{1}{k!} \sum_{j=0}^{k} \binom{k}{j} A^j B^{k-j} = \sum_{k=0}^{\infty} \sum_{j=0}^{k} \frac{1}{j!\,(k-j)!}\, A^j B^{k-j}$$

$$= \sum_{q=0}^{\infty} \sum_{p=0}^{\infty} \frac{1}{q!} \frac{1}{p!}\, A^q B^p = \left[\sum_{q=0}^{\infty} \frac{1}{q!} A^q \right] \left[\sum_{p=0}^{\infty} \frac{1}{p!} B^p \right] = e^A e^B$$

It then follows that $e^A e^B = e^{A+B} = e^{B+A} = e^B e^A$. With $B = -A$, we have $I = e^0 = e^{A-A} = e^A e^{-A}$. With $B = A$, we have $e^{2A} = e^{A+A} = e^A e^A = \left[e^A\right]^2$; a straightforward induction gives the general case. □

Although commutativity is a *sufficient* condition for the identities $e^{A+B} = e^A \cdot e^B = e^B \cdot e^A$ to hold, it is not *necessary*. The following exercises show that for noncommuting A, $B \in M_2$ it is possible to have $e^A \cdot e^B = e^B \cdot e^A = e^{A+B}$, or $e^A \cdot e^B = e^B \cdot e^A \neq e^{A+B}$, or even $e^A \cdot e^B \neq e^B \cdot e^A = e^{A+B}$.

Exercise. Consider

$$A = \begin{bmatrix} 0 & 0 \\ 0 & 2\pi i \end{bmatrix} \quad \text{and} \quad B = \begin{bmatrix} 0 & 1 \\ 0 & 2\pi i \end{bmatrix}$$

Verify that A and B do not commute, and that $e^A = e^B = e^{A+B} = I$. Conclude that $e^A \cdot e^B = e^B \cdot e^A = e^{A+B}$. *Hint:* B is diagonalizable, so $B = S \Lambda S^{-1}$ with $\Lambda = \text{diag}(0, 2\pi i)$ and $e^B = S e^\Lambda S^{-1} = SIS^{-1} = I$. The same argument works for $A + B$.

Exercise. Consider

$$A = \begin{bmatrix} \pi i & 0 \\ 0 & -\pi i \end{bmatrix} \quad \text{and} \quad B = \begin{bmatrix} 0 & 1 \\ 0 & 0 \end{bmatrix}$$

Verify that A and B do not commute, and that $e^A = e^{A+B} = -I$ and $e^B = \begin{bmatrix} 1 & 1 \\ 0 & 1 \end{bmatrix}$. Conclude that $e^A \cdot e^B = e^B \cdot e^A \neq e^{A+B}$. *Hint:* Use the method in the preceding exercise for e^A and e^{A+B}, and use (6.2.7) for e^B.

Exercise. Consider

$$A = \begin{bmatrix} 0 & 1 \\ 0 & 0 \end{bmatrix} \quad \text{and} \quad B = \begin{bmatrix} 0 & 0 \\ 0 & \zeta \end{bmatrix}$$

for a given $\zeta \neq 0$. Verify that A and B do not commute, and that

$$e^A = \begin{bmatrix} 1 & 1 \\ 0 & 1 \end{bmatrix} \quad \text{and } e^B = \begin{bmatrix} 1 & 0 \\ 0 & e^\zeta \end{bmatrix}$$

Show that

$$(A + B)^k = \begin{bmatrix} 0 & \zeta^{k-1} \\ 0 & \zeta^k \end{bmatrix} \quad \text{for } k = 1, 2, \ldots$$

and use the power series for e^t to compute

$$e^{A+B} = \begin{bmatrix} 1 & (e^\zeta - 1)/\zeta \\ 0 & e^\zeta \end{bmatrix}$$

Now let ζ be a nonzero root of the equation $e^z - z = 1$, for example, $\zeta = 2.08843\ldots + i \cdot 7.461489\ldots$, so that $e^{A+B} = \begin{bmatrix} 1 & 1 \\ 0 & e^\zeta \end{bmatrix}$; conclude that $e^A \cdot e^B \neq e^B \cdot e^A = e^{A+B}$ with this choice of ζ. See Problem 18 for another way to calculate this exponential.

The presence of nonreal complex numbers in the preceding exercises is inessential (see Problem 19 for equivalent real 4-by-4 examples), but the presence of a transcendental number like π is no accident. It is known that if all the entries of A, $B \in M_n$ are algebraic numbers and $n \geq 2$, then $e^A \cdot e^B = e^B \cdot e^A$ if and only if $AB = BA$. An *algebraic number* is a root of a polynomial equation with rational coefficients. Lindemann's famous 1882 theorem on the transcendence of π asserts that π is not an algebraic number; in 1873, Hermite proved that e is not an algebraic number.

Use of the Kronecker sum instead of the ordinary sum can sometimes permit one to overcome the problem of commutativity in establishing functional equations for primary matrix functions; See Problems 40 - 42.

Problems

1.   Show that the formulae of Sylvester (6.1.35-36), Buchheim (6.1.37), and Schwerdtfeger (6.1.38-39) extend directly to sufficiently smooth functions $f(t)$ that are not necessarily polynomials. Let $A \in M_n$ be given, and let

$$g(t) = (t - \lambda_1)^{s_1} \cdots (t - \lambda_\mu)^{s_\mu}$$ be a monic polynomial that annihilates A,

where $\lambda_1, \ldots, \lambda_\mu$ are distinct and all $s_i \geq 1$. The points $\{\lambda_i\}$ need not all be eigenvalues of A, but all the eigenvalues of A are in $\{\lambda_i\}$ and their multiplicities as zeroes of the minimal polynomial of A are not greater than the corresponding exponents in $g(t)$. Assume that $f(t)$ satisfies the domain and differentiability hypotheses in Definition (6.2.4) and let $f(A)$ be the primary matrix function with stem function $f(t)$. Verify:

(a) *Sylvester's formula*

$$f(A) = \sum_{i=1}^{k} f(\lambda_i) A_i \tag{6.2.39}$$

when A is diagonalizable; in this case, the *Frobenius covariants* A_i of A are defined in (6.1.36).

(b) *Buchheim's formula*

$$f(A) = \sum_{i=1}^{\mu} \left[\sum_{u=0}^{s_i-1} \frac{1}{u!} \varphi_i^{(u)}(\lambda_i)(A - \lambda_i I)^u \right] \prod_{\substack{j=1 \\ j \neq i}}^{\mu} (A - \lambda_j I)^{s_j} \tag{6.2.40}$$

where $\varphi_i(t) \equiv f(t)(t - \lambda_i)^{s_i} / g(t)$.

(c)   *Schwerdtfeger's formula*

$$f(A) = \sum_{i=1}^{\mu} A_i \sum_{k=0}^{s_i-1} \frac{1}{k!} f^{(k)}(\lambda_i)(A - \lambda_i I)^k \qquad (6.2.41a)$$

where the *Frobenius covariants* A_i are polynomials in A that are defined
in (6.1.40) for a general $A \in M_n$. If λ_i is *not* an eigenvalue of A, then
$A_i = 0$. If λ_i *is* an eigenvalue of A and the Jordan canonical form of A is
$A = SJS^{-1}$, then $A_i = SD_iS^{-1}$, where D_i is a block diagonal matrix that
is conformal with J; every block of J that has eigenvalue λ_i corresponds
to an identity block in D_i and all other blocks of D_i are zero. Notice
that $[A_i(A - \lambda_i I)]^k = A_i(A - \lambda_i I)^k$ is the kth power of a nilpotent
matrix corresponding to the λ_i-Jordan structure of A. Show that the
refinement (6.1.41) of Schwerdtfeger's formula

$$f(A) = \sum_{\substack{i=1,\dots,\mu \\ \lambda_i \in \sigma(A)}} A_i \sum_{k=0}^{r_i-1} \frac{1}{k!} f^{(k)}(\lambda_i)(A - \lambda_i I)^k \qquad (6.2.41b)$$

is also correct in this case. The outer sum in (6.2.41b) is over only those
indices i such that λ_i is an eigenvalue of A; for such an index i, r_i is the
multiplicity of λ_i as a zero of the minimal polynomial of A. Notice that
(6.2.41b) is exactly the same as (6.2.41a) when $g(\cdot)$ is the minimal polynomial of A. Explain how (6.2.41b) may be viewed as a way of gluing together
the formulae (6.2.7). In particular, for $f(t) = 1/(s-t)$, show that

$$(sI - A)^{-1} = \sum_{i=1}^{\mu} A_i \sum_{k=0}^{r_i-1} \frac{1}{(s-\lambda_i)^{k+1}} (A - \lambda_i I)^k \qquad (6.2.41c)$$

whenever $s \notin \sigma(A)$ and the minimal polynomial of A is $q(t) = (t-\lambda_1)^{r_1} \cdots$
$(t-\lambda_\mu)^{r_\mu}$, $\lambda_i \neq \lambda_j$ when $i \neq j$. Notice that this formula glues together the
expressions for the inverses of Jordan blocks obtained in Example (6.2.13).
This formula plays an important role in Section (6.6), where we develop a
chain rule formula for the derivative of $f(A(t))$.

In each of the preceding three cases, assume that $f(t)$ has whatever derivatives may be required at the roots of $g(t) = 0$ in order to evaluate the right-
hand side of each formula.

2. Suppose $A \in M_n$ is given and $f(A)$ is any matrix function (not necessarily a primary matrix function) that has the property (6.2.9(c)). If $B \in M_n$ commutes with A and is nonsingular, show that $B^{-1}f(A)B = f(A)$, and hence $f(A)B = Bf(A)$. Use a limit argument to show that $f(A)$ commutes with every $B \in M_n$ that commutes with A and use (4.4.19) to conclude from this that $f(A)$ must be a polynomial in A.

3. Let $f(t)$ be a given scalar-valued stem function, and let $A, B \in M_n$ be such that the primary matrix functions $f(A)$ and $f(B)$ can be defined by (6.2.4). If A and B commute, show that $f(A)$ and $f(B)$ commute. If $f(A)$ and $f(B)$ commute, must A and B commute?

4. Use Definition (6.2.4) to show that $\det e^A = e^{\operatorname{tr} A}$ for every $A \in M_n$. Use the power series for e^t to come to the same conclusion.

5. Use the fact that every matrix is similar to its transpose to show that $f(A^T)$ is similar to $f(A)$, where $f(A)$ and $f(A^T)$ are primary matrix functions. If A is symmetric, show that $f(A)$ must also be symmetric.

6. Consider the primary matrix functions e^A, $\sin A$, and $\cos A$ on M_n. Show that $e^{iA} = \cos A + i \sin A$. If $A \in M_n(\mathbb{R})$ or, more generally, if $A \in M_n$ and $\operatorname{tr} A$ is real, show that $|\det e^{iA}| = 1$. Show that $\cos(A + B) = \cos A \cos B - \sin A \sin B$ if $A, B \in M_n$ commute, but that this identity does not hold in general. For any given matrix norm $\|\|\cdot\|\|$ on M_n, show that

$$\|\|\cos A - \cos B\|\| \leq \cosh(\|\|B\|\| + \|\|A - B\|\|) - \cosh\|\|B\|\|$$

$$\leq \|\|A - B\|\| \cosh(\|\|B\|\| + \|\|A - B\|\|)$$

and deduce from these inequalities that the primary matrix function $\cos A$ is continuous on M_n.

7. Let $A \in M_n$ have the minimal polynomial $q_A(t) = (t + 1)^2(t - 2)^3$. Express e^A as a polynomial in A.

8. Let $A \in M_n$ have the minimal polynomial $q_A(t) = (t - \lambda_1)(t - \lambda_2)^2$, and let $f(t)$ be any given function that is defined at λ_1 and in a neighborhood of λ_2, and is differentiable at λ_2. Suppose $f(\lambda_1)$, $f(\lambda_2)$, and $f'(\lambda_2)$ are given. Find an explicit polynomial $r(t)$ of degree two such that $r(A) = f(A)$, a primary matrix function, for any A and $f(\cdot)$ that meet these conditions.

9. Show that $(e^A)^* = e^{A^*}$. Show that e^A is Hermitian if A is Hermitian and that it is unitary if A is skew-Hermitian.

10. Show that every unitary matrix $U \in M_n$ can be written as $U = e^{iA}$ for some Hermitian matrix $A \in M_n$.

11. Show that if $A \in M_n$ is skew-Hermitian, then $I + A$ is nonsingular and $B = (I - A)(I + A)^{-1}$ is unitary. Conversely, if $B \in M_n$ is a unitary matrix such that $I + B$ is nonsingular, show that $A = (I - B)(I + B)^{-1}$ is skew-Hermitian. What happens if A (or B) is real? This transformation between skew-Hermitian and unitary matrices is known as a *Cayley transform*.

12. Let the scalar-valued function $f(t)$ have domain $D \subset \mathbb{C}$.
  (a) Show that the primary matrix function $f(A)$ is normal for every normal $A \in M_n$ with $\sigma(A) \subset D$.
  (b) If $D \subset \mathbb{R}$, show that the primary matrix function $f(A)$ is Hermitian for every Hermitian $A \in M_n$ with $\sigma(D) \subset D$ if and only if $f(t)$ is real-valued on D.

13. If $A \in M_n$ and $I_m \in M_m$, use the mixed-product property in Lemma (4.2.10) to show that $(A \otimes I_m)^k = A^k \otimes I_m$ and $(I_m \otimes A)^k = I_m \otimes A^k$ for all $k = 1, 2, \ldots$. Show that if the primary matrix function $f(A)$ is defined, then so are the primary matrix functions $f(A \otimes I)$ and $f(I \otimes A)$. Show that $f(A \otimes I_m) = f(A) \otimes I_m$ and $f(I_m \otimes A) = I_m \otimes f(A)$.

14. Let $I_n, A \in M_n$ and $I_m, B \in M_m$ and recall from Theorem (4.4.5) that $A \otimes I_m$ and $I_n \otimes B$ commute.
  (a) Even though $e^{A+B} \neq e^A e^B$ in general, use Problem 13 to show that

$$e^{A \otimes I_m + I_n \otimes B} = e^A \otimes e^B$$

always. Thus, use of the Kronecker sum and product instead of the ordinary matrix sum and product can help to preserve the formal validity of some two-variable scalar functional equations for matrix functions.
  (b) Explain why the ordinary differential equation $X'(t) = AX(t) + X(t)B$, $X(0)$ given, $X(t) \in M_{n,m}$, is equivalent to

$$\frac{d}{dt} \operatorname{vec} X(t) = (I_m \otimes A + B^T \otimes I_n) \operatorname{vec} X(t)$$

Show that the latter differential equation has the solution

$$\text{vec } X(t) = e^{t(I_m \,\bullet\, A + B^T \,\bullet\, I_n)}\text{vec } X(0) = (e^{tB^T} \bullet\, e^{tA})\text{vec } X(0)$$

and explain why this is equivalent to $X(t) = e^{tA} X(0) e^{tB}$.

15. Show that

$$\sin(A \bullet I_m + I_n \bullet B) = (\sin A) \bullet (\cos B) + (\cos A) \bullet (\sin B)$$

for all $A \in M_n$, $B \in M_m$.

16. Consider the absolute value function $f(z) = |z|$, $z \in \mathbb{C}^n$. Discuss if or how $f(A)$ can be defined for $A \in M_n$. Can it be a primary matrix function? What about $f(A)$ for Hermitian $A \in M_n$? What about setting $f(A) = (AA^*)^{\frac{1}{2}}$, $(A^*A)^{\frac{1}{2}}$, $(A\overline{A})^{\frac{1}{2}}$, $(\overline{A}A)^{\frac{1}{2}}$, or Σ (where $A = V\Sigma W^*$ is a singular value decomposition of A)? Which of the fundamental properties in Theorem (6.2.9) do these satisfy?

17. Show that $e^{(A+B)t} - e^{At}e^{Bt} = (BA - AB)t^2/2 +$ higher-order terms in t. Conclude that there is some $\epsilon > 0$ such that $e^{(A+B)t} = e^{At}e^{Bt}$ for all $t \in [0,\epsilon)$ if and only if A and B commute.

18. Show that

$$e^{\begin{bmatrix} a & c \\ 0 & b \end{bmatrix}} = \begin{bmatrix} e^a & d \\ 0 & e^b \end{bmatrix}$$

for any $a, b, c \in \mathbb{C}$, where $d = c(e^b - e^a)/(b - a)$ if $b \neq a$, and $d = ce^a$ if $b = a$. Use this formula to verify the evaluations of all the matrix exponentials e^A, e^B, and e^{A+B} in the three exercises following Theorem (6.2.38).

19. Verify that the complex 2-by-2 examples in the three exercises following Theorem (6.2.38) can all be converted to real 4-by-4 examples by identifying each complex entry $a + ib$ with a real matrix $\begin{bmatrix} a & -b \\ b & a \end{bmatrix}$. For example, in the first exercise, identify

$$A = \begin{bmatrix} 0 & 0 \\ 0 & 2\pi i \end{bmatrix} \text{ with } A \equiv \begin{bmatrix} 0 & 0 \\ 0 & X \end{bmatrix} \in M_4(\mathbb{R})$$

and identify

$$B = \begin{bmatrix} 0 & 1 \\ 0 & 2\pi i \end{bmatrix} \text{ with } B \equiv \begin{bmatrix} 0 & I \\ 0 & X \end{bmatrix} \in M_4(\mathbb{R})$$

where 0, I, $X \in M_2(\mathbb{R})$ and $X \equiv \begin{bmatrix} 0 & -2\pi \\ 2\pi & 0 \end{bmatrix}$. Verify that A and B do not commute, but $e^A \cdot e^B = e^B \cdot e^A = e^{A+B}$. What are the corresponding identifications and conclusions in the other two exercises?

20. Let $A \in M_n$ and $\lambda \in \mathbb{C}$ be given. In the Jordan canonical form of A, show that the number of Jordan blocks of all sizes with eigenvalue λ is equal to $n - \text{rank}(A - \lambda I)$.

21. Let $k \geq 3$ be a given positive integer and let $\lambda \in \mathbb{C}$ be given. Suppose $f(\cdot)$ is a given function whose domain includes a neighborhood of λ, suppose $f(t)$ is $(k-1)$-times differentiable at λ, and suppose $f'(\lambda) = f''(\lambda) = 0$ and either $k = 3$ or $f^{(3)}(\lambda) \neq 0$. Show by a direct calculation that the Jordan canonical form of the primary matrix function $f(J_k(\lambda))$ has exactly three blocks and that they are:

$$J_{k/3}(f(\lambda)) \oplus J_{k/3}(f(\lambda)) \oplus J_{k/3}(f(\lambda)) \text{ if } k \equiv 0 \bmod 3$$

$$J_{(k+2)/3}(f(\lambda)) \oplus J_{(k-1)/3}(f(\lambda)) \oplus J_{(k-1)/3}(f(\lambda)) \text{ if } k \equiv 1 \bmod 3$$

$$J_{(k+1)/3}(f(\lambda)) \oplus J_{(k+1)/3}(f(\lambda)) \oplus J_{(k-2)/3}(f(\lambda)) \text{ if } k \equiv 2 \bmod 3$$

Verify that the formula in Theorem (6.2.25) gives the same Jordan canonical form for $f(J_k(\lambda))$.

22. Let $A, B \in M_n$. Even though $e^{A+B} \neq e^A e^B$ in general, show that $\det e^{A+B} = \det e^A \det e^B$ always.

23. The bound on the modulus of continuity of the matrix exponential in Corollary (6.2.32) relies on the bound (6.1.13) on the modulus of continuity of a matrix monomial. Show that if the bound (6.1.31) for the monomial is used instead, then the resulting bound for the exponential is

$$\||e^{A+E} - e^A\|| \leq \||E\|| \exp(\||E\||) \exp(\||A\||) \tag{6.2.42}$$

Show that this upper bound is always poorer than the one in Corollary (6.2.32).

24. Let $A, B \in M_n$ be Hermitian.
   (a) Show that $\{e^{itA}: t \in \mathbb{R}\}$ is a subgroup of the group of unitary matrices in M_n.
   (b) Show that e^{itA} commutes with e^{isB} for all $t, s \in \mathbb{R}$ if and only if there

is some $\epsilon > 0$ such that e^{itA} commutes with e^{isB} for all $t,\, s \in (-\epsilon, \epsilon)$.

(c) If $f(\cdot)$ is an absolutely integrable complex-valued function on $\mathbb{R}$, denote its Fourier transform by

$$\hat{f}(u) \equiv \int_{-\infty}^{\infty} e^{iut} f(t)\, dt$$

Show in detail that

$$\hat{f}(A) \equiv \int_{-\infty}^{\infty} e^{itA} f(t)\, dt$$

where both $\hat{f}(A)$ and e^{itA} are primary matrix functions and the integral has the usual meaning of n^2 entrywise ordinary integrals.

(d) Now let $f(\cdot)$ and $g(\cdot)$ be absolutely integrable functions on $\mathbb{R}$ and assume that e^{itA} commutes with e^{isB} for all $t,\, s \in \mathbb{R}$. Show that $\hat{f}(A)$ commutes with $\hat{g}(B)$.

(e) Suppose that all the eigenvalues of both A and B lie in the finite real interval (a, b). It is a fact that there is an absolutely integrable function $f(\cdot)$ such that $\hat{f}(u) \equiv u$ for all $u \in [a, b]$. Use this fact to show that e^{itA} commutes with e^{isB} for all $t,\, s \in \mathbb{R}$ (or, equivalently, just for all $t,\, s \in (-\epsilon, \epsilon)$ for some $\epsilon > 0$) if and only if A commutes with B.

25. Let $I \in M_2$ and let $R = \begin{bmatrix} 1 & 0 \\ 0 & -1 \end{bmatrix}$. Show that R is a square root of I. Why is there no primary matrix function $f(\cdot)$ such that $f(I) = R$?

26. Let $f(A)$ denote the function that takes a positive definite matrix into its unique positive definite square root, that is, $f(A)$ is positive definite and $f(A)^2 = A$ for every positive definite A. Show that $f(A)$ is a primary matrix function and use Theorem (6.2.37) to show that it is a continuous function on the open cone of positive definite matrices in M_n.

27. Using the notation in Example (6.2.16), let $A \in M_n$ be given, and suppose $A = SJS^{-1}$ is its Jordan canonical form. If D_θ is the complex plane slit from 0 to ∞ in such a way that $\sigma(e^A) \subset D_\theta$, and if $\log t$ is a given branch of the logarithm on D_θ, show that the primary matrix function $\log e^A$ is defined and that $\log e^A = A + SGS^{-1}$, where G is a diagonal matrix whose diagonal entries are integral multiples of $2\pi i$.

28. Examples (6.2.15-16) and Problem 27 illustrate the difficulties involved in defining $\log A$ as a primary function for nonsingular $A \in M_n$. Show that these difficulties disappear if one restricts the domain of e^A to $\mathbf{Z}_n \equiv \{A \in M_n: \sigma(A) \subset \mathbb{R}\}$. Show that $\mathbf{Z}_n$ is a closed subset of M_n. Show that $\log A$ can be defined as a primary matrix function on $\mathbf{Z}_n^+ \equiv \{A \in M_n: \sigma(A) \subset (0, \infty)\}$, and that $e^{\log A} = A$ for all $A \subset \mathbf{Z}_n^+$, $e^A \in \mathbf{Z}_n^+$ for all $A \in \mathbf{Z}_n$, and $\log e^A = A$ for all $A \in \mathbf{Z}_n$. As usual, e^A may be evaluated either as a primary matrix function or via its power series.

29. Provide details for the proof of the final assertion about Lipschitz continuity of $f(A)$ in Theorem (6.2.28).

30. Let $A \in M_n$ and the scalar-valued function $f(t)$ satisfy the hypotheses of Definition (6.2.4). Let A have minimal polynomial $q_A(t) = (t - \lambda_1)^{r_1} \cdots (t - \lambda_\mu)^{r_\mu}$, where $\lambda_1, \ldots, \lambda_\mu$ are distinct and all $r_i \geq 1$, and let A have Jordan canonical form $A = SJS^{-1}$. Let $\|\|\cdot\|\|_\infty$ denote the maximum absolute row sum matrix norm on M_n. Show that

$$\|\|f(A)\|\|_\infty \leq \kappa(S) \max_{1 \leq i \leq \mu} \sum_{k=0}^{r_i - 1} \frac{1}{k!} |f^{(k)}(\lambda_i)| \qquad (6.2.43)$$

where $\kappa(S) \equiv \|\|S\|\|_\infty \|\|S^{-1}\|\|_\infty$ is the condition number of S with respect to the norm $\|\|\cdot\|\|_\infty$. For the stem function $f(t) = e^t$, conclude that

$$\|\|e^A\|\|_\infty \leq \kappa(S) \max \{e^{\mathrm{Re}(\lambda)+1}: \lambda \in \sigma(A)\} \leq \kappa(S) \, e^{\rho(A)+1}$$

for every $A \in M_n$.

31. Let D be either a simply connected open subset of $\mathbb{C}$ or an open real interval. Let n be a given positive integer, and suppose $f(t)$ is a scalar-valued function that is n-times continuously differentiable on D, that is, $f'(t) \in A_n(D)$. If $A \in M_n$ is such that $tA \in \mathcal{D}_n(D)$ for all t in an open neighborhood $\mathcal{N}(t_0)$ of a given scalar t_0, use Theorem (6.2.34) and the fundamental theorem of calculus to prove the following identity for primary matrix functions:

$$f(tA) = f(t_0 A) + \int_{t_0}^{t} Af'(sA)\, ds \qquad \text{for all } t \in \mathcal{N}(t_0)$$

If, in addition, $f(t)$ is $(m + n)$-times continuously differentiable on D, that is, $f^{(m+1)} \in A_n(D)$, integrate by parts and show that

$$f(tA) = \sum_{k=0}^{m} \frac{1}{k!}(t-t_0)^k f^{(k)}(t_0 A) A^k + \frac{1}{m!}\int_{t_0}^{t}(t-s)^m f^{(m+1)}(sA)A^{m+1}\,ds$$

$$(6.2.44)$$

for all $t \in \mathcal{N}(t_0)$, where all the matrix functions are primary matrix functions. What does this give when $f(t)$ is a polynomial of degree $\le m$? What does this give when $f(t) = \log t$, $t_0 = 1$, and $m = 0$? In the special case of the stem function $f(t) = e^t$ and its associated primary matrix function $f(A) = e^A$, show that the integral remainder term in (6.2.44) tends to zero uniformly for all t in any given compact set as $m \to \infty$ and conclude that the primary matrix function e^{tA} can be represented as

$$e^{tA} = \sum_{k=0}^{\infty} \frac{1}{k!} t^k A^k$$

for every $A \in M_n$ and all $t \in \mathbb{C}$, where the convergence of the partial sums of this series is uniform for all t in any given compact set.

32. Starting with the entire analytic function $f(t) = \sin t$ as a stem function, use an argument similar to that used for the exponential function in Problem 31 to derive a power series representation for the primary matrix function $f(tA) = \sin tA$. Show carefully that the integral remainder term in (6.2.44) tends to zero for each fixed A uniformly for all t in any given compact set. Let $\xi \in \mathbb{C}^n$ be given and set $x(t) \equiv (\sin tA)\xi$. Show that $x''(t) = -A^2 x(t)$, $x(0) = 0$, and $x'(0) = A\xi$.

33. Use the explicit form of the inverse of a Jordan block given in Example (6.2.13) and the Cauchy integral theorem to show directly that, under the hypotheses of Theorem (6.2.28), the integral formula (6.2.29) agrees with the value of the primary matrix function $f(A)$ given by (6.2.4). Use this to show that the primary matrix function $f(A)$ is a continuous function of $A \in \mathcal{D}_n(D)$ whenever $f(t)$ is a complex-valued analytic function on a simply connected open set $D \subset \mathbb{C}$.

34. Provide details for the assertions about the Pythagorean identity for the sin and cos functions of complex and matrix arguments in Example (6.2.18).

35. Provide details for the calculations leading to Leibniz's rule in Example (6.2.18).

36. Provide details for the calculations leading to Faà di Bruno's rule in Example (6.2.21).

37. Let $D = (0,1)$ and $f(t) = t^{3/2}$. Show that $f(t)$ is continuously differentiable and has a bounded derivative on D, but the primary matrix function $f(A)$ is not Lipschitz continuous on $\mathcal{D}_2(D)$.

38. Suppose $f(t)$ is analytic in the disk $D_R = \{t \in \mathbb{C}: \ |t| < R\}$. Use the power series for $f(t)$ in D_R to prove (6.2.35) under appropriate hypotheses. Deduce (6.2.36) from this.

39. Consider the second-order initial-value problem

$$x''(t) = -Ax(t); \quad x(0) = \xi, \, x'(0) = \eta \text{ are given vectors in } \mathbb{C}^n$$

Suppose $A \in M_n$ is nonsingular and B is a square root of A, that is, $B^2 = A$. Consider the primary matrix functions $\cos Bt$ and $\sin Bt$, and use Theorem (6.2.34) to show that $x(t) \equiv (\cos Bt)\xi + (\sin Bt)B^{-1}\eta$ is a solution to the initial-value problem. Since $\cos u$ and $(\sin u)/u$ are both *even* analytic functions of u, that is, they are functions of u^2, show that the solution to the initial-value problem can be written as

$$x(t) = \left[\sum_{k=0}^{\infty} \frac{(-1)^k t^{2k}}{(2k)!} A^k \right] \xi + \left[\sum_{k=0}^{\infty} \frac{(-1)^k t^{2k+1}}{(2k+1)!} A^k \right] \eta$$

for *any* $A \in M_n$, singular or not. Show that this same conclusion is obtained by considering the equivalent first-order initial-value problem

$$y'(t) = Cy(t); \quad y(0) = [\xi^T \ \eta^T]^T \in \mathbb{C}^{2n}; \quad C \equiv \begin{bmatrix} 0 & I \\ -A & 0 \end{bmatrix}$$

where $y(t) \equiv [x(t)^T \ x'(t)^T]^T$. In this case, we always have $y(t) = e^{tC}y(0)$. Compute the explicit 2-by-2 block form of e^{tC}.

40. Let $A \in M_n$ be given and have eigenvalues $\lambda_1, \ldots, \lambda_n$ (including multiplicities), and let $f_{ij}(t)$, $i, j = 1, \ldots, n$, be n^2 given functions that are smooth enough to define $f_{ij}(A)$ as primary matrix functions. Show that the set of eigenvalues of the n^2-by-n^2 block matrix $B = [f_{ij}(A)]$ is the union (including multiplicities) of the sets of eigenvalues of the n matrices $[f_{ij}(\lambda_k)]$, $k = 1,\ldots, n$.

41. Let $A_0, A_1,\ldots \in M_n$ be a given infinite sequence of matrices such that for some matrix norm $|||\cdot|||$ on M_n either

(a) all $|||A_k||| \leq K/R^k$ for some finite positive constant K and some $R > 0$, or

(b) $A_k \neq 0$ for all sufficiently large k and $\lim_{k\to\infty} \dfrac{|||A_k|||}{|||A_{k+1}|||} = R > 0$, or

(c) $\lim_{k\to\infty} \sqrt[k]{|||A_k|||} = \dfrac{1}{R} > 0$.

Show that the infinite series

$$f(t) = \sum_{k=0}^{\infty} A_k t^k \tag{6.2.45}$$

converges with respect to the norm $|||\cdot|||$ for all $t \in \mathbb{C}$ such that $|t| < R$. Show that both of the infinite series

$$f(A) \equiv \sum_{k=0}^{\infty} A_k A^k \text{ and } \bar{f}(A) \equiv \sum_{k=0}^{\infty} A^k A_k \tag{6.2.46}$$

converge with respect to any norm $\|\cdot\|$ on M_n for all $A \in M_n$ such that $\rho(A) < R$.

42. Consider the functions $A \to \bar{A}$, $A \to A^T$, $A \to A^*$, $A \to \text{Re } A$, $A \to \text{H}(A) \equiv \frac{1}{2}(A + A^*)$, $A \to (A^*A)^{\frac{1}{2}}$, $A \to A^{-1}$ (nonsingular A). Which of these is a primary matrix function on M_n? On the diagonalizable matrices in M_n? On the normal matrices in M_n? On the Hermitian matrices in M_n?

43. The inverse function theorem, a result from analysis, was used to prove Corollary (6.2.12), while the proof of Theorem (6.2.25) is purely algebraic. Use Theorem (6.5.25) to prove the existence portion of Corollary (6.2.12): Show that, under the stated hypotheses on A and $f(\cdot)$ in Corollary (6.2.12), there is some X_0 such that $f(X_0) = A$, and describe how to construct X_0. When you construct X_0, what must you do to ensure that it is a primary matrix function of A?

44. Let $A, B \in M_n$ be given. If A commutes with B, show that A commutes with $f(B)$ for every stem function $f(t)$ for which the primary matrix function $f(B)$ is defined.

45. Suppose $f(t)$ is analytic in a simply connected open set $D \subset \mathbb{C}$ and let $A \in M_n$ be a given matrix whose eigenvalues lie in D. Use the integral representation (6.2.29) to prove (6.2.35).

46. Let $A \in M_n$ be given.

(a) If A is positive semidefinite, show that it has a unique positive semidefinite square root B, that is, $B^2 = A$.

(b) If A is positive definite, show that it has a unique Hermitian logarithm B, that is, $e^B = A$.

(c) If A is Hermitian, does it have a unique Hermitian arctan, that is, a B such that $\tan B = A$?

Further Readings: The problem of finding a continuous (or, more generally, a bounded) extension to not-necessarily diagonalizable matrices of a function $f(\cdot)$ defined by (6.2.1) on the diagonalizable matrices is discussed in P. C. Rosenbloom, Bounds on Functions of Matrices, *Amer. Math. Monthly* 74 (1967), 920–926. For more information on primary matrix functions, see [Cul], [Fer], and Chapter V of [Gan 59, Gan 86]. For a comprehensive historical survey of various ways to define $f(A)$, see R. F. Rinehart, The Equivalence of Definitions of a Matric Function, *Amer. Math. Monthly* 62 (1955), 395–413. The term *primary matrix function* for the value of $f(A)$ given in Definition (6.2.4) seems to be due to R. F. Rinehart in Elements of a Theory of Intrinsic Functions on Algebras, *Duke Math. J.* 27 (1960), 1–19. See Chapter IX of [Mac] for numerous references and some information about "multiple-valued analytic functions" of a matrix. For example, one can consider extending the meaning of (6.2.6-7) by allowing $f(\cdot)$ to take different values on different blocks corresponding to the same eigenvalue; this is discussed by M. Cipolla, Sulle Matrici Espressioni Analitiche de Un'Altra, *Rend. Circ. mat. Palermo* 56 (1932), 144–154. [Bel] discusses the matrix exponential with many examples. See Problem 33 in Section (6.5) for a bound on $||| e^{A+E} - e^A |||$ that is better than the one in Corollary (6.2.32). Our discussion of the possibilities for e^A, e^B, and e^{A+B} for noncommuting A and B is taken from E. M. E. Wermuth, Two Remarks on Matrix Exponentials, *Linear Algebra Appl.* 117 (1989), 127–132, which contains references to the literature on this fascinating special topic. For an interesting generalization of Theorem (6.2.38) on exponentials of commuting matrices, see Problem 35 in Section (6.5).

6.3   Hadamard matrix functions

In this section we consider a notion of "function of a matrix" that is entirely different from the one we have been discussing in Sections (6.1) and (6.2). If n^2 functions $f_{ij}(t)$ are given for $i, j = 1, 2,..., n$ with appropriate domains, and if $A = [a_{ij}] \in M_n$, the function $f: A \to f(A) = [f_{ij}(a_{ij})]$ is called a *Hadamard function* to distinguish it from the usual notion of matrix function.

If each function $f_{ij}(t)$ is a polynomial

$$f_{ij}(t) = a_{k_{ij}}(i,j)t^{k_{ij}} + a_{k_{ij}-1}(i,j)t^{k_{ij}-1} + \cdots + a_0(i,j) \qquad (6.3.1)$$

then $f(A) = [f_{ij}(a_{ij})] = [a_{k_{ij}}(i,j)\, a_{ij}^{k_{ij}} + \cdots]$ and we can write the function $f(t)$ as a matrix polynomial that formally looks like (6.1.2):

$$f(t) = A_m t^m + A_{m-1}t^{m-1} + \cdots + A_0,$$
$$m \equiv \max \{k_{ij}: i, j = 1,..., n\} \qquad (6.3.2)$$

Then

$$f(A) \equiv (A_m \circ A^{(m)}) + (A_{m-1} \circ A^{(m-1)}) + \cdots + A_0 \qquad (6.3.3a)$$

in which $A^{(k)} = [a_{ij}^k]$ is a Hadamard power and $A_k \circ A^{(k)}$ is a Hadamard product. Addition of terms in (6.3.3a) is ordinary matrix addition. Notice that because Hadamard multiplication is commutative, (6.3.3a) is identical to

$$f(A) = (A^{(m)} \circ A_m) + (A^{(m-1)} \circ A_{m-1}) + \cdots + A_0 \qquad (6.3.3b)$$

If each function $f_{ij}(t)$ is an analytic function with radius of convergence $R_{ij} > 0$, then the Hadamard function $f: A \to f(A) = [f_{ij}(a_{ij})]$ can be written as a power series

$$f(A) = \sum_{k=0}^{\infty} A_k \circ A^{(k)} \qquad (6.3.4a)$$

in which all the powers $A^{(k)}$ are Hadamard powers and all the products $A_k \circ A^{(k)}$ are Hadamard products. We must assume that $|a_{ij}| < R_{ij}$ for all $i, j = 1, 2, \ldots, n$.

Notationally and conceptually, the simplest case of (6.3.3a) and (6.3.4a) occurs when all the functions $f_{ij}(t)$ are the same, so $f(A) = [f(a_{ij})]$. If $f(t) = \Sigma_k a_k t^k$, then (6.3.4a) becomes

$$f(A) = \sum_{k=0}^{\infty} (a_k J) \circ A^{(k)} = \sum_{k=0}^{\infty} a_k A^{(k)} \qquad (6.3.4b)$$

where $J \in M_n$ has all entries $+1$ (this is the "identity matrix" with respect to Hadamard multiplication). A simple example of this situation is $f(t) = e^t$, for which

$$f(A) = [e^{a_{ij}}] = \sum_{k=0}^{\infty} \frac{1}{k!} A^{(k)} = \sum_{k=0}^{\infty} \frac{1}{k!} [a_{ij}^k]$$

Hadamard functions of a matrix arise naturally in the study of discretizations of integral operators, integral equations, and integral quadratic forms. Since positive definite integral operators are common in physical problems, many known results about Hadamard matrix functions concern Hadamard functions of positive definite matrices.

6.3.5   Theorem. Let $A \in M_n$ be positive semidefinite. If all the coefficient matrices A_k in (6.3.4a) are positive semidefinite, then the Hadamard function $f(A)$ defined by (6.3.4a) is positive semidefinite. In particular, if all the coefficients a_k in (6.3.4b) are nonnegative, then the Hadamard function $f(A)$ defined by (6.3.4b) is positive semidefinite.

Proof: The Schur product theorem (5.2.1) guarantees that each integer Hadamard power $A^{(k)}$ is positive semidefinite and that each Hadamard product $A_k \circ A^{(k)}$ is positive semidefinite. The result follows from the fact that a sum of positive semidefinite matrices is positive semidefinite.   ☐

Since $f(t) = e^t$ is an entire function with positive power series coefficients, the Hadamard function $e^{\circ A} = [e^{a_{ij}}]$ is positive semidefinite whenever $A = [a_{ij}]$ is positive semidefinite. Actually, a slightly stronger result holds in the special case of the exponential function. A matrix can be something less than positive semidefinite and still have a positive semidefinite Hadamard exponential.

6.3.6   **Theorem.** If a Hermitian matrix $A = [a_{ij}] \in M_n$ has the property that $x^* A x \geq 0$ for all $x = [x_i] \in \mathbf{C}^n$ such that $x_1 + x_2 + \cdots + x_n = 0$, then the Hadamard exponential $e^{\circ A} \equiv [e^{a_{ij}}]$ is positive semidefinite.

Proof: Let $e = [1, 1, ..., 1]^T \in \mathbf{C}^n$, and observe that for any $x \in \mathbf{C}^n$, the vector $\hat{x} \equiv x - \frac{1}{n} ee^* x$ has the property that the sum of its components is zero ($e^* \hat{x} = 0$). Thus for any $x \in \mathbf{C}^n$,

$$0 \leq \hat{x}^* A \hat{x} = (x - \tfrac{1}{n} ee^* x)^* A (x - \tfrac{1}{n} ee^* x)$$

$$= x^* \{ A - \tfrac{1}{n} ee^* A - \tfrac{1}{n} Aee^* + (\tfrac{1}{n})^2 (e^* Ae) ee^* \} x$$

This says that there is a positive semidefinite matrix

$$B \equiv A - \tfrac{1}{n} ee^* A - \tfrac{1}{n} Aee^* + (\tfrac{1}{n})^2 (e^* Ae) ee^* = [b_{ij}]$$

and a vector $y \equiv \frac{1}{n} Ae - \frac{1}{2}(\frac{1}{n})^2 (e^* Ae) e = [y_i]$ such that $A = B + ye^* + ey^*$, that is, $a_{ij} = b_{ij} + y_i + \bar{y}_j$. But then

$$e^{\circ A} = \left[e^{b_{ij} + y_i + \bar{y}_j} \right] = \left[e^{b_{ij}} e^{y_i} e^{\bar{y}_j} \right]$$

and for any $z = [z_i] \in \mathbf{C}^n$ we have

$$z^* e^{\circ A} z = \sum_{i,j=1}^{n} e^{b_{ij}} \left[z_i e^{y_i} \right] \left[\overline{z_j e^{y_j}} \right] \geq 0$$

by Theorem (6.3.5), since B is positive definite.   ☐

Remark. A Hermitian matrix $A \in M_n$ that satisfies the hypothesis of Theorem (6.3.6) is said to be *conditionally positive semidefinite*, or, sometimes, *almost positive semidefinite*.

Considered as Hadamard functions, $f(t) = e^t$ and $f(t) = t^k$ for any $k =$

1, 2,... have the property that they take positive semidefinite matrices into positive semidefinite matrices. What other functions have this property? A complete answer to this question is not known, but it is not difficult to develop some fairly strong necessary conditions.

6.3.7   Theorem. Let $f(\cdot)$ be an $(n-1)$-times continuously differentiable real-valued function on $(0, \infty)$, and suppose that the Hadamard function $f(A) \equiv [f(a_{ij})]$ is positive semidefinite for every positive semidefinite matrix $A = [a_{ij}] \in M_n$ that has positive entries. Then $f^{(k)}(t) \geq 0$ for all $t \in (0, \infty)$ and all $k = 0, 1,..., n-1$.

Proof: If $n = 1$, the assumptions guarantee trivially that $f(t) \geq 0$ for all $t > 0$. We assume that $n \geq 2$ and proceed by induction, observing that the hypotheses guarantee that $f(B)$ is positive semidefinite whenever $B \in M_m$ is positive semidefinite and has positive entries, $1 \leq m \leq n$. Thus, we may assume that $f^{(k)}(t) \geq 0$ for all $t > 0$ and all $k = 0, 1,..., n-2$, and we must prove that $f^{(n-1)}(t) \geq 0$ for all $t > 0$.

Let $a > 0$ and $\alpha = [\alpha_i]_{i=1}^n \in \mathbb{R}^n$ so that for all sufficiently small $t > 0$, the matrix $A(t) \equiv [a + t\alpha_i\alpha_j]_{i,j=1}^n$ is positive semidefinite and has positive entries. But then by hypothesis $f(A(t)) = [f(a + t\alpha_i\alpha_j)]_{i,j=1}^n$ is positive semidefinite so $\Delta(t) \equiv \det f(A(t)) \geq 0$. Because $f(A(0))$ has all columns identical, $\Delta(0) = 0$. Furthermore,

$$\frac{d^k}{dt^k} \Delta(t) \bigg|_{t=0} \equiv \Delta^{(k)}(0) = 0$$

for all $k = 0, 1,..., (n(n-1)/2) - 1$ because in each case there is at least one pair of proportional columns in each of the n^k determinant terms arising from the k-fold differentiation. Let $N \equiv n(n-1)/2$ and conclude from an explicit calculation that

$$0 \leq \lim_{t \to 0} \frac{\Delta(t)}{t^N} = \Delta^{(N)}(0)$$

$$= \binom{N}{1} \binom{N-1}{2} \binom{N-3}{3} \binom{N-6}{4} \cdots \binom{2n-3}{n-2} V^2(\alpha) f(a) f'(a) \cdots f^{(n-1)}(a)$$

where $\binom{m}{l} \equiv \dfrac{m!}{l!(m-l)!}$ is the binomial coefficient and

$$V(\alpha) \equiv \det \begin{bmatrix} 1 & \alpha_1 & \alpha_1^2 & \cdots & \alpha_1^{n-1} \\ 1 & \alpha_2 & \alpha_2^2 & \cdots & \alpha_2^{n-1} \\ \vdots & \vdots & \vdots & \cdots & \vdots \\ 1 & \alpha_n & \alpha_n^2 & \cdots & \alpha_n^{n-1} \end{bmatrix} = \prod_{1 \le j < i \le n} (\alpha_i - \alpha_j)$$

is the Vandermonde determinant (see (0.9.11) in [HJ]); notice that $V^2(\alpha) > 0$ if all $\alpha_i \neq \alpha_j$.

Therefore, we find, in general, that $f(a)f'(a) \cdots f^{(n-1)}(a) \ge 0$ is a necessary condition, and our induction hypothesis is that $f(a)$, $f'(a)$, ..., $f^{(n-2)}(a)$ are all nonnegative. Since we cannot conclude directly from this that $f^{(n-1)}(a) \ge 0$, recall that $g(t) \equiv t^n$ has the property asserted for $f(t)$ and so the function $f(t) + \tau g(t)$ has the same property for all $\tau \ge 0$. For $\tau \ge 0$, we have just shown that

$$p(\tau) \equiv (f(a) + \tau a^n)(f'(a) + \tau n a^{n-1}) \cdots (f^{(n-1)}(a) + \tau n! a) \ge 0$$

Now, $p(\tau)$ is a polynomial with at most n real zeroes, so for some $\epsilon > 0$, $p(\tau) > 0$ when $0 < \tau < \epsilon$. Thus, because we know $f(a)$, $f'(a)$,..., $f^{(n-2)}(a)$ are all nonnegative, we find $(f^{(n-1)}(a) + \tau n! a) > 0$ for all sufficiently small $\tau > 0$. We conclude that $f^{(n-1)}(a) \ge 0$ for all $a > 0$.   ☐

Remark. The smoothness assumption on $f(\cdot)$ in Theorem (6.3.7) is not necessary. If $f(\cdot)$ is merely a continuous function, one can use the argument of the proof to show that all the divided differences of $f(\cdot)$ up to order $n-1$ are nonnegative. From this alone it follows that $f(\cdot)$ must be $(n-3)$-times continuously differentiable and $f^{(n-3)}(t)$ must be nondecreasing and convex. Thus, considerable smoothness of $f(\cdot)$ follows from the property that it leaves invariant (as a Hadamard function) the cone of positive semidefinite matrices.

We know that all the integer powers $f(t) = t^k$ leave invariant (as Hadamard functions) the positive semidefinite matrices, but what about noninteger powers $f(t) = t^\alpha$? The necessary condition in Theorem (6.3.7) permits us to exclude some values of α at once. If $a_{ij} \ge 0$, then by a_{ij}^α we always mean the nonnegative choice of a_{ij}^α.

6.3.8   Corollary. Let $0 < \alpha < n-2$, α not an integer. There is some positive semidefinite matrix $A = [a_{ij}] \in M_n$ with positive entries such that the Hadamard power $A^{(\alpha)} \equiv [a_{ij}^\alpha]$ is not positive semidefinite.

Proof: Apply the theorem to $f(t) = t^\alpha$ and observe that if $\alpha < n-2$ is not an integer then not all the derivatives $f'(t) = \alpha t^{\alpha-1}$, $f''(t) = \alpha(\alpha-1)t^{\alpha-2},...,$ $f^{(n-1)}(t) = \alpha(\alpha-1)(\alpha-2)\cdots(\alpha-n+2)t^{\alpha-n+1}$ are nonnegative for $t > 0$. ☐

Although this corollary shows that the Schur Product Theorem (5.2.1) cannot be generalized to cover *all* fractional Hadamard powers of a positive semidefinite matrix with positive entries, it does identify precisely the values of α for which the fractional Hadamard powers $A^{(\alpha)}$ are not always positive semidefinite.

6.3.9   Theorem. Let $A = [a_{ij}] \in M_n$ be a positive semidefinite matrix with nonnegative entries. If $\alpha \geq n-2$, then the Hadamard power $A^{(\alpha)} \equiv [a_{ij}^\alpha]$ is positive semidefinite. Furthermore, the lower bound $n-2$ is, in general, the best possible.

Proof: If $n = 2$ and

$$A = \begin{bmatrix} a_{11} & a_{12} \\ a_{12} & a_{22} \end{bmatrix}$$

is nonsingular, we have $a_{11} > 0$ and $\det A = a_{11}a_{22} - (a_{12})^2 > 0$, or $a_{11}a_{12} > (a_{22})^2$. For any $\alpha > 0$ we also have $(a_{11})^\alpha > 0$ and $(a_{11})^\alpha(a_{12})^\alpha > (a_{12})^{2\alpha}$, or $(a_{11})^\alpha(a_{12})^\alpha - (a_{12})^{2\alpha} > 0$. By the determinant criterion in Theorem (7.2.5) in [HJ] we conclude that $A^{(\alpha)}$ is positive definite for all $\alpha > n-2 = 2-2 = 0$. The result for a singular A follows from a limiting argument.

Now let $n \geq 3$ and assume that the theorem has been proved for all matrices of order at most $n-1$ that satisfy the hypotheses. Partition A as

$$A = \begin{bmatrix} B & \omega \\ \omega^T & a_{nn} \end{bmatrix}$$

where $B = [a_{ij}]_{i,j=1}^{n-1} \in M_{n-1}$ and $\omega^T = [a_{n,1}, a_{n,2},..., a_{n,n-1}]$. Let $\eta^T = [\xi^T \ z]$, where $\xi \in \mathbb{C}^{n-1}$ and $z \in \mathbb{C}$, and observe that

$$\eta^* A \eta = \xi^* B \xi + 2\text{Re}(z\xi^* \omega) + a_{nn}|z|^2 \geq 0 \qquad (6.3.10)$$

for all $\xi \in \mathbb{C}^{n-1}$ and all $z \in \mathbb{C}$ because A is positive semidefinite. If $a_{nn} = 0$, this inequality shows that A cannot be positive semidefinite unless $\omega = 0$ and the last row and column of A vanish. In this event, $A^{(\alpha)}$ is positive semidefinite for all $\alpha \geq (n-1) - 2 = n - 3$ by the induction hypothesis.

We may therefore assume that $a_{nn} > 0$, and set $z = -\omega^T \xi / a_{nn}$ in (6.3.10) to obtain $\eta^* A \eta = \xi^* (B - \omega \omega^T / a_{nn}) \xi \geq 0$. Since $\xi \in \mathbb{C}^{n-1}$ is arbitrary, we conclude that $B - \omega \omega^T / a_{nn}$ is positive semidefinite. Notice that if we set

$$\zeta \equiv [\omega^T \ a_{nn}]^T / \sqrt{a_{nn}} = [a_{1n}, a_{2n}, ..., a_{nn}]^T / \sqrt{a_{nn}}$$

then

$$A - \zeta \zeta^T = \begin{bmatrix} B - \omega \omega^T / a_{nn} & 0 \\ 0 & 0 \end{bmatrix} \tag{6.3.11}$$

which we have just seen is positive semidefinite.

An elementary identity now permits us to conclude the proof. If $\alpha \geq 1$ and if $a, c \geq 0$, explicit evaluation of the integral shows that

$$a^\alpha = c^\alpha + \alpha \int_0^1 (a - c)\{ta + (1-t)c\}^{\alpha-1} dt$$

Entrywise application of this identity yields the matrix identity

$$[a_{ij}^\alpha] \equiv A^{(\alpha)} = (\zeta \zeta^T)^{(\alpha)} + \alpha \int_0^1 \left[A - \zeta \zeta^T \right] \circ \left[tA + (1-t)\zeta \zeta^T \right]^{(\alpha-1)} dt \tag{6.3.12}$$

for all $\alpha \geq 1$. By the Schur product theorem (5.2.1), and because the last row and column of $A - \zeta \zeta^T$ vanish, the integrand will be positive semidefinite if the upper left $(n-1)$-dimensional principal submatrix of $tA + (1-t)\zeta \zeta^T$ has the property that its $(\alpha-1)$th Hadamard power is positive semidefinite. But by the induction hypothesis, this is the case if $\alpha - 1 \geq (n-1) - 2 = n - 3$, that is, if $\alpha \geq n - 2$. The integral term is the limit of Riemann sums that are nonnegative linear combinations of positive semidefinite matrices and hence is positive semidefinite. The first term on the right-hand side of (6.3.12) is positive semidefinite because it is of the form $(\zeta \zeta^T)^{(\alpha)} = (\zeta_i^\alpha \zeta_j^\alpha) = yy^T$, $y_i = \zeta_i^\alpha$, which is trivially positive semidefinite. Corollary (6.3.8) shows that $n - 2$ is the best possible lower bound.   ☐

Although noninteger Hadamard powers $A^{(\alpha)}$ of a positive semidefinite matrix $A \in M_n$ need not be positive semidefinite if $\alpha < n-2$, there are matrices for which this is the case. For example, if $A \in M_n$ is conditionally positive semidefinite (satisfies the hypotheses of Theorem (6.3.6)) and if $\alpha > 0$, then αA is conditionally positive semidefinite and the Hadamard exponential $e^{\circ \alpha A} = \left[e^{\alpha a_{ij}} \right] = \left[e^{a_{ij}} \right]^{(\alpha)}$ is positive semidefinite by (6.3.6). This example is typical of all positive matrices that have the property that all their positive Hadamard powers are positive semidefinite.

6.3.13   Theorem. Let $A = [a_{ij}] \in M_n(\mathbb{R})$ be a symmetric matrix with positive entries. Then the Hadamard power $A^{(\alpha)} = [a_{ij}^\alpha]$ is positive semidefinite for all $\alpha > 0$ if and only if the Hadamard logarithm $B = \log^\circ(A) \equiv [\log a_{ij}]$ is conditionally positive definite, that is, $x^* B x \geq 0$ for all $x = [x_i] \in \mathbb{C}^n$ such that $x_1 + x_2 + \cdots + x_n = 0$.

Proof: Let $J \in M_n$ be the matrix with all entries equal to $+1$. If $A^{(\alpha)}$ is positive semidefinite for all $\alpha > 0$ and if $x \in \mathbb{C}^n$ is such that $x_1 + x_2 + \cdots + x_n = 0$, then

$$x^* \frac{1}{\alpha} A^{(\alpha)} x = x^* \left[\frac{1}{\alpha} A^{(\alpha)} - \frac{1}{\alpha} J \right] x = x^* \left[\frac{1}{\alpha} \left[A^{(\alpha)} - J \right] \right] x \geq 0$$

If we let $\alpha \to 0$ in this inequality and use the fact that $\lim_{\alpha \to 0} (a^\alpha - 1)/\alpha = \log a$ for any $a > 0$, the necessity of the condition follows. For sufficiency, use Theorem (6.3.6) to show that $A^{(\alpha)} = \left[e^{\alpha \log a_{ij}} \right]$ is positive semidefinite.   ☐

The result of Theorem (6.3.13) can be generalized to cover matrices $A \in M_n$ with nonnegative real entries (see Problem 6), and even to Hermitian matrices if care is taken to define the fractional powers a_{ij}^α consistently.

A positive semidefinite matrix $A \in M_n$ such that $A^{(\alpha)}$ is positive semidefinite for all $\alpha > 0$ is said to be *infinitely divisible*. The term comes from the theory of *infinitely divisible characteristic functions* in probability, which are continuous functions $\phi : \mathbb{R} \to \mathbb{C}$ with $\phi(0) = 1$ such that the kernel $K^\alpha(s,t) = \phi^\alpha(s-t)$ is positive semidefinite for all $\alpha > 0$.

Problems

1.   Consider the Hadamard functions $f_\alpha(A) \equiv [|a_{ij}|^\alpha]$, $\alpha > 0$. Show that $f_\alpha(A)$ is positive semidefinite whenever $A \in M_n$ is positive semidefinite

provided that $\alpha = 2, 4, 6,..., 2n-4$ or $\alpha \geq 2n-4$, but this need not be true for $\alpha = 1$.

2.   Denote the *incidence matrix of* $A = [a_{ij}] \in M_{m,n}$ by $M(A) \equiv [\mu_{ij}]$, where

$$\mu_{ij} = \begin{cases} 1 \text{ if } a_{ij} \neq 0 \\ 0 \text{ if } a_{ij} = 0 \end{cases}$$

If $A \in M_n$, $A \geq 0$, and A is infinitely divisible, show that $M(A)$ is positive semidefinite.

3.   If $A = [a_{ij}] \in M_n(\mathbb{R})$ has nonnegative entries and if A is infinitely divisible, show that $x^*[M(A) \circ \log^\circ(A)]x \geq 0$ for all $x \in \mathbb{C}^n$ such that $M(A)x = 0$. In the Hadamard product, $\log^\circ(A) \equiv [\log a_{ij}]$, and we agree that $0\log 0 \equiv 0$.

4.   If $A \in M_n$ is such that $M(A)$ is symmetric, show that $M(A)$ is positive semidefinite if and only if its graph $\Gamma(M(A))$ has the property that if any two nodes are connected by a path in $\Gamma(M(A))$ then they must be connected by an arc in $\Gamma(M(A))$.

5.   Show that the condition on $\Gamma(M(A))$ in Problem 4 is equivalent to the existence of a permutation matrix $P \in M_n$ such that $PM(A)P^T = B_1 \oplus B_2 \oplus \cdots \oplus B_k$ where each $B_i \in M_{n_i}$ has all entries $+1$.

6.   Use Theorem (6.3.13) and the result of Problem 5 to show that the necessary conditions in Problems 2 and 3 are also sufficient: If $A = [a_{ij}]$ is a nonnegative symmetric matrix, if $M(A)$ is positive semidefinite, and if $x^*[M(A) \circ \log^\circ(A)]x \geq 0$ for all $x \in \mathbb{C}^n$ such that $M(A)x = 0$, then A is infinitely divisible.

7.   Let

$$f(t) = a_0 + a_1 t + a_2 t^2 + \cdots + a_{n-2} t^{n-2} + \int_{\alpha=n-2}^{\infty} t^\alpha \, d\mu(\alpha)$$

where all $a_i \geq 0$, the measure $d\mu$ is nonnegative (you may take $d\mu(\alpha) = g(\alpha)d\alpha$ for $g(\alpha) \geq 0$ if you wish), and the integral is convergent for all $t \in [0, K]$ for some $K > 0$. Show that the Hadamard function $f(A)$ is positive semidefinite for all positive semidefinite $A = [a_{ij}] \in M_n(\mathbb{R})$ with $0 \leq a_{ij} \leq K$.

8.   Show that a Hermitian matrix $B = [b_{ij}] \in M_n$ is conditionally positive semidefinite if and only if the matrix

$$\Delta B \equiv \left[b_{ij} + b_{i+1,j+1} - b_{i,j+1} - b_{i+1,j} \right]_{i,j=1}^{n-1} \in M_{n-1}$$

is positive semidefinite.

9. Use Problem 8 to show that a continuously differentiable kernel $K(s,t)$ defined on $(a,b) \times (a,b) \subset \mathbb{R}^2$ has the property that

$$\int_a^b \int_a^b K(s,t)\, \bar\phi(s)\phi(t)\, dsdt \geq 0$$

for all continuous functions $\phi(\cdot)$ with compact support in (a,b) such that

$$\int_a^b \phi(t)\, dt = 0$$

if and only if

$$\int_a^b \int_a^b \frac{\partial^2}{\partial s \partial t} K(s,t)\bar\phi(s)\phi(t)\, dsdt \geq 0 \qquad (6.3.14)$$

for all continuous functions $\phi(\cdot)$ with compact support in (a,b).

10. Let $\phi: \mathbb{R} \to \mathbb{C}$ be a given infinitely divisible characteristic function. Use Problem 2 to show that $\phi(t) \neq 0$ for all $t \in \mathbb{R}$.

11. Let $A = [a_{ij}] \in M_n$ be given with Re $a_{ij} > 0$ for all $i, j = 1,\dots, n$, and suppose $-A$ is conditionally positive semidefinite. Let $f(t) = 1/t$ and show that the Hadamard reciprocal $f(A) \equiv [1/a_{ij}]$ is positive semidefinite. If A has real entries, show that $f(A)$ is infinitely divisible.

12. Let $B \equiv [\max \{i,j\}] \in M_n$, and let $A \equiv [1/\max \{i,j\}]$ be the Hadamard reciprocal of B. Use Problems 8 and 11 to show that A is positive definite and infinitely divisible.

13. Consider the Hilbert matrix $A = [a_{ij}] \equiv [1/(i+j-1)] \in M_n$. Use Theorem (6.3.13) to show that A is positive definite and infinitely divisible.

14. If $A \in M_n(\mathbb{R})$ is symmetric and has positive entries, a theorem of Marcus and Newman guarantees that there is a positive diagonal $D \in M_n(\mathbb{R})$ such that the positive symmetric matrix $DAD \equiv B$ is doubly stochastic. Use this

fact and Problem 11 to prove a theorem of R. Bapat: Show that if $A = [a_{ij}]$ $\in M_n(\mathbb{R})$ is symmetric and has positive entries, and if A has exactly one positive eigenvalue, then its Hadamard reciprocal $[1/a_{ij}]$ is infinitely divisible.

Further Readings: For more information about noninteger Hadamard powers and infinitely divisible matrices see On Infinitely Divisible Matrices, Kernels, and Functions, *Z. Wahrscheinlichkeitstheorie* 8 (1967), 219-230 by R. A. Horn; The Theory of Infinitely Divisible Matrices and Kernels, *Trans. Amer. Math.Soc.* 136 (1969), 269-286 by R. A. Horn; On Fractional Hadamard Powers of Positive Definite Matrices, *J. Math. Analysis Appl.* 61 (1977), 633-642 by C. H. FitzGerald and R. A. Horn. The results mentioned in Problem 14 are in M. Marcus and M. Newman, The Permanent of a Symmetric Matrix, *Notices Amer. Math. Soc.* 8 (1961), 595; R. Sinkhorn, A Relationship Between Arbitrary Positive Matrices and Doubly Stochastic Matrices, *Ann. Math. Statist.* 35 (1964), 876-879; and R. B. Bapat, Multinomial Probabilities, Permanents, and a Conjecture of Karlin and Rinott, *Proc. Amer. Math. Soc.* 102 (1988), 467-472.

6.4 Square roots, logarithms, and nonlinear matrix equations

We have been discussing how to define and evaluate $f(A)$ for $A \in M_n$; we now consider solving matrix equations of the form $f(X) = A$ for $X \in M_n$. All matrix functions in this section are the kinds of ordinary matrix functions discussed in Sections (6.1-2); Hadamard functions play no role in this section.

Many examples of linear matrix equations such as $AX + XB + C = 0$ have been considered in Chapter 4. In this section we are interested in finding or characterizing all matrices $X \in M_n$ that satisfy an equation such as

(a) $X^2 - I = 0$
(b) $X^2 - 4X + 4I = 0$
(c) $e^X - 5X = 0$
(d) $\cos X - I = 0$
(e) $BX^2 + CX + D = 0,$ $B, C, D \in M_n$ given
(f) $X^2 = A,$ $A \in M_n$ given
(g) $e^X = A,$ $A \in M_n$ given

These examples suggest four general types of matrix equations, not all mutually exclusive:

6.4.1 $p(X) = 0$, where $p(t)$ is an ordinary polynomial with scalar coefficients. Examples (a) and (b) are of this type with $p(t) = t^2 - 1$ and $p(t) = t^2 - 4t + 4$.

6.4.2 $f(X) = 0$, where $f(t)$ is a scalar-valued stem function, perhaps analytic, and $f(X)$ is a primary matrix function. Examples (c) and (d) are of this form with $f(t) = e^t - 5t$ and $f(t) = \cos t - 1$.

6.4.3 $P(X) = 0$, where $P(t)$ is a polynomial with matrix coefficients (6.1.2) and $P(X)$ is given by (6.1.3a or b). Examples (e) and (f) are of this type with $P(t) = Bt^2 + Ct + D$ and $P(t) = It^2 - A$.

6.4.4 $f(X) = A$, where $f(t)$ is a scalar-valued stem function (perhaps a polynomial or analytic function) and $f(X)$ is a primary matrix function. Examples (f) and (g) are of this form with $f(t) = t^2$ and $f(t) = e^t$.

Of course, equations of types (6.4.1-2) are subsumed under type (6.4.4), but it is convenient to analyze them as separate special cases. An equation of type (6.4.3) can be fundamentally different from the other three types because it might not arise from a primary matrix function associated with a scalar-valued stem function. As a consequence, it might not enjoy the important similarity invariance property (6.2.9(c)) that is the key to our approach to the other three types of equations.

If only a diagonalizable solution X is sought, some of these problems may become more tractable (even trivial) if one substitutes $X = S\Lambda S^{-1}$ into the given equation, simplifies, and then looks for diagonal solutions Λ. For example, (b) becomes $\Lambda^2 - 4\Lambda + 4I = 0$, and the only diagonal entries of Λ that satisfy this equation are $\lambda_i = 2$. Thus, $\Lambda = 2I$ is the only solution and hence all diagonalizable solutions are of the form $X = S\,2IS^{-1} = 2I$. However, any matrix whose Jordan canonical form is a direct sum of 2-by-2 blocks $J_2(2)$ and 1-by-1 blocks $J_1(2) = [2]$, in any combinations, will also be a solution. The moral is that while one might start by looking for a diagonalizable solution of a given matrix equation, there may be nondiagonalizable solutions that may require a more careful analysis to discover.

Suppose a given matrix equation can be written in the form $f(X) = A$, where $f(t)$ is a scalar-valued stem function and $f(X)$ is its associated primary matrix function. Then Corollary (6.2.12) gives a useful sufficient condition for the existence of a solution that is itself a primary matrix function of A. But there may be many solutions that are not primary matrix functions of A. How can they be found?

There are some simple general observations, described in Theorem (6.2.9(c,f,g)), that lead to a systematic way to discover whether there are any solutions to $f(X) = A$ and, if there are, to characterize and enumerate them. Suppose A has Jordan canonical form $A = SJS^{-1}$ with $J = J_{n_1}(\lambda_1) \oplus \cdots \oplus J_{n_s}(\lambda_s)$, and suppose one wants to know whether $f(X) = A$ has any solutions. Consider the following four reductions to successively simpler and more highly structured problems:

(a) $f(X) = A = SJS^{-1}$ if and only if $f(S^{-1}XS) = S^{-1}f(X)S = J$, so it suffices to consider whether $f(X) = J$ has any solutions.

(b) In fact, we do not have to solve $f(X) = J$ exactly. It suffices to consider only whether there is an X such that the Jordan canonical form of $f(X)$ is equal to J; if there is, then there is some nonsingular T such that $J = T^{-1}f(X)T = f(T^{-1}XT)$.

(c) Suppose the Jordan canonical form of X is J_X. Since the Jordan canonical form of $f(X)$ is the same as that of $f(J_X)$, it suffices to consider only whether there is a Jordan matrix J_X such that the Jordan canonical form of $f(J_X)$ is the same as J.

(d) Finally, think of J_X as a direct sum $J_X = J_{m_1}(\mu_1) \oplus \cdots \oplus J_{m_r}(\mu_r)$ with unknown direct summands $J_{m_i}(\lambda_i)$ and $m_1 + \cdots + m_r = n$. Since the Jordan canonical form of $f(J_X)$ is the direct sum of the Jordan canonical forms of $f(J_{m_i}(\mu_i))$, $i = 1, \ldots, r$, it suffices to consider only whether there are choices of μ and m such that $J = J_{n_1}(\lambda_1) \oplus \cdots \oplus J_{n_s}(\lambda_s)$ is the direct sum of the Jordan canonical forms of matrices of the form $f(J_m(\mu))$. An explicit formula for the Jordan canonical form of $f(J_{m_i}(\mu_i))$ is given in Theorem (6.2.25).

Each distinct scalar μ for which $f(\mu) = \lambda_i$ gives rise to a different possible solution to $f(X) = J_{n_i}(\lambda_i)$. If $f(\mu) = \lambda_i$, $n_i > 1$, and $f'(\mu) \neq 0$, then the Jordan canonical form of $f(J_m(\mu))$ is $J_m(\lambda_i)$, and only a Jordan block of size $m = n_i$ can be transformed by $f(\cdot)$ into a matrix similar to $J_{n_i}(\lambda_i)$ in this case. If $f'(\mu) = 0$, however, then the Jordan canonical form of $f(J_m(\mu))$ splits into two or more smaller blocks, whose sizes are given by Theorem (6.2.25), and hence one must choose a suitable value of $m > n_i$ to have $J_{n_i}(\lambda_i)$ as a direct summand in the Jordan canonical form of $f(J_m(\mu))$. In the latter case, for there to be any solution at all, Jordan blocks of A must be available to match up with *all* of the Jordan blocks of $f(J_m(\mu))$ that are split off by $f(\cdot)$.

The general principles involved in identifying and enumerating the solutions of $f(X) = A$ are simple in concept but can be complicated in practice, and an ad hoc approach tailored to the specific problem at hand may be the most efficient way to proceed. We illustrate with several examples.

If $p(t)$ is a given scalar polynomial

$$p(t) = t^m + a_{m-1}t^{m-1} + \cdots + a_1 t + a_0 = \prod_{i=1}^{k}(t - t_i)^{r_i} \qquad (6.4.5)$$

with zeroes t_i and multiplicities r_i, then the matrix equation of type (6.4.1)

$$p(X) = X^m + a_{m-1}X^{m-1} + \cdots + a_1 X + a_0 I = 0 \qquad (6.4.6)$$

for $X \in M_n$ is just the statement that $p(t)$ annihilates X. Observe that (6.4.6) determines whole similarity classes of solutions, since $p(SXS^{-1}) = S^{-1}p(X)S = 0$ whenever $p(X) = 0$. The minimal polynomial of X must divide any polynomial that annihilates X, so the set of solutions to (6.4.6) is precisely the set of matrices $X \in M_n$ such that the minimal polynomial $q_X(t)$ divides the given polynomial $p(t)$. The set of monic divisors of $p(t)$ of degree n or less is

$$\left\{ q(t) \colon q(t) = \prod_{i=1}^{k}(t - t_i)^{s_i}, \, 0 \le s_i \le r_i, \, s_1 + s_2 + \cdots + s_k \le n \right\} \qquad (6.4.7)$$

and the set of matrices $X \in M_n$ with minimal polynomial of this form is easily described in terms of the Jordan canonical form of X.

If $q(t) = (t - t_1)^{s_1}(t - t_2)^{s_2} \cdots (t - t_k)^{s_k}$ is a given polynomial of the form (6.4.7), and if $q(t)$ is the minimal polynomial of $X \in M_n$, then the Jordan canonical form of X must contain a Jordan block of the form

$$J_{s_i}(t_i) = \begin{bmatrix} t_i & 1 & & 0 \\ & \ddots & \ddots & \\ & & \ddots & 1 \\ 0 & & & t_i \end{bmatrix} \in M_{s_i}$$

for every nonzero exponent s_i. If $s_1 + s_2 + \cdots + s_k = n$, this is the only possible Jordan form for X (permutations of the blocks do not result in different Jordan canonical forms). If $s_1 + s_2 + \cdots + s_k < n$, the defect may be filled by any combination of diagonal blocks $J_{n_{i_j}}(t_i)$, repeated or not, so long as all orders $n_{i_j} \leq s_i$ and the total size of the Jordan form is n, that is,

$$s_1 + s_2 + \cdots + s_k + \sum_{i=1}^{k} \sum_j n_{i_j} = n$$

There are only finitely many such possibilities, and there are only finitely many polynomials in the set (6.4.7), so we can make in this way a complete list of all the finitely many similarity classes of matrices X that satisfy (6.4.6).

Consider example (a), for which $p(t) = t^2 - 1 = (t+1)(t-1) = 0$. The set (6.4.7) of possible minimal polynomials is just

$$\{(t-1)^{s_1}(t+1)^{s_2} : 0 \leq s_1 \leq 1, 0 \leq s_2 \leq 1, \text{ and } s_1 + s_2 \leq n\}$$

If $n = 1$, the only solutions are $X = \pm[1]$. If $n \geq 2$, the constraint $s_1 + s_2 \leq n$ imposes no restriction and there are exactly three choices for the minimal polynomial: $(t-1)$, $(t+1)$, and $(t-1)(t+1)$. In all three cases the roots are distinct, so all solutions X are diagonalizable and it suffices to consider the form of $\Lambda = \text{diag}(\lambda_1, \ldots, \lambda_n)$ if $X = S\Lambda S^{-1}$. If $q_X(t) = t \pm 1$, then $\Lambda = \mp I$ and $X = \mp I$. If $q_X(t) = (t-1)(t+1)$, then at least one λ_i must be +1, at least one must be -1, and the remaining $n-2$ diagonal entries may be assigned the values +1 or -1 in any order desired.

Example (b) has $p(t) = t^2 - 4t + 4 = (t-2)^2$, and the set of possible minimal polynomials is just $\{(t-2), (t-2)^2\}$ if $n \geq 2$. If $q_X(t) = t - 2$, $X = 2I$. If $q_X(t) = (t-2)^2$, then the Jordan canonical form of X must contain at least one block $J_2(2)$; the rest of the Jordan canonical form (of size $n-2$) may be filled with 2-by-2 blocks $J_2(2)$ or 1-by-1 blocks $J_1(2) = [2]$ in any number and order. Again, this construction determines finitely many equivalence classes of solutions of (b) under similarity.

If we have an equation of the type (6.4.2) with a primary matrix function $f(X)$, then we can consider as possible solutions only those matrices $X \in M_n$ whose eigenvalues are in the domain of $f(\cdot)$ and whose Jordan blocks are not too large with respect to the number of derivatives that $f(\cdot)$ has at the eigenvalues. Indeed, since we want $f(X) = 0$ it is clear that the only eigenvalues of X are at (ordinary scalar) roots of $f(t) = 0$ and that the size of the associated Jordan block is limited by the number of derivatives of $f(\cdot)$ that also vanish at the root.

To be precise, suppose one is given a function $f(t)$ whose domain is an open set D in $\mathbb{R}$ or $\mathbb{C}$ such that:

1. $f(t) = 0$ in D if and only if $t \in \{t_i\} \subset D$.

2. At each zero t_i, $f(\cdot)$ has d_i derivatives (possibly $d_i = +\infty$) and either $f^{(k)}(t_i) = 0$ for all $k = 1, 2, \ldots$, in which case we set $r_i = +\infty$, or there is a finite r_i such that $f(t_i) = f'(t_i) = f^*(t_i) = \cdots = f^{(r_i-1)}(t_i) = 0$, and either $f^{(r_i)}(t_i)$ does not exist or $f^{(r_i)}(t_i) \neq 0$.

This number r_i is the *multiplicity* of the root t_i of $f(t) = 0$. The *domain* of the primary matrix function $f(X)$ is the set of all matrices $X \in M_n$, all of whose eigenvalues $\{\lambda_j\}$ are in D, and such that if λ_j corresponds to a Jordan block of X of size s_j, then $f(t)$ has at least $s_j - 1$ derivatives at λ_j. By a solution of $f(X) = 0$ we mean a matrix X in the domain of $f(\cdot)$ such that $f(X) = 0$. If the function $f(t)$ is analytic in some open set D of the complex plane, then all the qualifications about the existence of derivatives in D disappear.

With this understanding of the meaning of a solution of $f(X) = 0$, one should start to solve the equation by first finding the roots t_i of $f(t) = 0$ in D, then finding the multiplicity r_i of each one. If $f(\cdot)$ has only finitely many zeroes, say k, and if each zero has finite multiplicity r_i, then by Theorem (6.2.9), $f(X) = 0$ if and only if $p(X) = 0$, where

$$p(t) = (t - t_1)^{r_1}(t - t_2)^{r_2} \cdots (t - t_k)^{r_k} \qquad (6.4.8)$$

and this is a problem of type (6.4.1), which we have solved already.

If there are infinitely many zeroes t_i (each of finite multiplicity) of $f(\cdot)$ in D, then the set of possible minimal polynomials of a solution X is not finite, but has the form

$$\{(t-t_1)^{s_1}(t-t_2)^{s_2} \cdots (t-t_k)^{s_k}: \ 0 \leq s_i \leq r_i, \ i = 1,\ldots, k,$$

$$s_1 + s_2 + \cdots + s_k \leq n, \ k = 1, 2,\ldots\}$$

In this case, the solutions are described as before, but there are no longer finitely many similarity classes of solutions.

If any zero t_i has infinite multiplicity $r_i = +\infty$, then one proceeds to solve as before except that there is no upper limit on the size of the Jordan block in X corresponding to the eigenvalue t_i; $J_k(t_i)$ is an acceptable Jordan block in X for every $k \leq n$.

If $f(t)$ is analytic and not identically zero in an open set $D \subset \mathbb{C}$, then every root of $f(t) = 0$ in D must have finite multiplicity and there can be only finitely many roots of $f(t) = 0$ in any compact subset of D. In the analytic case, most of the possible pathology of a general $f(\cdot)$ disappears.

As an example, consider the matrix equation $e^X - 5X = 0$, for which $f(t) = e^t - 5t$. Since this is an entire function, we take $D = \mathbb{C}$; the domain of $f(X)$ is all of M_n. Each root t_0 of $e^t - 5t = 0$ (for example, the real root at $t_0 = 2.5427...$) has multiplicity one since $f'(t_0) = e^{t_0} - 5 = 5t_0 - 5 \neq 0$ ($t_0 = 1$ is not a root). Thus, the only solutions to $e^X - 5X = 0$ are diagonalizable matrices whose eigenvalues are roots of $e^t - 5t = 0$.

To solve $\cos X - I = 0$ we must consider the entire function $f(t) = \cos t - 1 = -t^2/2! + t^4/4! - \cdots$, which has zeroes of multiplicity two at $t = 0$, $\pm 2\pi$, $\pm 4\pi,\ldots$ Thus, the solutions of $f(X) = 0$ have Jordan blocks of size one or two with eigenvalues $2k\pi$, $k = 0, \pm 1, \pm 2,\ldots$ All combinations of such blocks (with total size n) are possible, so there are infinitely many different similarity classes of solutions.

Problems of type (6.4.3) can also be reduced to problems of type (6.4.1), but it may not always be efficient to do so. If $P(t)$ is a given polynomial with matrix coefficients

$$P(t) = A_m t^m + A_{m-1} t^{m-1} + \cdots + A_1 t + A_0, A_k \in M_n \tag{6.4.9}$$

then the equation $P(X) = 0$, $X \in M_n$, has at least two possible interpretations.

$$P_l(X) = A_m X^m + A_{m-1} X^{m-1} + \cdots + A_1 X + A_0 = 0 \tag{6.4.10a}$$

or

$$P_r(X) = X^m A_m + X^{m-1} A_{m-1} + \cdots + X A_1 + A_0 = 0 \qquad (6.4.10b)$$

These two equations may have different solutions, of course. We choose to discuss the former; there is an exactly parallel development of the theory for the latter.

The Euclidean algorithm can be used on $P(t)$ with $tI - X$ as the divisor, and one can divide either on the left or on the right, with respective remainders $P_r(X)$ and $P_l(X)$:

$$P(t) = (tI - X)\tilde{Q}(t) + P_r(X) = Q(t)(tI - X) + P_l(X)$$

where $\tilde{Q}(t)$ and $Q(t)$ are polynomials of degree $m-1$ with matrix coefficients:

$$\tilde{Q}(t) = A_m t^{m-1} + (A_{m-1} + X A_m) t^{m-2} + \cdots$$

and

$$Q(t) = A_m t^{m-1} + (A_{m-1} + A_m X) t^{m-2} + \cdots$$

If $P_l(X) = 0$, this identity says that $P(t) = Q(t)(tI - X)$, and hence det $P(t)$ = det $Q(t)$ det$(tI - X) = p_X(t)$det $Q(t)$, where $p_X(t)$ is the characteristic polynomial of X. The function det $P(t)$ is a scalar polynomial of degree at most mn in t, so if we set $\phi(t) \equiv$ det $P(t)$ we have the identity

$$\phi(t) = (\det Q(t)) \, p_X(t)$$

if $X \in M_n$ is a solution to $P_l(X) = 0$. But $p_X(X) = 0$ by the Cayley-Hamilton Theorem, so we conclude that $\phi(X) = 0$.

The point of this argument is to show that every solution X to the matrix polynomial equation $P_l(X) = 0$ is also a solution to the scalar polynomial equation $\phi(X) = 0$, where $\phi(t) =$ det $P(t)$. This is an equation of type (6.4.1), and we know that it has finitely many similarity classes of solutions. If $\{J_1, J_2, ..., J_k\}$ are the finitely many different Jordan matrices that solve $\phi(X) = 0$, then all solutions to $\phi(X) = 0$ are of the form $X = S J_i S^{-1}$ for some $i = 1, 2, ..., k$ and some nonsingular $S \in M_n$. Not every such matrix is necessarily a solution to the original equation (6.4.10a), but all solutions to $P_l(X) = 0$ can be represented in this form.

If we choose some Jordan matrix solution J_i to $\phi(X) = 0$ and compute $P_l(S J_i S^{-1})$, we obtain the linear homogeneous equation

$$P_l(SJ_iS^{-1})S = A_mSJ_i^m + A_{m-1}SJ_i^{m-1} + \cdots + A_1SJ_i + A_0S = 0 \ (6.4.11)$$

for the unknown similarity S; all the terms A_j and J_i are known. Every nonsingular solution S to (6.4.11) yields a solution SJ_iS^{-1} to the original equation (6.4.10a), and every solution to (6.4.10a) is of this form. If (6.4.11) has no nonsingular solution S for any $i = 1,\ldots, k$, then we must conclude that (6.4.10a) has no solution $X \in M_n$.

Even though we have a general procedure to reduce all matrix polynomial equations (6.4.3) to scalar polynomial equations (6.4.1) and linear matrix equations (for which we have developed an extensive theory in Chapter 4), it is often of little practical use in solving a specific problem. For example, the simple polynomial equation $X^2 - A = 0$ with matrix coefficients has $P(t) = t^2I - A$, so if we use the general procedure, our first step is to compute $\phi(t) = \det P(t) = \det(t^2I - A) = p_A(t^2)$, where $p_A(t)$ is the characteristic polynomial of A. If $\lambda_1,\ldots, \lambda_k$ are the distinct eigenvalues of A with algebraic multiplicities $s_1,\ldots, s_k$, we have

$$\phi(t) = (t^2 - \lambda_1)^{s_1} \cdots (t^2 - \lambda_k)^{s_k}$$

$$= (t - \sqrt{\lambda_1})^{s_1}(t + \sqrt{\lambda_1})^{s_1} \cdots (t - \sqrt{\lambda_k})^{s_k}(t + \sqrt{\lambda_k})^{s_k}$$

where $\sqrt{\lambda_i}$ denotes either square root of the complex number λ_i; of course, the same choice of square root must be used in each pair of factors

$$(t - \sqrt{\lambda_i})^{s_i}(t + \sqrt{\lambda_i})^{s_i}$$

The next step is to consider all monic polynomial divisors of $\phi(t)$, write down all the Jordan forms J_i that have each divisor as a minimal polynomial, and then try to solve $SJ_i^2S^{-1} - A = 0$, or $SJ_i^2 - AS = 0$ for a nonsingular $S \in M_n$. Although the general theory covers this problem, it is more efficient and illuminating to recast it as an inhomogeneous equation $f(X) = A$ with the primary matrix function $f(X) = X^2$.

Any solution of the equation $X^2 = A$ gives a "square root" of $A \in M_n$. When $n = 1$, we know that every complex number has a square root, and every nonzero complex number has two distinct square roots. In particular, the "square root" is not a function on $\mathbb{C}$. For matrices, the situation is even more complicated.

Let $A = SJS^{-1}$ be the Jordan canonical form of the given matrix A, so that if $X^2 = A = SJS^{-1}$, then $S^{-1}X^2S = (S^{-1}XS)^2 = J$. It suffices, therefore, to solve the equation $X^2 = J$. But if X is such that the Jordan canonical form of X^2 is equal to J, then there is some nonsingular T such that $J = TX^2T^{-1} = (TXT^{-1})^2$. Thus, it suffices to find an X such that the Jordan canonical form of X^2 is equal to J. If the Jordan canonical form of X itself is J_X, then the Jordan canonical form of X^2 is the same as that of $(J_X)^2$, so it suffices to find a Jordan matrix J_X such that the Jordan canonical form of $(J_X)^2$ is equal to J. Finally, if $J_X = J_{m_1}(\mu_1) \oplus \cdots \oplus J_{m_r}(\mu_r)$, then the Jordan canonical form of $(J_X)^2$ is the same as the direct sum of the Jordan canonical forms of $J_{m_i}(\mu_i)^2$, $i = 1,..., r$. Thus, to solve $X^2 = A$, it suffices to consider only whether there are choices of scalars μ and positive integers m such that the given Jordan canonical form J is the direct sum of the Jordan canonical forms of matrices of the form $J_m(\lambda)^2$. If $\mu \neq 0$, we know from Theorem (6.2.25) that the Jordan canonical form of $J_m(\mu)^2$ is $J_m(\mu^2)$, so every nonsingular Jordan block $J_k(\lambda)$ has a square root; in fact, it has square roots that lie in two distinct similarity classes with Jordan canonical forms $J_k(\pm\sqrt{\lambda})$. If necessary, these square roots of Jordan blocks can be computed explicitly using (6.2.7), as in Example (6.2.14).

Thus, every nonsingular matrix $A \in M_n$ has a square root, and it has square roots in at least 2^μ different similarity classes if A has μ distinct eigenvalues. It has square roots in at most 2^ν different similarity classes if the Jordan canonical form of A is the direct sum of ν Jordan blocks; there are exactly 2^ν similarity classes if all the Jordan blocks with the same eigenvalue have different sizes, but there are fewer than 2^ν if two or more blocks of the same size have the same eigenvalue, since permuting blocks does not change the similarity class of a Jordan canonical form. Some of these nonsimilar square roots may not be "functions" of A, however. If each of the ν blocks has a different eigenvalue (that is, A is nonderogatory), a Lagrange-Hermite interpolation polynomial (6.1.16) can always be used to express *any* square root of A as a polynomial in A. If the same eigenvalue λ occurs in two or more blocks, however, polynomial interpolation is possible only if the *same* choice is made for $\lambda^{\frac{1}{2}}$ for *all* of them; if different choices are made in this case, one obtains a square root of A that is not a "function" of A in the sense that it cannot be obtained as a polynomial in A, and therefore cannot be a primary matrix function $f(A)$ with a *single-valued* function $f(\cdot)$. Some authors have studied extending the notion of "function of a matrix" to allow multiple-valued $f(\cdot)$ in (6.2.4), with possibly different values of $f(\lambda)$ in different Jordan blocks with the same eigenvalue.

Exercise. Consider $A = I \in M_2$ and

$$X = \begin{bmatrix} 2 & -1 \\ 3 & -2 \end{bmatrix}$$

Show that $X^2 = I$ and that X cannot be a polynomial in A. Explain why there are exactly three similarity classes of square roots of A, with respective Jordan canonical forms

$$I = \begin{bmatrix} 1 & 0 \\ 0 & 1 \end{bmatrix}, \, -I = \begin{bmatrix} -1 & 0 \\ 0 & -1 \end{bmatrix}, \text{ and } \begin{bmatrix} 1 & 0 \\ 0 & -1 \end{bmatrix}$$

Explain why the first two can be achieved by a "function" of A, and the last one cannot; note that X is in the similarity class of the last one.

What happens if A is singular? The example $A = \begin{bmatrix} 1 & 0 \\ 0 & 0 \end{bmatrix}$ shows that A might have a square root, while $A = \begin{bmatrix} 0 & 1 \\ 0 & 0 \end{bmatrix}$ shows that A might not have a square root (see Problem 7).

Since each nonsingular Jordan block of A has a square root, it suffices to consider the direct sum of all the singular Jordan blocks of A. If A has a square root, then this direct sum is the Jordan canonical form of the square of a direct sum of singular Jordan blocks. Which direct sums of singular Jordan blocks can arise in this way?

Let $k > 1$. We know from Theorem (6.2.25) that the Jordan canonical form of $J_k(0)^2$ consists of exactly two Jordan blocks $J_{k/2}(0) \oplus J_{k/2}(0)$ if $k > 1$ is even, and it consists of exactly two Jordan blocks $J_{(k+1)/2}(0) \oplus J_{(k-1)/2}(0)$ if $k > 1$ is odd.

If $k = 1$, $J_1(0)^2 = [0]$ is a 1-by-1 block, and this is the only Jordan block that is similar to the square of a singular Jordan block.

Putting together this information, we can determine whether or not a given singular Jordan matrix J has a square root as follows: Arrange the diagonal blocks of J by decreasing size, so $J = J_{k_1}(0) \oplus J_{k_2}(0) \oplus \cdots \oplus J_{k_p}(0)$ with $k_1 \geq k_2 \geq k_3 \geq \cdots \geq k_p \geq 1$. Consider the differences in sizes of successive pairs of blocks: $\Delta_1 = k_1 - k_2$, $\Delta_3 = k_3 - k_4$, $\Delta_5 = k_5 - k_6$, etc., and suppose J is the Jordan canonical form of the square of a singular Jordan matrix $\bar{J}$. We have seen that $\Delta_1 = 0$ or 1 because either $k_1 = 1$ [in which case $J_1(0) \oplus J_1(0)$

corresponds to $(J_1(0) \oplus J_1(0))^2$ or to $J_2(0)^2]$ or $k_1 > 1$ and $J_{k_1}(0) \oplus J_{k_2}(0)$ corresponds to the square of the largest Jordan block in $\tilde{J}$, which has size $k_1 + k_2$. The same reasoning shows that $\Delta_3, \Delta_5, \ldots$ must all have the value 0 or 1 and an acceptable square root corresponding to the pair $J_{k_i}(0) \oplus J_{k_{i+1}}(0)$ is $J_{k_i+k_{i+1}}(0)$, $i = 1, 3, 5, \ldots$. If p (the total number of singular Jordan blocks in J) is odd, then the last block $J_{k_p}(0)$ is left unpaired in this process and must therefore have size 1 since it must be the square of a singular Jordan block. Conversely, if the successive differences (and k_p, if p is odd) satisfy these conditions, then the pairing process described constructs a square root for J.

Exercise. Suppose a given $A \in M_{20}$ has exactly four singular Jordan blocks with sizes 4, 4, 3, and 3. Show that A has a square root whose two singular Jordan blocks have sizes 8 and 6, and another (necessarily nonsimilar) square root whose two singular Jordan blocks both have size 7. The point is that the pairing algorithm in the preceding paragraph might not be the *only* way to construct a square root of the nilpotent part of A.

Exercise. Let $A = J_2(0) \oplus J_2(0) \in M_4$. Show that $B = J_4(0)$ is the only possible Jordan canonical form of a square root of A in M_4. Show that there is no polynomial $p(t)$ such that $B = p(A)$, and conclude that B cannot be a primary matrix function of A.

Suppose $A \in M_n$ is singular and suppose there is a polynomial $r(t)$ such that $B = r(A)$ is a square root of A. Then $r(0) = 0$, $r(t) = tg(t)$ for some polynomial $g(t)$, and $A = B^2 = A^2 g(A)^2$, which is clearly impossible if rank $A^2 < $ rank A. Thus, rank $A = $ rank A^2 in this case, which means that every singular Jordan block of A is 1-by-1. Conversely, if A is singular and has minimal polynomial $q_A(t) = t(t-\lambda_1)^{r_1} \cdots (t-\lambda_\mu)^{r_\mu}$ with distinct nonzero $\lambda_1, \ldots, \lambda_\mu$ and all $r_i \geq 1$, let $g(t)$ be a polynomial that interpolates the function $f(t) = 1/\sqrt{t}$ and its derivatives at the (necessarily nonzero) roots of the polynomial $q_A(t)/t = 0$, and let $r(t) \equiv tg(t)$. For each nonsingular Jordan block $J_{n_i}(\lambda_i)$ of A, $g(J_{n_i}(\lambda_i)) = [J_{n_i}(\lambda_i)]^{-\frac{1}{2}}$ and hence Corollary (6.2.10(d)) guarantees that $r(J_{n_i}(\lambda_i)) = J_{n_i}(\lambda_i) [J_{n_i}(\lambda_i)]^{-\frac{1}{2}} = J_{n_i}(\lambda_i)^{\frac{1}{2}}$. Since all the singular Jordan blocks of A are 1-by-1 and $r(0) = 0$, we conclude from (6.1.6) that $r(A)$ is a square root of A. Thus, a given singular $A \in M_n$ has a square root that is a polynomial in A if and only if rank $A = $ rank A^2.

Since this latter condition is trivially satisfied if A is nonsingular (in which case we already know that A has a square root that is a polynomial in A), we conclude that a given $A \in M_n$ has a square root that is a polynomial in A if and only if rank A = rank A^2.

If we agree that a "square root" of a matrix $A \in M_n$ is any matrix $B \in M_n$ such that $B^2 = A$, we can summarize what we have learned about the solutions of the equation $X^2 - A = 0$ in the following theorem:

6.4.12   Theorem. Let $A \in M_n$ be given.

(a)   If A is nonsingular and has μ distinct eigenvalues and ν Jordan blocks in its Jordan canonical form, it has at least 2^μ and at most 2^ν nonsimilar square roots. Moreover, at least one of its square roots can be expressed as a polynomial in A.

(b)   If A is singular and has Jordan canonical form $A = SJS^{-1}$, let $J_{k_1}(0) \oplus J_{k_2}(0) \oplus \cdots \oplus J_{k_p}(0)$ be the singular part of J with the blocks arranged in decreasing order of size: $k_1 \geq k_2 \geq \cdots \geq k_p \geq 1$. Define $\Delta_1 = k_1 - k_2$, $\Delta_3 = k_3 - k_4$, ... Then A has a square root if and only if $\Delta_i = 0$ or 1 for $i = 1, 3, 5, \ldots$ and, if p is odd, $k_p = 1$. Moreover, A has a square root that is a polynomial in A if and only if $k_1 = 1$, a condition that is equivalent to requiring that rank A = rank A^2.

(c)   If A has a square root, its set of square roots lies in finitely many different similarity classes.

Since the sizes and numbers of the Jordan blocks $J_k(\lambda)$ of a matrix A can be inferred from the sequence of ranks of the powers $(A - \lambda I)^k$, $k = 1, 2, \ldots$, the necessary and sufficient condition on the sizes of the singular Jordan blocks of A in part (b) of the preceding theorem can be restated in terms of ranks of powers. Let $A \in M_n$ be a given singular matrix, and let $r_0 = n$, $r_k = $ rank A^k, $k = 1, 2, \ldots$. The sequence $r_0, r_1, r_2, \ldots$ is decreasing and eventually becomes constant. If $r_{k_1-1} > r_{k_1} = r_{k_1+1} = \cdots$, then the largest singular Jordan block in A has size k_1, which is the *index of the matrix with respect to the eigenvalue* $\lambda = 0$. The difference $r_{k_1-1} - r_{k_1}$ is the number of singular Jordan blocks of size k_1. If this number is even, the blocks of size k_1 can all be paired together in forming a square root. If this number is odd, then one block is left over after the blocks are paired and A can have a

square root only if either $k_1 = 1$ (so that no further pairing is required), or
there is at least one singular Jordan block of size $k_1 - 1$ available to be paired
with it; this is the case only if $r_{k_1-2} - r_{k_1-1} > r_{k_1-1} - r_{k_1}$ since $r_{k_1-2} - r_{k_1-1}$
equals the total number of singular Jordan blocks of sizes k_1 and $k_1 - 1$. This
reasoning is easily continued backward through the sequence of ranks r_k. If
all the differences $r_i - r_{i+1}$ are even, $i = k_1 - 1, k_1 - 3,...$, then A has a square
root. If any difference $r_i - r_{i+1}$ is odd, then $r_{i-1} - r_i$ must have a larger value
if A is to have a square root. Since $r_0 - r_1$ is the total number of singular
Jordan blocks of all sizes, if $r_0 - r_1$ is odd we must also require that there be
at least one block of size 1, that is, $1 \leq$ (# of singular blocks of all sizes ≥ 1) -
(# of singular blocks of all sizes ≥ 2) $= (r_0 - r_1) - (r_1 - r_2) = r_0 - 2r_1 + r_2$.
Notice that $r_k \equiv n$ if A is nonsingular, so all the successive differences
$r_i - r_{i+1}$ are zero and A trivially satisfies the criteria for a square root in this
case.

6.4.13   Corollary. Let $A \in M_n$, and let $r_0 = n$, $r_k = \text{rank } A^k$ for $k = 1, 2,...$.
Then A has a square root if and only if the sequence $\{r_k - r_{k+1}\}$, $k = 0, 1, 2,...$
does not contain two successive occurrences of the same odd integer and, if
$r_0 - r_1$ is odd, $r_0 - 2r_1 + r_2 \geq 1$.

Although we now know exactly when a given complex matrix has a
complex square root, one sometimes needs to answer a slightly different
question: When does a given *real* matrix $A \in M_n(\mathbb{R})$ have a *real* square root?
The equivalent criteria in Theorem (6.4.12) and Corollary (6.4.13) are still
necessary, of course, but they do not guarantee that any of the possible
square roots are real. The crucial observation needed here is that if one looks
at the Jordan canonical form of a real matrix, then the Jordan blocks with
nonreal eigenvalues occur only in conjugate pairs, that is, there is an *even*
number of Jordan blocks of each size for each nonreal eigenvalue. Moreover,
a given complex matrix is similar to a real matrix if and only if the nonreal
blocks in its Jordan canonical form occur in conjugate pairs; for more
details, see the discussion of the real Jordan canonical form in Section (3.4)
of [HJ].

If there is some $B \in M_n$ such that $B^2 = A$, then any Jordan block of A
with a negative eigenvalue corresponds to a Jordan block of B of the same
size with a purely imaginary eigenvalue. If B is real, such blocks must occur
in conjugate pairs, which means that the Jordan blocks of A with negative
eigenvalues must also occur in pairs, just like the nonreal blocks of A.

Conversely, let J be the Jordan canonical form of the real matrix $A \in M_n(\mathbb{R})$, suppose all of the Jordan blocks in J with negative eigenvalues occur in pairs, and suppose A satisfies the rank conditions in Corollary (6.4.13). Form a square root for J using the process leading to Theorem (6.4.12) for the singular blocks, and using the primary-function method in Example (6.2.14) for each individual nonsingular Jordan block, but be careful to choose conjugate values for the square root for the two members of each pair of blocks with nonreal or negative eigenvalues; blocks or groups of blocks with nonnegative eigenvalues necessarily have real square roots. Denote the resulting (possibly complex) block diagonal upper triangular matrix by C, so $C^2 = J$. Each diagonal block of C is similar to a Jordan block of the same size with the same eigenvalue, so C is similar to a real matrix R because of the conjugate pairing of its nonreal Jordan blocks. Thus, the real matrix R^2 is similar to $C^2 = J$, and J is similar to the real matrix A, so R^2 is similar to A. Recall that two real matrices are similar if and only if they are similar via a real similarity, since they must have the same real Jordan canonical form, which can always be attained via a real similarity. Thus, there is a real nonsingular $S \in M_n(\mathbb{R})$ such that $A = SR^2S^{-1} = SRS^{-1}SRS^{-1} = (SRS^{-1})^2$ and the real matrix SRS^{-1} is therefore a real square root of A.

In the preceding argument, notice that if A has any pairs of negative eigenvalues, the necessity of choosing conjugate purely imaginary values for the square roots of the two members of each pair precludes any possibility that a real square root of A could be a polynomial in A or a primary matrix function of A. We summarize our observations as the following theorem.

6.4.14   Theorem. Let $A \in M_n(\mathbb{R})$ be a given real matrix. There exists a real $B \in M_n(\mathbb{R})$ with $B^2 = A$ if and only if A satisfies the rank condition in Corollary (6.4.13) and has an even number of Jordan blocks of each size for every negative eigenvalue. If A has any negative eigenvalues, no real square root of A can be a polynomial in A or a primary matrix function of A.

See Problem 18 for a reformulation of the latter condition in terms of ranks of powers.

The same reasoning that we have used to analyze the equation $X^2 = A$ can be used to analyze $X^m = A$ for $m = 3, 4, \ldots$. Every nonsingular $A \in M_n$ has an mth root, in fact, a great many of them, and the existence of an mth root of a singular matrix is determined entirely by the sequence of sizes of its singular Jordan blocks. See Problems 10-13.

Other inhomogeneous matrix equations of the form $f(X) = A$ for a primary matrix function $f(X)$ can be analyzed in a similar way. For example, when $f(t) = e^t$, a solution to the equation $e^X = A$ may be thought of as a "logarithm" of the given matrix A.

Since $\det e^X = e^{\operatorname{tr} X} \neq 0$ for all $X \in M_n$ [see Problem 4 in Section (6.2)], an obvious necessary condition for solvability of the equation $e^X = A$ is that A be nonsingular, and we saw in Example (6.2.15) that this condition is sufficient as well.

Following the argument in our analysis of the equation $X^2 = A$, we begin by analyzing the possible Jordan block structure of e^X. If $X = SJS^{-1}$ with $J = J_{k_1}(\lambda_1) \oplus \cdots \oplus J_{k_p}(\lambda_p)$, then

$$e^X = e^{SJS^{-1}} = Se^J S^{-1} = S e^{J_{k_1}(\lambda_1) \oplus \cdots \oplus J_{k_p}(\lambda_p)} S^{-1}$$

$$= S\left[e^{J_{k_1}(\lambda_1)} \oplus \cdots \oplus e^{J_{k_p}(\lambda_p)} \right] S^{-1}$$

so the Jordan canonical form of e^X is determined by the Jordan canonical forms of $e^{J_{k_i}(\lambda_i)}$. But the derivative of e^t is never zero, so Theorem (6.2.25) guarantees that the exponential preserves the size of every Jordan block; the Jordan canonical form of $e^{J_k(\lambda)}$ is $J_k(e^\lambda)$.

This analysis shows that every nonsingular Jordan block $J_k(\lambda)$ is similar to a matrix of the form $e^{J_k(\log \lambda)}$, where $\log \lambda = \log |\lambda| + i \arg(\lambda) + 2j\pi i$, $j = 0, \pm 1, \pm 2,...$, is any one of infinitely many branches of the complex logarithm. Except for the singular ones, there are no impossible Jordan canonical forms for e^X. If the Jordan canonical form of the given matrix A is $J_{k_1}(\lambda_1) \oplus \cdots \oplus J_{k_p}(\lambda_p)$, then every solution $X \in M_n$ to $e^X - A = 0$ will have Jordan canonical form $J_{k_1}(\log \lambda_1) \oplus \cdots \oplus J_{k_p}(\log \lambda_p)$. Since each value $\log \lambda_i$ may be chosen in infinitely many ways, there are infinitely many nonsimilar logarithms of each nonsingular matrix $A \in M_n$. The same reasoning used to prove Theorem (6.4.14) shows that a real nonsingular matrix has a real logarithm if and only if the Jordan blocks corresponding to every negative eigenvalue of A are paired; see Problems 18-19.

6.4.15   Theorem. Let $A \in M_n$ be given.

(a)   If A is nonsingular, there are solutions $X \in M_n$ to the equation $e^X = A$ that lie in infinitely many different similarity classes; at least one of these solutions is a polynomial in A.

(b) If A is singular, there is no $X \in M_n$ such that $e^X = A$.

(c) Suppose A is real. There is a real $X \in M_n(\mathbb{R})$ such that $e^X = A$ if and only if A is nonsingular and has an even number of Jordan blocks of each size for every negative eigenvalue. If A has any negative eigenvalues, no real solution of $e^X = A$ can be a polynomial in A or a primary matrix function of A.

One interesting application of the square root primary matrix function is to prove the following analog of the classical polar decomposition.

6.4.16 **Theorem.** Let $A \in M_n$ be given.

(a) If A is nonsingular, then $A = GQ$, where $Q \in M_n$ is complex orthogonal ($QQ^T = I$), $G \in M_n$ is complex symmetric ($G = G^T$), and G is a polynomial in AA^T.

(b) Suppose $A = GQ$, where $G = G^T$ and $QQ^T = I$. Then $G^2 = AA^T$. If G commutes with Q, then A commutes with A^T. Conversely, if A commutes with A^T and if G is a polynomial in AA^T, then G commutes with Q.

(c) If A is real and nonsingular, there are real factors G and Q satisfying the conditions in (a).

Proof: If $A = GQ$ for some symmetric G and orthogonal Q, then $AA^T = GQQ^TG^T = GG^T = G^2$, so G must be a symmetric square root of AA^T. Since A is nonsingular, AA^T is nonsingular and has nonsingular square roots. Among the square roots of AA^T, use Theorem (6.4.12(a)) to choose one that is a polynomial in AA^T and denote it by G. Then G must be symmetric since AA^T is symmetric and any polynomial in a symmetric matrix is symmetric. Now define $Q \equiv G^{-1}A$ and compute $QQ^T = G^{-1}A(G^{-1}A)^T = G^{-1}AA^T(G^T)^{-1} = G^{-1}G^2G^{-1} = I$, so Q is orthogonal and $GQ = A$.

If G commutes with Q, then G commutes with Q^{-1}, which is a polynomial in Q. Since $Q^{-1} = Q^T$, we have $A^TA = Q^TG^2Q = G^2Q^TQ = G^2 = AA^T$. Conversely, if $AA^T = A^TA$, then $G^2 = AA^T = A^TA = (GQ)^T(GQ) = Q^TG^2Q = Q^TAA^TQ$, so $QAA^T = AA^TQ$, that is, Q commutes with AA^T. If G is a polynomial in AA^T, Q must also commute with G.

If A is real, Corollary (7.3.3) in [HJ] guarantees that $A = PU$ for a unique real Hermitian (real symmetric) positive semidefinite matrix $P = (AA^*)^{\frac{1}{2}} = (AA^T)^{\frac{1}{2}}$ and a real unitary (real orthogonal) matrix U. P is always a polynomial in AA^T; one may use the Lagrange interpolating polynomial that takes the values $+\sqrt{\lambda_i}$ at the (necessarily nonnegative) eigenvalues λ_i of AA^T. Setting $G = P$ and $Q = U$ verifies assertion (c).   ☐

Although there is a formal analogy between the symmetric-orthogonal factorization in Theorem (6.4.16) and the positive definite-unitary factorization in the classical polar decomposition, there are some important differences. If A is nonsingular, the positive definite factor P in the ordinary polar decomposition $A = PU$ is always uniquely determined by A, but the symmetric factor G in the decomposition $A = GQ$ is never uniquely determined; it may be taken to be any primary matrix function square root of AA^T, of which there are at least two that are not similar. If A is singular, it always has a polar decomposition $A = PU$, but it might not have a factorization of the form $A = GQ$. For example, if

$$A = \begin{bmatrix} 0 & i \\ 0 & 1 \end{bmatrix}$$

then

$$AA^T = \begin{bmatrix} -1 & i \\ i & 1 \end{bmatrix} \text{ and } A^TA = 0$$

so these two matrices are not similar. But if $A = GQ$, then $AA^T = G^2$ would be similar to $A^TA = Q^TG^2Q$, so such a factorization is impossible.

The following theorem, which we present without proof, characterizes the possibly singular square matrices for which the orthogonal-symmetric factorization in Theorem (6.4.16) can be achieved.

6.4.17   Theorem. Let $A \in M_n$ be given. There exists a complex orthogonal $Q \in M_n$ and a complex symmetric $G \in M_n$ such that $A = GQ$ if and only if rank $(AA^T)^k =$ rank $(A^TA)^k$ for $k = 1,..., n$.

If A is nonsingular, then the rank conditions in Theorem (6.4.17) are trivially satisfied; no conditions need to be imposed to be able to write $A = GQ$ in the nonsingular case. The rank conditions in Theorem (6.4.17) are equivalent to assuming that AA^T and A^TA are similar; see Problem 26. See Problem 34 for an extension of the theorem to the non-square case.

As a consequence of Theorem (6.4.16), we have the following generalization of the classical fact (see (4.4.13) in [HJ]) that two complex symmetric matrices are similar if and only if they are orthogonally similar (see Theorem (4.4.13) in [HJ] for a special case). It may also be thought of as an analog of the fact that two normal matrices are similar if and only if they are unitarily similar (see Problem 31 in Section (2.5) of [HJ]).

6.4.18   Corollary. Let $A, B \in M_n$ be given matrices and suppose there is a single polynomial $p(t)$ such that $A^T = p(A)$ and $B^T = p(B)$. Then A and B are similar if and only if they are complex orthogonally similar.

Proof: If there is a nonsingular $S \in M_n$ such that $A = S^{-1}BS$, then $A^T = p(A) = p(S^{-1}BS) = S^{-1}p(B)S = S^{-1}B^TS$, so $A = S^TB(S^T)^{-1}$ and hence $S^{-1}BS = A = S^TB(S^T)^{-1}$. Thus, $B(SS^T) = (SS^T)B$. Now use (6.4.17) to write $S = GQ$ with $G = G^T$, $QQ^T = I$, and $G = g(SS^T)$ for some polynomial $g(t)$. Since B commutes with SS^T, it also commutes with $G = g(SS^T)$. Thus, $A = S^{-1}BS = (GQ)^{-1}B(GQ) = Q^TG^{-1}BGQ = Q^TG^{-1}GBQ = Q^TBQ$, so A is complex orthogonally similar to B via Q.   □

6.4.19   Corollary. Let $A, B \in M_n$ be given matrices that are either both complex symmetric, both complex skew-symmetric, or both complex orthogonal. Then A and B are similar if and only if they are complex orthogonally similar.

Proof: Using Corollary (6.4.18), we need only show that in each case there is a polynomial $p(t)$ such that both $A^T = p(A)$ and $B^T = p(B)$. In the symmetric case, take $p(t) = t$; in the skew-symmetric case, take $p(t) = -t$. In the orthogonal case, we can invoke Theorem (6.2.9(a)) and choose for $p(t)$ a polynomial that interpolates the function $f(t) = 1/t$ and its derivatives on the spectrum of A, which is the same as the spectrum of B.   □

In proving both the classical polar decomposition $A = PU$ (P positive semidefinite and U unitary) and the analogous factorization $A = GQ$ (G symmetric and Q orthogonal) in Theorem (6.4.16), a key step is to show that the positive definite matrix AA^* has a positive definite square root and that the symmetric matrix AA^T has a symmetric square root. Both of these factorizations are natural generalizations to matrices of the polar form $z = re^{i\theta}$ for complex scalars, but there is another, even more natural, generalization that may not be as familiar: $A = RE$, where R is real and $E\overline{E} = I$. A square matrix satisfying the latter condition is said to be *coninvolutory* or *circular.*

If such a factorization, of the form $A = RE$, is possible for a given nonsingular matrix A, then $\overline{A}^{-1}A = (\overline{RE})^{-1}(RE) = \overline{E}^{-1}R^{-1}RE = E^2$ is the square of a coninvolutory matrix. A simple calculation shows that any matrix of the form $\overline{A}^{-1}A$ is coninvolutory; conversely, Theorem (4.6.11) in [HJ] ensures that every coninvolutory matrix $E \in M_n$ is condiagonalizable and is consimilar to the identity, so $E = \overline{B}^{-1}B$ for some $B \in M_n$. Thus, if we wish to study the factorization $A = RE$, we are forced to ask whether an arbitrary coninvolutory matrix has a coninvolutory square root. It does, and we can use simple properties of the integral representation (6.2.29) for a primary matrix function logarithm to prove a useful stronger result. Recall from Definition (6.2.26) that $\mathcal{D}_n(D) = \{A \in M_n : \sigma(A) \subset D \subset \mathbb{C}\}$.

6.4.20   Theorem. Let $R > 1$ and $\theta \in [0, 2\pi)$ be given. Let $D_{R,\theta}$ denote the (simply connected) region in the complex plane between the circles $|z| = R$ and $|z| = 1/R$, excluding the ray segment $\{z = re^{i\theta} : 1/R \le r \le R\}$. Let Γ denote the boundary of $D_{R,\theta}$, traversed in the positive sense. For $A \in \mathcal{D}_n(D_{R,\theta})$, define

$$\text{Log } A \equiv \frac{1}{2\pi i}\oint_\Gamma (\log z)(zI - A)^{-1}\, dz, \quad \Gamma = \partial D_{R,\theta} \qquad (6.4.21)$$

where the principal branch of the scalar logarithm is used, that is, $\log z$ is real for $z > 0$. Then

(a)   Log A is a primary matrix function that is continuous on $\mathcal{D}_n(D_{R,\theta})$, and $e^{\text{Log } A} = A$ for all $A \in \mathcal{D}_n(D_{R,\theta})$.

(b)   For all $A \in \mathcal{D}_n(D_{R,\theta})$ we have

      (1)   Log $A^T = (\text{Log } A)^T$

      (2)   Log $\overline{A}^{-1} = -(\overline{\text{Log } A})$

      (3)   Log$(A^{-1})^* = \text{Log}(A^*)^{-1} = -(\text{Log } A)^*$

(c)   If $A \in \mathcal{D}_n(D_{R,\pi})$, that is, if A has no negative eigenvalues, and if we compute Log A with $\theta = \pi$ in (6.4.21), then

      (1)   Log $\overline{A} = (\overline{\text{Log } A})$

      (2)   Log $A^{-1} = -\text{Log } A$

Proof: The assertions in (a) are known from Theorem (6.2.28) and are recorded here for convenience. Notice that (b)(1) follows from the fact that Log A is a polynomial in A, and (b)(3) is a consequence of (b)(1) and (b)(2). To prove (b)(2), make the change of variable $z \to (\bar{z})^{-1}$ in (6.4.21) and notice that $\Gamma \to \overline{\Gamma}^{-1} = -\Gamma$ when we do this. Thus,

$$\text{Log } A = \frac{-1}{2\pi i} \oint_{-\Gamma} [\log (\bar{z})^{-1}][(\bar{z})^{-1} I - A]^{-1} \, d\bar{z}/(\bar{z})^2$$

and hence

$$\overline{\text{Log } A} = \frac{1}{2\pi i} \oint_{-\Gamma} (\log z^{-1})(z^{-1} I - \overline{A})^{-1} \, dz/z^2$$

$$= \frac{-1}{2\pi i} \oint_{\Gamma} (-\log z)(I - z\overline{A})^{-1} \, dz/z$$

$$= \frac{1}{2\pi i} \oint_{\Gamma} (\log z)(I - z\overline{A})^{-1} \, dz/z$$

$$= \frac{1}{2\pi i} \oint_{\Gamma} (\log z)[z^{-1} I - (zI - \overline{A}^{-1})^{-1}] \, dz$$

$$= -\frac{1}{2\pi i} \oint_{\Gamma} (\log z)(zI - \overline{A}^{-1})^{-1}] \, dz = -\text{Log } \overline{A}^{-1}$$

For the last equality, we use the fact that $(\log z)/z$ is a single-valued analytic function in $D_{R,\theta}$, and hence its integral around Γ is zero.

To prove (c)(1), make the change of variable $z \to \bar{z}$ and notice that $\Gamma \to \overline{\Gamma} = -\Gamma$ when we do this; although $D_{R,\pi}$ is symmetric with respect to the x-axis, its boundary Γ is traversed in the negative sense when $z \to \bar{z}$. Thus,

$$\text{Log } A = \frac{1}{2\pi i} \oint_{\Gamma} (\log \bar{z})(\bar{z} I - A)^{-1} \, d\bar{z}$$

$$= \frac{1}{2\pi i} \oint_{-\Gamma} (\log \bar{z})(\bar{z} I - A)^{-1} \, d\bar{z}$$

$$= \frac{-1}{2\pi i} \oint_\Gamma (\log \bar{z})(\bar{z}I - A)^{-1}\, d\bar{z}$$

and hence

$$\overline{\text{Log }A} = \frac{1}{2\pi i} \oint_\Gamma (\log z)(zI - \overline{A})^{-1}\, dz = \text{Log }\overline{A}$$

as asserted. Finally, (c)(2) is a consequence of (b)(2) and (c)(1).   ∎

6.4.22   Corollary. Let $E \in M_n$ be given. The following are equivalent:

(a) $E\overline{E} = I$, that is, E is coninvolutory.

(b) $E = e^{iS}$, where $S \in M_n(\mathbb{R})$ is real.

(c) $E = F^2$, where $F \in M_n$ is coninvolutory.

For a coninvolutory E, the real matrix S in (b) may be taken to be $S = (\text{Log }E)/i$ (using (6.4.21) to compute Log E), and the coninvolutory matrix F in (c) may be taken to be $F = e^{i\frac{1}{2}S}$. Thus, both S and F may be taken to be polynomials in E.

Proof: Suppose E is coninvolutory, so $\overline{E}^{-1} = E$. Choose $R > 0$ so that $R > \max \{\rho(E), 1/\rho(E^{-1})\}$ and choose $\theta \in [0, 2\pi)$ so that no eigenvalues of E lie on the ray segment $\{z = re^{i\theta} : 1/R \le r \le R\}$. Use these values of R and θ to construct the domain $D_{R,\theta}$, use (6.4.21) to compute Log E, and use the identity $E^{-1} = \overline{E}$ and (6.4.20(b)(2)) to write

$$\text{Log }E = \text{Log }\overline{E}^{-1} = -(\overline{\text{Log }E})$$

Thus, Log E is purely imaginary, so if we set $S \equiv (\text{Log }E)/i$ we have the representation (b). Assuming (b), (c) follows with $F = e^{i\frac{1}{2}S}$. Finally, (c) implies (a) since $E\overline{E} = FF\overline{F}\overline{F} = F I \overline{F} = I$. The final assertions follow from the fact that Log E as defined in Theorem (6.4.20) is a primary matrix function.   ∎

We now have all the tools needed to establish a third polar decomposition theorem.

6.4.23 **Theorem.** Let $A \in M_n$ be a given nonsingular matrix.

(a) $A = RE$, where $E \in M_n$ is coninvolutory ($E\overline{E} = I$), $R \in M_n(\mathbb{R})$ is real and nonsingular, and E is a polynomial in $\overline{A}^{-1}A$.

(b) Suppose $A = RE$, where $R, E \in M_n$, R is real and nonsingular, and $E\overline{E} = I$. Then $E^2 = \overline{A}^{-1}A$. If R commutes with E, then A commutes with $\overline{A}$ (that is, $A\overline{A}$ is real). Conversely, if A commutes with $\overline{A}$ and if E is a polynomial in $\overline{A}^{-1}A$, then R commutes with E.

Proof: Since $\overline{A}^{-1}A$ is coninvolutory, Corollary (6.4.22) guarantees that it has a coninvolutory square root E, that is, $\overline{A}^{-1}A = E^2$ and $E\overline{E} = I$; moreover, E may be taken to be a polynomial in $\overline{A}^{-1}A$. Since $\overline{A}^{-1}A = E^2$, we have $A = \overline{A}E^2 = (\overline{A}E)E$ and hence $\overline{AE} = (\overline{A}E)E\overline{E} = \overline{A}E = \overline{A}\overline{E}$. Thus, $R \equiv A\overline{E}$ is real and $A = RE$.

Suppose $A = RE$. If R commutes with E, then $A\overline{A} = RER\overline{E} = R\overline{E}RE = \overline{A}A = \overline{A\overline{A}}$, so A commutes with $\overline{A}$, that is, $A\overline{A}$ is real. Conversely, if A commutes with $\overline{A}$, then $RER\overline{E} = A\overline{A} = \overline{A}A = R\overline{E}RE$, so $ER\overline{E} = \overline{E}RE$ and hence $E^2R = E(ER\overline{E})E = E(\overline{E}RE)E = RE^2$, so R commutes with $E^2 = \overline{A}^{-1}A$. If E is a polynomial in $\overline{A}^{-1}A$, it follows that R commutes with E. □

6.4.24 **Theorem.** Let $A \in M_{m,n}$ be given with $m \geq n$. There exists a real matrix $R \in M_{m,n}(\mathbb{R})$ and a coninvolutory matrix $E \in M_n$ ($E\overline{E} = I$) such that $A = RE$ if and only if there exists a nonsingular $S \in M_n$ such that $A = \overline{A}S$, that is, if and only if A and $\overline{A}$ have the same range.

Proof: First observe that if A has a factorization of the desired form ($A = RE$), and if $S \in M_n$ is nonsingular, then $AS = RES = R(ES)$ and $ES \in M_n$ is nonsingular, so Theorem (6.4.23) permits us to write $ES = R'E'$ with R' real and E' coninvolutory. Thus, $AS = R(R'E') = (RR')E'$ is a factorization of the desired form for AS. Also, if $S, S_1 \in M_n$ are nonsingular, $A = \overline{A}S$, and $A_1 \equiv AS_1$, then $A_1 = \overline{A}SS_1 = (\overline{A}S_1)(\overline{S_1}^{-1}SS_1) = \overline{A_1}S_2$, where $S_2 \equiv \overline{S_1}^{-1}SS_1$ is nonsingular. Thus, the asserted equivalent conditions are both invariant under right multiplication by any nonsingular matrix. Since there is a permutation matrix $P \in M_n$ such that $AP = [A_1 \ A_2]$ with rank $A = $ rank $A_1 = r$ and $A_1 \in M_{m,r}$, we may assume that $A = [A_1 \ A_2]$ with $A_1 \in M_{m,r}$ and rank $A = $ rank A_1. Since we can then write $A_2 = A_1 B$ for some $B \in M_{r,n-r}$, we have

$$A = [A_1 \ A_2] = [A_1 \ A_1 B] = [A_1 \ 0] \begin{bmatrix} I & B \\ 0 & I \end{bmatrix}$$

and the indicated block matrix is nonsingular. Thus, we may further assume that $A = [A_1 \ 0]$ with a full rank $A_1 \in M_{m,r}$. Now assume that $A = \overline{A}S$ for some nonsingular $S \in M_n$ and write this identity in block form as

$$A = [A_1 \ 0] = [\overline{A}_1 S_{11} \ \overline{A}_1 S_{12}] = [\overline{A}_1 \ 0] \begin{bmatrix} S_{11} & S_{12} \\ S_{21} & S_{22} \end{bmatrix} = \overline{A}S$$

that is, $0 = \overline{A}_1 S_{12}$ and $A_1 = \overline{A}_1 S_{11} = (\overline{A}_1 S_{11}) S_{11} = A_1 (\overline{S}_{11} S_{11})$, or $A_1 (I - \overline{S}_{11} S_{11}) = 0$. Since A_1 and $\overline{A}_1$ have full rank, we conclude that $S_{12} = 0$ and $\overline{S}_{11} S_{11} = I$, that is, S_{11} is nonsingular and coninvolutory. Now use Corollary (6.4.22) to write $S_{11} = E^2$ for a coninvolutory E. Then $A_1 = \overline{A}_1 S_{11} = \overline{A}_1 E^2$, so $A_1 \overline{E} = \overline{A}_1 E = \overline{(A_1 \overline{E})} \equiv R$ is a real matrix and $A_1 = RE$. Finally,

$$A = [A_1 \ 0] = [RE \ 0] = [R \ 0] \begin{bmatrix} E & 0 \\ 0 & I \end{bmatrix}$$

is a factorization of the desired form. The converse implication is immediate, for if $A = RE$, then $A = RE = (\overline{A}E)E = \overline{A}E^2$ and E^2 is nonsingular. ☐

Problems

1. How many distinct similarity classes of solutions $X \in M_n$ are there to $p(X) = X^2 - I = 0$?

2. Describe the set of all solutions $X \in M_n$ to the equation $X^2 = X$. How does this generalize to the equation $X^k = X$, $k > 2$?

3. Describe the set of all solutions $X \in M_n$ to the equation $X^k = 0$ for given

$k > 1$.

4.   Let $p(t)$ be a given polynomial with scalar coefficients, and think of $f(t) \equiv p(t)$ as an entire analytic function. Analyze the solutions to $p(X) = 0$ as an example of a problem of type (6.4.2) and compare it with the analysis given in the text for equations of type (6.4.1).

5.   Show that every solution to the matrix polynomial equation (6.4.10b) $P_R(X) = 0$, $X \in M_n$, is also a solution to the scalar polynomial equation $\phi(X) = 0$, where $\phi(t) = \det P(t)$ and $P(t)$ is given by (6.4.9).

6.   Show that every solution X to (6.4.10b) is of the form $X = SJS^{-1}$, where J is a Jordan matrix that is a solution to $\phi(X) = 0$ and $\phi(t)$ is as in Problem 5. If $\phi(J) = 0$, show that $X = SJS^{-1}$ is a solution to (6.4.10b) if and only if S is a nonsingular solution to the linear homogeneous equation $J^m SA_m + J^{m-1}SA_{m-1} + \cdots + JSA_1 + SA_0 = 0$.

7.   Show by direct calculation that there is no matrix $A = \begin{bmatrix} a & b \\ c & d \end{bmatrix} \in M_2$ such that $A^2 = \begin{bmatrix} 0 & 1 \\ 0 & 0 \end{bmatrix}$.

8.   Use the criteria in Theorem (6.4.12) and Corollary (6.4.13) to show that $A = \begin{bmatrix} 0 & 1 \\ 0 & 0 \end{bmatrix}$ has no square root.

9.   Let $A \in M_n$ be given, and let $f(X)$ be a primary matrix function. If $Y \in M_n$ is in the domain of $f(\cdot)$ and if $f(Y) = A$, show that $f(SYS^{-1}) = A$ for any nonsingular $S \in M_n$ that commutes with A. What does this imply if the given right-hand side A is a scalar matrix, in particular, if $A = 0$ or I?

10.   Use the formulae in Theorem (6.2.25) to analyze the Jordan structure of the cube of a singular Jordan block $J_k(0)$. Verify that if $k \equiv 0 \bmod 3$, then $(J_k(0))^3$ has three Jordan blocks of size $k/3$; if $k \equiv 1 \bmod 3$, then $(J_k(0))^3$ has one Jordan block of size $(k+2)/3$ and two blocks of size $(k-1)/3$; if $k \equiv 2 \bmod 3$, then $(J_k(0))^3$ has two Jordan blocks of size $(k+1)/3$ and one block of size $(k-2)/3$.

11.   Let $A \in M_n$ be a singular matrix, and let $J_{k_1}(0) \oplus J_{k_2}(0) \oplus J_{k_3}(0) \oplus \cdots \oplus J_{k_p}(0)$ be the singular part of the Jordan canonical form of A with $k_1 \geq k_2 \geq k_3 \geq \cdots \geq k_p$. Let $\Delta_1 = (k_1 - k_2, k_2 - k_3)$, $\Delta_4 = (k_4 - k_5, k_5 - k_6)$, ..., $\Delta_i = (k_i - k_{i+1}, k_{i+1} - k_{i+2})$, $i = 1, 4, 7, \dots$. Show that A has a cube root if and only if each Δ_i is of the form $(0,0)$ or $(1,0)$ or $(0,1)$. If p is not an integer multiple of 3, we must also require: $k_p = 1$ if $p \equiv 1 \bmod 3$, or $k_p = k_{p-1} = 1$ if

$p \equiv 2 \mod 3$. Show that if A has a cube root then it has cube roots in only finitely many different similarity classes.

12. What are the analogs of Corollary (6.4.13) and Theorem (6.4.14) for cube roots?

13. Formulate and prove an analog of Theorem (6.4.12) for m th roots.

14. In the discussion preceding Theorem (6.4.15), it was shown that if the Jordan canonical form of the nonsingular matrix $A \in M_n$ is $J = J_{k_1}(\lambda_1) \oplus \cdots \oplus J_{k_p}(\lambda_p)$, then every solution X to $e^X = A$ has the Jordan canonical form $J_{k_1}(\log \lambda_1) \oplus \cdots \oplus J_{k_p}(\log \lambda_p)$. Although this determines the similarity class of X, it does not give an explicit value for X. Show that if $S_i \in M_{k_i}$ is a nonsingular matrix such that $S_i e^{J_{k_i}(\log \lambda_i)} S_i^{-1} = J_{k_i}(\lambda_i)$ (Why is there such an S_i?), and if $A = SJS^{-1}$ is the Jordan canonical form of A, then

$$X = S\left[S_1 J_{k_1}(\log \lambda_1) S_1^{-1} \oplus \cdots \oplus S_p J_{k_p}(\log \lambda_p) S_p^{-1} \right] S^{-1}$$

is a solution to $e^X = A$.

15. Analyze the equation $\sin X - A = 0$ for $X, A \in M_n$ following the analysis of $e^X - A = 0$ given in the text. For any complex number z, $\sin z = (e^{iz} - e^{-iz})/2i$ and $\cos z = (e^{iz} + e^{-iz})/2$.

16. For $X \in M_4$, describe explicitly the complete set of solutions of the equation $\cos X - I = 0$; do the same for the equation $\sin X - I = 0$. What is the most striking difference between these two sets of solutions and why?

17. Consider the stem function $f(t) \equiv e^{-1/t^2}$ for real $t \neq 0$, $f(0) \equiv 0$, which has continuous derivatives of all orders on all of $\mathbb{R}$ but is not analytic in any neighborhood of $t = 0$. Every derivative of $f(t)$ vanishes at $t = 0$. Let $f(X)$ denote the associated primary matrix function on $\mathcal{D}_n(\mathbb{R})$. What are the solutions to $f(X) = 0$ in $\mathcal{D}_n(\mathbb{R})$, $n = 1, 2, \ldots$? This is a case in which $f(t)$ has a zero of infinite multiplicity.

18. Let $A \in M_n$ be given and let λ be an eigenvalue of A. Let $r_0(\lambda) = n$, $r_k(\lambda) = \text{rank}(A - \lambda I)^k$, $k = 1, 2, \ldots, n$. Show that the number of Jordan blocks of A with size k and eigenvalue λ is equal to $[r_{k-1}(\lambda) - r_k(\lambda)] - [r_k(\lambda) - r_{k+1}(\lambda)] = r_{k-1}(\lambda) - 2r_k(\lambda) + r_{k+1}(\lambda)$, $k = 1, 2, \ldots$. Use this observation to reformulate the Jordan block size conditions in Theorems (6.4.14-15) in

terms of these second differences of ranks of powers.

19. Modify the reasoning used to prove Theorem (6.4.14), as necessary, to give a detailed proof for the assertions about a real logarithm of a real matrix in Theorem (6.4.15). Observe that the square root and the logarithm are both *typically real* analytic functions: There is always a choice of arguments for which $f(\bar{z}) = \overline{f(z)}$; explain why this observation is important in this context.

20. Suppose $A \in M_n$ is nonsingular and has μ distinct eigenvalues $\lambda_1, ..., \lambda_\mu$. Among the Jordan blocks of A with eigenvalue λ_i, suppose there are k_i blocks with *different* sizes, $i = 1, ..., \mu$, and let $\kappa \equiv k_1 + \cdots + k_\mu$. Show that A has square roots in at least 2^κ different similarity classes. When is this lower bound exact? Describe how to compute the exact number of similarity classes that contain square roots of A.

21. Consider

$$G \equiv \begin{bmatrix} 1 & 0 \\ 0 & -1 \end{bmatrix}, \quad Q \equiv 2^{-\frac{1}{2}} \begin{bmatrix} 1 & 1 \\ -1 & 1 \end{bmatrix}, \text{ and } A = 2^{-\frac{1}{2}} \begin{bmatrix} 1 & 1 \\ 1 & -1 \end{bmatrix}$$

Show that Q is orthogonal, G is symmetric, $A = GQ$, and A commutes with A^T, but G does not commute with Q. Does this contradict Theorem (6.4.16(b))? Why?

22. The singular value and polar decompositions are equivalent in the sense that each may be derived from the other and every matrix has both, but the situation is more complicated for symmetric-orthogonal analogs of these classical results. Let $A \in M_n$ be given. If there are complex orthogonal $Q_1, Q_2 \in M_n$ and a diagonal $\Lambda \in M_n$ such that $A = Q_1 \Lambda Q_2^T$, show that A can be written as $A = GQ$, as in Theorems (6.4.16-17). Conversely, consider the symmetric matrix $B = \begin{bmatrix} 1 & i \\ i & -1 \end{bmatrix}$, which has a trivial symmetric-orthogonal decomposition $B = BI$. Compute $B^T B = B^2$ and explain why this B cannot be written as $B = Q_1 \Lambda Q_2^T$. In general, it is known that a given $A \in M_n$ can be written as $A = Q_1 \Lambda Q_2^T$ for some orthogonal Q_1, Q_2 and some diagonal Λ if and only if $A^T A$ is diagonalizable and rank $A = $ rank $A^T A$.

23. Let $A \in M_n$ be a given complex symmetric matrix, $A = A^T$. If A has a square root B, show that it has a symmetric square root that is similar to B.

24. If $A \in M_n$ can be written as $A = GQ$ for some complex symmetric $G \in$

M_n and a complex orthogonal $Q \in M_n$, show that AA^T is similar to A^TA. Use this to show that the rank conditions in Theorem (6.4.17) are necessary.

25. Let $A \in M_n$ be given. If A is nonsingular, use Theorem (6.4.16) to show that A^TA is complex orthogonally similar to AA^T. Consider $\begin{bmatrix} 0 & i \\ 0 & 1 \end{bmatrix}$ to show that this need not be true if A is singular. In general, use Theorem (6.4.17) to show that AA^T is complex orthogonally similar to A^TA if and only if rank $(AA^T)^k = $ rank $(A^TA)^k$ for $k = 1, 2,..., n$. Since analogies are often a fruitful source of conjectures, it is interesting to note that the ordinary polar decomposition shows that AA^* is always unitarily similar to A^*A without any further conditions.

26. Let $A \in M_n$ be given. Prove—without using Theorem (6.4.17)—that the following conditions are equivalent:

(a)   rank $(AA^T)^k = $ rank $(A^TA)^k$ for $k = 1,..., n$.

(b) AA^T is similar to A^TA.

(c) AA^T is orthogonally similar to A^TA.

(d)   There exists an orthogonal $P \in M_n$ such that $\hat{A} \equiv AP$ commutes with $\hat{A}^T$.

27. Verify the following representations of various classes of matrices in terms of exponentials and products of matrices.

27a. $U \in M_n$ is unitary if and only if $U = e^{iH}$ for some Hermitian $H \in M_n$; one may choose H to be a polynomial in U, and H may be taken to be positive definite.

27b. Every $A \in M_n$ can be represented as $A = Pe^{iH}$, where $P, H \in M_n$ are Hermitian and P is positive semidefinite; H may be taken to be positive definite. The factor P is uniquely determined as the unique positive semidefinite square root of AA^*.

27c. $U \in M_n$ is both symmetric and unitary if and only if there is a real symmetric $B \in M_n(\mathbb{R})$ such that $U = e^{iB}$; one may choose B to be a polynomial in U, and B may be taken to be positive definite.

27d. $U \in M_n$ is unitary ($U^*U = I$) if and only if there is a real orthogonal $Q \in M_n(\mathbb{R})$ and a real symmetric $B \in M_n(\mathbb{R})$ such that $U = Qe^{iB}$; one may choose B to be a polynomial in U^TU. Thus, every unitary matrix is the product of a real orthogonal matrix and a complex symmetric unitary coninvolutory matrix;   in particular, this is a simultaneous polar decomposition of U as a product of a real matrix and a coninvolutory matrix as well as a product of a complex symmetric matrix and an orthogonal matrix.   Also conclude that every unitary $U \in M_n$ can be written as $U = Q_1 D Q_2$, where Q_1, $Q_2 \in M_n(\mathbb{R})$ are real orthogonal matrices and $D \in M_n$ is a diagonal unitary matrix.

27e. $P \in M_n$ is both positive definite and orthogonal if and only if there is a real skew-symmetric $B \in M_n(\mathbb{R})$ such that $P = e^{iB}$; one may choose B to be a polynomial in P.

27f. $P \in M_n$ is complex orthogonal $(P^T P = I)$ if and only if there is a real orthogonal $Q \in M_n(\mathbb{R})$ and a real skew-symmetric $B \in M_n(\mathbb{R})$ such that $P = Q e^{iB}$; one may choose B to be a polynomial in $P^* P$. Thus, every complex orthogonal matrix is the product of a real orthogonal matrix and a positive definite Hermitian orthogonal coninvolutory matrix; in particular, this is a simultaneous polar decomposition of P as a product of a unitary matrix and a positive definite matrix as well as a product of a real matrix and a coninvolutory matrix.

27g. $E \in M_n$ is coninvolutory $(E\overline{E} = I)$ if and only if there is a real $B \in M_n(\mathbb{R})$ such that $E = e^{iB}$; one may choose B to be a polynomial in E.

27h. Every nonsingular $A \in M_n$ can be written as $A = R e^{iB}$, where R, $B \in M_n(\mathbb{R})$; one may choose B to be a polynomial in $\overline{A}^{-1} A$.

28.   Problems 27d and 27f help to explain the perhaps mysterious similarities and differences between unitary and complex orthogonal matrices.   Both can be written in the form $A = Q e^{iB}$ with Q, $B \in M_n(\mathbb{R})$ and Q orthogonal, but a unitary A corresponds to a symmetric B and an orthogonal A corresponds to a skew-symmetric B.   What does this say when $n = 1$?

29.   Consider $A = \begin{bmatrix} 0 & 1 \\ 1 & 0 \end{bmatrix}$, $R = \begin{bmatrix} 1 & 0 \\ 0 & -1 \end{bmatrix}$, and $E = \begin{bmatrix} 0 & 1 \\ -1 & 0 \end{bmatrix}$.   Show that $A = RE$, R is real, E is coninvolutory, and A commutes with $\overline{A}$, but R does not commute with E.   Does this contradict Theorem (6.4.23(b))?   Explain.

30.   Use Theorem (6.4.20) to show that if $A \in M_n(\mathbb{R})$ is real and has no negative eigenvalues, then $A = e^B$ for some real $B \in M_n(\mathbb{R})$.   Contrast this result with Theorem (6.4.15(c)).

31.   If $B \in M_n$ is skew-symmetric, show that $P \equiv e^{iB}$ is complex orthogonal. Use Theorem (6.4.20) to show that if $P \in M_n$ is complex orthogonal and has no negative eigenvalues, then $P = e^{iB}$ for some skew-symmetric $B \in M_n$.

However, consider $n = 1$, $P = -1$ to show that not every complex orthogonal matrix $P \in M_n$ can be written as $P = e^{iB}$ with a skew-symmetric $B \in M_n$.

32. Use the representation in Problem 27f to show that $P \in M_n$ is both orthogonal and Hermitian if and only if there are $Q, B \in M_n(\mathbb{R})$ such that $P = Qe^{iB}$, where Q is real orthogonal, symmetric, and idempotent ($QQ^T = I$, $Q^T = Q$, $Q^2 = I$), B is real skew-symmetric ($B^T = -B$), and Q commutes with B. Show that one may take $Q = I$ if and only if P is positive definite, so this representation generalizes Problem 27e.

33. Let $A \in M_n$ be given. It is always true that $A = XA^TY$ for some nonsingular $X, Y \in M_n$, but suppose $A = XA^TX$ for some nonsingular $X \in M_n$. Use Problem 23 to show that there is a symmetric $S \in M_n$ such that $S^2 = A^TA$ and rank S = rank A.

34. Let $A \in M_{m,n}$ be given. When $m \geq n$, we wish to write $A = QS$ with $Q \in M_{m,n}$, $S = S^T \in M_n$, and $Q^TQ = I \in M_n$. If $m \leq n$ we wish to write $A = SQ$ with $Q \in M_{m,n}$, $S = S^T \in M_m$, and $QQ^T = I \in M_m$.
   (a) When $m \geq n$, show that the desired factorization can be achieved if and only if $\tilde{A} \equiv [A\ 0] \in M_m$ can be written as $\tilde{A} = QS$ with $Q, S \in M_m$, $S = S^T$, and $Q^TQ = I$.
   (b) When $m \geq n$, use (a) and Theorem (6.4.17) to show that A has a factorization of the desired form if and only if AA^T is similar to $A^TA \oplus 0_{m-n}$, and that this condition is equivalent to requiring that rank$(A^TA)^k$ = rank$(AA^T)^k$ for all $k = 1, \ldots, n$.
   (c) State and verify the analogs of (a) and (b) when $m \leq n$.
   (d) In all cases, show that A has the desired factorization if and only if rank$(A^TA)^k$ = rank$(AA^T)^k$, $k = 1, \ldots, n$.

35. Let $A, B \in M_n$ be given. Show that range A = range B if and only if there is a nonsingular $S \in M_n$ such that $A = BS$.

36. Restate and verify Theorem (6.4.24) when $m \leq n$.

37. The purpose of this problem is to show that a square complex matrix A with a positive spectrum has a unique square root with a positive spectrum; moreover, this square root is a polynomial in A. See Problem 20 in Section (4.4) for a different approach to this conclusion via an important property of the centralizer of A.
   (a) Let $J_m(\lambda) \in M_m$ be a Jordan block with positive eigenvalue λ. Suppose $K \in M_m$ is an upper triangular Toeplitz matrix with positive main diagonal.

Show that $K^2 = J_m(\lambda)^2$ if and only if $K = J_m(\lambda)$.

(b) Let $J \in M_n$ be a direct sum of Jordan blocks with positive eigenvalues. Suppose $K \in M_n$ is a block diagonal matrix conformal to J, and suppose that each diagonal block of K is an upper triangular Toeplitz matrix with positive main diagonal. Use (a) to show that $K^2 = J^2$ if and only if $K = J$.

(c) Suppose $A \in M_n$ has a positive spectrum. Show that there is a primary matrix function square root of A that has a positive spectrum, that is, there is a $B \in M_n$ such that B is a polynomial in A, $B^2 = A$, and B has a positive spectrum. Moreover, use (b) to show that this matrix B is the unique square root of A with a positive spectrum, that is, if $C \in M_n$ has a positive spectrum and $C^2 = A = B^2$, then $C = B$. Conclude that any matrix that commutes with A must commute with any square root of A that has a positive spectrum.

(d) Explain why "square root" may be replaced by "kth root" in (c) for each positive integer k.

(e) Compare the results in (c) and (d) with Theorem (7.2.6) in [HJ], which concerns positive semidefinite matrices.

Further Readings. See Chapter VIII of [Gan 59, Gan 86] for more information about matrix equations, logarithms, and square roots. See also Square Roots of Complex Matrices, *Linear Multilinear Algebra* 1 (1974), 289-293 by G. W. Cross and P. Lancaster, and On Some Exponential and Polar Representations of Matrices, *Nieuw Archief voor Wiskunde* (3) III (1955), 20-32, by N. G. De Bruijn and G. Szekeres; the latter paper also discusses the existence of a real square root of a real matrix. Some partial generalizations of Theorem (6.4.16) to the possibly singular case, numerous examples, and a conjectured general result are in D. Choudhury and R. A. Horn, A Complex Orthogonal-Symmetric Analog of the Polar Decomposition, *SIAM J. Alg. Disc. Meth.* 8 (1987), 219-225; the conjecture is resolved by Theorem (6.4.17), which is proved in I. Kaplansky, Algebraic Polar Decomposition, *SIAM J. Matrix Analysis Appl.* 11 (1990), 213-217. At the end of Problem 22 there is a characterization stated for the matrices that have an orthogonal-diagonal-orthogonal factorization; for a proof, see D. Choudhury and R. A. Horn, An Analog of the Singular Value Decomposition for Complex Orthogonal Equivalence, *Linear Multilinear Algebra* 21 (1987), 149-162. The results on orthogonal-symmetric and real-coninvolutory factorizations in Problem 34 and Theorem (6.4.24) are due to Dennis Merino.

6.5   Matrices of functions

The usual analytic notions of continuity, differentiability, integrability, etc., all carry over to vector-valued and matrix-valued functions of a scalar $f: t \rightarrow A(t) = [a_{ij}(t)]$ if we interpret them entrywise. Thus, $A(t)$ is said to be continuous, differentiable, or integrable if each entry $a_{ij}(t)$ has the respective property.

If $A(t)$, $B(t)$, $A_i(t) \in M_n$ are differentiable, the following formulae are easily verified:

$$\frac{d}{dt}(aA(t) + bB(t)) = a\frac{d}{dt}A(t) + b\frac{d}{dt}B(t) \text{ for all } a, b \in \mathbb{C} \qquad (6.5.1)$$

$$\frac{d}{dt}[A(t)B(t)] = \left[\frac{d}{dt}A(t)\right]B(t) + A(t)\left[\frac{d}{dt}B(t)\right] \qquad (6.5.2)$$

$$\frac{d}{dt}\left[A_1(t)A_2(t)\cdots A_k(t)\right] = \left[\frac{d}{dt}A_1(t)\right]A_2(t)\cdots A_k(t) \qquad (6.5.3)$$

$$+ A_1(t)\left[\frac{d}{dt}A_2(t)\right]A_3(t)\cdots A_k(t) + \cdots$$

$$+ A_1(t)\cdots A_{k-2}(t)\left[\frac{d}{dt}A_{k-1}(t)\right]A_k(t)$$

$$+ A_1(t)\cdots A_{k-1}(t)\left[\frac{d}{dt}A_k(t)\right]$$

$$\frac{d}{dt}C = 0 \text{ if } C \in M_n \text{ is a constant matrix} \qquad (6.5.4)$$

$$\frac{d}{dt}[A(t)]^m = \left[\frac{d}{dt}A(t)\right]A^{m-1}(t) + A(t)\left[\frac{d}{dt}A(t)\right]A^{m-2}(t) + \cdots$$

$$+ A^{m-2}(t)\left[\frac{d}{dt}A(t)\right]A(t) + A^{m-1}(t)\left[\frac{d}{dt}A(t)\right] \quad (6.5.5)$$

If $A(t)$ commutes with $\frac{d}{dt}A(t)$, then

$$\frac{d}{dt}[A(t)]^m = mA^{m-1}(t)\left[\frac{d}{dt}A(t)\right] = m\left[\frac{d}{dt}A(t)\right]A^{m-1}(t) \qquad (6.5.5a)$$

If

$$f(t) = \sum_{m=0}^{\infty} a_m t^m \text{ and } g(t) = \sum_{m=0}^{\infty} b_m t^m$$

are analytic, if $\rho(A(t))$ is less than the radius of convergence of $g(t)$, if $\rho(g(A(t))$ is less than the radius of convergence of $f(t)$, and if $\frac{d}{dt}A(t)$ commutes with $A(t)$, then (6.5.5a) can be used to show that $\frac{d}{dt}A(t)$ commutes with $g'(A(t))$ and $f'(g(A(t)))$, $g'(A(t))$ commutes with $f'(g(A(t)))$, and

$$\frac{d}{dt}f(g(A(t))) = f'(g(A(t)))g'(A(t))\frac{d}{dt}A(t) \qquad (6.5.6)$$

where the factors on the right can be multiplied in any order.

If $A(t)$ is differentiable and nonsingular at some point t, then

$$0 = \frac{d}{dt}I = \frac{d}{dt}\left[A(t)A^{-1}(t)\right] = \left[\frac{d}{dt}A(t)\right]A^{-1}(t) + A(t)\left[\frac{d}{dt}A^{-1}(t)\right]$$

so the formula

$$\frac{d}{dt}A^{-1}(t) = -A^{-1}(t)\left[\frac{d}{dt}A(t)\right]A^{-1}(t) \qquad (6.5.7)$$

follows from (6.5.2) and (6.5.4). The trace function is a linear function, so

$$\frac{d}{dt}\text{tr } A(t) = \text{tr }\frac{d}{dt}A(t) \qquad (6.5.8)$$

Since the determinant function is not linear, its derivative is a bit more complicated. If

$$A(t) = \begin{bmatrix} a_{11}(t) & a_{12}(t) \\ a_{21}(t) & a_{22}(t) \end{bmatrix} \in M_2$$

then

$$\frac{d}{dt}\det A(t) = \frac{d}{dt}(a_{11}a_{22} - a_{12}a_{21})$$

$$= a'_{11}a_{22} - a'_{12}a_{21} - a'_{21}a_{12} + a'_{22}a_{11}$$

$$= \det\begin{bmatrix} a'_{11} & a'_{12} \\ a_{21} & a_{22} \end{bmatrix} + \det\begin{bmatrix} a_{11} & a_{12} \\ a'_{21} & a'_{22} \end{bmatrix}$$

Using the Lagrange expansion by cofactors, one shows inductively that

$$\frac{d}{dt}\det A(t) = \sum_{i=1}^{n}\det A_{(i)}(t) = \sum_{i=1}^{n}\det A^T_{(i)}(t) \qquad (6.5.9)$$

where $A_{(i)}(t)$ is the matrix that coincides with the matrix $A(t)$ except that every entry in the ith row (equivalently, columns could be used) is differentiated with respect to t. As a corollary to this formula, observe that

$$\frac{d}{dt}\det(tI - A) = \sum_{i=1}^{n} \det(tI - A)_{(i)}$$

$$= \text{sum of all the } (n-1)\text{-by-}(n-1) \text{ principal minors of } (tI - A)$$

because each term $(tI - A)_{(i)}$ has as its ith row the basis vector e_i^T. Thus, using the matrix adj X of cofactors of X, a compact expression for the derivative of the characteristic polynomial is given by the formula

$$\frac{d}{dt}p_A(t) = \text{tr adj}(tI - A) \qquad (6.5.10)$$

As an example of the utility of some of these formulae, consider the function

$$\log(1 - t) = -\sum_{k=1}^{\infty} \frac{1}{k} t^k$$

which has radius of convergence 1. We know that if $A \in M_n$ and $\rho(A) < 1$, then the series

$$\log(I - A) \equiv -\sum_{k=1}^{\infty} \frac{1}{k} A^k \qquad (6.5.11)$$

is convergent. We know from the theory of primary matrix functions [see Example (6.2.15)] that $e^{\log(I - A)} = I - A$ under these conditions. Let us see how this important identity follows from direct calculation with power series.

If $A \in M_n$ is given, then $\rho(tA) < 1$ if $|t| < 1/\rho(A)$ and $\rho(A) > 0$, or $\rho(tA) < 1$ for all $t \in \mathbb{C}$ if $\rho(A) = 0$. For t in this range, we can consider the matrix function

$$B(t) \equiv \log(I - tA) = -\sum_{k=1}^{\infty} \frac{t^k}{k} A^k \qquad (6.5.12)$$

Notice that

$$\tfrac{d}{dt}B(t) = -\sum_{k=1}^{\infty} \frac{k t^{k-1}}{k} A^k = -\sum_{k=1}^{\infty} t^{k-1}A^k = -A\sum_{k=0}^{\infty} t^k A^k = -A(I-tA)^{-1}$$

and $B(0) = 0$. Now consider the matrix function $f(t) \equiv e^{\log(I-tA)}$, for which $f(0) = e^{\log(I)} = e^0 = I$. Using (6.5.6), we compute

$$f'(t) = \tfrac{d}{dt}e^{\log(I-tA)} = -e^{\log(I-tA)}A(I-tA)^{-1}, \quad f'(0) = -A$$

Now use (6.5.7), (6.5.6), and (6.5.3) to obtain

$$\begin{aligned}
f''(t) &= \tfrac{d}{dt}\Big[-e^{\log(I-tA)}A(I-tA)^{-1}\Big]\\
&= \Big[e^{\log(I-tA)}A(I-tA)^{-1}\Big]A(I-tA)^{-1}\\
&\quad - e^{\log(I-tA)}A\Big[(I-tA)^{-1}A(I-tA)^{-1}\Big]\\
&= 0
\end{aligned}$$

Thus, $f'(t) = \text{constant} = f'(0) = -A$ and $f(t) = -tA + \text{constant} = I - tA$, since $f(0) = I$. We conclude that if $\log(I-tA)$ is defined by (6.5.12) and if e^X is defined by its power series, then

$$e^{\log(I-tA)} = \exp\left[-\sum_{k=1}^{\infty} \frac{t^k}{k}A^k\right] = I-tA \tag{6.5.13}$$

for all $|t| < 1/\rho(A)$ if $\rho(A) > 0$, and for all $t \in \mathbb{C}$ if $\rho(A) = 0$. Thus, the formula (6.5.12) produces a "log" of $I-tA$ for all suitably small t.

We already know [see Example (6.2.15)] that the identity $e^{\log(I-tA)} = I-tA$ for *primary* matrix functions holds whenever $I-tA$ is nonsingular, a condition much less restrictive than requiring that $\rho(tA) < 1$. The point of (6.5.13) is that a direct analytic argument, independent of the general theory of primary matrix functions, can be given to show that the asserted *power series* identity holds under the stated hypotheses. This identity can be very useful in computations.

A similar argument shows that the role of the exponential and logarithm as inverse matrix functions can be reversed under suitable hypotheses. If $A \in M_n$ is given, then every eigenvalue of $I-e^{tA}$ is of the form $1-e^{\lambda t}$, where λ is an eigenvalue of A. Thus, $|1-e^{\lambda t}| \le \exp(|\lambda t|) - 1$ by Corollary

(6.2.32), so $\rho(I-e^{tA}) < 1$ for all $|t| < (\ln 2)/\rho(A)$ if $\rho(A) > 0$ or for all $t \in \mathbb{C}$ if $\rho(A) = 0$. For t in this range, consider the matrix function

$$g(t) \equiv \log e^{tA} \equiv \log[I-(I-e^{tA})] \equiv -\sum_{k=1}^{\infty} \frac{1}{k}(I-e^{tA})^k$$

for which $g(0) = 0$. Now use this series and (6.5.6) to compute

$$g'(t) = \sum_{k=1}^{\infty} (I-e^{tA})^{k-1} e^{tA} A = [I-(I-e^{tA})]^{-1} e^{tA} A$$

$$= (e^{tA})^{-1} e^{tA} A = A$$

With one integration it now follows that $g(t) = tA + g(0) = tA$. Our conclusion is that if $\log[I-(I-e^{tA})]$ is defined by the power series (6.5.11), then

$$\log e^{tA} \equiv \log[I-(I-e^{tA})] = -\sum_{k=1}^{\infty} \frac{1}{k}(I-e^{tA})^k = tA \qquad (6.5.14)$$

for all $|t| < (\ln 2)/\rho(A)$ if $\rho(A) > 0$, and for all $t \in \mathbb{C}$ if $\rho(A) = 0$.

We already know [see Example (6.2.16)] that a branch of the logarithm can be chosen so that the identity $\log e^{tA} = tA$ for *primary* matrix functions holds whenever the spectrum of tA lies in an open horizontal strip with height 2π. In particular, this condition certainly holds if $\rho(A) > 0$ and $|t| < \pi/\rho(A)$, which is less restrictive than requiring $|t| < (\ln 2)/\rho(A)$. The point of (6.5.14), as with (6.5.13), is that the asserted *power series* identity can be proved with a direct analytic argument that does not rely on the general theory of primary matrix functions.

One consequence of (6.5.13) is that the familiar limit $e^{-t} = \lim_{m \to \infty}(1-\frac{t}{m})^m$ has an analog for matrix functions. If $A \in M_n$ is given, then for all $m > \rho(A)$ we have, using Theorem (6.2.38) and the power series (6.5.11),

$$(I-\tfrac{1}{m}A)^m = \left[e^{\log(I-A/m)} \right]^m = e^{m\log(I-A/m)}$$

$$= e^{-m\,[A/m + A^2/(2m^2) + A^3/(3m^3) + \cdots\,]}$$

$$= e^{-A} + O(1/m) = e^{-A} e^{O(1/m)}$$

where the term $O(\frac{1}{m}) \to 0$ as $m \to \infty$. Thus, continuity of the exponential primary matrix function [for quantitative bounds see Corollary (6.2.32)] ensures that

$$\lim_{m \to \infty} (I - \frac{1}{m}A)^m = e^{-A} \qquad (6.5.15)$$

for any $A \in M_n$.

Another consequence of (6.5.13) is a very useful matrix version of a famous formula from the classical theory of Lie groups. If $A, B \in M_n$ are given, and if $|||\cdot|||$ is a given matrix norm on M_n, then Corollary (6.2.32) gives the bound

$$||| e^{tA} e^{tB} - I ||| = ||| (e^{tA} - I) + (e^{tB} - I) + (e^{tA} - I)(e^{tB} - I) |||$$

$$\leq ||| e^{tA} - I ||| + ||| e^{tB} - I ||| + ||| e^{tA} - I ||| \; ||| e^{tB} - I |||$$

$$\leq [\exp(|t| \; ||| A |||) - 1] + [\exp(|t| \; ||| B |||) - 1]$$

$$+ [\exp(|t| \; ||| A |||) - 1] [\exp(|t| \; ||| B |||) - 1]$$

$$= \exp[|t| \, (||| A ||| + ||| B |||)] - 1$$

and hence

$$\rho(e^{tA} e^{tB} - I) \leq ||| e^{tA} e^{tB} - I ||| < 1$$

for all $|t| < (\ln 2)/(||| A ||| + ||| B |||)$ if either A or B is nonzero, or for all $t \in \mathbb{C}$ if $A = B = 0$. For all t in this range, (6.5.13) guarantees that

$$e^{tA} e^{tB} = e^{\log[I - (I - e^{tA} e^{tB})]} = \exp\left[-\sum_{k=1}^{\infty} \frac{1}{k} (I - e^{tA} e^{tB})^k \right] \qquad (6.5.16)$$

Thus, for each small t, there is some $C(t; A, B) \in M_n$ such that $e^{tA} e^{tB} = e^{C(t; A, B)}$, and

$$C(t;A,B) = -\sum_{k=1}^{\infty} \frac{1}{k}(I - e^{tA} e^{tB})^k$$

$$= -\sum_{k=1}^{\infty} \frac{1}{k}\left[I - \left[\sum_{j=0}^{\infty} \frac{t^j}{j!}A^j \right] \left[\sum_{m=0}^{\infty} \frac{t^m}{m!}B^m \right] \right]^k$$

$$= \sum_{k=1}^{\infty} \frac{(-1)^{k+1}}{k}\left[\sum_{m=1}^{\infty} \frac{t^m}{m!}\left[\sum_{j=0}^{m} \binom{m}{j} A^j B^{m-j} \right] \right]^k$$

$$\equiv \sum_{k=1}^{\infty} C_k(A,B)t^k$$

$$= (A + B)t + O(t^2) \tag{6.5.17}$$

as $t \to 0$. The matrix coefficients $C_k(A,B)$ in this power series for $C(t;A,B)$ depend on A and B in a complicated way, but the first coefficient, $C_1(A,B)$ $= A + B$, is easily computed; see Problem 18 for some higher-order coefficients. Substituting (6.5.17) into (6.5.16) gives

$$e^{tA} e^{tB} = e^{t(A+B)} + O(t^2) \tag{6.5.18}$$

as $t \to 0$. Now let $t = 1/m$ and take m th powers of this identity to obtain

$$\left[e^{A/m} e^{B/m} \right]^m = \left[e^{(A+B)/m} + O(1/m^2) \right]^m = e^{(A+B)} + O(1/m)$$

as $m \to \infty$. Continuity of the matrix exponential as guaranteed in Corollary (6.2.32) now gives the important *Lie product formula*:

$$\lim_{m \to \infty} \left[e^{A/m} e^{B/m} \right]^m = e^{A+B} \tag{6.5.19}$$

which holds for any $A, B \in M_n$. The same argument results in a similar formula for $e^{A_1 + \cdots + A_k}$; see Problem 27.

This formula has the advantage of expressing e^{A+B} (whose value in terms of e^A, e^B, and $e^A e^B$ is, in general, unknown) in terms of a limit of ordinary products of exponentials of A and B. There are many interesting applications of the Lie product formula; we consider first some inequalities for matrix exponentials that arise in statistical mechanics, population biology, and quantum mechanics.

6.5.20   Theorem. Let $f: M_n \to \mathbb{C}$ be a continuous function such that

   (a) $f(XY) = f(YX)$ for all $X, Y \in M_n$, and

   (b) $|f(X^{2k})| \le f([XX^*]^k)$ for all $X \in M_n$ and all $k = 1, 2, \dots$.

Then

   (1) $f(XY) \ge 0$ for all positive semidefinite $X, Y \in M_n$; in particular, $f(e^A) \ge 0$ whenever $A \in M_n$ is Hermitian.

   (2) $|f(e^{A+B})| \le f(e^{(A+A^*+B+B^*)/2}) \le f(e^{(A+A^*)/2}\, e^{(B+B^*)/2})$ for all $A, B \in M_n$.

   (3) $|f(e^A)| \le f(e^{(A+A^*)/2})$ for all $A \in M_n$.

   (4) $0 \le f(e^{A+B}) \le f(e^A e^B)$ for all Hermitian $A, B \in M_n$.

Proof: If X and Y are positive semidefinite, they have positive semidefinite square roots $X^{\frac{1}{2}}$ and $Y^{\frac{1}{2}}$. Using conditions (a) and (b), compute $f(XY) = f(X^{\frac{1}{2}} X^{\frac{1}{2}} Y) = f(X^{\frac{1}{2}} YX^{\frac{1}{2}}) = f([X^{\frac{1}{2}} Y^{\frac{1}{2}}][X^{\frac{1}{2}} Y^{\frac{1}{2}}]^*) \ge |f([X^{\frac{1}{2}} Y^{\frac{1}{2}}]^2)| \ge 0$, which verifies assertion (1).

Let $X, Y \in M_n$ be given. For any integer $k \ge 1$ we have

$$f([X^*X\,YY^*]^{2^{k-1}}) = f(X^*[X\,YY^*X^*] \cdots [XYY^*X^*]X\,YY^*)$$

$$= f([X\,YY^*X^*] \cdots [X\,YY^*X^*]X\,YY^*X^*) \qquad \text{[by (a)]}$$

$$= f([X\,YY^*X^*]^{2^{k-1}}) = f([(XY)(XY)^*]^{2^{k-1}})$$

$$\ge |f([XY]^{2^k})| \qquad\qquad\qquad\qquad\qquad \text{[by (b)]}$$

Iterating this inequality gives

$$|f([XY]^{2^k})| \le f([(X^*X)(YY^*)]^{2^{k-1}})$$

$$\le f([(X^*X)^2(YY^*)^2]^{2^{k-2}})$$

$$\leq \cdots \leq f([(X^*X)^{2^{m-1}}(YY^*)^{2^{m-1}}]^{2^{k-m}})$$

$$\leq \cdots \leq f([(X^*X)^{2^{k-1}}(YY^*)^{2^{k-1}}]^{2^{k-k}})$$

$$= f([X^*X]^{2^{k-1}}[YY^*]^{2^{k-1}}) \qquad (6.5.21)$$

For given $A, B \in M_n$, set $X = e^{2^{-k}A}$ and $Y = e^{2^{-k}B}$ in (6.5.21) to obtain

$$|f([e^{2^{-k}A} e^{2^{-k}B}]^{2^k})| \leq f([e^{2^{-k}A^*} e^{2^{-k}A}]^{2^{k-1}} [e^{2^{-k}B} e^{2^{-k}B^*}]^{2^{k-1}})$$

$$= f([e^{\frac{1}{2}A^*/2^{k-1}} e^{\frac{1}{2}A/2^{k-1}}]^{2^{k-1}} [e^{\frac{1}{2}B/2^{k-1}} e^{\frac{1}{2}B^*/2^{k-1}}])$$

Now let $k \to \infty$ and use the Lie product formula (6.5.19) and the continuity of $f(\cdot)$ to conclude that

$$|f(e^{A+B})| \leq f(e^{(A+A^*)/2} e^{(B+B^*)/2})$$

This verifies the outer inequalities in (2).

The outer inequalities in (2) imply both (3) (take $B = 0$) and (4) (take A and B to be Hermitian). The inner inequality $f(e^{(A+A^*+B+B^*)/2}) \leq f(e^{(A+A^*)/2} e^{(B+B^*)/2})$ in (2) now follows from (4) with A replaced by $(A + A^*)/2$ and B replaced by $(B + B^*)/2$. The other inner inequality

$$|f(e^{A+B})| \leq f(e^{(A+A^*+B+B^*)/2})$$

in (2) follows from (3) with A replaced by $A + B$.   ☐

The motivating special case of the preceding theorem that arose in statistical mechanics was (6.5.20(4)) for the function $f(X) = \operatorname{tr} X$, for which condition (a) is trivial. In the process of verifying condition (b) for the trace function, we can obtain an interesting inequality for unitarily invariant norms as well.

6.5.22   Corollary. For any $X \in M_n$, let $\{\lambda_i(X)\}$ denote the eigenvalues of X ordered by decreasing absolute value $(|\lambda_1(X)| \geq |\lambda_2(X)| \geq \cdots \geq |\lambda_n(X)|)$. For $m = 1, \ldots, n$, define

$$f_m(X) \equiv |\lambda_1(X)| + \cdots + |\lambda_m(X)|$$

and

$$\varphi_m(X) \equiv |\lambda_1(X)\,\lambda_2(X) \cdots \lambda_m(X)|$$

Then

(1)   Each function $f_m(\cdot)$ satisfies the inequalities (6.5.20(1–4)). In particular, the spectral radius function $f_1(\cdot) = \rho(\cdot)$ satisfies these inequalities.

(2)   Each function $\varphi_m(\cdot)$ satisfies the inequalities (6.5.20(1–4)).

(3)   The trace and determinant functions satisfy the inequalities (6.5.20(1–4)).

(4) $\| e^{A+B} \| \le \| e^A e^B \|$ for every unitarily invariant norm $\|\cdot\|$ on M_n and all Hermitian $A, B \in M_n$.

Proof: To prove assertions (1) and (2), it suffices to verify that the functions $f_m(X)$ and $\varphi_m(X)$ satisfy the hypotheses of Theorem (6.5.20).

The functions $f_m(X)$ and $\varphi_m(X)$ are continuous on M_n since the eigenvalues of a matrix are continuous functions of its entries; they satisfy condition (a) of the theorem since XY and YX have the same eigenvalues.

To verify condition (b) for $f_m(\cdot)$, use the additive eigenvalue–singular value inequalities of Weyl and A. Horn (3.3.13,14) to compute

$$f_m(X^{2k}) = \sum_{i=1}^{m} |\lambda_i(X^{2k})| \le \sum_{i=1}^{m} \sigma_i(X^{2k}) \le \sum_{i=1}^{m} [\sigma_i(X)]^{2k}$$

$$= \sum_{i=1}^{m} [\sigma_i(XX^*)]^k = \sum_{i=1}^{m} [\lambda_i(XX^*)]^k = \sum_{i=1}^{m} \lambda_i([XX^*]^k)$$

$$= f_m([XX^*]^k)$$

where $\sigma_1(Z) \ge \sigma_2(Z) \ge \cdots \ge \sigma_n(Z)$ denote the ordered singular values of Z. The last equality uses the fact that XX^* is positive semidefinite, so all of its eigenvalues are nonnegative.

A similar calculation with the product form of the Weyl-Horn inequalities (3.3.2,4) verifies condition (b) for the functions $\varphi_m(\cdot)$:

$$\varphi_m(X^{2k}) = \prod_{i=1}^{m} |\lambda_i(X^{2k})| \le \prod_{i=1}^{m} \sigma_i(X^{2k}) \le \prod_{i=1}^{m} \sigma_i(X)^{2k}$$

$$= \prod_{i=1}^{m} \sigma_i(XX^*)^k = \prod_{i=1}^{m} \lambda_i(XX^*)^k = \prod_{i=1}^{m} \lambda_i([XX^*]^k)$$

$$= \varphi_m([XX^*]^k)$$

Both the trace and determinant are continuous functions on M_n that satisfy condition (a) since they are functions only of the eigenvalues of their arguments. The determinant satisfies condition (b) since $\det(XY) = \det(X)\det(Y)$ for all X, $Y \in M_n$. It is easy to see that the trace function satisfies condition (b) by using the fact that $|\text{tr}(X)| = |\lambda_1(X) + \cdots + \lambda_n(X)| \le f_n(X)$ for all $X \in M_n$, and $f_n(\cdot)$ has already been shown to satisfy condition (b): $|\text{tr}(X^{2k})| \le f_n(X^{2k}) \le f_n([XX^*]^k) = \text{tr}([XX^*]^k)$.

Finally, assertion (4) follows from (1): Notice that

$$\sigma_1(e^{A+B}) + \cdots + \sigma_m(e^{A+B}) = f_m(e^{A+B}) \le f_m(e^A e^B)$$

$$\le \sigma_1(e^A e^B) + \cdots + \sigma_m(e^A e^B)$$

and this outer inequality for $m = 1, 2,..., n$ is exactly what one needs to prove (4) by Corollary (3.5.9)   ∎

See Problems 25, 28, and 29 for upper bounds on, and a refinement of, the trace inequalities in (6.5.22(3)). For an interesting application of the inequalities (6.5.22(2)) for products of eigenvalues, see the majorization inequalities in Problem 41.

As a second application of the Lie product formula, we use it to improve the bound in Corollary (6.2.32) on the modulus of continuity of the matrix exponential function in an important special case. First observe that

$$\frac{d}{dt} e^{(1-t)A} e^{tB} = -Ae^{(1-t)A} e^{tB} + e^{(1-t)A} Be^{tB}$$

$$= e^{(1-t)A} (B-A) e^{tB}$$

for any A, $B \in M_n$, and hence

$$e^B - e^A = \int_0^1 \frac{d}{dt} \left[e^{(1-t)A} e^{tB} \right] dt = \int_0^1 e^{(1-t)A} (B-A) e^{tB} dt \quad (6.5.23)$$

For any matrix norm $|||\cdot|||$, the triangle inequality and submultiplicativity now give

$$||| e^B - e^A ||| \leq ||| B - A ||| \int_0^1 ||| e^{(1-t)A} ||| \, ||| e^{tB} ||| \, dt \qquad (6.5.24)$$

The triangle inequality again, and the power series for the exponential, give the general bound

$$||| e^X ||| = \left|\left|\left| \sum_{k=0}^{\infty} \frac{1}{k!} X^k \right|\right|\right| \leq \sum_{k=0}^{\infty} \frac{1}{k!} ||| X |||^k = \exp ||| X ||| \qquad (6.5.25)$$

for any $X \in M_n$.   Finally, the Lie product formula (6.5.19) permits us to write

$$e^{tB} = \lim_{m \to \infty} \left[e^{t(B-A)/m} e^{tA/m} \right]^m$$

and hence

$$||| e^{tB} ||| = \lim_{m \to \infty} \left|\left|\left| \left[e^{t(B-A)/m} e^{tA/m} \right]^m \right|\right|\right|$$

$$\leq \lim_{m \to \infty} \sup ||| e^{t(B-A)/m} e^{tA/m} |||^m$$

$$\leq \lim_{m \to \infty} \sup ||| e^{t(B-A)/m} |||^m \, ||| e^{tA/m} |||^m$$

$$\leq \lim_{m \to \infty} \sup [\exp(|t| \, ||| B - A |||/m)]^m \, ||| e^{tA/m} |||^m$$

$$= \exp(|t| \, ||| B - A |||) \lim_{m \to \infty} \sup ||| e^{tA/m} |||^m \qquad (6.5.26)$$

So far, $A, B \in M_n$ have been unrestricted and $|||\cdot|||$ has been any matrix norm. We now assume that A is normal and that $|||\cdot|||$ is the spectral norm. For any normal $X \in M_n$ we have $||| X |||_2 = \rho(X)$ (the spectral radius) and

$\||| e^X \||_2 = \rho(e^X) = e^{\mu(X)}$, where $\mu(X) \equiv \max \{ \text{Re } \lambda : \lambda \in \sigma(X) \}$, and hence

$$\||| e^{sX} \||_2 = e^{\mu(sX)} = e^{s\mu(X)} = [e^{\mu(X)}]^s = \||| e^X \||_2^s, \quad s \geq 0 \qquad (6.5.27)$$

Applying this bound to (6.5.26), we have

$$\||| e^{tB} \||_2 \leq \exp(|t| \, \||| B - A \||_2) \lim_{m \to \infty} \sup \||| e^{tA/m} \||_2^m$$

$$= \exp(|t| \, \||| B - A \||_2) \, \||| e^A \||_2^t \qquad (6.5.28)$$

when A is normal. Combining these observations, we have the following improvement in the bound in Corollary (6.2.32) when A is normal.

6.5.29   Theorem. Let $A, E \in M_n$ be given with A normal, and let $\||| \cdot \||_2$ denote the spectral norm on M_n. Then

$$\||| e^{A+E} - e^A \||_2 \leq [\exp(\||| E \||_2) - 1] \, \||| e^A \||_2 \qquad (6.5.30)$$

Proof: Since the asserted bound is trivially correct for $E = 0$, assume that $E \neq 0$, let $B = A + E$, and use (6.5.24, 27, 28) to write

$$\||| e^{A+E} - e^A \||_2 \leq \||| E \||_2 \int_0^1 \||| e^{(1-t)A} \||_2 \, \||| e^{tB} \||_2 \, dt$$

$$\leq \||| E \||_2 \int_0^1 \||| e^A \||_2^{1-t} [\exp(t \||| E \||_2)] \, \||| e^A \||_2^t \, dt$$

$$= \||| E \||_2 \, \||| e^A \||_2 \int_0^1 \exp(t \||| E \||_2) \, dt$$

$$= [\exp(\||| E \||_2) - 1] \, \||| e^A \||_2 \qquad \qquad \square$$

It is frequently necessary to compute the matrix function e^{tA}, and we know how to do so, in principle, using either its power series or the definition of a primary matrix function (6.2.4). The power series definition is useful if A is nilpotent or has a simple annihilating polynomial, but, in general, it does not give a finite procedure for determining e^{tA} explicitly.

Although it is easy to evaluate $e^{tJ_k(\lambda)}$, using the Jordan canonical form to compute e^{tA} as a primary matrix function can be inconvenient because the preliminary reduction of A to Jordan canonical form may not be easily accomplished. In principle, we know that for each $A \in M_n$, every power A^k for $k \geq n$ can be expressed as a linear combination of $I, A, A^2, ..., A^{n-1}$ and hence, given A, it must be possible to express

$$e^{tA} = \sum_{k=0}^{\infty} \frac{t^k}{k!} A^k$$

$$= f_0(t)I + f_1(t)A + f_2(t)A^2 + \cdots + f_{n-1}(t)A^{n-1} \qquad (6.5.31)$$

for some functions $\{f_i(t)\}$ that depend on A. It is interesting, and useful, to know that there are ways to compute these functions $\{f_i(t)\}$ that do not depend on summing the exponential series or knowing the Jordan canonical form of A.

The general theory of differential equations guarantees that the first-order initial-value problem

$$\tfrac{d}{dt}X(t) = F(X,t), \ X(0) \text{ given} \qquad (6.5.32)$$

has a unique solution $X(t)$ in a neighborhood of $t = 0$ provided that the right-hand side $F(X,t)$ is continuous in t and is Lipschitz continuous in X in some neighborhood of the given initial values $(X(0), 0)$. The unknown function $X(t)$ could be a scalar, vector, or matrix function, and $F(X,t)$ is a function with the appropriate domain and range. The function $F(X,t)$ is Lipschitz continuous in X on a given set of pairs (X,t) if there is some positive real number L such that

$$\|F(X,t) - F(Z,t)\| \leq L\|X - Z\|$$

for all pairs (X,t), (Z,t) in the given set; $\|\cdot\|$ is the ordinary absolute value function if A and $F(X,t)$ are scalars, and is some given vector or matrix norm if X and $F(X)$ are vectors or matrices.

For our purposes here, we are interested only in initial-value problems that have a right-hand side of the form $F(X,t) = A(t)X + Y(t)$, where $A(t) \in M_n$ is a given continuous matrix function and $Y(t)$ is a given continuous function that is vector-valued if $X(t)$ is a vector or is matrix-valued if

$X(t)$ is a matrix. In this event, $F(X,t) - F(Z,t) = A(t)(X - Z)$, so in the vector-valued case we have

$$\| F(X,t) - F(Z,t) \| = \| A(t)(X - Z) \| \leq \| A(t) \| \| X - Z \| \leq L \| X - Z \|$$

where $\| \cdot \|$ is a matrix norm that is compatible with the vector norm $\| \cdot \|$; the constant L may be taken to be the maximum of the continuous function $\| A(t) \|$ in a suitable neighborhood of $t = 0$. In the matrix-valued case we have, similarly,

$$\| F(X,t) - F(Z,t) \| = \| A(t)(X - Z) \| \leq \| A(t) \| \| X - Z \| \leq L \| X - Z \|$$

In either case, $F(X,t)$ is Lipschitz continuous.

Our conclusion is that a first-order initial-value problem of the form

$$\tfrac{d}{dt}X(t) = A(t)X(t) + Y(t), \qquad X(0) \text{ given}$$

has a unique solution on a real interval (a,b) containing 0 if the given functions $A(t)$ and $Y(t)$ are continuous on (a,b). In particular, these criteria are certainly met for $(a,b) = (-\infty,\infty)$ if $A(t) = A \in M_n$ is a given constant matrix and $Y(t) \equiv 0$. Thus, the initial-value problem

$$\tfrac{d}{dt}X(t) = AX(t), \; X(0) = I, \; X(t) \in M_n \tag{6.5.33}$$

has a unique solution $X(t)$ for all $t \in \mathbb{R}$.

Since we already know that $X(t) = e^{tA}$ is *a* solution, and hence is the *only* solution, to (6.5.33), our procedure will be to construct a function of the form

$$X(t) = f_0(t)I + f_1(t)A + f_2(t)A^2 + \cdots + f_{n-1}(t)A^{n-1} \tag{6.5.34}$$

such that $X(0) = I$ and $\tfrac{d}{dt}X(t) = AX(t)$ for all $t \in \mathbb{R}$. By uniqueness, we shall then be able to conclude that $X(t) = e^{tA}$. There are two algorithms due to Putzer for finding such a function $X(t)$. They both rely on knowledge of an annihilating polynomial $p(t)$ for A; one could use the characteristic or minimal polynomial, but any annihilating polynomial will do.

6.5.35   Theorem. (Putzer's algorithm #1) Let $A \in M_n$ be given, and let $p(t) = (t - \lambda_1)(t - \lambda_2) \cdots (t - \lambda_k)$ be a given polynomial of degree k such that

$p(A) = 0$. Define

$$P_0 \equiv I, \quad P_j \equiv \prod_{i=1}^{j} (A - \lambda_i I), \quad j = 1, 2, \dots, k$$

and let $r_1(t), r_2(t), \dots, r_k(t)$ be the unique solutions of the first-order linear scalar differential equations

$$\tfrac{d}{dt} r_1(t) = \lambda_1 r_1(t), \quad r_1(0) = 1$$

$$\tfrac{d}{dt} r_j(t) = \lambda_j r_j(t) + r_{j-1}(t), \; r_j(0) = 0, \, j = 2, 3, \dots, k \tag{6.5.36}$$

Then

$$e^{tA} = \sum_{j=0}^{k-1} r_{j+1}(t) P_j \tag{6.5.37}$$

Proof: Set

$$X(t) \equiv \sum_{j=0}^{k-1} r_{j+1}(t) P_j$$

and observe that

$$X(0) = \sum_{j=0}^{k-1} r_{j+1}(0) P_j = r_1(0) P_0 = I$$

so it suffices to show that $\tfrac{d}{dt} X(t) = A X(t)$ for all $t \in \mathbb{R}$. If we define $r_0(t) \equiv 0$, we can use the differential equations for $r_j(t)$ and the definition of P_j to compute

$$\tfrac{d}{dt} X(t) = \sum_{j=0}^{k-1} \tfrac{d}{dt} r_{j+1}(t) P_j = \sum_{j=0}^{k-1} [\lambda_{j+1} r_{j+1}(t) + r_j(t)] P_j$$

$$= \sum_{j=0}^{k-1} \lambda_{j+1} r_{j+1}(t) P_j + \sum_{j=0}^{k-2} r_{j+1}(t) P_{j+1}$$

$$= \sum_{j=0}^{k-2} [P_{j+1} + \lambda_{j+1} P_j] r_{j+1}(t) + \lambda_k r_k(t) P_{k-1}$$

$$= \sum_{j=0}^{k-2} [(A - \lambda_{j+1} I) P_j + \lambda_{j+1} P_j] r_{j+1}(t) + \lambda_k r_k(t) P_{k-1}$$

$$= \sum_{j=0}^{k-2} A P_j r_{j+1}(t) + \lambda_k r_k(t) P_{k-1}$$

$$= A \sum_{j=0}^{k-1} r_{j+1}(t) P_j - A P_{k-1} r_k(t) + \lambda_k r_k(t) P_{k-1}$$

$$= A X(t) - (A - \lambda_k I) P_{k-1} r_k(t) = A X(t) - P_k r_k(t)$$

$$= A X(t)$$

where we use at last the fact that $P_k = p(A) = 0$. $\qquad$ □

The functions $r_j(t)$ defined by (6.5.36) are very easy to compute since only a first-order linear differential equation must be solved at each step, but the expression (6.5.37) for e^{tA} is presented as a linear combination of polynomials in A rather than in the form (6.5.34), to which it can, of course, be reduced by multiplying out all the products P_j and collecting like powers of A. Putzer has a second algorithm that gives an expression for e^{tA} directly in terms of the powers of A, but the internal computations are more involved. See Problem 8.

As an example, suppose $A \in M_n$ has only two distinct eigenvalues λ_1, λ_2 and that $p(t) = (t - \lambda_1)^2 (t - \lambda_2)$ annihilates A. Then A might be diagonalizable [if $q(t) = (t - \lambda_1)(t - \lambda_2)$ also annihilates A] or not, and if it is not and if n is large, A could have any one of many different Jordan canonical forms. Nevertheless, Putzer's algorithm says that if we solve

$$\tfrac{d}{dt} r_1(t) = \lambda_1 r_1(t), \qquad\qquad r_1(0) = 1$$

$$\tfrac{d}{dt} r_2(t) = \lambda_1 r_2(t) + r_1(t), \qquad\qquad r_2(0) = 0$$

$$\tfrac{d}{dt} r_3(t) = \lambda_2 r_3(t) + r_2(t), \qquad\qquad r_3(0) = 0$$

for the coefficient functions

$$r_1(t) = e^{\lambda_1 t}$$

$$r_2(t) = t e^{\lambda_1 t}$$

$$r_3(t) = \left[[(\lambda_1 - \lambda_2)t - 1]e^{\lambda_1 t} + e^{\lambda_2 t} \right] / (\lambda_1 - \lambda_2)^2$$

then

$$e^{tA} = e^{\lambda_1 t} I + t e^{\lambda_1 t}(A - \lambda_1 I) + r_3(t)(A - \lambda_1 I)^2 \qquad (6.5.38)$$

independently of the value of n or the Jordan canonical form of A. See Problem 12.

Problems

1.   Provide details for a proof of (6.5.6).

2.   Show that $e^{\log(I - J_k(0))} = I - J_k(0)$ for all $k = 1, 2, \ldots$.

3.   Show that, for each $A \in M_n$ and for all $t \in \mathbf{C}$ such that $0 < |t| < 1/\rho(A)$ if $\rho(A) > 0$, or for all nonzero $t \in \mathbf{C}$ if $\rho(A) = 0$,

$$t^n p_A(\tfrac{1}{t}) = \exp\left[-\sum_{k=1}^{\infty} \frac{t^k}{k} \operatorname{tr} A^k \right]$$

where $p_A(t)$ is the characteristic polynomial of A.

4a.   Let $A(t) = [a_{ij}(t)]$ be a continuous n-by-n matrix-valued function of $t \in [0, t_0]$. The general theory of differential equations guarantees that there exists a unique solution $X(t) \in M_n$ to the homogeneous first-order linear system of differential equations

$$\tfrac{d}{dt}X(t) = A(t)X(t), \quad X(0) = I, \; t \in [0, t_0) \qquad (6.5.39)$$

Use (6.5.9) to prove *Jacobi's identity*

$$\det X(t) = \exp\left[\int_0^t \operatorname{tr} A(s)\, ds\right] \quad \text{for all } t \in [0, t_0) \qquad (6.5.40)$$

and deduce that $X(t)$ is nonsingular for all $t \in [0, t_0)$. What does this say when $A(t) = A = \text{constant}$? Compare with Problem 4 in Section (6.2).

4b. Let $X(t)$ denote the solution to the homogeneous equation (6.5.39). The *Green's matrix* of (6.5.39) is the function $G(t,s) \equiv X(t)X^{-1}(s)$. Show that $G(t,t) = I$ and $\frac{d}{dt}G(t,s) = A(t)G(t,s)$, and deduce that if $y(t) \in \mathbb{C}^n$ is a given continuous function on $[0, t_0)$, then

$$x(t) \equiv X(t)x(0) + \int_0^t G(t,s)y(s)\, ds$$

is the solution to the inhomogeneous first-order initial-value problem

$$\tfrac{d}{dt}x(t) = A(t)x(t) + y(t), \quad t \in (0, t_0), \quad x(0) \in \mathbb{C}^n \text{ given} \qquad (6.5.41)$$

4c. Now suppose $A \in M_n$ is a constant matrix. Show that the Green's matrix of (6.5.39) is $G(t,s) = e^{A(t-s)}$, and that the solution to the inhomogeneous first-order initial-value problem

$$\tfrac{d}{dt}x(t) = Ax(t) + y(t), \quad t \in (0, t_0), \quad x(0) \in \mathbb{C}^n \text{ given}$$

is

$$x(t) = e^{tA}x(0) + e^{tA}\int_0^t e^{-sA}y(s)\, ds$$

4d. Suppose that $x(0) \geq 0$ and that $y(t) \geq 0$ and $e^{tA} \geq 0$ for all $t \in [0, t_0)$, and let $x(t)$ be the solution to (6.5.41). Show that $x(t) \geq 0$ for *all* such initial conditions $x(0)$ and forcing terms $y(t)$ and all $t \in [0, t_0)$. These are all entrywise inequalities.

4e. Show that $e^{tA} \geq 0$ for all $t > 0$ if and only if $a_{ij} \geq 0$ for all $i \neq j$. If A satisfies these inequalities, use Problem 4d to show that the solution $x(t)$ to (6.5.41) (with $A(t) = A$) has nonnegative entries if $x(0) \geq 0$ and if $y(t) \geq 0$ for all $t \in [0, t_0)$.

5a. Let $A, B, C \in M_n$ be given. Verify that $X(t) = e^{-tA}Ce^{-tB}$ is a solution to the problem

$$\tfrac{d}{dt}X(t) = -AX(t) - X(t)B, \ X(0) = C \tag{6.5.42}$$

(see Problem 14 in Section (6.2) for a derivation). Use the results from the general theory of differential equations discussed in this section to explain why this problem has a unique solution.

5b. Let $A, B, C \in M_n$ be given and suppose

$$\lim_{t\to\infty} e^{-tA}Ce^{-tB} = 0 \tag{6.5.43}$$

and

$$\int_0^\infty e^{-tA}Ce^{-tB}\, dt \ \text{ is convergent} \tag{6.5.44}$$

Use Problem 5a to show that the matrix

$$Y \equiv \int_0^\infty e^{-tA}Ce^{-tB}\, dt \tag{6.5.45}$$

is a solution to the linear matrix equation $AY + YB = C$.

5c.   For a given $A \in M_n$, define $\tau(A) \equiv \min\{\text{Re}\,\lambda : \lambda \text{ is an eigenvalue of } A\}$. Suppose $\tau(A) > 0$, that is, A is positive stable. For a given norm $\|\cdot\|$ on M_n, show that there exists a positive constant K (that depends on A) such that $\|e^{-tA}\| \le Ke^{-\tau(A)t}$ for all $t \ge 0$.

5d.   If $A, B \in M_n$ are positive stable, use the norm bound in Problem 5c to show that the two convergence conditions (6.5.43–44) are satisfied for any $C \in M_n$, and conclude that the integral (6.5.45) gives a solution to the linear matrix equation $AY + YB = C$ for any $C \in M_n$.

5e.   Let $A \in M_n$ be positive stable. According to Theorem (4.4.6), the Lyapunov equation $GA + A^*G = H$ has a unique solution for each given $H \in M_n$. Use Problem 5d to show that the solution can be represented as

$$G = \int_0^\infty e^{-tA^*}He^{-tA}\, dt$$

Use this representation to deduce that the solution G is Hermitian if the right-hand side H is Hermitian, and that G is positive definite if H is positive definite. Compare with Theorem (2.2.3).

6.   Give an example of a matrix function $A(t) \in M_n$ for which $\frac{d}{dt}A(t)$ commutes with $A(t)$, and another example for which it does not.

7.   In applying Putzer's algorithm (6.5.35), one must solve repeatedly a simple linear first-order differential equation of the form $y'(t) = ay(t) + z(t)$, $y(0)$ given, where a is a given scalar and $z(t)$ is a known function. Verify that

$$y(t) = e^{at} \int_{s=0}^{t} e^{-as}z(s)\,ds + y(0)e^{at} \tag{6.5.46}$$

gives the (unique) solution to this problem.

8.   (Putzer's algorithm #2) Let $A \in M_n$ and let $p(t) = t^k + a_{k-1}t^{k-1} + \cdots + a_1 t + a_0$ be a polynomial such that $p(A) = 0$. Construct the (scalar) solution $z(t)$ to the differential equation

$$z^{(k)}(t) + a_{k-1}z^{(k-1)}(t) + \cdots + a_1 z'(t) + a_0 z(t) = 0$$

$$z(0) = z'(0) = \cdots = z^{(k-2)}(0) = 0 \text{ and } z^{(k-1)}(0) = 1$$

Let

$$Z(t) = \begin{bmatrix} z(t) \\ z'(t) \\ \vdots \\ z^{(k-1)}(t) \end{bmatrix} \text{ and } C = \begin{bmatrix} a_1 & a_2 & \cdots & a_{k-1} & 1 \\ a_2 & a_3 & \cdots & 1 & 0 \\ \vdots & & \ddots & & \\ a_{k-1} & 1 & 0 & & \\ 1 & 0 & & & 0 \end{bmatrix} \in M_k$$

and define $Q(t) = \begin{bmatrix} q_0(t) \\ q_1(t) \\ \vdots \\ q_{k-1}(t) \end{bmatrix}$ by $Q(t) \equiv CZ(t)$. Then show that

$$e^{tA} = \sum_{j=0}^{k-1} q_j(t)A^j$$

9.   If $A \in M_n$ is nilpotent of index $k \leq n$, compute e^{tA} explicitly using its power series. Now carry out Putzer's first algorithm (6.5.35) and Putzer's second algorithm in Problem 8. Compare the results. Verify that each

solution satisfies (6.5.33). What are the possible Jordan canonical forms for A? Discuss what would be necessary if (6.2.4) were used to compute e^{tA} as a primary matrix function.

10. Same as Problem 9, but assume that $A \in M_n$ is idempotent, $A^2 = A$.

11. Same as Problem 9, but assume that $A \in M_n$ is tripotent, $A^3 = A$.

12. Write the solution (6.5.38) in the form (6.5.34) and show that it satisfies (6.5.33).

13. If $A \in M_n$ is positive stable, use Problem 5b to show that

$$A^{-1} = \int_0^\infty e^{-tA}\, dt$$

14. If $A, B \in M_n$ are skew-Hermitian and C is unitary, use Problem 5a to show that the (unique) solution $X(t)$ to

$$\tfrac{d}{dt}X(t) = AX(t) + X(t)B, \ X(0) = C$$

is unitary for all $t \in \mathbb{R}$.

15. Use Corollary (6.5.22(4)) to show that $\| e^{A+B} \| \le e^{\lambda_{max}(A)} \| e^B \|$ for all Hermitian $A, B \in M_n$ and any unitarily invariant norm $\|\cdot\|$ on M_n.

16. Show that every M-matrix $A \in M_n(\mathbb{R})$ has an mth root that is an M-matrix for every $m = 2, 3, \ldots$.

17. Use the Lie product formula (6.5.19) to show that $\det(e^A e^B) = \det(e^{A+B})$ for all $A, B \in M_n$. Also give an elementary proof using the fact that $\det e^X = e^{\operatorname{tr} X}$.

18. (a) Take the determinant of both sides of (6.5.16) and show that $t\operatorname{tr}(A+B) = \operatorname{tr} C(t;A,B)$ for all sufficiently small t, where $e^{tA}e^{tB} = e^{C(t;A,B)}$ and $C(t;A,B)$ is given by (6.5.17).
(b) Now use (6.5.17) to show that the matrix coefficients $C_k(A,B)$ in the power series

$$C(t;A,B) = (A+B)t + \sum_{k=2}^{\infty} C_k(A,B)t^k$$

satisfy tr $C_k(A,B) = 0$ for all $k = 2, 3,.....$ Conclude from Theorem (4.5.2) that every coefficient $C_k(A,B)$ is a commutator, that is, for each $k = 2, 3,...$ there are X_k, $Y_k \in M_n$ such that $C_k(A,B) = X_k Y_k - Y_k X_k$. It is often convenient to use the *Lie bracket* abbreviation for a commutator: $[X,Y] \equiv XY - YX$.

(c) Calculate the coefficient of the second-order term in the series (6.5.17) and show that

$$\log[I - (I - e^{tA} e^{tB})] = t(A + B) + \tfrac{1}{2} t^2 (AB - BA) + O(t^3) \quad (6.5.47)$$

for any A, $B \in M_n$ as $t \to 0$. Thus, the second coefficient $C_2(A,B) = \tfrac{1}{2}[A,B]$ is a rational multiple of the commutator of A and B. Verify that the next two coefficients can be written as

$$C_3(A,B) = \frac{1}{12}[A - B,[A,B]] = -\frac{1}{12}[[A,B],A - B]$$

$$= \frac{1}{12}\Big[(A - B)(AB - BA) - (AB - BA)(A - B)\Big]$$

and

$$C_4(A,B) = -\frac{1}{24}[A,[B,[A,B]]] = -\frac{1}{24}[[[A,B],B],A]$$

The appearance of these commutators of commutators involving A and B is no accident; the famous Campbell-Baker-Hausdorff theorem asserts that *every* coefficient $C_k(A,B)$ in the power series (6.5.17) for $e^{tA} e^{tB}$ is a rational linear combination of iterated commutators of A and B.

19. Let A, $B \in M_n$ be given. Show that

$$e^{tA} e^{tB} e^{-tA} e^{-tB} = e^{t^2(AB - BA)} + O(t^3) \quad (6.5.48)$$

as $t \to 0$. Use this identity to prove that

$$\lim_{m \to \infty} \Big[e^{A/m} e^{B/m} e^{-A/m} e^{-B/m} \Big]^{m^2} = e^{AB - BA} \quad (6.5.49)$$

for any A, $B \in M_n$.

20. A natural conjecture to generalize the exponential trace inequality in Corollary (6.5.22(2)) is

$$|\operatorname{tr}(e^A\, e^B e^C)| \le \operatorname{tr}(e^{A\,+\,B\,+\,C}) \qquad (6.5.50)$$

for Hermitian A, B, $C \in M_n$, but this is not true in general. Provide details for the following counterexample in M_2: Let

$$S_1 = \begin{bmatrix} 0 & 1 \\ 1 & 0 \end{bmatrix}, \ S_2 = \begin{bmatrix} 0 & -i \\ i & 0 \end{bmatrix}, \ S_3 = \begin{bmatrix} 1 & 0 \\ 0 & -1 \end{bmatrix}$$

(these are the *Pauli spin matrices* that arise in quantum mechanics), show that $(a_1 S_1 + a_2 S_2 + a_3 S_3)^2 = a^2 I$ for all $a_1, a_2, a_3 \in \mathbb{R}$, where $a \equiv (a_1^2 + a_2^2 + a_3^2)^{\frac{1}{2}}$, and show that

$$e^{a_1 S_1 + a_2 S_2 + a_3 S_3} = (\cosh a)\, I + \frac{\sinh a}{a}(a_1 S_1 + a_2 S_2 + a_3 S_3)$$

Now let

$$A = t S_1, \quad B = t S_2, \quad C = t(S_3 - S_2 - S_1)$$

and verify that

$$\operatorname{tr}(e^{A\,+\,B\,+\,C}) = 2 \cosh t$$

while

$$|\operatorname{tr}(e^A e^B e^C)| = 2|\cosh^2 t \cosh 3^{\frac{1}{2}} t - 2 \cdot 3^{-\frac{1}{2}} \cosh t \sinh t \sinh 3^{\frac{1}{2}} t$$
$$+\ i\, 3^{-\frac{1}{2}} \sinh^2 t \sinh 3^{\frac{1}{2}} t|$$

$$= 2\left[1 - \frac{t^4}{12} + O(t^6)\right] \cosh t$$

as $t \to 0$. Thus, the conjectured inequality (6.5.50) is violated for all sufficiently small nonzero t.

21. If A, $B \in M_n$ are positive semidefinite and $\alpha \in [0,1]$, show that

$$\operatorname{tr}(A^\alpha B^{1-\alpha}) \le (\operatorname{tr} A)^\alpha (\operatorname{tr} B)^{1-\alpha} \qquad (6.5.51)$$

22. Use Problem 21 and Corollary (6.5.22(2)) to show that the function $f(X) \equiv \log \operatorname{tr} e^X$ is a convex mapping on the Hermitian matrices in M_n, that is,

$$\mathrm{tr}(e^{\,\alpha A + (1-\alpha)B}) \le (\mathrm{tr}\, e^A)^\alpha (\mathrm{tr}\, e^B)^{1-\alpha} \tag{6.5.52}$$

for any Hermitian $A, B \in M_n$ and any $\alpha \in [0,1]$.

23. Let $A(t) \in M_n$ be differentiable and nonsingular. Use (6.5.9) to prove that

$$\frac{d}{dt} \log \det A(t) = \mathrm{tr}\left[A(t)^{-1}\frac{d}{dt}A(t)\right] \tag{6.5.53}$$

Use this to show that if $A(t)$ is positive definite and $dA(t)/dt$ is positive semidefinite, then $\det A(t)$ is a strictly increasing function of t.

24. Let $X, Y \in M_n$ be given.
   (a) Use the Cauchy-Schwarz inequality for the Frobenius norm to show that

$$|\mathrm{tr}\, XY^*| \le (\mathrm{tr}\, XX^*)^{\frac12}(\mathrm{tr}\, YY^*)^{\frac12} \le \tfrac12 \mathrm{tr}\,(X^*X + Y^*Y) \tag{6.5.54a}$$

   (b) Let $\sigma_1(X) \ge \cdots \ge \sigma_n(X) \ge 0$ denote the ordered singular values of X. Show that

$$|\mathrm{tr}\, XY^*| \le \sum_{i=1}^n \sigma_i(X)\sigma_i(Y) \le \left[\sum_{i=1}^n \sigma_i(X)^p\right]^{1/p}\left[\sum_{i=1}^n \sigma_i(Y)^q\right]^{1/q}$$

$$\le \frac1p\left[\sum_{i=1}^n \sigma_i(X)^p\right] + \frac1q\left[\sum_{i=1}^n \sigma_i(Y)^q\right] \tag{6.5.54b}$$

   whenever $p, q > 0$ and $(1/p) + (1/q) = 1$. Explain why (6.5.54a) is a special case of (6.5.54b).

25. Let $A, B \in M_n$ be given. Use (6.5.54a,b) to show that the inequalities in Theorem (6.5.20(2)) for the trace function have the following upper bounds:

$$|\mathrm{tr}\, e^{A+B}| \le \mathrm{tr}(e^{(A+A^*+B+B^*)/2})$$

$$\le \mathrm{tr}\, e^{(A+A^*)/2} e^{(B+B^*)/2} \le (\mathrm{tr}\, e^{A+A^*})^{\frac12}(\mathrm{tr}\, e^{B+B^*})^{\frac12}$$

$$\le \tfrac12 \mathrm{tr}(e^{A+A^*} + e^{B+B^*}) \tag{6.5.55a}$$

and, more generally,

$$\left|\operatorname{tr} e^{A+B}\right| \leq \operatorname{tr}\left(e^{(A+A^*+B+B^*)/2}\right)$$

$$\leq \left[\operatorname{tr} e^{p(A+A^*)/2}\right]^{1/p}\left[\operatorname{tr} e^{q(B+B^*)/2}\right]^{1/q}$$

$$\leq \frac{1}{p}\operatorname{tr} e^{p(A+A^*)/2} + \frac{1}{q}\operatorname{tr} e^{q(B+B^*)/2} \qquad (6.5.55b)$$

whenever $p, q > 0$ and $(1/p) + (1/q) = 1$. If A and B are Hermitian, deduce that $\operatorname{tr} e^{A+B} \leq \operatorname{tr} e^A e^B \leq (\operatorname{tr} e^{pA})/p + (\operatorname{tr} e^{qB})/q$.

26. For any $X \in M_n$ and all $m = 1, 2, \ldots$, show that

$$\left|\operatorname{tr} X^{2m}\right| \leq \operatorname{tr}[(X^*)^m X^m] \leq \operatorname{tr}(X^*X)^m \qquad (6.5.56)$$

27. Extend the argument used to prove the Lie product formula (6.5.19) and show that

$$\lim_{m\to\infty}\left[e^{A_1/m}e^{A_2/m}\cdots e^{A_k/m}\right]^m = e^{A_1+A_2+\cdots+A_k} \qquad (6.5.57)$$

for any $A_1, \ldots, A_k \in M_n$ and all $k = 1, 2, \ldots$.

28. Use Problems 26 and 27 to prove the following refinement of the inequalities in Theorem (6.5.20(2)) for the trace function: For all $A, B \in M_n$,

$$\left|\operatorname{tr} e^{A+B}\right| \leq \operatorname{tr}\left(e^{(A+B)/2}e^{(A+B)^*/2}\right) \leq \operatorname{tr} e^{(A+A^*+B+B^*)/2} \qquad (6.5.58)$$

29. For any $A \in M_n$, use (6.5.58) to show that

$$\left|\operatorname{tr} e^{2A}\right| \leq \operatorname{tr}(e^A e^{A^*}) \leq \operatorname{tr} e^{A+A^*} \leq \left[n\operatorname{tr} e^{2(A+A^*)}\right]^{\frac{1}{2}}$$

$$\leq \tfrac{1}{2}n + \tfrac{1}{2}\operatorname{tr} e^{2(A+A^*)} \qquad (6.5.59)$$

30. Show that the bound (6.5.30) holds for any pair of commuting matrices $A, E \in M_n$ without any other assumptions.

31. Why is the bound in (6.5.30) better than the bound in Corollary (6.2.32)? Does it make much difference? Verify with several numerical

examples.

32. Suppose $A, E \in M_n$ are both normal. Modify the proof of Theorem (6.5.29) to show that

$$\||| e^{A+E} \||_2 \leq \||| e^A \||_2 \||| e^E \||_2 \tag{6.5.60}$$

and deduce that

$$\||| e^{A+E} - e^A \||_2 \leq \rho(E) \frac{e^{\mu(E)} - 1}{\mu(E)} \||| e^A \||_2 \tag{6.5.61}$$

where $\mu(E) \equiv \max \{\text{Re } \lambda : \lambda \in \sigma(E)\}$ and $\rho(E) \equiv \max \{|\lambda| : \lambda \in \sigma(E)\}$. Show that the bound (6.5.30) becomes

$$\||| e^{A+E} - e^A \||_2 \leq (e^{\rho(E)} - 1) \||| e^A \||_2 \tag{6.5.62}$$

in this case. Show that (6.5.61) is a better bound than (6.5.62) for all normal $A, E \in M_n$.

33. Let $\||| \cdot \|||$ be a given matrix norm on M_n. Show that the general bound (6.5.24) implies the bound

$$\||| e^{A+E} - e^A \||| \leq \||| E \||| \frac{\exp(\||| A + E \||| - \||| A \|||) - 1}{\||| A + E \||| - \||| A \|||} \exp(\||| A \|||) \tag{6.5.63}$$

for all $A, E \in M_n$, where the ratio is understood to have the value 1 if $\||| A + E \||| - \||| A \||| = 0$. Show that the bound in (6.5.63) is always at least as good as the bound in Corollary (6.2.32).

34. Use the formula (6.2.36) for $\frac{d}{dt} e^{tA}$ and some of the identities (6.5.1-7) to show that $e^{-A} = (e^A)^{-1}$ and $e^{mA} = (e^A)^m$ for every $A \in M_n$ and every integer $m = \pm 1, \pm 2, \ldots$.

35. For $X, Y \in M_n$, let $[X, Y] \equiv XY - YX$ and notice that X commutes with Y if and only if $[X, Y] = 0$. The objective of this problem is to prove *Weyl's identity* (6.5.64), which may be thought of as a generalization of the assertion in Theorem (6.2.38) that $e^A e^B = e^{A+B}$ whenever A and B commute. Let $A, B \in M_n$ be given. Define $g(t) \equiv e^{-t(A+B)} e^{tA} e^{tB}$ and $h(t) \equiv e^{tA} B e^{-tA}$.

(a) Show that $g'(t) = e^{-t(A+B)}[e^{tA}, B] e^{tB}$ and $g(0) = I$.

(b) Show that $h'(t) = [A, h(t)] = e^{tA}[A,B]e^{-tA}$, $h''(t) = [A, h'(t)]$, $h(0)$
$= B$, and $h'(0) = [A,B]$.

(c) Suppose A commutes with $[A,B]$. Show that A commutes with
$h'(t)$, $h''(t) \equiv 0$, and $e^{tA}Be^{-tA} = h(t) = B + t[A,B]$. Conclude that
$[e^{tA}, B] = t[A,B]e^{tA}$ in this case.

(d) Now suppose both A and B commute with $[A,B]$. Show that $g'(t)$
$= t[A,B]g(t)$. Define $\gamma(t) \equiv e^{t^2[A,B]/2}$; verify that $\gamma'(t) = t[A,B]\gamma(t)$
and $\gamma(0) = I$.

(e)   Conclude that

$$e^A e^B = e^{A + B + \frac{1}{2}[A,B]} \qquad (6.5.64)$$

whenever A and B both commute with $[A,B]$.

(f) If A commutes with $[A,B]$, show that $[A,B]$ is nilpotent. Let
$T_{n-1}(t) \equiv 1 + t + t^2/2! + \cdots + t^{n-1}/(n-1)!$. Show that

$$e^A e^B - e^{A+B} = e^A e^B(I - T_{n-1}(-\tfrac{1}{2}[A,B])) = -e^{A+B}(I - T_{n-1}(\tfrac{1}{2}[A,B]))$$

whenever A and B both commute with $[A,B]$.

36. Let $P(t) \in M_n$ be a given differentiable matrix-valued function for $t \in$
$(-a,a) \subset \mathbb{R}$, $a > 0$. Suppose there is a positive integer $m \geq 2$ such that $P(t)^m =$
$P(t)$ for all $t \in (-a,a)$; when $m = 2$, each $P(t)$ is therefore a *projection*. Show
that $P(t)$ commutes with $P'(t)$ for all $t \in (-a,a)$ if and only if $P(t)$ is constant,
that is, $P(t) \equiv P(0)$ for all $t \in (-a,a)$.

37. Let a given function $f \colon M_n \to \mathbb{C}$ satisfy condition (b) of Theorem (6.5.20)
for $k = 1$. Show that $|f(e^A)| \leq f(e^{\frac{1}{2}A}e^{\frac{1}{2}A^*})$ for any $A \in M_n$, and that $|f(e^A)|$
$\leq f(e^{(A+A^*)/2})$ if A is normal.

38. If $f(\cdot)$ satisfies the hypotheses of Theorem (6.5.20), extend the argument given for (6.5.20(1)) to show that $f([XY]^k) \geq 0$ whenever $X, Y \in M_n$
are positive semidefinite.

39. Let $A \in M_n$ be given. Show that $|\mathrm{tr}\, e^{A+B}| \leq \mathrm{tr}\, e^{(A+A^*)/2}$ and $\rho(e^{A+B})$
$\leq \rho(e^{(A+A^*)/2})$ for every skew-Hermitian $B \in M_n$. What does this say when
$n = 1$?

40. Let $A \in M_n$ be given. We know from a simple field of values argument
[see (1.5.8)] that $\lambda_{max}(\frac{1}{2}(A + A^*)) = \max\{\mathrm{Re}\,\alpha \colon \alpha \in F(A)\} \geq \max\{\mathrm{Re}\,\lambda \colon$
$\lambda \in \sigma(A)\}$. Use the inequality (6.5.20(3)) with the spectral radius function
to give another proof that

$$\max \{\text{Re } \lambda : \lambda \in \sigma(A)\} \leq \lambda_{max}(\tfrac{1}{2}(A + A^*)) \tag{6.5.65}$$

Explain why this inequality implies the useful bound

$$\rho(A) \leq \lambda_{max}(\tfrac{1}{2}(A + A^T)) \quad \text{for } A \geq 0 \tag{6.5.66}$$

41. For $X \in M_n$, let $\{\hat{\lambda}_i(X)\}$ denote the eigenvalues of X ordered by decreasing real part, so $\text{Re } \hat{\lambda}_1(X) \geq \text{Re } \hat{\lambda}_2(X) \geq \cdots \geq \text{Re } \hat{\lambda}_n(X)$; if X is Hermitian, this gives the usual decreasing ordering of its eigenvalues $\lambda_{max}(X) = \hat{\lambda}_1(X) \geq \cdots \geq \hat{\lambda}_n(X) = \lambda_{min}(X)$. Let $A \in M_n$ be given. Use the inequality (6.5.20(3)) with the absolute eigenvalue product functions $\varphi_m(X)$ defined in Corollary (6.5.22) to prove the majorization

$$\sum_{i=1}^{m} \text{Re } \hat{\lambda}_i(A) \leq \sum_{i=1}^{m} \hat{\lambda}_i(\tfrac{1}{2}(A + A^*)), \quad m = 1,\ldots, n \tag{6.5.67}$$

with equality for $m = n$. Notice that the inequality (6.5.65) is the case $m = 1$ of this majorization.

42. Let $\{\beta_i\} \subset \mathbb{C}$ and $\{\eta_i\} \subset \mathbb{R}$ be two given sets of n scalars each, and assume that they are arranged so that $\text{Re } \beta_1 \geq \text{Re } \beta_2 \geq \cdots \geq \text{Re } \beta_n$ and $\eta_1 \geq \eta_2 \geq \cdots \geq \eta_n$. Show that there is an $A \in M_n$ such that $\sigma(A) = \{\beta_i\}$ and $\sigma(\tfrac{1}{2}[A + A^*]) = \{\eta_i\}$ if and only if

$$\sum_{i=1}^{m} \text{Re } \beta_i \leq \sum_{i=1}^{m} \eta_i \quad \text{for } m = 1,\ldots, n \tag{6.5.68}$$

with equality for $m = n$.

43. Using the notation of Problem 41, modify the proof of (6.5.67) to show that, for any $A, B \in M_n$,

$$\sum_{i=1}^{m} \text{Re } \hat{\lambda}_i(A + B) \leq \sum_{i=1}^{m} [\hat{\lambda}_i(\tfrac{1}{2}(A + A^*)) + \hat{\lambda}_i(\tfrac{1}{2}(B + B^*))], \, m = 1,\ldots, n \tag{6.5.69}$$

What does this majorization say when both A and B are Hermitian? Show that (6.5.69) implies Theorem (4.3.27) in [HJ]. Show that (6.5.69) for any $A, B \in M_n$ follows from (6.5.67) and Theorem (4.3.27) in [HJ].

44. Show that $\| e^{A+B} \| \le e^{\lambda_{max}(A)} \| e^{B} \|$ for all Hermitian $A, B \in M_n$ and any unitarily invariant norm $\| \cdot \|$ on M_n.

Further Reading: See [Bel] and [Rog] for many results about matrix calculus. The simple bounds leading up to (6.5.16) show that the power series in t for $\log[I - (I - e^{tA} e^{tB})]$ converges absolutely for $t = 1$ provided $\|| A \||$ and $\|| B \||$ are less than $\frac{1}{4}\log 2 = 0.346...$, where $\|| \cdot \||$ is any given matrix norm, but one can do much better; for a proof that absolute convergence actually holds whenever $\|| A \||$ and $\|| B \||$ are less than 1, and for references to the literature, see M. Newman, W. So, and R. C. Thompson, Convergence Domains for the Campbell-Baker-Hausdorff Formula, *Linear Multilinear Algebra* 24 (1989), 301-310. There is a short proof of the Campbell-Baker-Hausdorff theorem (referred to in Problem 18) and references for six other proofs in D. Ž. Đoković, An Elementary Proof of the Baker-Campbell-Hausdorff-Dynkin Formula, *Math. Zeit.* 143 (1975), 209-211. For more about inequalities for the exponential of the type discussed in Theorem (6.5.20) and Corollary (6.5.22), as well as references to the physics literature, see C. J. Thompson, Inequalities and Partial Orders on Matrix Spaces, *Ind. Univ. Math. J.* 21 (1971), 469-480. Thompson also proves a theorem of A. Lenard: *Every* elementary symmetric function of the eigenvalues (not just the trace and determinant) satisfies the inequalities in Theorem (6.5.20). For other results of this type, many applications, and extensive references to the literature, see J. E. Cohen, S. Friedland, T. Kato, and F. P. Kelly, Eigenvalue Inequalities for Products of Matrix Exponentials, *Linear Algebra Appl.* 45 (1982), 55-95. The proof given for Theorem (6.5.29) is in E. M. E. Wermuth, Two Remarks on Matrix Exponentials, *Linear Algebra Appl.* 117 (1989), 127-132; the bound in Problem 33 is also due to Wermuth. See Chapter XI of [Gan 59, Gan 86] for the representations of various classes of matrices in terms of exponentials and products of matrices, which arise in the theory of Lie groups. Putzer's algorithms for computing e^{tA} are in E. J. Putzer, Avoiding the Jordan Canonical Form in the Discussion of Linear Systems With Constant Coefficients, *Amer. Math. Monthly* 73 (1966), 2-7. For a detailed discussion and comparison of various ways to compute a numerical approximation to e^A, see the survey article by C. Moler and C. F. Van Loan, Nineteen Dubious Ways to Compute the Exponential of a Matrix, *SIAM Review* 20 (1978), 801-836. The authors warn: "In principle, the exponential of a matrix could be computed in many ways....In practice, consideration of computational stability and efficiency indicates that some of the methods are preferable to others, but that none are completely satisfactory."

6.6   A chain rule for functions of a matrix

In the preceding section we have considered various special cases of the problem of evaluating $\frac{d}{dt} f(A(t))$, and we have focused on polynomials and on analytic functions such as $f(t) = e^t$ and $f(t) = \log(1-t)$. In this section we are interested in a general formula for the derivative of the primary matrix function $f(A(t))$, where $f(\cdot)$ is a sufficiently differentiable function with suitable domain and $A(t) \in M_n$ is a differentiably parametrized family of matrices. For example, if $A, B \in M_n$ are positive definite and $f(t) = \sqrt{t}$, we want to be able to evaluate $\frac{d}{dt}(tA + (1-t)B)^{\frac{1}{2}}$ in a useful way for $t \in (0,1)$.

Since the primary matrix function $f(A(t))$ is defined for each t as

$$f(A(t)) = S(t) f(J(t)) S(t)^{-1} \tag{6.6.1}$$

where $A(t)$ has Jordan canonical form $A(t) = S(t)J(t)S(t)^{-1}$, one straightforward approach to evaluating $\frac{d}{dt} f(A(t))$ would be to use the product rule to differentiate the right-hand side of (6.6.1). However, this may not be possible, even for the identity function $f(z) = z$. The difficulty is that even if $A(t)$ is a very smooth function of t, there may be no continuous Jordan canonical form factorization (6.6.1); see Problem 1 for an example in which $A(t)$ is a C^∞ family of 2-by-2 real symmetric matrices. However, even though the individual factors in the right-hand side of (6.6.1) might not be differentiable, under reasonable hypotheses their product is differentiable and there is a useful formula for the derivative.

We begin with the special case of a function $f(\cdot)$ that is analytic in a domain $D \subset \mathbb{C}$ and a given family $A(t) \in M_n$ that is defined on an open real interval (a,b), is differentiable at $t = t_0 \in (a,b)$, and has $\sigma(A(t)) \subset D$ for all $t \in (a,b)$. We may use Theorem (6.2.28) to write

$$f(A(t)) = \frac{1}{2\pi i} \oint_\Gamma f(s)[sI - A(t)]^{-1} \, ds$$

for all t in a neighborhood of $t = t_0$, where $\Gamma \subset D$ is any simple closed rectifiable curve that strictly encloses all the eigenvalues of $A(t_0)$. Using (6.5.7), we compute

$$\frac{d}{dt} f(A(t))\Big|_{t=t_0} = \frac{1}{2\pi i} \oint_\Gamma f(s) \frac{d}{dt}[sI - A(t)]^{-1}\Big|_{t=t_0} ds$$

$$= \frac{1}{2\pi i} \oint_\Gamma f(s) \left[[sI - A(t_0)]^{-1} \frac{d}{dt} A(t)\Big|_{t=t_0} [sI - A(t_0)]^{-1} \right] ds \quad (6.6.2)$$

Thus, $f(A(t))$ is differentiable at $t = t_0$, and this formula makes it clear that if $A(t)$ is continuously differentiable in a neighborhood of $t = t_0$, then so is $f(A(t))$.

Since we wish to obtain a formula that is applicable when $f(\cdot)$ is not necessarily analytic, we make two observations directed toward elimination of the line integral from (6.6.2).

First, recall from (6.2.41b,c) that Schwerdtfeger's formula can be used to write $[sI - A(t)]^{-1}$ in a form that decouples its dependence on the parameter s from the Jordan structure of A:

$$[sI - A(t)]^{-1} = \sum_{j=1}^{\mu(t)} A_j(t) \sum_{l=0}^{r_j(t)-1} \frac{1}{(s - \lambda_j(t))^{l+1}} [A(t) - \lambda_j(t)I]^l \quad (6.6.3)$$

Here, we assume that $s \notin \sigma(A(t))$, $A(t)$ has $\mu(t)$ distinct eigenvalues $\lambda_1(t),\dots,$ $\lambda_{\mu(t)}(t)$, and their multiplicities as zeroes of the minimal polynomial of $A(t)$ are $r_1(t),\dots, r_{\mu(t)}$, respectively. The Frobenius covariants $\{A_j(t)\}_{j=1,\dots,\mu(t)}$ are a family of projections onto the generalized eigenspaces of $A(t)$ that have the properties (6.1.40.1-9); in particular, each projection $A_j(t)$ is a polynomial in $A(t)$, $A_j(t)A_k(t) = 0$ if $j \neq k$, and $A_j(t)^m = A_j(t)$ for all $m = 1, 2,\dots$. If $A(t) = S(t)J(t)S(t)^{-1}$ is the Jordan canonical form of $A(t)$, each Frobenius covariant $A_j(t)$ may be visualized easily as $A_j(t) = S(t)D_j(t)S(t)^{-1}$, where $D_j(t)$ is a block diagonal matrix conformal with $J(t)$; every block of $J(t)$ that has eigenvalue $\lambda_j(t)$ corresponds to an identity block of the same size in the same position in $D_j(t)$ and all other blocks of $D_j(t)$ are zero.

Second, observe that

$$l!\,m! \frac{1}{2\pi i} \oint_\Gamma \frac{f(s)\, ds}{(s - \lambda_j)^{l+1}(s - \lambda_k)^{m+1}} = \frac{\partial^{l+m}}{\partial \lambda_j^l\, \partial \lambda_k^m} \frac{1}{2\pi i} \oint_\Gamma \frac{f(s)\, ds}{(s - \lambda_j)(s - \lambda_k)}$$

$$= \frac{\partial^{l+m}}{\partial \lambda_j^l\, \partial \lambda_k^m} \left(\frac{f(\lambda_j) - f(\lambda_k)}{\lambda_j - \lambda_k} \right) = \frac{\partial^{l+m}}{\partial u^l\, \partial v^m} \left(\frac{f(u) - f(v)}{u - v} \right)\Bigg|_{\substack{u = \lambda_j \\ v = \lambda_k}}$$

$$= \frac{\partial^{l+m}}{\partial u^l \ \partial v^m} \Delta f(u,v) \Big|_{\substack{u=\lambda_j \\ v=\lambda_k}} \tag{6.6.4}$$

where the difference quotient $\Delta f(u,v) = [f(u) - f(v)]/(u-v)$ is to be interpreted as $f'(u)$ when $u = v$. If we now substitute (6.6.3) into (6.6.2), we obtain

$$\tfrac{d}{dt} f(A(t)) \Big|_{t=t_0} = \frac{1}{2\pi i} \oint_\Gamma f(s) \{ [sI - A(t_0)]^{-1} \tfrac{d}{dt} A(t) \Big|_{t=t_0} [sI - A(t_0)]^{-1} \} \, ds$$

$$= \sum_{j,\,k=1}^{\mu(t)} \sum_{l=0}^{r_j(t)-1} \sum_{m=0}^{r_k(t)-1} A_j(t)[A(t) - \lambda_j(t)I]^l \left[\tfrac{d}{dt} A(t) \right] A_k(t)[A(t) - \lambda_k(t)I]^m \Big|_{t=t_0}$$

$$\times \frac{1}{2\pi i} \oint_\Gamma \frac{f(s) \, ds}{[s - \lambda_j(t_0)]^{l+1}[s - \lambda_k(t_0)]^{m+1}}$$

Finally, (6.6.4) permits us to rewrite this identity as

$$\tfrac{d}{dt} f(A(t))$$

$$= \sum_{j,\,k=1}^{\mu(t)} \sum_{l=0}^{r_j(t)-1} \sum_{m=0}^{r_k(t)-1} A_j(t)[A(t) - \lambda_j(t)I]^l \left[\tfrac{d}{dt} A(t) \right] A_k(t)[A(t) - \lambda_k(t)I]^m$$

$$\times \frac{1}{l!\,m!} \frac{\partial^{l+m}}{\partial u^l \ \partial v^m} \Delta f(u,v) \Big|_{\substack{u=\lambda_j(t) \\ v=\lambda_k(t)}} \tag{6.6.5}$$

for all $t \in (a,b)$ at which $A(t)$ is differentiable.

Notice that the indicated derivatives of the difference quotient in (6.6.5), evaluated at the indicated eigenvalues of $A(t)$, are uniquely determined by the values of $\{ f^{(k)}(\lambda_j(t)): \ k = 0, 1, \ldots, 2r_j(t) - 1, \ j = 1, 2, \ldots, \mu(t) \}$. Thus, if $t_0 \in (a,b)$ is given and $g(\cdot)$ is *any* analytic function on D such that $f^{(k)}(\lambda_j(t)) = g^{(k)}(\lambda_j(t))$ for $k = 0, 1, \ldots, 2r_j(t) - 1$ and $j = 1, 2, \ldots, \mu(t)$, then replacement of $f(\cdot)$ by $g(\cdot)$ does not change the last expression in (6.6.5) for $t = t_0$. This means that

$$\frac{d}{dt} f(A(t))\Big|_{t=t_0} = \frac{d}{dt} g(A(t))\Big|_{t=t_0} \qquad (6.6.6)$$

under these conditions. In particular, choose $g(t) = r_{A(t_0)\bullet A(t_0)}(t)$, where $r_{A(t_0)\bullet A(t_0)}(t)$ is the Newton polynomial defined by (6.1.29a,b) that interpolates $f(\cdot)$ and its derivatives at the zeroes of the characteristic polynomial of $A(t_0) \bullet A(t_0)$. Then $f^{(k)}(\lambda_j(t_0)) = r_{A(t_0)\bullet A(t_0)}{}^{(k)}(\lambda_j(t_0))$ for $k = 0, 1,\ldots,$ $2s_j(t_0) - 1$ and $j = 1,\ldots, \mu(t)$, where $s_1(t_0),\ldots, s_{\mu(t_0)}(t_0)$ are the respective algebraic multiplicities of the distinct eigenvalues $\lambda_1(t_0),\ldots, \lambda_{\mu(t_0)}(t_0)$ of $A(t_0)$. Since $s_j(t_0) \geq r_j(t_0)$ for $j = 1,\ldots, \mu(t)$, it follows that

$$\frac{d}{dt} f(A(t))\Big|_{t=t_0} = \frac{d}{dt} r_{A(t_0)\bullet A(t_0)}(A(t))\Big|_{t=t_0} \qquad (6.6.7)$$

a formula that reminds us of the basic identity

$$f(A(t))\Big|_{t=t_0} = r_{A(t_0)}(A(t))\Big|_{t=t_0}$$

for any primary matrix function.

So far, we have assumed that $f(\cdot)$ is analytic on a suitable domain, but if the family $A(t)$ has only real eigenvalues (for example, if each $A(t)$ is Hermitian), and if $f(\cdot)$ is defined on a suitable real interval and has sufficiently many derivatives there, then both the Newton polynomial $r_{A(t_0)\bullet A(t_0)}(t)$ and the difference quotient expressions in (6.6.5) are meaningful, and it is not unreasonable to hope that (6.6.5) and (6.6.7) might be valid in this case.

Let D be a given open real interval, let $A(t) \in M_n$ be a given family with $\sigma(A(t)) \subset D$ for all $t \in (a,b)$, and suppose $A(t)$ is continuous for $t \in (a,b)$. Recall from Theorem (6.2.27) that if $f(\cdot)$ is a given $(n-1)$-times continuously differentiable scalar-valued function on D, then $f(A(t))$ is defined and continuous for all $t \in (a,b)$. Now suppose that $A(t)$ is continuously differentiable for $t \in (a,b)$ and that $f(t)$ is $(2n-1)$-times continuously differentiable on D. We shall show that $f(A(t))$ is continuously differentiable for $t \in (a,b)$ and that both (6.6.5) and (6.6.7) are valid.

Let $t_0 \in (a,b)$ be given. Notice that for $t \in (a,b)$ and $t \neq t_0$, the polynomial $r_{A(t_0)\bullet A(t)}(\cdot)$ interpolates $f(\cdot)$ at each eigenvalue of $A(t_0)$ and $A(t)$ to an order equal to one less than the sum of its algebraic multiplicities as zeroes of the characteristic polynomials of $A(t)$ and $A(t_0)$, respectively. By

Theorem (6.2.9(e)), we know that $f(A(t_0)) = r_{A(t_0)\bullet A(t)}(A(t_0))$ and $f(A(t)) = r_{A(t_0)\bullet A(t)}(A(t))$. In order to verify the formula (6.6.7), it suffices to show that

$$\lim_{t \to t_0} \left[\frac{1}{t-t_0}[f(A(t)) - f(A(t_0))] \right.$$

$$\left. - \frac{1}{t-t_0}[r_{A(t_0)\bullet A(t_0)}(A(t)) - r_{A(t_0)\bullet A(t_0)}(A(t_0))] \right] = 0$$

For t in a neighborhood of t_0, $t \neq t_0$, we compute

$$\frac{1}{t-t_0}\left[[f(A(t)) - f(A(t_0))] - [r_{A(t_0)\bullet A(t_0)}(A(t)) - r_{A(t_0)\bullet A(t_0)}(A(t_0))] \right]$$

$$= \frac{1}{t-t_0}\left[[r_{A(t_0)\bullet A(t)}(A(t)) - r_{A(t_0)\bullet A(t)}(A(t_0))] \right.$$

$$\left. - [r_{A(t_0)\bullet A(t_0)}(A(t)) - r_{A(t_0)\bullet A(t_0)}(A(t_0))] \right]$$

$$= \frac{1}{t-t_0}\int_{\tau=t_0}^{t} \frac{d}{d\tau}\left[r_{A(t_0)\bullet A(t)}(A(\tau)) - r_{A(t_0)\bullet A(t_0)}(A(\tau)) \right] d\tau \qquad (6.6.8)$$

Thus, it is sufficient to show that the integrand in (6.6.8) tends to zero uniformly in τ as $t \to t_0$, that is, for every $\epsilon > 0$ it suffices to show that there exists some $\delta > 0$ and some norm $\|\cdot\|$ on M_n such that

$$\left\| \frac{d}{d\tau}\left[r_{A(t_0)\bullet A(t)}(A(\tau)) - r_{A(t_0)\bullet A(t_0)}(A(\tau)) \right] \right\| \leq \epsilon$$

for all τ such that $|t_0 - \tau| \leq \delta$.

For any polynomial $g(\cdot)$, we may use (6.6.2) to write

$$\frac{d}{d\tau}g(A(\tau)) = \frac{1}{2\pi i}\oint_\Gamma g(s)\left[[sI - A(\tau)]^{-1}\left[\frac{d}{d\tau}A(\tau) \right][sI - A(\tau)]^{-1} \right] ds \qquad (6.6.9)$$

for all τ in a neighborhood of t_0, where Γ is a fixed (that is, Γ is independent of τ) simple closed rectifiable curve that encloses D. Let $|||\cdot|||$ be any given matrix norm and observe that continuous differentiability of $A(t)$ implies that $|||\frac{d}{d\tau}A(\tau)||| \leq |||A(t_0)||| + 1$ for all τ in a suitable neighborhood of t_0. Similarly, $\{ |||[sI - A(\tau)]^{-1}||| : s \in \Gamma \}$ is bounded for τ in some neighborhood of t_0. Thus, for each τ in a neighborhood of t_0 we have the bound

$$|||\tfrac{d}{d\tau}g(A(\tau))||| \leq \frac{1}{2\pi}|\Gamma| \max_{s \in \Gamma}|g(s)| \; |||\tfrac{d}{d\tau}A(\tau)||| \max_{s \in \Gamma}|||[sI - A(\tau)]^{-1}|||^2$$

$$\leq K \max_{s \in \Gamma}|g(s)| \tag{6.6.10}$$

where $|\Gamma|$ denotes the length of Γ and K is a finite positive constant. Applying this bound to the integrand in (6.6.8) gives

$$|||\tfrac{d}{d\tau}\left[r_{A(t_0)\bullet A(t)}(A(\tau)) - r_{A(t_0)\bullet A(t_0)}(A(\tau)) \right]|||$$

$$\leq K \max_{s \in \Gamma}|r_{A(t_0)\bullet A(t)}(s) - r_{A(t_0)\bullet A(t_0)}(s)|$$

Corollary (6.1.28(1)) now ensures that all of the coefficients of $r_{A(t_0)\bullet A(t)}(\cdot)$ tend to the respective coefficients of $r_{A(t_0)\bullet A(t_0)}(\cdot)$ as $t \to t_0$, so

$$\max_{s \in \Gamma}|r_{A(t_0)\bullet A(t)}(s) - r_{A(t_0)\bullet A(t_0)}(s)| \to 0 \text{ as } t \to t_0$$

and we are done. Notice that it is the smoothness hypothesis in Corollary (6.1.28), applied to $A(t_0) \oplus A(t_0) \in M_{2n}$, that requires us to assume that $f(\cdot)$ has $2n - 1$ continuous derivatives.

In fact, we have actually proved a little more than the assertion (6.6.7): Using (6.6.9), we have

$$\tfrac{d}{dt}f(A(t))\Big|_{t=t_0} = \tfrac{d}{dt}r_{A(t_0)\bullet A(t_0)}(A(t))\Big|_{t=t_0}$$

$$= \frac{1}{2\pi i}\oint_\Gamma r_{A(t_0)\bullet A(t_0)}(s)\,\{[sI - A(t_0)]^{-1}\tfrac{d}{dt}A(t)\Big|_{t=t_0}[sI - A(t_0)]^{-1}\}\,ds$$

$$\tag{6.6.11}$$

Since $A(t)$ and $\frac{d}{dt}A(t)$ are both assumed to be continuous in a neighborhood of t_0, Corollary (6.1.28(1)) ensures that the last expression in (6.6.11), and hence the first also, depends continuously on the parameter t_0. The conclusion is that the stated hypotheses are sufficient to ensure that $f(A(t))$ is *continuously* differentiable in a neighborhood of t_0, and

$$\frac{d}{dt} f(A(t))\Big|_{t=\tau} = \frac{d}{dt} r_{A(\tau)\bullet A(\tau)}(A(t))\Big|_{t=\tau} \qquad (6.6.12)$$

for all τ in a neighborhood of t_0.

Finally, it is easy to connect (6.6.7) to the formula (6.6.5). Since $r_{A(t_0)\bullet A(t_0)}(t)$ is a polynomial in t, and hence is an analytic function of t, (6.6.5) can be applied to give an expression for the right-hand side of (6.6.7); it is just the right-hand side of (6.6.5) in which $f(\cdot)$ is replaced by $r_{A(t_0)\bullet A(t_0)}(\cdot)$. Each term

$$\frac{\partial^{l+m}}{\partial u^l\, \partial v^m}\left(\frac{r_{A(t_0)\bullet A(t_0)}(u) - r_{A(t_0)\bullet A(t_0)}(v)}{u-v}\right)\Bigg|_{\substack{u=\lambda_j(t_0)\\ v=\lambda_k(t_0)}} \qquad (6.6.13)$$

in this expression involves various derivatives of $r_{A(t_0)\bullet A(t_0)}(t)$ with respect to t, evaluated at eigenvalues of $A(t_0)$. But since the polynomial $r_{A(t_0)\bullet A(t_0)}(\cdot)$ interpolates $f(\cdot)$ and its derivatives at the eigenvalues of $A(t_0) \oplus A(t_0)$, each of these derivatives is equal to the corresponding derivative of $f(\cdot)$. As a consequence, each term of the form (6.6.13) may be replaced with a corresponding term

$$\frac{\partial^{l+m}}{\partial u^l\, \partial v^m}\left(\frac{f(u) - f(v)}{u-v}\right)\Bigg|_{\substack{u=\lambda_j(t_0)\\ v=\lambda_k(t_0)}}$$

and hence we have the desired formula

$$\frac{d}{dt} f(A(t))\Big|_{t=t_0} = \frac{d}{dt} r_{A(t_0)\bullet A(t_0)}(A(t))\Big|_{t=t_0}$$

$$= \sum_{j,\,k=1}^{\mu(t)} \sum_{l=0}^{r_j(t)-1} \sum_{m=0}^{r_k(t)-1} A_j(t)[A(t)-\lambda_j(t)I]^l\Big[\tfrac{d}{dt}A(t)\Big]A_k(t)[A(t)-\lambda_k(t)I]^m \Bigg|_{t=t_0}$$

$$\times\,\frac{1}{l!\,m!}\,\frac{\partial^{l+m}}{\partial u^l\,\partial v^m}\,\Delta^r A(t_0)\bullet A(t_0)(u,v)\Bigg|_{\substack{u=\lambda_j(t_0)\\v=\lambda_k(t_0)}}$$

$$= \sum_{j,\,k=1}^{\mu(t)} \sum_{l=0}^{r_j(t)-1} \sum_{m=0}^{r_k(t)-1} A_j(t)[A(t)-\lambda_j(t)I]^l\Big[\tfrac{d}{dt}A(t)\Big]A_k(t)[A(t)-\lambda_k(t)I]^m \Bigg|_{t=t_0}$$

$$\times\,\frac{1}{l!\,m!}\,\frac{\partial^{l+m}}{\partial u^l\,\partial v^m}\,\Delta f(u,v)\Bigg|_{\substack{u=\lambda_j(t_0)\\v=\lambda_k(t_0)}}$$

The same reasoning shows that (6.6.6) is valid in this case, provided that $g(\cdot)$ is $(2n-1)$-times continuously differentiable on the interval D.

The following theorem summarizes what we have proved so far.

6.6.14   Theorem. Let $n \geq 1$ be a given positive integer, and let $D \subset \mathbf{C}$ be either

(a)   a given simply connected open set, or
(b)   a given open real interval.

Let $A(t) \in M_n$ be a given family of matrices that is continuously differentiable for t in a given open real interval (a,b), let $A'(t) \equiv \tfrac{d}{dt} A(t)$, and suppose $\sigma(A(t)) \subset D$ for all $t \in (a,b)$. For each $t \in (a,b)$, let $\lambda_1(t),\dots,\lambda_{\mu(t)}$ denote the $\mu(t)$ distinct eigenvalues of $A(t)$, and let $r_1(t),\dots,r_{\mu(t)}$ denote their respective multiplicities as zeroes of the minimal polynomial of $A(t)$. Let $f\colon D \to \mathbf{C}$ be a given function; in case (a) assume that $f(\cdot)$ is analytic on D and in case (b) assume that $f(\cdot)$ is $(2n-1)$-times continuously differentiable on D. Then

(1) $f(A(t))$ is continuously differentiable for $t \in (a,b)$.

(2)   In case (a), let $t_0 \in (a,b)$ be given and let $\Gamma \subset D$ be a simple closed rectifiable curve that strictly encloses $\sigma(A(t_0))$. Then

$$\tfrac{d}{dt} f(A(t)) = \frac{1}{2\pi i}\oint_\Gamma f(s)\left[[sI - A(t)]^{-1} A'(t) [sI - A(t)]^{-1} \right] ds \quad (6.6.15)$$

for all t in a neighborhood of t_0.

(3) Let $t_0 \in D$ be given and let $r_{A(t_0) \bullet A(t_0)}(\cdot)$ denote the Newton polynomial [defined by (6.1.29a)] that interpolates $f(\cdot)$ and its derivatives at the zeroes of the characteristic polynomial of $A(t_0) \bullet A(t_0)$. Then

$$\frac{d}{dt} f(A(t))\Big|_{t=t_0} = \frac{d}{dt} r_{A(t_0) \bullet A(t_0)}(A(t))\Big|_{t=t_0} \tag{6.6.16}$$

(4) More generally, let $t_0 \in D$ be given and let $g: D \to \mathbb{C}$ be a given function that satisfies the same smoothness conditions as $f(\cdot)$ in each case (a) and (b). If $f^{(k)}(\lambda_j(t_0)) = g^{(k)}(\lambda_j(t_0))$ for $k = 0, 1,..., 2r_j(t_0) - 1$ and $j = 1, 2,..., \mu(t_0)$, then

$$\frac{d}{dt} f(A(t))\Big|_{t=t_0} = \frac{d}{dt} g(A(t))\Big|_{t=t_0} \tag{6.6.17}$$

(5) For each $t \in (a,b)$ we have

$$\frac{d}{dt} f(A(t)) = \sum_{j,k=1}^{\mu(t)} \sum_{l=0}^{r_j(t)-1} \sum_{m=0}^{r_k(t)-1} \frac{1}{l!\,m!} \frac{\partial^{l+m}}{\partial u^l\, \partial v^m} \Delta f(u,v)\Big|_{\substack{u=\lambda_j(t) \\ v=\lambda_k(t)}}$$

$$\times A_j(t)[A(t) - \lambda_j(t)I]^l\, A'(t)\, A_k(t)[A(t) - \lambda_k(t)I]^m \tag{6.6.18}$$

where $\{A_j(t)\}_{j=1,...,\mu(t)}$ are the Frobenius covariants of $A(t)$ defined in (6.1.40) and $\Delta f(u,v) \equiv [f(u) - f(v)]/(u-v)$ for $u \neq v$ with $\Delta f(u,u) \equiv f'(u)$.

One immediate consequence of this theorem is the following generalization of Theorem (6.2.34) and the identity (6.5.5a).

6.6.19 Corollary. With the notation and hypotheses of Theorem (6.6.14), let $t_0 \in (a,b)$ be given and suppose $A'(t_0)$ commutes with $A(t_0)$. Then

$$\frac{d}{dt} f(A(t))\Big|_{t=t_0} = A'(t_0)\, f'(A(t_0)) = f'(A(t_0))\, A'(t_0)$$

Proof: We know that

$$\tfrac{d}{dt} f(A(t))\big|_{t=t_0} = \tfrac{d}{dt}\, r_{A(t_0)\bullet A(t_0)}(A(t))\big|_{t=t_0}$$

and it follows from (6.5.5a) that

$$\tfrac{d}{dt}\, r_{A(t_0)\bullet A(t_0)}(A(t))\big|_{t=t_0} = A'(t_0)\, r'_{A(t_0)\bullet A(t_0)}(A(t_0))$$

$$= r'_{A(t_0)\bullet A(t_0)}(A(t_0))\, A'(t_0)$$

since $A'(t_0)$ commutes with $A(t_0)$. Because

$$f^{(k)}(\lambda_i(t_0)) = r^{(k)}_{A(t_0)\bullet A(t_0)}(\lambda_i(t_0)) \text{ for } k = 0, 1, \dots, 2r_i(t_0) - 1$$

the derivatives of $f'(\cdot)$ and $r'_{A(t_0)\bullet A(t_0)}(\cdot)$ agree at each eigenvalue $\lambda_i(t_0)$ up to order $2r_i(t_0) - 2 \geq r_i(t_0) - 1$, and it follows from Theorem (6.2.9(e)) that $r'_{A(t_0)\bullet A(t_0)}(A(t_0)) = f'(A(t_0))$. □

We have considered in detail the problem of calculating the first derivative of $f(A(t))$ under various conditions, but the methods we have used also yield in a straightforward way increasingly complex formulae for the second and higher derivatives. For the second derivative one obtains the following:

6.6.20   Theorem. With the notation and hypotheses of Theorem (6.6.14) assume, in addition, that $A(t)$ is twice continuously differentiable for $t \in (a,b)$ and let $A''(t) \equiv \tfrac{d^2}{dt^2} A(t)$. In case (b) assume that $f(\cdot)$ is $(3n-1)$-times continuously differentiable on D. Then

(1) $f(A(t))$ is twice continuously differentiable for $t \in (a,b)$.

(2) In case (a), let $t_0 \in (a,b)$ be given and let $\Gamma \subset D$ be a simple closed rectifiable curve that strictly encloses $\sigma(A(t_0))$. Then

$$\tfrac{d^2}{dt^2} f(A(t))\big|_{t=t_0}$$

$$= \frac{2}{2\pi i} \oint_\Gamma f(s) \{[sI - A(t)]^{-1} A'(t) [sI - A(t)]^{-1} A'(t) [sI - A(t)]^{-1}\} \, ds$$

$$+ \frac{1}{2\pi i} \oint_\Gamma f(s) \{[sI - A(t)]^{-1} A''(t) [sI - A(t)]^{-1}\} \, ds \quad (6.6.21)$$

for all t in a neighborhood of t_0.

(3) Let $t_0 \in D$ be given and let $r_{A(t_0) \bullet A(t_0) \bullet A(t_0)}(\cdot)$ denote the Newton polynomial [defined by (6.1.29a)] that interpolates $f(\cdot)$ and its derivatives at the zeroes of the characteristic polynomial of $A(t_0) \bullet A(t_0) \bullet A(t_0)$. Then

$$\frac{d^2}{dt^2} f(A(t))\Big|_{t=t_0} = \frac{d^2}{dt^2} r_{A(t_0) \bullet A(t_0) \bullet A(t_0)}(A(t))\Big|_{t=t_0} \quad (6.6.22)$$

(4) More generally, let $t_0 \in D$ be given and let $g \colon D \to \mathbf{C}$ be a given function that satisfies the same smoothness conditions as $f(\cdot)$ in each case (a) and (b). If $f^{(k)}(\lambda_j(t_0)) = g^{(k)}(\lambda_j(t_0))$ for $k = 0, 1, \ldots, 3r_j(t_0) - 1$ and $j = 1, 2, \ldots, \mu(t_0)$, then

$$\frac{d^2}{dt^2} f(A(t))\Big|_{t=t_0} = \frac{d^2}{dt^2} g(A(t))\Big|_{t=t_0} \quad (6.6.23)$$

(5) For each $t \in (a, b)$ we have

$$\frac{d^2}{dt^2} f(A(t)) = 2 \sum_{j,k,p=1}^{\mu(t)} \sum_{l=0}^{r_j(t)-1} \sum_{m=0}^{r_k(t)-1} \sum_{q=0}^{r_p(t)-1} \left(\frac{1}{l! \, m! \, q!} \right.$$

$$\times \frac{\partial^{l+m+q}}{\partial u^l \, \partial v^m \, \partial w^q} \Delta^2 f(u, v, w)\Big|_{\substack{u=\lambda_j(t) \\ v=\lambda_k(t) \\ w=\lambda_p(t)}}$$

$$\times A_j(t)[A(t) - \lambda_j(t)I]^l A'(t) A_k(t)[A(t) - \lambda_k(t)I]^m$$

$$\times A'(t) A_p(t)[A(t) - \lambda_p(t)I]^p \Big)$$

$$+ \sum_{j,k=1}^{\mu(t)} \sum_{l=0}^{r_j(t)-1} \sum_{m=0}^{r_k(t)-1} \frac{1}{l!\,m!} \frac{\partial^{l+m}}{\partial u^l\,\partial v^m} \Delta f(u,v)\Big|_{\substack{u=\lambda_j(t)\\ v=\lambda_k(t)}}$$

$$\times A_j(t)[A(t)-\lambda_j(t)I]^l\, A^\bullet(t)\, A_k(t)[A(t)-\lambda_k(t)I]^m \quad (6.6.24)$$

where the divided differences $\Delta f(u,v)$ and $\Delta^2 f(u,v,w)$ are defined in (6.1.17a,b).

If $A(t)$ is diagonalizable for some $t \in (a,b)$, then all of the multiplicities $r_i(t)$ are one and the formulae (6.6.18,24) simplify considerably, under the same hypotheses, to

$$\frac{d}{dt} f(A(t)) = \sum_{j,k=1}^{\mu(t)} \Delta f(\lambda_j(t),\lambda_k(t)) A_j(t)\, A'(t)\, A_k(t) \quad (6.6.25)$$

and

$$\frac{d^2}{dt^2} f(A(t)) = 2 \sum_{j,k,p=1}^{\mu(t)} \Delta^2 f(\lambda_j(t),\lambda_k(t),\lambda_p(t))\, A_j(t)\, A'(t)\, A_k(t)\, A'(t)\, A_p(t)$$

$$+ \sum_{j,k=1}^{\mu(t)} \Delta f(\lambda_j(t),\lambda_k(t))\, A_j(t)\, A^\bullet(t)\, A_k(t) \quad (6.6.26)$$

For the given point $t \in (a,b)$ at which $A(t)$ is diagonalizable, write $A(t) = S(t)\Lambda(t)S(t)^{-1}$, where $\Lambda(t)$ is diagonal. Then each Frobenius covariant $A_j(t)$ may be written as $A_j(t) = S(t) D_j(t) S(t)^{-1}$, where the diagonal matrix $D_j(t)$ has a 1 in position i,i if the i,i entry in $\Lambda(t)$ is $\lambda_j(t)$, $i=1,...,n$; all other entries of $D_j(t)$ are zero. Using this notation, (6.6.25) becomes

$$\frac{d}{dt} f(A(t)) = \sum_{j,k=1}^{\mu(t)} \Delta f(\lambda_j(t),\lambda_k(t))\, S(t) D_j(t) S(t)^{-1} A'(t) S(t) D_k(t) S(t)^{-1}$$

$$= S(t)\left[\sum_{j,k=1}^{\mu(t)} \Delta f(\lambda_j(t),\lambda_k(t))\, D_j(t)\, [S(t)^{-1}A'(t)S(t)]\, D_k(t)\right] S(t)^{-1} \quad (6.6.27)$$

Each term in the preceding sum has the effect of multiplying every entry in a certain submatrix of $S(t)^{-1}A'(t)S(t)$ by a certain difference quotient, so the overall effect of the sum is to perform a Hadamard product of the matrix $S(t)^{-1}A'(t)S(t)$ with a matrix of difference quotients. The form of this Hadamard product may be more apparent if we make two notational changes: Express each $D_j(t)$ as a sum of elementary matrices E_{ii}, where the i,i entry of E_{ii} is one and all other entries are zero;   write $A(t) = S(t)\,\Lambda(t)\,S(t)^{-1}$ with $\Lambda(t) = \mathrm{diag}(\lambda_1(t),...,\lambda_n(t))$, that is, $\lambda_1(t),...,\lambda_n(t)$ *now denote all the eigenvalues of* $A(t)$, *including their algebraic multiplicities.* With this new notation we have

$$\tfrac{d}{dt}f(A(t)) = S(t)\left[\sum_{i,\,j=1}^{n}\Delta f(\lambda_i(t),\lambda_j(t))\,E_{ii}\,[S(t)^{-1}A'(t)S(t)]\,E_{jj}\right]S(t)^{-1}$$

$$= S(t)\left[[\Delta f(\lambda_i(t),\lambda_j(t))]_{i,j=1}^{n}\circ [S(t)^{-1}A'(t)S(t)]\right]S(t)^{-1} \quad (6.6.28)$$

Notice that, under the domain and smoothness hypotheses on $f(\cdot)$ and $A(\cdot)$ in Theorem (6.6.14), this formula is valid at *each* point $t \in (a,b)$ at which $A(t)$ is diagonalizable; it is not necessary to assume that $A(t)$ is diagonalizable for *all t*.

If one proceeds in exactly the same way to rewrite the formula for the second derivative in terms of Hadamard products, the second sum in (6.6.26) achieves a form identical to (6.6.28) with $A'(t)$ replaced by $A^{\bullet}(t)$. The first sum becomes

$$\sum_{i,j,k=1}^{n}\Delta^2 f(\lambda_i(t),\lambda_j(t),\lambda_k(t))$$

$$\times\; S(t)\,E_{ii}\,S(t)^{-1}A'(t)\,S(t)\,E_{jj}\,S(t)^{-1}A'(t)\,S(t)\,E_{kk}\,S(t)^{-1}$$

$$= S(t)\left[\sum_{k=1}^{n}\sum_{i,j=1}^{n}\Delta^2 f(\lambda_i(t),\lambda_j(t),\lambda_k(t))\right.$$

$$\left.\times\;[E_{ii}\,S(t)^{-1}A'(t)\,S(t)]\,E_{kk}\,[S(t)^{-1}A'(t)\,S(t)\,E_{jj}]\right]S(t)^{-1}$$

$$= S(t) \left[\sum_{k=1}^{n} [\Delta^2 f(\lambda_i(t), \lambda_j(t), \lambda_k(t))]_{i,j=1}^{n} \right.$$

$$\left. \circ [C_k(S(t)^{-1} A'(t) S(t)) \, R_k(S(t)^{-1} A'(t) S(t))] \right] S(t)^{-1}$$

We have used the symmetry of the second divided difference $\Delta^2 f(u,v,w)$ in its three arguments and have written $C_k(X) \in M_{n,1}$ (respectively, $R_k(X) \in M_{1,n}$) to denote the kth column (respectively, the kth row) of $X \in M_n$.

Thus, the complete Hadamard product version of the expression (6.6.26) for the second derivative at a point t at which $A(t) = S(t) \Lambda(t) S(t)^{-1}$ is diagonalizable and $\Lambda(t) = \mathrm{diag}(\lambda_1(t), \ldots, \lambda_n(t))$ is

$$\frac{d^2}{dt^2} f(A(t)) = S(t) \left[2 \sum_{k=1}^{n} \left[[\Delta^2 f(\lambda_i(t), \lambda_j(t), \lambda_k(t))]_{i,j=1}^{n} \right. \right.$$

$$\left. \circ [C_k(S(t)^{-1} A'(t) S(t)) \, R_k(S(t)^{-1} A'(t) S(t))] \right]$$

$$+ [\Delta f(\lambda_i(t), \lambda_j(t))]_{i,j=1}^{n} \circ [S(t)^{-1} A^\bullet(t) S(t)] \bigg] S(t)^{-1} \quad (6.6.29)$$

If $A(t)$ is Hermitian for all $t \in (a,b)$, then $A(t)$ is diagonalizable for all $t \in (a,b)$ and the formulae (6.6.25-29) apply, but we can improve considerably on the underlying smoothness hypotheses on $f(\cdot)$ that they inherit from Theorems (6.6.14,20). The new argument available in this case relies on the fact that all $A(t)$ are *unitarily* diagonalizable, so there is a uniform upper bound available on the norms of the diagonalizing similarities.

6.6.30 Theorem. Let $n \geq 1$ be a given positive integer, let D be a given open real interval, let $A(t) \in M_n$ be a given family of Hermitian matrices for t in a given open real interval (a,b), and suppose $\sigma(A(t)) \subset D$ for all $t \in (a,b)$. For each $t \in (a,b)$, let $A(t) = U(t) \Lambda(t) U(t)^*$, where $U(t) \in M_n$ is unitary and $\Lambda(t) = \mathrm{diag}(\lambda_1(t), \ldots, \lambda_n(t))$. Let $f: D \to \mathbb{C}$ be a given function.

(1) If $A(\cdot)$ is continuously differentiable on (a,b) and $f(\cdot)$ is continuously differentiable on D, then $f(A(t))$ is continuously differentiable for $t \in (a,b)$ and

$$\frac{d}{dt} f(A(t))$$

$$= U(t) \left[[\Delta f(\lambda_i(t), \lambda_j(t))]_{i,j=1}^n \circ [U(t)^* A'(t) U(t)] \right] U(t)^* \qquad (6.6.31)$$

for all $t \in (a,b)$.

(2) If $A(t)$ is twice continuously differentiable on (a,b) and $f(\cdot)$ is twice continuously differentiable on D, then $f(A(t))$ is twice continuously differentiable for $t \in (a,b)$ and

$$\frac{d^2}{dt^2} f(A(t))$$

$$= U(t) \left[2 \sum_{k=1}^n \left[[\Delta^2 f(\lambda_i(t), \lambda_j(t), \lambda_k(t))]_{i,j=1}^n \right. \right.$$

$$\left. \circ [C_k(U(t)^* A'(t) U(t)) \, C_k(U(t)^* A'(t) U(t))^*] \right]$$

$$\left. + [\Delta f(\lambda_i(t), \lambda_j(t))]_{i,j=1}^n \circ [U(t)^* A^{\cdot}(t) U(t)] \right] U(t)^* \qquad (6.6.32)$$

where $C_k(X) \in M_{n,1}$ denotes the kth column of $X \in M_n$.

Proof: For any continuously differentiable function $\phi(\cdot)$ on (a,b), define

$$\mathcal{D}(\phi;t) \equiv U(t) \left[[\Delta\phi(\lambda_i(t), \lambda_j(t))]_{i,j=1}^n \circ [U(t)^* A'(t) U(t)] \right] U(t)^*$$

and notice that this expression is linear in ϕ. By (6.6.28) we know that $\frac{d}{dt} p(A(t)) = \mathcal{D}(p;t)$ for any polynomial $p(\cdot)$. It suffices to show that $f(A(t))$ is continuously differentiable and $\frac{d}{dt} f(A(t)) = \mathcal{D}(f;t)$ for all t in the interior of an arbitrary compact subset of D, so let $\alpha < \beta$ be given with $[\alpha,\beta] \subset D$. Using the fact that the Frobenius norm $\|\cdot\|_2$ is both ordinary and Hadamard product submultiplicative on M_n and assigns norm $\sqrt{n}$ to every unitary matrix in M_n, compute

$$\|\mathcal{D}(\phi;t)\|_2 = \left\| U(t) \left[[\Delta\phi(\lambda_i(t), \lambda_j(t))]_{i,j=1}^n \circ [U(t)^* A'(t) U(t)] \right] U(t)^* \right\|_2$$

$$\leq \| U(t)\|_2 \, \|[\Delta\phi(\lambda_i(t),\lambda_j(t))]_{i,j=1}^n\|_2 \, \| U(t)^*\|_2 \, \| A'(t)\|_2 \, \| U(t)\|_2 \, \| U(t)^*\|_2$$

$$= n^2 \, \|[\Delta\phi(\lambda_i(t),\lambda_j(t))]_{i,j=1}^n\|_2 \, \| A'(t)\|_2$$

$$\leq n^{5/2} \max \{|\Delta\phi(u,v)|: \; u, v \in [\alpha,\beta]\} \max \{\| A'(\tau)\|_2: \; \tau \in [\alpha,\beta]\}$$

$$= K \max \{|\phi'(\tau)|: \; \tau \in [\alpha,\beta]\}$$

where $K \equiv n^{5/2} \max \{\| A'(\tau)\|_2: \; \tau \in [\alpha,\beta]\}$ is finite and independent of ϕ.

Now let $\{\phi_k(\cdot)\}_{k=1,2,\dots}$ be a sequence of polynomials such that $\phi_k(t) \to f(t)$ and $\phi_k'(t) \to f'(t)$ uniformly in $t \in [\alpha,\beta]$ as $k \to \infty$. Then

$$\| \mathcal{D}(f;t) - \mathcal{D}(\phi_k;t) \|_2 = \| \mathcal{D}(f - \phi_k;t) \|_2$$

$$\leq K \max \{|f'(\tau) - \phi_k'(\tau)|: \; \tau \in [\alpha,\beta]\} \to 0$$

uniformly in $t \in [\alpha,\beta]$ as $k \to \infty$. Since Theorem (6.6.14(1)) ensures that $\mathcal{D}(\phi_k;t) = \frac{d}{dt}\phi_k(A(t))$ is continuous in t and $\mathcal{D}(\phi_k;t) \to \mathcal{D}(f;t)$ uniformly on $[\alpha,\beta]$, it follows that $\mathcal{D}(f;t)$ is a continuous function of $t \in (\alpha,\beta)$. Finally, let $t \in (\alpha,\beta)$ be given, let $\Delta t \neq 0$ be small enough so that $t + \Delta t \in [\alpha,\beta]$, and use the pointwise convergence of $\phi_k(t)$ to $f(t)$ and the uniform convergence of $\mathcal{D}(\phi_k;t)$ to $\mathcal{D}(f;t)$ to compute

$$\frac{1}{\Delta t}[f(A(t+\Delta t)) - f(A(t))] = \lim_{k\to\infty} \frac{1}{\Delta t}[\phi_k(A(t+\Delta t)) - \phi_k(A(t))]$$

$$= \lim_{k\to\infty} \frac{1}{\Delta t} \int_t^{t+\Delta t} \frac{d}{d\tau}\phi_k(A(\tau)) \, d\tau = \lim_{k\to\infty} \frac{1}{\Delta t} \int_t^{t+\Delta t} \mathcal{D}(\phi_k;\tau) \, d\tau$$

$$= \frac{1}{\Delta t} \int_t^{t+\Delta t} \lim_{k\to\infty} \mathcal{D}(\phi_k;\tau) \, d\tau = \frac{1}{\Delta t} \int_t^{t+\Delta t} \mathcal{D}(f;\tau) \, d\tau$$

Finally, let $\Delta t \to 0$ and conclude that $\frac{d}{dt} f(A(t)) = \mathcal{D}(f;t)$ for all $t \in (\alpha,\beta)$.

A similar argument shows that (6.6.32) follows from (6.6.29); the form of the second factor in the Hadamard product in the sum in (6.6.32) results from the fact that $R_k(X) = C_k(X)^*$ for a Hermitian X. ☐

The preceding characterizations of derivatives of a primary function of a parametrized family of matrices have an interesting application to primary matrix function analogs of monotone and convex real-valued functions. It is convenient to denote the set of n-by-n Hermitian matrices with spectrum in an open real interval (a,b) by $H_n(a,b)$. Recall the positive semidefinite partial order on $H_n(a,b)$, sometimes called the *Loewner partial order*: $B \succeq A$ if A and B are Hermitian matrices of the same size and $B - A$ is positive semidefinite; we write $B \succ A$ if $B - A$ is positive definite.

6.6.33   Definition. Let (a,b) be a given open real interval and let $n \geq 1$ be a given positive integer. A real-valued function $f(\cdot)$ on (a,b) is said to be a *monotone matrix function on* $H_n(a,b)$ if the primary matrix function $f(\cdot)$ on $H_n(a,b)$ has the property that

$$f(B) \succeq f(4) \text{ whenever } A, B \in H_n(a,b) \text{ and } B \succeq A \qquad (6.6.34)$$

A monotone matrix function $f(\cdot)$ on $H_n(a,b)$ is said to be *strictly monotone* if

$$f(B) \succ f(A) \text{ whenever } A, B \in H_n(a,b) \text{ and } B \succ A \qquad (6.6.35)$$

Once one has the formula (6.6.31), it is easy to give a useful characterization of the continuously differentiable functions that are monotone matrix functions.

6.6.36   Theorem. Let (a,b) be a given open real interval, let $n \geq 1$ be a given positive integer, and let $f: (a,b) \to \mathbb{R}$ be a given continuously differentiable function. Then

(1) $f(\cdot)$ is a monotone matrix function on $H_n(a,b)$ if and only if

$$[\Delta f(t_i, t_j)]_{i,j=1}^{n} \succeq 0 \text{ for all } t_1, \ldots, t_n \in (a,b) \qquad (6.6.37)$$

(2) Let $n \geq 2$ be given and suppose $f(\cdot)$ is a monotone matrix function on $H_n(a,b)$. Then $f'(t) \geq 0$ for all $t \in (a,b)$. Moreover, $f'(t_0) = 0$ for some $t_0 \in (a,b)$ if and only if $f(\cdot)$ is a constant function on (a,b), that is, $f(t) = \alpha$ for some $\alpha \in \mathbb{R}$ and all $t \in (a,b)$.

(3) Let $n \geq 2$ be given and suppose $f(\cdot)$ is a monotone matrix function on $H_n(a,b)$. If $f(\cdot)$ is not a constant function on (a,b), then it is a strictly

monotone matrix function on $H_n(a,b)$, that is, $f(B) \succ f(A)$ whenever $A, B \in H_n(a,b)$ and $B \succ A$.

Proof: Let $A, B \in H_n(a,b)$ be given with $B \succeq A$, set $A(t) \equiv (1-t)A + tB$ with $t \in [0,1]$, and suppose $A(t) = U(t) \Lambda(t) U(t)^*$, where $\Lambda(t) = \text{diag}(\lambda_1(t),...,\lambda_n(t))$ and $U(t) \in M_n$ is unitary. Then $A'(t) = B - A \succeq 0$ and (6.6.31) gives the identity

$$f(B) - f(A) = \int_0^1 \tfrac{d}{dt} f(A(t))\, dt$$

$$= \int_0^1 U(t) \left([\Delta f(\lambda_i(t),\lambda_j(t))]_{i,j=1}^n \circ [U(t)^* (B-A) U(t)] \right) U(t)^*\, dt$$

The Schur product theorem (5.2.1) now makes evident the sufficiency of the condition in (1) since the preceding integral is the limit of sums of Hadamard products of positive semidefinite matrices.

Conversely, let $\xi \in \mathbb{C}^n$ be given, let $\Lambda = \text{diag}(\lambda_1,...,\lambda_n)$ be any given diagonal matrix in $H_n(a,b)$, and let $\epsilon > 0$ be small enough so that both $A \equiv \Lambda - \epsilon\xi\xi^*$ and $B \equiv \Lambda + \epsilon\xi\xi^*$ are in $H_n(a,b)$. Let $A(t) \equiv (1-t)A + tB$, so that $A(\tfrac{1}{2}) = \Lambda$, $A'(t) = 2\epsilon\xi\xi^* \succeq 0$, and $A(s) \succeq A(t)$ if $s, t \in [0,1]$ and $s \geq t$. If $f(\cdot)$ is a monotone matrix function on $H_n(a,b)$, then

$$\tfrac{d}{dt} f(A(t)) \Big|_{t=\frac{1}{2}} = \lim_{\substack{h \to 0 \\ h > 0}} \tfrac{1}{h} [f(A(\tfrac{1}{2} + h)) - f(A(\tfrac{1}{2}))]$$

is positive semidefinite since it is the limit of positive semidefinite matrices. But (6.6.31) gives

$$\tfrac{d}{dt} f(A(t)) \Big|_{t=\frac{1}{2}} = [\Delta f(\lambda_i,\lambda_j)]_{i,j=1}^n \circ (2\epsilon\xi\xi^*)$$

so the Hadamard product $[\Delta f(\lambda_i,\lambda_j)]_{i,j=1}^n \circ (\xi\xi^*)$ is positive semidefinite. Now take $e^T \equiv [1, ..., 1]^T \in \mathbb{R}^n$ and compute

$$e^T \{ [\Delta f(\lambda_i,\lambda_j)]_{i,j=1}^n \circ (\xi\xi^*) \} e = \xi^T ([\Delta f(\lambda_i,\lambda_j)]_{i,j=1}^n) \bar{\xi} \geq 0$$

Since $\xi \in \mathbf{C}^n$ is arbitrary, we conclude that $[\Delta f(\lambda_i, \lambda_j)]_{i,j=1}^n$ is positive semi-definite for all $\lambda_1, ..., \lambda_n \in (a, b)$.

The second assertion follows from the first and the fact that a zero main diagonal entry in a positive semidefinite matrix must lie in an entire row of zero entries. If $n \geq 2$ and $f(\cdot)$ is a monotone matrix function on $H_n(a, b)$, then

$$\begin{bmatrix} \Delta f(t_0, t_0) & \Delta f(t_0, t) \\ \Delta f(t, t_0) & \Delta f(t, t) \end{bmatrix} = \begin{bmatrix} f'(t_0) & \Delta f(t_0, t) \\ \Delta f(t_0, t) & f'(t) \end{bmatrix} \geq 0$$

for all $t_0, t \in (a, b)$; in particular, $f'(t) \geq 0$ for all $t \in (a, b)$. If $f'(t_0) = 0$, then $\Delta f(t_0, t) = 0$ so $f(t) = f(t_0)$ for all $t \in (a, b)$.

The last assertion follows from the preceding integral representation for $f(B) - f(A)$, (6.6.37), (2), and the assurance of Theorem (5.2.1) that the Hadamard product of a positive definite matrix and a positive semidefinite matrix with positive main diagonal entries is positive definite.   □

Exercise. For $f(t) = -1/t$, show that $\Delta f(s, t) = 1/st$ for all nonzero s, t, so the matrix of difference quotients in (6.6.37) is a rank one positive semidefinite matrix when all $t_i \neq 0$. Conclude that $f(t) = -1/t$ is a monotone matrix function on $H_n(0, \infty)$ for all $n = 1, 2, ...$ For a direct proof of this fact, see Corollary (7.7.4) in [HJ].

Exercise. Although $f(t) = t^2$ is a monotone scalar function on $(0, \infty)$, show that $\Delta f(s, t) = s + t$ and the determinant of the matrix of difference quotients in (6.6.37) is $-(t_1 - t_2)^2$ when $n = 2$. Conclude that $f(t) = t^2$ is not a monotone matrix function on $H_n(a, b)$ for any $n \geq 2$ and any $a < b$.

One can significantly strengthen the conclusion of (6.6.36(3)) in return for a stronger, but very natural, hypothesis. If $n \geq 2$, $t_1 \geq \cdots \geq t_n$, and some values of t_i are repeated [this corresponds to the occurrence of multiple eigenvalues of $A(t) = (1 - t)A + tB$ for some $t \in (0, 1)$], the positive semidefinite matrix $[\Delta f(t_i, t_j)]$ is structurally singular because it contains principal (and nonprincipal) submatrices in which all the entries are equal. On the other hand, if $A, B \in H_n(a, b)$ have a common eigenvector for which they have the same eigenvalue, $g(B) - g(A)$ is singular for every primary matrix function $g(\cdot)$ defined on (a, b); see Problem 34. The following theorem describes a type of matrix monotonicity that is stronger than strict monotonicity. It ensures that if the matrix of difference quotients $[\Delta f(t_i, t_j)]$ is positive definite whenever that is structurally possible, then $f(B) \succ f(A)$ whenever $B \succeq A$ and it is structurally possible for $f(B) - f(A)$ to be nonsingular.

6.6.38   Theorem. Let (a,b) be a given open real interval and let $n \geq 2$ be a given integer. Let $f(\cdot)$ be a given continuously differentiable real-valued function on (a,b), and suppose that

(a) $[\Delta f(t_i, t_j)]^m_{i,j=1} \succ 0$ whenever $t_1,\ldots, t_m \in (a,b)$ are distinct and $1 \leq m \leq n$.

Then

(1) $f(\cdot)$ is a monotone matrix function on $H_n(a,b)$; and

(2) If $A, B \in H_n(a,b)$ and $A \succeq B$, and if there is no common eigenvector of A and B for which they have the same eigenvalue, then $f(B) \succ f(A)$.

Proof: Let $A(t) \equiv (1-t)A + tB = U(t)\, \Lambda(t)\, U(t)^*$ for each $t \in (0,1)$, where $U(t)$ is unitary, $\Lambda(t) = \mathrm{diag}(\lambda_1(t),\ldots, \lambda_n(t))$, and $\lambda_1(t) \geq \cdots \geq \lambda_n(t)$. It suffices to show for each $t \in (0,1)$ that

$$[\Delta f(t_i, t_j)]^n_{i,j=1} \circ [U(t)^*(B-A)U(t)] \succ 0$$

Fix $t \in (0,1)$, set $U \equiv U(t)$, and write $L \equiv \Lambda(t) = \lambda_1 I_{n_1} \oplus \cdots \oplus \lambda_p I_{n_p} \equiv [L_{ij}]^p_{i,j=1}$, where $1 \leq p \leq n$, $\lambda_1 > \cdots > \lambda_p$ are the distinct eigenvalues of $A(t)$, $n_1 + \cdots + n_p = n$, each $L_{ij} \in M_{n_i, n_j}$, each $L_{ii} = \lambda_i I_{n_i}$, and $L_{ij} = 0$ if $i \neq j$. Then $A(t) = (1-t)A + tB = ULU^*$, so

$$B = t^{-1}[ULU^* - (1-t)A]$$

$$B - A = t^{-1}(ULU^* - A)$$

and

$$U^*(B-A)U = t^{-1}(L - U^*AU)$$

540   **Matrices and functions**

Notice that

$$[\Delta f(\lambda_i(t),\lambda_j(t))]^n_{i,j=1} = [\Delta f(\lambda_i,\lambda_j)J_{n_i,n_j}]^p_{i,j=1}$$

and hence we wish to show that

$$[\Delta f(\lambda_i,\lambda_j)J_{n_i,n_j}]^p_{i,j=1} \circ (L - U^*AU) \succ 0$$

Partition $A \equiv U^*AU = [A_{ij}]^p_{i,j=1}$ conformally to the partition of L. Using Theorem (5.3.6) and the assumption that $[\Delta f(\lambda_i,\lambda_j)]^p_{i,j=1} \succ 0$, it suffices to show that $L_{ii} - A_{ii} \succ 0$ for each $i = 1,...,p$. Since $B - A \succeq 0$, we have $tU^*(B-A)U = L - U^*AU = L - A \succeq 0$, and hence every diagonal block $L_{ii} - A_{ii}$ of $L - A$ is positive semidefinite for $i = 1,...,p$. If some diagonal block $L_{ii} - A_{ii}$ is singular, there is a nonzero $x \in \mathbb{C}^{n_i}$ such that $0 = (L_{ii} - A_{ii})x = (\lambda_i I_{n_i} - A_{ii})x = \lambda_i x - A_{ii}x$, so x is an eigenvector of A_{ii} and $A_{ii}x = \lambda_i x$. Define $\xi \in \mathbb{C}^n$ by $\xi^T \equiv [0 \ ... \ 0 \ x^T \ 0 \ ... \ 0]^T$ (x in the ith block position of ξ, zeroes elsewhere), and compute $\xi^*(L-A)\xi = x^*(L_{ii}-A_{ii})x = 0$. Since $L - A \succeq 0$, ξ must be in the nullspace of $L - A$, so $0 = (L-A)\xi = \lambda_i \xi - A\xi$, $\lambda_i \xi = A\xi = U^*AU\xi$, and $A(U\xi) = \lambda_i(U\xi)$. But

$$B(U\xi) = t^{-1}[ULU^* - (1-t)A](U\xi)$$

$$= t^{-1}[UL\xi - (1-t)A(U\xi)]$$

$$= t^{-1}[\lambda_i(U\xi) - (1-t)\lambda_i(U\xi)] = \lambda_i(U\xi)$$

so $U\xi$ is a common eigenvector of A and B for which both have the same eigenvalue λ_i. Since this is precluded by hypothesis, it follows that all the diagonal blocks $L_{ii} - A_{ii}$ of $L - A$ are positive definite, as desired. $\quad\Box$

6.6.39   Definition. Let (a,b) be a given open real interval. A monotone matrix function $f(\cdot)$ on $H_n(a,b)$ is said to be *strongly monotone* if $f(B) \succ f(A)$ whenever $A, B \in H_n(a,b)$, $B \succeq A$, and there is no common eigenvector of A and B for which they have the same eigenvalue.

We have shown that the condition (6.6.38(a)) is sufficient to guarantee that a continuously differentiable monotone matrix function is strongly monotone. If $f(\cdot)$ is a monotone matrix function on $H_m(a,b)$ for every $m = 1, 2, \ldots$, then it is known that $f(\cdot)$ satisfies the condition in (6.6.38(a)) for a given positive integer n if and only if $f(\cdot)$ is not a rational function of degree at most $n-1$. This observation implies that $f(t) = \log t$ and $f_\alpha(t) = t^\alpha$ for $\alpha \in (0,1)$ are strongly monotone; see Problems 17, 18, and 20. The rational function $f(t) = -1/t$ is a strictly monotone matrix function on $H_n(0,\infty)$ for all $n = 1, 2, \ldots$, but it is not strongly monotone on $H_n(a,b)$ for any $n \geq 2$ and any $b > a > 0$; see Problem 37.

The matrix of difference quotients in (6.6.37) is often called the *Loewner matrix* of $f(\cdot)$ because it plays a key role in C. Loewner's seminal 1934 paper on monotone matrix functions. Loewner showed that if for some $n \geq 2$ and some open real interval (a,b) a given real-valued function $f(\cdot)$—not assumed to be differentiable or even continuous—is a monotone matrix function on $H_n(a,b)$, then it has at least $2n-3$ continuous derivatives on (a,b), its odd-order derivatives up to order $2n-3$ are all nonnegative, and $f^{(2n-3)}(\cdot)$ is convex on (a,b). Thus, a monotone matrix function on $H_2(a,b)$ must be continuously differentiable on (a,b), so the hypothesis of continuous differentiability in Theorem (6.6.36) is needed only for the scalar case $n = 1$ of (6.6.36(1)) and to show that (6.6.37) is a sufficient condition for matrix monotonicity. Loewner also showed that $f(\cdot)$ is a monotone matrix function on $H_n(a,b)$ for *every* $n = 1, 2, \ldots$ if and only if $f(\cdot)$ is the restriction to (a,b) of an analytic function on the upper half-plane $\{z \in \mathbb{C}: \operatorname{Im} z > 0\}$ that maps the upper half-plane into itself and has finite real boundary values

$$f(t) \equiv \lim_{\epsilon \downarrow 0} f(t + i\epsilon) \text{ for } t \in (a,b)$$

In particular, such a function must be real-analytic on (a,b). This *mapping criterion* can be very useful in deciding whether a given analytic function is a monotone matrix function; see Problems 20-22.

If $\alpha, \beta \in \mathbb{R}$ with $\alpha \geq 0$, and if $d\mu$ is a positive Borel measure on $\mathbb{R}$ that has no mass in the interval (a,b) and for which the integral

$$\int_{-\infty}^{\infty} \frac{d\mu(u)}{1+u^2}$$

is convergent, then any function of the form

$$f(z) = \alpha z + \beta + \int_{-\infty}^{\infty} \left[\frac{1}{u-z} - \frac{u}{u^2+1} \right] d\mu(u), \quad \text{Im } z > 0 \qquad (6.6.40)$$

maps the upper half-plane into itself and has finite real boundary values on (a,b) and therefore is a monotone matrix function on $H_n(a,b)$ for all $n = 1, 2,\dots$. Conversely, if $f(\cdot)$ is a monotone matrix function on $H_n(a,b)$ for every $n = 1, 2,\dots$, then it has a representation of the form (6.6.40) for $z \in (a,b)$, and this formula gives the analytic continuation of $f(\cdot)$ to the upper half-plane.

Simple examples of monotone matrix functions can be constructed with the formula (6.6.40) and measures of the form $d\mu(t) = m(t) \, dt$, where $m(t) \geq 0$ for all $t \in \mathbb{R}$ and $m(t) \equiv 0$ for all $t \in (a,b)$. For example, if $m(t) = 1$ for $t < 0$ and $m(t) = 0$ for $t \geq 0$, the resulting function is $f(z) = \alpha z + \beta + \log z$. If $m(t) = \sqrt{-t}/\pi$ for $t < 0$ and $m(t) = 0$ for $t > 0$, the resulting function is $f(z) = \alpha z + \beta - 1/\sqrt{2} + \sqrt{z}$. If $d\mu$ has point mass 1 at $t = 0$, the resulting function is $f(z) = \alpha z + \beta - 1/z$. Functions that can be represented in the form (6.6.40) are sometimes called *functions of positive type* or *Pick functions*.

It is easy to show that a continuously differentiable real-valued function $f(\cdot)$ on (a,b) has the property that the Loewner matrices in (6.6.37) are positive semidefinite for all $n = 1, 2,\dots$ if and only if the difference quotient kernel

$$K_f(s,t) = \frac{f(s) - f(t)}{s - t}, \qquad K_f(t,t) \equiv f'(t)$$

on $(a,b) \times (a,b)$ is positive semidefinite, that is

$$\int_a^b \int_a^b K_f(s,t) \overline{\phi}(s) \phi(t) \, ds dt \geq 0 \qquad (6.6.41)$$

for all continuous functions $\phi(t)$ with compact support in (a,b). A function $\phi(\cdot)$ on (a,b) is said to have *compact support* if there is a compact set $D \subset (a,b)$ such that $\phi(t) \equiv 0$ if $t \in (a,b)$ and $t \notin D$.

Thus, the condition (6.6.41) must be equivalent to the existence of a representation of the form (6.6.40) for $f(z)$. Since the condition (6.6.41) involves knowledge only of the behavior of the given function $f(\cdot)$ on the real interval (a,b), it may be thought of as a test that gives a necessary and sufficient condition for a function $f(\cdot)$ on (a,b) to possess an analytic continuation into the upper half-plane that maps the upper half-plane into itself.

It is certainly fair to say that the connection between positive semidefiniteness of the Loewner matrices in (6.6.37) and the representation (6.6.40) is not obvious. That there *is* a connection is a very pretty piece of matrix analysis that starts with the simple question of how to generalize the notion of a monotone increasing scalar function to a function of Hermitian matrices.

As a final comment about monotone matrix functions, we note that it may be of interest to know not only whether a given function $f(\cdot)$ on (a,b) can be continued analytically to give a function that takes the upper half-plane into itself, but also to know whether this analytic continuation is a one-to-one function, sometimes called a *univalent* or *schlicht* function. It turns out that this is the case if and only if the kernel $K_f(s,t)$ not only satisfies (6.6.41), but also satisfies the stronger condition

$$\int_a^b \int_a^b K_f^\alpha(s,t)\bar{\phi}(s)\phi(t)\, ds dt \geq 0 \quad \text{for all } \alpha > 0 \tag{6.6.42}$$

for all continuous functions $\phi(t)$ with compact support in (a,b). In terms of the Loewner matrix in (6.6.37), this is equivalent to having

$$\left[\left[\frac{f(t_i) - f(t_j)}{t_i - t_j} \right]^\alpha \right]_{i,j=1}^n \tag{6.6.43}$$

be positive semidefinite for all $\alpha > 0$, all choices of the points $t_1,\ldots, t_n \in (a,b)$, and all $n = 1, 2,\ldots$, that is, the difference quotient kernel $K_f(s,t)$ is *infinitely divisible*. Theorem (6.3.13) and the criterion in (6.3.14) can be very useful in showing that $K_f(s,t)$ is infinitely divisible.

A continuously differentiable function $f: (a,b) \to \mathbb{R}$ is convex if and only if its derivative is monotone increasing on (a,b). An analog of this familiar condition is valid for a natural notion of convex matrix function, but the exact condition is perhaps not what one might first guess.

6.6.44 **Definition.** Let l, m, and n be given positive integers and let $C \subset M_{l,m}$ be a given convex set. A Hermitian matrix-valued function $f: C \to H_n(-\infty,\infty)$, not necessarily a primary matrix function, is said to be a *convex function on C* if

$$(1-\alpha)f(A) + \alpha f(B) \succeq f((1-\alpha)A + \alpha B) \qquad (6.6.45)$$

for all $A, B \in C$ and all $\alpha \in [0,1]$; $f(\cdot)$ is said to be *concave on* C if $-f(\cdot)$ is convex on C. A convex function $f(\cdot)$ on C is said to be *strictly convex* if

$$(1-\alpha)f(A) + \alpha f(B) \succ f((1-\alpha)A + \alpha B) \qquad (6.6.46)$$

for all $A, B \in C$ such that $A - B$ has full rank and all $\alpha \in (0,1)$; a concave function $f(\cdot)$ on C is said to be *strictly concave* if $-f(\cdot)$ is strictly convex.

We are particularly interested in convex and concave primary matrix functions, which arise by taking $C = H_n(a,b)$ in the preceding definition.

6.6.47   Definition. Let (a,b) be a given open real interval and let $n \geq 1$ be a given positive integer. A real-valued function $f(\cdot)$ on (a,b) is said to be a *convex* (respectively, *strictly convex*) *matrix function on* $H_n(a,b)$ if the primary matrix function $f: H_n(a,b) \to H_n(-\infty,\infty)$ is a convex (respectively, strictly convex) function on $H_n(a,b)$.

Let $f(\cdot)$ be a given real-valued function on (a,b). Set $A(t) \equiv (1-t)A + tB$ and write $\phi(t;\xi;A,B) \equiv \xi^* f(A(t))\xi$ for $t \in [0,1]$, $\xi \in \mathbb{C}^n$, and $A, B \in H_n(a,b)$. It is easy to show that $f(\cdot)$ is a convex matrix function on $H_n(a,b)$ if and only if $\phi(t;\xi;A,B)$ is a convex function of t for every $\xi \in \mathbb{C}^n$; see Problem 24. Now assume that $f(\cdot)$ is twice continuously differentiable, so $\phi(t;\xi;A,B)$ is twice continuously differentiable in t. Since a twice continuously differentiable function $g: (a,b) \to \mathbb{R}$ is convex if and only if $g''(t) \geq 0$ for all $t \in (a,b)$, and is strictly convex if $g''(t) > 0$ for all $t \in (a,b)$, we see that $\phi(t;\xi;A,B)$ is convex in t if and only if

$$\phi''(t;\xi;A,B) = \xi^* \frac{d^2}{dt^2} f(A(t))\xi \geq 0 \text{ for all } t \in (0,1)$$

and it is strictly convex if

$$\phi''(t;\xi;A,B) = \xi^* \frac{d^2}{dt^2} f(A(t))\xi > 0 \text{ for all } t \in (0,1)$$

Since $\xi \in \mathbb{C}^n$ is arbitrary, we conclude that $f(\cdot)$ is a convex matrix function on $H_n(a,b)$ if and only if

$$\frac{d^2}{dt^2} f((1-t)A + tB) \succeq 0 \text{ for all } t \in (0,1) \text{ and all } A, B \in H_n(a,b) \quad (6.6.48)$$

and $f(\cdot)$ is a strictly convex matrix function on $H_n(a,b)$ if

$$\frac{d^2}{dt^2} f((1-t)A + tB) \succ 0 \text{ for all } t \in (0,1) \text{ and all } A, B \in H_n(a,b)$$
$$\text{such that } A - B \text{ is nonsingular} \quad (6.6.49)$$

Now let $A(t) = U(t) \Lambda(t) U(t)^*$ with $\Lambda(t) = \mathrm{diag}(\lambda_1(t),\dots,\lambda_n(t))$ and a unitary $U(t) \in M_n$, and use (6.6.32) to compute

$$\frac{d^2}{dt^2} f((1-t)A + tB) = 2\, U(t) \left(\sum_{k=1}^{n} \left[[\Delta^2 f(\lambda_i(t),\lambda_j(t),\lambda_k(t))]_{i,j=1}^{n} \right. \right.$$

$$\left. \left. \circ \left[C_k(U(t)^* (B-A) U(t)) \, C_k(U(t)^* (B-A) U(t))^* \right] \right] \right) U(t)^* \quad (6.6.50)$$

The second term in (6.6.32) is absent because $A'(t) = 0$. Since each of the rank one matrices

$$C_k(U(t)^* (B-A) U(t)) \, C_k(U(t)^* (B-A) U(t))^*$$

is positive semidefinite, the Schur product theorem (5.2.1) ensures that $\frac{d^2}{dt^2} f((1-t)A + tB) \succeq 0$ for every $A, B \in H_n(a,b)$ and all $t \in (0,1)$, and hence $f(\cdot)$ is a convex matrix function, if

$$[\Delta^2 f(t_i,t_j,t_k)]_{i,j=1}^{n} \succeq 0 \text{ for all } t_1,\dots, t_n \in (a,b) \text{ and all } k = 1,\dots, n$$

In fact, it is sufficient to require that these conditions hold for *any one* value of k. To see this, suppose $n \geq 2$ and assume that

$$[\Delta^2 f(t_i,t_j,t_1)]_{i,j=1}^{n} \succeq 0 \text{ for all } t_1,\dots, t_n \in (a,b) \quad (6.6.51)$$

Let k be a given integer between 2 and n and let $t_1' = t_k$, $t_k' = t_1$, and $t_i' = t_i$ for $i = 2,\dots, k-1, k+1,\dots, n$. Let $Q = [q_{ij}] \in M_n$ denote the permutation matrix that permutes rows and columns 1 and k, that is, $q_{1k} = q_{k1} = 1$, $q_{ii} = 1$ for $i = 2,\dots, k-1, k+1,\dots, n$, and all other entries are zero. Then

$[\Delta^2 f(t'_i, t'_j, t'_1)]^n_{i,j=1} \succeq 0$ and hence

$$[\Delta^2 f(t_i, t_j, t_k)]^n_{i,j=1} = Q \, [\Delta^2 f(t'_i, t'_j, t'_1)]^n_{i,j=1} \, Q^T \succeq 0$$

Thus, the condition (6.6.51) is sufficient to ensure that $f(\cdot)$ is a convex matrix function on $H_n(a,b)$.

To show that (6.6.51) is also a necessary condition, let $\Lambda = \mathrm{diag}(\lambda_1, ..., \lambda_n) \in H_n(a,b)$ be given and let X be any given Hermitian matrix. Choose $\epsilon > 0$ small enough so that $A \equiv \Lambda - \epsilon X$ and $B \equiv \Lambda + \epsilon X$ are both in $H_n(a,b)$. Then for $A(t) \equiv (1-t)A + tB$ we have $A(\tfrac{1}{2}) = \Lambda$, $A'(t) = 2\epsilon X$, and

$$\frac{d^2}{dt^2} f(A(t)) \Big|_{t=\frac{1}{2}} = 4\epsilon \sum_{k=1}^n [\Delta^2 f(\lambda_i, \lambda_j, \lambda_k)]^n_{i,j=1} \circ [C_k(X)C_k(X)^*]$$

Now assume that $f(\cdot)$ is a convex matrix function on $H_n(a,b)$, so

$$\sum_{k=1}^n \left[[\Delta^2 f(t_i, t_j, t_k)]^n_{i,j=1} \circ [C_k(X)C_k(X)^*] \right] \succeq 0$$

for all Hermitian $X \in M_n$ and all choices of $t_1, ..., t_n \in (a,b)$. Finally, let $L > 0$, let $X = [x_{ij}] \in M_n$ have $x_{11} = L$, $x_{1j} = x_{j1} = 1$ for $i = 2, ..., n$, and $x_{ij} = 0$ for $i, j = 2, ..., n$. For any given $\xi = [\xi_i] \in \mathbb{C}^n$, set $\eta \equiv [\xi_1/L, \xi_2, ..., \xi_n]^T$ and compute

$$0 \le \eta^* \left(\sum_{k=1}^n \left[[\Delta^2 f(t_i, t_j, t_k)]^n_{i,j=1} \circ [C_k(X)C_k(X)^*] \right] \right) \eta$$

$$= \sum_{i,j=1}^n \Delta^2 f(t_i, t_j, t_1) x_{i1} x_{1j} \bar{\eta}_i \eta_j$$

$$+ \sum_{k=2}^n \sum_{i,j=1}^n \Delta^2 f(t_i, t_j, t_k) x_{ik} x_{kj} \bar{\eta}_i \eta_j$$

$$= \sum_{i,j=1}^n \Delta^2 f(t_i, t_j, t_1) \bar{\xi}_i \xi_j + \sum_{k=2}^n \Delta^2 f(t_1, t_1, t_k) |\xi_1|^2 / L^2$$

$$= \xi^*[\Delta^2 f(t_i, t_j, t_1)]_{i,j=1}^n \xi + \sum_{k=2}^n \Delta^2 f(t_1, t_1, t_k) |\xi_1|^2 / L^2$$

Now let $L \to \infty$ and conclude that $[\Delta^2 f(t_i, t_j, t_1)]_{i,j=1}^n \succeq 0$ for every choice of $\{t_1, ..., t_n\} \subset (a,b)$.

Thus, the condition (6.6.51) is both necessary and sufficient for $f(\cdot)$ to be a convex matrix function on $H_n(a,b)$. If this condition is satisfied, and if $n \geq 2$, notice that $[\Delta^2 f(t_i, t_j, t_1)]_{i,j=2}^n \succeq 0$ for every choice of $t_2, ..., t_n \in (a,b)$ and all $t_1 \in (a,b)$, which is equivalent to saying that $\Delta f(\cdot, t_1)$ is a monotone matrix function on $H_{n-1}(a,b)$ for every $t_1 \in (a,b)$. The latter condition then implies that $f(\cdot)$ is a convex matrix function on $H_{n-1}(a,b)$. In particular, $f(\cdot)$ is a convex matrix function on $H_n(a,b)$ for *every* $n = 1, 2, ...$ if and only if $\Delta f(\cdot, t_0)$ is a monotone matrix function on $H_n(a,b)$ for *every* $n = 1, 2, ...$ for all $t_0 \in (a,b)$.

What we have learned can be summarized as follows:

6.6.52   Theorem. Let (a,b) be a given open real interval, let $n \geq 1$ be a given positive integer, and let $f: (a,b) \to \mathbb{R}$ be a given twice continuously differentiable function. Then

(1) $f(\cdot)$ is a convex matrix function on $H_n(a,b)$ if and only if

$$[\Delta^2 f(t_i, t_j, t_1)]_{i,j=1}^n \succeq 0 \text{ for all } t_1, ..., t_n \in (a,b) \qquad (6.6.53)$$

(2) Let $n \geq 2$ be given and suppose $f(\cdot)$ is a convex matrix function on $H_n(a,b)$. Then $f''(t) \geq 0$ for every $t \in (a,b)$. Moreover, $f''(t_0) = 0$ for some $t_0 \in (a,b)$ if and only if $f(\cdot)$ is a linear function on (a,b), that is, $f(t) = \alpha + \beta t$ for some $\alpha, \beta \in \mathbb{R}$ and all $t \in (a,b)$.

(3) If $f(\cdot)$ is a convex matrix function on $H_n(a,b)$, then $\Delta f(\cdot, t_0)$ is a monotone matrix function on $H_{n-1}(a,b)$ for every $t_0 \in (a,b)$.

(4) $f(\cdot)$ is a convex matrix function on $H_n(a,b)$ for every $n = 1, 2, ...$ if and only if $\Delta f(\cdot, t_0)$ is a monotone matrix function on $H_n(a,b)$ for every $n = 1, 2, ...$ and every $t_0 \in (a,b)$.

(5) $f(\cdot)$ is a strictly convex matrix function on $H_n(a,b)$ if

$$\sum_{k=1}^{n} \left[[\Delta^2 f(t_i,t_j,t_k)]_{i,j=1}^{n} \circ [C_k(X)C_k(X)^*] \right] \succ 0 \qquad (6.6.54)$$

for every nonsingular Hermitian $X \in M_n$ and all $t_1,\dots, t_n \in (a,b)$, where $C_k(X) \in M_{n,1}$ denotes the kth column of $X \in M_n$.

Proof: Only assertions (2) and (5) still require some comments. To prove (2), use (1) and proceed as in the proof of (6.6.36(2)): For any distinct $t_1, t_2 \in (a,b)$ we have

$$\begin{bmatrix} \Delta^2 f(t_1,t_1,t_1) & \Delta^2 f(t_1,t_2,t_1) \\ \Delta^2 f(t_2,t_1,t_1) & \Delta^2 f(t_2,t_2,t_1) \end{bmatrix} \succeq 0$$

so $\Delta^2 f(t_1,t_1,t_1) = \tfrac{1}{2} f''(t_1) \geq 0$ for all $t_1 \in (a,b)$. If $f''(t_1) = 0$, then $0 = \Delta^2 f(t_1,t_2,t_1) = (f'(t_1) - \Delta f(t_2,t_1))/(t_1 - t_2)$, so $f'(t_1) - \Delta f(t_2,t_1) = 0$ for all $t_2 \in (a,b)$, which says that $f(t) = f(t_1) + (t - t_1)f'(t_1)$ for all $t \in (a,b)$.

Since every Hermitian matrix is a limit of nonsingular Hermitian matrices, a limiting argument shows that the condition (6.6.54) implies that $f(\cdot)$ is a convex matrix function on $H_n(a,b)$. For strict convexity it is sufficient to have $\frac{d^2}{dt^2} f((1-t)A + tB) \succ 0$ for every $A, B \in H_n(a,b)$ such that $A - B$ is nonsingular and all $t \in (0,1)$; (6.6.50) shows that this is the case under the conditions stated in (5). ☐

Although some ingenuity may be required to verify from first principles that a given function is a convex or strictly convex matrix function, the criteria (6.6.52(1),(4)) can be powerful tools for this purpose.

Exercise. For $f(t) = t^2$, show that $\Delta f(t,t_0) = t + t_0$, which is a monotone matrix function on $H_n(-\infty,\infty)$ for all $n = 1, 2,\dots$ and all $t_0 \in \mathbb{R}$. Conclude that $f(t) = t^2$ is a convex matrix function on $H_n(-\infty,\infty)$ for all $n = 1, 2,\dots$. Show that $\Delta^2 f(s,t,u) \equiv 1$ and that the sum in (6.6.54) reduces to

$$\sum_{k=1}^{n} C_k(X)C_k(X)^* = X^2$$

in this case. Since $X^2 \succ 0$ whenever X is Hermitian and nonsingular, conclude that $f(t) = t^2$ is a strictly convex matrix function on the set of

n-by-n Hermitian matrices for all $n = 1, 2, \ldots$. For a direct proof of these results, see Problem 25. Notice that $f'(t) = 2t$ is a monotone matrix function in this case.

Exercise. For $f(t) = 1/t$, show that $\Delta f(t, t_0) = -1/tt_0$, which is a monotone matrix function on $H_n(0, \infty)$ for all $n = 1, 2, \ldots$ and all $t_0 > 0$. Conclude that $f(t) = 1/t$ is a convex matrix function on $H_n(0, \infty)$ for all $n = 1, 2, \ldots$. Show that $\Delta^2 f(s, t, u) = (stu)^{-1}$ and that the sum in (6.6.54) reduces to

$$\sum_{k=1}^{n} \left[[(t_i t_j)^{-1}]_{i,j=1}^{n} \circ [\, t_k^{-1} C_k(X) C_k(X)^* \,] \right]$$

$$= [(t_i t_j)^{-1}]_{i,j=1}^{n} \circ [X \operatorname{diag}(1/t_1, \ldots, 1/t_n) X]$$

in this case. Invoke Theorem (5.2.1) to show that this matrix is positive definite whenever $t_1, \ldots, t_n > 0$ and X is Hermitian and nonsingular. Conclude that $f(t) = 1/t$ is a strictly convex matrix function on the set of n-by-n positive definite matrices for all $n = 1, 2, \ldots$. For a direct proof of these results, see Problem 26. In this case, $f'(t) = -1/t^2$ is *not* a monotone matrix function; see Problem 32.

Exercise. For $f(t) = -\sqrt{t}$, show that $\Delta f(t, t_0) = -(\sqrt{t} + \sqrt{t_0})^{-1}$, which is a monotone matrix function on $H_n(0, \infty)$ for all $n = 1, 2, \ldots$ and all $t_0 > 0$ (see Problem 14). Conclude that $f(t) = -\sqrt{t}$ is a convex matrix function on $H_n(0, \infty)$ for all $n = 1, 2, \ldots$. Show that

$$\Delta^2 f(s, t, u) = (\sqrt{s} + \sqrt{t})^{-1}(\sqrt{s} + \sqrt{u})^{-1}(\sqrt{t} + \sqrt{u})^{-1}$$

and that, in this case, the sum in (6.6.54) reduces to

$$\sum_{k=1}^{n} [(\sqrt{t_i} + \sqrt{t_j})^{-1}(\sqrt{t_i} + \sqrt{t_k})^{-1}(\sqrt{t_j} + \sqrt{t_k})^{-1}] \circ [C_k(X) C_k(X)^*]$$

$$= Z \circ (Z \circ X)^2$$

where $Z \equiv [(\sqrt{t_i} + \sqrt{t_j})^{-1}]_{i,j=1}^{n}$, $t_1 \geq \cdots \geq t_n > 0$, and X is Hermitian and nonsingular. Problem 13 and Theorem (5.2.1) show that $Z \circ (Z \circ X)^2$ is positive semidefinite, so $f(t) = -\sqrt{t}$ is a convex matrix function on $H_n(0, \infty)$

for all $n = 1, 2,...$. Since both Z and $Z \circ X$ can be singular (see Problem 40), Theorem (5.2.1) cannot be used as in the preceding exercise to show that $f(t) = -\sqrt{t}$ is strictly convex. Fortunately, a result stronger than Theorem (5.2.1) is available and is just what is needed here. Use Corollary (5.3.7) to show that $Z \circ (Z \circ X)^2 \succ 0$ and conclude that $f(t) = -\sqrt{t}$ is a strictly convex matrix function on the set of n-by-n positive definite matrices for all $n = 1, 2,...$. See Problem 17 for a different proof of a more general result.

In 1936, F. Kraus showed that if a given real-valued function—not assumed to be twice differentiable or even continuous—is a convex matrix function on $H_2(a,b)$ for some $a < b$, then $f(\cdot)$ must be twice continuously differentiable on (a,b). Thus, the hypothesis of twice continuous differentiability in Theorem (6.5.52) is actually unnecessary for most of its assertions. Because of criterion (6.6.52(3)) and Loewner's discoveries about smoothness properties of monotone matrix functions, we see that if $n \geq 3$ and $f(\cdot)$ is a convex matrix function on $H_n(a,b)$, then it has at least $2(n-1)-3 = 2n-5$ continuous derivatives on (a,b). Since functions that are monotone matrix functions on $H_n(a,b)$ for all $n = 1, 2,...$ are analytic, have an integral representation (6.6.40), and map the upper half-plane into itself, it follows that a function $f(\cdot)$ on (a,b) that is a convex matrix function on $H_n(a,b)$ for all $n = 1, 2,...$ is analytic on (a,b), has an integral representation obtained from (6.6.40) (see Problem 27), and has an analytic continuation into the upper half-plane for which $\text{Im}[(f(z) - f(t_0))/(z - t_0)] > 0$ whenever $\text{Im}(z) > 0$ and $t_0 \in (a,b)$.

Problems

1. Consider the real symmetric family $A(t) \in M_2$ defined for all $t \in \mathbb{R}$ by

$$A(t) = e^{-1/t^2} \begin{bmatrix} \cos(2/t) & \sin(2/t) \\ \sin(2/t) & -\cos(2/t) \end{bmatrix} \quad \text{for } t \neq 0, \; A(0) \equiv 0$$

Show that $A(t) \in C^{\infty}(\mathbb{R})$ but $A(t)$ is not analytic at $t = 0$. Show that the eigenvalues of $A(t)$ are $\{\pm e^{-1/t^2}\}$ for $t \neq 0$ and are $\{0,0\}$ for $t = 0$, so it is possible to choose the eigenvalues $\{\lambda_1(t), \lambda_2(t)\}$ of $A(t)$ as C^{∞} (but not analytic) functions of $t \in \mathbb{R}$. For all $t \neq 0$, show that the normalized real eigenvectors corresponding to the eigenvalues e^{-1/t^2} and $-e^{-1/t^2}$ are

$$\pm \begin{bmatrix} \cos(1/t) \\ \sin(1/t) \end{bmatrix} \text{ and } \pm \begin{bmatrix} \sin(1/t) \\ -\cos(1/t) \end{bmatrix}$$

Although $A(t)$ can be real orthogonally diagonalized for each $t \in \mathbb{R}$, show that there is no way to choose a continuous real orthogonal family $Q(t)$ so that $A(t) = Q(t) \Lambda(t) Q(t)^T$ for all $t \in \mathbb{R}$, with $\Lambda(t) = \text{diag}(\lambda_1(t), \lambda_2(t))$.

2. Using the methods in the proof of Theorem (6.6.14), give a detailed proof of Theorem (6.6.20).

3. Using Theorems (6.6.14,20) as a guide, state and verify the analog of Theorem (6.6.20) for the third derivative.

4. Using the methods in the proof of Theorem (6.6.30(1)), give a detailed proof of Theorem (6.6.30(2)).

5. Use the formula (6.6.18) and properties (6.1.40.1-9) of the Frobenius covariants to give an alternate proof of Corollary (6.6.19).

6. In the spirit of Corollary (6.6.19), generalize the identity (6.5.6) from analytic functions represented by a power series to primary matrix functions. Under suitable assumptions on smoothness, domain, and range of $f(\cdot)$ and $g(\cdot)$ (what are they?), show that

$$\frac{d}{dt} f(g(A(t))) = f'(g(A(t))) \, g'(A(t)) \, A'(t)$$

if $A(t)$ commutes with $A'(t)$.

7. Let $n \geq 1$ be a given integer; suppose $A(t) \in M_n$ is continuously differentiable in a neighborhood of $t = 0$ and suppose $A(0) = I$. If $A(t)$ has only real eigenvalues in a neighborhood of $t = 0$ [case (i)], assume that a given scalar-valued function $f(\cdot)$ is $(2n-1)$-times continuously differentiable in a real neighborhood of 1; otherwise [case (ii)], assume that $f(\cdot)$ is analytic in a neighborhood of 1. Use (6.6.28) to show that

$$\frac{d}{dt} f(A(t)) \Big|_{t=0} = f'(1) \, A'(0)$$

If, in addition, $A(t)$ is twice continuously differentiable in a neighborhood of $t = 0$ and $f(\cdot)$ has $3n-1$ continuous derivatives in a real neighborhood of $t = 1$ [case (i)] or is analytic in a neighborhood of $t = 1$ [case (ii)], as required by the eigenvalues of $A(t)$ near $t = 0$, use (6.6.29) to show that

$$\frac{d^2}{dt^2} f(A(t)) \Big|_{t=0} = f^{\bullet\bullet}(1)[A'(0)]^2 + f'(1)A^{\bullet\bullet}(0)$$

In particular, if $A \in M_n$ is given, state conditions on $f(\cdot)$ under which one can conclude that

$$\frac{d}{dt} f(e^{At}) \Big|_{t=0} = f'(1) A$$

and

$$\frac{d^2}{dt^2} f(e^{At}) \Big|_{t=0} = [f^{\bullet\bullet}(1) + f'(1)] A^2$$

8.   What do the formulae (6.6.15,18,21,24,29) say when $n = 1$?

9.   Let $f(\cdot)$ be a given continuously differentiable function on (a,b), and suppose $\pm f(\cdot)$ is a monotone matrix function on $H_n(a,b)$ for some $n \geq 2$. Use the criterion (6.6.37) to show that

$$f'(s)f'(t)(s-t)^2 \geq [f(s) - f(t)]^2 \text{ for all } s,\, t \in (a,b) \qquad (6.6.55)$$

In particular, deduce that $f'(t_0) = 0$ for some $t_0 \in (a,b)$ if and only if $f(\cdot)$ is a constant function on (a,b).

10.   Suppose $f(\cdot)$ is a monotone matrix function on $H_n(a,b)$, $g(\cdot)$ is a monotone matrix function on $H_n(c,d)$, and the range of $g(\cdot)$ on (c,d) is contained in (a,b). Show that the composition $f(g(t))$ is a monotone matrix function on $H_n(c,d)$.

11.   Suppose $f(\cdot)$ is a monotone matrix function on $H_n(a,b)$ for some $n \geq 1$. Show that $f(\cdot)$ is a monotone matrix function on $H_m(a,b)$ for all $m = 1,\dots,n$. In particular, $f(\cdot)$ is an ordinary monotone increasing function on (a,b). State and prove a similar result for a convex matrix function on $H_n(a,b)$.

12.   Let (a,b) be a given open real interval and let $n \geq 1$ be a given integer. Show that the set of monotone matrix functions on $H_n(a,b)$ and the set of convex matrix functions on $H_n(a,b)$ are both convex cones, that is, they are closed under convex combinations and multiplication by positive real numbers. Moreover, show that both sets contain all the real constant functions.

13.   Use the identity

$$\frac{1}{\sqrt{s} + \sqrt{t}} = \int_0^\infty e^{-(\sqrt{s} + \sqrt{t})\tau} \, d\tau \qquad (6.6.56)$$

to show that $[1/(\sqrt{t_i} + \sqrt{t_j})]_{i,j=1}^n \succeq 0$ for all $t_1,..., t_n \in (0,\infty)$. Now use (6.6.37) to show that $f(t) = \sqrt{t}$ is a monotone matrix function on $H_n(0,\infty)$ for all $n = 1, 2,...$.

14. Use the results in Problems 10, 12, and 13 to show without computations that $f(t) = -1/(\sqrt{t} + c)$ is a monotone matrix function on $H_n(0,\infty)$ for all $n = 1, 2,...$, and all $c \geq 0$. Now compute $\Delta f(s,t)$ and use the criterion (6.6.37) to come to the same conclusion.

15. Use the criterion (6.6.42), Theorem (6.3.13), and the criterion (6.3.14) to show that the function $f(t) = \sqrt{t}$ on $(0,\infty)$ possesses an analytic continuation onto the upper half-plane that maps it one-to-one into itself.

16. Use the criteria (6.6.37,53,54) to show for all $u \geq 0$ that $f_u(t) \equiv t/(t + u)$ is a monotone matrix function on $H_n(-u,\infty)$ for all $n = 1, 2,...$ and $-f_u(t)$ is a strictly convex matrix function on $H_n(-u,\infty)$ for all $n = 1, 2,...$.

17. Let $f_\alpha(t) \equiv t^\alpha$ for $t \in [0,\infty)$ and $\alpha \in (0,1)$. Use the identity

$$t^\alpha = \frac{\sin(\alpha\pi)}{\pi} \int_0^\infty \frac{tu^{\alpha-1}}{t + u} \, du, \quad 0 < \alpha < 1, \; t \geq 0 \qquad (6.6.57)$$

to show that $f_\alpha(\cdot)$ is a monotone and strictly concave matrix function on $H_n(0,\infty)$ for all $n = 1, 2,...$.

18. Let (a,b) be a given open real interval. Show for all $u \leq 0$ that $f_u(t) \equiv 1/(u - t)$ is a monotone matrix function on $H_n(u,\infty)$ for all $n = 1, 2,...$ and $-f_u(\cdot)$ is a strictly convex matrix function on $H_n(u,\infty)$ for all $n = 1, 2,...$. Verify that

$$\lim_{T\to\infty} \int_{-T}^0 \left[\frac{1}{u-t} - \frac{u}{u^2 + 1} \right] du = \int_{-\infty}^0 \frac{1 + ut}{(u - t)(u^2 + 1)} \, du = \log t \qquad (6.6.58)$$

for all $t > 0$. Use this identity and Problem 12 to show that $f(t) = \log t$ is a monotone and strictly concave matrix function on $H_n(0,\infty)$ for all $n = 1, 2,...$ and that $g(t) \equiv t \log t$ is strictly convex on $H_n(0,\infty)$ for all $n = 1, 2,...$.

19. Consider the linear fractional transformation $f(t) = (\alpha t + \beta)/(\gamma t + \delta)$ with $\alpha, \beta, \gamma, \delta \in \mathbb{R}$ and $|\gamma| + |\delta| \neq 0$. If $\alpha\delta - \beta\gamma \geq 0$, show that $f(t)$ is a

monotone matrix function on $H_n(a,\infty)$ for all $n = 1, 2,...$ for some $a \in \mathbb{R}$; one may take $a = -\infty$ if $\gamma = 0$, or $a = -\delta/\gamma$ if $\gamma \neq 0$. What happens if $\alpha\delta - \beta\gamma < 0$?

20. Show that the function $f(z) = \log z = \log |z| + i \arg z$ (use the principal value of the argument function, for which $\arg z = 0$ when z is real and positive) maps the upper half-plane onto an infinite strip $\{z: 0 < \operatorname{Im} z < \pi\}$ in the upper half-plane and that $f(t)$ is real if $t > 0$. Show that the function $f(z) = z^\alpha$, $0 < \alpha \leq 1$, maps the upper half-plane onto a wedge with vertex at the origin, one side on the positive real axis, and the other side making an angle of $\pi\alpha$ with the real axis, and that $f(t)$ is real if $t > 0$. What region does $f(z) = -1/z$ map the upper half-plane into? Sketch these three regions and explain why, if $A, B \in M_n$ are positive definite matrices such that $A - B$ is positive semidefinite, then $\log A - \log B$, $(A)^\alpha - (B)^\alpha$ $(0 < \alpha \leq 1)$, and $B^{-1} - A^{-1}$ must all be positive semidefinite; if $A - B$ is positive definite, explain why each of these differences must be positive definite.

21. Let $(a,b) \subset \mathbb{R}$ be a given nonempty open interval. Use Loewner's mapping criterion to show that $f(t) = t^2$ is not a monotone matrix function on $H_n(a,b)$ for all $n = 1, 2,....$ Use the necessary condition (6.6.55) to give an elementary proof of a stronger result: $f(t) = t^2$ is not a monotone matrix function on $H_2(a,b)$. Why does this imply that $f(t) = t^2$ is not a monotone matrix function on $H_n(a,b)$ for *any* $n \geq 2$?

22. Same as Problem 21, but for $f(t) = e^t$.

23. If $A, B \in H_n(a,b)$, show that $(1-t)A + tB \in H_n(a,b)$ for all $t \in [0,1]$.

24. Let $C \subset M_{l,m}$ be a given convex set and let $f: C \to H_n(-\infty,\infty)$ be a given function, not necessarily a primary matrix function. Use Definition (6.6.44) to show that $f(\cdot)$ is convex if and only if the real-valued function $\phi(t;\xi;A,B) \equiv \xi^* f((1-t)A + tB)\xi$ is convex on $t \in [0,1]$ for each $\xi \in \mathbb{C}^m$ and all $A, B \in C$.

25. Let $A, B \in M_n$ and $\alpha \in \mathbb{C}$ be given. Verify that

$$\alpha A^2 + (1-\alpha)B^2 - [\alpha A + (1-\alpha)B]^2 = \alpha(1-\alpha)(A-B)^2 \qquad (6.6.59)$$

and use this identity to show that $f(t) = t^2$ is a strictly convex matrix function on $H_n(-\infty,\infty)$ for all $n = 1, 2,....$

26. Let $A, B \in M_n$ be given nonsingular matrices and let $\alpha \in \mathbb{C}$. Verify that

$$\alpha A^{-1} + (1 - \alpha)B^{-1} - [\alpha A + (1 - \alpha)B]^{-1}$$

$$= \alpha(1 - \alpha)A^{-1}(B - A)B^{-1}[\alpha B^{-1} + (1 - \alpha)A^{-1}]^{-1}A^{-1}(B - A)B^{-1}$$
$$(6.6.60)$$

and use this identity to show that $f(t) = 1/t$ is a strictly convex matrix function on $H_n(0,\infty)$ for all $n = 1, 2,\dots$. This fact is useful in the theory of linear-optimal statistical designs.

27.   Use (6.6.52(4)) and (6.6.40) to obtain integral representations for a function that is a convex matrix function on $H_n(a,b)$ for all $n = 1, 2,\dots$.

28.   Use the criterion (6.6.52(5)) to show that $f(t) = 1/\sqrt{t}$ is a strictly convex matrix function on $H_n(0,\infty)$ for all $n = 1, 2,\dots$.

29.   Let $(a,b) \subset \mathbb{R}$ be a given open interval and suppose a given twice continuously differentiable function $f(\cdot)$ is a convex or concave matrix function on $H_n(a,b)$ for some $n \geq 2$. Use the criterion (6.6.53) to show that

$$f''(s)[f(s) - f(t) - (s - t)f'(t)](s - t)^2 \geq 2[f(s) - f(t) - (s - t)f'(s)]^2 \quad (6.6.61)$$

for all $s, t \in (a,b)$. In particular, deduce that $f''(t_0) = 0$ for some $t_0 \in (a,b)$ if and only if $f(\cdot)$ is a linear function on all of (a,b). Moreover, deduce that there are distinct points $t_0, t_1 \in (a,b)$ for which $f(t_1) = f(t_0) + (t_1 - t_0)f'(t_0)$ if and only if $f(\cdot)$ is a linear function on all of (a,b). What does this mean geometrically? If $f(\cdot)$ is matrix convex on $H_n(a,b)$ and is not a linear function on (a,b), conclude that $[\Delta^2 f(t_i,t_j,t_k)]_{i,j=1}^n$ is a positive semidefinite matrix with positive entries for all $k = 1,\dots, n$.

30.   Let $(a,b) \subset \mathbb{R}$ be a given nonempty open interval. Use the criterion (6.6.53) to show that $f(t) = t^3$ is not a convex or concave matrix function on $H_n(a,b)$ for any $n \geq 2$. Also use the necessary condition (6.6.61) to show that $f(t) = t^3$ is not a convex or concave matrix function on $H_2(a,b)$.

31.   Let $(a,b) \subset \mathbb{R}$ be a given nonempty open interval. Show that $f(t) = e^t$ is not a convex matrix function on $H_n(a,b)$ for any $n \geq 2$.

32.   Consider the function $f(t) = 1/t$, which is a strictly convex matrix function on $H_n(0,\infty)$ for all $n \geq 1$. Use the criterion (6.6.55) to show that $f'(\cdot)$ is not a monotone matrix function on $H_n(0,\infty)$ for any $n \geq 2$. Evaluate $f'(-1 + i)$ and use Loewner's mapping criterion for monotone matrix functions to come to the same conclusion.

33. Consider the following examples of convex matrix-valued functions on a convex set of matrices that are not primary matrix functions: Let $C \in M_n$ be given and consider $\phi: M_{m,n} \to M_m$ given by $\phi(A) \equiv A^*CA$. Verify the identity

$$\alpha\phi(A) + (1-\alpha)\phi(B) - \phi(\alpha A + (1-\alpha)B) = \alpha(1-\alpha)(A-B)^*C(A-B)$$
(6.6.62)

Show that the function $\phi(A) \equiv A^*CA$ on $M_{m,n}$ $(C \in M_m)$ is convex if C is positive semidefinite, and it is strictly convex if C is positive definite and $m \geq n$. What is this for $n = 1$? What is this for $C = I$ and $m = n$? What is this if $C = I$, $m = n$, and the domain of $\phi(\cdot)$ is restricted to Hermitian matrices?

34. Let (a,b) be a given open real interval, let $n \geq 1$ be a given positive integer, and let $f: (a,b) \to \mathbb{R}$ be a given continuously differentiable function. If $t_1,..., t_n \in (a,b)$ and $t_i = t_j$ for some $i \neq j$, show that the Loewner matrix $[\Delta f(t_i, t_j)]_{i,j=1}^n$ is singular. If $A, B \in H_n(a,b)$ have a common eigenvector for which A and B have the same eigenvalue, and if $f: H_n(a,b) \to H_n(-\infty,\infty)$ is a primary matrix function, show that $f(B) - f(A)$ is singular.

35. Show that a strongly monotone matrix function on $H_n(a,b)$ is also a strictly monotone matrix function there. What is the relationship between these two concepts when $n = 1$?

36. Use the representation (6.6.56) to show that $f(t) = \sqrt{t}$ is a strongly monotone matrix function on $H_n(0,\infty)$ for all $n \geq 1$.

37. Let $A, B \in M_n$ be nonsingular and let $(a,b) \subset (0,\infty)$ be any given nonempty open interval. Verify that

$$A^{-1} - B^{-1} = A^{-1}(B-A)B^{-1}$$
(6.6.63)

Use this identity, and the fact that $f(t) = -1/t$ is a monotone matrix function on $H_n(0,\infty)$ for all $n = 1, 2,...$, to show that $f(t) = -1/t$ is a strictly monotone matrix function on $H_n(a,b)$ for all $n = 1, 2,...$ that is not strongly monotone for any $n \geq 2$.

38. Let $f(\cdot)$ be a continuous real-valued function on an open real interval (a,b). If the primary matrix function $f: H_n(a,b) \to H_n(-\infty,\infty)$ satisfies the condition (6.6.35), show that it also satisfies (6.6.34) and hence is a monotone matrix function on $H_n(a,b)$ that is strictly monotone. Similarly, if $f(\cdot)$ satisfies the condition (6.6.46), show that it also satisfies (6.6.45) and hence is a convex matrix function on $H_n(a,b)$ that is strictly convex.

39. Let (a,b) be a given open real interval, let $n \geq 1$ be a given integer, and suppose a given continuously differentiable real-valued function $f(\cdot)$ on (a,b) is a monotone matrix function on $H_n(a,b)$. For any $A, B \in H_n(a,b)$ define the closed interval

$$I(A,B) \equiv [\min \{\lambda_{min}(A), \lambda_{min}(B)\}, \max \{\lambda_{max}(A), \lambda_{max}(B)\}] \subset (a,b)$$

where $\lambda_{min}(X)$ and $\lambda_{max}(X)$ denote the smallest and largest eigenvalues of a Hermitian matrix X. Use the integral representation

$$f(B) - f(A) = \int_0^1 \tfrac{d}{dt} f((1-t)A + tB) \, dt \tag{6.6.64}$$

employed in the proof of Theorem (6.6.36) to show that

$$\||\, f(B) - f(A) \,||_2 \leq \max \{f'(t): t \in I(A,B)\} \, \||\, B - A \,||_2 \tag{6.6.65}$$

Conclude that the monotone matrix function $f(\cdot)$ is Lipschitz continuous on $H_n(\alpha,\beta)$ for any choice of α, β such that $a < \alpha < \beta < b$. Compare and contrast this result with Theorem (6.2.30). If $A, B \in H_n(a,b)$ and $B \succeq A$, show that

$$\lambda_{min}(f(B) - f(A)) \geq \min \{f'(t): t \in [\lambda_{min}(A), \lambda_{max}(B)]\} \, \lambda_{min}(B - A)$$

and deduce that

$$\||\, [f(B) - f(A)]^{-1} \,||_2 \leq \left[\min \{f'(t): t \in [\lambda_{min}(A), \lambda_{max}(B)]\}\right]^{-1} \||\, (B - A)^{-1} \,||_2$$

if $B \succ A$ and $f(\cdot)$ is not a constant function on (a,b). For the square root function $f(t) = \sqrt{t}$ use these bounds to show that

$$\||\, B^{\frac{1}{2}} - A^{\frac{1}{2}} \,||_2 \leq \tfrac{1}{2} [\max \{\||\, A^{-1} \,||_2, \||\, B^{-1} \,||_2\}]^{\frac{1}{2}} \, \||\, B - A \,||_2$$

for any positive definite A, B of the same size. If, in addition, $B \succeq A \succ 0$, show that

$$\lambda_{min}(B^{\frac{1}{2}} - A^{\frac{1}{2}}) \geq \tfrac{1}{2} ||| B |||_2^{-\frac{1}{2}} \lambda_{min}(B - A)$$

Finally, if $B \succ A \succ 0$, show that

$$||| (B^{\frac{1}{2}} - A^{\frac{1}{2}})^{-1} |||_2 \leq 2 ||| (B - A)^{-1} |||_2 \, (||| B |||_2)^{-\frac{1}{2}}$$

40. Let $t_1 = 1$, $t_2 = 1/4$, and $X = \begin{bmatrix} 1 & 3 \\ 3 & 8 \end{bmatrix}$. Calculate $Z \equiv [(\sqrt{t_i} + \sqrt{t_j})^{-1}]_{i,j=1}^2$, $K \equiv [(t_i t_j)^{-1}]_{i,j=1}^2$, and $D \equiv \mathrm{diag}(t_1^{-1}, t_2^{-1})$. Using these explicit numerical matrices, show that

(a) X is nonsingular and Hermitian;

(b) $(Z \circ X)^2$ and $K \circ [(Z \circ X)D(Z \circ X)]$ are positive semidefinite and singular and have positive main diagonal entries; and

(c) $Z \circ (Z \circ X)^2$ and $K \circ [Z \circ (Z \circ X)^2]$ are positive definite.

Explain why the results in (c) should be expected.

41. The purpose of this problem is to study a sufficient condition for strict matrix convexity that is simpler, but much weaker, than (6.6.54).
(a) Let m be a given positive integer, let $A_1,..., A_m \in M_n$ be given positive definite matrices, and let $\zeta_1,..., \zeta_m \in \mathbb{C}^n$ be given vectors such that, for any given $x \in \mathbb{C}^n$, $x \circ \zeta_k = 0$ for all $k = 1,..., m$ if and only if $x = 0$. Show that

$$\sum_{k=1}^m A_k \circ (\zeta_k \zeta_k^*) \succ 0$$

(b) Let n be a given positive integer and let $f(\cdot)$ be a given twice continuously differentiable function on an open real interval (a,b). Show that $f(\cdot)$ is a strictly convex matrix function on $H_n(a,b)$ if it satisfies the following strong form of (6.6.53):

$$[\Delta f(t_i, t_j, t_1)]_{i,j=1}^n \succ 0 \text{ for all } t_1,..., t_n \in (a,b)$$

(c) However, show that the criterion in (b) fails to ensure that $f(t) = t^2$ is a strictly convex matrix function on $H_2(0,\infty)$, and that it would still fail to do so even if we could narrow the criterion by requiring that the points $t_1,..., t_n$

be distinct.

42. Let l, m, n, and p be given positive integers, let $C \subset M_{l,m}$ be a given convex set, let $g: C \to H_n(-\infty,\infty)$ be a given convex function on C, let $K \subset H_n(-\infty,\infty)$ be a convex set such that $g(C) \subset K$, and let $f: K \to H_p(-\infty,\infty)$ be a convex function on K that is also monotone (respectively, strictly monotone) in the sense that whenever A, $B \in K$ and $B \succeq A$ (respectively, $B \succ A$), then $f(B) \succeq f(A)$ (respectively, $f(B) \succ f(A)$).
(a) Use the definitions to show that the composition $h(\cdot) \equiv f(g(\cdot))$: $C \to H_p(-\infty,\infty)$ is a convex function on C that is strictly convex if $g(\cdot)$ is strictly convex on C and $f(\cdot)$ is strictly monotone on K.
(b) Use this principle to show that $h_1(A) \equiv \operatorname{tr} A^{-1}$, $h_1(A) \equiv -\operatorname{tr} A^{\frac{1}{2}}$, and $h_3(A) \equiv \operatorname{tr} A^{-\frac{1}{2}}$ are all strictly convex real-valued functions on the positive definite matrices of any given size, and that $h_4(A) \equiv \operatorname{tr} A^2$ is a strictly convex real-valued function on the Hermitian matrices of any given size.
(c) Suppose $K \subset H_n(0,\infty)$ and take $f(t) = -1/t$. Use (a) to show that $1/g(\cdot)$ is a concave function on C that is strictly concave if $g(\cdot)$ is strictly convex.

43. Let A, $B \in M_n$ be given. Suppose $\cup \{\sigma((1-t)A + tB): 0 \leq t \leq 1\} \subset D$, where $D \subset \mathbb{C}$ is a simply connected open set if $\sigma((1-t)A + tB)$ is not entirely real for some $t \in [0,1]$ [case (a)], and D is an open real interval if $\sigma((1-t)A + tB) \subset \mathbb{R}$ for all $t \in [0,1]$ [case (b)]. Let $f: D \to \mathbb{C}$ be a given function that is analytic on D in case (a) and is $(2n-1)$-times continuously differentiable on the real interval D in case (b). Let $\|\|\cdot\|\|$ be a given matrix norm on M_n. If A and B commute, show that

$$\|\|f(B) - f(A)\|\| \leq \|\|B - A\|\| \max_{0 \leq t \leq 1} \|\|f'((1-t)A + tB)\|\| \qquad (6.6.66)$$

where $f(A)$, $f(B)$, and $f'((1-t)A + tB)$ are all primary matrix functions.

Notes and Further Readings: For a detailed discussion of the basic formulae (6.6.18,24,31,32) with extensions to the infinite dimensional case, see Ju. L. Daleckiĭ and S. G. Kreĭn, Integration and Differentiation of Functions of Hermitian Operators and Applications to the Theory of Perturbations, and Ju. L. Daleckiĭ, Differentiation of Non-Hermitian Matrix Functions Depending on a Parameter, pages 1-30 and 73-87, respectively, in American Mathematical Society Translations, Series 2, Vol. 47, American Mathematical Society, Providence, R. I., 1965. For a thorough exposition of the theory of monotone matrix functions and many references to the literature see

[Don]; the original paper is K. Löwner (C. Loewner), Über monotone Matrixfunctionen, *Math. Zeit.* 38 (1934), 177-216. Monotone matrix functions appear naturally in the study of electrical networks, and have come up in the theory of quantum mechanical particle interactions. For a concise modern discussion of monotone and convex matrix functions with numerous examples and counterexamples and references to the literature, see C. Davis, Notions Generalizing Convexity for Functions Defined on Spaces of Matrices, in *Convexity: Proceedings of Symposia in Pure Mathematics*, Vol. VII, American Mathematical Society, Providence, R. I., 1963, pp. 187-201; also see Section 16E of [MOl]. The original paper on convex matrix functions is by a student of Loewner's, F. Kraus, Über konvexe Matrixfunctionen, *Math. Zeit.* 41 (1936), 18-42. There is one more paper (by another student of Loewner's) in this original series of investigations of generalizations to matrix functions of classical properties of scalar functions: O. Dobsch, Matrixfunctionen beschränkter Schwankung, *Math. Zeit.* 43 (1937), 353-388. Dobsch showed that matrix functions of bounded variation are, in some sense, not very interesting because every ordinary scalar function of bounded variation on $[a,b]$ is a matrix function of bounded variation on $H_n(a,b)$ for every order $n = 1, 2, \ldots$. For details about analytic continuation to a univalent mapping of the upper half-plane into itself, see On Boundary Values of a Schlicht Mapping, *Proc. Amer. Math. Soc.* 18 (1967), 782-787 by R. A. Horn. Many additional properties of monotone matrix functions are discussed in R. Horn, The Hadamard Product in [Joh], pp. 87-169; in particular, there is a different proof of Theorem (6.6.38). The original proof of Theorem (6.6.38) is due to R. Mathias; see his paper Equivalence of Two Partial Orders on a Convex Cone of Positive Semidefinite Matrices, *Linear Algebra Appl.* 151 (1991), 27-55.

Hints for problems

Section (1.0)

2. $\rho(A + B) \le \|\| A + B \|\|_2$, where the matrix norm $\|\|\cdot\|\|_2$ is the spectral norm (see Section (5.6) of [HJ]). How are $\rho(A)$ and $\|\| A \|\|_2$ related when A is normal?

Section (1.2)

4. Consider $A = \begin{bmatrix} 0 & 1 \\ 1 & 0 \end{bmatrix}$ and $B = \begin{bmatrix} 1 & 0 \\ 0 & -1 \end{bmatrix}$.

15. Show by example that they are not the same, but by (1.2.12) they both must be "angularly" the same. Thus, the only difference can be stretching or contracting along rays from the origin.

17. The orthogonal complement of the intersection of the nullspaces of A and A^* is the span of the union of the column spaces of A and A^*. Let $U \equiv [P \ P'] \in M_n$ be unitary, let $U^* x \equiv \begin{bmatrix} \xi \\ \eta \end{bmatrix}$ with $\xi \in \mathbf{C}^k$, and show that $x^* A x = \xi^*(P^* A P) \xi$.

20. Permute the basis to $\{e_1, e_4, e_3, e_2\}$ and use (1.2.10).

22. $x^* A^{-1} x = z^* A^* z$ for $z \equiv A^{-1} x$.

23. (a) If $A x = \lambda x$, then $x^* A x = \lambda x^* x$ and $x^* A^* x = \bar{\lambda} x^* x$, add.

25. (a) Let $B = V \Sigma W^*$ be a singular value decomposition of B and show that $|x^* B y| \le \sigma_1(B)\|x\|_2 \|y\|_2$, with equality when x and y are the first columns of V and W, respectively.

561

Section (1.3)

1. This may be done by applying the Gram-Schmidt procedure to a basis of $\mathbb{C}^n$ whose first two vectors are x and y if x and y are independent, or whose first vector is just one of them if they are dependent and one is nonzero. There is nothing to do if both are zero. Householder transformations can also be used. Try both approaches.

4. Work back from the special form, using the transformations that produced it.

13. Reduce A to upper triangular form by a unitary similarity and use (1.3.6).

14. Use (1.3.6).

17. (c) Direct computation or recall the Newton identities in Section (1.2) of [HJ].

Section (1.4)

1. (a) For the first four, let $F(A) \equiv \mathrm{Co}(\sigma(A))$.
(b) For all but (1.2.4), let $F(A) \equiv \{z: z \in F(A) \text{ and } \mathrm{Re}\, z \text{ is a minimum}\}$
(c) For all but (1.2.3), let $F(A) \equiv \mathrm{Co}(F(A) \cup \{\mathrm{tr}\, A\})$ or, even simpler, $F(A) = \phi$
(d) For all but (1.2.2), let $F(A)$ be the boundary of $F(A)$
   (e) For the last four, let $F(A)$ be the interior of $F(A)$, when $F(A)$ has an interior, and let $F(A) = F(A)$ otherwise.

Section (1.5)

5. Consider 2-by-2 examples of the form $\begin{bmatrix} 0 & a \\ b & 0 \end{bmatrix}$.

15. For each $x = [x_1, ..., x_n]^T \in \mathbb{C}^n$, define $x^{(i)} \equiv [x_1, ..., x_{i-1}, x_{i+1}, ..., x_n]^T \in \mathbb{C}^{n-1}$ and consider $(x^{(1)*} A_1 x^{(1)} + \cdots + x^{(n)*} A_n x^{(n)})/n$.

16. Let a denote the smallest, and b the largest, eigenvalue of A, and show that the area ratio in (1.5.17) is exactly $(\beta - \alpha)/(a - b)$; we have $\beta \geq \alpha > 0$ and we want $a > 0$.

23. (1) $|x^* A x| \leq |x|^T |A| |x|$
(m) $|x^* A x| \leq |x|^T A |x| = |x|^T \mathrm{H}(A) |x|$

(o) Use (g) and (d).

24. If $x \in \mathbf{C}^n$ is a unit vector such that $\||A|\|_2 = |x^*Ax|$, use the Cauchy-Schwarz inequality and look carefully at the case of equality.

29. Notice that $J_{2k}(0)$ is a principal submatrix of $J_{2k+1}(0)$, and that $J_{2k}(0)^k = \begin{bmatrix} 0 & I \\ 0 & 0 \end{bmatrix}$ with $I \in M_k$. Use Problem 25 in Section (1.2) and the power inequality in Problem 23(d).

Section (1.6)

1. Use the unitary similarity invariance property to reduce to the case of an upper triangular matrix and take the two eigenvectors into account.

2. Consider the eigenvalues of $H(A)$ and $S(A)$. Write the eigenvalues λ_1, λ_2 in terms of the entries a_{ij} using the quadratic formula and observe that, upon performing a unitary similarity to triangular form, the absolute value of the off-diagonal entry is $(\Sigma_{i,j} |a_{ij}|^2 - |\lambda_1|^2 - |\lambda_2|^2)^{\frac{1}{2}}$.

6. Consider (1.6.8) and think geometrically.

8. Consider the proof of (1.6.3).

12. Eigenvectors of A^* can be thought of as left eigenvectors of A, while those of A are right eigenvectors.

21. (b) $(I - P^2)^{\frac{1}{2}}$ and P are simultaneously (unitarily) diagonalizable and hence they commute. Alternatively, use the singular value decomposition for A.
(c) Use the singular value decomposition of A.
(f) See the 1982 paper by Thompson and Kuo cited at the end of the section for an outline of a calculation to verify the general case.
(g) If $U = \begin{bmatrix} A & B \\ * & * \end{bmatrix}$ is unitary, then $B \in M_{k,n-k}$ and $B^*B = I - A^*A$, so rank $B = \text{rank}(I - A^*A)$.

23. See Problem 11 of Section (1.2) and explicitly compute both sides of the asserted identity.

26. Suppose $p(\cdot)$ has degree k and let $B \in M_{(k+1)n}$ be unitary and such that $B^m = \begin{bmatrix} A^m & * \\ * & * \end{bmatrix}$ for $m = 1, 2, ..., k$. Then we have $p(B) = \begin{bmatrix} p(A) & * \\ * & * \end{bmatrix}$. But $p(B)$ is a contraction by the spectral theorem, so $\||p(A)|\|_2 \le \||p(B)|\|_2 \le 1$.

27. If A is row stochastic, show that every power A^m is row stochastic and hence $\{A^m:\ m = 1, 2,...\}$ is uniformly bounded, which is not possible if the Jordan canonical form of A contains any Jordan blocks $J_k(1)$ with $k > 1$.

28. If all the points in $\sigma(A)$ are collinear, then either all of the line segment $L \equiv \text{Co}(\sigma(A))$ lies in the boundary of $F(A)$ or the relative interior of L lies in the interior of $F(A)$.

30. Use (3.1.13) in [HJ], the direct sum property (1.2.10), and (1.6.6).

31. Use the preceding problem and the characterization of inner products given in Problem 14 of Section (7.2) of [HJ].

32. Consider $A = I_{n-2} \oplus \begin{bmatrix} 0 & 2 \\ 0 & 0 \end{bmatrix}$. Show that a matrix in M_2 is spectral if and only if it is radial if and only if it is normal. Show that for every $n \geq 3$ there are spectral matrices in M_n that are not radial.

35. Use the fact that $\rho(A) = \lim \| A^k \|^{1/k}$ as $k \to \infty$ for any norm $\|\cdot\|$ on M_n and the power inequality $r(A^k) \leq r(A)^k$; see Problem 23 in Section 1.5.

40. (c) If $Q = \begin{bmatrix} A & B \\ * & * \end{bmatrix}$ is complex orthogonal, then $B \in M_{k,n-k}$ and $B^T B = I - A^T A$, so rank $B \geq \text{rank}(I - A^T A)$.

Section (1.7)

2. Show that $[H(A)]^2 - H(A^2)$ is positive semidefinite and hence $\lambda_{max}[H(A^2)] \leq \lambda_{max}[H(A)]^2$.

11. Use (1.5.2).

14. Use Schur's unitary triangularization theorem (2.3.1) in [HJ] to show that one may assume without loss of generality that A is upper triangular. Use the results of the exercises after (2.4.1) in [HJ] to show that it is then sufficient to consider the 2-by-2 case.

17. Use (1.7.11) and pick t_0 so that $F(A + t_0 I)$ lies to the right of the imaginary axis.

18. Recall from Theorem (2.5.5) in [HJ] that a commuting family of normal matrices is simultaneously unitarily diagonalizable.

20. Use (1.7.9).

24. (a) implies that $(e^{i\theta}S)A^* = A(e^{i\theta}S)$ and $A(e^{i\theta}S)^* = (e^{i\theta}S)^* A^*$ so

$AH(e^{i\theta}S) = H(e^{i\theta}S)A^*$. Now use (1.3.5) to choose $\theta \in \mathbb{R}$ so that $P \equiv H(e^{i\theta}S)$ is positive definite, show that $K = P^{-\frac{1}{2}}AP^{\frac{1}{2}} = P^{\frac{1}{2}}A^*P^{-\frac{1}{2}}$ is Hermitian, and note that $A = P(P^{-\frac{1}{2}}KP^{-\frac{1}{2}})$.

25. See Problem 24. When is a normal matrix similar to a Hermitian matrix?

26. Use $A = U^*A^*U$ to show that A commutes with U^2. Let $\{\lambda_1,..., \lambda_n\} = \sigma(U)$. Since $\sigma(e^{i\theta}U) \subset RHP$ for some $\theta \in \mathbb{R}$, the interpolation problem $p(\lambda_i^2) = \lambda_i$, $i = 1,..., n$, has a polynomial solution $p(t)$ (of degree at most $n-1$), so $U = p(U^2)$ commutes with A.

27. (a) If $\zeta \in F(B)$ then $|\zeta| \leq r(B) \leq ||| B |||_2$.

28. Perform a simultaneous unitary similarity of A, B, and C by U that diagonalizes A and C; the diagonal entries of UBU^* are in $F(B)$.

Section (2.1)

3. Use induction. If $p(t) = t^k + \gamma_1 t^{k-1} + \gamma_2 t^{k-2} + \cdots + \gamma_{k-1}t + \gamma_k$ is a real polynomial with $(-1)^m\gamma_m > 0$, $m = 1,..., k$, show that the coefficients of the products $p(t)(t-\lambda)$ and $p(t)(t^2 - at + b)$ both have the same strict sign-alternation pattern if λ, a, $b > 0$.

5. Use the result of Problem 2 to construct examples in $M_2(\mathbb{R})$.

7. Let $z_0 = re^{i\theta}$ be in the complement of W_n and consider the value of the characteristic polynomial $p_A(z_0) = z_0^n - E_1(A)z_0^{n-1} + E_2(A)z_0^{n-2} - E_3(A)z_0^{n-3} + \cdots$. What is $p_A(0)$? If $r > 0$, what is the sign of $Im[(-1)^kE_k(A)z_0^{n-k}]$?

8. Perform a congruence on A via $\begin{bmatrix} I & -B^{-1}C \\ 0 & I \end{bmatrix}$.

Section (2.2)

2. (b) Consider (a) with $z_i(t) = \exp(\lambda_i t)$, $a = -\infty$, $b = 0$.

3. Use Problem 2 and Theorem (5.2.1).

5. Consider a small perturbation A_ϵ of A that is positive stable and diagonalizable.

9. Consider $\int_0^\infty \frac{d}{dt}P(t)\, dt$

Section (2.3)

2.   See Problem 2 of Section (2.1).

3.   If A satisfies (a) or (b), then so does DA; explain why it suffices to show that (a) and (b) imply that A is positive stable. Use (2.3.2); check that $E_1(A)E_2(A) - E_3(A) > 0$ if and only if $(2xyz - ac\beta - a\gamma b) + (x + y)(xy - aa) + (x + z)(xz - b\beta) + (y + z)(yz - c\gamma) > 0$.

Section (2.4)

2.   The equation $GA + A^*G = I$ has a positive definite solution G, so $(GH)K + K(GH)^* = I$.

7.   If Hermitian G, H, with H positive definite, satisfy (2.4.1), show that G^2 is positive definite and that G is a Lyapunov solution for each $\alpha G + (1-\alpha)A, 0 \leq \alpha \leq 1$. Conclude that $i(\alpha G + (1-\alpha)A)$ is constant, $0 \leq \alpha \leq 1$, since no $\alpha G + (1-\alpha)A$ may have an eigenvalue with zero real part, and thus that $i(A)$, $\alpha = 0$, is the same as $i(G)$, $\alpha = 1$.

8.   Carefully follow the development (2.4.1) through (2.4.6).

9.   Give a diagonal Lyapunov solution with positive semidefinite right-hand side and use (2.4.7) or Problem 8.

10.   Show that there is a positive diagonal Lyapunov solution with positive definite right-hand side.

11.   If B is positive or negative definite, the assertion follows immediately from (1.7.8). If B is nonsingular and indefinite, let α be a given point in $F(A)$, let $A(t) \equiv (1 - t)\alpha I + tA$, and observe that $F(A(t)) \subset F(A) \subset RHP$ for $0 \leq t \leq 1$. Use (1.7.8) to argue that $\sigma[A(t)B]$ varies continuously with t and lies in the union of two fixed disjoint angular cones, so the number of eigenvalues of $A(t)B$ in each cone is constant as t varies from 0 to 1. Reduce the singular case to the nonsingular case as in the proof of (2.4.15).

12.   Set $G \to H^{-1}$ and $A \to HA$.

Section (2.5)

3.   Use (2.5.3.17).

5.   Write $\alpha A + (1 - \alpha)B = \alpha B[B^{-1}A + (\alpha^{-1} - 1)I]$.

6.   (a) Use Theorem 2.5.4(a) and (2.5.3.12); $A = B - R$ with $R \geq 0$, so $B^{-1}A = I - B^{-1}R \in Z_n$. If $x > 0$ and $Ax > 0$, then $B^{-1}Ax > 0$. Use Problem 5.

9.   Use the fact that, for an index set $\alpha \subseteq \{1, 2, ..., n\}$ and a nonsingular matrix $A \in M_n$, $\det A^{-1}(\alpha) = \det A(\alpha')/\det A$ (see (0.8.4) in [HJ]).

10.   Use (2.5.3.13) and (2.5.12). Prove the same implication if A is the inverse of an H-matrix.

11.   Use the result of Problem 10.

12.   (c) Use (2.5.3.13) and (2.5.12) to find a positive diagonal $D = \operatorname{diag}(d_1, ..., d_n)$ such that BD is strictly row diagonally dominant and $(BD)^{-1}$ is strictly diagonally dominant of its column entries.   Now take $E \equiv \operatorname{diag}(\beta_{11}/d_1, ..., \beta_{nn}/d_n)$ and check that $(EBD)^{-1}$ satisfies the conditions in (a).   Then $\det[A \circ (EBD)^{-1}] \geq \det A$. Use Hadamard's inequality for an inverse M-matrix in Problem 9.

15.   Partition A^{-1} conformally with A and use the presentation of the partitioned inverse given in (0.7.3) in [HJ]; the Schur complement of A_{11} is the inverse of the block $(A^{-1})_{22}$.   If A is an M-matrix, then $A^{-1} \geq 0$. Conversely, use the alternate presentation for the block

$$(A^{-1})_{11} = (A_{11})^{-1} + (A_{11})^{-1}A_{12}[A_{22} - A_{21}(A_{11})^{-1}A_{12}]^{-1}A_{21}(A_{11})^{-1}$$

to show that if A_{11} and its Schur complement are M-matrices then $A^{-1} \geq 0$.

17.   (a) Using the notation of the proof of Theorem (2.5.12), first suppose A is strictly row diagonally dominant and all $a_{ii} > 0$, $\det A_{ii} > 0$. Show that $r_2(A)\det A_{11} + \epsilon\det A_{12} \geq 0$ by incorporating the factor $r_2(A)$ into the first column of A_{11} in the same way ϵ was incorporated into the first column of A_{12}.

18.   Use the Laplace expansion for $\det A$ (see (0.3.1) in [HJ]) and (2.5.6.1).

20.   (b) Show that there is some $\epsilon > 0$ such that $(1 - \epsilon)I - \epsilon A \geq 0$ and, for such an ϵ, define $B_\epsilon \equiv [(1 - \epsilon)I - \epsilon A]/[(1 - \epsilon) - \epsilon\tau(A)]$. Show that $B_\epsilon \geq 0$, $\rho(B_\epsilon) = 1$, and $\lambda_\epsilon \equiv [(1 - \epsilon) - \epsilon\lambda]/[(1 - \epsilon) - \epsilon\tau(A)] \in \sigma(B_\epsilon)$ whenever $\lambda \in \sigma(A)$. Use the result of Problem 22c of Section (1.5) (a consequence of the Kellogg-Stephens theorem) to show that λ_ϵ lies in the closed wedge L_n defined by (1.5.19) and deduce that $\lambda - \tau(A) = re^{i\theta}$ with $r \geq 0$ and $\theta \leq \pi/2 - \pi/n$.

22.   Let $\lambda(t) \equiv 1 + t - t\lambda$, and show that one may choose $t > 0$ small enough

so that $\lambda(t) - 1 = t(1 - \lambda)$ lies in the closed polygon whose n vertices are the nth roots of unity $\{e^{2\pi i k/n}: k = 1,..., n\}$. Then write $\lambda(t) = \Sigma_k \alpha_k e^{2\pi i k/n}$ with all $\alpha_k \geq 0$ and $\alpha_1 + \cdots + \alpha_n = 1$. Use Problem 21 to conclude that $\lambda(t)$ is an eigenvalue of a doubly stochastic (circulant) matrix $B \in M_n$ whose eigenvalues lie in the wedge L_n defined by (1.5.19). Set $A \equiv [(1 + t)I - B]/t$ and show that A is an M-matrix with $\tau(A) = 1$ and $\lambda \in \sigma(A)$.

23. If $\lambda > 0$, consider λI. Suppose λ is nonreal and define $A(\gamma) \equiv [a_{ij}(\gamma)] \in M_n$ by $a_{i,i+1} = 1$ for $i = 1,..., n-1$, $a_{n,1} = \gamma$, and all other entries are zero. If n is *odd*, let $\gamma = +1$ and show that the characteristic polynomial of $\beta[A(1) + \alpha I]$ is $\beta^n[(t - \alpha)^n - 1]$. Consider the coefficients and show that if $\alpha, \beta > 0$, $\beta[A(1) + \alpha I]$ is a P-matrix with $\beta[e^{i(n\pm 1)\pi/n} + \alpha]$ in its spectrum. If $\pm \text{Im } \lambda > 0$, show that one can choose $\alpha, \beta > 0$ so that $\lambda = \beta[e^{i(n\pm 1)\pi/n} + \alpha]$. If n is *even*, let $\gamma = -1$, show that the characteristic polynomial of $\beta[A(-1) + \alpha I]$ is $\beta^n[(t - \alpha)^n + 1]$, show that $\beta[A(-1) + \alpha I]$ is a P-matrix with the points $\beta[e^{i(n\pm 1)\pi/n} + \alpha]$ in its spectrum, and conclude that λ can be represented as such a point for some $\alpha, \beta > 0$.

25. See Problem 3 of Section (2.1) for the necessity of this coefficient condition. For the sufficiency, consider the polynomial $p_\epsilon(t) \equiv p(t + \epsilon) = t^n + b_1(\epsilon)t^{n-1} + \cdots + b_{n-1}(\epsilon)t + b_n(\epsilon)$ for $\epsilon > 0$ small enough that $(-1)^k b_k(\epsilon) > 0$ for all $k = 1,..., n$. Consider the matrix $A_\epsilon \equiv [a_{ij}] \in M_n(\mathbb{R})$ with $a_{i,i+1} = 1$ for $i = 1,..., n-1$, $a_{ii} = \epsilon$ for $i = 1,..., n-1$, $a_{n,n-i+1} = -b_i(\epsilon)$ for $i = 1,..., n-1$, $a_{nn} = -b_1(\epsilon) + \epsilon$, and all other entries zero. Show that A_ϵ is a P-matrix whose characteristic polynomial is $p(t)$; see Problem 11 of Section (3.3) of [HJ] regarding the *companion matrix*.

28. Use Theorem (2.5.4(b)) and Problem 19.

29. Provide justification for the following identities and estimate:

$$\tau(DA) = \rho(A^{-1}D^{-1})^{-1} \geq \left[\rho\left[A^{-1}(\max_{1 \leq i \leq n} d_i^{-1})I\right]\right]^{-1}$$

$$= \left[\rho(A^{-1}) \max_{1 \leq i \leq n} d_i^{-1}\right]^{-1} = \tau(A) \min_{1 \leq i \leq n} d_i$$

31. Construct a diagonal unitary matrix D such that the main diagonal entries of DB are positive, let $A = \alpha I - P$ with $\alpha > \rho(P)$, and set $R \equiv \alpha I - DB$. Check that $|R| \leq P$, so $\rho(R) \leq \rho(|R|) \leq \rho(P) < \alpha$. Then $|(DB)^{-1}| = |\alpha^{-1}(I - \alpha^{-1}R)^{-1}| = |\Sigma_k \alpha^{-k-1} R^k| \leq \Sigma_k \alpha^{-k-1} P^k = A^{-1}$. The induction argument for Theorem (2.5.4(c)) works here, too; the crucial inequality

is $(\det A_{11})/\det A = (A^{-1})_{nn} \geq |(B^{-1})_{nn}| = |\det B_{11}/\det B|$.

33. If $\mathrm{Re}\,\lambda < 0$, then $|\lambda + t|$ is a strictly decreasing function of t for small $t > 0$, but $\det(A + tI) = |\lambda_1 + t| \cdots |\lambda_n + t|$ is a strictly increasing function of t for all $t > 0$.

34. (a) Write $B = \alpha I - P$ as in Problem 19, and use Fan's theorem to show that the spectrum of A is contained in the union of discs of the form $\{|z - a_{ii}|: \rho(P) - p_{ii}\}$. Then show that $\rho(P) - p_{ii} = b_{ii} - \tau(B)$.
(b) Since $\tau(B) > 0$, the hypotheses ensure that the discs in (a) with $\mathrm{Re}\,a_{ii} < 0$ lie to the left of the line $\mathrm{Re}\,z = -\tau(B)$, and those with $\mathrm{Re}\,a_{ii} > 0$ lie to the right of the line $\mathrm{Re}\,z = \tau(B)$; use the argument in Theorem (6.1.1) in [HJ].

35. Use induction on n.

36. Use Sylvester's determinant identity (0.8.6) in [HJ].

Section (3.0)

4. (b) Using the singular value decomposition, it suffices to consider the case in which $A = \begin{bmatrix} S & 0 \\ 0 & 0 \end{bmatrix}$ with a positive diagonal $S \in M_r$, with $r = \mathrm{rank}\,A$.
Write $B = \begin{bmatrix} B_{11} & B_{12} \\ B_{21} & B_{22} \end{bmatrix}$ with blocks conformal to those of A, and use the criterion to show that $B_{12} = 0$, $B_{21} = 0$, $SB_{11}^* = B_{11}S$, and $B_{11}^* S = SB_{11}$. Conclude that B_{11} is Hermitian and commutes with S.

Section (3.1)

5. (f) Using the notation of (3.1.8(b)(3)), consider E_r and $\frac{1}{2}[(I_r \oplus I_{n-r}) + (I_r \oplus (-I_{n-r}))]$, where $I_k \in M_k$ is an identity matrix.

6. $|x^* A y| \leq \|x\|_2 \|Ay\|_2 \leq \|x\|_2 \|y\|_2\, \sigma_1(A)$. Use the singular value decomposition $A = V\Sigma W^*$ to find vectors x, y for which this inequality is an equality.

7. Let $A = V\Sigma W^*$ be a singular value decomposition and consider $C_r = WI_r V^*$, where $I_r = \mathrm{diag}\,(1, ..., 1, 0, ..., 0) \in M_n$ has r ones.

8. (a) If $Ax = x$, then $\|x - A^* x\|_2^2 = \|x - A^* A x\|_2^2 = \|x\|_2^2 - 2\|Ax\|_2^2 + \|A^* Ax\|_2^2 \leq 0$.

19. Use the reduction algorithm in the proof of Theorem (2.3.1) in [HJ] to construct a unitary triangularization of A; use the argument in the proof of Theorem $(3.1.1)$ to show that the resulting upper triangular matrix is actually diagonal.

25. Write $A = I + iB$ for some Hermitian $B \in M_n$ and compute $A^* A$.

26. For any $z \in \mathbb{C}^n$ with $\|x\|_2 = 1$, show that $4 = 4\| Ux \|_2^2 = \| Ax \|_2^2 + 2x^*H(A^*B)x + \| Bx \|_2^2 \leq 4$. Conclude that A and B are unitary and $H(A^*B) = I$. Use Problem 25.

27. (a) If $A = V\Sigma W^*$ is a singular value decomposition of $A \in B_n$ and if $\sigma_k(A) < 1$, choose $\epsilon > 0$ so $B = V(\Sigma + \epsilon E_{kk})W^*$ and $C = V(\Sigma - \epsilon E_{kk})W^*$ are both in B_n, where $E_{kk} \in M_n$ has entry 1 in position k,k and all other entries are zero. Then $A = \frac{1}{2}(B + C)$, so A is not an extreme point of B_n.

28. Use the discussion preceding Theorem $(3.1.2)$ and the Cauchy-Schwarz inequality to show that $\sigma_i(A) = |x_i^* A x_i| \leq \| Ax_i \|_2 \leq \sigma_i(A)$ and hence $x_i = c_i(Ax_i)$ with $|c_i| = 1$.

31. Compute $(A + \alpha I)^*(A + \alpha I)$ and use Corollary $(4.3.3)$ in [HJ].

32. See Problem 26 in Section (4.4) of [HJ].

33. (a) Write $U = [U_{ij}]$ as a block matrix conformal with D. Notice that

$$\| U_{11} \|_2^2 + \| U_{21} \|_2^2 + \cdots + \| U_{r1} \|_2^2 = n$$
$$= \| U_{11} \|_2^2 + \| U_{12} \|_2^2 + \cdots + \| U_{1r} \|_2^2$$
$$= \| U_{11} \|_2^2 + |d_2/d_1|^2 \| U_{21} \|_2^2 + \cdots + |d_r/d_1|^2 \| U_{r1} \|_2^2$$

(b) If $B = S\Lambda S^{-1}$, $\Lambda = \operatorname{diag}(\lambda_1,\ldots,\lambda_n)$, require $p(\lambda_i)^2 = \lambda_i$, $i = 1,\ldots, n$.

37. (e) Let $U \in M_n$ be a unitary matrix such that $U(A + B) \geq 0$ and compute $|A + B| = H(U(A + B))$.

39. Compute $\operatorname{tr} A^* A$ with $A = H(A) + S(A)$.

45. Using the notation of Theorem $(4.5.9)$ in [HJ], $\lambda_k(AA^*) = \lambda_k(S[BS^{-1}S^{-*}B^*]S^*) = \theta_k \lambda_k(BS^{-1}S^{-*}B^*) = \theta_k \lambda_k(S^{-*}[B^*B]S^{-1})$.

Section (3.2)

4. (c) Write $[s_1 \ \ldots \ s_n]^T = Q[\sigma_1 \ \ldots \ \sigma_n]^T$, where Q is doubly substochastic and hence is a finite convex combination of generalized permutation matrices.

Section (3.3)

4. First generalize Theorem (3.3.4).

8. The largest singular values of a matrix are unaffected if zero rows are deleted.

9. Using the notation of the proof of (3.3.2), apply (3.3.4) to the product $U_k^* A U_k = \Delta_k$.

10. (a) $|\text{tr } X| = |\Sigma_i \lambda_i(X)| \le \Sigma_i |\lambda_i(X)|$.

12. The lower bounds are trivial if B is singular; if not, $A = (AB)B^{-1}$.

14. See Problem 19 in Section (3.1).

18. (c) Expand $[A - B, A - B]_F$ and use (b).

19. (c) Use the argument given to prove (3.3.17).

20. (a) Let $A = U\Delta U^*$ be a unitary upper triangularization of A, so $\text{H}(A) = U\text{H}(\Delta)U^*$. What are the main diagonal entries of $\text{H}(\Delta)$? Apply the Schur majorization theorem (4.3.26) in [HJ] to $\text{H}(\Delta)$, whose eigenvalues are the same as those of $\text{H}(A)$.
(b) Use Theorem (4.3.32) in [HJ] to construct $B = B^T = [b_{ij}] \in M_n(\mathbb{R})$ with $b_{ii} = \text{Re } \lambda_i$, $i = 1, ..., n$, and eigenvalues $\{\eta_i\}$. Let $Q^T BQ = \text{diag}(\text{Re } \lambda_1, ..., \text{Re } \lambda_n)$ for $Q \in M_n(\mathbb{R})$, $Q^T = Q^{-1}$. Let $C = [c_{ij}] \in M_n$ have $c_{ii} = \lambda_i$, $c_{ij} = 2b_{ij}$ if $j > i$, and $c_{ij} = 0$ if $j < i$, and take $A \equiv Q^T CQ$.

21. (a) Let D be a diagonal unitary matrix such that DA has nonnegative main diagonal entries, and let $B_k \in M_k$ be the principal submatrix of DA corresponding to the main diagonal entries $|a|_{[1]}, ..., |a|_{[k]}$. Argue that $|\text{tr } B_k| \le \sigma_1(B_k) + \cdots + \sigma_k(B_k) \le \sigma_1(A) + \cdots + \sigma_k(A)$.

26. (c) Use induction and Theorem (3.3.14); Lemma (3.3.8); $p = 2$.

30. Use Theorem (3.3.21) and (3.3.18).

Section (3.4)

7. (a) Simultaneously unitarily diagonalize the family $\{P_i\}$ with $P_i = U\Lambda_i U^*$, Λ_i a direct sum of zero blocks and an identity matrix. Do the same with the family $\{Q_i\}$, $Q_i = VD_i V^*$, D_i a direct sum of zero blocks and an identity matrix. Then $U^* \hat{A} V = \Lambda_1(U^* A V)D_1 + \cdots + \Lambda_m(U^* A V)D_m$.

Use Problem 6.

8. $A + \alpha I$, $B + \beta I$, and $A + B + (\alpha + \beta)I$ are all positive definite for some $\alpha, \beta \geq 0$.

9. (b) Use successive interchanges of pairs of entries to transform Py to y, and invoke (a) at each step.

Section (3.5)

3. Note that $E_{11}^2 = E_{11}$ and use Corollary (3.5.10).

7. Write $A = B^*B$ and partition $B = [B_1 \ B_2]$, $B_1 \in M_{n,k}$, $B_2 \in M_{n,n-k}$. Now write A and det $A_{\|\cdot\|}$ in terms of B_1 and B_2 and apply (3.5.22) to show that det $A_{\|\cdot\|} \geq 0$. Problem 6 in Section (7.5) of [HJ] gives a counterexample in the block 4-by-4 case (the blocks are 1-by-1 matrices). For a 3-by-3 block counterexample, partition this same matrix so that $A_{11} = A_{33} = [10]$ and $A_{22} = \text{diag}(10, 10)$. Compute the 3-by-3 norm compression obtained with the spectral norm and show that it has a negative eigenvalue.

8. Use the majorization relation between the main diagonal entries and the eigenvalues of AA^*.

13. Use Corollaries (3.1.3) and (3.5.9).

14. In this case, $\sigma(A) \geq \sigma(B)$ entrywise.

16. To prove the triangle inequality for $\nu(\cdot)$, use Problem 14 and the matrix-valued triangle inequality (3.1.15).

20. (c) Use the fact (Section (5.1) of [HJ]) that a norm $\nu(\cdot)$ on a real or complex vector space V is derived from an inner product if and only if it satisfies the parallelogram identity $\nu(x)^2 + \nu(y)^2 = \frac{1}{2}[\nu(x + y) + \nu(x - y)]$ for all $x, y \in V$. Define $g(x) = \| \text{diag}(x) \|$ for all $x \in \mathbb{R}^n$ and use (b).

22. (a) Let $f(t_1, t_2) = t_1 + t_2$ in Corollary (3.5.11) and use (3.5.8).
(b) Write $A = U\Lambda U^*$ with $\Lambda = \text{diag}(\lambda_1, ..., \lambda_n)$, $\lambda_1 \geq \cdots \geq \lambda_n \geq 0$, and a unitary $U \in M_n$. For a given $k \in \{1, ..., n\}$, let $\Lambda_1 \equiv \text{diag}(\lambda_1, ..., \lambda_k, 0, ..., 0)$, $\Lambda_2 \equiv \text{diag}(0, ..., 0, \lambda_{k+1}, ..., \lambda_n)$, $B \equiv U\Lambda_1 U^*$, $C \equiv U\Lambda_2 U^*$, and partition $B = [B_{ij}]_{i,j=1}^2$ and $C = [C_{ij}]_{i,j=1}^2$ conformally with A. Then $A = B + C$ and

$$N_k(A) = N_k(B) = \text{tr } B = \text{tr } B_{11} + \text{tr } B_{22} = N_k(B_{11}) + N_k(B_{22})$$

$$\leq N_k(B_{11} + C_{11}) + N_k(B_{22} + C_{22}) = N_k(A_{11}) + N_k(A_{22})$$

Note that tr $B_{22} = \lambda_1(B_{22}) + \cdots + \lambda_r(B_{22}) = N_k(B_{22})$ for $r \equiv$ rank $B_{22} \leq k$ = rank B, and recall the monotonicity theorem, Corollary (4.3.3) in [HJ].
(c) See Problem 7 in Section (3.4). In fact, the same argument shows that $\|\hat{X}\| \leq \|X\|$ for any unitarily invariant norm $\|\cdot\|$ on M_n and any pinching $\hat{X}$ of $X \in M_n$, even if X is not positive semidefinite.

Section (3.6)

1.   If (b) holds, let Δ be a real upper triangular matrix with eigenvalues $\sqrt{\lambda_i}$ and singular values $\sqrt{\sigma_i}$; take $A = \Delta^T \Delta$. If (c) holds, let $A = \Delta^* \Delta$ and show that diag $\Delta = (\sqrt{\lambda_1},..., \sqrt{\lambda_n})$ and $\sigma_i(\Delta) = \sigma_i(A)^{\frac{1}{2}}$.

Section (3.7)

3.   Apply Theorem (6.4.1) in [HJ] to the block matrix in Problem 46 of Section (3.1).

Section (3.8)

2.   For each choice of indices j, l, m, p one has $E_{jl} \bullet E_{mp} = \Sigma_{i,q} \gamma_{jlmpiq} E_{iq}$. Show that $E_{ij} A E_{lm} B E_{pq} = a_{jl} b_{mp} E_{iq}$ if $A = \Sigma_{j,l} a_{jl} E_{jl}$, $B = \Sigma_{m,p} b_{mp} E_{mp}$.

Section (4.2)

10.   If the ith row of A is nonzero, consider the ith diagonal block $(AA^*)_{ii} BB^* = (A^*A)_{ii} B^*B$ from the identity $AA^* \bullet BB^* = A^*A \bullet B^*B$.

13.   Use the formula for rank$(A \bullet B)$ given in Theorem (4.2.15).

15.   Apply (4.2.13) repeatedly.

16.   Realize that $\Pi(A)$ is a principal submatrix of $A^{\bullet n}$.

19.   Use the same strategy as in the proof of Theorem (4.2.12).

24.   Consider $(A - B) \bullet C + B \bullet (C - D)$.

29.   Consider the Ky Fan 2-norm $N_2(X) \equiv \sigma_1(X) + \sigma_2(X)$ and $\frac{1}{2}N_2(X)$, $A = \text{diag}(2,1)$, $B = \text{diag}(3,2)$.

Section (4.3)

5. Express the coefficient matrix $K(T)$ of this linear transformation, as in Lemma (4.3.2), in Kronecker form (4.3.1b).

14. To describe the Jordan structure of $A \oplus B$ associated with a zero eigenvalue, it suffices to determine rank $(A \oplus B)^k = (\text{rank } A^k)(\text{rank } B^k)$, and these values are easily determined from the Jordan structures of A and B.

16. Use Lemma (4.3.1).

22. Use (4.3.9b).

Section (4.4)

8. Convert this equation to one involving the appropriate Kronecker sum and examine its eigenvalues.

9. Use the formula in (4.4.14).

13. First use the Schur triangularization theorem to reduce A to upper triangular form with the respective eigenvalues λ_i grouped together.

18. (d) $A \oplus I_n$ and $I_n \oplus A$ are normal and commute; Theorem (4.4.5).

20. (a) $\dim C(B) = \dim C(B^k)$, $k = 1, 2, \ldots$ when all the eigenvalues of B are positive.
(b) If $B^2 = A = C^2$ and all the eigenvalues of B and C are positive, notice that the sizes of the Jordan blocks of A, B, and C are the same; in particular, B and C are similar. If $C = SBS^{-1}$ and $B^2 = C^2$, then S commutes with B^2 and hence with B.

Section (4.5)

2. See Problem 12 of Section (4.2).

6. If $B \in M_n$ and tr $B = n$, then $||| B ||| \geq \rho(B) \geq 1$.

10. Use Theorem (4.5.4) to write $A = BC$ in which B and C have distinct positive spectra; B and C are products of two positive definites (Problem 9 in Section (7.6) of [HJ]). If $P_1 P_2 P_3 P_4 = \alpha I$ and all P_i are positive definite, then $P_1 P_2 = \alpha P_3^{-1} P_4^{-1}$ and $\alpha > 0$ (each product has positive spectrum). If $A = \alpha I$ and α is not positive, $A = (\alpha P^{-1})P$ for a non-scalar positive definite P.

11. If $\det A = 1$, use (4.5.4) to write $A = BC$ with $\sigma(B) = \{1, b_1, b_1^{-1},...,$ $b_k, b_k^{-1}\}$ when $n = 2k + 1$ and analogously for C.

12. Use (4.5.4) again.

15. (b) (γ) If (β), let $X_1 = XY$, $X_2 = X^{-1}$, $X_3 = Y^{-1}B$.
(c) $B^{-1}A = XYX^{-1}Y^{-1}$.
(d) (γ) If (β), then $A = (Y^{-1}B)(B^{-1}YX)B(B^{-1}Y)(X^{-1}Y^{-1}B)$.
(d) (ϵ) If (δ), let $X_2 = C^{-1}AB = DBC^{-1}$.

Section (5.1)

5. $\|\|C\|\|_2 = \max\{|x^*Cy|: x \text{ and } y \text{ are unit vectors}\}$. Recall that $\|\|C\|\|_2$ is an upper bound for any Euclidean row or column length of C.

Section (5.2)

6. $x^*(A \circ B)x$ is the sum of the entries of $(D_x^* A D_x) \circ B$.

10. $(a_{11} \cdots a_{nn})\det B \geq \det AB = \det(A \circ B) \geq (a_{11} \cdots a_{nn})\det B$.

13. Let $B = VMV^*$ with $M = \text{diag}(\mu_1,..., \mu_n)$ and $V = [v_{ij}]$ unitary, and let $x = [x_i] \in \mathbb{C}^n$. Show that $x^*(A \circ B)x = \Sigma_{r,s} \lambda_r \mu_s |\Sigma_i x_i u_{ir} v_{is}|^2 \leq x^*(|A| \circ |B|)x$ and use the monotonicity theorem (4.3.3) in [HJ].

14. $\max\{\lambda_1(A \circ B), -\lambda_n(A \circ B)\} = \max\{|x^*(A \circ B)x|: x \in \mathbb{C}^n, \|x\|_2 = 1\}$ $\leq \max\{x^*(\text{abs}(A) \circ \text{abs}(B))x: \|x\|_2 = 1\}$.

15. (a) If $\text{tr } AB = 0$, then $\text{tr}[(UAU^*)(UBU^*)] = 0$ for every unitary U. Choose U so that $UAU^* = \Lambda_1 \oplus 0$ for a positive diagonal $\Lambda_1 \in M_r$. If $\text{tr}[(\Lambda_1 \oplus 0)(UBU^*)] = 0$, the first r main diagonal entries of the positive semidefinite matrix UBU^* are zero, so $UBU^* = 0 \oplus B_2$ with $B_2 \in M_{n-r}$.
(b) To show that $(2) \Rightarrow (1)$, apply (a) to $(D_z^* A D_z)$ and B^T.
(c) $(A \circ B)z = 0 \Rightarrow (D_z^* A D_z)B^T = 0 \Rightarrow (D_z^* A D_z) = 0 \Rightarrow z = 0$.

Section (5.3)

2.   Partition S by columns and use the fact that the spectral norm $\|\|\cdot\|\|_2$ on M_n is compatible with the Euclidean norm $\|\cdot\|_2$ on $\mathbb{C}^n$. You may wish to refer to facts in Chapter 5 of [HJ].

3.   Let $R = (A^{\frac{1}{2}}B^{\frac{1}{2}})^{-1}$ and $S = A^{\frac{1}{2}}B^{\frac{1}{2}}B^{-\frac{1}{2}}D_z B^{\frac{1}{2}T}$.

Section (5.4)

4.   Use (5.4.2b(i)) and (5.5.1).

Section (5.5)

8.   (b) If $A = U \circ \overline{U}$, write $U = [u_{ij}] = [u_1 \ u_2 \ u_3]$ with $u_i \in \mathbb{C}^3$ and show that $u_1 \perp u_2$ implies $u_{21} u_{22} = 0$, which contradicts $a_{21} a_{22} \neq 0$.
(d) A convex combination of unitary matrices is a contraction.

Section (5.7)

2.   Consider the role of the irreducible components, determine the case of equality in Lemma (5.7.8), and use (5.7.10).

5.   Either (1) mimic the proof for positive definite matrices in [HJ], noting the differences in this case, or (2) apply the monotonicity of det for M-matrices [Theorem (2.5.4(c))] and the fact that an M-matrix may be scaled so that its inverse is diagonally dominant of its column entries, [(2.5.3.13) and (2.5.12)].

7.   Apply the weighted arithmetic-geometric mean inequality (Appendix B of [HJ]) to the product

$$\prod_{\substack{i,\,j=1 \\ p_{ij}>0}}^{n} \left[\frac{q_{ij} v_i u_j / v^T Q u}{p_{ij} y_i x_j / y^T P x} \right]^{\frac{p_{ij} y_i x_j}{y^T P x}}$$

and simplify, as in the proof of Lemma (5.7.28).

10. Use the result of Problem 7.

15. (b) $A \circ B \geq A \star B$.   Use Problems 31 and 19 in Section (2.5) and Corollary (5.7.4.1).

Section (6.1)

2. $P(0) = 0$ implies $A_0 = 0$, $P'(0) = 0$ implies $A_1 = 0$, and so forth.

9.   If $P(\lambda)x = 0$ and $x \neq 0$, consider $x^* P(\lambda)x$. What can you say about the roots of $at^2 + bt + c = 0$ if a, b, and c are all positive?

15. Use (6.1.41) to evaluate $p(s,B)$. Use it again with each of the polynomials $p_{0,l}(s,\eta_j)$, $l = 0, 1,\ldots, \beta_j$, $j = 1,\ldots, \nu$, to evaluate $p_{0,l}(A,\eta_j)$. Then use (6.1.40.4) and the commutativity hypothesis.

22. Calculate $\Delta^2 f(a,b,(1-t)a + tb)$ and use (6.1.44). For the converse, use (6.1.46) with $a = \xi - h$, $b = \xi + h$; divide by h^2 and let $h \to 0$.

Section (6.2)

11. Consider the scalar-valued stem function $f(t) = (1-t)/(1+t)$. What is $|f(it)|$ if $t \in \mathbb{R}$? What is Re $f(t)$ if $|t| = 1$?

18. If $b \neq a$, explicitly diagonalize the matrix and use (6.2.1). To handle the case in which $b = a$, either pass to the limit, use (6.2.7), or use the power series for e^t to compute

$$e^{\begin{bmatrix} a & c \\ 0 & a \end{bmatrix}} = e^{aI + \begin{bmatrix} 0 & c \\ 0 & 0 \end{bmatrix}} = e^a e^{\begin{bmatrix} 0 & c \\ 0 & 0 \end{bmatrix}} = e^a \left[I + \begin{bmatrix} 0 & c \\ 0 & 0 \end{bmatrix} \right]$$

22. $\det e^X = e^{\operatorname{tr} X}$ for any $X \in M_n$; see Problem 4.

37. Define $f(J_2(0))$ in the obvious way by continuity and show that if $f(J_2(\lambda))$ were a Lipschitz continuous function of $\lambda \in (0,1)$ with respect to a given norm $\|\cdot\|$, then $\|f(J_2(\lambda)) - f(J_2(0))\| \leq L\lambda$ for some $L > 0$.

40. Let $A = SJS^{-1}$ be a Jordan canonical form for A. Show that B is similar to $C = [f_{ij}(J)]$ via the similarity $S \bullet \cdots \bullet S$. Construct a permutation matrix P such that the diagonal blocks of $P^T C P$ are $[f_{ij}(\lambda_1)],\ldots, [f_{ij}(\lambda_n)]$ and observe that $P^T C P$ is block upper triangular.

45. Use (6.5.7).

Section (6.3)

1. See the problems at the end of Section (7.5) of [HJ].

2. Consider $\lim\limits_{\alpha \to 0} A^{(\alpha)}$.

10. Let $a \equiv \inf\{t > 0: \phi(t) = 0\}$ and consider $A \equiv [\phi(t_i - t_j)] \in M_3$ with $t_1 = a$, $t_2 = a/2$, and $t_3 = 0$.

11. $1/t^\alpha = \int_0^\infty e^{-ts} s^{\alpha-1} \, ds/\Gamma(\alpha)$ for any $\alpha, t > 0$.

12. Show that $\Delta B = -I$ and that $A = CC^T$, where $C = [c_{ij}]$ is lower triangular and $c_{ij} = 1/i$ for $j = 1, \ldots, i$. Show that $\det A = 1/(n!)^2$.

13. Show that $a_{ij} = i^{-1}j^{-1}/(1 - \alpha_{ij})^{-1}$ with $\alpha_{ij} = (i-1)(j-1)/ij$. Then $\log \alpha_{ij} = -\log i - \log j + \sum_{k=1}^\infty (\alpha_{ij})^k/k$.

14. The doubly stochastic matrix B has exactly one positive eigenvalue, which is its Perron root, and $Be = \rho(B)e$. If $x \in \mathbb{R}^n$ and $x \perp e$, then x is a linear combination of eigenvectors of B corresponding to nonpositive eigenvalues, so $x^T Bx \leq 0$.

Section (6.4)

7. Compute A^2 and try to solve the four equations. It simplifies the calculation to notice that all the eigenvalues of A are zero (since all the eigenvalues of A^2 are zero), so $\operatorname{tr} A = a + d = 0$. What is the 1,2 entry of A^2?

13. What are the allowed forms for the $(m-1)$-tuples $\Delta_1 \equiv (k_1 - k_2, k_2 - k_3, \ldots, k_{m-1} - k_m), \ldots$?

23. B is similar to a symmetric matrix C (Theorem (4.4.9) in [HJ]), so the symmetric matrices A and C^2 are similar. Use Corollary (6.4.19).

26. Use the argument in the proof of Theorem (1.3.20) in [HJ] to show that
$$\operatorname{rank} \begin{bmatrix} AA^T & 0 \\ A^T & 0 \end{bmatrix}^k = \operatorname{rank} \begin{bmatrix} 0 & 0 \\ A^T & A^TA \end{bmatrix}^k, \quad k = 1, \ldots, n.$$ Conclude that the nonsingular Jordan blocks of AA^T and A^TA are identical. What does condition (a) say about the singular Jordan blocks in AA^T and A^TA? Use Corollary (6.4.19).

27a. Calculate $C \equiv \operatorname{Log} U$ by (6.4.21) and use Theorem (6.4.20) to show that $C^* = -C$; take $B = C/i$. One can also use the spectral theorem.

27b. Use the classical polar decomposition (see (7.3.3) in [HJ]) to write $A = PU$ and use Problem 27a.

27c. Calculate $C \equiv \operatorname{Log} U$ by (6.4.21) and use Theorem (6.4.20) to show that $C = C^T$ and C is purely imaginary.

27d. Use Problem 27c to determine a real symmetric matrix B such that $U^T U = e^{2iB}$, and set $Q = U e^{-iB}$. Show that Q is unitary as well as orthogonal, and conclude that it is real.

27e. Calculate $C \equiv \text{Log } P$ by (6.4.21) with $\theta = \pi$ and use Theorem (6.4.20) to show that $C = -C^T$ and C is purely imaginary.

27f. Use Problem 27e to determine a real skew-symmetric B such that $P^* P = e^{2iB}$, and set $Q = P e^{-iB}$. Show that Q is both unitary and orthogonal, and conclude that it is real.

32. Use Problem 27f to write $P = Q e^{iB}$, where $Q, B \in M_n(\mathbb{R})$, $QQ^T = I$, $B^T = -B$, and B is a polynomial in $P^* P$. Since P is normal, use Theorem (7.3.4) in [HJ] to show that Q commutes with e^{iB} and compute $P = P^*$ to show that $Q^T = Q$. Then B is a polynomial in $P^2 = (e^{iB})^2$, so Q commutes with B. Let $C \equiv e^{iB}$, which is positive definite, and write $P = C^{\frac{1}{2}} [C^{-\frac{1}{2}} Q C^{\frac{1}{2}}] C^{\frac{1}{2}}$, so P is congruent to a matrix that is similar to Q. Since P and Q have the same signature, $Q = I$ if P is positive definite.

33. Set $B = A^T X$, so rank $B = $ rank A, $A = XB$, $A^T = BX^{-1}$, and $A^T A = B^2$.

34. (a) If $[A \ \ 0] = QS$, write $Q = [Q_1 \ \ Q_2]$ with $Q_1 \in M_{m,n}$ and $S = \begin{bmatrix} S_{11} & S_{12} \\ S_{12}^T & S_{22} \end{bmatrix}$ with $S_{11} \in M_m$. Then $A = Q_1 S_{11} + Q_2 S_{12}^T$, $0 = Q_1 S_{12} + Q_2 S_{22}$, $Q_1^T Q_1 = I$, and $Q_1^T Q_2 = 0$. Now show that $S_{12} = 0$, so $A = Q_1 S_{11}$.

35. If range $A = $ range B (that is, the column spaces are the same), then rank $A = $ rank $B = r$ and there are permutation matrices $P, Q \in M_n$, matrices $C_1, C_2 \in M_{r,n-r}$, and a nonsingular $S_1 \in M_r$ such that $AP = [A_1 \ \ A_2]$, $BQ = [B_1 \ \ B_2]$ with $A_1, B_1 \in M_{n,r}$ having full rank r, $A_2 = A_1 C_1$, $B_2 = B_1 C_2$, $B_1 = A_1 S_1$. Notice that $[A_1 \ \ A_1 C_1] = [A_1 \ \ 0] \begin{bmatrix} I & C_1 \\ 0 & I \end{bmatrix}$.

37. (a) Denote the entries in the first row of K by $\alpha_0, \alpha_1, ..., \alpha_{m-1}$. Explicitly compute the entries in the first rows of $K^2 = J_m(\lambda)^2$, and solve sequentially for $\alpha_0, \alpha_1, ..., \alpha_{m-1}$.
(c) See Example (6.2.14). If $C = SJS^{-1}$, then $C^2 = SJ^2 S^{-1} = A$, so $B = p(A) = Sp(J^2)S^{-1}$. Note that $K \equiv p(J^2)$ satisfies the hypotheses in (b). But $B^2 = SK^2 S^{-1} = SJ^2 S^{-1}$, so $K = J$ and $B = C$.

Section (6.5)

4a. Let $X^T(t) = [x_1(t)\ x_2(t)\ \cdots\ x_n(t)]$, where each $x_i(t) \in \mathbb{C}^n$. Note that $\det X(t) = \det X^T(t)$ and show that

$$\tfrac{d}{dt}\det X(t) = \tfrac{d}{dt}\det X^T(t) = \det\left[\frac{dx_1}{dt}\ x_2\ \cdots\ x_n\right] + \cdots$$

$$= \det\left[[a_{11}x_1 + a_{12}x_2 + \cdots + a_{1n}x_n]\ x_2\ \cdots\ x_n\right] + \cdots$$

$$= \det[a_{11}x_1\ x_2\ \cdots\ x_n] + \cdots = a_{11}(t)\det X^T(t) + \cdots$$

$$= \operatorname{tr} A(t)\ \det X^T(t) = \operatorname{tr} A(t)\ \det X(t)$$

Show that this first-order ordinary differential equation has the solution (6.5.40).

4e. Use the power series for the matrix exponential to show necessity. For sufficiency, show that $aI + A \geq 0$ for some $a > 0$, so that $e^A = e^{-aI + aI + A} = (e^{-aI})e^{aI + A} = e^{-a}e^{aI + A} \geq 0$.

5b. Integrate (6.5.42) to show that

$$\int_0^t \tfrac{d}{ds}X(s)\ ds = X(t) - C = -A\left[\int_0^t X(s)\ ds\right] - \left[\int_0^t X(s)\ ds\right]B$$

Now let $t \to \infty$.

5c. Use (6.2.6) to express e^{-tA} and show that all of the Jordan blocks of $e^{\tau(A)t}e^{-tA}$ remain bounded as $t \to \infty$.

8. As in Theorem (6.5.35), show that $X(0) = I$ and $\tfrac{d}{dt}X(t) = AX(t)$ if $X(t) = q_0(t)I + q_1(t)A + \cdots + q_{k-1}(t)A^{k-1}$.

15. See Problem 18 in Section (7.3) of [HJ].

16. Use (2.5.3.2) to write $A = \alpha I - P$ with $P \geq 0$ and $\alpha > \rho(P)$. Then

$$\left[\tfrac{1}{\alpha}A\right]^{1/m} = \left[I - \tfrac{1}{\alpha}P\right]^{1/m} = \sum_{k=0}^{\infty}(-1)^k\left[\genfrac{}{}{0pt}{}{\frac{1}{m}}{k}\right]\left[-\tfrac{1}{\alpha}P\right]^k = I - Q$$

where $Q \geq 0$ and $\rho(Q) < 1$, so $A^{1/m} = \alpha^{1/m}\left[\tfrac{1}{\alpha}A\right]^{1/m}$ is an M-matrix. It may be helpful to use (5.6.10) in [HJ] to guarantee that there is a matrix norm $|||\cdot|||$ for which $|||\frac{1}{\alpha}P||| < 1$; recall that $\binom{t}{k} = t(t-1)(t-2)\cdots(t-k+1)/k!$, even when t is not a positive integer.

19. $e^{tA}e^{tB}e^{-tA}e^{-tB} = e^{\log[I-(I-e^{tA}e^{tB}e^{-tA}e^{-tB})]} = \cdots$ as in (6.5.16–17).

21. It suffices to consider only diagonal $A = \Lambda = \text{diag}(\lambda_1,\ldots,\lambda_n)$. Why? Compute $\text{tr}(\Lambda^\alpha B^{1-\alpha})$ and use Hölder's inequality.

26. Use (6.5.54a) and see the proof of Corollary (6.5.22).

28. Replace X by $e^{A/2m} e^{B/2m}$ in (6.5.56) and use (6.5.19) and Problem 27.

29. Write $\text{tr}\, e^{A+A^*} = e^{\lambda_1} + \cdots + e^{\lambda_n}$.

30. Show that (6.5.23) becomes $e^B - e^A = (B-A)\, e^A \int_0^1 e^{t(B-A)}\, dt$ when A and B commute.

33. Apply (6.5.25) to (6.5.24) and integrate.

34. Differentiate $A(t) \equiv e^{-tmA}(e^{tA})^m$.

35. (f) See Problem 12 in Section (2.4) of [HJ].

36. Differentiate $P(t)^m = P(t)$ and left-multiply by $P(t)$.

42. See the preceding problem for necessity. For sufficiency, use Theorem (4.3.32) in [HJ] to construct a real symmetric $B = [b_{ij}] \in M_n(\mathbb{R})$ with main diagonal entries $b_{ii} = \text{Re}\,\beta_i$ and $\sigma(B) = \{\eta_i\}$. Let $B = PHP^T$ with a real orthogonal P and a diagonal $H = [\eta_{ij}]$ with $\eta_{ii} = \eta_i$. Form an upper triangular $\Delta = [d_{ij}]$ with $d_{ii} = \beta_i$ and $d_{ij} = 2b_{ij}$ for all $j > i$. Check that $A \equiv P\Delta P^T$ has the asserted properties.

43. Use (6.5.20(2)) with the absolute eigenvalue product functions $\varphi_m(\cdot)$ defined in Corollary (6.5.22). Notice that

$$\prod_{i=1}^m \lambda_i(e^{\frac{1}{2}(A+A^*)}\, e^{\frac{1}{2}(B+B^*)}) \leq \prod_{i=1}^m \sigma_i(e^{\frac{1}{2}(A+A^*)}\, e^{\frac{1}{2}(B+B^*)})$$

$$\leq \prod_{i=1}^m \sigma_i(e^{\frac{1}{2}(A+A^*)})\sigma_i(e^{\frac{1}{2}(B+B^*)}) = \prod_{i=1}^m \lambda_i(e^{\frac{1}{2}(A+A^*)})\lambda_i(e^{\frac{1}{2}(B+B^*)})$$

for $m = 1, ..., n$.

44. See Corollaries (6.5.22) and (3.5.10).

Section (6.6)

6. Use Corollary (6.2.11)

9. Consider $\det [\Delta f(t_i, t_j)]_{i,j=1}^2$.

10. Use the definition (6.6.33).

17. Using the result in Problem 12, it suffices to consider only the functions $g_u(t) \equiv t/(t + u)$ for $u \geq 0$.

22. Use a Taylor expansion to show that $h^2 e^h \geq (e^h - 1)^2$ is false for all sufficiently small $h > 0$.

28. Show that $\Delta^2 f(s,t,u) = (stu)^{-\frac{1}{2}}(\sqrt{s} + \sqrt{u})^{-1}(\sqrt{t} + \sqrt{u})^{-1} + (st)^{-\frac{1}{2}}(\sqrt{s} + \sqrt{u})^{-1}(\sqrt{t} + \sqrt{u})^{-1}(\sqrt{s} + \sqrt{t})^{-1}$, then show that the sum in (6.6.54) reduces to $K \circ [(Z \circ X)D(Z \circ X)] + K \circ Z \circ (Z \circ X)^2$, where $K \equiv [(t_i t_j)^{-\frac{1}{2}}]_{i,j=1}^n$, $Z \equiv [(\sqrt{t_i} + \sqrt{t_j})^{-1}]_{i,j=1}^n$, $D = \text{diag}(t_1^{-1}, ..., t_n^{-1})$, $t_1 \geq \cdots \geq t_n > 0$, and X is nonsingular and Hermitian. Although both terms are positive semidefinite, the first can be singular; see Problem 40. Use Corollary (5.3.7) [as in the third exercise following Theorem (6.6.52)] and Theorem (5.2.1) to show that the term $K \circ [Z \circ (Z \circ X)^2]$ is always positive definite.

29. Consider $\det [\Delta^2 f(t_i, t_j, t_1)]_{i,j=1}^2$. Note (6.1.44).

31. Use the criterion (6.6.61), then use a Taylor series expansion to show that $e^h(e^h - 1 - h)h^2 \geq 2(e^h - 1 - he^h)^2$ is false for all sufficiently small $h > 0$.

37. Consider

$$A = \begin{bmatrix} \lambda & 0 \\ 0 & \lambda - \epsilon \end{bmatrix} \text{ and } B = \begin{bmatrix} \lambda + \epsilon & \sqrt{2}\epsilon \\ \sqrt{2}\epsilon & \lambda + \epsilon \end{bmatrix}$$

for $\lambda \in (a,b)$ and small $\epsilon > 0$.

39. Use an upper bound from Theorem (5.5.18) and a lower bound from Theorem (5.3.4). Recall that $\xi^* X \xi \geq \lambda_{min}(X)$ for any Hermitian matrix X and unit vector ξ.

40. Use Corollary (5.3.7) and Theorem (5.2.1).

43. Use the identity (6.6.64) and Corollary (6.6.19).

References

The following reference list is ordered alphabetically by *author*. This ordering does not always coincide with an alphabetical ordering by *mnemonic code*.

Ait   A. C. Aitken. *Determinants and Matrices*. 9th ed. Oliver and Boyd, Edinburgh, 1956.

AmMo   A. R. Amir-Moéz. *Extreme Properties of Linear Transformations*, Polygonal Publishing House, Washington, N.J., 1990.

And   T. Ando. "Totally Positive Matrices," *Linear Algebra Appl.* 90 (1987), 165-219.

BGR   J. Ball, I. Gohberg, and L. Rodman. *Interpolation of Rational Matrix Functions*. Birkhäuser, Basel, 1990.

Bar 75   S. Barnett. *Introduction to Mathematical Control Theory*. Clarendon Press, Oxford, 1975.

Bar 79   S. Barnett. *Matrix Methods for Engineers and Scientists*. McGraw-Hill, London, 1979.

Bar 83   S. Barnett. *Polynomials and Linear Control Systems*. Dekker, New York, 1983.

Bar 84   S. Barnett. *Matrices in Control Theory*, revised edition. Krieger, Malabar, Fl., 1984.

BSt   S. Barnett and C. Storey. *Matrix Methods in Stability Theory*. Barnes & Noble, New York, 1970.

BB   E. F. Beckenbach and R. Bellman. *Inequalities*. Springer-Verlag, New York, 1965.

BeLy   G. R. Belitskii and Y. I. Lyubich. *Matrix Norms and Their Applications*. Birkhäuser, Basel, 1988.

Bel   R. Bellman. *Introduction to Matrix Analysis*. 2nd ed. McGraw-

Hill, New York, 1970.

BPl   A. Berman and R. Plemmons. *Nonnegative Matrices in the Mathematical Sciences.* Academic Press, New York, 1979.

Bha   R. Bhatia. *Perturbation Bounds for Matrix Eigenvalues.* Pitman Research Notes in Mathematics Series, No. 162, Longman Scientific and Technical, Harlow, 1987.

Boa   R. P. Boas, Jr. *A Primer of Real Functions.* 2nd ed. Carus Mathematical Monographs, No. 13. Mathematical Association of America, Washington, D.C., 1972.

Bru   R. Brualdi, D. Carlson, B. Datta, C. Johnson, and R. Plemmons, eds. *Linear Algebra and Its Role in Systems Theory* (Proceedings of the AMS-IMS-SIAM Joint Summer Research Conference, July 29-August 4, 1984), Vol. 47 in the Contemporary Mathematics series, American Mathematical Society, Providence, R. I., 1985.

CaLe   J. A. Carpenter and R. A. Lewis. *KWIC Index for Numerical Algebra.* U.S. Dept. of Commerce, Springfield, Va. Microfiche and printed versions available from National Technical Information Service, U.S. Dept. of Commerce, 5285 Port Royal Road, Springfield, VA 22161.

Chi   L. Childs. *A Concrete Introduction to Higher Algebra.* Springer, New York, 1983.

Cul   C. G. Cullen. *Matrices and Linear Transformations.* 2nd ed. Addison-Wesley, Reading, Mass., 1972.

Dat   B. N. Datta, ed. *Linear Algebra in Signals, Systems, and Control.* SIAM, Philadelphia, Penn., 1988.

Dei   A. S. Deif. *Advanced Matrix Theory for Scientists and Engineers.* Abacus Press, Tunbridge Wells & London, Kent, England, 1982.

Don   W. F. Donoghue, Jr. *Monotone Matrix Functions and Analytic Continuation.* Springer-Verlag, Berlin, 1974.

Fad   V. N. Faddeeva. Trans. C. D. Benster. *Computational Methods of Linear Algebra.* Dover, New York, 1959.

Fan   Ky Fan. *Convex Sets and Their Applications.* Lecture Notes, Applied Mathematics Division, Argonne National Laboratory, Summer 1959.

Fer   W. L. Ferrar. *Finite Matrices.* Clarendon Press, Oxford, 1951.

Fie 75   M. Fiedler. *Spectral Properties of Some Classes of Matrices.* Lecture Notes, Report No. 75.01R, Chalmers University of Technology and the University of Göteborg, 1975.

Fie 86   M. Fiedler. *Special Matrices and Their Applications in Numerical*

586   References

Mathematics. Nijhof, Dordrecht, 1986.

Fra   J. Franklin. *Matrix Theory*. Prentice-Hall, Englewood Cliffs, N.J., 1968.

Gan 59   F. R. Gantmacher. *The Theory of Matrices*. 2 vols. Chelsea, New York, 1959.

Gant   F. R. Gantmacher. *Applications of the Theory of Matrices*. Interscience, New York, 1959.

Gan 86   F. R. Gantmakher. *Matrizentheorie*. Springer, Berlin, 1986.

GKr   F. R. Gantmacher and M. G. Krein. *Oszillationsmatrizen, Oszilla-tionskerne, und kleine Schwingungen mechanische Systeme*. Akademie-Verlag, Berlin, 1960.

Gel   A. O. Gel'fand. *The Calculus of Finite Differences*, GITTL, Moscow, 1952; 2nd ed., Fitzmatgiz, Moscow, 1959 (Russian). *Differenzenrechnung*, Deutscher Verlag der Wissenschaften, Berlin, 1958. *Calcul des Differences Finies*, Dunod, Paris, 1963.

GLj   I. M. Glazman and L. J. Ljubovič. *Finite Dimensional Linear Analysis: A Systematic Presentation*. The MIT Press, Cambridge, Mass., 1974.

GoKr   I. Gohberg and M. G. Kreïn. *Introduction to the Theory of Linear Nonselfadjoint Operators*. Translations of Mathematical Monographs, Vol. 18, American Mathematical Society, Providence, R.I., 1969.

GLR 82   I. Gohberg, P. Lancaster, and L. Rodman. *Matrix Polynomials*. Academic Press, New York, 1982.

GLR 83   I. Gohberg, P. Lancaster, and L. Rodman. *Matrices and Indefinite Scalar Products*. Birkhäuser-Verlag, Boston, 1983.

GLR 86   I. Gohberg, P. Lancaster, and L. Rodman. *Invariant Subspaces of Matrices with Applications*. Wiley, New York, 1986.

GVl   G. Golub and C. VanLoan. *Matrix Computations*. 2nd ed. Johns Hopkins University Press, Baltimore, 1989.

GoBa   M. J. C. Gover and S. Barnett, eds. *Applications of Matrix Theory* (Proceedings of the IMA Conference at Bradford), Oxford University Press, Oxford, 1989.

Grah   A. Graham. *Kronecker Products and Matrix Calculus*. Ellis Horwood, West Sussex, 1981.

Gray   F. A. Graybill. *Matrices with Applications to Statistics*. 2nd ed. Wadsworth, Belmont, Calif., 1983.

Gre   W. H. Greub. *Multilinear Algebra*. 2nd ed. Springer-Verlag, New York, 1978.

Hal 58   P. R. Halmos. *Finite-Dimensional Vector Spaces*. Van Nostrand, Princton, N.J., 1958.

Hal 67   P. R. Halmos. *A Hilbert Space Problem Book*. Van Nostrand, Princeton, N.J., 1967.

Hel   J. W. Helton. *Operator Theory, Analytic Functions, Matrices and Electrical Engineering*. American Mathematical Society, Providence, R. I., 1987.

HeWi   I. N. Herstein and D. J. Winter. *Matrix Theory and Linear Algebra*. Macmillan, New York, 1988.

HSm   M. W. Hirsch and S. Smale. *Differential Equations, Dynamical Systems, and Linear Algebra*. Academic Press, New York, 1974.

HKu   K. Hoffman and R. Kunze. *Linear Algebra*. 2nd ed. Prentice-Hall, Englewood Cliffs, N.J., 1971.

HJ   R. A. Horn and C. R. Johnson. *Matrix Analysis*. Cambridge University Press, Cambridge, 1985.

Hou 65   A. S. Householder. *The Theory of Matrices in Numerical Analysis*. Blaisdell, New York, 1964.

Hou 72   A. S. Householder. *Lectures on Numerical Algebra*. Mathematical Association of America, Buffalo, N.Y., 1972.

Jac   N. Jacobson. *The Theory of Rings*. American Mathematical Society, New York, 1943.

Joh   C. R. Johnson, ed. *Matrix Theory and Applications*. Proceedings of Applied Mathematics, Vol. 40, American Mathematical Society, Providence, R. I., 1990.

Kap   I. Kaplansky. *Linear Algebra and Geometry: A Second Course*. 2nd ed. Chelsea, New York, 1974.

Kar   S. Karlin. *Total Positivity*. Stanford University Press, Stanford, Calif., 1968.

Kat   T. Kato. *A Short Introduction to Perturbation Theory for Linear Operators*. Springer-Verlag, New York, 1982.

Kel   R. B. Kellogg. *Topics in Matrix Theory*. Lecture Notes, Report No. 71.04, Chalmers Institute of Technology and the University of Göteborg, 1971.

Kow   H. Kowalsky. *Lineare Algebra*. 4th ed. deGruyter, Berlin, 1969.

LaTi   P. Lancaster and M. Tismenetsky. *The Theory of Matrices With Applications*. 2nd ed. Academic Press, New York, 1985.

Lang   S. Lang. *Linear Algebra*. 3rd ed. Springer-Verlag, New York, 1987.

LaH   C. Lawson and R. Hanson. *Solving Least Squares Problems*.

Prentice-Hall, Englewood Cliffs, N.J., 1974.

Mac   C. C. MacDuffee. *The Theory of Matrices*. J. Springer, Berlin, 1933; Chelsea, New York, 1946.

Mar   M. Marcus. *Finite Dimensional Multilinear Algebra*. 2 vols. Dekker, New York, 1973-5.

MMi   M. Marcus and H. Minc. *A Survey of Matrix Theory and Matrix Inequalities*. Allyn and Bacon, Boston, 1964.

MOl   A. W. Marshall and I. Olkin. *Inequalities: Theory of Majorization and Its Applications*. Academic Press, New York, 1979.

McA   T. J. McAvoy. *Interaction Analysis*. Instrument Society of America, Research Triangle Park, 1983.

Meh   M. L. Mehta. *Elements of Matrix Theory*. Hindustan Publishing Corp., Delhi, 1977.

Mir   L. Mirsky. *An Introduction to Linear Algebra*. Clarendon Press, Oxford, 1963; Dover Publications, New York, 1982.

Mui   T. Muir. *The Theory of Determinants in the Historical Order of Development*. 4 vols. Macmillan, London, 1906, 1911, 1920, 1923; Dover, New York, 1966. *Contributions to the History of Determinants, 1900-1920*. Blackie, London, 1930.

Ner   E. Nering. *Linear Algebra and Matrix Theory*. 2nd ed. Wiley, New York, 1963.

New   M. Newman. *Integral Matrices*. Academic Press, New York, 1972.

NoDa   B. Noble and J. W. Daniel. *Applied Linear Algebra*. 3rd ed. Prentice-Hall, Englewood Cliffs, N.J., 1988.

Oue   D. Ouellette, "Schur Complements and Statistics," *Linear Algebra Appl.* 36 (1981), 187-295.

Par   B. Parlett. *The Symmetric Eigenvalue Problem*. Prentice-Hall, Englewood Cliffs, N.J., 1980.

Pau   V. Paulsen. *Completely Bounded Maps and Dilations*. Pitman Research Notes in Mathematics Series, No. 146, Longman Scientific and Technical, Harlow, 1986.

Per   S. Perlis. *Theory of Matrices*. Addison-Wesley, Reading, Mass., 1958.

Pie   A. Pietsch. *Eigenvalues and s-Numbers*. Cambridge University Press, Cambridge, 1987.

Rod   L. Rodman. *An Introduction to Operator Polynomials*. Birkhäuser, Basel, 1989.

Rog   G. S. Rogers. *Matrix Derivatives*. Lecture Notes in Statistics,

Vol. 2. Dekker, New York, 1980.

Rud W. Rudin. *Principles of Mathematical Analysis*. 3rd ed. McGraw-Hill, New York, 1976.

Schn H. Schneider. *Recent Advances in Matrix Theory*. University of Wisconsin Press, Madison, Wis., 1964.

Schw H. Schwerdtfeger. *Introduction to Linear Algebra and the Theory of Matrices*. 2nd ed. P. Noordhoff N.V., Groningen, 1961.

Sen E. Seneta. *Nonnegative Matrices*. Springer-Verlag, New York, 1981.

Ste G. W. Stewart. *Introduction to Matrix Computations*. Academic Press, New York, 1973.

StSu G. W. Stewart and J.-g. Sun. *Matrix Perturbation Theory*. Academic Press, Boston, 1990.

Str G. Strang. *Linear Algebra and Its Applications*. 3rd ed. Harcourt Brace Jovanovich, San Diego, 1988.

STy D. A. Suprenenko and R. I. Tyshkevich. *Commutative Matrices*. Academic Press, New York, 1968.

Tod J. Todd (ed.). *Survey of Numerical Analysis*. McGraw-Hill, New York, 1962.

Tur H. W. Turnbull. *The Theory of Determinants, Matrices and Invariants*. Blackie, London, 1950.

TuA H. W. Turnbull and A. C. Aitken. *An Introduction to the Theory of Canonical Matrices*. Blackie, London, 1932.

UhGr F. Uhlig and R. Grone, eds. *Current Trends in Matrix Theory*. North-Holland, New York, 1987.

Val F. A. Valentine. *Convex Sets*. McGraw-Hill, New York, 1964.

Var R. S. Varga. *Matrix Iterative Analysis*. Prentice-Hall, Englewood Cliffs, N.J., 1962.

Wed J. H. M. Wedderburn. *Lectures on Matrices*. American Mathematical Society Colloquium Publications XVII. American Mathematical Society, New York, 1934.

Wie H. Wielandt. *Topics in the Analytic Theory of Matrices*. Lecture Notes prepared by R. Meyer. Department of Mathematics, University of Wisconsin, Madison, Wis., 1967.

Wil J. H. Wilkinson. *The Algebraic Eigenvalue Problem*. Clarendon Press, Oxford, 1965.

Notation

$\mathbb{R}$	the real numbers
$\mathbb{R}^n$	vector space (over $\mathbb{R}$) of real n-vectors thought of as columns, $M_{n,1}(\mathbb{R})$
$\mathbb{R}^n_+\!\downarrow$	decreasingly ordered nonnegative n-vectors, 204
$\mathbb{C}$	the complex numbers
$\mathbb{C}^n$	vector space (over $\mathbb{C}$) of complex n-vectors thought of as columns, $M_{n,1}(\mathbb{C})$
$\mathbb{F}$	usually a field (usually $\mathbb{R}$ or $\mathbb{C}$)
$\mathbb{F}^n$	vector space (over the field $\mathbb{F}$) of n-vectors with entries from $\mathbb{F}$ thought of as columns, $M_{n,1}(\mathbb{F})$
$\mathbb{Q}$	the quaternions, 86
$\mathbb{Q}^n$	module (over $\mathbb{Q}$) of quaternion n-vectors thought of as columns, $M_{n,1}(\mathbb{Q})$, 86
$M_{m,n}(\mathbb{F})$	m-by-n matrices with entries from $\mathbb{F}$
$M_{m,n}$	m-by-n complex matrices, $M_{m,n}(\mathbb{C})$
M_n	n-by-n complex matrices, $M_{n,n}(\mathbb{C})$
$A, B, \ldots$	matrices, $A = [a_{ij}] \in M_{m,n}(\mathbb{F}), \ldots$
$x, y, \ldots$	column vectors, $x = [x_i] \in \mathbb{F}^n, \ldots$
I_n	identity matrix in $M_n(\mathbb{F})$, 268
I	identity matrix when the dimension is implicit in the context, (0.2.4) in [HJ]
0	zero scalar, vector, or matrix
E_{ij}	matrix with entry 1 in position i,j and zeroes elsewhere
$\overline{A}$	entrywise conjugate of $A \in M_{m,n}(\mathbb{C})$ or $M_{m,n}(\mathbb{Q})$
A^T	transpose, $A^T \equiv [a_{ji}]$ is n-by-m if $A = [a_{ij}]$ is m-by-n, (0.2.5) in [HJ]

A^*	Hermitian adjoint of $A \in M_{m,n}(\mathbb{C})$ or $M_{m,n}(\mathbb{Q})$, $\overline{A}^T$, (0.2.5) in [HJ]
A^{-1}	inverse of a nonsingular $A \in M_n(\mathbb{F})$, (0.5) in [HJ]
$A^{\frac{1}{2}}$	a square matrix such that $(A^{\frac{1}{2}})^2 = A$
$|A|$	either the entrywise absolute value $[|a_{ij}|]$ or $(A^*A)^{\frac{1}{2}}$, the unique positive semidefinite square root, depending upon the context, 124, 211, 311
$A^{\dagger}$	Moore-Penrose generalized inverse of $A \in M_{m,n}$, Problem 7 in Section (7.3) of [HJ]
e_i	i th standard basis vector in $\mathbf{F}^n$ (usually)
e	vector of all ones, $e = e_1 + \cdots + e_n \in \mathbf{F}^n$ (usually), 451
e	base of natural logarithms, 2.718...
$A \geq B$	entrywise inequality for $A, B \in M_{m,n}(\mathbb{R})$, $a_{ij} \geq b_{ij}$, 112
$A \succeq B$	Loewner partial order, $A - B$ is positive semidefinite, 160
adj A	adjugate (classical adjoint) of $A \in M_n(\mathbb{F})$, the transposed matrix of cofactors, (0.8.2) in [HJ]
$A \bullet B$	bilinear product of matrices, 232
$A \bullet_L B, A \bullet_R B$	associated bilinear products of matrices, 232
$\binom{n}{k}$	binomial coefficient, $n(n-1)\cdots(n-k+1)/k(k-1)\cdots 2 \cdot 1$
∂S	boundary of the set S, 50
$C(A)$	centralizer of $A \in M_n$, 274-5
$p_A(t)$	characteristic polynomial of $A \in M_n(\mathbb{F})$, $\det(tI - A)$, (1.2.3) in [HJ]
$C_k(X)$	k th column of $X \in M_{m,n}(\mathbb{F})$, $C_k(X) \in M_{m,1}$, 533, 548
$c_i(A)$	i th (decreasingly ordered) Euclidean column length of $A \in M_{m,n}$, 333
$[X,Y]$	commutator (Lie bracket) of $X, Y \in M_n(\mathbb{F})$, $XY - YX$, 288
$M(A)$	comparison matrix, 123; *see also* incidence matrix, 457
S^c	complement of the set S, 26
$\kappa(A)$	condition number (for inversion, with respect to a given matrix norm) of a nonsingular $A \in M_n$, 331
$\kappa_2(A)$	spectral condition number of a nonsingular $A \in M_n$ (condition number for inversion, with respect to the spectral norm), 193
$\text{Co}(S)$	Convex hull of the set S, 11
$\delta(A; \|\cdot\|)$	defect from normality of $A \in M_n$ with respect to a given unitarily invariant norm $\|\cdot\|$, 192
$d_i(A)$	i th (decreasingly ordered) row deficit of $A \in M_{m,n}$, 228

$C_i'(A)$	i th deleted absolute column sum of $A \in M_{m,n}$, 31
$R_i'(A)$	i th deleted absolute row sum of $A \in M_{m,n}$, 31
$g_i(A)$	$\frac{1}{2}[R_i'(A) + C_i'(A)]$, 31
det A	determinant of $A \in M_n(\mathbb{F})$, (0.3.1) in [HJ]
diag(A)	vector whose entries are the main diagonal entries of the matrix A, 169
diag(x)	diagonal matrix whose main diagonal entries are the entries of the vector x, 305
D_x	diag(x), 305
⊕	direct sum, 12
$\Gamma(A)$	directed graph of $A \in M_n(\mathbb{F})$, (6.2.11) in [HJ]
$\Delta^i f(t_1,\ldots,t_{i+1})$	i th divided difference of the function f, 391-4
$\|\cdot\|^D$	dual norm of a vector norm $\|\cdot\|$, 203
$f^D(\cdot)$	dual norm of a pre-norm $f(\cdot)$, (5.4.12) in [HJ]
λ	eigenvalue of $A \in M_n$ (usually)
$\lambda_{max}(A)$	algebraically largest eigenvalue if $\sigma(A) \subset \mathbb{R}$, 312
$\lambda_{min}(A)$	algebraically smallest eigenvalue if $\sigma(A \subset \mathbb{R}$, 312
$\{\lambda_i(A)\}$	set of eigenvalues (spectrum) of $A \in M_n$
$n!$	factorial, $n(n-1)\cdots2\cdot1$
$A \star B$	Fan product of $A, B \in M_{m,n}$, 357
$F(A)$	field of values (numerical range) of $A \in M_n$, 5
$F'(A)$	angular field of values of $A \in M_n$, 6
$F_{In}(A,\Theta)$	constructive inner approximation for field of values, 37
$F_{Out}(A,\Theta)$	constructive outer approximation for field of values, 37
$G(A)$	Geršgorin region of $A \in M_n$, 31
$G_F(A)$	Geršgorin-type inclusion region for $F(A)$, 31
$GL(n,\mathbb{F})$	group of nonsingular matrices in M_n, (0.5) in [HJ]
$G^h(A)$	induced norm (via the norm $G(\cdot)$) of $X \to A \circ X$ (5.5.14), 339
$A \circ B$	Hadamard product of $A, B \in M_{m,n}(\mathbb{F})$, $[a_{ij}b_{ij}]$, 298
H(A)	Hermitian part of $A \in M_n$, $\frac{1}{2}(A + A^*)$, 9
$H_n(a,b)$	all n-by-n Hermitian matrices with spectra in $(a,b) \subset \mathbb{R}$, 536
$M(A)$	incidence matrix, 457; *see also* comparison matrix, 123
$i(A)$	inertia of $A \in M_n$, $(i_+(A), i_-(A), i_0(A))$, 91-2
$J_k(\lambda)$	Jordan block of size k with eigenvalue λ, 25, 385
$N_k(A)$	Ky Fan k-norm of $A \in M_{m,n}$, 198
⊗	Kronecker (tensor) product, 243
LHP	open left half-plane in $\mathbb{C}$, 9
$l_i(A)$	i th (decreasingly ordered) absolute line sum of $A \in M_{m,n}$, 224

$q_A(t)$	minimal polynomial of $A \in M_n(\mathbf{F})$, (3.3.2) in [HJ]
$\tau(A)$	min $\{\text{Re } \lambda: \ \lambda \in \sigma(A)\}$, minimum eigenvalue of a Z-matrix, M-matrix, or general matrix, 128-9, 359
$r_A(A)$	Newton polynomial for $A \in M_n$, 396-7
$\|\cdot\|_1$	l_1 norm (absolute sum norm) on $\mathbf{C}^n$ or $M_{m,n}$, (5.2.2) in [HJ] and Section (5.6) of [HJ]
$\|\cdot\|_2$	l_2 (Euclidean) norm on $\mathbf{C}^n$; Frobenius, Schur, or Hilbert-Schmidt norm on $M_{m,n}$, (5.2.1) in [HJ] and Section (5.6) of [HJ]
$\|\cdot\|_\infty$	l_∞ (maximum absolute value) norm on $\mathbf{C}^n$ or $M_{m,n}$, (5.2.3) in [HJ] and Section (5.7) of [HJ]
$\|\cdot\|_p$	l_p norm on $\mathbf{C}^n$ or $M_{m,n}$, (5.2.4) in [HJ]
$\|\cdot\|_{tr}$	trace norm (sum of singular values), 215
$\|\|\cdot\|\|_1$	maximum absolute column sum norm on $M_{m,n}$, (5.6.4) in [HJ]
$\|\|\cdot\|\|_2$	spectral norm on $M_{m,n}$, induced by Euclidean norm on vectors, (5.6.6) in [HJ]
$\|\|\cdot\|\|_\infty$	maximum absolute row sum norm on $M_{m,n}$, (5.6.5) in [HJ]
$\|A\|_\alpha$	$\alpha_1 \sigma_1(A) + \alpha_2 \sigma_2(A) + \cdots$ for $A \in M_{m,n}$, all $\alpha_i > 0$, 204
$r(A)$	numerical radius of $A \in M_n$ (usually), 7
$\perp$	orthogonal complement, (0.6.6) in [HJ]
$p_i(A)$	decreasingly ordered main diagonal entries of $(AA^*)^{\frac{1}{2}}$, 342
$P(A)$	algebra of all polynomials in $A \in M_n$, 275
per A	permanent of $A \in M_n(\mathbf{F})$, (0.3.2) in [HJ]
$\Phi(A)$	$A \circ A^{-1}$ for nonsingular $A \in M_n(\mathbf{F})$, 322
$\Phi_T(A)$	$A \circ (A^{-1})^T$ for nonsingular $A \in M_n$, 322
$q_i(A)$	decreasingly ordered main diagonal entries of $(A^*A)^{\frac{1}{2}}$, 342
rank A	rank of $A \in M_{m,n}(\mathbf{F})$, (0.4.1) in [HJ]
RHP	open right half-plane in $\mathbf{C}$, 9
RHP_0	closed right half-plane in $\mathbf{C}$, 9
$r_i(A)$	ith (decreasingly ordered) Euclidean row length of $A \in M_{m,n}$, 333
$R_k(X)$	kth row of $X \in M_{m,n}(\mathbf{F})$, $R_k(X) \in M_{1,n}$, 533
sgn	signum of a permutation or sign of a real number (± 1 or 0), (0.3.2) in [HJ]
$\{\sigma_i(A)\}$	singular values of $A \in M_{m,n}$, $\sigma_1 \geq \sigma_2 \geq \cdots \geq \sigma_{\min\{m,n\}} \geq 0$, 146
$\sigma_1(A)$	largest singular value of $A \in M_{m,n}$, $\|\|A\|\|_2$, 146

Span S	span of a subset S of a vector space, (0.1.3) in [HJ]
spread(A)	maximum distance between eigenvalues of A, 264
S(A)	skew-Hermitian part of $A \in M_n$, $\frac{1}{2}(A - A^*)$, 9
$\rho(A)$	spectral radius of $A \in M_n$, 5
$\sigma(A)$	spectrum (set of eigenvalues) of $A \in M_n$, 1
$A(\alpha,\beta)$	submatrix of $A \in M_{m,n}(\mathbf{F})$ determined by the index sets α, β, (0.7.1) in [HJ]
$A(\alpha)$	principal submatrix $A(\alpha,\alpha)$, (0.7.1) in [HJ]
tr A	trace of $A \in M_n(\mathbf{F})$, (1.2.5) in [HJ]
UHP	open upper half-plane in $\mathbf{C}$, 9
UHP_0	closed upper half-plane in $\mathbf{C}$, 27
$P(m,n)$	vec-permutation matrix in M_{mn}, 259
vec A	vector of stacked columns of $A \in M_{m,n}(\mathbf{F})$, 244
Z_n	set of real n-by-n matrices with nonpositive off-diagonal entries, 113

Index

595